Series Title: Mucins – Potential Regulators of Cell Functions

Volume Title: Gel-Forming and Soluble Mucins

Authored By

Joseph Z. Zaretsky and Daniel H. Wreschner

Department of Cell Research and Immunology
George S. Wise Faculty of Life Sciences
Tel Aviv University
Israel

Bentham Science Publishers
Executive Suite Y - 2
PO Box 7917, Saif Zone
Sharjah, U.A.E.
subscriptions@benthamscience.org

Bentham Science Publishers
P.O. Box 446
Oak Park, IL 60301-0446
USA
subscriptions@benthamscience.org

Bentham Science Publishers
P.O. Box 294
1400 AG Bussum
THE NETHERLANDS
subscriptions@benthamscience.org

CONTENTS

Foreword *i*

Preface *ii*

CHAPTERS

PART I: MUCINS: GENERAL CHARACTERISTICS

1. General Properties and Functions of Mucus and Mucins 3

2. Secreted and Membrane-Bound Mucins: Similarities and Differences 11

3. Secreted Mucins 29

PART II: GEL-FORMING MUCINS

4. Gel-Forming Mucin MUC2 44

5. Gel-Forming Mucin MUC5AC 145

6. Gel-Forming Mucin MUC5B 246

7. Gel-Forming Mucin MUC6 316

8. Gel-Forming Mucin MUC19 387

PART III: SOLUBLE MUCINS

9. Mucin MUC7 398

10. Mucin MUC8 418

11. Mucin MUC9 425

PART IV: SECRETED MUCINS-REGULATORS OF CELL FUNCTIONS

12.	Secreted Mucin Multifunctionality: Overt Functions	452
13.	Secreted Mucins: Covert Functions	547
	Index	569

FOREWORD

It is a great pleasure to write the Foreword for the manuscript "Mucins – potential regulators of cell functions" prepared by internationally known scientists Drs. Joseph Z. Zaretsky and Daniel H. Wreschner. It is a very comprehensive publication providing an excellent insight into the properties and functions of mucin glycopoteins. In recent years, major breakthroughs have taken place in the field. Significant progress has been made in understanding of structure and biosynthesis of mucins as well as in prediction and identification of their functions. Although a lot of literature is available that covers various aspects of mucin studies, there is an obvious requirement for a complete comprehensive analysis of recently obtained data which could change previous knowledge of the biological role of mucins. Such book is indeed the need of the hour. The book covers not only general concepts but also presents details which help to understand the role of mucins in human development, cell differentiation and defense, innate immunity and regulation of cell functions. The authors have taken efforts to focus attention on basic as well as advanced aspects of the research of secreted mucins. All aspects of the mucin problems are presented in a straightforward fashion, with insightful sifting and appraisal of evidences, and in a logical and scientific manner that would help beginners and experts alike to enjoy reading and simplify understanding. The book would be very useful not only for the "muciners", as Isabelle Van Seunengen called researches studying mucins, but also for biologists working in adjacent sections of the life sciences. It is useful as a handbook containing comprehensive information on structure and properties of both gel-forming and soluble mucins. It would serve also as an excellent reference book for students and investigators interested in molecular biology, biochemistry, immunology, genetics, embryology, histology and clinical medicine. I strongly recommend the book to students and researchers studying the rapidly progressing branch of modern biology - "mucinology". I am confident that essential new ideas presented in the book would stimulate further studies in this important area.

Professor Roald Nezlin

Weizmann Institute of Science
Rehavot
Israel

PREFACE

Mucus is a viscous colloid gel developed in the course of evolution by live systems as one of the major instruments of cell defense. Mucus protects cells from mechanical and chemical stresses, hydrates cells and organisms, lubricates epithelial surfaces, and enables exchange of water, chemicals, metabolites, nutrients, gases, odorants, hormones and gametes. It has the ability to trap and immobilize pathogens and small particles before they come into contact with epithelial surfaces. These important functions are determined by the properties of specific glycoproteins known as mucins.

The term "mucin" was coined for large multifunctional glycoproteins that are secreted by epithelial cells into extracellular space and have specific structural domains that perform specific functions. This group of mucin glycoproteins is called "secreted mucins" in contrast to non-secreted "membrane-bound mucins". Research in the past several years has shown that secreted mucins are polyvalent proteins involved in multiple cell processes, with active roles in maintenance of homeostasis under physiological conditions and development of disease under pathological conditions.

The past three decades have witnessed the rapid development of a new area of science called mucinology – the study of the properties and functions of mucins. The rapid advances in this field are reflected by the number of articles published on the subject at various times in the past 30 years: 180 papers from 1980 to 1982, some 700 from 1989 to 1990, and more than 2200 articles between 2010 and 2011. The latest monograph on mucins appeared in 2008, and already the field has burgeoned with a wealth of new data waiting to be analyzed and integrated. This book is meant to meet this need.

A total of 21 mucin genes have been cloned and sequenced and their polypeptide products studied to varying extents. The secreted mucins are encoded by 8 of these genes and 13 other encode the membrane-bound mucins. The secreted mucins are the subject of this book: five are gel-forming (MUC2, MUC5AC, MUC5B, MUC6 and MUC19) and three are soluble (MUC7, MUC8 and MUC9).

As follows from the title of the book, the properties and functions of the gel-forming and soluble mucins and the corresponding genes attest to their multifunctional character. Functions that have been discovered and documented to date, so called overt functions, are described in detail. Another whole set of potential functions, the covert functions, hinted at by indirect evidence, mainly bioinformatics data, and await experimental verification. They too are addressed.

The 13 chapters of the book are collected into four parts. The first part (Chapters 1-3) presents the general characteristics of mucins and mucin classification (Chapter 1), a comparison of secreted and membrane-bound mucin properties (Chapter 2), and the structural and evolutionary aspects of the secreted mucins (Chapter 3). The second part (Chapters 4-8) presents detailed information about the structure, and the biochemical, biophysical and genetic properties of the gel-forming mucins and the corresponding genes, including their promoters and regulatory mechanisms: MUC2 (Chapter 4), MUC5AC (Chapter 5), MUC5B (Chapter 6), MUC6 (Chapter 7) and MUC19 (Chapter 8). Also described is the expression of these genes at transcriptome and proteome levels under physiological conditions, including embryonic and fetal development, and in pathology. The third part of the book (Chapters 9-11) covers the properties and expression of genes encoding the soluble mucins MUC7 (Chapter 9), MUC8 (Chapter 10) and MUC9 (Chapter 11), including the structure and biochemical properties of individual glycoproteins comprising this group of mucins. In the fourth part, the overt and covert functions of the gel-forming and soluble mucins are analyzed. Chapter 12 contains information about experimentally-proven mucin functions and the involvement of the secreted mucins in the fundamental processes of cell physiology and pathology, including oncogenesis. Chapter 13 summarizes bioinformatics data that point to the potential of the secreted mucin glycoproteins to interact with various protein partners and thereby contribute to the regulation of cell functions.

COMPETING INTERESTS

The authors have declared that no competing interests associated with this work exist.

ACKNOWLEDGEMENTS

The authors thank Drs. Itay Barnea and Edward Nemirovsky for help in bioinformatics analysis and computer design.

Joseph Z. Zaretsky

Department of Cell Research and Immunology
George S. Wise Faculty of Life Sciences, Tel Aviv University
Israel
E-mail: josephz@post.tau.ac.il

CHAPTER 1

General Properties and Functions of Mucus and Mucins

Abstract: Mucus, a viscous colloid gel, is an important element of cell defense developed in the course of evolution by live systems. Mucins, the main components of mucus, are glycoproteins characterized by specific structure and functions. Two subfamilies of the large mucin superfamily have been identified: secreted mucin glycoproteins and membrane-bound mucins. The secreted mucins are further subdivided into two groups: insoluble gel-forming mucins and soluble mucins. All gel-forming mucins share several features, such as specific domain structures, glycosylation patterns and biosynthetic pathways that differ from those of the membrane-bound mucins. Several classifications of the mucin glycoproteins have been proposed, but no one is universal. Further studies of the mucins are needed for development of an appropriate classification system.

Keywords: Mucus, mucins, structure, classification.

1.1. MUCUS: PROPERTIES, ROLE IN EVOLUTION AND FUNCTIONS

Mucus, a viscous colloid gel, has been developed in the course of evolution by live systems as one of the ingenious instruments of cell defense. It protects cells from mechanical and chemical stresses, hydrates cells and organisms, lubricates epithelial surfaces, and filters nutrients. Mucus is a dynamic semi-permeable barrier that enables exchange of water, chemicals, metabolites, nutrients, gases, odorants, and interaction of hormones and gametes. At the same time it is impermeable to most pathogens under physiological conditions. The functions and "characteristics of mucus gel may vary from one organism to another, from one tissue to another, and even within a tissue may vary depending on the physiological conditions" [1].

In primitive organisms like gastropod mollusk (class *Gastropoda*), the mucus serves a protection function, facilitates movement, and participates in communication. Fish mucus is known to contain many biologically active peptides and proteins that enable several biological functions such as respiration, ionic and osmotic regulation, reproduction, excretion, disease resistance, communication, parental feeding, and nest building [2-8]. In vertebrates, mucus covers all mucous membranes, especially the epithelial surfaces. In mammals, it

Joseph Z. Zaretsky and Daniel H. Wreschner

protects epithelial cells of the respiratory, gastrointestinal, urogenital, visual, and auditory systems; in amphibians, mucus film covers the epidermis, and in fish, protects the gills. It is a key component of innate defense against pathogens such as bacteria, viruses and fungi [9, 10].

Mucus has evolved to have robust barrier mechanisms that trap and immobilize pathogens and small particles before they come into contact with epithelial surfaces [11]. Epithelial cells constantly secret mucus. The thickness of the mucus "blanket" is determined by the balance between the rate of mucus secretion and the rate of its degradation or shedding. Under physiological conditions, mucus must be movable. Cilia, located on the surface of epithelial cells, can transport the mucus if it possesses appropriate viscoelasticity: it has to be high enough to prevent gravitational flow but low enough to enable rapid ciliary transport and clearance [11, 12]. The mucus viscoelasticity depends strongly on mucins, gigantic glycoprotein molecules that determine the biophysical and biochemical properties of mucus and fulfill, or at least, participate in, most mucus-mediated functions. Many other factors also contribute to mucus viscoelasticity, including secreted lipids, trefoil proteins, non-mucin glycoproteins as well as salts and pH [13].

1.2. MUCINS: GENERAL CHARACTERISTICS

As pointed out by Theodoropoulos *et al.* [14], it is important to differentiate the term "mucus", which refers to an aggregate secretion consisting of water, ions, inorganic salts, cell debris, and various peptides and proteins, from the term "mucin", which refers to specific glycoprotein molecules within this mucous secretion. Historically, the term "mucin" was coined in reference to large multifunctional glycoproteins secreted by epithelial cells into extracellular space and characterized by the presence of specific structural domains that fulfill specific functions [15]. This group of mucus glycoproteins is called "secreted mucins" in contrast to non-secreted "membrane-bound mucins". Physically, molecules of the secreted mucins look like "long flexible strings", the central region(s) of which are densely coated with negatively charged glycans of different lengths. The glycosylated and highly hydrophilic regions of secreted mucins are separated by relatively hydrophobic non-glycosylated regions, so called "naked" regions [11]. The presence of the alternating hydrophobic and hydrophilic regions in the mucin structure determines flexibility of the molecule. The ability of secreted mucins to form elastic gel is associated with the cysteine residues present in the cysteine-rich domains that are involved in disulfide-bound formation between different mucin molecules. This mechanism is responsible for the occurrence of the gigantic macromolecular gel-forming polymers. The mucin-based gel is able to trap and immobilize pathogens, particles, and nutrients. One of the main functions of secreted mucins – lubrication of the epithelial surfaces – is also associated with their ability to develop dynamic gel structures. Among the known secreted mucins, five belong to a structurally homogeneous group of insoluble gel-forming mucins while the three others belong to a group of soluble mucins.

As noted above, in addition to secreted mucins, there is a group of mucin glycoproteins tethered to the cell membrane – the so-called "membrane-bound mucins". These mucins also function as cell "defenders", but fulfill many other functions as well, including epithelial cell renewal, differentiation, signal transduction, cell adhesion and intercellular communications. The typical membrane-bound mucin contains three structurally and functionally different elements: N-terminal extracellular domain, transmembrane region, and C-terminal intracellular cytoplasmic domain. The extracellular part of the molecule functions as a sensor of extracellular insults, and also participates in intercellular communications, and cell adhesion and repulsion. The functions of the transmembrane domain have not been delineated, as opposed to the cytoplasmic region of the membrane-bound mucin, which has been extensively studied. It participates in multiple signal transduction pathways, thereby exerting control over numerous cell functions and homeostatic parameters. Most of the membrane-bound mucins are cell receptors: they accept extracellular signals and transfer them inside the cell where they activate various intracellular processes.

Under physiological conditions, both the secreted and membrane-bound mucins exhibit highly coordinated organ-, tissue- and cell-specific expression. However, environmental factors that affect cellular integrity may cause alterations in "mucin homeostasis" resulting in the development of pathological states such as cancer and inflammation. Therefore, as pointed out by Andrianifahanana *et al.* [15], "it is crucial to comprehend the underlying basis of molecular mechanisms controlling mucin production in order to design and implement adequate therapeutic strategies for combating these diseases".

1.3. MUCIN CLASSIFICATION

The simplest mucin classification divides them into two main groups: secreted and membrane-bound mucins. So far, a total of 21 human mucins have been identified and "more are likely awaiting discovery" [15]. The group of secreted mucins contains 8 mucin glycoproteins, and the group of membrane-bound mucins contains 13 mucins (Fig. **1**). Although mucins have being studied for many decades, "there is still no clear definition of a "mucin" and the increasing number of genes with the symbol *MUC* is unfortunately not helping" [16]. Thus, the classification of mucins is not straightforward and has been a subject of many discussions [17-22]. The difficulty in assigning a newly discovered mucin protein to a specific mucin group has led to a large group of so-called mucin-like proteins.

Many investigators define "mucin" as a glycoprotein molecule whose central polymorphic domain contains a variable number of tandem repeats (VNTR) enriched in proline, threonine and serine residues heavily glycosylated with O-glycans of different lengths [14, 19]. According to Chaturverdi *et al.* [23], these features are "a hallmark of the mucin family". Although this definition appears to be widely accepted by the scientific community [20, 24, 25], it does not take into consideration many of the structural features found in the mucin glycoproteins isolated from different species. While the symbol "*MUC*" has been assigned to mucin genes in accordance with the definition, many genes encoding O-glycoproteins containing VNTR have been given this designation, but "only some of these genes encode true mucins while others encode non-mucin adhesion O-glycoproteins" [21]. Several classifications of mucin proteins have been suggested, but none meets all the criteria. Rose and Voynow [22] suggested defining mucin as a glycoprotein containing tandem repeat (TR) domain(s) enriched in proline, threonine and serine residues, whereas protein molecules with significant amounts of O-glycosylated serine and threonine, but without TR, would be defined as mucin-like glycoproteins. According to these criteria, three genes, *MUC14, MUC15* and *MUC18,* designated in GenBank as mucins, should be attributed to a group of mucin-like proteins as they do not contain TR although they do have numerous serine and threonine residues [26].

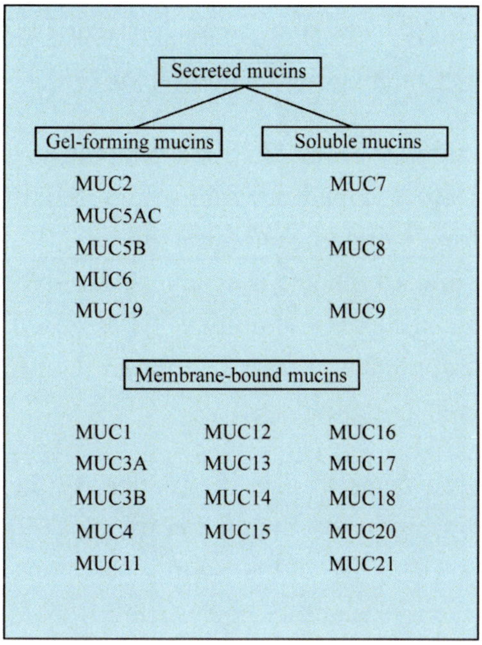

Figure 1: Mucin glycoprotein classification (based on the data reported in [17-23]).

The problems with mucin classification are associated not only with the interpretation of the TR and O-glycosylation patterns [26-29], but also with the ambiguity of some domains found in mucins. For instance, several domains considered typical for mucins – SEA, VWC, VWD, CYS, EGF, NIDO and AMOP – are found in non-mucin proteins as well. On the other hand, some mucins do not contain domain(s) present in other members of the mucin family [30-34]. The evolutionary histories of the domains found in mucins are not clear in many cases. Some mucins have evolved from different ancestors while others have a common progenitor [35], making it difficult in some cases to explain the presence or absence of a particular domain in a particular mucin molecule. For example, the membrane-bound MUC4 mucin has no SEA domain, which is characteristic of all other membrane-tethered mucins. One could assume that *MUC4* gene originated from the SEA domain-containing ancestor common to all membrane-bound mucins, but in the course of evolution the domain was lost by the *MUC4* gene. On the other hand, the absence of the SEA domain in the *MUC4* gene may indicate that this mucin originated from a unique ancestor not common to other membrane-bound mucins. This possibility is strengthened by the finding that MUC4 mucin has EGF, NIDO, AMOP and VWD domains not present in MUC1, MUC16 and other membrane-bound mucins [35].

As pointed out by Duraisamy *et al.* [35], "in contrast to most protein families, *MUC* family members are grouped according to a biophysical structure rather than having evolved from common ancestral genes". These and other authors emphasize that "classifying all mucins in one gene family is not justified because of the lack of common sequence homology" [17, 35]. Duraisamy *et al.* [35] suggested an alternative approach to classification of the *MUC* family: phylogenic analysis based on sequence homology.

Several attempts had been made to classify mucins. In the HUGO classification [36], 17 proteins were assigned to mucin family (*MUC* genes) [13], although, according to Lang *et al.* [19], "this family contains proteins that differ considerably". The Mucin Database constructed by Mucin Biology Group (University of Gothenberg) [37] contains 21 mucin genes identified in human, most but not all of which were also found in mouse and rat, and some in chicken. This database also includes genes coding for mucins isolated from fish (*Takifugu* and *Zebrafish*), frog (*Xenopus tropicalis*), fly (*Drosophila melanogaster*), ascidian (*Cione intestinalis*), lancelet (*Branchiostoma floridae),* worm (*Ceanorhabditis elegans*), and sea urchin (*Strongylocentrotus purpuratus*). It should be noted that although this classification takes into consideration most of the known mucin genes, it does not contain mucin and "mucin-like" genes discovered in recent years, including mucins identified in yeast (*Saccharomyces cerevisiae*) [38] and in protozoan and metazoan parasites [14, 39, 40].

In summary, more studies based on phylogenic analysis of nucleotide and amino acid sequences as well as comparison of domain composition and structural and functional pecularities are needed to better understand the origin of the mucin genes and to create an appropriate classification for this group of glycoproteins.

REFERENCES

[1] Desseyn JL. Mucin CYS domains are ancient and highly conserved modules that evolved in concert. Mol Phylogenet Evol 2009;52:284-92.
[2] Jakowska S. Mucus secretion in fish--a note. Ann N Y Acad Sci 1963;106:458-62.
[3] Negus VE. The function of mucus. Acta Otolaryngol 1963;56:204-14.

[4] Ingram RH, Jr. Mechanical aids to lung expansion. Am Rev Respir Dis 1980;122:23-4.
[5] Ellis RJ. Protein transport across membranes: an introduction. Biochem Soc Symp 1981:223-34.
[6] Aranishi F, Nakane M. Epidermal proteinases in the European eel. Physiol Zool 1997;70:563-70.
[7] Chong K, Joshi S, Jin LT, Shu-Chien AC. Proteomics profiling of epidermal mucus secretion of a cichlid (Symphysodon aequifasciata) demonstrating parental care behavior. Proteomics 2006;6:2251-8.
[8] Bragadeeswaran S, Priyadharshini S, Prabhu K, Rani SR. Antimicrobial and hemolytic activity of fish epidermal mucus Cynoglossus arel and Arius caelatus. Asian Pac J Trop Med 4:305-9.
[9] Rogan MP, Geraghty P, Greene CM, *et al.* Antimicrobial proteins and polypeptides in pulmonary innate defence. Respir Res 2006;7:29.
[10] Thornton DJ, Rousseau K, McGuckin MA. Structure and function of the polymeric mucins in airways mucus. Annu Rev Physiol 2008;70:459-86.
[11] Cone RA. Barrier properties of mucus. Adv Drug Deliv Rev 2009;61:75-85.
[12] Tarran R, Grubb BR, Gatzy JT, Davis CW, Boucher RC. The relative roles of passive surface forces and active ion transport in the modulation of airway surface liquid volume and composition. J Gen Physiol 2001;118:223-36.
[13] Thornton DJ, Sheehan JK. From mucins to mucus: toward a more coherent understanding of this essential barrier. Proc Am Thorac Soc 2004;1:54-61.
[14] Theodoropoulos G, Hicks SJ, Corfield AP, Miller BG, Carrington SD. The role of mucins in host-parasite interactions: Part II - helminth parasites. Trends Parasitol 2001;17:130-5.
[15] Andrianifahanana M, Moniaux N, Batra SK. Regulation of mucin expression: mechanistic aspects and implications for cancer and inflammatory diseases. Biochim Biophys Acta 2006;1765:189-222.
[16] Desseyn JL, Tetaert D, Gouyer V. Architecture of the large membrane-bound mucins. Gene 2008;410:215-22.
[17] Dekker J, Rossen JW, Buller HA, Einerhand AW. The MUC family: an obituary. Trends Biochem Sci 2002;27:126-31.
[18] Desseyn JL, Aubert JP, Porchet N, Laine A. Evolution of the large secreted gel-forming mucins. Mol Biol Evol 2000;17:1175-84.
[19] Lang T, Hansson GC, Samuelsson T. Gel-forming mucins appeared early in metazoan evolution. Proc Natl Acad Sci U S A 2007;104:16209-14.
[20] Moniaux N, Escande F, Porchet N, Aubert JP, Batra SK. Structural organization and classification of the human mucin genes. Front Biosci 2001;6:D1192-206.
[21] Porchet N, Aubert JP. [MUC genes: mucin or not mucin? That is the question]. Med Sci (Paris) 2004;20:569-74.
[22] Rose MC, Voynow JA. Respiratory tract mucin genes and mucin glycoproteins in health and disease. Physiol Rev 2006;86:245-78.
[23] Chaturvedi P, Singh AP, Batra SK. Structure, evolution, and biology of the MUC4 mucin. Faseb J 2008;22:966-81.
[24] Strous GJ, Dekker J. Mucin-type glycoproteins. Crit Rev Biochem Mol Biol 1992;27:57-92.
[25] Gendler SJ, Spicer AP. Epithelial mucin genes. Annu Rev Physiol 1995;57:607-34.
[26] Meezaman D, Charles P, Daskal E, *et al.* Cloning and analysis of cDNA encoding a major airway glycoprotein, human tracheobronchial mucin (MUC5). J Biol Chem 1994;269:12932-9.

[27] Gum JR, Jr., Ho JJ, Pratt WS, *et al.* MUC3 human intestinal mucin. Analysis of gene structure, the carboxyl terminus, and a novel upstream repetitive region. J Biol Chem 1997;272:26678-86.

[28] Higuchi T, Orita T, Nakanishi S, *et al.* Molecular cloning, genomic structure, and expression analysis of MUC20, a novel mucin protein, up-regulated in injured kidney. J Biol Chem 2004;279:1968-79.

[29] Nollet S, Moniaux N, Maury J, *et al.* Human mucin gene MUC4: organization of its 5'-region and polymorphism of its central tandem repeat array. Biochem J 1998;332 (Pt 3):739-48.

[30] Hollingsworth MA, Swanson BJ. Mucins in cancer: protection and control of the cell surface. Nat Rev Cancer 2004;4:45-60.

[31] Macao B, Johansson DG, Hansson GC, Hard T. Autoproteolysis coupled to protein folding in the SEA domain of the membrane-bound MUC1 mucin. Nat Struct Mol Biol 2006;13:71-6.

[32] Perez-Vilar J, Hill RL. The structure and assembly of secreted mucins. J Biol Chem 1999;274:31751-4.

[33] Rousseau K, Byrne C, Kim YS, *et al.* The complete genomic organization of the human MUC6 and MUC2 mucin genes. Genomics 2004;83:936-9.

[34] Brunelli R, Papi M, Arcovito G, *et al.* Globular structure of human ovulatory cervical mucus. Faseb J 2007;21:3872-6.

[35] Duraisamy S, Ramasamy S, Kharbanda S, Kufe D. Distinct evolution of the human carcinoma-associated transmembrane mucins, MUC1, MUC4 AND MUC16. Gene 2006;373:28-34.

[36] HUGO classification, www.genenames.org.

[37] Mucin Database, http://www.medkem.gu.se/mucinbiology/databases/.

[38] Cullen PJ. Signaling mucins: the new kids on the MAPK block. Crit Rev Eukaryot Gene Expr 2007;17:241-57.

[39] Hicks SJ, Theodoropoulos G, Carrington SD, Corfield AP. The role of mucins in host-parasite interactions. Part I-protozoan parasites. Parasitol Today 2000;16:476-81.

[40] Roger E, Gourbal B, Grunau C, *et al.* Expression analysis of highly polymorphic mucin proteins (Sm PoMuc) from the parasite Schistosoma mansoni. Mol Biochem Parasitol 2008;157:217-27.

Send Orders of Reprints at reprints@benthamscience.net
Mucins – Potential Regulators of Cell Functions, 2013, 11-28

CHAPTER 2

Secreted and Membrane-Bound Mucins: Similarities and Differences

Abstract: Two main subfamilies of the mucin glycoproteins have been identified: secreted and membrane-bound. The secreted mucins can be further divided into insoluble gel-forming mucins, including MUC2, MUC5AC, MUC5B, MUC6 and MUC19, and soluble mucins, including MUC7, MUC8 and MUC9. Evolutionary studies showed that the gel-forming mucins are more ancient than the membrane-bound mucins. The evolutionary separation of these two subfamilies is partially reflected in the chromosomal localization of the genes encoding each of mucins. The differences between secreted and membrane-bound mucins are also reflected in the composition of their structural domains, in biosynthesis of their precursors and in posttranslational modifications. Despite some differences, the common features of mucin glycoproteins, such as the structure of the mucin specific domain with its tandem repeats and associated functions, relate them to the same protein family.

Keywords: Mucins, gel-forming, soluble, membrane-bound, evolution, structure, biosynthesis, proteolytic modification.

2.1. MUCIN GENES

Based on their evolution, structure, biosynthesis, cell topology and functions, mucins were divided into two main groups: secreted and membrane-bound. The secreted mucins can be further classified as gel-forming or soluble (non-gel-forming) [1-9]. The gel-forming mucins are large glycoproteins encoded by the *MUC2, MUC5AC, MUC5B, MUC6* and *MUC19* genes [9]; the soluble non-gel-forming mucins – only three identified to date – are encoded by the *MUC7, MUC8* and *MUC9* genes [10]. The group of membrane-bound mucins includes glycoproteins produced by the *MUC1, MUC3A, MUC3B, MUC4, MUC11-18, MUC20 and MUC21* genes [11-13]. Mucins are multifunctional proteins that are involved in defense shields, cell communication network and signal transduction systems. Being important constituents of saliva, they play a special role in speech [4, 9, 11, 13].

2.2. EVOLUTION OF SECRETED AND MEMBRANE-BOUND MUCINS

The different evolutionary ages of the gel-forming and membrane-bound mucins suggest different evolutionary histories. While the gel-forming mucins have been

Joseph Z. Zaretsky and Daniel H. Wreschner

found in very primitive life systems, membrane-bound mucins occurred late in evolution and are observed only in vertebrates. Moreover, expression of a relatively young gene, *MUC1,* is detected only in mammals [5, 14]. According to the current view, the origin of gel-forming mucins can be traced to lower *Metazoa* such as *N. vectensis* [5]. Gel-forming mucins and mucin-related proteins were identified in a variety of organisms belonging to different evolutionary branches: in lower animals such as *C. intestinalis, B. floridae* and *S. purpurants* (*Chordata* class), as well as in vertebrates including insects, fishes, amphibians, birds, and mammals [5]. The evolutionary distribution of the gel-forming and membrane-bound mucin glycoproteins is illustrated in Fig. **1**.

As it emerges from the Fig. **1**, gel-forming mucin related proteins are found in organisms with radial symmetry such as *N. Vectensis* (sea anemone), in the members of Ehinoderma such as *S. Purpuratus* (sea urchin) and in the members of Cephalo- and Urochordata [examples *C. intestinalis* (tunicate) and *B. floridae* (lancelot), respectively]. Gel-forming mucins MUC2 and MUC5 have been found in Actinopterygii (*D.Rerio* (fish)), Amphibia [*X. Tropicalis* (frog)), Aves [*G. galus* (chicken)) and Mammalia (*Homo sapiens* (human being) and *Mus Musculus* (rodent)). The MUC6 mucin was also detected in Amphibia (*X. Tropicalis* (frog)), Aves (*G. galus* (chicken)) and Mammalia (*Homo sapiens* (human being) and *Mus Musculus* (rodent)), but not in Actinopterygii, while gel-forming mucin MUC19 has been found till now only in Mammalia (*Homo sapiens* (human being) and *Mus Musculus* (rodent)). It has to be noted that MUC19 was identified only recently and more bioinformatics and experimental research are needed to study its association with the representatives of different evolutionary branches.

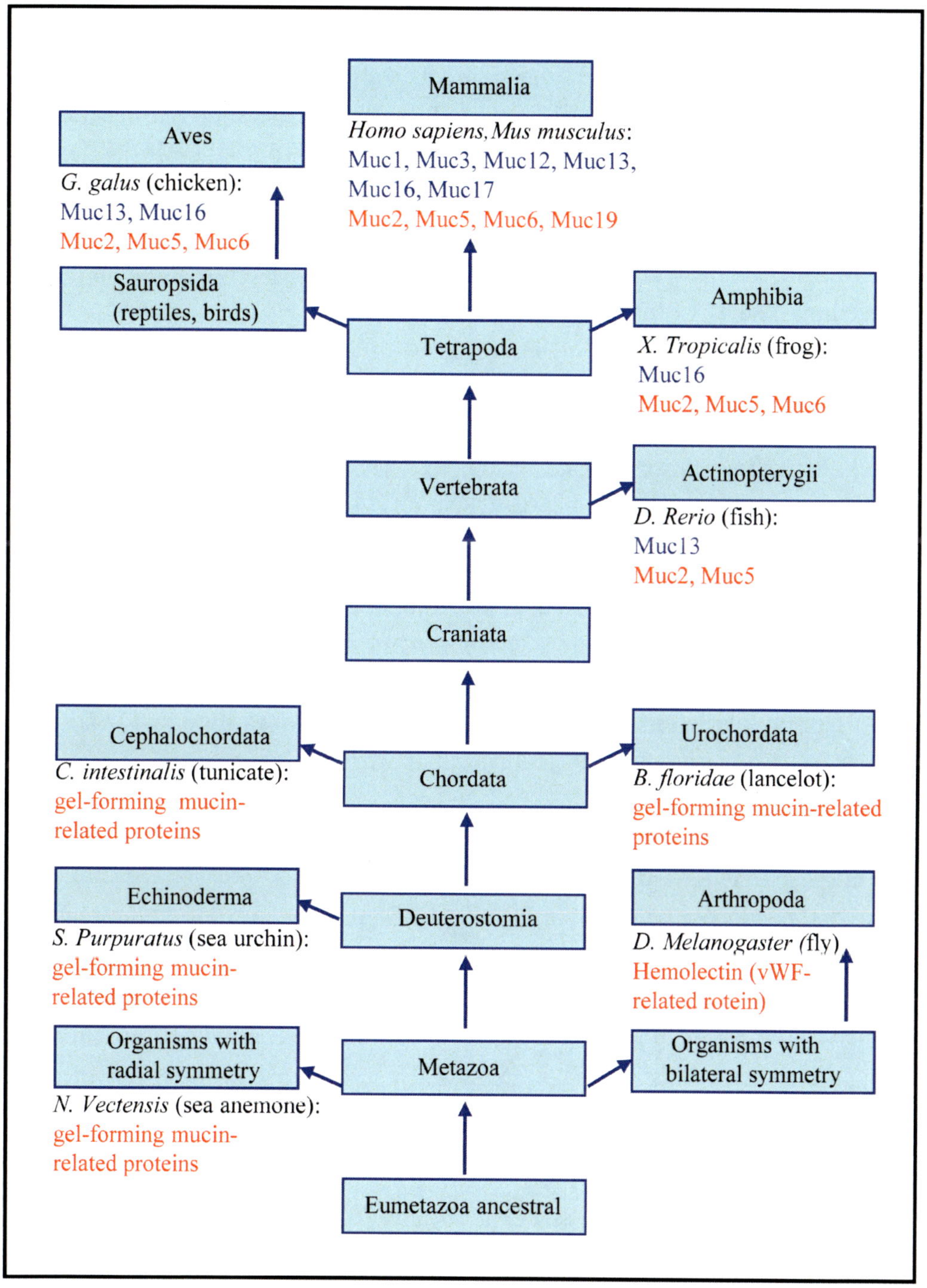

Figure 1: Evolution of the gel-forming and membrane-bound mucins (based on the data reported in [1-9, 14]).

The membrane-bound mucins, which appeared later in evolution than gel-forming mucins, are a heterogeneous collection of subgroups with different genetic backgrounds. The MUC1 mucin is found only in mammals, although genetically it demonstrates connection to the gel-forming mucin MUC5B detected at the early stages of evolution [14]. Duraisamy *et al.* [14] established that the *HSPG2* gene, encoding a large single-chain polypeptide found in mammals, amphibians, fishes, insects, worms and sea urchins, is an ancestor of a cluster of membrane-bound mucin-coding genes including *MUC1, MUC3, MUC12, MUC13* and *MUC17. MUC3* evolved from *MUC13* and branched to *MUC12* and *MUC17*. The MUC4 mucin has two evolutionary ancestors: one is shared with the nidogen protein and the other one is genetically close to the Sushi-domain-containing protein [14]. These ancestors differ from those of *MUC1* and *MUC16*. The *MUC16* evolved from the *agrin* gene and represents a member of a separate evolutionary group [14].

The evolutionary histories of the gel-forming and membrane-bound mucins are reflected in their different molecular and biochemical properties. For instance, alternative splicing, a relatively new evolutionary mechanism ensuring biological diversity, is actively utilized for production of the membrane-bound mucins [15-27] but is rarely exploited by the more ancient gel-forming mucins [23-25]. The following is a review of other features that differentiate the gel-forming and membrane-bound mucins.

2.3. CHROMOSOMAL CLUSTERING OF MUCIN GENES

Genetic analysis has shown that at least some of the gel-forming and membrane-bound mucin genes tend to cluster distribution [28-30]. For example, four of the five gel-forming genes – *MUC6, MUC2, MUC5AC* and *MUC5B* – are clustered on human chromosome 11p15 (Fig. **2**). Their mouse orthologs tend to cluster on the syntenic mouse chromosome 7 F5 [28-30]. In contrast to these mucin genes, *MUC19*, the fifth member of the gel-forming mucin group, is not part of 11p15 cluster: in human, it is located on chromosome 12q12, and in mouse on chromosome 15 E3 [30-32].

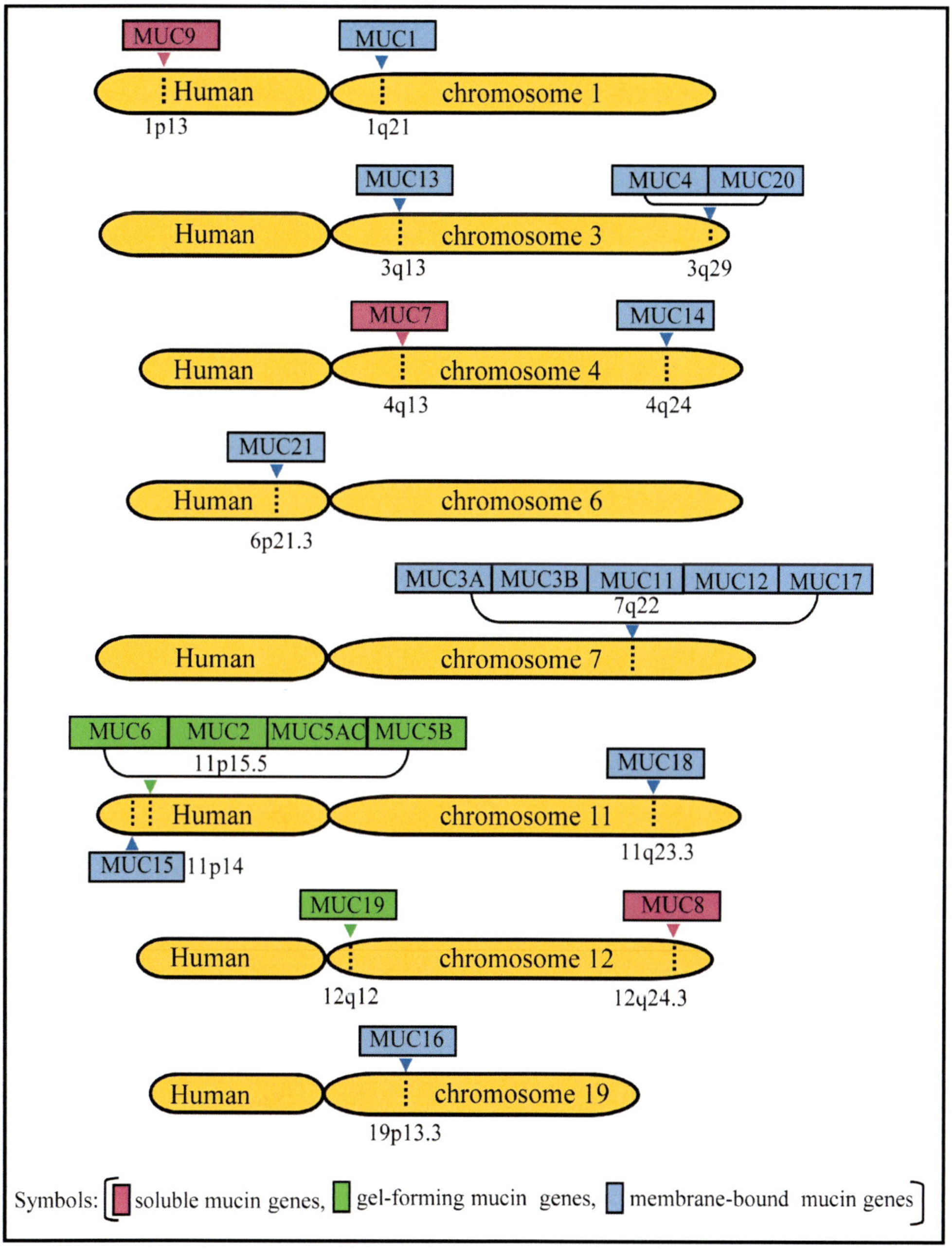

Figure 2: Chromosomal localization of the mucin genes (based on data extracted from [28-30, 33-37]).

Two conserved clusters of membrane-bound mucin loci are present in human and mouse. The first one, comprised of *MUC3A, MUC3B, MUC11, MUC12* and

MUC17, is located on human chromosome 7q22 [33-37]. Among these human genes, only *MUC3* has an ortholog in mice, *Muc3*, which was identified at syntenic mouse chromosome 5G2. Interestingly, the mouse *Muc3* exhibits a higher degree of sequence homology with the rat *Muc3* and chimpanzee *Muc17* genes than with the human genes *MUC3A, MUC3B, MUC11 and MUC12* that constitute one cluster [10]. Two other membrane-bound mucin genes, *MUC4* and *MUC20*, make up the second cluster found in human on chromosome 3q29 and in mouse on the syntenic chromosome 16B2 [10, 26, 38]. The rest of the membrane-bound mucin genes – *MUC1, MUC13, MUC14, MUC15, MUC16, MUC18* and *MUC21* – are distributed separately on human chromosomes 1q21, 3q13.3, 4q24, 11p14.3, 19q13.2, 11q23.3 and 6, respectively. Their mouse orthologs are located on syntenic mouse chromosomes 3F1, 16B2, 3G3, 2E3, 9A2 and 9A5.2, respectively [10, 39-43]. *MUC17* was found on chromosome 6 in human, although its precise location on the chromosome has not been established and its mouse ortholog has not been detected.

No cluster distribution was found in the soluble mucin genes. The genes comprising this group, *MUC7, MUC8* and *MUC9*, are located on chromosomes 4q13.3, 12q24.3 and 1p13, respectively [44-46].

The cluster distribution of some of the mucin genes may indicate a common evolutionary history and/or common mechanisms of transcriptional regulation. Dispersion of other mucin genes between different chromosomes may reflect non-related origin of these genes.

2.4. STRUCTURE OF MUCIN GLYCOPROTEINS

A mucin glycoprotein is made up of 15-20% polypeptide component and 80% carbohydrates, mostly O-linked glycans [1, 47]. N-glycans also appear in all the mucins studied, although in much smaller proportions [48-52]. Structurally, the polypeptide backbone of a mucin molecule can be divided into three regions: N-terminal, central and C-terminal. The N- and C-terminal regions are sparsely O-glycosylated, while the central region is heavily glycosylated with O-glycans attached to the serine and threonine residues, the most abundant amino acids of the tandemly repeated (TR) sequences [47]. The central region is also called "mucin" domain.

2.4.1. Structure of Gel-Forming Mucins

The N- and C-terminal regions of gel-forming and membrane-bound mucins have major structural differences. The N-terminal region of the gel-forming mucins contain cysteine-rich D1-, D2-, D'- and D3-domains, similar to the corresponding domains of von Willebrand factor (Fig. **3**).

The C-terminal region of the gel-forming mucins contains a cystine-knot (CK) domain [47, 53-58]. Although D-domains and CK-domain are the main sources of cysteine residues involved in the formation of disulfide bonds, some of the gel-forming mucins also contain "multiple copies of "naked" cysteine-enriched domains (CYS-domains) that interrupt or are adjacent to the mucin domains" [59, 60]. The cysteine residues of the CK- domain are involved in dimerization of the monomeric mucin molecules by forming disulfide-bond linkage between monomers; the cysteine residues of D1-D3 domains participate in the next stage of mucin polymerization [47]. Interestingly, the CYS- domains are implicated in reversible mucin-mucin interactions that play a central role in changing mucus viscoelastic properties by transforming a globular mucin structure into a fibrous web. The changing of physico-chemical properties of cervical mucus during the menstrual cycle is an example of such transition mediated by MUC5B gel-forming mucin [61].

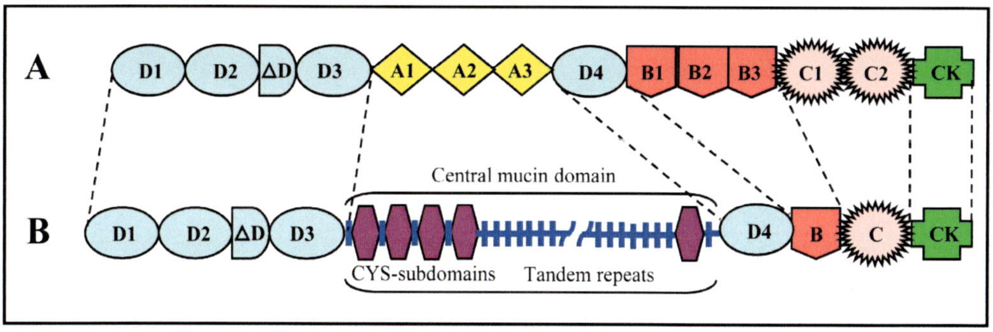

Figure 3: Domain structure of a pro-gel-forming mucin and von Willebrand factor (data extracted from [47, 53-58]; A – von Willebrand factor, B- pro-gel-forming mucin).

The presence of the CK structure in the gel-forming mucins links them to a superfamily of proteins containing cystine-knot. This superfamily includes numerous proteins, including TGFβ, PDGF-like proteins, glycoprotein hormones,

Norrie disease protein, von Willebrand factor, bone morphogenetic protein antagonists, slit-like protein, *etc.* As noted by Vitt *et al.* [58], "phylogenic analysis revealed the ancient evolution of these proteins and the relationship between hormones (*e.g.* TGFβ) and extracellular matrix proteins (*e.g.* mucins). The cystine-knots are absent in the unicellular yeast genome but present in nematode, fly, and higher species, indicating that the cystine knot structure evolved in extracellular signaling molecules of multicellular organisms". The presence of CK in gel-forming mucin molecules suggests that these mucins may fulfill signaling functions in addition to participation in mechanical defense and lubrication.

In contrast to gel-forming mucins, membrane-bound mucins do not contain cysteine-rich D- and CYS-domains or CK motifs, which probably accounts for their not forming multimers as gel-forming mucins do. Of note, all membrane-bound mucins except MUC1 and MUC16 contain cysteine-rich EGF-domains, which mean they could potentially form disulfide-bond linkage. In addition to EGF-domains, MUC4 mucin contains a cysteine-rich AMOP-domain. The ability of these domains to create disulfide-bridges is sufficient to form only homo- and/or hetherodimers but not multimers [62, 63].

The central part of a gel-forming mucin is defined as the "mucin domain" (Fig. **3**). It consists of one or more tandem repeat-containing regions, flanked by and/or interspersed with the "naked" CYS-subdomains [7, 64-68]. The tandem repeats are abundant in proline, threonine and serine residues, giving this domain the name of PTS-domain. Because the number of tandem repeats in a mucin domain is a matter of allele-specific polymorphism, this region is also called VNTR (variable number of tandem repeats) domain. Many studies have shown that serine and threonine residues of the PTS-domain are heavily O-glycosylated [69-71]. The impact of O-glycans on the physico-chemical properties and functions of mucins is difficult to overestimate: it extends and stiffens the mucin molecule [72], resulting in the large volume mucin occupies in solution, which is important for the formation of the defensive mucus gel [13, 73].

2.4.2. Structure of the Membrane-Bound Mucins

While D-, PTS- and CK-domains can be considered the hallmarks of the gel-forming mucins, the PTS-, SEA-, TM- (transmembrane) and CT- (cytoplasmic) domains are the hallmarks of the membrane-bound mucins. These domains facilitate multiple specific functions.

Structurally, the most simply organized membrane-bound mucin is MUC1, which contains canonical PTS-, SEA-, TM- and CT-domains. Qualitatively, but not quantitatively, the same domain composition is observed in MUC16, but while the MUC1 molecule contains only one SEA-domain, the MUC16 glycoprotein contains 16 such modules [41, 74]. The precise domain structure of other membrane-bound mucins will be discussed in the second volume of this series. We describe here only the general structure of these proteins.

The PTS-domain of a membrane-bound mucin exhibits the same biochemical characteristics as the corresponding domain of the gel-forming mucins. However, although this domain is a part of membrane-bound molecule, in some membrane-bound mucins (*e.g.* MUC1, MUC4), it can also be secreted into medium by shedding the extracellular part of the molecule (α-subunit) that is non-covalently bound to the rest of membrane-tethered mucin molecule (β-subunit) [2, 75].

The uniqueness of the SEA-domain is its ability to undergo auto-proteolysis. As pointed out by Cone [76], "the SEA-domain appears to have evolved to break apart in response to mechanical stress, shedding the mucin without disrupting the membrane". The SEA self-cleave generates two non-equal subunits, α and β, which form a heterodimer by non-covalent bonds [20, 75] that supply a well-regulated tensile breaking strength [76]. As a heterodimer, the subunits fulfill important and complex functions of communication between extra- and intra-cellular compartments. However, after separation of the two subunits and shedding of α-subunit, they may function autonomously. Two hydrophobic regions on the external surface of the SEA-domain that may interact with other proteins or other hydrophobic molecules [76] further increase the multifunctionality of the originally synthesized molecule.

The ectodomain (α-subunit) of the membrane-tethered mucin is responsible mainly for extracellular functions [14, 77, 78], whereas the β-subunit, consisting mainly of transmembrane and cytoplasmic domains, is an important player in intracellular processes, where it functions as a scaffold for various signaling molecules and regulators [11, 20, 79]. The cytoplasmic domain of the membrane-bound mucins contains a number of potential tyrosine and serine phosphorylation sites. By phosphorylating specific sites, the cytoplasmic domain acquires the ability to interact with different molecules and become an active participant in basic cell processes [11, 20, 80]. The cytoplasmic domain of a membrane-bound mucin appears to be a crossroads of numerous intracellular signaling pathways. The ability of membrane-bound mucins to connect different pathways and direct signaling to specific molecular substrates is yet another indication of the enormous multifunctionality of these molecules.

2.5. SIMILARITIES AND DIFFERENCES IN BIOSYNTHESIS OF GEL-FORMING AND MEMBRANE-BOUND MUCINS

2.5.1. Biosynthesis of Mucin Polypeptide Precursors

Biosynthesis of both gel-forming and membrane-bound mucins occurs on polyribosomes in the endoplasmic reticulum. Early studies established that less than 1 minute is necessary for synthesis of the full length polypeptide precursors of the gel-forming mucins [69, 81]. The dimers of the MUC2, MUC5AC and MUC6 mucins occur in the rough endoplasmic reticulum within the first 30 min of biosynthesis, whereas the first dimmers of the MUC5B mucin take about 4 hours to form [51, 52, 69]. The next stage of the gel-forming mucin biosynthesis, N-glycosylation, is likely to occur co-translationally [81]. N-glycans were shown to be necessary for efficient oligomerization [51, 52]. Moreover, the N-glycans expressed on mucin molecules interact with chaperones, calnexin and calreticulin, which modulate mucin biosynthesis during the folding and oligomerization stages in the endoplasmic reticulum [82, 83]. O-glycosylation of the mucin precursor occurs within the first hour of protein core biosynthesis after apomucin dimerization. It has been shown that N-glycosylation, C-mannosylation, folding, and dimerization of the primary peptide occur in endoplasmic reticulum, while O-glycosylation, sulphation, oligomerization and proteolysis of immature mucin molecules take place during their transit through the Golgi complex. Importantly, apomucins lacking N-glycans are degraded in the endoplasmic reticulum and cannot be transported to the Golgi apparatus [82, 84].

Like in the gel-forming mucins, the membrane-bound mucin apoproteins are also synthesized rapidly within the first several minutes of biosynthesis. The nascent polypeptide chain undergoes co-translational self-cleavage followed by non-covalent binding of the N-terminal α-subunit and the C-terminal β-subunit of the peptide precursor. In all membrane-bound mucins except the MUC4, the auto-cleavage proteolysis takes place at the SEA-domain [75, 85-88]. Interestingly, N-glycosylation plays an important role in this reaction, evidenced by the finding that N-glycosylation pattern of the SEA-domain of the rodent Muc3 constitutes a control point for modulation of the proteolytic cleavage targeted to the SEA-module [49]. After the relatively short phase of apomucin biosynthesis and initial structural rearrangements, both gel-forming and membrane-bound mucins undergo a relatively long process of protein core glycosylation.

2.5.2. Glycosylation of Mucin Polypeptide Backbone

Despite the structural differences between secreted and membrane-bound mucins, they share the following common features in the glycosylation of their core backbones. N-linked core glycosylation is the first event in the addition of sugar moieties to polypeptide chains, a process that occurs co-translationally in the lumen of the endoplasmic reticulum. The sequence, Asn-X-Ser/Thr, where X can be any amino acid except proline [82, 89], is the N-glycosylation target common for the gel-forming and membrane-bound mucins. N-glycans are responsible for appropriate folding of the newly synthesized polypeptide molecules. In the membrane-bound mucins, N-glycans also control apical cell membrane targeting [50]. In secreted gel-forming mucins, N-linked oligosaccharides play a role in the disulphide-dependent dimerization [51, 52, 69, 90]. Usually, the N-glycan structures of membrane-bound mucins are represented by the sialylated hybrid-type N-glycans [89, 91, 92]. The chemical content of N-glycosides of secreted mucins has not been fully identified [4].

O-glycosylation of serine and threonine residues is one of the characteristic steps in biosynthesis of both the secreted and membrane-bound mucins. The incorporation of O-linked oligosaccharides into secreted gel-forming mucins begins after the completion of N-glycosylation and dimer formations [51, 52, 82, 93]. Initial O-glycosylation of gel-forming mucins takes place in the *cis*-Golgi compartments [94, 95]. The main O-glycosylation process occurs in the *medial*-Golgi and comes to completion in *trans*-Golgi cisterns [93, 96]. The fully glycosylated, mature mucin molecules are stored in the secretory granules for further release into extracellular environment in response to mucin secretagogues [4, 97].

In membrane-bound mucins, the initial O-glycosylation occurs in *trans*-Golgi. Then, partially glycosylated molecules shuttle between the Golgi compartments and the cell membrane until O-glycosylation is completed by addition of sialic acid residues. The resultant mature mucin molecule is then re-translocated to the cell apical membrane where it fulfills its native receptor function [98, 99].

2.5.3. Proteolytic Modifications of Gel-Forming and Membrane-Bound Mucins

One of the important steps in biosynthesis of both the secreted and membrane-bound mucins is proteolytic modification of immature precursors. However, the nature of the proteolytic reactions involved in processing and maturation of secreted and membrane-tethered mucins is different. Most of the membrane-bound mucins undergo auto-catalytic proteolysis in the SEA-domain that yields two unequal subunits, which, as noted above, develop heterodimers by non-covalent bonds [75, 86, 87]. In addition to auto-cleavage, the membrane-bound mucins undergo proteolysis by metallo-proteases and γ-secretase that results in release of the ectodomain and mobilization of the cytoplasmic tail, respectively. These proteolytic reactions are crucial for activation of the membrane-bound mucin-mediated signal transduction functions [100, 101].

Proteolytic events involved in processing the gel-forming mucins are associated with N- and C-terminal regions of the mucin molecules. The C-terminal cleavage is formed by an auto-catalytic mechanism triggered by low pH in the late secretory pathway. The cleavage site is located in the D4-domain between the Asp and Pro residues in the GD^PH sequence [102]. The cleavage produces a new C-terminus that has the potential to link the cleaved mucin to other proteins [102, 103]. Remarkably, the GDPH sequence is found not only in the gel-forming mucins but in other mucins and non-mucin proteins as well [104, 105]; on the other hand, of the gel-forming mucins, only MUC2 and MUC5AC, but not MUC5B and MUC6, have the GDPH sequence at their C-terminal regions [102]. Nevertheless, although the GDPH site was not found in the MUC5B protein, the C-terminal cleavage of this mucin was also described [106].

It appears that the GDPH sequence-associated cleavage is an important step in the processing of many proteins. Notably, the GDPH sequences are found in human and rat MUC4 mucins, the only membrane-bound mucins that do not have the SEA domain responsible for auto-cleavage in other membrane-bound mucins. It is likely that in the absence of the SEA-domain, its function is carried out by the GDPH sequence. This possibility is strengthened by the finding that two subunits of the MUC4 mucin were produced by auto-cleavage of the Asp-Pro bond at the GD^PH site [107, 108]. Thus, proteolytic modifications of gel-forming mucins and membrane-bound mucins have some features in common. Because the gel-forming mucins are more ancient than membrane-bound mucins in evolution terms [5, 14], cleavage at the GDPH site appears to be an older mechanism than SEA-domain mediated proteolysis.

As noted above, the N-terminally located D1-D3 domains of MUC2, MUC5AC and MUC5B mucins display a high degree of similarity with corresponding domains of the von Willibrand factor [53, 109, 110]. This similarity is reflected also in proteolytic processing of the two entities associated with the N-terminal region. Some gel-forming mucins undergo proteolytic cleavage at the N-terminal D'-domain followed by a second cleavage at the D3-domain [109, 111].

The analysis of the main structural features and potential functions of the gel-forming and membrane-bound mucins further attests to their multistructurality and multifunctionality. This chapter focused on the main structural and functional aspects of gel-forming and membrane-bound mucins The next chapters of this volume will focus on the secreted mucins, with special attention to their potentials as regulators of cell functions.

REFERENCES

[1] Gendler SJ, Spicer AP. Epithelial mucin genes. Annu Rev Physiol 1995;57:607-34.

[2] Carraway KL, Price-Schiavi SA, Komatsu M, *et al.* Muc4/sialomucin complex in the mammary gland and breast cancer. J Mammary Gland Biol Neoplasia 2001;6:323-37.

[3] Desseyn JL, Tetaert D, Gouyer V. Architecture of the large membrane-bound mucins. Gene 2008;410:215-22.

[4] Rose MC, Voynow JA. Respiratory tract mucin genes and mucin glycoproteins in health and disease. Physiol Rev 2006;86:245-78.

[5] Lang T, Hansson GC, Samuelsson T. Gel-forming mucins appeared early in metazoan evolution. Proc Natl Acad Sci USA 2007;104:16209-14.

[6] Carraway KL, 3rd, Funes M, Workman HC, Sweeney C. Contribution of membrane mucins to tumor progression through modulation of cellular growth signaling pathways. Curr Top Dev Biol 2007;78:1-22.

[7] Desseyn JL, Aubert JP, Porchet N, Laine A. Evolution of the large secreted gel-forming mucins. Mol Biol Evol 2000;17:1175-84.
[8] Rousseau K, Byrne C, Kim YS, *et al.* The complete genomic organization of the human MUC6 and MUC2 mucin genes. Genomics 2004;83:936-9.
[9] Voynow JA, Rubin BK. Mucins, mucus, and sputum. Chest 2009;135:505-12.
[10] Williams OW, Sharafkhaneh A, Kim V, Dickey BF, Evans CM. Airway mucus: From production to secretion. Am J Respir Cell Mol Biol 2006;34:527-36.
[11] Hattrup CL, Gendler SJ. Structure and function of the cell surface (tethered) mucins. Annu Rev Physiol 2008;70:431-57.
[12] Crawley SC, Gum JR, Jr., Hicks JW, *et al.* Genomic organization and structure of the 3' region of human MUC3: alternative splicing predicts membrane-bound and soluble forms of the mucin. Biochem Biophys Res Commun 1999;263:728-36.
[13] Thornton DJ, Rousseau K, McGuckin MA. Structure and function of the polymeric mucins in airways mucus. Annu Rev Physiol 2008;70:459-86.
[14] Duraisamy S, Kufe T, Ramasamy S, Kufe D. Evolution of the human MUC1 oncoprotein. Int J Oncol 2007;31:671-7.
[15] Smorodinsky N, Weiss M, Hartmann ML, *et al.* Detection of a secreted MUC1/SEC protein by MUC1 isoform specific monoclonal antibodies. Biochem Biophys Res Commun 1996;228:115-21.
[16] Baruch A, Hartmann M, Yoeli M, *et al.* The breast cancer-associated MUC1 gene generates both a receptor and its cognate binding protein. Cancer Res 1999;59:1552-61.
[17] Wreschner DH, Hareuveni M, Tsarfaty I, *et al.* Human epithelial tumor antigen cDNA sequences. Differential splicing may generate multiple protein forms. Eur J Biochem 1990;189:463-73.
[18] Levitin F, Baruch A, Weiss M, *et al.* A novel protein derived from the MUC1 gene by alternative splicing and frameshifting. J Biol Chem 2005;280:10655-63.
[19] Williams CJ, Wreschner DH, Tanaka A, *et al.* Multiple protein forms of the human breast tumor-associated epithelial membrane antigen (EMA) are generated by alternative splicing and induced by hormonal stimulation. Biochem Biophys Res Commun 1990;170:1331-8.
[20] Chaturvedi P, Singh AP, Batra SK. Structure, evolution, and biology of the MUC4 mucin. Faseb J 2008;22:966-81.
[21] Choudhury A, Moniaux N, Ringel J, *et al.* Alternate splicing at the 3'-end of the human pancreatic tumor-associated mucin MUC4 cDNA. Teratog Carcinog Mutagen 2001;21:83-96.
[22] Escande F, Lemaitre L, Moniaux N, *et al.* Genomic organization of MUC4 mucin gene. Towards the characterization of splice variants. Eur J Biochem 2002;269:3637-44.
[23] Escande F, Porchet N, Aubert JP, Buisine MP. The mouse Muc5b mucin gene: cDNA and genomic structures, chromosomal localization and expression. Biochem J 2002;363:589-98.
[24] Das B, Cash MN, Hand AR, Shivazad A, Culp DJ. Expression of Muc19/Smgc gene products during murine sublingual gland development: cytodifferentiation and maturation of salivary mucous cells. J Histochem Cytochem 2009;57:383-96.
[25] Sternberg LR, Byrd JC, Hansson GC, Liu KF, Bresalier RS. Alternative splicing of the human MUC2 gene. Arch Biochem Biophys 2004;421:21-33.
[26] Higuchi T, Orita T, Nakanishi S, *et al.* Molecular cloning, genomic structure, and expression analysis of MUC20, a novel mucin protein, up-regulated in injured kidney. J Biol Chem 2004;279:1968-79.
[27] Zaretsky JZ, Barnea I, Aylon Y, *et al.* MUC1 gene overexpressed in breast cancer: structure and transcriptional activity of the MUC1 promoter and role of estrogen receptor alpha (ERalpha) in regulation of the MUC1 gene expression. Mol Cancer 2006;5:57.
[28] Escande F, Porchet N, Bernigaud A, *et al.* The mouse secreted gel-forming mucin gene cluster. Biochim Biophys Acta 2004;1676:240-50.

[29] Lang T, Hansson GC, Samuelsson T. An inventory of mucin genes in the chicken genome shows that the mucin domain of Muc13 is encoded by multiple exons and that ovomucin is part of a locus of related gel-forming mucins. BMC Genomics 2006;7:197.
[30] Evans CM, Koo JS. Airway mucus: the good, the bad, the sticky. Pharmacol Ther 2009;121:332-48.
[31] Chen Y, Zhao YH, Kalaslavadi TB, *et al.* Genome-wide search and identification of a novel gel-forming mucin MUC19/Muc19 in glandular tissues. Am J Respir Cell Mol Biol 2004;30:155-65.
[32] Culp DJ, Latchney LR, Fallon MA, *et al.* The gene encoding mouse Muc19: cDNA, genomic organization and relationship to Smgc. Physiol Genomics 2004;19:303-18.
[33] Fox MF, Lahbib F, Pratt W, *et al.* Regional localization of the intestinal mucin gene MUC3 to chromosome 7q22. Ann Hum Genet 1992;56:281-7.
[34] Gum JR, Hicks JW, Swallow DM, *et al.* Molecular cloning of cDNAs derived from a novel human intestinal mucin gene. Biochem Biophys Res Commun 1990;171:407-15.
[35] Gum JR, Jr., Crawley SC, Hicks JW, Szymkowski DE, Kim YS. MUC17, a novel membrane-tethered mucin. Biochem Biophys Res Commun 2002;291:466-75.
[36] Pratt WS, Crawley S, Hicks J, *et al.* Multiple transcripts of MUC3: evidence for two genes, MUC3A and MUC3B. Biochem Biophys Res Commun 2000;275:916-23.
[37] Williams SJ, McGuckin MA, Gotley DC, *et al.* Two novel mucin genes down-regulated in colorectal cancer identified by differential display. Cancer Res 1999;59:4083-9.
[38] Porchet N, Nguyen VC, Dufosse J, *et al.* Molecular cloning and chromosomal localization of a novel human tracheo-bronchial mucin cDNA containing tandemly repeated sequences of 48 base pairs. Biochem Biophys Res Commun 1991;175:414-22.
[39] Middleton-Price H, Gendler S, Malcolm S. Close linkage of PUM and SPTA within chromosome band 1q21. Ann Hum Genet 1988;52:273-8.
[40] Swallow DM, Gendler S, Griffiths B, *et al.* The human tumour-associated epithelial mucins are coded by an expressed hypervariable gene locus PUM. Nature 1987;328:82-4.
[41] Yin BW, Lloyd KO. Molecular cloning of the CA125 ovarian cancer antigen: identification as a new mucin, MUC16. J Biol Chem 2001;276:27371-5.
[42] Itoh Y, Kamata-Sakurai M, Denda-Nagai K, *et al.* Identification and expression of human epiglycanin/MUC21: a novel transmembrane mucin. Glycobiology 2008;18:74-83.
[43] Williams SJ, Wreschner DH, Tran M, *et al.* Muc13, a novel human cell surface mucin expressed by epithelial and hemopoietic cells. J Biol Chem 2001;276:18327-36.
[44] Bobek LA, Tsai H, Biesbrock AR, Levine MJ. Molecular cloning, sequence, and specificity of expression of the gene encoding the low molecular weight human salivary mucin (MUC7). J Biol Chem 1993;268:20563-9.
[45] Shankar V, Pichan P, Eddy RL, Jr., *et al.* Chromosomal localization of a human mucin gene (MUC8) and cloning of the cDNA corresponding to the carboxy terminus. Am J Respir Cell Mol Biol 1997;16:232-41.
[46] Lapensee L, Paquette Y, Bleau G. Allelic polymorphism and chromosomal localization of the human oviductin gene (MUC9). Fertil Steril 1997;68:702-8.
[47] Offner GD, Troxler RF. Heterogeneity of high-molecular-weight human salivary mucins. Adv Dent Res 2000;14:69-75.
[48] Yu SY, Khoo KH, Yang Z, Herp A, Wu AM. Glycomic mapping of O- and N-linked glycans from major rat sublingual mucin. Glycoconj J 2008;25:199-212.
[49] He Y, Li Y, Peng Z, *et al.* Role of N-glycosylation of the SEA module of rodent Muc3 in posttranslational processing of its carboxy-terminal domain. Glycobiology 2009;19:1094-102.
[50] Ho JJ, Jaituni RS, Crawley SC, *et al.* N-glycosylation is required for the surface localization of MUC17 mucin. Int J Oncol 2003;23:585-92.

[51] Asker N, Axelsson MA, Olofsson SO, Hansson GC. Dimerization of the human MUC2 mucin in the endoplasmic reticulum is followed by a N-glycosylation-dependent transfer of the mono- and dimers to the Golgi apparatus. J Biol Chem 1998;273:18857-63.
[52] Asker N, Axelsson MA, Olofsson SO, Hansson GC. Human MUC5AC mucin dimerizes in the rough endoplasmic reticulum, similarly to the MUC2 mucin. Biochem J 1998;335 (Pt 2):381-7.
[53] Gum JR, Jr., Hicks JW, Toribara NW, Siddiki B, Kim YS. Molecular cloning of human intestinal mucin (MUC2) cDNA. Identification of the amino terminus and overall sequence similarity to prepro-von Willebrand factor. J Biol Chem 1994;269:2440-6.
[54] Offner GD, Nunes DP, Keates AC, Afdhal NH, Troxler RF. The amino-terminal sequence of MUC5B contains conserved multifunctional D domains: implications for tissue-specific mucin functions. Biochem Biophys Res Commun 1998;251:350-5.
[55] Keates AC, Nunes DP, Afdhal NH, Troxler RF, Offner GD. Molecular cloning of a major human gall bladder mucin: complete C-terminal sequence and genomic organization of MUC5B. Biochem J 1997;324 (Pt 1):295-303.
[56] Toribara NW, Ho SB, Gum E, *et al.* The carboxyl-terminal sequence of the human secretory mucin, MUC6. Analysis Of the primary amino acid sequence. J Biol Chem 1997;272:16398-403.
[57] Perez-Vilar J, Eckhardt AE, DeLuca A, Hill RL. Porcine submaxillary mucin forms disulfide-linked multimers through its amino-terminal D-domains. J Biol Chem 1998;273:14442-9.
[58] Vitt UA, Hsu SY, Hsueh AJ. Evolution and classification of cystine knot-containing hormones and related extracellular signaling molecules. Mol Endocrinol 2001;15:681-94.
[59] Desseyn JL. Mucin CYS domains are ancient and highly conserved modules that evolved in concert. Mol Phylogenet Evol 2009;52:284-92.
[60] Thornton DJ, Howard M, Khan N, Sheehan JK. Identification of two glycoforms of the MUC5B mucin in human respiratory mucus. Evidence for a cysteine-rich sequence repeated within the molecule. J Biol Chem 1997;272:9561-6.
[61] Brunelli R, Papi M, Arcovito G, *et al.* Globular structure of human ovulatory cervical mucus. Faseb J 2007;21:3872-6.
[62] Ciccarelli FD, Doerks T, Bork P. AMOP, a protein module alternatively spliced in cancer cells. Trends Biochem Sci 2002;27:113-5.
[63] Moniaux N, Nollet S, Porchet N, *et al.* Complete sequence of the human mucin MUC4: a putative cell membrane-associated mucin. Biochem J 1999;338 (Pt 2):325-33.
[64] Allen A, Hutton DA, Pearson JP. The MUC2 gene product: a human intestinal mucin. Int J Biochem Cell Biol 1998;30:797-801.
[65] Toribara NW, Gum JR, Jr., Culhane PJ, *et al.* MUC-2 human small intestinal mucin gene structure. Repeated arrays and polymorphism. J Clin Invest 1991;88:1005-13.
[66] van de Bovenkamp JH, Hau CM, Strous GJ, *et al.* Molecular cloning of human gastric mucin MUC5AC reveals conserved cysteine-rich D-domains and a putative leucine zipper motif. Biochem Biophys Res Commun 1998;245:853-9.
[67] Desseyn JL, Guyonnet-Duperat V, Porchet N, Aubert JP, Laine A. Human mucin gene MUC5B, the 10.7-kb large central exon encodes various alternate subdomains resulting in a super-repeat. Structural evidence for a 11p15.5 gene family. J Biol Chem 1997;272:3168-78.
[68] Desseyn JL, Buisine MP, Porchet N, Aubert JP, Laine A. Genomic organization of the human mucin gene MUC5B. cDNA and genomic sequences upstream of the large central exon. J Biol Chem 1998;273:30157-64.
[69] van Klinken BJ, Einerhand AW, Buller HA, Dekker J. The oligomerization of a family of four genetically clustered human gastrointestinal mucins. Glycobiology 1998;8:67-75.
[70] Dekker J, Strous GJ. Covalent oligomerization of rat gastric mucin occurs in the rough endoplasmic reticulum, is N-glycosylation-dependent, and precedes initial O-glycosylation. J Biol Chem 1990;265:18116-22.

[71] Lamblin G, Degroote S, Perini JM, *et al.* Human airway mucin glycosylation: a combinatory of carbohydrate determinants which vary in cystic fibrosis. Glycoconj J 2001;18:661-84.
[72] Jentoft N. Why are proteins O-glycosylated? Trends Biochem Sci 1990;15:291-4.
[73] Forstner JF, Forstner GG, editors. Physiology of the gastrointestinal tract. New York: Raven Press; 1994.
[74] Maeda T, Inoue M, Koshiba S, *et al.* Solution structure of the SEA domain from the murine homologue of ovarian cancer antigen CA125 (MUC16). J Biol Chem 2004;279:13174-82.
[75] Levitin F, Stern O, Weiss M, *et al.* The MUC1 SEA module is a self-cleaving domain. J Biol Chem 2005;280:33374-86.
[76] Cone RA. Barrier properties of mucus. Adv Drug Deliv Rev 2009;61:75-85.
[77] Rhodes JM, Campbell BJ, Yu LG. Lectin-epithelial interactions in the human colon. Biochem Soc Trans 2008;36:1482-6.
[78] Linden SK, Florin TH, McGuckin MA. Mucin dynamics in intestinal bacterial infection. PLoS One 2008;3:e3952.
[79] Singh PK, Hollingsworth MA. Cell surface-associated mucins in signal transduction. Trends Cell Biol 2006;16:467-76.
[80] Zrihan-Licht S, Baruch A, Elroy-Stein O, Keydar I, Wreschner DH. Tyrosine phosphorylation of the MUC1 breast cancer membrane proteins. Cytokine receptor-like molecules. FEBS Lett 1994;356:130-6.
[81] McGuckin MA, Devine PL, Ward BG. Early steps in the biosynthesis of MUC2 epithelial mucin in colon cancer cells. Biochem Cell Biol 1996;74:87-93.
[82] Andrianifahanana M, Moniaux N, Batra SK. Regulation of mucin expression: mechanistic aspects and implications for cancer and inflammatory diseases. Biochim Biophys Acta 2006;1765:189-222.
[83] McCool DJ, Okada Y, Forstner JF, Forstner GG. Roles of calreticulin and calnexin during mucin synthesis in LS180 and HT29/A1 human colonic adenocarcinoma cells. Biochem J 1999;341 (Pt 3):593-600.
[84] Perez-Vilar J. Mucin granule intraluminal organization. Am J Respir Cell Mol Biol 2007;36:183-90.
[85] Wang R, Khatri IA, Forstner JF. C-terminal domain of rodent intestinal mucin Muc3 is proteolytically cleaved in the endoplasmic reticulum to generate extracellular and membrane components. Biochem J 2002;366:623-31.
[86] Ligtenberg MJ, Kruijshaar L, Buijs F, *et al.* Cell-associated episialin is a complex containing two proteins derived from a common precursor. J Biol Chem 1992;267:6171-7.
[87] Macao B, Johansson DG, Hansson GC, Hard T. Autoproteolysis coupled to protein folding in the SEA domain of the membrane-bound MUC1 mucin. Nat Struct Mol Biol 2006;13:71-6.
[88] Davies JR, Kirkham S, Svitacheva N, Thornton DJ, Carlstedt I. MUC16 is produced in tracheal surface epithelium and submucosal glands and is present in secretions from normal human airway and cultured bronchial epithelial cells. Int J Biochem Cell Biol 2007;39:1943-54.
[89] Parry S, Hanisch FG, Leir SH, *et al.* N-Glycosylation of the MUC1 mucin in epithelial cells and secretions. Glycobiology 2006;16:623-34.
[90] Bell SL, Xu G, Khatri IA, *et al.* N-linked oligosaccharides play a role in disulphide-dependent dimerization of intestinal mucin Muc2. Biochem J 2003;373:893-900.
[91] Matsushita T, Sadamoto R, Ohyabu N, *et al.* Functional neoglycopeptides: synthesis and characterization of a new class of MUC1 glycoprotein models having core 2-based O-glycan and complex-type N-glycan chains. Biochemistry 2009;48:11117-33.
[92] Pallesen LT, Pedersen LR, Petersen TE, Rasmussen JT. Characterization of carbohydrate structures of bovine MUC15 and distribution of the mucin in bovine milk. J Dairy Sci 2007;90:3143-52.

[93] Perez-Vilar J, Hill RL. The structure and assembly of secreted mucins. J Biol Chem 1999;274:31751-4.
[94] Roth J, Wang Y, Eckhardt AE, Hill RL. Subcellular localization of the UDP-N-acetyl-D-galactosamine: polypeptide N-acetylgalactosaminyltransferase-mediated O-glycosylation reaction in the submaxillary gland. Proc Natl Acad Sci U S A 1994;91:8935-9.
[95] Deschuyteneer M, Eckhardt AE, Roth J, Hill RL. The subcellular localization of apomucin and nonreducing terminal N-acetylgalactosamine in porcine submaxillary glands. J Biol Chem 1988;263:2452-9.
[96] Van den Steen P, Rudd PM, Dwek RA, Opdenakker G. Concepts and principles of O-linked glycosylation. Crit Rev Biochem Mol Biol 1998;33:151-208.
[97] Sperber K, Shim J, Mehra M, *et al.* Mucin secretion in inflammatory bowel disease: comparison of a macrophage-derived mucin secretagogue (MMS-68) to conventional secretagogues. Inflamm Bowel Dis 1998;4:12-7.
[98] Pimental RA, Julian J, Gendler SJ, Carson DD. Synthesis and intracellular trafficking of Muc-1 and mucins by polarized mouse uterine epithelial cells. J Biol Chem 1996;271:28128-37.
[99] Litvinov SV, Hilkens J. The epithelial sialomucin, episialin, is sialylated during recycling. J Biol Chem 1993;268:21364-71.
[100] Julian J, Dharmaraj N, Carson DD. MUC1 is a substrate for gamma-secretase. J Cell Biochem 2009;108:802-15.
[101] Thathiah A, Brayman M, Dharmaraj N, *et al.* Tumor necrosis factor alpha stimulates MUC1 synthesis and ectodomain release in a human uterine epithelial cell line. Endocrinology 2004;145:4192-203.
[102] Lidell ME, Johansson ME, Hansson GC. An autocatalytic cleavage in the C terminus of the human MUC2 mucin occurs at the low pH of the late secretory pathway. J Biol Chem 2003;278:13944-51.
[103] Fahim RE, Forstner GG, Forstner JF. Heterogeneity of rat goblet-cell mucin before and after reduction. Biochem J 1983;209:117-24.
[104] Lidell ME, Hansson GC. Cleavage in the GDPH sequence of the C-terminal cysteine-rich part of the human MUC5AC mucin. Biochem J 2006;399:121-9.
[105] Pagani A, Silvestri L, Nai A, Camaschella C. Hemojuvelin N-terminal mutants reach the plasma membrane but do not activate the hepcidin response. Haematologica 2008;93:1466-72.
[106] Wickstrom C, Davies JR, Eriksen GV, Veerman EC, Carlstedt I. MUC5B is a major gel-forming, oligomeric mucin from human salivary gland, respiratory tract and endocervix: identification of glycoforms and C-terminal cleavage. Biochem J 1998;334 (Pt 3):685-93.
[107] Sheng ZQ, Hull SR, Carraway KL. Biosynthesis of the cell surface sialomucin complex of ascites 13762 rat mammary adenocarcinoma cells from a high molecular weight precursor. J Biol Chem 1990;265:8505-10.
[108] Sheng Z, Wu K, Carraway KL, Fregien N. Molecular cloning of the transmembrane component of the 13762 mammary adenocarcinoma sialomucin complex. A new member of the epidermal growth factor superfamily. J Biol Chem 1992;267:16341-6.
[109] Wickstrom C, Carlstedt I. N-terminal cleavage of the salivary MUC5B mucin. Analogy with the Van Willebrand propolypeptide? J Biol Chem 2001;276:47116-21.
[110] Desseyn JL, Aubert JP, Van Seuningen I, Porchet N, Laine A. Genomic organization of the 3' region of the human mucin gene MUC5B. J Biol Chem 1997;272:16873-83.
[111] Thornton DJ, Gray T, Nettesheim P, *et al.* Characterization of mucins from cultured normal human tracheobronchial epithelial cells. Am J Physiol Lung Cell Mol Physiol 2000;278:L1118-28.

CHAPTER 3

Secreted Mucins

Abstract: The group of secreted mucins has eight members: five genes encode the gel-forming mucins (MUC2, MUC5AC, MUC5B, MUC6 and MUC19) and three genes encode the soluble mucins (MUC7, MUC8 and MUC9). Gel-forming mucins share structural and evolutionary features with von Willibrand Factor. Soluble mucins differ from gel-forming mucins in that the former all do not have the von Willibrand Factor specific domains while mucin specific tandem repeat-containing domain is a common feature of all secreted mucins. Genes encoding the soluble mucins are located at different chromosomes, whereas all gel-forming mucin genes except *MUC19* are clustered on chromosome locus 11p15.5 and the *MUC19* gene is located on chromosome 12. Structural aspects and functions of the secreted mucins are discussed.

Keywords: Secreted mucins, classification, MUC2, MUC5AC, MUC5B, MUC6, MUC19, MUC7, MUC8, MUC9, chromosomal localization.

3.1. GENERAL CHARACTERISTICS

The group of secreted mucins contains eight glycoproteins: five gel-forming mucins and three soluble (non-gel-forming) mucins (Fig. **1**).

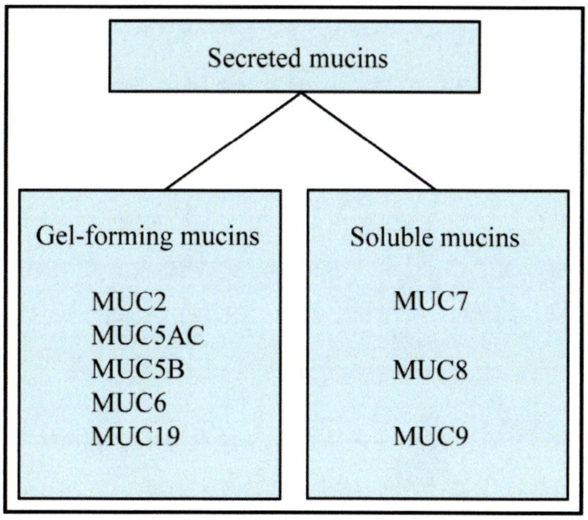

Figure 1: Classification of the secreted mucins (the data reported in [3-12]).

These glycoproteins play various important roles in normal cell physiology and pathology. They form a protective mucus barrier between epithelia and harmful

Joseph Z. Zaretsky and Daniel H. Wreschner

exogenous agents that threaten the lumen of the respiratory, gastrointestinal and genitourinary tracts as well as of the visual and auditory systems [1]. Secreted mucins determine structure and rheological properties of the mucus gel and mucosal fluids. Defense of the underlying epithelial cells is thought to be the main function of the gel-forming and soluble mucins, although careful comparison of their structural characteristics with those of other proteins suggests additional functions [2] (see Chapters 12 and 13).

The soluble mucins are differed from the gel-forming mucins in many aspects. The only structural feature common to both the gel-forming and soluble mucins is the presence of the mucin specific VNTR-containing domain rich in highly glycosylated serine and threonine residues. Functionally, soluble mucins differ from gel-forming mucins. Of note, the soluble mucins also differ from each other whereas the gel-forming mucins have many features in common.

3.2. GEL-FORMING MUCINS: STRUCTURE, CHROMOSOMAL LOCALIZATION AND EVOLUTION

The molecular structures of all gel-forming mucins are essentially the same and exhibit similarity to the von Willebrand factor (vWF) and have common evolutionary and regulatory mechanisms [3-12]. Like vWF, the N-terminal region of the gel-forming mucins consists of four D-domains: three (D1, D2 and D3) full-length vWF D-domains and one (D') truncated. D' is located between D2- and D3-domains, as follows: D1-D2-D'-D3. The central part of the gel-forming mucin molecule has a domain containing variable numbers of tandem repeats of different lengths that are enriched in proline, serine and threonine. The C-terminal regions of these molecules are similar, but not identical, to the C-terminus of the vWF, and contain the D4-, B- and C-domains and the CK-module [4, 6] (Fig. **2**).

The basic composition and order of the main domains in the vWF and in the MUC2, MUC5AC and MUC5B mucins are the same. The only difference is in the central region, where the vWF molecule contains three A-domains (A1-A2-A3) and the molecules of gel-forming mucins have the non-identical mucin domains instead of A-domains. In addition to the structural elements described, there are several CYS-subdomains in both vWF and gel-forming mucins, differing in number and position in each mucin. It should be noted that, in contrast to the MUC2, MUC5AC and MUC5B mucins, MUC6 and MUC19 contain neither the D4- nor the B-domains (Fig. **2**). *MUC6* also differs from other gel-forming genes by chromosomal orientation [13].

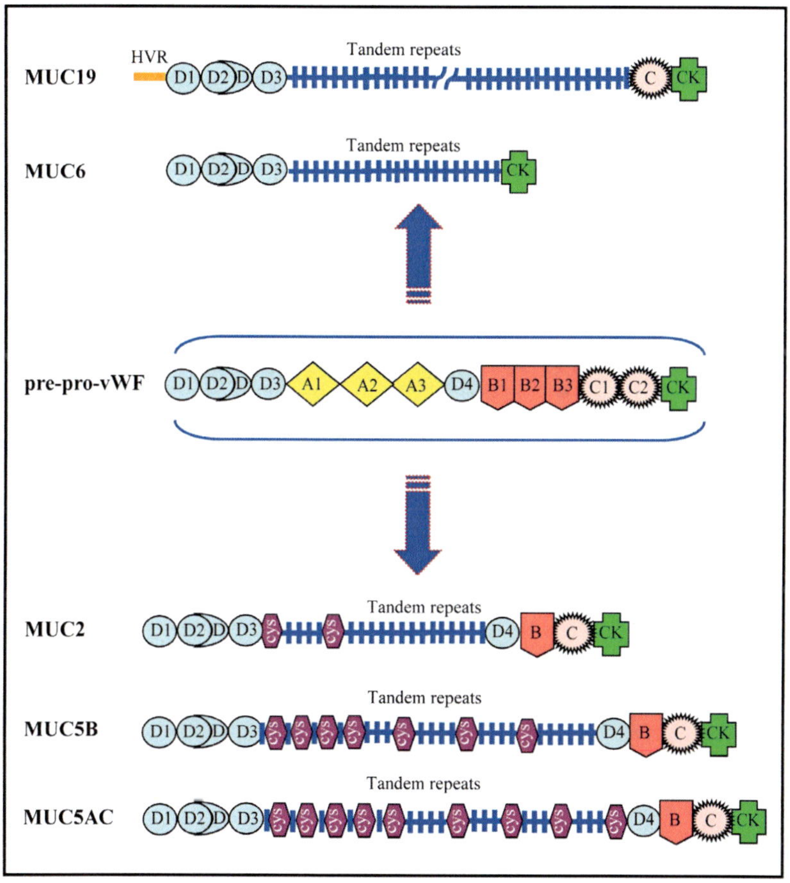

Figure 2: Domain structure of the gel-forming mucins and von Willebrand factor (vWF) (based on the data reported in [3-12]).

The structural similarity of the gel-forming mucin polypeptides suggests a definite order of events during their evolution. However, the *MUC6* and *MUC19* genes introduce some disorder in this tidy organized evolution. These two genes exhibit different phylogenetic pathways from the evolutionary route of other gel-forming mucins, although there is a definite degree of similarity between all gel-forming mucin genes, including *MUC6* and *MUC19*. On the one hand, the N-termini of the

MUC6 and *MUC19* genes are similar to those of other gel-forming mucin genes and the *vWF* gene. On the other hand, the C-termini of *MUC6* and *MUC19* are partially alike but markedly different from the C-terminal regions of the *MUC2*, *MUC5AC* and *MUC5B* genes and the *vWF* gene [9, 12]. In contrast to the other gel-forming mucin genes, which are located on the 11th chromosome as a cluster, the human *MUC19* gene resides at chromosome locus 12q12, close to the *vWF* gene (12p13), and the *MUC6* gene, structurally very similar to *MUC19*, is an integral part of the 11p15.5 cluster [12] (Fig. **3**).

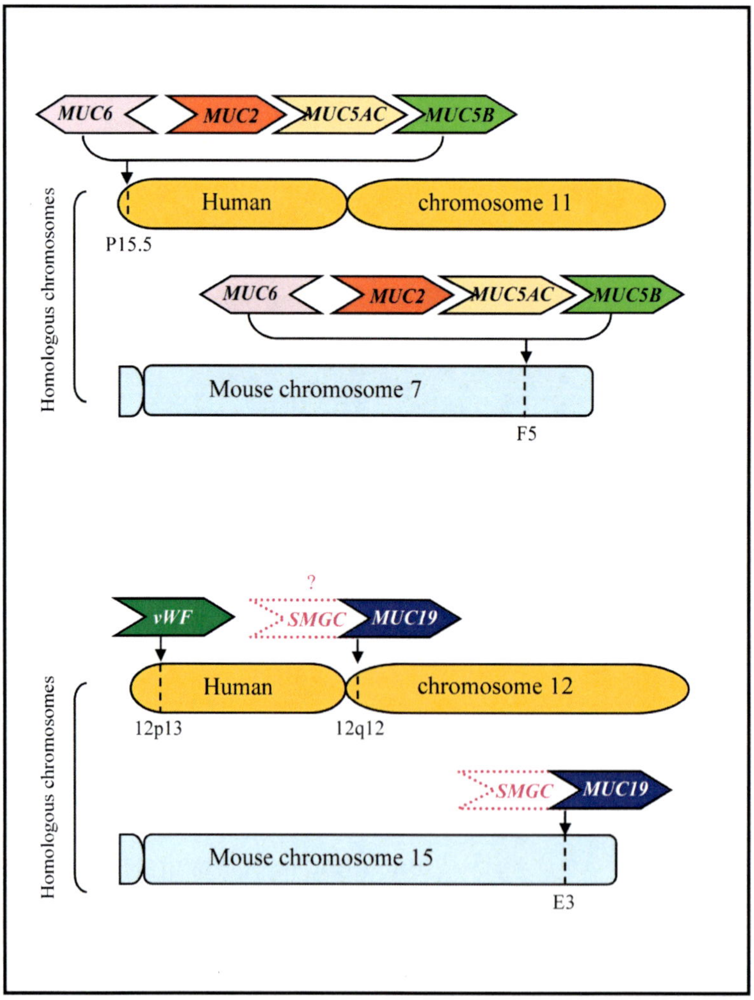

Figure 3: Chromosomal localization of vWF, SMGC and gel-forming mucin genes (based on the data reported in [5, 12]).

Despite the above-mentioned differences, the gel-forming mucins have a common evolutionary history that is partially reflected in the cluster-type distribution of their genes: four out of five gel-forming mucin genes – *MUC6*, *MUC2*, *MUC5AC* and *MUC5B* – are located on the same chromosome locus (11p15.5 locus in human, and 7 F5 locus in mouse) [5]. This cluster organization implies common evolution from the ancestral gene by gene duplication mechanisms. Indeed, computational and phylogenic analyses performed by Desseyn *et al.* [6] point to the evolutionary history of the human *MUC6*, *MUC2*, *MUC5AC* and *MUC5B* mucin genes from an ancestral gene common to the *vWF* gene. The recent studies by Lang *et al.* [7] confirmed observations made by others [6, 8] regarding the origin of the gel-forming mucin genes. The genomic organization of mouse mucin genes is similar to their human homologs, pointing to the high degree of evolutionary conservation. This conservation is reflected not only at the genomic level, but can also be seen at the transcriptome and proteome levels [9-11].

Interestingly, Chen and co-workers [12] consider the *MUC19* gene evolutionarily much closer to *MUC2, MUC5AC* and *MUC5B* than to *MUC6*, although *MUC6,* as mentioned above, belongs to the same chromosomal cluster as *MUC2*, *MUC5AC* and *MUC5B* [13]. According to these authors [12], "*MUC19* shares a similar ancestor with the other gel-forming mucins and branched out evolutionarily later than *MUC6*" (Fig. **4**). In contrast to Chen *et al.* [12], Kawahara and Nishida [14] found that *MUC19* is an ortholog of the fish *spiggin* gene, pointing to a common ancestor gene. This finding suggests that *MUC19* shares the common ancestor gene with the fish *spiggin* genes, but not with the gel-forming genes – a finding that underscores the peculiarity of the evolution of the *MUC19* gene.

The difference between the *MUC19* gene and other gel-forming mucin genes is seen also at the transcriptional level. In mouse, fusion of *Smgc* with *Muc19* genes, occurred during evolution, resulted in common regulation of transcription of both genes followed by developmentally regulated splicing. The *Smgc/Muc19* gene is composed of 60 exons: the Smgc protein is encoded by exons 1-18 while the Muc19 transcript incorporates exon 1 and then skips to exons 19-60 [9, 15].

The structural characteristics and evolutionary history as well as biochemical properties and functional activities of the gel-forming mucins are disucced in Chapters 4-8.

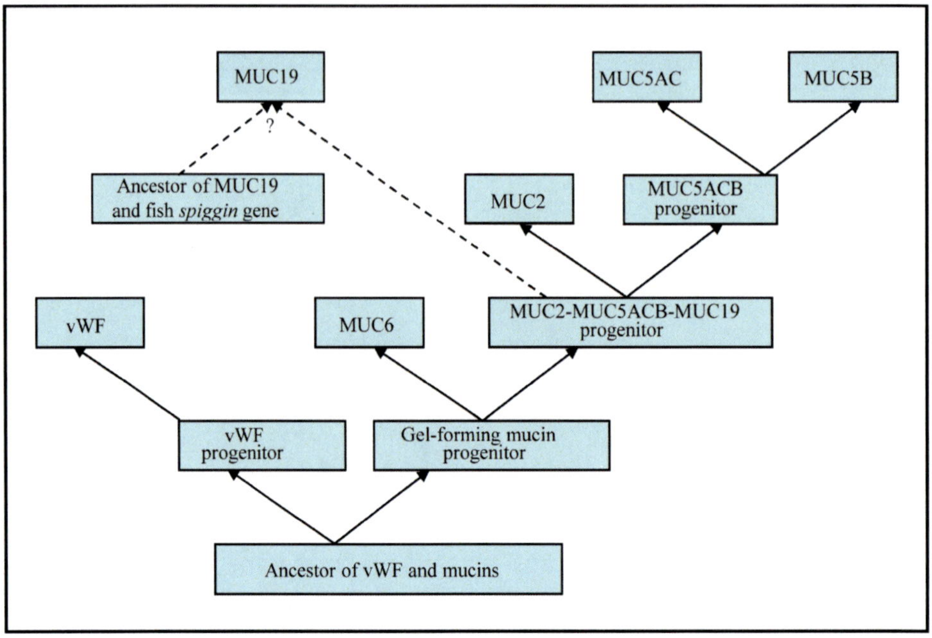

Figure 4: Evolution of the gel-forming genes (based on the data reported in [6, 12, 14]).

3.3. SOLUBLE MUCINS: STRUCTURAL AND FUNCTIONAL CHARACTERISTICS

While the group of soluble mucins is small, it consists of highly multifunctional glycoproteins that play important roles in both cell physiology and pathology. Despite the importance of their functions, these proteins have been studied much less than gel-forming mucins, and some aspects of their structure and functions can only be outlined.

In contrast to gel-forming mucin genes, most of which are located at chromosome locus 11p15.5 as a cluster, the genes encoding soluble mucins reside on different chromosomes. *MUC7* is located at chromosome locus 4q13, *MUC8* at chromosome locus 12q24.3, and *MUC9* at chromosome locus 1p13. This distribution suggests the absence of evolutionary relationship between them.

3.3.1. Mucin MUC7

The *MUC7* gene encodes a small, soluble salivary protein consisting of 377 amino acid residues, part of which are highly glycosylated and contains variable amounts of fucose and sialic acid [16-23]. Analysis of the MUC7 mucin amino acid sequence revealed five distinct structural domains in this apomucin molecule: 1) N-terminal histatin-like domain, 2) moderately glycosylated proline-rich domain, 3) highly glycosylated tandem repeat-containing domain, 4) mouse Muc1 and human MUC2-like domains, and 5) C-terminal leucine zipper-containing module [22]. The presence in the human *MUC7* gene of sequences homologous to the sequences of the mouse membrane-bound mucin *Muc1* gene and human gel-forming *MUC2* gene indicates a possible crossing point in their evolutionary histories [22]. Importantly, each domain of the MUC7 mucin possesses its own biological function(s), pointing to the multifunctionality of this mucin [24, 25].

As follows from its name, the histatin-like domain expresses candidacidal activity [26-33]. The proline moiety of the second domain assists in extending the structure of the MUC7 mucin, contributing to its role in defense of epithelial surfaces. The central mucin-like domain fulfills several mucin-specific functions including lubrication and hydration. The C-terminal domains are apparently involved in self-association of mucin polypeptide important for protective barrier function [24].

The mechanisms controlling transcription of the *MUC7* gene have not been extensively studied, but several transcription factor binding sites have been identified in the promoter region of the *MUC7* gene, including *cis*-elements specific for NFκB, c/EBPβ, FOXD3, Oct, TCF11 and STATs [16, 34]. Various cytokines as well as growth factors and bacterial lipopolysaccharides have been shown to up-regulate expression of *MUC7* [35-38], indicating the involvement of diverse signaling pathways in the regulation of *MUC7* transcription [38].

The MUC7 mucin plays important roles in both normal cell physiology and pathology. It was thought to be involved in clearance of bacteria, viruses and fungi in the oral cavity and to aid in mastication, speech and swallowing. It has been shown to fulfill "sanitary inspection" of the oral cavity by preventing oral infections, participate in mastication, food bolus formation and speaking [16, 25, 33, 39-42]. It lubricates oral surfaces, protects them from mechanical and chemical damage, and participates in formation of impermeable defensive barrier [25, 43-48]. Recently, *MUC7* expression was detected also in the respiratory and genitourinary tracts, in the lachrymal gland, and in the middle ear epithelium [35, 41-43]. It is expressed in inflamed airways [49, 50], in bladder cancer [43, 51], in pyelonephritic kidney [52], and in ampullary carcinoma [53]. Taken together, these findings show that MUC7 mucin is an important player in many fundamental physiological processes maintaining cell homeostasis and in the pathogenesis of various diseases.

3.3.2. Mucin MUC8

The MUC8 mucin is one of the less studied members of the mucin superfamily. Some unique features are however taking shape. Like all other mucins, MUC8 contains a mucin-specific domain characterized by a variable number of tandem repeats. While all other mucins contain only one type of a consensus repeat, the MUC8 polypeptide contains two types: one composed of 13 amino acids and another of 41 amino acids [54].

Under physiological conditions, *MUC8* gene is widely expressed in many epithelial tissues, ranging from the airways [54, 55] to the stomach [56], the male and female reproductive tract [57, 58], and the middle ear [59]. It serves as a specific marker of ciliated cell differentiation [60]. In pathology, it has been implicated in the pathogenesis of mucus hypersecretion in chronic airway inflammatory diseases [61], and in the pathogenesis of ampullary and endometrial carcinomas [53, 58]. The "immune-type" agglutination of motile sperm is also thought to be induced by MUC8 specific antibodies [58]. These sporadic investigations give an idea of the functional activity of the MUC8 mucin; clearly, more systemic studies are needed to gain further insight into the functional aspects of the MUC8 glycoprotein.

The few studies on the mechanisms responsible for regulation of *MUC8* transcription have established that inflammatory mediators such as TNFα, IL-1β, IL-4, LPS and PAF up-regulate *MUC8* expression [61-66], whereas suppressors of cytokine signaling (SOCS) may inhibit transcriptional activity of the gene [67]. Several transcription factors have been implicated in *MUC8* gene regulation, among which CREB and AP2α appear to be of particular importance [61, 68-71].

Taken together, the available information shows that in spite of the important contribution of MUC8 mucin to normal cell physiology (differentiation of ciliated cells) and pathology (airway inflammation [54, 55, 60, 62, 72], endometrial cancer [58]), study of its molecular, biochemical and physiological characteristics is still at the very beginning. More research is needed to elucidate the main parameters of this apparently important mucin glycoprotein.

3.3.3. Mucin MUC9

The glycoprotein MUC9 is the third soluble mucin. It differs from MUC7 and MUC8 in several ways. The MUC9 mucin is a relatively large epithelial glycoprotein containing, like other mucins, numerous tandem repeats that are highly O-glycosylated at the threonine and serine residues [73-77]. However, unlike the other mucins, MUC9 belongs simultaneously to two different protein families: the mucin family and the family of 18 glycosyl hydrolases [78]. In addition to a mucin-specific domain, it contains a chitinase (glycosyl hydrolase)-specific domain [79-82]. While MUC9 does not express chitinase activity, which has been lost during evolution as a result of mutation(s) [81], the presence of a chitinase domain in the MUC9 structure may indicate a common ancestor for the two genes [83].

In contrast to the widely expressed MUC7 and MUC8 mucins, the expression of MUC9 glycoprotein is restricted mainly to the epithelial cells of the oviduct. Thus, the MUC9 mucin is also known as the oviductal glycoprotein 1, or oviductin [80]. In accordance with this name, it was considered to be expressed exclusively in the oviductal microenvironment [76, 77, 84] until its expression was recently and unexpectedly detected in the middle ear epithelium [41, 85]. The MUC9 mucin is an active and important participant in the processes of biological reproduction [76, 77, 86-90]. It interacts directly with oocytes, spermatozoa and embryos [80, 91] – interactions that have positive effects on sperm mobility and viability, capacitation, sperm-ovum binding, ovum penetration, fertilization and early embryo development [76, 80, 86, 88, 89, 92-97]. MUC9 acts both on the surface of gametes and intracellularly [2, 98, 99], and defends gametes and embryos against maternal immune attack [100]. The multiple amino acid motifs present in the MUC9 polypeptide, which allow interactions of the MUC9 molecule with a number of partner-proteins, may enlist the MUC9 to various intracellular signaling pathways [2].

In summary, three glycoproteins, MUC7, MUC8 and MUC9, comprising a group of soluble mucins, share very few features at the structural and functional levels, but are unified into one group thanks to the presence of mucin domains in their structures, the ability to stay in a soluble form, and the power to provide cells with defense against mechanical and chemical stresses, infections and immunological attacks. These properties demonstrate the importance of soluble mucins for cell physiology and pathology and merit further research. The individual properties of MUC7, MUC8 and MUC9 mucins are described in Chapters 9, 10 and 11, respectively.

REFERENCES

[1] Allen A, Hutton DA, Pearson JP. The MUC2 gene product: a human intestinal mucin. Int J Biochem Cell Biol 1998;30:797-801.

[2] Kadam KM, D'Souza SJ, Natraj U. Identification of cellular isoform of oviduct-specific glycoprotein: role in oviduct tissue remodeling? Cell Tissue Res 2007;330:545-56.

[3] Desseyn J, Gouyer V, Tetaert D. Architecture of the gel-forming mucins. In: Van Seuningen I, editor. The epithelial mucins:structure/function Roles in cancer and inflammatory diseases. Kerala: Research Signpost; 2008. p. 1-16.

[4] Thornton DJ, Rousseau K, McGuckin MA. Structure and function of the polymeric mucins in airways mucus. Annu Rev Physiol 2008;70:459-86.

[5] Evans CM, Koo JS. Airway mucus: the good, the bad, the sticky. Pharmacol Ther 2009;121:332-48.

[6] Desseyn JL, Aubert JP, Porchet N, Laine A. Evolution of the large secreted gel-forming mucins. Mol Biol Evol 2000;17:1175-84.

[7] Lang T, Hansson GC, Samuelsson T. Gel-forming mucins appeared early in metazoan evolution. Proc Natl Acad Sci USA 2007;104:16209-14.

[8] Escande F, Porchet N, Bernigaud A, *et al.* The mouse secreted gel-forming mucin gene cluster. Biochim Biophys Acta 2004;1676:240-50.

[9] Culp DJ, Latchney LR, Fallon MA, *et al.* The gene encoding mouse Muc19: cDNA, genomic organization and relationship to Smgc. Physiol Genomics 2004;19:303-18.

[10] Zinzen KM, Hand AR, Yankova M, Ball WD, Mirels L. Molecular cloning and characterization of the neonatal rat and mouse submandibular gland protein SMGC. Gene 2004;334:23-33.

[11] Das B, Cash MN, Hand AR, Shivazad A, Culp DJ. Expression of Muc19/Smgc gene products during murine sublingual gland development: cytodifferentiation and maturation of salivary mucous cells. J Histochem Cytochem 2009;57:383-96.

[12] Chen Y, Zhao YH, Kalaslavadi TB, *et al.* Genome-wide search and identification of a novel gel-forming mucin MUC19/Muc19 in glandular tissues. Am J Respir Cell Mol Biol 2004;30:155-65.

[13] Pigny P, Guyonnet-Duperat V, Hill AS, *et al.* Human mucin genes assigned to 11p15.5: identification and organization of a cluster of genes. Genomics 1996;38:340-52.

[14] Kawahara R, Nishida M. Extensive lineage-specific gene duplication and evolution of the spiggin multi-gene family in stickleback. BMC Evol Biol 2007;7:209.
[15] Das B, Cash MN, Hand AR, *et al.* Tissue Distibution of Murine Muc19/Smgc Gene Products. J Histochem Cytochem 2010;58:141-56.
[16] Bobek LA, Liu J, Sait SN, *et al.* Structure and chromosomal localization of the human salivary mucin gene, MUC7. Genomics 1996;31:277-82.
[17] Bobek LA, Tsai H, Biesbrock AR, Levine MJ. Molecular cloning, sequence, and specificity of expression of the gene encoding the low molecular weight human salivary mucin (MUC7). J Biol Chem 1993;268:20563-9.
[18] Ramasubbu N, Reddy MS, Bergey EJ, *et al.* Large-scale purification and characterization of the major phosphoproteins and mucins of human submandibular-sublingual saliva. Biochem J 1991;280 (Pt 2):341-52.
[19] Reddy MS, Bobek LA, Haraszthy GG, Biesbrock AR, Levine MJ. Structural features of the low-molecular-mass human salivary mucin. Biochem J 1992;287 (Pt 2):639-43.
[20] Levine MJ, Reddy MS, Tabak LA, *et al.* Structural aspects of salivary glycoproteins. J Dent Res 1987;66:436-41.
[21] Loomis RE, Prakobphol A, Levine MJ, Reddy MS, Jones PC. Biochemical and biophysical comparison of two mucins from human submandibular-sublingual saliva. Arch Biochem Biophys 1987;258:452-64.
[22] Gururaja TL, Ramasubbu N, Venugopalan P, *et al.* Structural features of the human salivary mucin, MUC7. Glycoconj J 1998;15:457-67.
[23] Shori DK, Genter T, Hansen J, *et al.* Altered sialyl- and fucosyl-linkage on mucins in cystic fibrosis patients promotes formation of the sialyl-Lewis X determinant on salivary MUC-5B and MUC-7. Pflugers Arch 2001;443 Suppl 1:S55-61.
[24] Satyanarayana J, Gururaja TL, Narasimhamurthy S, Naganagowda GA, Levine MJ. Synthesis and conformational features of human salivary mucin C-terminal derived peptide epitope carrying Thomsen-Friedenreich antigen: implications for its role in self-association. Biopolymers 2001;58:500-10.
[25] Amerongen AV, Bolscher JG, Veerman EC. Salivary mucins: protective functions in relation to their diversity. Glycobiology 1995;5:733-40.
[26] Oppenheim FG, Xu T, McMillian FM, *et al.* Histatins, a novel family of histidine-rich proteins in human parotid secretion. Isolation, characterization, primary structure, and fungistatic effects on Candida albicans. J Biol Chem 1988;263:7472-7.
[27] Raj PA, Edgerton M, Levine MJ. Salivary histatins: dependence of sequence, chain length and helical conformation for candidacidal activity. J Biol Chem 1990;265:3898-905.
[28] Xu T, Levitz SM, Diamond RD, Oppenheim FG. Anticandidal activity of major human salivary histatins. Infect Immun 1991;59:2549-54.
[29] Tsai H, Bobek LA. Human salivary histatins: promising anti-fungal therapeutic agents. Crit Rev Oral Biol Med 1998;9:480-97.
[30] Liu B, Rayment SA, Gyurko C, *et al.* The recombinant N-terminal region of human salivary mucin MG2 (MUC7) contains a binding domain for oral Streptococci and exhibits candidacidal activity. Biochem J 2000;345 Pt 3:557-64.
[31] Liu B, Rayment SA, Soares RV, *et al.* Interaction of human salivary mucin MG2, its recombinant N-terminal region and a synthetic peptide with Actinobacillus actinomycetemcomitans. J Periodontal Res 2002;37:416-24.

[32] Situ H, Bobek LA. *In vitro* assessment of antifungal therapeutic potential of salivary histatin-5, two variants of histatin-5, and salivary mucin (MUC7) domain 1. Antimicrob Agents Chemother 2000;44:1485-93.

[33] Gururaja TL, Levine JH, Tran DT, *et al.* Candidacidal activity prompted by N-terminus histatin-like domain of human salivary mucin (MUC7)1. Biochim Biophys Acta 1999;1431:107-19.

[34] Li S, Bobek LA. Functional analysis of human MUC7 mucin gene 5'-flanking region in lung epithelial cells. Am J Respir Cell Mol Biol 2006;35:593-601.

[35] Li S, Intini G, Bobek LA. Modulation of MUC7 mucin expression by exogenous factors in airway cells *in vitro* and *in vivo*. Am J Respir Cell Mol Biol 2006;35:95-102.

[36] Woenckhaus M, Klein-Hitpass L, Grepmeier U, *et al.* Smoking and cancer-related gene expression in bronchial epithelium and non-small-cell lung cancers. J Pathol 2006;210:192-204.

[37] Gray T, Koo JS, Nettesheim P. Regulation of mucous differentiation and mucin gene expression in the tracheobronchial epithelium. Toxicology 2001;160:35-46.

[38] Kim H, Seo JH, Kim KH. The effect of p38 mitogen-activated protein kinase on mucin gene expression and apoptosis in Helicobacter pylori-infected gastric epithelial cells. Ann N Y Acad Sci 2003;1010:90-4.

[39] Wei GX, Campagna AN, Bobek LA. Factors affecting antimicrobial activity of MUC7 12-mer, a human salivary mucin-derived peptide. Ann Clin Microbiol Antimicrob 2007;6:14.

[40] Wei GX, Bobek LA. *In vitro* synergic antifungal effect of MUC7 12-mer with histatin-5 12-mer or miconazole. J Antimicrob Chemother 2004;53:750-8.

[41] Kerschner JE. Mucin gene expression in human middle ear epithelium. Laryngoscope 2007;117:1666-76.

[42] Gipson IK. Distribution of mucins at the ocular surface. Exp Eye Res 2004;78:379-88.

[43] Okegawa T, Kinjo M, Horie S, Nutahara K, Higashihara E. Detection of mucin 7 gene expression in exfoliated cells in urine from patients with bladder tumor. Urology 2003;62:182-6.

[44] Fisher SJ, Prakobphol A, Kajisa L, Murray PA. External radiolabelling of components of pellicle on human enamel and cementum. Arch Oral Biol 1987;32:509-17.

[45] Al-Hashimi I, Levine MJ. Characterization of *in vivo* salivary-derived enamel pellicle. Arch Oral Biol 1989;34:289-95.

[46] Jensen JL, Lamkin MS, Oppenheim FG. Adsorption of human salivary proteins to hydroxyapatite: a comparison between whole saliva and glandular salivary secretions. J Dent Res 1992;71:1569-76.

[47] Nieuw Amerongen AV, Oderkerk CH, Driessen AA. Role of mucins from human whole saliva in the protection of tooth enamel against demineralization *in vitro*. Caries Res 1987;21:297-309.

[48] Nieuw Amerongen AV, Oderkerk CH, Veerman EC. Influence of phytate on the adsorption of human salivary mucins onto hydroxyapatite. J Biol Buccale 1988;16:203-8.

[49] Sharma P, Dudus L, Nielsen PA, *et al.* MUC5B and MUC7 are differentially expressed in mucous and serous cells of submucosal glands in human bronchial airways. Am J Respir Cell Mol Biol 1998;19:30-7.

[50] Martinez-Anton A, Debolos C, Garrido M, *et al.* Mucin genes have different expression patterns in healthy and diseased upper airway mucosa. Clin Exp Allergy 2006;36:448-57.

[51] Kinjo M, Okegawa T, Horie S, Nutahara K, Higashihara E. Detection of circulating MUC7-positive cells by reverse transcription-polymerase chain reaction in bladder cancer patients. Int J Urol 2004;11:38-43.

[52] Lehmann J, Suttmann H, Gerber M, *et al.* [Expression of antimicrobial peptide MUC7 in kidneys with pyelonephritis]. Urologe A 2006;45:1300, 2-7.

[53] Paulsen FP, Varoga D, Paulsen AR, Corfield A, Tsokos M. Prognostic value of mucins in the classification of ampullary carcinomas. Hum Pathol 2006;37:160-7.

[54] Shankar V, Gilmore MS, Elkins RC, Sachdev GP. A novel human airway mucin cDNA encodes a protein with unique tandem-repeat organization. Biochem J 1994;300 (Pt 2):295-8.

[55] Leikauf GD, Borchers MT, Prows DR, Simpson LG. Mucin apoprotein expression in COPD. Chest 2002;121:166S-82S.

[56] Shankar V, Pichan P, Eddy RL, Jr., *et al.* Chromosomal localization of a human mucin gene (MUC8) and cloning of the cDNA corresponding to the carboxy terminus. Am J Respir Cell Mol Biol 1997;16:232-41.

[57] D'Cruz OJ, Dunn TS, Pichan P, Hass GG, Jr., Sachdev GP. Antigenic cross-reactivity of human tracheal mucin with human sperm and trophoblasts correlates with the expression of mucin 8 gene messenger ribonucleic acid in reproductive tract tissues. Fertil Steril 1996;66:316-26.

[58] Hebbar V, Damera G, Sachdev GP. Differential expression of MUC genes in endometrial and cervical tissues and tumors. BMC Cancer 2005;5:124.

[59] Choi JY, Kim JY, Kim CW, *et al.* IL-1beta promotes the ciliogenesis of human middle ear epithelial cells: possible linkage with the expression of mucin gene 8. Acta Otolaryngol 2005;125:260-5.

[60] Kim CH, Kim HJ, Song KS, *et al.* MUC8 as a ciliated cell marker in human nasal epithelium. Acta Otolaryngol 2005;125:76-81.

[61] Seong JK, Koo JS, Lee WJ, *et al.* Upregulation of MUC8 and downregulation of MUC5AC by inflammatory mediators in human nasal polyps and cultured nasal epithelium. Acta Otolaryngol 2002;122:401-7.

[62] Song KS, Seong JK, Chung KC, *et al.* Induction of MUC8 gene expression by interleukin-1 beta is mediated by a sequential ERK MAPK/RSK1/CREB cascade pathway in human airway epithelial cells. J Biol Chem 2003;278:34890-6.

[63] Bernatchez PN, Allen BG, Gelinas DS, Guillemette G, Sirois MG. Regulation of VEGF-induced endothelial cell PAF synthesis: role of p42/44 MAPK, p38 MAPK and PI3K pathways. Br J Pharmacol 2001;134:1253-62.

[64] Waetzig GH, Seegert D, Rosenstiel P, Nikolaus S, Schreiber S. p38 mitogen-activated protein kinase is activated and linked to TNF-alpha signaling in inflammatory bowel disease. J Immunol 2002;168:5342-51.

[65] Plath KE, Grabbe J, Gibbs BF. Calcineurin antagonists differentially affect mediator secretion, p38 mitogen-activated protein kinase and extracellular signal-regulated kinases from immunologically activated human basophils. Clin Exp Allergy 2003;33:342-50.

[66] Cho KN, Choi JY, Kim CH, *et al.* Prostaglandin E2 induces MUC8 gene expression *via* a mechanism involving ERK MAPK/RSK1/cAMP response element binding protein activation in human airway epithelial cells. J Biol Chem 2005;280:6676-81.

[67] Song KS, Kim K, Chung KC, Seol JH, Yoon JH. Interaction of SOCS3 with NonO attenuates IL 1beta dependent MUC8 gene expression. Biochem Biophys Res Commun 2008;377:946-51.

[68] Hennig G, Lowrick O, Birchmeier W, Behrens J. Mechanisms identified in the transcriptional control of epithelial gene expression. J Biol Chem 1996;271:595-602.

[69] Batsche E, Muchardt C, Behrens J, Hurst HC, Cremisi C. RB and c-Myc activate expression of the E-cadherin gene in epithelial cells through interaction with transcription factor AP-2. Mol Cell Biol 1998;18:3647-58.

[70] Anttila MA, Kellokoski JK, Moisio KI, *et al.* Expression of transcription factor AP-2alpha predicts survival in epithelial ovarian cancer. Br J Cancer 2000;82:1974-83.

[71] Moon UY, Kim CH, Choi JY, *et al.* AP2alpha is essential for MUC8 gene expression in human airway epithelial cells. J Cell Biochem 2010;110:1386-98.

[72] Lopez-Ferrer A, Curull V, Barranco C, *et al.* Mucins as differentiation markers in bronchial epithelium. Squamous cell carcinoma and adenocarcinoma display similar expression patterns. Am J Respir Cell Mol Biol 2001;24:22-9.

[73] Agarwal A, Yeung WS, Lee KF. Cloning and characterization of the human oviduct-specific glycoprotein (HuOGP) gene promoter. Mol Hum Reprod 2002;8:167-75.

[74] Arias EB, Verhage HG, Jaffe RC. Complementary deoxyribonucleic acid cloning and molecular characterization of an estrogen-dependent human oviductal glycoprotein. Biol Reprod 1994;51:685-94.

[75] Lapensee L, Paquette Y, Bleau G. Allelic polymorphism and chromosomal localization of the human oviductin gene (MUC9). Fertil Steril 1997;68:702-8.

[76] Verhage HG, Mavrogianis PA, O'Day-Bowman MB, *et al.* Characteristics of an oviductal glycoprotein and its potential role in the fertilization process. Biol Reprod 1998;58:1098-101.

[77] Buhi WC, Alvarez IM, Kouba AJ. Secreted proteins of the oviduct. Cells Tissues Organs 2000;166:165-79.

[78] Jaffe RC, Arias EB, O'Day-Bowman MB, *et al.* Regional distribution and hormonal control of estrogen-dependent oviduct-specific glycoprotein messenger ribonucleic acid in the baboon (Papio anubis). Biol Reprod 1996;55:421-6.

[79] Buhi WC, Alvarez IM, Choi I, Cleaver BD, Simmen FA. Molecular cloning and characterization of an estrogen-dependent porcine oviductal secretory glycoprotein. Biol Reprod 1996;55:1305-14.

[80] Buhi WC. Characterization and biological roles of oviduct-specific, oestrogen-dependent glycoprotein. Reproduction 2002;123:355-62.

[81] DeSouza MM, Murray MK. An estrogen-dependent secretory protein, which shares identity with chitinases, is expressed in a temporally and regionally specific manner in the sheep oviduct at the time of fertilization and embryo development. Endocrinology 1995;136:2485-96.

[82] Lagow E, DeSouza MM, Carson DD. Mammalian reproductive tract mucins. Hum Reprod Update 1999;5:280-92.

[83] Watanabe T, Kobori K, Miyashita K, *et al.* Identification of glutamic acid 204 and aspartic acid 200 in chitinase A1 of Bacillus circulans WL-12 as essential residues for chitinase activity. J Biol Chem 1993;268:18567-72.

[84] Hendrix E, Hewetson A, Mansharamani M, Chilton BS. Oviductin (Muc9) is expressed in rabbit endocervix. Endocrinology 2001;142:2151-4.

[85] Takeuchi K, Yagawa M, Ishinaga H, *et al.* Mucin gene expression in the effusions of otitis media with effusion. Int J Pediatr Otorhinolaryngol 2003;67:53-8.

[86] Yong P, Gu Z, Luo JP, Wang JR, Tso JK. Antibodies against the C-terminal peptide of rabbit oviductin inhibit mouse early embryo development to pass 2-cell stage. Cell Res 2002;12:69-78.

[87] Boatman DE, Magnoni GE. Identification of a sperm penetration factor in the oviduct of the golden hamster. Biol Reprod 1995;52:199-207.

[88] O'Day-Bowman MB, Mavrogianis PA, Reuter LM, *et al.* Association of oviduct-specific glycoproteins with human and baboon (Papio anubis) ovarian oocytes and enhancement of human sperm binding to human hemizonae following *in vitro* incubation. Biol Reprod 1996;54:60-9.

[89] Kouba AJ, Abeydeera LR, Alvarez IM, Day BN, Buhi WC. Effects of the porcine oviduct-specific glycoprotein on fertilization, polyspermy, and embryonic development *in vitro*. Biol Reprod 2000;63:242-50.

[90] Boatman DE. Responses of gametes to the oviductal environment. Hum Reprod 1997;12:133-49.

[91] Akatsuka K, Yoshida-Komiya H, Tulsiani DR, *et al.* Rat zona pellucida glycoproteins: molecular cloning and characterization of the three major components. Mol Reprod Dev 1998;51:454-67.

[92] Killian GJ. Evidence for the role of oviduct secretions in sperm function, fertilization and embryo development. Anim Reprod Sci 2004;82-83:141-53.

[93] McCauley TC, Buhi WC, Wu GM, *et al.* Oviduct-specific glycoprotein modulates sperm-zona binding and improves efficiency of porcine fertilization *in vitro*. Biol Reprod 2003;69:828-34.

[94] Anderson SH, Killian GJ. Effect of macromolecules from oviductal conditioned medium on bovine sperm motion and capacitation. Biol Reprod 1994;51:795-9.

[95] Nancarrow CD, Hill JL. Co-culture, oviduct secretion and the function of oviduct-specific glycoproteins. Cell Biol Int 1994;18:1105-14.

[96] Sendai Y, Komiya H, Suzuki K, *et al.* Molecular cloning and characterization of a mouse oviduct-specific glycoprotein. Biol Reprod 1995;53:285-94.

[97] Satoh T, Abe H, Sendai Y, Iwata H, Hoshi H. Biochemical characterization of a bovine oviduct-specific sialo-glycoprotein that sustains sperm viability *in vitro*. Biochim Biophys Acta 1995;1266:117-23.

[98] Kan FW, Roux E, Bleau G. Immunolocalization of oviductin in endocytic compartments in the blastomeres of developing embryos in the golden hamster. Biol Reprod 1993;48:77-88.

[99] Buhi WC, O'Brien B, Alvarez IM, Erdos G, Dubois D. Immunogold localization of porcine oviductal secretory proteins within the zona pellucida, perivitelline space, and plasma membrane of oviductal and uterine oocytes and early embryos. Biol Reprod 1993;48:1274-83.

[100] Oliphant G, Cabot C, Ross P, Marta J. Control of the humoral immune system within the rabbit oviduct. Biol Reprod 1984;31:205-12.

Mucins – Potential Regulators of Cell Functions, 2013, 44-144

CHAPTER 4

Gel-Forming Mucin MUC2

Abstract: The MUC2 glycoprotein is one of the most extensively studied secreted gel-forming mucins. The gene encoding this protein has been cloned, sequenced and analyzed both structurally and functionally. The exon/intron composition of the *MUC2* gene has been established, and its transcription investigated. The transcriptional activity of the gene is controlled by promoter regulatory elements and by the epigenetic mechanism. Biosynthesis of the MUC2 protein precursor and its processing and maturation associated with posttranslational modifications, such as N- and O-glycosylation, sulfation, sialylation and posttranslational proteolysis, have been thoroughly studied. Expression of *MUC2* gene has been analyzed in different organs, tissues and cells under physiological conditions, including embryogenesis and fetal development, and in different forms of pathology. The molecular biology aspects of the MUC2 glycoprotein functioning, as well as its biosynthesis, posttranslational modifications and expression are discussed.

Keywords: MUC2, domain structure, biosynthesis, promoter, transcriptional regulation, expression, development, pathology.

The group of gel-forming mucins, consisting of the MUC2, MUC5AC, MUC5B, MUC6 and MUC19 glycoproteins, represents a subfamily of the large mucin superfamily. They have attracted much attention because of the important functions they perform in physiology and pathology of epithelial cells. Some of the mucins have been studied thoroughly while others much less. MUC2 is the best studied and MUC19 the least. This chapter presents a detailed analysis of the MUC2 mucin; other gel-forming mucins are covered in the subsequent chapters.

4.1. MUC2 MUCIN: DOMAIN STRUCTURE

Among the gel-forming mucin genes, the *MUC2* was the first to be cloned and completely sequenced [1-3]. Located at chromosome locus 11p15.5, the human *MUC2* gene, containing 49 exons, encodes the core protein comprised of 5179 amino acid residues in its commonest allelic form [2-5]. The N-terminal D1-D3 domains comprise 1400 amino acids including multiple cysteine residues (Fig. **1**). This region is followed by a short, 347 amino acid fragment consisting of irregular heavily O-glycosylated repeats. Downstream to these repeats is a large 2300 amino acid-containing PTS domain that includes regular repeats, each

Joseph Z. Zaretsky and Daniel H. Wreschner

containing 23 amino acids. The number of repeats varies between 50 and 115 copies in different alleles (VNTR). The PTS-domain is densely O-glycosylated at serine and threonine residues. Between the 347 amino acid-containing domain and the PTS-domain, there is a short cysteine-rich bridge composed of 148 amino acids. The C-terminal region of the MUC2 glycoprotein is formed by the cysteine-rich D4-, B-, C- and CK-domains, and consists of 984 amino acids (Fig. **1**) The very C-end of the MUC2 mucin contains heparin binding site [1].

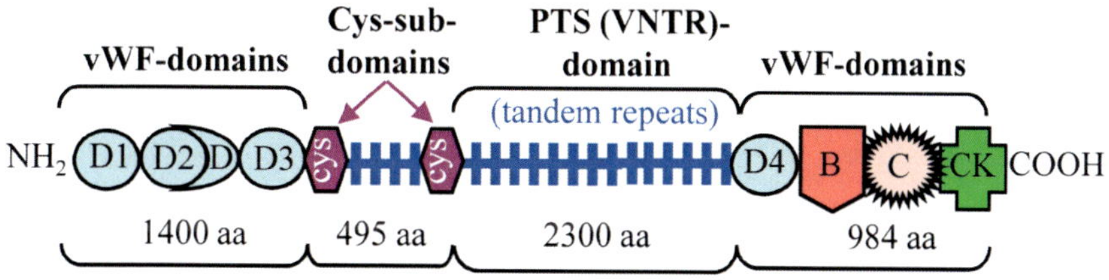

Figure 1: Domain structure of the MUC2 mucin glycoprotein (based on the data reported in [2-5]).

4.2. REGULATION OF *MUC2* GENE TRANSCRIPTION

Expression of the *MUC2* gene is regulated by both genetic and epigenetic mechanisms that carry out their respective tasks in a highly coordinated manner. The peculiarities of the *MUC2* gene's promoter structure and functions as well as the details of its epigenetic regulation are discussed below.

4.2.1. *MUC2* promoter

The *MUC2* gene is expressed mostly in the goblet cells in the small and large intestine [6, 7]. Other tissues and organs of the gastro-intestinal tract, such as stomach, esophagus and gallbladder, do not synthesize significant quantities of the MUC2 glycoprotein [8, 9]. Nevertheless, it is expressed under specific conditions in other tissues as well: for example, low-level expression of *MUC2* is detected in the epithelium and glands of endocervix and in airways [10-12]. While it is highly expressed in normal colon epithelium, its expression in colon cancer is drastically decreased [13, 14], suggesting that *MUC2* expression is highly associated with the level of differentiation characteristic of normal intestinal epithelium [8]. A number of biologically active molecules regulate the *MUC2* gene expression in various cell types, including cytokines, growth factors, differentiation stimulating agents, bacterial and viral products, and many other factors [15]. These data point to the complex organization of the *MUC2* promoter that enables it to support specific combinations of *cis*-elements with the potential to bind different transcription factors in cell- and tissue-specific manner.

The regulatory region of the *MUC2* gene is represented by the promoter which, according to Velcich *et al.* [16], is "stretched out" by approximately 12 kb. This promoter contains a typical TATA box located at position -31/-25 upstream to the transcription initiation site [15]. In addition to the canonical TATA box, there are several TATA box-like sites throughout the promoter; one, a TAATAAT sequence, is located far upstream (-3390/-3384) to the canonical transcription start site. In this context, the 8 noncanonical transcription start sites found in the *MUC2* promoter by Velcich *et al.* [16] are of great importance.

The presence of repetitive sequences of different lengths is characteristic of the human *MUC2* promoter. Several repetitive sequences have been found: the ATCC motif is repeated 101 times in the region between nucleotides -5232 and -4330; the pentanucleotide CCTGC is repeated 26 times in the region between bases -3450 and -1; a 90-nucleotide-containing segment is repeated twice with almost 90% similarity; three direct repeats of the TCCTGCC sequence with only one nucleotide substitution are located near the canonical transcription start site. In addition, there are multiple Alu repeats [8].

The number and distribution of different transcription factor binding *cis*-elements throughout the promoter region of a given gene is specific to that gene and determine the expression of the gene in various cells and tissues. Thus, analysis of the *cis*-elements in a given promoter is important for evaluation of a given gene transcriptional regulation. Fig. **2** illustrates the partial *cis*-element map of the *MUC2* promoter. The role of each of the identified *cis*-elements and corresponding transcription factors in the regulation of *MUC2* gene transcription is discussed below.

<u>Transcription factors Sp1/Sp3</u>: The *MUC2* promoter region encompassing the first 150 bases upstream to the canonical TATA-box contains a number of *cis*-

elements specific for several transcription factors (Fig. **2**). Sequence analysis of the human *MUC2* promoter revealed the presence of several GC boxes, the putative binding sites for the Sp family of transcription factors suggested to play a role in the regulation of the *MUC2* gene. The region comprising the bases from -91 to -73 (relative to the transcription start point) contains CACCC motif specific for transcription factors of Sp1 family. This region is important for basal promoter activity in the all cell lines tested [8, 16-18]. A comparison of the mouse and human *MUC2* promoters showed a high degree of sequence conservation including the presence, location and composition of GC boxes. This conservation of promoter sequences suggests functional relevance of the *Muc2* and *MUC2* expression in the intestine of mouse and human both *in vivo* and *in vitro* [17, 19].

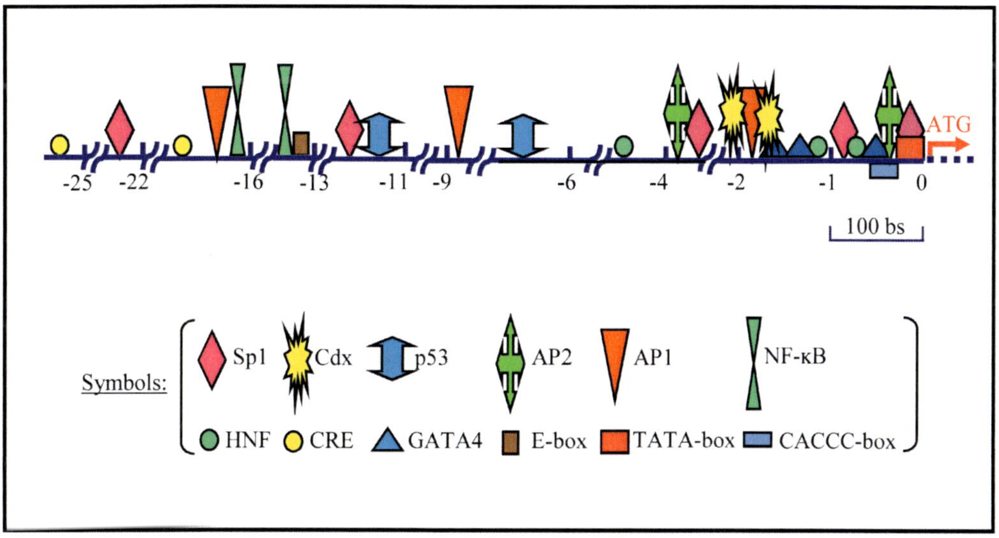

Figure 2: Transcription factor *cis*-element map of the *MUC2* promoter (constructed on the data reported in [8, 16-18, 38, 49-52, 70]).

The importance of Sp1/Sp3 transcription factors and the corresponding Sp-binding sites for expression of the *MUC2* gene was demonstrated in several studies. These transcription factors are involved in different signaling cascades that participate in regulation of *MUC2* expression. Growth factors play an important role in the physiology and pathology of many cell types including goblet cells, the main producers of MUC2 mucin [20]. EGF and TGFα up-regulate *MUC2* transcription through activation of Ras-Raf-MEK1/2-ERK1/2 pathway. This cascade leads to phosphorylation of the Sp1 transcription factor, resulting in the binding of the Sp1 protein to the Sp1-specific *cis*-element located in the -2627/-2097 region of the *MUC2* promoter. This binding activates *MUC2* transcription [21].

The Sp1 transcription factor is also involved in adenosine-mediated up-regulation of *MUC2* gene expression. Adenosine, which is accumulated extracellularly in tissues under oxidative or metabolic stresses, was shown by McNamara *et al.* [22] to affect mucin production *via* the epithelial cell signaling pathway initiated at the adenosine receptor. The activated receptor transduces signals through a Ca^{2+} ion channel to EGFR, followed by activation of the Ras-Raf -MEK1/2-ERK1/2 canonical signal transduction pathway, resulting in binding of the phosphorylated Sp1 transcription factor to the *MUC2* promoter.

The importance of the Sp1/Sp3 transcription factors for expression of *MUC2* was clearly demonstrated by the use of specific inhibitors of their activity. Aslam *et al.* [17] showed that mithramycin, an inhibitor of Sp1/Sp3 binding, blocked the expression of *MUC2* in human adeno-carcinoma cell line HT29. The Sp1 transcription factor was shown to function as a main regulator of basic *MUC2* promoter activity in intestinal goblet cells. Importantly, interaction of Sp1 with other transcription factors may activate *MUC2* transcription in other cells and tissues, for example in the rat airway epithelium [18]. In this context, interaction of Sp1 with Ap2 transcription factor, whose binding site is adjusted to the Sp1-specific *cis*-element in murine and rat *Muc2* genes, is of particular interest [17]. Interestingly, Sp1 and Sp3 transcription factors belong to the same family but may function differently: Sp1 is a powerful activator of *MUC2* transcription while Sp3 is a strong inhibitor of *MUC2* expression [21]. The Sp1 transcription factor is involved also in epigenetic transcriptional regulation of the *MUC2* gene by prevention of CpG methylation, a process that leads to *MUC2* gene silencing [17, 23, 24].

<u>*Transcription factor Cdx2*</u>: Functional segmentation of the *MUC2* promoter has been established by several research groups [8, 16, 17, 25-29]. *Cis*-elements located between nucleotides -91 and -73 participate mostly in regulation of basic *MUC2* transcription [16, 17]. The segment located between the bases -228 and -171 controls mainly the cell-type specific transcription of the *MUC2* gene [8]. This

segment contains two binding sites specific for the Cdx2 transcription factor, which is an intestine-specific homeodomain protein and a principal regulator of the intestinal genes including *MUC2* [25-29]. These two Cdx2-binding sites participate in concert in *MUC2* transcriptional regulation: only simultaneous mutations in the two sites results in inhibition of Cdx2-mediated *trans*-activation of the *MUC2* promoter [29].

Although both Cdx homeodomain proteins, Cdx1 and Cdx2, are important for intestinal epithelium differentiation [30], only Cdx2 interacts with the corresponding *cis*-elements in the *MUC2* promoter activating transcription of the *MUC2* gene, and thereby plays an important role in goblet cell differentiation [28]. Mesquita *et al.* [29] showed that Cdx1 had no significant effect on the *MUC2* promoter activity in several gastric carcinoma cell lines tested, whereas simultaneous expression of both the Cdx1 and Cdx2 transcription factors resulted in *MUC2* expression in intestinal metaplastic cells and in gastric carcinoma cells [31]. Importantly, the *MUC2* promoter could be *trans*-activated in gastric carcinoma cell lines by ectopical over-expression of Cdx2 transcription factor, showing that Cdx2 may drive transcription of the *MUC2* gene in both intestinal and nonintestinal epithelial cells.

The synergism between homeodomain proteins and zinc finger transcription factors has been repeatedly described as an important mechanism of gene regulation [32, 33]. Such cooperation has not been observed, however, with the *MUC2* gene. Sp1 protein, a member of the zinc finger transcription factor family, and Cdx2, a homeodomain transcription factor, did not cooperate in cell-specific Cdx2-mediated regulation of *MUC2* promoter [29]. According to Mesquita *et al.* [29], this is consistent with the data showing that Sp1 is a general ubiquitous transcription factor involved mainly in basic, but not cell-specific, regulation of transcription. This opinion contradicts, however, the finding of Nogami *et al.* [18] that Sp1 does participate in regulation of the *MUC2* cell-specific expression in the airway epithelia.

Transcription factors GATA: GATA proteins, members of the zinc finger family of transcription factors, have been shown to be expressed in the gastro-intestinal epithelium. Moreover, transcription factors GATA-4/-5/-6 are involved in gastrointestinal cell differentiation [34]. The *MUC2* promoter contains several GATA-4 protein binding *cis*-elements in close proximity to the Cdx2 binding sites (Fig. **2**) – a topology that suggests their possible cooperation in regulation of *MUC2* transcription. Cooperation between GATA and Cdx in the regulation of the *MUC2* gene has not been reported, but synergism in activities of these transcription factors in transcriptional regulation of other intestine-specific genes has been described [35, 36]. One such study reports cooperation between Cdx2, GATA-4 and HNF-1α (hepatocyte nuclear factor also known as forkhead box A, FOXA) for the intestine-specific sucrase-isomaltase gene [37]. Of interest, the binding sites for HNF (FOXA) are present also in the *MUC2* gene promoter (Fig. **2**). Moreover, a comprehensive study by Van Seuningen's group [38] suggests that the murine *Muc2* mucin gene is transcriptionally regulated by both the GATA-4 and HNF-3α transcription factors. This study showed that the GATA-4 transcription factor is expressed in the intestinal goblet cells in parallel with the Muc2 mucin. Moreover, the *Muc2* promoter contains several cognate GATA-4 binding sites located between nucleotides -168 and -96. These *cis*-elements directly bind molecules of GATA-4 transcription factor, resulting in activation of the *Muc2* promoter. The HNF-3α (Foxa1) and HNF-3β (Foxa2) transcription factors also bind to the *Muc2* promoter at their cognate *cis*-elements (-108/-104), located next to the GATA-4 *cis*-element (-99/-96), and up-regulate transcription of the endogenous *Muc2* gene and an artificial construct directed by the *Muc2* promoter [39]. Although these data suggest close cooperation of GATA, HNF-3α (Foxa1) and HNF-3β (Foxa2) transcription factors in transcriptional regulation of the murine *Muc2* gene, Van der Sluis *et al.* [38, 39] could not detect synergism between them in regulation of the human *MUC2* gene.

A definite parallelism between expression of the MUC2 mucin, on the one hand, and GATA-4,-5,-6 and HNF (Foxa) transcription factors, on the other hand, has been observed during embryonic and fetal development. The MUC2 mucin is expressed early in embryonic development of the intestine, and high expression of this gene is observed also in differentiated intestinal epithelia [40, 41]. The GATA-5 and GATA-6 proteins are also expressed during embryonic development of the intestine and in differentiated intestinal cells, supporting transcription of several intestine-specific genes [36, 42-45]. Like MUC2 protein and GATA transcription factors, the HNFs (Foxa1 and Foxa2) are also involved in morphogenesis of mammalian gut endoderm, lineage development, and intestinal cell differentiation [25, 46]. As pointed out by Van der Sluis *et al.* [38], "all these data are in favor of an important role for GATA-4 factor in Muc2 spatio-temporal expression pattern observed in embryonic, fetal, and adult small intestine, and identifies the *Muc2*, a gene that is a marker of goblet cells, as a direct target of transcription factors involved in intestinal development and cell differentiation". The importance of these findings is also evidenced by the fact that GATA-4 controls the Muc2 mucin expression not only in the intestine goblet cells, but ectopically in other tissues, where both the *GATA-4* and *MUC2* genes are co-expressed [47, 48].

Transcription factor p53: The p53 protein is one of the transcription factors that regulate *MUC2* transcription. It has been shown to mediate stress-induced *Muc2 trans*-activation in a colon cancer cell line *via* binding to the p53 cognate *cis*-elements located at positions -1131/-1100 and - 676/- 650 in mouse *Muc2* promoter [49]. Both sites were found to contribute to stimulation of the *Muc2* promoter activity in response to stress factors such as actinomycin D, UV and X-ray radiation. Importantly, a spectrum of cells expressing *MUC2* under control of p53 is not limited to intestinal cells: Ookawa *et al.* [49] reported up-regulation of the *MUC2* gene in osteosarcoma, hepatome, breast cancer and lung cancer cells by exogenous p53. These data show that p53 is a powerful and probably universal activator of the *MUC2* gene expression.

Transcription factor NF-κB: Different stress factors activate different transcription factors, which, in turn, interact with different but specific *cis*-elements in the *MUC2* promoter and up-regulate *MUC2* gene transcription. NF-κB transcription factor appears to be the main player in the cell's defense against bacterial stress. The *MUC2* promoter contains at least two NFκB binding sites, located between the nucleotides -1628 and -1400 (Fig. **2**). The NF-κB transcription factor binds to its cognate *cis*-element at the -1458/-1430 position, resulting in up-regulation of *MUC2* transcription in airway epithelial cells in response to *Pseudomonas aeruginosa (P.aeruginosa)* infection [50]. These authors established that bacterial lipopolysaccharide, LPS, the major component of the Gram-negative bacteria outer membrane and a potent activator of the host defense response [51], is the main factor mediating *P. aeruginosa*-induced up-regulation of *MUC2* transcription. LPS was shown to activate the signal transduction cascade consisting of LPS-LBP-Ras-Raf-1-MEK1/2-ERK1/2-pp90rsk-NF-κB(p50/p65). At the last stage of this cascade, two subunits of the NF-κB transcription factor bind to its cognate *cis*-element in the *MUC2* promoter, activating MUC2 mucin expression [50-52].

The described pathway is not the only one that up-regulates *MUC2* transcription in response to Gram-negative bacterial infection. For instance, when the non-typeable *Haemophilus influenzae* bacterium (NTHi) infects the airway epithelial cells, it interacts with the cell surface in a way different from that of *P. aeruginoza.* NTHi interacts with cell membrane through two cell membrane receptors: the Toll-like receptor 2 (TLR2) and the TGFβ receptor type I/II (TβRI/II). The TLR2 receptor directs the NTHi-induced signal to the cascade TLR2->MyD->TAK->NIK->IKKβ/γ->I*k*Bα/NF-*k*B(p50/p65)-> NF*k*B(p50/p65). In other words, the signal obtained by TLR2 from NTHi is transmitted to MyD, and, *via* a TAK1-NIK-IKKβ/γ-dependent phosphorylation of the I*k*Bα, activates NF*k*B, which then is translocated to the nucleus, where it binds to the cognate *cis*-element in the *MUC2* promoter [53]. NHTi may also interact directly with the TβRII receptor. This results in phosphorylation of Smad3/4 and activation of TβRI/II-Smad3/4 signaling pathway through the following steps: (1) the phosphorylated TβRII receptor forms a heterodimer with the TβRI receptor; (2) the heterodimer TβRI/II phosphorylates the Smad3 protein; (3) the phosphorylated Smad3 is translocated to the nucleus, where (4) it forms a complex with the Smad4. However, the Smad3/4 complex cannot bind directly to the *MUC2* promoter, as the latter does not contain Smad-binding *cis*-elements. Instead, (5) Smad3/4 interacts with NF*k*Bp65/p50, after which (6) the NFκBp65/p50:Smad3/4 complex binds to the NF*k*B-specific *cis*-element in the *MUC2* promoter, resulting in up-regulation of *MUC2* transcription. Thus, cooperation between the TGFβ-Smad signaling pathway and NF-κB mediates NTHi-induced transcription of the *MUC2* mucin gene [53].

Recently, NFκB was also implicated in a signaling pathway that activates *MUC2* gene transcription in response to leukotriene D_4 (LTD_4). LTD_4 participates in the host-defense reactions and in pathogenesis of such conditions as immediate hypersensitivity and inflammation [54, 55]. This pathway is initiated by interaction of LTD_4 with its cognate cysteinyl-leukotriene receptor (CysLT1), followed by activation of the cascade G-protein, PKC, MEK, ERK and NFκB. The last one interacts with the *MUC2* promoter, thereby activating the *MUC2* transcription.

In addition to LTD_4 and bacterial stress, NFκB responds to other stress factors such as the bile acids. The bile acids are cytotoxic to esophageal epithelial cells and induce a strong cell-protective response by enhanced production of gel-forming mucins, particularly the MUC2 mucin. Deoxycholic acid up-regulates *MUC2* transcription by activation of NF-*k*B, but in this case activation of the *MUC2* gene occurs *via* PKC, MAPK - independent pathway [56]. The presented examples show that NF-κB transcription factor is a multi-potent element of gene regulation machinery that can be activated by different signaling pathways. NF-κB may functionally interact with multiple adaptor proteins, resulting in transcriptional up-regulation of multiple genes including *MUC2*.

NF-*k*B transcription factor is not only one of the transcription factors that mediate activation of *MUC2* to protect the cell from bacterial stress, although it is an important one. For instance, *Helicobacter pylori (H.pylori)* infection stimulates production of the MUC2 mucin by a mechanism that involves down-regulation of the *Sox2* gene. The contact of *H. pylori* with epithelial cells induces the immune response associated with production of the pro-inflammatory cytokines, including INF-γ, TNF-α, IL-1β and IL-4 [57, 58]. INF-γ inhibits synthesis of the Sox2 transcription factor [59, 60]. The down-regulation of the *Sox2* gene up-regulates Cdx2 protein production, which, in turn, activates *MUC2* transcription.

<u>*Transcription factor CREB*</u>: The *MUC2* promoter also contains the binding sites specific for the CREB and AP-1 transcription factors (Fig. **2**). The CREB family of transcription factors plays a critical role in controlling cell growth, cell cycle progression, and differentiation of many cell types including intestinal and airway epithelial cells [61-64]. The CREB proteins control gene transcription through binding to the specific CRE *cis*-element (5'-TGACGTCA-3') in the promoters of the relevant genes.

The CREB transcription factor has the potential to control *MUC2* expression by a number of signaling cascades. For instance, the *MUC2* gene can be activated in human intestinal epithelial cells by the vasoactive intestinal peptide (VIP), which triggers signal transmission from the cell surface through the VIP-receptor to cAMP ->PKA-> MEK-> ERK1/2-> MSK-> CREB/ATF1 pathway [61]. The activated CREB/ATF1 binds to the CRE *cis*-element of the *MUC2* promoter located at -2571/-2563 nucleotides, resulting in up-regulation of *MUC2* gene expression. The activating signal from PKA may also be transmitted to the p38 and then to MSK->CREB/ATF1. Importantly, VIP can also induce reciprocal inhibition of *MUC2* gene expression by a mechanism involving activation of JNK, which, through interaction with c-Jun, blocks binding of the activated CREB/ATF1 to the *MUC2* promoter.

Regulation of *MUC2* gene activity by the CREB transcription factor can also occur *via* a non-classical retinoic acid (RA) signaling pathway [62]. In normal human tracheo-bronchial cells, RA mediates its activity not in the classical way through RAR/RXR receptors, but through activation of PKCα followed by transmission of the activated signal *via* Raf-> MEK-> ERK-> RSK-> CREB signaling. The activated CREB binds to the CRE-*cis*-elements in human *MUC2* promoter at positions -1047/-1039, -422/-414 and -194/-186, leading to up-regulation of *MUC2* gene expression.

Transcription factor AP-1: The 5'-untranslated region of the *MUC2* gene harbors three AP-1 binding sites located at positions corresponding to the -1871, -818 and -169 nucleotides (Fig. **2**). One of these AP-1sites (-1871) is involved in up-regulation of *MUC2* gene expression by galectin-3 [65]. This pleiotropic member of the β-galactoside-binding protein family [66-68] modulates MUC2 mucin expression in human cancer cells at the transcriptional level *via* direct interaction of the nuclear galectin-3 with the c-Jun/Fra-1 transcription complex, followed by its binding to the AP-1 *cis*- element (-1871) in the *MUC2* promoter [69]. Importantly, the MUC2 mucin is a ligand of galactin-3 [69]. The ability of galactin-3 to up-regulate production of its own ligand may explain the effect of galactin-3 and MUC2 proteins on colon cancer cells and metastasis.

Another AP-1 site of the *MUC2* promoter is located at the nucleotides -818/-808 and is involved in regulation of *MUC2* expression by short-chain fatty acids (SCFA) [70]. SCFAs, and especially butyrate, are known to affect intestinal-specific gene expression [70-73]. Low concentration (1 mM) of butyrate activates MUC2 mucin expression by stimulation of specific binding of the c-Fos/c-Jun complex to the -818/-808 AP-1 binding site; high concentrations of butyrate (5-15 mM) induce repression of *MUC2* expression [70]. It appears that the concentration-dependent effects of butyrate on *MUC2* transcription result from butyrate's multifunctionality. Butyrate may affect *MUC2* gene transcription also through an epigenetic mechanism by changing the "histone code"; butyrate is a well-known inhibitor of histone deacetylases (HDAC) [74], which participate in epigenetic regulation of *MUC2* gene activity [70, 75] (see section 4.2.2).

4.2.2. *MUC2* Gene: Epigenetic Regulation

The term "epigenetics" literally means "outside conventional genetics" and is used to describe the mechanisms that regulate gene expression by posttranslational modifications of either the DNA itself or of the proteins that are "intimately associated with DNA as the key mediators" [76]. It should be pointed out that reversible posttranslational modifications of histone proteins in combination with reversible methylation of specific CpG sites in genomic DNA, especially in a gene promoter, are an important part of gene regulation. These epigenetic mechanisms determine the accessibility of the promoter *cis*-elements to the corresponding transcription factors, thereby facilitating the dynamic regulation of gene transcription [77].

MUC2 as well as other gel-forming mucin genes are located in "a methylation hot spot" of the genome [4]. Repression of the *MUC2* gene observed in many cancer cells is a result of the site-specific methylation within its promoter, and/or specific modifications of histones composing "a regulative histone code" [24]. The term "histone code", introduced into the scientific lexicon by Jenuwein and Allis [78], refers to the concept that "chromatin, the physiological template of all eukaryotic genetic information, is subject to a diverse array of posttranslational modifications that largely impinge on histone amino termini, thereby regulating access to the underlying DNA. Distinct histone amino-terminal modifications can generate synergistic or antagonistic interaction affinities for chromatin-associated proteins, which in turn dictate dynamic transition between transcriptionally active or transcriptionally silent chromatine states. The combinatorial nature of histone amino-terminal modifications thus reveals a "histone code" that considerably extends the information potential of the genetic code".

The first indication of epigenetic regulation of the *MUC2* gene was obtained by Hanski *et al.* [79, 80], who showed that suppression of *MUC2* in colon carcinoma cells is associated with methylation of its promoter region; and that the expression of the *MUC2* gene could be enhanced by treating the MUC2-expressing colon carcinoma cells with 5-aza-2'-deoxycytidine, an inhibitor of DNA-methylation. These results were confirmed and extended by Siedow *et al.* [23] and Ho *et al.* [81], who showed that expression of the *MUC2* gene in pancreatic cancer cells correlates well with the methylation status of the proximal region of the *MUC2* gene promoter: the *MUC2* gene promoter is highly methylated in the MUC2-non-producing cell line PANC-1 and not in the MUC2-expressing cell line BxPC-3. The *de novo* expression of the *MUC2* gene in pancreatic MUC2-negative cells was associated with promoter demethylation triggered by treating the cells with an inhibitor of DNA methyltransferase.

Gratchev *et al.* [82] investigated the methylation status of the nine CpG sites located between nucleotides -350 and +20 of the *MUC2* promoter in the clones of the colon carcinoma cell line COLO 205, characterized by different levels of the *MUC2* gene expression. Five CpG sites in the MUC2-positive clone, were not methylated at all and the level of methylation of the four other CpG sites was relatively low. In contrast, all nine CpG sites in the MUC2 non-expressing clone were highly methylated.

Hamada *et al.* [83] took the CpG methylation of the *MUC2* promoter a step further by mapping the methylation patterns of the 59 CpG sites present in a larger region (~1900 bp) of the *MUC2* promoter. The *MUC2* promoter in the PANC-1 cell line, which, as noted above, does not express MUC2 mucin, appeared to be highly methylated: 28 out of 59 CpG sites were completely methylated, and the average methylation level was 87%. In the MUC2-positive BxPC3 cell line, the promoter region contained only 2 completely methylated sites, and the average level of methylation was 43%. In MUC2-positive normal colon crypts, no CpG site was completely methylated and the average level of methylation reached only 33%. The authors further showed that CpG sites of the *MUC2* promoter are unevenly methylated, confirming earlier observations by Gratchev *et al.* [82] and Mesquita *et al.* [84] that only certain CpG sites in the *MUC2* promoter region, including those in the AP-2 and SP1 binding motifs, are important in *MUC2* transcriptional regulation by DNA-methylation.

Silvester *et al.* [85] analyzed the influence of methylation of the CpG sites located distal (between the -3299 and -1791 bases of the *MUC2* promoter) to those studied by Hamada *et al.* [83] on *MUC2* expression. They found that methylation of these sites is cell dependent and strongly correlated with the stages of cell proliferation and differentiation. DNA methylation strongly suppressed *MUC2* gene expression in proliferative epithelial cells, although this effect was lost once cells became confluent and differentiated. Demethylation of three specific sites in the *MUC2* promoter, -2481, -2347 and -2331, was observed during the transit of colonic cells HT-29 STD and HT-29 5F7 from proliferation stage to differentiation. The decreased methylation of the promoter region can explain the increase in gel-forming mucin expression concomitant with cell differentiation described by Silvester *et al.* [85]. The strong influence of epigenetic DNA methylation on the *MUC2* promoter activity is also seen in the finding that methylation of *MUC2* promoter dramatically impaired activation of the *MUC2* gene by Sp1 transcription factor (50-100% loss). The authors stress that the discovered correlation between the level of methylation of specific CpG sites in the *MUC2* promoter and the degree of cell differentiation points to MUC2 as a marker of differentiated mucus-secreting cells; and that screening for methylation of the key CpG sites in the *MUC2* promoter may be a useful diagnostic and prognostic tool to identify cancer cells undergoing de-differentiation.

To better understand the role of DNA methylation and histone modification in regulation of *MUC2* gene expression, Yamada *et al.* [86] performed methylation-specific PCR and chromatin immunoprecipitation assays in MUC2-negative PANC-1 cells and BxPC3 cells highly expressing MUC2 mucin. They uncovered the epigenetic mechanism that regulates expression of the *MUC2* gene by a combination of highly coordinated DNA methylation and histone modification associated with the *MUC2* promoter. Important conclusions emerged from this study. 1) Once again it was shown that the CpG sites of the *MUC2* promoter are differentially involved in regulation of *MUC2* gene expression: the CpG sites located relatively close to the TATA-box between nucleotides -300 and -1 are more critical to the methylation-related *MUC2* gene silencing than the sites located more upstream. The level of methylation of these sites in MUC2-non-producing cells is higher than the corresponding index in cells actively expressing

MUC2 mucin. 2) There is a significant correlation between histone H3-K9 methylation level and DNA methylation level in MUC2-positive and MUC2-negative cells. 3) Changes in DNA methylation have little effect on histone H3-K9/K27 acetylation status, whereas histone modification strongly affects the *MUC2* DNA methylation landscape. These results unambiguously demonstrated the importance of coordinated changes in DNA methylation and histone modification for epigenetic regulation of the *MUC2* gene.

The next important step in the study of epigenetic mechanisms in the regulation of gel-forming mucin gene expression was made by Van Seuningen's group [24]. Performing a comparative study for the first time, they demonstrated the impact of DNA methylation and histone modifications in transcriptional regulation of the clustered *MUC6*, *MUC2*, *MUC5AC* and *MUC5B* genes in epithelial cancer cells [24] and found that epigenetic regulation of the 11p15 mucin genes is complex and gene- and cell-specific. Although all four genes are located in the same chromosomal locus, known as "a hot methylation spot" [87], only *MUC2* and *MUC5B* are strongly regulated by the DNA methylation and histone modifications. *MUC5AC* is rarely affected by epigenetic regulatory mechanisms, and *MUC6* appears to be completely insensitive to this type of regulation.

These and other studies established that DNA methylation and histone modification are partners in transcriptional repression of the *MUC2* gene. Chromatin immunoprecipitation assay revealed that repression of *MUC2* by DNA methylation in PANC1 cells (MUC2-negative cells) was associated with histone H3 deacetylation and simultaneous H3K9 methylation. In contrast, in the MUC2-positive LS174T cells, high level of *MUC2* expression correlated with hyperacetylation of histones H3 and H4 in the *MUC2* promoter region. The characteristic configuration of the modified histones and associated proteins creates "regulatory histone code" specific for a given gene (*e.g.* *MUC2*) expressed in a given cell type under the specific conditions. The dynamic nature of this code can be deciphered by specific transcription patterns of the *MUC2* gene. The authors conclude that the active or repressive state of the *MUC2* gene in a given cell system depends on the balance between methylation and demethylation of specific CpG sites in the *MUC2* promoter and expression of its histone code.

The recent discovery of small non-coding RNAs (miRNAs) that contribute to gene silencing add a new layer to the complex mechanism of epigenomic regulation [88-90]. However, no information about the role of miRNAs in regulation of the *MUC2* gene activity has been reported.

4.3. MUC2 MRNA: SPLICED AND NON-SPLICED ISOFORMS

In this section, the alternative splicing of the MUC2 primary transcript is discussed. In contrast to membrane-bound mucins, which actively use the alternative splicing mechanisms for generation of multiple mRNA isoforms, these mechanisms are utilized to a much lesser degree by gel-forming mucins. For a long time it was not clear whether the MUC2 apoprotein is translated from a single MUC2 mRNA, or there are several MUC2 mRNA isoforms that may be translated into different MUC2 protein isoforms. The conflicting results obtained in the studies in which expression of the MUC2 glycoprotein was detected by the antibodies developed against tandem repeats [85, 91-93] suggested that both options are relevant. When detected by anti-TR antibodies, the levels of MUC2 expression in colorectal adenocarcinomas were decreased compared with the normal colorectal epithelium, and were high in mucinous carcinomas [85, 91-93]. One explanation for these results is that all the cell lines studied have the same amount of MUC2 apomucin molecules which were differently glycosylated in different cancer cells. The highly glycosylated patterns of the MUC2 mucin can not be detected by anti-TR antibodies, which would show artificially low levels of MUC2 mucin in, for instance, the colorectal cancer cell line. In cells, containing less glycosylated MUC2 glycoprotein, an equal amount of the same antibodies will detect more TR-containing molecules compared with the cells producing highly glycosylated MUC2 protein.

The reported results may also be explained by the translation in different cells of the differently spliced MUC2 mRNAs. Differently spliced MUC2 mRNAs may direct synthesis of different MUC2 protein isoforms in mucinous carcinomas and in colorectal adenocarcinoma. The mucinous carcinoma cells may produce more TR-containing mRNA, whereas in the MUC2 mRNA from colorectal adenocarcinoma, the TR-coding sequences could be spliced off, at least in a part of the MUC2 mRNAs. As a result, the colorectal adenocarcinoma cells may synthesize fewer TR-containing MUC2 mucin polypeptides compared with the mucinous carcinoma cells, although the total amount of MUC2 mucin molecules can be the same in both types of cells. The combination of two mechanisms entailing glycosylation and alternative splicing can not be ruled out.

Sternberg *et al.* [94] showed that the MUC2 primary transcript indeed undergoes alternative splicing, at least in the colon cancer cell line HM7. This event results in two mRNA isoforms, designated MUC2 mRNA and MUC2.1 mRNA. MUC2 mRNA is translated into full-length MUC2 protein, while MUC2.1 mRNA directs synthesis of MUC2 protein isoform that does not contain TRII domain composed of 23 aa tandemly repeated up to 115 times, depending on the allele variations [95]. The full-length MUC2 mRNA consists of 15720 bp, whereas MUC2.1 mRNAs (GenBank Accession No. NM_002457) lacks 6958 bp coding for TRII epitope. At present, this spliced variant is the only experimentally identified MUC2 isoform; others may be discovered. The *MUC2* mucin gene contains multiple cryptic donor and acceptor splice sequences (J. Zaretsky, unpublished data), which, in the appropriate cells and under specific conditions may be recognized and utilized by the splicing machine. This supposition is strengthened by the fact that *MUC2* contains several exonic minisatellite sequences [96], which are known to induce alternative splicing by activation of cryptic splice sites [97-101].

4.4. MUC2 MUCIN: BIOSYNTHESIS, PROCESSING AND SECRETION

Translation of the MUC2 mRNA occurs on polysomes of the endoplasmic reticulum, resulting in synthesis of the high molecular weight (~600-650 kDa) precursor. The synthesis of the precursor polypeptide was shown to be fast, taking less than 1 minute –the first precursor molecules could be detected within 1 minute by the pulse-chase assay [102]. The process of MUC2 glycoprotein maturation includes several stages: 1) initial N-glycosylation; 2) precursor dimerization, 3) O-glycosylation, and 4) proteolytic modifications accompanied by 5) oligo- and multimerization. The mature MUC2 multimers undergo packaging into vesicules followed by constitutive or induced secretion [16, 103, 104] (Fig. **3**).

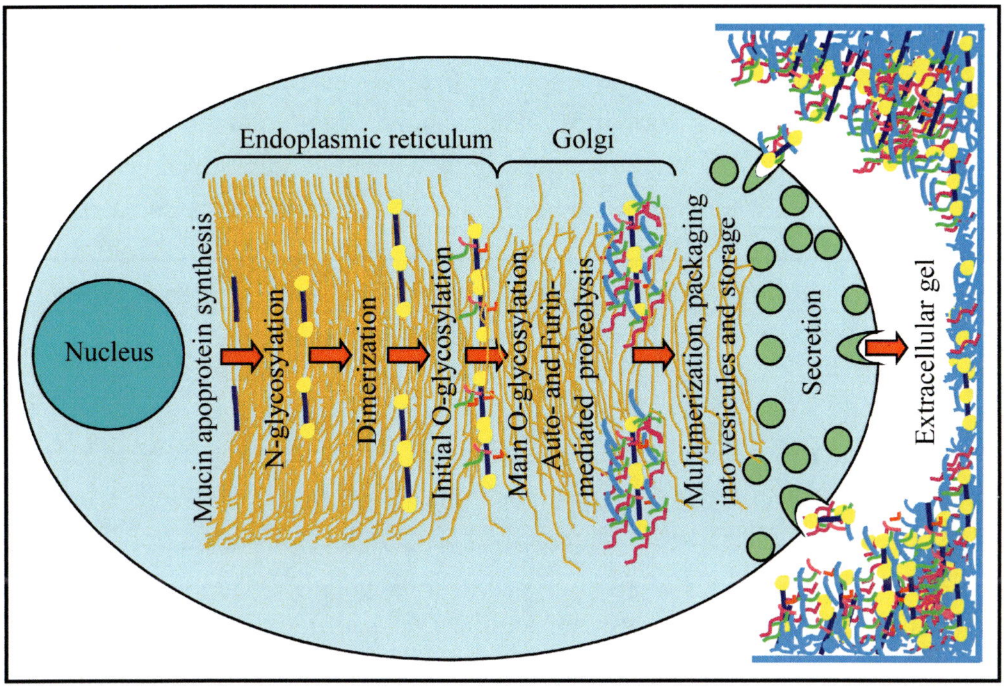

Figure 3: Biosynthesis, processing and secretion of MUC2 mucin (based on the data reported in [102, 105, 107-109]).

4.4.1. N-Glycosylation and Dimerization of MUC2 Precursor

N-glycosylation of the nascent MUC2 polypeptide occurs co-translationally, concomitant with the internalization of the growing polypeptide chains into rough endoplasmic reticulum (RER), where dimerization of the monomers occurs [15]. Asker *et al.* [105] showed that initial dimerization of the MUC2 precursor monomers begins immediately after translocation of the apomucin molecules into RER, as detected by calnexin reactivity. Inhibition of N-glycosylation and site-specific mutation of the N-glycan-binding sites in the cysteine knot region of the MUC2 molecule demonstrated the importance of N-glycosylation for the MUC2 mucin dimerization and secretion [106]. Tunicamycin, an inhibitor of N-glycosylation, induces retardation of rat and human MUC2 oligomer formation [107-109]. MUC2 has been shown to bind also calreticulin, an N-glycan-specific chaperone, indicating that the interaction of N-glycosylated mucin with chaperone is part of the normal MUC2 maturation process [110]. However, different N-glycans linked to the MUC2 molecule play different roles in mucin dimerization. Bell *et al.* [106] showed that two N-glycans, N9 and N10, located near the first and the second cysteines of the cysteine knot, play different roles in the process of disulfide-dependent dimer formation. They showed that mutation of the N10 glycan did not induce any major changes in synthesis, maturation or secretion of the MUC2 mucin, whereas mutation of the N9 glycan was associated with incorrectly assembled MUC2 dimers and secretion of the unstable mucin molecules. It would be important to study the role of other N-glycans present in the fully glycosylated MUC2 molecule in mucin processing.

Like N-glycans, the cysteine residues of the cysteine knot also influence MUC2 dimer formation and secretion. Deletion of cysteine knot from the rat Muc2 mucin resulted in inability of the truncated molecule to form dimers. The structural integrity of the cysteine knot, maintained by intramolecular bonds, appears to be important for mucin dimerization. Mutations of the cysteine residues of the cysteine knot – comprised of three intramolecular disulfide bonds (Cys-1=Cys-4, Cys-2=Cys-5, Cys-3=Cys-6) – lead to impaired dimerization, although the extent and character of the changes associated with mutation of a given cysteine residue are different. Bell *et al.* [111] showed that unpaired Cys-X residue plays a key role in the dimerization process, as substitution of the Cys-X residue for Ala completely abolished dimerization of the rat Muc2 mucin. The correctly dimerized MUC2 molecules are transferred to the Golgi, where the next steps of processing and maturation take place.

N-glycans play an important role not only in MUC2 dimerization, but also in targeting of dimers to the Golgi [105, 106, 110]. Importantly, MUC2 oligomerization itself is not affected by inhibition of RER-to-Golgi transport and is independent of medial- and trans-Golgi functions [109]. However, normal functioning of the Golgi apparatus is of great importance for further maturation of the MUC2 mucin associated with O-glycosylation in the Golgi cisternaes [102, 107, 109].

4.4.2. O-Glycosylation of MUC2 Apomucin

O-glycosylation of the MUC2 apoprotein is a step-by-step process mediated by specific carbohydrate-transferases: N-acetyl-galactosaminyl-, N-acetyl-glucosaminyl-, galactosyl-, sialyl-, fucosyl- and sulfo-transferases [112]. The first step in this process involves the transfer of a GalNAc residue to serine or threonine residues. At the second stage, the core structures containing Gal and Glc carbohydrates are added, followed by poly-N-acetyllactosamine elongation of O-glycan chains. The terminal stage of O-glycosylation includes sialylation, fucosylation and sulfation of the pre-synthesized carbohydrate chains. It should be pointed out that every step of O-glycosylation is mediated by multiple transferase isoforms that demonstrate strong, but in some cases overlapping, tissue specificity [112-119]. Different activity and cell specificity of the transferases involved in O-glycosylation of the MUC2 apomucin are reflected, for example, in differences of MUC2 glycosylation along the proximal-to-distal axis in rat colon. Whereas biosynthesis of the rat Muc2 precursor polypeptide appeared to be constant in all regions of rat colon with respect to both amount and molecular mass, a gradient in sulphation of mature mucin molecules was observed, increasing from proximal to distal regions [120]. This difference in sulphation may have implications for the local function of mucin observed in animal models and humans [121-123].

It was recently shown that the activity and specificity of the transferases involved in MUC2 O-glycosylation are dependent on the amino acid sequences of the protein substrate and the nearby carbohydrate moiety. Brockhausen *et al.* [124] examined the site-directed O-glycosylation of mucin type proteins by several transferases, using a series of synthetic MUC2 VNTR-derived peptides and glycopeptides as acceptors. They found that specific threonine residues were the preferred sites for the GalNAc addition. Notably, proline in the +3 position especially enhanced primary threonine glycosylation. The inverse relationship was found between the size of adjacent glycans and the rate of GalNAc addition. The four studied enzymes (polypeptide Gal-NAc-transferase, core 1β3-Gal-transferase, core 2β6-GlcNAc-transferase, and β4-Gal-transferase) could distinguish between substrates having different amino acid sequences and O-glycan specific sites. The authors came to the conclusion that activities of these enzymes assembling the extended core 2 structure on the MUC2 tandem repeat-derived peptides are affected by the amino acid sequence and presence of carbohydrates on nearby residues in the acceptor glycopeptides. For more detailed information on chemical and enzymatic properties of participants in the O-glycasylation process, the readers are refereed to Brockhausen's paper [124] and to several recent excellent reviews [112, 125].

The study of the dynamics of MUC2 biosynthesis and processing has yielded the following time table and consequences. Within the first minutes of biosynthesis, both the MUC2 monomeric molecules and dimers appear. O-glycosylation of MUC2 core begins within 0.5-1 hour of MUC2 precursor biosynthesis, when the major population of mucin molecules is represented by dimers. Secretion of mature mucin molecules begins after the first hour and is completed within 4 hours of MUC2 biosynthesis [102]. Thus, it appears that MUC2 dimers, but not monomers, are the main substrates for MUC2-specific O-glycosylation. This conclusion contradicts the data obtained in some studies. Axelsson *et al.* [126] found two populations of O-glycosylated MUC2 molecules in colon adenocarcinoma LS 174T cells. One population contained large water-insoluble oligomers formed by non-reducible intermolecular bonds (non-disulfide bonds).These molecules contained O-linked glycans although they occurred early in biosynthesis before transferring of the dimers to the Golgi, where, as noted above, the main O-glycosylation processes take place. The second population of MUC2 molecules was represented mainly by classical (glycosylated in the Golgi) water-soluble disulfide bond-mediated dimers. Importantly, this set of MUC2 molecules also contained O-glycosylated monomers. Thus, on the one hand, some MUC2 molecules were O-glycosylated in the "wrong" place and time before being transferred to the Golgi where O-glycosylation usually proceeds; and, on the other hand, monomers that are usually present in the RER but absent in the Golgi, were found in this compartment in the O-glycosylated form. Interestingly, Asker *et al.* [105] also found both O-glycosylated monomers and dimers in the Golgi compartments.

The discrepancy between these studies may be explained by different extents of O-glycosylation of MUC2 mucin precursors in different cell compartments. Studies of the rat gastric mucin showed that initial O-glycosylation is a continuous process: it begins on the nascent polypeptides in the RER, but with the addition of the major site of core GalNAc in the late endoplasmic reticulum compartment [127], or in the Golgi [107]. Dekker and Strous [107] found that 5% of the gastric mucin precursor monomer contained O-linked GalNAc carbohydrate and considered it the initial glycosylation step in the multistep-process of mucin O-glycosylation. McCool *et al.* [108] showed that about 13% of monomeric

MUC2 molecules in the human colonic cell line LS180 were partially O-glycosylated and contained only short carbohydrate structure limited to GalNAc but not longer O-linked core Galβ1,3GalNAc sugar. The authors interpreted these results as evidence of limited O-glycosylation of MUC2 monomers that may occur before oligomerization, *i.e.* very soon after completion of the peptide biosynthesis. They considered that "early O-glycosylation of some serine and threonine residues can help to stabilize and extend the conformation in the center of the nascent mucin chains so that the cysteine-rich N- and C-terminal domains are kept well separated". The main conclusion from this study is that oligomerization may follow some core O-glycosylation with GalNAc, but precedes elongation of oligosaccharide chains.

The aforementioned studies illustrate the complexity of the MUC2 glycosylation process and suggest the possibility of variable changes in the MUC2 properties at each stage of this process, which, in turn, may influence cell physiology and pathology.

4.4.3. Proteolytic Modifications of MUC2 Mucin Precursor

The striking similarity between the N- and C-termini of MUC2 and the von Willebrand factor makes it likely that posttranslational proteolytic modifications of these two glycoproteins also follow similar pathways. However, this is not the case. While the proteolytic events resulting in mature vWF proteins are associated with the N-terminal region, proteolysis of MUC2 precursor is associated with the two cleavages within both the N- and the C-termini of the molecule [128-130].

Some 40 years ago investigators debated the existence of a discrete "link" mucin-derived peptide thought to join together the high molecular weight mucin molecules to form polymers by intermolecular disulfide bond bridges [131, 132]. Later on, Xu *et al.* [133] established the nature of the "link" peptide, which in the rat cell system was a C-terminal 689 amino acid-containing glycopeptides (118-kDa) resulting from a hypothetical proteolytic cleavage of the rat Muc2 mucin. Lidell *et al.* [134], working with human MUC2 mucin, confirmed the results of Xu *et al.* [133] and showed that the "hypothetic proteolysis" is a real auto-catalytical cleavage occurring in the C-terminus of the MUC2 mucin at the low pH of the late secretory pathway. They showed that the cleavage site specific for this reaction is represented by the

^{4485}GDPH^{4488} sequence, and proteolysis of the peptide bond occurs between Asp (D) and Pro (P) residues. Importantly, the GDPH sequence detected in many mucin and non-mucin glycoproteins was found to undergo auto-cleavage [135-138], indicating the nature of the proteolytic reaction associated with this sequence. This posttranslational modification may have a number of functional consequences, including those leading to conformational changes that may affect viscosity of the cleaved MUC2 mucin. Another scenario suggests interaction of the cleaved mucin with other proteins [134] that may incorporate the MUC2 mucin into different metabolic and/or signaling pathways.

The auto-cleavage described above is not the only proteolytic reaction in post-translational proteolysis of the MUC2 mucin. Xu *et al.* [139] found at the very C-terminus of the rat Muc2 and human MUC2 mucins a sequence corresponding to the consensus target recognition site (R-X-R/K-R^X) specific for endoprotease furin [140]. Further analysis showed that the protease furin indeed cleaves rat and human MUC2 mucins at this site, cutting off *in vivo* the positively charged C-terminal 9 aa-containing peptide RTRR^**SSPRLLGRK**. The cationic C-tail region of the MUC2 mucin plays an important role in correct folding of the primary monomers, ensuring formation of mucin dimers in the ER. The short C-terminal fragment of the MUC2 mucin is removed prior to the passage of mucin into apical storage granule mass of goblet cells, thus regulating packaging, routing and storage of mucin molecules in granules [139].

The two posttranslational proteolytic modifications of the MUC2 mucin discussed above are apparently extremely important for the biochemistry of this protein, but the consequence of these events in the processing of the MUC2 mucin is not clear. Does the furin protease-mediated proteolysis influence the auto-cleavage reaction at the GDPH-site of the MUC2 molecule, and *vice versa*? Do these reactions occur simultaneously, or there is a time interval between them? These and other questions important for understanding the role of posttranslational proteolysis in mucin physiology and pathology await clarification.

4.4.4. Multimerization of MUC2 Mucin

The physiological defense barrier provided by mucin gel results from multimerization of mucin dimers. Multimer formation differs from dimer formation in several aspects. Perez-Vilar *et al.* [141] found that disruption of the Golgi complex by brefeldin A inhibits multimer formation of the porcine homolog of the human MUC2 mucin [142]. Agents that increase the pH of the *trans*-Golgi compartments also inhibit formation of multimers, but do not affect formation of mucin dimers. Perez-Vilar *et al.* [141] established that interchain disulfide bonds that give rise to multimerization are formed at a slightly acidic pH in the *trans*-Golgi complex through cysteine residues of the N-terminal D-domains. Thus, the assembly of gel-forming mucin involves two separate steps: dimerization in the endoplasmic reticulum and multimerization in the *trans*-Golgi compartments. While the molecular mechanisms that determine specific compartmentalization of mucin polymerization are not known, the CGLCG motifs found in the D1- and D3-domains seem to play a critical role in this process [141, 143]. Interestingly, this motif in the D3-domain prevents multimerization of mucin in endoplasmic reticulum and *cis/medial*-Golgi compartments, while the same sequence in D1-domain determines multimer formation in the *trans*-Golgi cisternae [141, 143].

In summary, the process of MUC2 biosynthesis and posttranslational modification is a multistep process. It includes synthesis of the polypeptide precursor in the rough endoplasmic reticulum, dimerization in the rough and smooth endoplasmic reticulum, glycosylation in the Golgi, and, ultimately, formation of mature multimeric molecules followed by their packaging and storage in secretory vesicules. After secretion, these multimers form mucin gel that apparently undergoes proteolytic changes in response to physiological or pathological signals.

4.5. MUC2 MUCIN: EXPRESSION AND FUNCTIONS

MUC2 and many other mucin genes were cloned less than two decades ago. Nevertheless, some fundamental data on MUC2 mucin properties and functions have been obtained. While mucins expressed in respiratory, gastrointestinal and urogenital tracts share many functions, the pattern of individual mucin expression is organ-, tissue- and cell-specific [10]. The MUC2 glycoprotein is considered a predominantly intestinal mucin [6], despite being expressed in embryogenesis and development in all parts of the gastrointestinal and respiratory tracts, but not in urogenital organs [144]. Importantly, when a given mucin is expressed in different tissues, it may fulfill different functions, which may even be opposites. For example, MUC2 mucin is constantly expressed in colon and prevents malignant transformation of colon epithelium, thereby functioning as a tumor suppressor [145]. It is never detected in normal stomach epithelium, but is always associated with intestinal metaplasia and dysplasia of stomach epithelium and with gastric adenocarcinoma [146-148], suggesting oncogenic potentials of this mucin.

MUC2 displays several types of functional activities. It is active in embryogenesis and development, in innate immunity, in mucus gel formation, in supplying a mechanical shield for epithelial cells, in renewal of epithelial cell populations throughout adult life of the organism, and in pathological states such as inflammation and carcinogeneis. In cancer, it may function as a tumor suppressor or as a tumor stimulating factor, depending on the type and location of the cancer. In addition, MUC2 may function as regulator of pH, hydration and lubrication [5, 15, 145]. The expression patterns of the MUC2 mucin and some associated functions are discussed below.

4.5.1. Expression of *MUC2* Gene in Embryogenesis and Development

The MUC2 glycoprotein is an active participant in the complex biochemical program that determines the development of epithelial organs from the primitive gut [149]. Busine *et al.* [150] showed that MUC2 as well as other gel-forming and membrane-bound mucins exhibit a complex spatio-temporal expression pattern that is organ- and tissue-specific, and therefore quite different in developing airways, stomach and intestine. *MUC2* expression in the embryonic lung cells differs from that observed in normal adult respiratory mucosae. In undifferentiated epithelial cells of human fetal lungs, MUC2 mRNA is present by 9-10 weeks of gestation (Table **1**).

Table 1: Expression of the *MUC2* Gene in Human Embryonic and Fetal Respiratory Tract.

Organ/Tissue/Cell	Gestation (weeks)	References
Embryonic and fetal trachea, bronchi, tubules and terminal sacs	9-10	[151]
Fetal respiratory mucosa	13	[153]
Goblet cells of bronchus and bronchioles	19	[151]
Goblet cells of bronchus, and submucosal glands	23, and in adults	[144]

In fetus older than 23 weeks and in adult organism, *MUC2* is strongly expressed in epithelial goblet, basal and suprabasal cells, and in submucosal glands. Although *MUC2* expression is detectable in embryonic lungs, the level of its expression is very low compared with other mucin genes such as *MUC1* or *MUC4* [151]. However, the dynamics and site-specificity of the *MUC2* expression in the embryonic airways [152] suggest specific functions of the MUC2 mucin in lung development.

In contrast to the respiratory tract, cells and tissues of the developing gastrointestinal organs express *MUC2* at high levels [151]. However, according to several studies [149, 154, 155], liver and pancreas are the exceptions to the rule, as no expression of MUC2 mRNA and protein could be detected in these organs throughout embryogenesis and development. Lopez-Ferrer *et al.* [156], on the other hand, did observe MUC2 apomucin expression in developing pancreas. *MUC2* is expressed at a high level in the fetal duodenum as early as 10 weeks of gestation (Table **2**) [149]. Its expression in the goblet cells of jejunum, ileum and colon was observed two weeks later, although it was already detected in non-differentiated epithelial cells by 9 weeks of gestation [151].

Table 2: Expression of the *MUC2* Gene in Human Embryonic and Fetal Gastrointestinal Tract.

Organ/Tissues/Cells	Gestation (weeks)	References
Intestinal epithelial cytodifferentiation	9	[40]
Duodenum (surface and crypt epithelial cells	10	[149]
Duodenum glands	26, and adults	[149]
Jejunum goblet cells	12	[151]
Ileum goblet cells	12	[151]
Colon goblet cells	12	[151]
Stomach antral glands	26	[150]
Stomach surface epithelium and fundic glands	N/E	[149]
Fetal gallbladder	18	[149]
Liver and Pancreas	N/E	[149, 151, 154]

Legend: N/E – not expressed.

The pattern of *MUC2* expression is different from that of other mucin genes, in particular *MUC1*gene. *MUC1* is expressed only in colon; it could not be detected in jejunum and ileum at any stage from 16 weeks of gestation; it is expressed at a moderate level in cryptic epithelial cells and in decreased amount in cells migrating along the villi. In contrast, *MUC2* is expressed in high levels in both crypt and in villus epithelium [151], and a high level of intestinal *MUC2* expression is constantly detectable in adults [149, 151].

MUC2 mRNA is detected in the fetal stomach, by 26 weeks of gestation; however, its expression is weak and limited to the antral glands. MUC2 expression is not detected in the surface epithelium of stomach and fundic glands at any stage of gestation. A striking feature of the pyloric and cardiac regions of the developing stomach is the presence of "intestinal-like" mucosa areas containing goblet cells. These "intestinal-like" regions have virtually disappeared at birth [157]. Their significance in the developing stomach is not clear, but one can suppose that these cells, which are the main producers of the MUC2 secreted mucin, may have physiological meaning in the context of specific cell differentiation in stomach. A specific role of *MUC2* gene expression in the development of gastrointestinal organs becomes obvious when compared with its total suppression during normal development of such organs as kidney, male genital ducts and mid-trimester testis (Table **3**) [151, 154, 155].

Table 3: Expression of the *MUC2* Gene in Human Embryonic and Fetal Urogenital Tract.

Organ/Tissue/Cell	**Gestation (weeks)**	**References**
Male genital ducts and testis	very low levels at all stages	[151]
Kidney	N/E	[151, 154, 155]

Legend: N/E – not expressed.

Importantly, the expression of *MUC2* in embryogenesis, development and normal adult organisms is strongly regulated and coordinated with the expression of other mucins. In pathological conditions its expression is disregulated and occurs in cells and tissues known as MUC2 mucin non-producing cells.

4.5.2. MUC2 Mucin Function in Innate Immunity

Innate immunity is a natural pre-existing and nonspecifc defense mechanism independent of prior sensitization to an antigen by infection or vaccination. Since it is not stimulated by specific antigens, innate immunity is generally nonspecific in contrast to acquired immunity. Mucins contribute to innate immunity by their ability to entrap microbes, particles or irritants for clearance from respiratory, gastrointestinal and urogenital tracts as well as from conjuctiva and middle ear [104]. Mucus gel contains a number of antibacterial peptides and multiple protective enzymes, providing extracellular surface with antimicrobial activity. In this context, the gel-forming mucins contribute to innate immunity as scaffolds for cell-protective agents.

The intestinal microbes stimulate development of both innate and acquired components of the mucosal immune system. As pointed out by Deplancke and Gaskins [158], "much of the structure and many of the functions of the mammalian intestine seem to have evolved to enable the host to tolerate the antigenic and chemical challenges associated with the permanent carriage of a complex microbiota". The mucus gel that overlies the intestinal epithelium is the anatomical site at which the host first encounters gut bacteria. It functions as a dynamic defense barrier responsive to signals received from intestinal microbes or parasites [159-161]. The MUC2 mucin forms a heterogeneous mucus gel containing two distinct layers: a "loose" outer layer pervious to bacteria, and an adherent inner layer that defends the underlying epithelium from direct contact with microorganisms [162].

The efficiency of the innate defense mechanism provided by Muc2 mucin was recently investigated by Hasnain *et al.* [163] in a mouse model of nematode *Trichris muris* infection. Worm infection induced hyperplasia of goblet cells and increased production of Muc2 mucin. An increase in Muc2 mucin expression correlated with worm expulsion, and absence of Muc2 expression resulted in a delay in worm elimination. The physical properties of the mucus barrier were also altered during infection, resulting in a less porous network with overall changes that had a direct effect on the viability of the whipworm. This study clearly showed that a Muc2-enriched mucus barrier is important for a well-coordinated response in the gut to worm infection. It further highlighted the functional dynamics and highly regulated nature of the mucus barrier during intestinal infection [163].

4.6. EXPRESSION OF *MUC2* GENE UNDER PHYSIOLOGICAL CONDITIONS

Numerous studies have demonstrated the importance of the gel-forming mucins for normal physiology of epithelial surfaces [1, 6, 10, 15, 144, 149, 152, 158, 163]. The physiological pattern of an individual mucin expression is tissue- and cell-specific, although epithelial surfaces of most parenchimatous organs generally express several types of mucins. The most pronounced expression of the MUC2 mucin is in the intestine [1, 164], with definite expression at much lower level in other organs and tissues both during embryogenesis (Tables **1-3**) and in adults [15, 165]. The following sections address the physiological expression of the MUC2 mucin in cells and tissues of different organs.

4.6.1. Expression of *MUC2* Gene in the Gastrointestinal Tract

MUC2 is highly expressed in goblet cells of normal duodenum and to a lesser extent in Brunner's gland [149]. The goblet cells of villi and crypts are the main locations of MUC2 mucin synthesis and secretion in the small and colorectal intestine [11, 166, 167]. At the same time, MUC2 is rarely detected in normal salivary glands [168, 169], esophagus, stomach [15], pancreas [16, 170] and hepatobiliary system [15, 16]. Interestingly, the serous and mucous acini of salivary glands do not express MUC2 mucin, while its expression in the intercalated and striated ducts could be detected by different antibodies [15]. Analogously, the acini and islets of Langerhans in the pancreas do not express the MUC2 apomucin, whereas its expression, albeit at a low level, was observed in the pancreatic ducts and ductules [15]. The normal surface epithelium of stomach is also generally MUC2-negative. Very low levels of MUC2 synthesis were detected in some areas of the stomach, including the pyloric glands and mucous neck cells of the stomach body [15, 16].

4.6.2. Expression of *MUC2* Gene in the Respiratory Tract

A relatively high level of *MUC2* expression is observed in the respiratory tract under physiological conditions. In normal respiratory organs, *MUC2* is substantially expressed in basal and goblet cells. Its expression, albeit at lower level, is also detected in mucous and some serous cells [171, 172]. MUC2 mRNA and protein expression was observed in normal nasal mucosa [173, 174], in normal ciliated epithelium and in serous and mucous glands in bronchus [15], and in bronchial washings from normal airways, but at a very low level [175]. The search for MUC2 expression in the lung alveoli yielded negative results [15]. It should be noted that relatively small numbers of studies have been performed in regard to evaluate expression of the *MUC2* gene in normal respiratory tract tissues. Further studies of this issue are needed.

4.6.3. Expression of *MUC2* Gene in the Urogenital Tract

The organs of the urogenital tract are less studied in regard of *MUC2* expression than cells and tissues of the gastrointestinal and respiratory tracts.

4.6.3.1. Male Urogenital Organs

Relatively little is known about mucin synthesis and secretion in the male urogenic and reproductive tracts. With regard to *MUC2* expression, the prostate has been studied more thoroughly than other male reproductive organs, however, with conflicting results. Durrant *et al.* [176] demonstrated the presence of the MUC2 mucin in normal prostate by immunohistochemical method, while Ho *et al.* [14] failed to detect MUC2-specific transcripts in normal prostate. According to Lagow *et al.* [177], the discrepancy between these two studies may be explained by assuming that MUC2 mucin is transported to the prostate from other sources. Zhang *et al.* [178] detected MUC2-positive normal prostate epithelium in 3 of 10 prostate samples, while Osunkoya *et al.* [179] stated that MUC2 "is not present in either normal prostate or the majority of conventional adenocarcinomas of this organ". The absence of *MUC2* expression was also documented in the testis by the RT-PCR method [180]. Conflicting results were obtained in studies of normal urothelium of the bladder: Retz *et al.* [181] observed expression of MUC2 in all samples of normal bladders studied, while Walsh *et al.* [182] found no MUC2 mucin in any of 11 normal urothelium specimens. No indication of *MUC2* expression in normal kidney was found in several studies [151, 154]. These studies demonstrate the need for further research to clarify *MUC2* gene activity in the male urogenital system.

4.6.3.2. Female Genital Organs

Relatively more information is available regarding *MUC2* expression in the female reproductive organs [177, 183-187]. It appears that mucins, including MUC2 glycoprotein, play important roles in the female reproductive tract. They provide a barrier to sperm and pathogen entrance into the endometrium and form protective covering to the vaginal epithelium. The amount and properties of the locally produced mucins are under control of the reproductive hormones and depend on the phase of the reproductive cycle [183]. Like expression of other mucins synthesized in the female reproductive system, the MUC2 expression also undergoes changes coordinated with the menstrual cycle. Variable levels of MUC2 expression have been detected by immunochemistry in endometrium and endocervix during the secretory phase, while no differences in MUC2 expression were observed in these tissues in the postmenopausal period [183]. Alameda *et al.* [184] also reported changes in MUC2 expression dependent on the endometrial cycle: the level of MUC2 expression increased during the secretory phase, suggesting a role for MUC2 in mucus secretion. However, other investigators did not observe MUC2 expression in normal endometrium and endocervix [15, 185, 186], or found its expression in those tissues at low to negligible levels [187]. The conflicting results of the different studies may be due to differences in the reproductive epithelium samples being taken at the different stages of the menstrual cycle. Expression of MUC2 was also not detected in the surface and follicular epithelium of the ovary [15, 165], the vagina, Fallopian tube, or endo- or ectocervix tissues [183].

4.6.4. Expression of *MUC2* Gene in the Breast, Eye and Ear

MUC2 expression was also studied in "nonmucinous organs" sucg as mammary gland, eye and middle ear. MUC2 expression was not detected in secretory tubule, acini and intralobular duct of the normal breast [15, 165]. In the eye, it was consistently expressed in the normal ocular surface epithelium and in tear fluid [188-190]. Apparently, biosynthesis and processing of the MUC2 mucin in ocular tissues have specific peculiarities. According to Berry *et al.* [191], MUC2 mucin expressed on the ocular surface is smaller in size than intracellular species. The authors consider that eye physiology dictates that the full-length MUC2 mucin undergo additional postsecretory cleavage in the pre-ocular compartments. *MUC2* expression was recently detected in human [192, 193] and animal [194] middle ear epithelium.

4.7. EXPRESSION OF *MUC2* GENE UNDER PATHOLOGICAL CONDITIONS

The data presented above show that the *MUC2* gene is expressed in normal physiological conditions in most epithelial tissues. And, although its level of expression is different in different organs, it is strongly controlled by complex regulatory mechanisms. In pathology, the *MUC2* gene is expressed aberrantly: cells that in physiological conditions produce MUC2 at very low levels or do not produce it at all synthesize and secrete MUC2 mucin in large quantities, while cells that in normal condition produce MUC2 mucin at high levels decrease or even stop synthesis of this glycoprotein. The expression and functions of the MUC2 mucin in the pathology of different organs are discussed below.

4.7.1. Expression of *MUC2* Gene in the Respiratory Tract

Enhanced secretion of mucus by human airway epithelial cells is an acute defense mechanism in response to inhaled stimulating substances or hazardous chemicals. Mucus hypersecretion in chronic inflammatory airway disorders such as asthma, chronic obstructive pulmonary disease (COPD), chronic bronchitis, allergic rhinitis and rhinosinusitis is one of the most important characteristics of these diseases [195]. Many studies have showed differences in the content of mucins, including MUC2, in airway tissues in physiological conditions and in pathology.

As shown above, the expression of *MUC2* in developing airways is restricted mainly to the epithelium of large bronchioles and occurs relatively late in development (19 weeks of gestation) [151]. The expression level of the MUC2 mRNA is very low in the developing lung and higher in the adult lung where it can be detected in goblet cells of the main bronchus and bronchioles [151]. Nevertheless, the levels of the MUC2 mRNA and protein in human lung and bronchial tissues are very low compared with those in human intestine and colon [15, 175]. Importantly, despite low expression of *MUC2* under physiological conditions, its expression in the pathologically changed airways (inflammatory and/or malignant diseases) is sharply up-regulated.

Inflammatory diseases: The early studies on airway inflammation performed in a rat model disclosed up-regulation of the *Muc2* gene expression compared with corresponding non-inflamed tissues [196]. According to Ho *et al.* [14], over-expression of *MUC2* is characteristic of chronic bronchitis and lung form of cystic fibrosis (CF), suggesting a key role of MUC2 mucin in the pathogenesis of these diseases. However, according to the data obtained by Chambers *et al.* [151], it is unlikely that MUC2 mucin plays an important role in the development of lung CF disease, at least in children, although later in life this mucin apparently contributes to CF-associated pathology of large airways. It has been reported that transcriptional activation of *MUC2* by LPS of *P. aeruginosa* may play a role in the pathogenesis of CF [52, 152]. As shown by Li *et al.* [197], the *MUC2* gene is expressed at three to four-fold higher levels in CF nasal mucosa than in non-CF nasal tissue. The *MUC2* gene is also expressed in idiopathic pulmonary fibrosis, although at a much lower level than expression of the *MUC5B* gene [198].

Several studies showed that up-regulation of the *MUC2* gene is characteristic of asthma and COPD [199-202]. Multiple data indicate that in asthmatic "irritated" airways, *MUC2* is significantly over-expressed in goblet cells, which do not express this mucin in the physiological state [199, 201, 202]. At the same time, MUC2 mucin has been described as an inconspicuous and irregular component in COPD sputum [175, 201], an observation confirmed by Caramori *et al.* [203], who also showed that MUC2 is not a major secretory mucin in the peripheral airways.

Thus, it appears that MUC2 mucin is indeed involved in the pathogenesis of the respiratory tract inflammatory and allergic diseases. Its contribution to the course of different types of airway pathology is not identical and is apparently disease-specific.

<u>Lung malignancies</u>: There is ample evidence of the involvement of MUC2 and other mucins in the progression of human carcinomas and promotion of tumor cell metastasis [152, 204]. The possible involvement of MUC2 glycoprotein in the pathogenesis of lung cancer has been demonstrated [205-209].

The WHO classification of lung cancer defines a typical adenomatous hyperplasia (AAH) as a premalignant lesion of lung [210]. This definition is supported by several studies that showed AAH to be a precursor of bronchioloalveolar carcinoma (BAC) [205, 206]. Progression from AAH through non-invasive BAC to invasive mixed subtype (MX) has been described [207-209]. Awaya *et al.* [207] showed that the expression levels of *MUC2*, *MUC5AC* and *MUC6* are significantly increased in the progression from AAH through BAC to MX. Alterations in the expression of these genes are associated with dedifferentiation of bronchial epithelium. Importantly, the levels of *MUC2, MUC5AC* and *MUC6* expression in highly aggressive mucinous BACs are higher than in non-mucinous BACs, which demonstrate non-aggressive behavior. This observation correlates with the data reported by Barsky *et al.* [211], who attributed different behavior of lung carcinomas to different expression of the secretory mucins. Higher aggressiveness of mucinous lung tumors is also in line with the findings of Nishiumi *et al.* [212] that small adenocarcinomas of lung, characterized by high expression of MUC2 and MUC6 mucins, showed a significantly higher incidence of nodal metastasis than those in which these mucins were not expressed. Moreover, cases with high MUC2 mucin expression had a significantly poorer prognosis than those without MUC2 expression. Interestingly, in gastrointestinal tumors, MUC2 expression is associated with noninvasive proliferation of tumors and a favorable outcome for the patients [213-215]. It appears that in the gastrointestinal tract, MUC2 mucin behaves as a tumor suppressor that restrains the tumorigenic process, whereas in the airway organs the mucin fulfills tumor-promoting functions.

An important observation was made by Stacher *et al.* [216], who found that the up-regulated expression of *MUC2* in premalignant atypical goblet cell hyperplasia in congenital cystic adenomatoid malformation (CCAM) is associated exclusively with the nuclear localization of the expressed MUC2 and IL-4Rα glycoproteins, while in normal cells these two proteins are usually expressed in the cytoplasm. The authors consider nuclear translocation and over-expression of MUC2 and IL-4Rα to be indicative of their involvement in neoplastic transformation, particularly in CCAM.

It is well known that primary pulmonary adenocarcinomas (PPAC) are very heterogeneous tumors that contain a wide variety of histological features including papillary, acinar, tubular, solid, bronchioloalveolar, and mucin-producing elements. Several variants of PPAC have been identified, including mucinous ("colloid") adenocarcinoma, mucinous cystadenocarcinoma, signet-ring cell adenocarcinoma, clear cell adenocarcinoma and pulmonary adenocarcinoma with enteric differentiation (PAED) [217-219]. It is clinically important to distinguish PAED from metastatic colorectal carcinoma (MCR) and other lung carcinomas, but their differentiation is difficult [220]. Inamura *et al.* [217] showed that PAED clearly has a separate phenotype of specific markers that allows distinguishing it from other pulmonary adenocarcinomas. Immunohistochemically, pulmonary adenocarcinomas are generally positive for CK7, TTF1 and Napsin A, and negative for MUC2, CDX2 and CK20 markers [221]. In contrast, the coordinated expression of MUC2 and CDX2 was observed in primary mucinous (so called "colloid") carcinomas of the lung [222]. As emphasized by Mesquita *et al.* [223], the coordinated expression of MUC2 and CDX2 in lung is not surprising and is determined by CDX2, a transcriptional regulator of *MUC2* gene expression. According to the above studies, some types of lung carcinomas express MUC2 mucin while others do not [224]. These conflicting findings highlight the need for further study of the relationship between *MUC2* expression and histological differentiation of lung carcinomas. Correlation between MUC2 mucin expression and expression of other cell and tumor specific markers also awaits clarification.

4.7.2. Expression of *MUC2* Gene in the Gastrointestinal Tract

Epithelial cells of the gastrointestinal tract are the main producers of the MUC2 glycoproteins. Abnormal expression of MUC2 mucin is an important element in the pathogenesis of many gastrointestinal diseases, including inflammatory diseases such as ulcerative colitis and Crohn's disease and different types of carcinomas. Table **4** summarizes the expression of *MUC2* in different organs of the gastrointestinal tract under pathological conditions.

Table 4: Expression of *MUC2* Gene in Pathology of Gastrointestinal Tract.

Organ	MUC2	References
Esophagus:		
Barrett's esophagus	+	[225-227]
Esophagus adenocarcinoma	+	[227]
Stomach:		
Gastric adenocarcinoma	+	[228, 229]

Table 4: cont….

Duodenum:		
Duodenal goblet cells	+	[230]
Brunner glands	-	[230]
Pancreas:		
Colloid adenocarcinoma	+	[165]
Ductal adenocarcinoma	-	[231]
Papillary mucinous carcinoma		[231]
Colon:		
Non-mucinous colorectal carcinoma	-	[79, 232, 233]
Mucinous colorectal carcinoma	+	[5, 234]

Esophagus: Normal esophageal mucosa is covered by a squamous epithelium, which does not express MUC2 mucin [225, 235, 236]. Transition from the normal epithelium to aggressive adenocarcinoma follows several stages, including a pathological state called Barrett's esophagus (BE). BE is defined as columnar metaplasia with goblet cells in the distal esophagus due to chronic duodeno-gastro-esophageal reflux [237]. It is a premalignant lesion and a major risk factor for esophageal adenocarcinoma. The latter develops through a dysplasia-carcinoma pathway [227]. *MUC2* expression in Barrett's esophagus has been extensively studied [238-240]. Glickman *et al.*'s comprehensive study [227] of MUC2 core polypeptide expression profile in the progression of neoplasia in Barrett's esophagus showed that at the nondysplastic stage of the disease, MUC2 was expressed in cytoplasm of both goblet and nongoblet columnar cells in 100% of the cases tested. All goblet cells demonstrated strong, diffused cytoplasmic staining in histochemical tests, while nongoblet cells displayed only focal staining.

The low- and high-grade dysplasia samples also showed a high rate of MUC2 expression: 90-100% of MUC2-positive cells in the analyzed specimens. A drastic decrease in MUC2 expression was seen in adenocarcinoma: only 40% of the biopsies were MUC2-positive with focal staining [227]. Decrease in *MUC2* expression during progression of Barrett's esophagus from less malignant to highly aggressive adenocarcinomas was observed in many studies [225, 226, 236, 241], confirming Glickman *et al.*'s main conclusion [227] that progression from metaplasia to low- and high-grade dysplasia and further to carcinoma is associated with a significant decrease in MUC2 expression and alterations in the staining pattern.

Mucin expression in adenocarcinoma arising in Barrett's esophagus may indicate a definite tumorigenesis pathway [242]. Histological analysis of Barrett's epithelium showed that it is not a homogenous cell layer, but a combination of heterogenous elements. As Szachnowicz *et al.* [242] noted, "Barrett's esophagus is a complex, mosaic of cell, gland, and architectural types, showing variable degrees of atrophy and maturation toward intestinal and gastric epithelium". Thus, of the described above dynamics in MUC2 expression may be relevant only to specific cells in the indicated histological complex. By using immunochemical detection of MUC2 and MUC5AC, markers specific for intestinal goblet cells and for gastric collumnal cells, respectively, Szachnowicz *et al.* [242] showed that specimens of adenocarcinoma in Barrett's esophagus were either intestinal (MUC2-positive) or gastric (MUC5AC-positive) type. The results of this study imply that mucin pattern reflects the origin of the adenocarcinoma in Barrett's esophagus. These data are consistent with the findings of Brown *et al.* [243] who distinguish between two types of Barrett's esophagus: adenomatous dysplasia (type 1) and foveolar hyperplastic dysplasia (type 2). Epithelium of type 1 expresses the intestinal differentiation markers MUC2, CDX2 and villin, but not MUC5AC, whereas epithelium of type 2 expresses MUC5AC mucin but not MUC2.

Several theories have been put forth to explain the origin of Barrett's esophagus [244-247]. The prevailing hypothesis, expounded by Peters and Avisar [248], is that "Barrett's esophagus occurs *via* abnormal differentiation of esophageal epithelial stem cells exposed to gastric juice by the chronic epithelial erosion and injury of gastro-esophageal reflux disease". This hypothesis is supported by the finding that the basal layer of the esophageal epithelium contains pluripotent stem cells that may differentiate into squamous epithelial cells after obtaining physiological differentiation-specific signals [249, 250]. However, epithelial erosion known to occur secondary to gastro-esophageal reflux may expose esophageal stem cells to nonphysiological chemical matter present in luminal contents, which, in turn, may activate various signaling in gastric or intestinal differentiation pathways in esophagus. Bile components present in duodenal content have been implicated in the pathogenesis of Barrett's esophagus by activation of specific transcription factors and signaling pathways [251, 252]. One such transcription factor is CDX2 [238]. Interestingly, the primary effect of the bile acids, at least in the rat model, is directed to activation of the *Cdx1* gene, which then up-regulates *Cdx2* [253]. CDX2 has been shown to be an important transcriptional regulator of embryonic differentiation and maintenance of normal adult epithelium in small and large intestines through regulation of *MUC2* gene activity [254, 255]. *CDX2* and *MUC2* are not expressed in normal esophageal epithelium, but are over-expressed in intestinal metaplastic mucosa in Barrett's esophagus [256-258]. Deoxycholic acid was found to up-regulate goblet-specific MUC2 mucin in concert with CDX2 protein in human esophageal cells [238]. This study also established that bile acid alone, or in combination with acid pH, induces *CDX2* expression through activation of the *EGFR* gene [259]. The activation of both *CDX2* and *MUC2* genes in response to bile acid exposure in esophageal cells observed in experiments conducted by Hu *et al.* [238] is consistent with the findings of Yamamoto *et al.* [28] and Mesquita *et al.* [29] that CDX2 transcription factor directly regulates transcription of the *MUC2* gene in gastric and colon adenocarcinoma cells.

In summary, the presented data show the up-regulation of the *EGFR, CDX2* and *MUC2* genes after exposure of esophageal mucosa to the bile acids, leading to intestinal metaplasia of esophageal epithelium. These data demonstrate the crucial role of the bile acids-EGFR-CDX2-MUC2 pathway in the pathogenesis of Barrett's esophagus. The effect of the bile acids on the esophageal epithelium could be accelerated by nitric oxide generated intraluminally through the entero-salivary recirculation of dietary nitrate in the presence of refluxed gastroduodenal contents [239]. Consistent with these results there are findings of Vaninetti *et al.* [260] who showed that, "although nitric oxide alone did not induce CDX2 expression in the normal esophageal cell line Het1A, it greatly enhanced bile-acid induced CDX2 expression up to 98-fold compared with bile acid alone, suggesting a direct pathway that links nitric oxide to a key transcription factor that controls the tissue phenotype".

The described mechanism of *MUC2* activation in esophageal epithelium is not the only one by which *MUC2* gene can be activated in response to bile acids. As shown by Wu *et al.* [56], deoxycholic acid may induce *MUC2* over-expression in human adenocarcinoma cells by activation of NFκB transcription through a process independent of MAPK, but involving PKC.

Thus, the consequent transition of the normal esophageal epithelium to Barrett's dysplastic epithelium and further to adenocarcinoma is associated with a signaling cascades that include interaction of bile acids and nitric oxide with esophageal stem cells, followed by activation of a long chain of signaling events leading to expression of the *EGFR*, *CDX1*, *CDX2*, NFκB and *MUC2* genes. The expression of these genes changes as the carcinogenic process progresses from initial dysplasia to aggressive carcinoma.

Stomach: The role of MUC2 mucin in stomach pathology has been thoroughly studied, however, the obtained results are often controversial. The available studies were focused on two main types of stomach pathology – gastric cancer and chronic infection by *Helicobacter pylori*. Below, the results of these studies will be discussed.

a) Stomach cancer: The role of MUC2 mucin in stomach cancer had been intensively investigated, with conflicting results [147, 228, 261, 262]. On the one hand, *MUC2* has been reported to be expressed in intestinal metaplasia and gastric carcinoma, but not in normal mucosa of the stomach [146, 147]; and on the other hand, almost 17% of gastric cancer specimens without histological signs of intestinal metaplasia also express MUC2 mRNA [263]. Gastric cancers constitute a highly heterogeneous group of tumors with respect to genetics, histopathology, biological behavior and mucin expression. Several classifications based on histopathological features have been proposed. Lauren [264] classified stomach adenocarcinomas into two histological types, intestinal and diffuse. Nakamura *et al.* [265] defined them as differentiated or undifferentiated. Muligan [266] divided gastric cancer into three groups: intestinal cell carcinomas, mucous cell carcinomas and pylorocardiac tumors. According to the WHO classification, there are five subgroups of gastric cancers: papillary, tubular, mucinous, signet-ring cell and undifferentiated [267]. Goseki *et al.* [268] divided all types of gastric cancers into four groups, according to the configuration of histological features. A recently suggested classification is based on the expression of the phenotypic markers, including human gastric mucin (HGM), gel-forming mucins MUC2 and MUC6, lymphocyte marker CD10, and staining with Con A. In this system, gastric carcinomas was classified as having either a gastric (G), gastric and intestinal mixed (M) or intestinal (I) phenotype [269-274]. Although phenotypic marker expression is known to generally reproduce the pattern of expression in the tissue of origin [275, 276], the nature of gastric cancer is still unknown. In spite of the numerous classifications, the histological content of stomach cancer has not been delineated, and the expression profile of MUC2 mucin and its role in gastric tumors remain lunclear. Nevertheless, attempts have been made to correlate MUC2 expression with histological types of gastric cancer.

Leteurtre *et al.* [277] analyzed the relationships between expression of gel-forming mucins, including MUC2, and histopathological phenotypes corresponding to different classifications in a series of gastric carcinomas. Expression of MUC2 was observed in all subtypes of studied tumors in all classifications (Lauren, Mulligan, WHO and Goseki), with the highest level of MUC2 expression significantly associated with the mucinous subtype in WHO classification and with high tubular differentiation and high mucin content in the group II of Goseki's classification. The absence of MUC2 mRNA was more often observed in the signet-ring subtype of gastric cancer (WHO classification) and in the diffuse subtype carcinomas (Lauren's classification). Surprisingly, in this study the expression of MUC2 was not significantly associated with the intestinal subtypes of Lauren's and Mulligan's classifications, in contrast to other investigations in which specimens with intestinal metaplasia were highly positive for MUC2 [146, 263, 278]. Interestingly, Conze *et al.* [279] observed relatively high levels of MUC2 mucin expression in gastric carcinomas of both intestinal and diffuse types, and low level of the MUC2 antigen in carcinomas of mixed type. Leteurtre *et al.* [277] noted that MUC2 expression was not restricted to a specific type of cells, but could be detected in cells characteristic of different histopathological types of gastric tumors. The relationship between MUC2 expression and the histology of the MUC2-producing cells became even more complex when Mitsuuchi *et al.* [263] reported MUC2 expressed not only by malignant tissue, but also by surrounding regenerative epithelium associated with chronic gastritis or peptic ulcer induced by *H. pylori* infection. As noted by the authors of this study, the MUC2-positive specimens did not contain goblet cells. These data are in agreement with Lee *et al.* [148], who found intestinal metaplasia regions adjacent to the tumor area and showed that metaplastic cells were highly MUC2-positive. Both Mitsuuchi *et al.* [263] and Lee *et al.* [148] suggested that increased MUC2 expression in intestinal metaplasia in the neighborhood of carcinomas is an adaptive process, a "mechanism for self-cure", which may provide additional defense against the underlying injurious factors. Such behavior of MUC2 mucin in gastric carcinoma is reminiscent of its tumor suppressor function in colon cancer [145].

Important information on the dynamics of MUC2 expression in stomach tumors was obtained by Yamazaki *et al.* [274]. They found that benign gastric adenomas accounted for the largest amount of MUC2-positive specimens (87.5%), while the early and advanced carcinomas accounted for progressively decreasing amounts of MUC2-positive tumors (73.5% and 61.5%, respectively). Interestingly, *MUC2* expression in gastric adenomas correlated with *CD10* expression, mutations in *APC* and intestinal phenotype, and was inversely associated with the expression of *HMG*, *MUC6* and mutated *p53* compared with early and advanced carcinoma. These findings demonstrate that gastric adenomas are substantially different from early and advanced carcinomas in terms of tumor differentiation phenotype and genetic alterations. A number of investigators consider the observed interdependence of MUC2 expression and the above phenotypic markers to reflect an underlying network of interactions between general regulators of embryonic development and differentiation (HMG proteins), and tissue- and cell-specific transcription factors such as SOX, Shh and CDX [280-283]. CDX2, a caudal-related homeobox transcription factor, is well known for its role in development and maintenance of gastrointestinal tract tissues. In normal adults, it is expressed in intestine but not in stomach and esophagus [284]. CDX2 expression has been implicated in the development of stomach cancer, although its role in gastric carcinogenesis is controversial [285-287]. CDX2 regulates *MUC2* expression by direct binding to the *MUC2* promoter [28]. Thus, the pattern of CDX2 expression in gastric tissues under normal and pathological conditions may influence the pattern and character of MUC2 mucin expression. Roessler *et al.* [288] showed that CDX2 and MUC2 were expressed in more than 80% of the specimens exhibiting intestinal metaplasia, while their expression was decreased in gastric carcinoma to 57% and 21%, respectively. Of note, the expression of neither of the two molecules correlated with WHO, Lauren and Goseki classifications (with the exception of a significantly stronger MUC2 expression in mucinous tumors). The same results were obtained by Liu *et al.* [285], who showed that CDX2 and MUC2 expressions are progressively decreased in human gastric intestinal metaplasia, dysplasia and carcinoma sequence, with the only difference being a parallel increase in MUC2 and CDX2 expression in mucinous tumors.

It is becoming increasingly clear that knowledge the factors controlling *CDX2* expression is required for a better understanding of the mechanisms regulating expression of *MUC2* in the stomach. Several factors are implicated in ectopic activation and regulation of the *CDX2* gene activity in gastric epithelial cells. One of them is physiological bile, or, more precisely, bile acids. Clinical and experimental data show that bile reflux into the stomach contributes to gastric carcinogenesis through induction of intestinal metaplasia [289, 290]. Xu *et al.* [291] showed that the bile acids induced *Cdx2* and *Muc2* expression in normal rat gastric epithelium in a dose-dependent manner through interaction with the farnesoid X receptor (FXR), a nuclear receptor for bile acids. In addition to FXR, bile acids may activate gastrointestinal genes also by interaction with the membrane EGFR receptor, promoting gastric carcinogenesis through activation of both nuclear and membrane receptors.

<u>b) Helicobacter pylori infection</u>: *Helicobacter pylori* (*H. pylori*) is thought to be a pathological factor that plays an important role in various types of stomach pathology, including gastric carcinogenesis [292, 293]. *H.pylori* causes gastritis, gastroduodenal ulcer, gastric atrophy, intestinal metaplasia and gastric cancer [294, 295]. In gastric epithelium, *H. pylori* infection leads to alterations in expression of mucins, transcription factors and trefoil proteins. Infection of human gastric carcinoma cells with *H. pylori* induces both activation of the *CDX2* and *MUC2* genes and inhibition of the *Sox2* gene in a dose-dependent manner [296]. These *in vitro* data are in agreement with the *in vivo* results, which also demonstrate up-regulation of *CDX2* and *MUC2* gene expression in response to *H. pylori* infection [297-299]. *H. pylori* up-regulates MUC2 mucin expression not only in cancerous cells, but also in pericancerous mucosa, and this aberrant expression of the *MUC2* gene leads to intestinal metaplasia [300]. Asonuma *et al.* [283] identified the pathway leading to development of intestinal metaplasia in stomach: *H. pylori* infection induces host immune response, which includes high expression of IL-4, an interleukin that has potential to inhibit activity of the *Sox2*

gene; the decrease in Sox2 expression leads to activation of the *CDX2* and *MUC2* genes and, as a result, to development of intestinal metaplasia in gastric mucosa. Mejias-Luque *et al.* [301] recently documented the direct effect of the *H. pylor*-induced immune reaction on MUC2 mucin expression. They showed that two types of inflammation induced by *H. pylori* infection, acute and chronic reactions, modulate *MUC2* gene expression in gastric tumors in two ways: Expression of *MUC2* was significantly higher in human gastric tumors with the features of chronic inflammation (lymphoplasmocytic cells) than in tumors with acute inflammation (polymorphonuclear cells). The high level of *MUC2* expression observed in this study could be attributed to chronic inflammation cytokines (TNF-α, IL-1β, IL-4, *etc.*) induced by *H. pylori*, which, in turn, linked *MUC2* expression to MAPK, Ras/MEK or NFκB pathways [302].

The role of inflammatory cells in up-regulation of *CDX2* and development of intestinal metaplasia in the *H. pylori*-infected gastric epithelium was clarified by the elegant studies of Bleuming *et al.* [303] and Barros *et al.* [304]. Inflammatory cells are known to produce the same morphogens as those that regulate epithelial cell fate decisions in the adult gastrointestinal tract [303, 305, 306]. Bleuming *et al.* [303] asked the question how the influx of inflammatory cells into the *lamina propria* of the *H. pylori*-infected stomach changes the relatively normal gastric morphogenetic landscape of the bone morphgenetic protein (BMP) pathway. BMP signaling in normal gastric mucosa is restricted to the epithelial cells located at both ends of the renewal gastric unit axis. In the *H. pylori*-infected stomach, the expression of BMP is increased mainly because of influx of BMP2-producing inflammatory cells. As a result, the activity of the BMP-pathway "shifts from the differentiated epithelial cells in the gastric pit cell region to the precursor cells in the isthmus", and the morphogenetic landscape in the inflammed stomach mucosa is altered by morphogens secreted from inflammatory cells [303]. Based on these data, Barros *et al.* [304] showed that BMP2 and BMP4, the key elements of the BMP/SMAD pathways, co-localize with CDX2 in the regions of intestinal metaplasia. Thus, the coordinated expression of BMP, CDX2 and MUC2 proteins promotes intestinal differentiation in human gastric epithelium.

Beside *H. pylori*, another infectious agent, Epstein-Barr virus (EBV), is also considered an etiological factor of stomach adenocarcinomas, although it associates only with 16% of gastric cancers [307, 308]. In contrast to *H. pylori*-induced carcinogenesis, associated as noted above with increased expression of the CDX2 transcription factor and MUC2 mucin, gastric cancer induced by EBV infection is negatively associated with CDX2 and MUC2 markers [309, 310]. The EBV(+) stomach cancers are characterized by low expression of intestinal phenotype markers [310] and have very specific profiles of differentiation markers: they show a low level of keratin expression [311], and have a null/gastric phenotype as determined by the expression pattern of gastric specific mucins MUC5AC and MUC6 [312]. According to Fukayama *et al.* [313], these findings suggest that "the targets of EBV infection and subsequent transformation are the precursor cells with intrinsic differentiation potential toward the gastric cell type but not the intestinal type". The nature of the "precursor cells" remains elusive. Recently, a case report of a patient with EBV(+) gastric carcinoma developed after bone marrow transplantation [314] raised the question whether gastric carcinoma may originate from the donor-derived stem cells through reprogramming or cell fusion with the recipient's stem cells [315, 316].

Studies of epithelial and hematopoietic cells, which continuously undergo cell renewal, suggest that cancer cells may originate from a stem cell compartment [317]. The current hypothesis that considers cancer stem cells (CSC) as progenitors of stomach tumors is the most popular and the most debated [317, 318]. According to the American Association for Cancer Research, CSCs are "cells within a tumor that possess the capacity for self-renewal and that can cause the heterologous lineages of cancer cells that constitute the tumor" [319]. CSCs have been already identified in many solid tumors [320-322]. The origin of human gastric CSCs has yet to be elucidated [323]. Data from experimental models, including a mouse model of *H. pylori*-induced gastric cancer, have implicated bone marrow stem cells (BMSC) as a potential candidate for the role of gastric cancer progenitor [324, 325]. More studies are needed to solve this problem, which is crucial for both experimental and clinical oncology.

4.7.3. Expression of *MUC2* Gene in the Pancreo-Hepato-Biliary System

Diseases of pancreas, liver, gallbladder and biliary ducts comprise a substantial part of human pathology. The main pathological conditions detected in these organs and associated with abnormal activity of mucin genes include such nosologies as pancreatic cancer, hepato- and cholangiocarcinomas, hepatolithiasis and infection. In the following sections, the role of MUC2 mucin in pathogenesis of these diseases are discussed.

4.7.3.1. Pancreas

The malignant lesions of pancreas are associated with multiple alterations in gene expression, including the *MUC2* gene. To better understand the role MUC2 mucin plays in the pancreatic cancer, it is important to discuss first the origin, histogenesis, genetic abnormalities, topology and differentiation patterns of pancreatic neoplasms.

<u>General characteristics of pancreatic neoplasms</u>: Pancreatic cancer can be of exocrine or endocrine types. Most pancreatic neoplasms are of the exocrine type [326, 327]. Infiltrating ductal carcinoma is the most common exocrine neoplasm [326]. The origin of human pancreatic cancer is not clearly defined. It has been postulated that the initiation and progression of pancreatic neoplasms result from oncogenic mutations in the normal adult pancreatic stem cells or their differentiated daughter cells (Fig. **4**). Cancer stem cells (CSCs) – genetically altered cells – comprise 0.1%-1% of cells in a given tumor [326, 328-330]. They acquire the capacity for self-renewal through deregulation of critical self-renewal pathways, including Hedgehog and Notch [326, 331, 332]. Two potential candidates for CSCs have emerged in the exocrine pancreas. They are differentiated acinar and centroacinar cells (Fig. **4**), which, under specific pathological conditions, may accumulate oncogenic mutations in the appropriate oncogenes (*KRAS2, BRAF*) and tumor suppressor genes (*TP53, CDKN2A/p16, SMAD4/DPC4*), thereby acquiring the properties of CSCs [333-335]. Recently, several groups isolated human pancreatic CSCs possessing several CSC-specific markers (Fig. **4**) [321, 336-340]. Gene expression analysis revealed several expression abnormalities specific for different stages of pancreatic cancer from benign adenoma to invasive adenocarcinoma. Among the abnormally expressed genes are mucin genes, including *MUC2*. Different types of pancreatic neoplasms express different mucins in stage- and type-specific manners. Expression of *MUC2* is tightly associated with specific types and stages of pancreatic carcinogenesis.

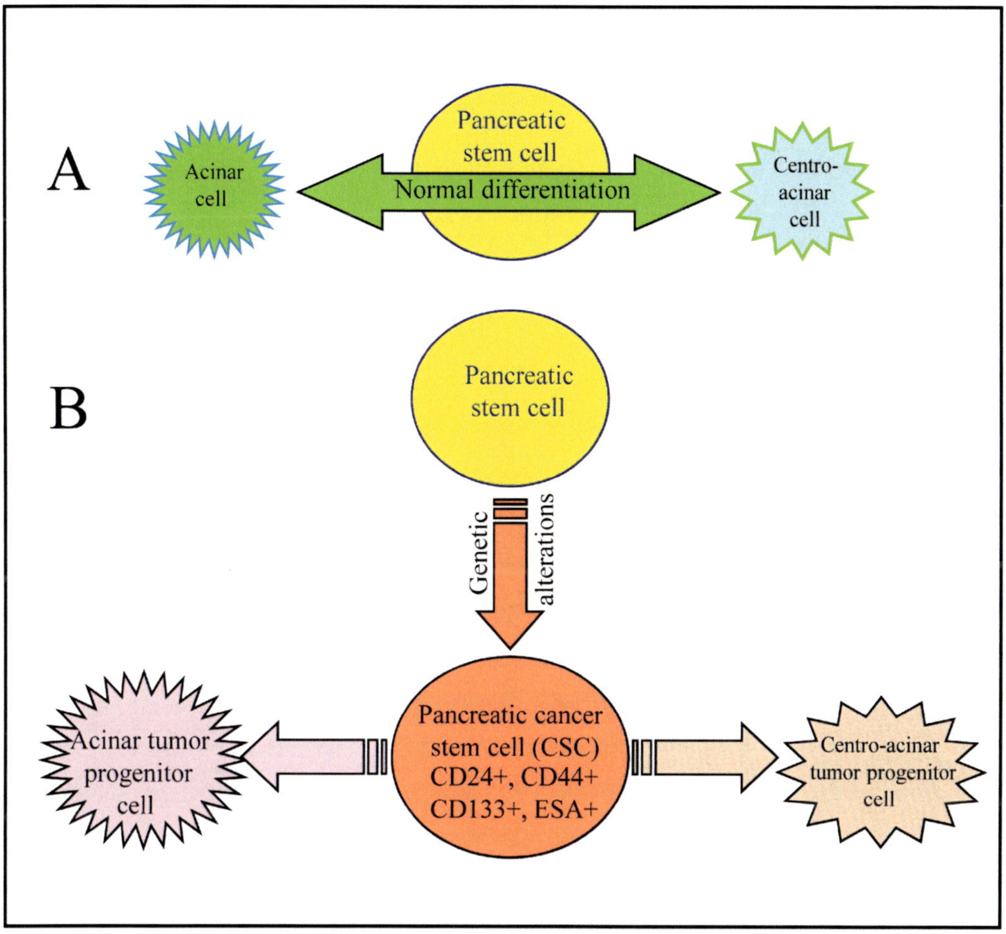

Figure 4: Transition of pancreatic stem cells into cancer stem and progenitor cells (based on the data reported in [326, 328-330]; A- normal differentiation, B malignant transformation).

Three histologically distinct lesions, known as precursors of ductal adenocarcinoma of pancreas, have been described. These include pancreatic intraepithelial neoplasias (PanIN), intraductal papillary mucinous neoplasms (IPMN), and mucinous cyctic neoplasms (MCN) [341-343]. Depending on the degree of cytologic and tissue atypia, PanIN can be subclassified into PanIN-1A, PanIN-1B, PanIN-2 and PanIN-3. The last one may develop into infiltrating adenocarcinoma [326, 344]. Recently, the foamy variant of PanIN was described [345]. Both noninvasive IPMNs and MCNs of the pancreas may progress to invasive adenocarcinoma over time [343] (Fig. **5**).

Mutations in the *KRAS2* gene are the early events detectable already at the PanIN-1A and PanIN-1B stages, whereas inactivating mutations in the *p16/CDKN2A* gene are associated more often with the intermediate stage (PanIN-2). Inactivation of the *SMAD4/TP53* and *BRCA2* genes usually occur in the late lesions (PanIN-3) (Fig. **5**) [343, 346].

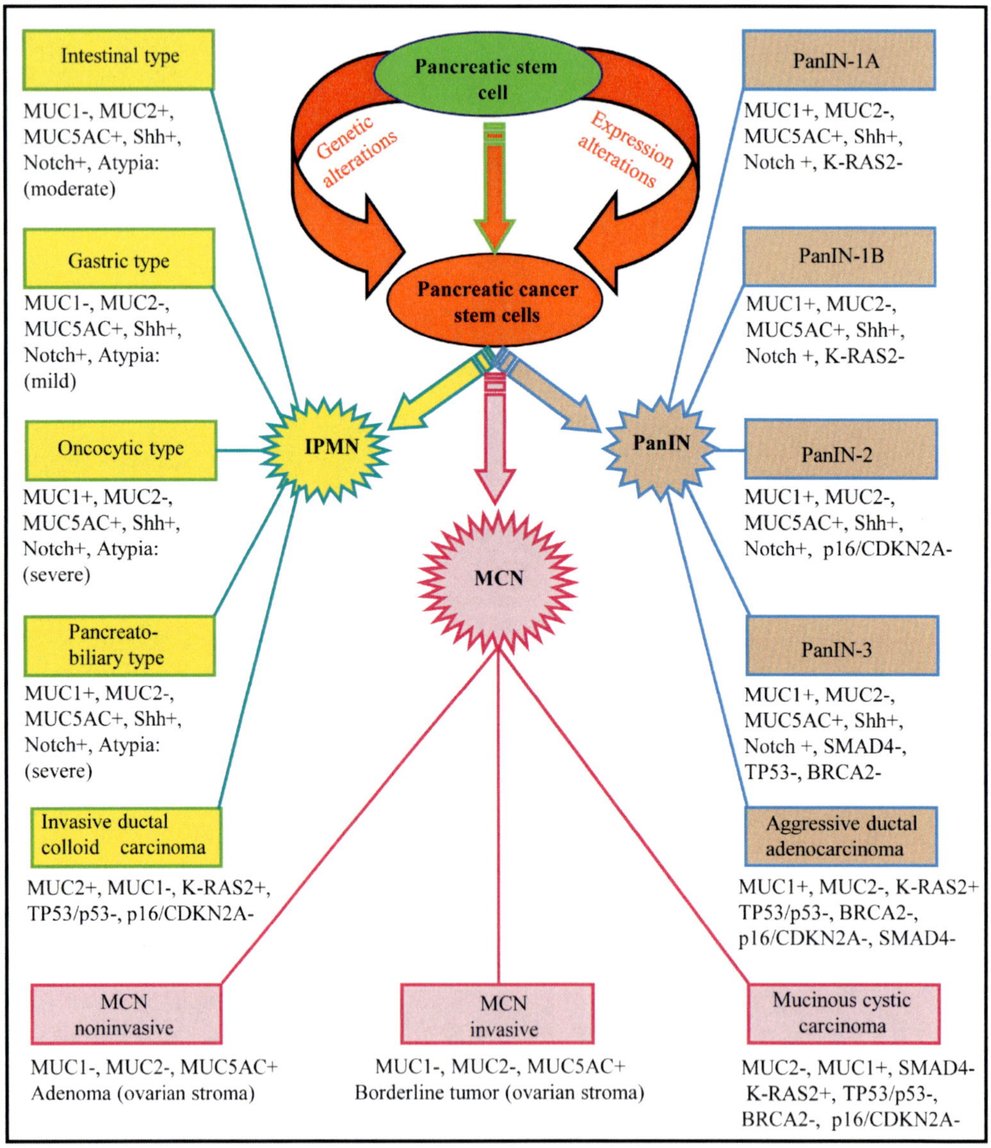

Figure 5: Gene expression in pancreatic neoplasms PanIN, IPMN and MCN (based on the data reported in [326-344, 348, 349, 352, 377, 378]).

IPMNs are pancreatic epithelial mucinous neoplasms and usually arise within the main pancreatic duct or its branches [347]. They demonstrate the same genetic alterations as PanIN neoplasms with minor variations [348, 349]. IPMNs consist of a spectrum of neoplasms characterized by morphological and immunohistochemical variations. Four histologically distinctive IPMN subtypes, including gastric type, intestinal type, pancreatobiliary type and oncocytic type, have been suggested [350, 351]. IPMNs may progress stepwise from adenomas to intraductal carcinomas and then to invasive colloid carcinoma (Fig. **5**) [352].

MCNs represent the third histologically distinct group of pancreatic neoplasms. The vast majority of MCNs arises in women and contains ovarian-type stroma [353]. The genetic alterations in MCN have not been extensively studied, but according to Maitra and Hruban [326], "they can have many abnormalities found in infiltrating ductal adenocarcinomas of the pancreas, but at a lower frequency" (Fig. **5**).

Besides the *cis*-genetic abnormalities mentioned above, epigenetic *trans*-factors also play important roles in pancreatic carcinogenesis. Epigenetic silencing is frequently observed in pancreatic cancers and involves tumor suppression genes and genes critical for homeostasis [326, 354]. On the other hand, genes involved in sustaining of renewal potentials of tissue stem cells and participating in developmental signaling pathways are often super-expressed in human malignancies, including pancreatic cancer [331-333, 355, 356].

Pancreatic intraepithelial (PanIN) and intraductal papillary mucinous (IPMN) neoplasms: The MUC2 mucin expression profile in pancreatic cancer is type specific. As shown in Fig. **5**, cells of the PanIN type of pancreatic neoplasia do not express MUC2 at any stage of tumor progression [342, 357, 358]. Nor was this mucin expressed in gastric, oncocytic and pancreatobiliary types of IPMNs. It was, however, expressed in intestinal IPMNs at a high level [23, 350, 359-361]. An important observation regarding MUC2 expression in IPMN and PanIN during progression of these neoplasms to colloid carcinoma and ductal invasive adenocarcinoma, respectively, was made by Adsay *et al.* [357]. The authors showed that 54% of IPMNs expressed MUC2, whereas none of the PanINs did. In contrast, PanINs, especially higher grade lesions, were often positive for MUC1 (61% of PanIN-3), whereas only 20% of IPMNs expressed this mucin. This dichotomy was further seen in the invasive carcinomas: transition of IPMNs to colloid carcinomas was associated with an increased frequency of MUC2 expression from 54% to 100%, and a decreased proportion of MUC1-positive lesions from 20% to 0%, respectively. On the other hand, transition of PanIN to ductal adenocarcinoma was not associated with changes in *MUC2* and/or *MUC1* gene expression. MUC1, a marker of aggressive phenotype, is expressed in pancreatic neoplasia in some higher-grade PanINs, and are consistently found in infiltrating conventional ductal carcinoma [362]. MUC2 mucin appears to be a marker of an indolent phenotype of pancreatic neoplasms. It is not expressed in the normal pancreas, PanINs or ductal adenocarcinomas, but is highly expressed in IPMNs and is consistently seen in colloid carcinomas [357]. Interestingly, like in Barrett's esophagus, *H. pylori* gastritis and gastric atrophy [31, 362-365] co-expression of MUC2 mucin and CDX2 transcription factor, the main regulator of normal intestinal differentiation and trigger of "intestinal reprogramming", have been observed also in IPMNs and colloid carcinomas of pancreas.

The data reported by Adsay *et al.* [357] and discussed above concur with the well known facts that colloid carcinoma is less aggressive than ductal adenocarcinoma [362, 366, 367], and that IPMNs with a high expression of MUC2 mucin have a lower invasion and metastatic potential than MUC2-negative neoplasms of pancreas [231, 361, 368]. However, some authors reported data that contradict the previously published results. For instance, Nakamura *et al.* [360] found that MUC2-positive IPMNs had a higher incidence of malignant transformation and invasive behavior than MUC2-negative tumors. In agreement with Nakamara's results, Sanada and Yoshida [369] described a case of benign IPMN of the pancreas with gradual transition from peripheral normal MUC2-negative pancreas tissue to MUC2-negative gastric-type adenoma, and further to MUC2-positive intestinal-type lesion, whose histology exhibited features intermittent between adenoma and carcinoma. These investigators concluded that the MUC2-negative gastric type lesion is a less malignant precursor of a more aggressive MUC2-positive intestinal type of IPMN.

The findings reported by Nakamura *et al.* [360] and Sanada and Yoshida [369], although important, need to be confirmed. At present, there is consensus that tumors expressing MUC2 are less aggressive and have better prognosis than

MUC2-negative neoplasms [370-372]. This conclusion is in agreement with the idea that MUC2 mucin is a tumor suppressor molecule. The ability of the MUC2 protein to function as an antitumorogenic factor, originally established by Velcich *et al.* [145], was further supported by the finding that the *MUC2* gene expression is triggered by p53, a known tumor suppressor and regulator of cell proliferation in response to stress signals [49, 373]. Interconnection between *MUC2* and *p53* expression suggests that *MUC2* gene is also part of a stress response program [49]. Moreover, Yuasa [48] and Levi *et al.* [374] consider that "MUC2 expression may not be an oncogenic event *per se*, but a protective mechanism, which is activated to prevent further exposure to carcinogens". This opinion is supported by the data of Ji *et al.* [375] showing that MUC2 mucin defends the attacked epithelium by formation of a protective gel barrier.

Thus, on the basis of MUC2 expression, the IPMNs may be classified into two distinct subtypes: MUC2-positive lesions including intestinal type of IPMNs and colloid carcinomas, and MUC2-negative tumors comprised of gastric, pancreato-biliary and oncocytic types of IPMNs. The first group of pancreatic tumors appears to be the result of a biologically indolent pathway of pancreatic carcinogenesis with intestinal lineage. As tumor suppressors and important molecules of "intestinal programming", MUC2 and CDX2 may be considered not only as diagnostic markers, but as active regulators of the intestinal metaplastic pathway [362].

<u>Mucinous cystic neoplasm (MCN)</u>: The mucinous cystic neoplasm (MCN) is the third histologically distinct type of cancer precursor lesions in the pancreas. According to Yoon *et al.* [376], MCNs comprise about 25% of cystic neoplasms of the exocrine pancreas. MCNs and IPMNs have many features in common [377, 378], which impedes their differential diagnosis. Histological differences between them can help in diagnostics. In contrast to IPMNs, which are cystic dilatations of preexisting ducts, MCNs form *de novo* cystic tumors with underlying ovarian type stroma. According to Adsay [379], "the presence of ovarian-type stroma has now almost become a requirement for the diagnosis of MCN". Malignant transformation of IPMN ranges from as low as 6% to as high as 92%, while malignant transformation of MCNs is seen in only 6-36% of the cases studied [380, 381]. Noninfiltrating MCNs can be distinguished from infiltrating MCN,

IPMNs and ductal adenocarcinomas by mucin expression profiles. Noninfiltrating MCNs express MUC5AC but not MUC1 and MUC2 mucins; infiltrating MCNs express MUC5AC and MUC1 but not MUC2 [352, 358].

Ampullary cancer: Among the pancreatic cancers, carcinoma of the papilla of Vater (ampullary cancer) attracts special attention, although it comprises only 5%-8% of all gastrointestinal cancers. The special interest is due to the unique location of Vater's valve on the border between two completely different types of mucosa [380, 382]. Ampulla of Vater is located at the junction of the main pancreatic and distal bile ducts, within the head of the pancreas, surrounded by the parenchyma of the pancreas and duodenum [383]. Ampullary tumors (mostly adenocarcinomas) may arise from epithelium of the confluence of the distal bile duct and the pancreatic duct as well as from the duodenal mucosa. Histologically they are divided into intestinal and pancreatobiliary subtypes [384]. The intestinal type usually expresses such markers as MUC2, CDX2 and cytokeratin 20 (CK20); the pancreatobiliary type are mostly negative for these markers, but positive for MUC1, MUC5AC and CK7 [382-386]. Chu *et al.* [382] showed that immunophenotype may differentiate between adenocarcinoma of duodenal papillary origin and ampullary carcinoma of pancreatobiliary origin: the combination of markers specific for the former is MUC1(-)/CK17(-)/MUC2(+)/CDX2(+); and for the latter is MUC1(+)/CK17(+)/MUC2(-)/CDX2(-). Intestinal type tumors are similar biologically to colon cancer, while tumors with pancreatobiliary type of differentiation are biologically close to ductal adenocarcinomas [383].

4.7.3.2. Liver and Gallbladder

Liver malignancies: Hepatocellular carcinoma (HCC) and intrahepatic cholangiocarcinoma (ICC) are two major primary liver carcinomas in adults [387]. The third type of primary liver tumor is a mixed hepatocellular cholangiocarcinoma (HC-CC), a rare malignancy containing unequivocal elements of both HCC and ICC [388]. The origin of human liver tumors is not fully understood. The traditional view that they arise from mature cell types has been challenged in recent years, with several studies suggesting that they can be derived from hepatic progenitor cells (HPCs) [389]. Animal models of hepatocarcinogenesis have shown that the hepatobiliary malignancies are often developed from the facultative bipotential progenitor cells called "oval cells", which possess properties of'stem cells' [390]. Evidence of the direct involvement of the mouse oval cells in histogenesis of HCC was brought by Dumble *et al.* [391]. Recent studies confirm the existence of hepatic progenitor cells also in human, although their role in human hepatocarcinogenesis has not been clearly established [392-395]. Expression of some progenitor cell specific markers including Muc1 mucin has been found on the surface of the rat oval cells [395]; however, there is no evidence at present of expression of other mucin genes, including *Muc2*.

The modern classification of primary liver tumors is based on both their histology and topology (Fig. **6**). HCC is a parenchymal hepatocellular cancer, while intrahepatic cholangiocarcinoma (ICC) and extrahepatic cholangiocarcinomas (ECC) are ductal malignancies. ICCs are subdivided into peripheral and hilar subtypes on the basis of origin. Peripheral cholangiocarcinoma (CC) arises from distal branches of the intrahepatic bile tree, while hilar ICC originates from the right or left hepatic ducts (Fig. **6**) [396]. Peripheral CCs are classified as ductular or duct carcinomas depending on the size of the duct [387]. In an early study, Sasaki *et al.* [397] showed that nondysplastic large intrahepatic bile ducts do not express MUC1 and MUC2 glycoproteins; MUC5/6 apomucins were expressed in only 14% of samples, whereas MUC3 was highly expressed in all samples. In biliary epithelial dysplasia, MUC1 and MUC2 were expressed in 29% of the samples; MUC3 in 88%; and MUC5/6 in 75%.

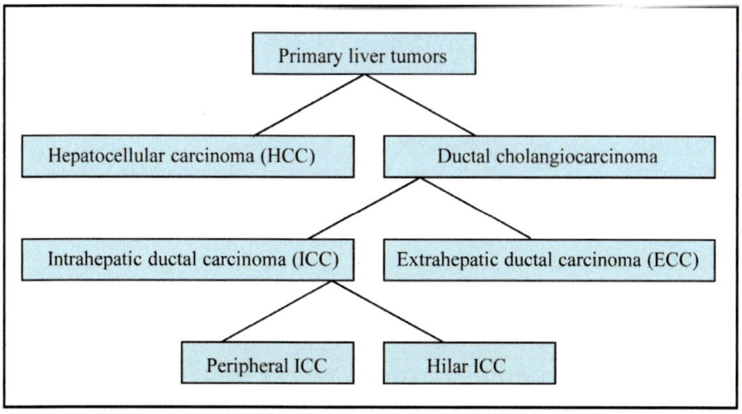

Figure 6: Classification of the primary liver tumors (based on the data reported in [387, 396]).

In cholangiocarcinomas, MUC1, MUC2, MUC3 and MUC5/6 were expressed in 100%, 21%, 68% and 79% of hilar carcinomas, and in 100%, 10%, 10% and 50% of peripheral tumors, respectively. According to this study, frequent and aberrant expression of "gastric type" MUC5/6 apomucin in biliary epithelial displasia, as well as in CC, suggests biliary epithelial cell transformation to gastric phenotypic metaplasia during the dysplasia-carcinoma progression.

A very similar expression profile of MUC1, MUC2 MUC5AC and MUC6 mucins was observed by Park *et al.* [398] in ICC and ECC as well as in adenocarcinomas of the gallbladder and pancreas. This and other studies [399, 400] point to a definite parallelism between the intraductal tumors of the pancreas and biliary system of the liver. In fact, Zen *et al.* [400] reported similarities between mucin and cytokeratine expression profiles in intraductal papillary mucinous neoplasia of the biliary tract and pancreatic papillary mucinous neoplasm. These authors found that the same specific combinations of mucin and cytokeratin expression patterns are characteristic of definite types of biliary and pancreatic neoplasms. According to their study, biliary intraepithelial neoplasia (BilIN) is characterized by the MUC1+/MUC2-/CK7+/CK20- expression pattern. This form of hepatic tumor is a counterpart of PanIN pancreatic cancer. Intraductal papillary neoplasm of the bile duct (IPN-B) is characterized by the intestinal phenotype of mucin expression pattern, MUC1-/MUC2+/CK7+/CK20+, corresponding to pancreatic IPMN. According to Zen *et al.* [400], transformation of BilPN and IPN-B into intrahepatic cholangio-carcinoma progresses *via* three pathways: 1) transition of BilPN to tubular adenocarcinoma associated with MUC1+/MUC2-/CK7+/CK2- expression profile; 2) transition of IPN-B to tubular carcinoma associated with MUC1+/MUC2+/CK7+/CK20+ pattern, and 3) transition of IPN-B to colloid carcinoma characterized by MUC1-/MUC2+/CK7+/CK20+ phenotype. Critical analysis of these data by Kloppel and Kosmahl [399] led them to conclude that there is "a conspicuous similarity between the intraductal tumors" of intrahepatic biliary system and pancreas.

Variations in expression of MUC1-6 mucins in ICCs arising in different types of ducts have been reported in numerous studies [215, 401-405]. ICCs of hilar type arising from large bile ducts frequently express MUC3, whereas its expression in peripheral type ICC is relatively rare. Different pathological conditions may influence mucin expression in ICC. ICC associated with chronic viral hepatitis or cirrhosis demonstrates lack of MUC3 expression even in large bile ducts [406]. Thanks to these numerous studies, prognosis of ICC is often associated with a mucin expression profile. Increased expression of MUC1, MUC4 and MUC5AC mucins has been related to aggressive behavior of carcinoma and poorer prognosis, while the expression of MUC2 and MUC6 indicates better prognosis [215, 401, 407-409]. Specific association of MUC2, MUC5AC and MUC6 mucin expression with definite types of ICC allowed categorization of ICC into intestinal (MUC2+) type, gastric (foveolar MUC5AC+) type, and pyloric gland (MUC6+) type, respectively [404].

The data presented above show that the participation of mucins in ICC carcinogenesis has been intensively studied. The role of mucin in HCC, on the other hand, has been much less investigated. Cao *et al.* [410, 411] studied the expression of *MUC1* and *MUC2* in hepatocellular carcinomas and preneoplastic hepatocellular lesions, and showed that *MUC1* was not expressed in normal hepatocytes but was expressed in 38% patients with HCC. Yuan *et al.* [412] found that 64% of HCC tumors expressed *MUC1*. Moreover, its expression was less pronounced in well differentiated tumors and became more evident with tumor progression to moderate and poor differentiated states.

In contrast to Cao *et al.* [410, 411] and Yuan *et al.* [412], Inagaki *et al.* [413] and Xu *et al.* [414] could not detect any *MUC1* expression in HCC. Among gel-forming mucins, only MUC2 expression has been analyzed in the context of HCC. Nevertheless, MUC2 expression was not found in either normal hepatocytes or HCC [411]. Clearly more studies are needed to clarify the role of mucins in HCC carcinogenesis.

Liver and gallbladder stone-associated and inflammatory diseases: Altered expression of mucin genes has been detected in various inflammatory diseases of the liver and gallbladder [404, 407, 415, 416], highlighting the important role of mucins in these types of hepatobiliary pathology.

a) Hepatolithiasis: Three factors are crucial for the pathogenesis of hepatolithiasis: bacterial infection, bile stasis, and alteration of the bile composition resulting from aberrant expression of mucins, TFF and DMBT1 proteins [417]. It has been shown that over-production of mucin glycoproteins plays a role in the initiation and progression of hepatolithiasis [418, 419]. Mucin hypersecretion in the gallbladder has been implicated in the pathogenesis of biliary sludge and in gallstone disease [420]. Mucins occur as structural components of gallstones that promote cholesterol crystal nucleation [421, 422]. Inhibition of mucin secretion by aspirin prevents gallstone formation in an animal model [423]. Mucins appear to be important molecules in the pathogenesis of cholesterol gallstones as they fulfill a matrix function in stone formation [424]. It is widely accepted that the increased expression of mucins in gallbladder is a prerequisite for gallstone development, and increased amounts of mucins are consistently observed in gallbladder bile of patients [425, 426]. High levels of the *de novo* expressed MUC2, MUC5AC and MUC6 mucins and increased expression of MUC3 and MUC5B glycoproteins in large bile ducts and peribiliary glands in human have been reported in association with hepatolithiasis [418, 427]. Lee and Liu [428] analyzed mucin gene expression in gallbladders containing cholesterol stones or calcium billirubinate stones, and in gallbladders without stones. The MUC2 mRNA was not detected in normal gallbladders free of stones, but was expressed in 14% of gallbladders containing cholesterol stones and in 25% of gallbladders with calcium billirubinate stones. The *de novo* expression of *MUC2* was associated with *de novo* transcription of the *MUC4* gene and up-regulation of *MUC1*, *MUC3*, *MUC5B* and *MUC6*.

Trefoil factors TTF1-3 are human mucin-associated proteins whose expression are augmented markedly in the biliary mucosa in hepatolithiasis together with gel-forming mucins [429, 430]. MUC2 and MUC5AC mucins interact with TTFs [431], which, in turn, interact with the DMBT1, a supposed receptor of TTFs [432], resulting in significant increase of mucin gel viscosity. According to Sasaki *et al.* [404], "it is conceivable that the increased expression of gel-forming MUC mucins, TFF peptides and DMBT1 may play a role cooperatively in lithogenesis".

b) Cholecystitis: Cholecystitis is an inflammatory gallbladder disease whose pathogenesis depends on two crucial factors: gallstones and infection. Mucin expression patterns in the inflamed gallbladder are significantly different from those observed in the normal gallbladder epithelium. No expression of MUC2 mucin was detected in normal gallbladder mucosa, which is characterized by expression of a unique pattern of other mucin genes consisting of high levels of *MUC3, MUC5AC, MUC5B and MUC6* expression [433]. This pattern is dramatically changed under inflammation conditions. Ho *et al.* [433] analyzed mucin gene expression in normal gallbladder and in surgical specimens with acute, mild and chronic cholecystitis and found mucin expression patterns significantly altered in specimens with cholecystitis compared with normal gallbladder. Greater degree of inflammation was associated with higher expression of *MUC2* and decreased expression of *MUC3*, *MUC5AC*, *MUC5B* and *MUC6*. In acute cholecystitis, the amount of MUC2-positive cells reached 50%. The MUC1 apomucin was infrequently found in the inflamed gallbladder and tended to occur in specimens with greater degrees of inflammation.

Taken together, these studies indicate that mucin genes are differentially expressed in the pathologically changed pancreas, liver and gallbladder. It appears that MUC2 mucin participates actively in the pathogenesis of diseases associated with these organs, albeit differently in each of the diseases. More studies are needed to delineate the functions of the MUC2 mucin in various diseases of pancreatohepatobiliary system.

4.7.4. Expression of *MUC2* Gene in the Colon

Colon cancer, ulcerative colitis (UC) and Crohn's disease (CD) are the main forms of pathology detected in the colon. The role of mucins in pathogenesis of these diseases have been intensively studied. Below, the information demonstrating the involvement of MUC2 mucin in development of these nosologies is presented.

<u>Colon cancer</u>: The role of *MUC2* gene in the pathogenesis of colorectal adenocarcinoma has been investigated by several laboratories [5, 232, 233]. Velcich *et al.* [145] were the first to show the involvement of MUC2 mucin in suppression of intestinal tumor development. The authors found that *MUC2-/-* mice displayed aberrant intestinal crypt morphology and altered cell maturation and migration; frequently developed adenomas in the small intestine and rectum that further progressed to invasive adenocarcinoma; and that the absence of MUC2 mucin leads to colorectal cancer, while its presence suppresses

development of intestinal tumors. Thus, the tumor suppressor nature of the *MUC2* gene was established [145].

The precise mechanism of MUC2 tumor suppression activity is not clear. It appears that MUC2 mucin-mediated tumor suppression occurs in a way different from that of the classical tumor suppressors such as p53. MUC2 may act through interaction with or activation of other proteins or pathways. In this context, an interesting observation has been made by Fijneman *et al.* [434] showing that carcinogenesis in *Muc2*-deficient mice can be prevented by artificial activation of phospholipase *Pla2g2α* gene. It is not known whether alterations in morphology, proliferation and migration of cells in *Muc2*-deficient mice are the primary responses to the absence of the Muc2 mucin, or are secondary reactions to the insufficient defense of the gut epithelium that is often observed in the inflamed intestine. Recently, Van der Sluis *et al.* [435] showed that *Muc2*-deficient mice develop chronic colon inflammation and ulcerative colotis, conditions that play an important part in intestinal tumorigenesis [436]. Other factors have been implicated to colon carcinogenesis, including the association of a decrease in *MUC2* expression in colon adenocarcinomas with hypermethylation of the *MUC2* promoter [5, 82], loss of p53 activity [49], and overexpression of SOX9 protein, a repressor of the *CDX2* and *MUC2* genes [437].

In contrast to the data demonstrating tumor suppression activity of the MUC2 glycoprotein, some studies indicate the oncogenic potential of this mucin [5, 14, 234, 359, 438-440]. In these studies, high levels of MUC2 mucin expression have been detected in various malignant neoplasms. A distinct subtype of colon cancer, a mucinous carcinoma producing a substantial level of MUC2, has been described [5, 234], and shown to be linked to microsatellite instability [91]. The mucinous tumors have been found in different organs that do not usually express *MUC2* in normal physiological conditions. Importantly, most, if not all, of these tumors are MUC2-positive [179, 359, 438-440]. Ho *et al.* [14] found expression of MUC2 glycoprotein in 90-94% of colon adenocarcinomas, with no mention of the type of colonic cancer. The expression of MUC2 glycoprotein in mucinous tumors raised doubt about the ability of the MUC2 mucin to function as a true tumor suppressor. Furthermore, activation of its expression in tumors originating from tissues normally not producing MUC2 suggests oncogenic activity associated with this mucin. There is no clear answer to this question at the present time, although the better survival prognosis of MUC2-positive tumors in comparison with MUC2-negative adenocarcinomas [229, 359] may be considered as evidence in favor of the tumor suppressor nature of the MUC2 glycoprotein. An alternative hypothesis suggests that, like many other tumor suppressors, the MUC2 mucin may possess dual functions: in specific cell types it functions as a tumor suppressor, while in others, its oncogenic potentials prevail.

Inflammatory diseases: Ulcerative colitis (UC) and Crohn's disease (CD) are the main chronic inflammatory bowel diseases (IBD) [441]. In the previous sections it was noted that the intestinal mucus barrier – consisting of two gel layers, an inner firmly adherent layer and outer loosely adherent layer, both containing gel-forming MUC2 mucin [442] – provides host defense against endogenous and exogenous noxious substances, mechanical stress and luminal pathogens [162, 443, 444]. Therefore, changes in expression or properties of the secreted MUC2 mucin may lead to diminished protection of the intestinal mucosa, and hence play a role in the pathogenesis of IBDs, including UC and CD.

In UC, goblet cells, the main producers of the MUC2 mucin, are reduced in number and size [443]. *Muc2*-deficient mice, which do not have goblet cells and do not express Muc2 glycoprotein, were shown to have deficient mucus layers with increased permeability and enhanced bacterial adhesion to epithelial cell surfaces. These mice spontaneously develop colitis, demonstrating that Muc2 mucin is a critical factor in colon protection and an important element in the pathophysiology of UC [435, 443, 445]. This observation correlates well with Strugala *et al.'s* report [446] that the mucus layer of patients with severe UC is decreased in thickness and has large areas with no mucus layer at all. At the same time, CD patients in this study showed no changes in the thickness of the adherent mucus layer in the rectum and no significant discontinuities. These findings are in agreement with those reported by Myllarneimi and Nickels [447], according to which the number of goblet cells in the colon and the level of mucin secretion in patients with CD are normal or even increased, and the crypt architecture and epithelial mucin are usually preserved, except in the acute state.

These findings are in accord also with the morphological data of McCormick *et al.* [448], who found that the amount of MUC2 in the mucosa of UC patients is less abundant than in normal mucosa, and not altered at all in the mucosa of CD patients. In contrast to these data, Hanski *et al.* [449] found that although the steady-state concentration of MUC2 mRNA was not affected by UC or CD, the amount of detectable MUC2 protein, assessed by Immunohistochemistry, was significantly increased in both diseases compared with the normal colonic mucosa. The authors attributed the obvious discrepancy between the unaltered mRNA expression and the increased level of the detectable MUC2 mucin to aberrant hypoglycosylation of the MUC2 core protein resulting in augmented binding of MUC2-specific antibodies to mucin backbone. This explanation correlates well with previous reports showing that in UC and CD, the sialylation rate of mucins, including MUC2, is increased and the level of sulfation is decreased [122, 444, 450-452]. Moreover, according to the combined data of Van Klinken *et al.* [444] and Tytgat *et al.* [450-452], synthesis of hyposulfated MUC2 is actually decreased twofold in the acute phase of UC, but its secretion is proportionally increased as a compensatory mechanism to maintain the integrity of the defense. It is noteworthy that the constant amount of sulfated MUC2 in the intestinal lumen is of great importance, as sulfate is thought to confer resistance to enzymatic degradation of the mucosa barrier [453].

In line with the data showing hypoglycosylation of the MUC2 protein in UC and CD [449], Shaoul *et al.* [454] found that MUC2 mucin is present in UC and CD in two different forms: mature MUC2, detected in the goblet cells, and immature MUC2, present in the secretory granules of cells that are not phenotypically goblet cells. Interestingly, two major patterns of MUC2 glycoprotein were also found in UC by Larsson *et al.* [455]: the normal pattern in controls and in patients with inactive UC, and the aberrant profile in patients with active phase of disease. The aberrant form was found to consist of an increase in a subset of the smaller glycans and a decrease of several complex glycans. Importantly, the aberrant expression profile was reversible upon remission. Furthermore, the magnitude of the glycan shift observed in UC was significantly correlated with both the degree of inflammatory process and the course of disease [455].

Larsson *et al.*'s data [455] are supported by findings that not only glycosylation but also synthesis of the MUC2 apomucin in UC appears to be related to the activity of the disease: both total amount of MUC2 mucin and the efficiency of

MUC2 precursor biosynthesis are usually decreased in the active phase of UC and return to control levels upon remission [451, 456]. The diminished synthesis of MUC2 precursor is "most likely due to ineffective translation of the available MUC2 mRNA" [451], which level, according to Weiss *et al.* [233], is constant and independent of disease activity. However, Longman *et al.* [457] found reduced content of MUC2 mRNA and reduced numbers of goblet cells in mucosa adjacent to ulceration in patients with severe UC, whereas the expression of MUC2 mRNA and protein levels appeared to be normal in areas distant from ulceration.

The level of Muc2 mucin has also been investigated in several animal models of experimental colitis. Renes *et al.* [121] analyzed alteration in Muc2 biosynthesis and secretion in dextran sulfate sodium (DSS) induced colitis and found that DSS-induced damage was associated with a decrease in the number of goblet cells in colon, accompanied by the maintenance or even increase in biosynthesis of Muc2 precursor, total Muc2 levels and total Muc2 secretion. The authors consider that since the amount of Muc2 mRNA is decreased and total level of the Muc2 protein is maintained or even elevated during the regenerative phase, Muc2 translation efficiency is apparently increased. Noting that the Muc2 glycoprotein becomes undersulfated, and the sulfated Muc2 is preferentially secreted during active disease and the regenerative phase, the authors suggest that the barrier functions of the mucus layer during DSS-induced disease are maintained or even elevated. Although the data in the DSS-induced mouse model of UC are of great importance, they contrast with the changes observed in the UC patients [451, 455 457]. The noted discrepancies raise question about the relevance of this model to human disease.

Two other mouse models of UC were recently developed and studied by Heazlewood *et al.* [458]. In this study, two random mutations, *Winnie* and *Eeyore*, in D3- and D4- domains of the Muc2 mucin, respectively, were introduced into the *Muc2* gene. The mutant mice showed aberrant Muc2 biosynthesis, less stored Muc2 mucin in goblet cells, a diminished mucus barrier, and increased susceptibility to colitis induced by a luminal toxin. The pathology that developed in mutant mice was accompanied by accumulation of the Muc2 precursor and impaired ability for glycosylation, oligomerization and targeting to cell compartments. As a result, the mutated Muc2 protein underwent misfolding and aberrant assembly, which led to accumulation of the misfolded molecules in endoplasmic reticulum causing substantial ER stress, goblet cell pathology, premature goblet cell apoptosis, and development of chronic intestinal inflammation. Importantly, in UC patients, a similar accumulation of nonglycosylated MUC2 precursor in goblet cells together with ultrastructural and biochemical evidence of ER stress detected even in morphologically noninflamed intestinal tissues were observed. The presence of aberrant goblet cells in nonaffected proximal colon of UC patients, as reported by Heazlewood *et al.* [458], suggests that this defect most probably precedes inflammation. As stressed by the authors of this study, comparison of the changes observed in the *Winnie* and *Eeyore* models with the alterations detected in human disease suggests that ER stress-related MUC2 depletion could be a fundamental component of the pathogenesis of human colitis.

The etiology of UC and CD is complex, undefined, and suggests familial clustering [459-461]. Since *MUC2* gene demonstrates direct involvement in the development of UC and CD, on the one hand, and the gene contains a high level of genetically determined length polymorphism, on the other hand [95, 462], the possible association of UC and/or CD with the definite *MUC2* allele(s) has been studied. No association was found between any *MUC2* allele and the occurrence of UC or CD [463] in the few studies conducted to date. The negative result obtained by Swallow *et al.* [463] seems to exclude the possibility that the short *MUC2* allele may predispose to UC. However, this result does not rule out the possibility that other variations in the *MUC2* gene, such as single nucleotide polymorphism (SNP) within a repeat, may play a role. Indeed, CD was recently found to be associated with a definite *MUC2* SNP variant (MUC2 V116M) [464]. Further genetic studies are needed to determine the possible connection between aberrant intestinal expression of mucin genes, including *MUC2*, and their allelic and/or SNP variants, associated with development of inflammatory bowel diseases.

Although CD and UC share many clinical features, they differ in some biological and biochemical parameters. The goblet cell phenotypes and secreted mucin profiles represent important differences between the two diseases [443, 458].

Typical features of UC include a reduction in the number of goblet cells, a decrease in MUC2 mucin production and sulfation, accumulation of MUC2 misfolded precursors, a reduction in secreted mucin and diminution of mucus layer thickness [444, 449, 465]. In contrast, in CD, the typical features are an increase in the number of goblet cells and in the thickness of the mucus layer [466, 467]. Gersemann *et al.* [468] described a principal difference between CD and UC in goblet cell differentiation in which a comparable level of inflammation induced activation of transcription factors Hath1 and KLF4 only in CD goblet cells and not in UC goblet cells. As noted by the authors [468], the association of inflammation with enhanced goblet cell differentiation in CD but not in UC might have "pathogenic importance".

Another difference between CD and UC is related to the topography of MUC2 mucin expression in the pathological intestine. In UC, accumulation of non-glycosylated MUC2 precursor in goblet and phenotypically nongoblet cells is observed in both ulceration areas and morphologically unaffected proximal regions. In CD, MUC2 expression is observed only in histologically unaffected mucosa, while its expression in the mucosa adjacent to the ulcer margins, in so-called ulcer associated cell lineage (UACL), is sharply decreased or can not be detected at all [469, 470].

Although CD is of great clinical importance, little is know about the associated genetic and biochemical processes. The role of *MUC2* in the pathogenesis of CD has been studied insufficiently, and more research is needed to elucidate functions of different mucins, including MUC2, in the development of CD. The limit of available information on the role of *MUC2* gene in UC and CD, suggests that alterations in MUC2 synthesis, glycosylation, oligomerization, targeting and secretion affect the integrity and protective capacities of the defensive mucus layer, resulting in induction and perpetuation of intestinal inflammation.

4.7.5. Expression of *MUC2* Gene in the Male Urogenital Tract

While the male urogenital tract is the least studied with regard to *MUC2* expression, nevertheless, the available information gives a general idea of the gene activity in the male urogenital organs.

Kidney: The development of kidney is a complex process that is thought to involve more than 300 genes including several mucin genes. Interestingly, the expression of *MUC2* in kidney during embryonic and fetal development has not been detected [155, 471, 472]. According to Leroy *et al.* [473], four mucin genes (*MUC1*, *MUC3*, *MUC4* and *MUC6*), but not *MUC2*, are substantially expressed in normal kidney. Abnormal fetal development causes an autosomal recessive polycystic kidney disease (ARPKD), in which only *MUC1* is expressed. No expression of *MUC2* or other secretory mucin genes, could be detected in ARKPD [155], indicating a negative regulation of these genes in this type of renal malformation. Interestingly, some of these genes are suppressed also in renal cell carcinoma [473]. When expression of several mucin genes, *MUC1*-7, was analyzed in different types of renal cell carcinomas, no types of renal neoplasms expressed *MUC2*, *MUC4*, *MUC5AC*, *MUC5B* and *MUC7*, whereas *MUC1*, *MUC3* and *MUC6* were substantially expressed [473]. The absence of *MUC2* gene expression in renal cell carcinomas was also documented by Lau *et al.* [474] and Hayashi *et al.* [475]. However, recently, Chu *et al.* [476] reported strong expression of *MUC2*, *CDX2* and *CK20* in the intestinal type of mucous borderline tumor arising from mixed epithelial and stromal tumor of kidney. The limited number of studies and conflicting results argue for further investigations of mucin gene expression in renal pathology.

Urinary bladder: Little information is available on the repertoire of mucin gene expression in normal and pathologically changed human urinary bladder. Cystitis glandularis (CG) – a metaplastic alteration of the bladder urothelium – is thought to be induced by chronic inflammation. Two types of CG have been identified: intestinal type (CGIT) and typical one (CGTP). They were shown to have different mucin profiles [477]: CGIT expresses *MUC2* and *MUC5AC*, while CGTP expresses *MUC1* and *CD10*. The importance of these data is highlighted by the suggestion that CG may progress to bladder adenocarcinoma [477].

Expression of mucins in different types of bladder carcinomas has been also studied. Two research groups examined *MUC2* expression in the signet-ring cell carcinoma component of urothelial bladder carcinoma and in primary mucinous adenocarcinoma of the bladder [478, 479]. In both studies, the signet cells were found to be MUC2-positive, while the urothelial carcinoma (UrCa) cell component of the same tumor was MUC2-negative. MUC2-negativity of UrCa was also reported by Lau *et al.* [474]. Other studies found the opposite: Hayashi *et al.* [475] showed that 35% of UCs tested were MUC2-positive with MUC2 staining observed in goblet cell-like vesicules and perinuclear regions of tumor cells. Approximately the same proportion (40%) of transitional cell carcinomas of the bladder was MUC2-positive in the study of Walsh *et al.* [182]. *MUC2* gene was highly expressed in all malignant tissue samples of papillary UrCa examined by Retz *et al.* [181]. In other studies, the expression of *MUC2* had differential mode with relatively high levels in the low-grade noninvasive tumors and low levels in the high-grade noninvasive and invasive UrCa [480, 481]. Recently, *MUC2* expression was observed in UrCa with abundant mixoid stroma [482].

Collectively, the available data show that the *MUC2* gene is actively expressed in most types of UrCa. However, the small numbers of the tested samples preclude drawing conclusions about the biological role of MUC2 mucin in the pathogenesis of UrCa. According to Marques *et al.* [479], the expression of markers (MUC2 and MUC5AC) specific for the mucin-producing tumors is very heterogeneous and observed in a wide variety of tumors. Therefore, immunohistochemical evaluation of these markers for diagnostic purpose is not usually reasonable. The role of these mucins in the biology of the bladder and in the pathogenesis of the urinary bladder diseases requires further investigation.

<u>Prostate</u>: The available information regarding *MUC2* expression in normal and malignant prostate is controversial. According to Cozzi *et al.* [483], both normal and malignant prostate tissues are MUC2-negative. These data correlate with the results reported by Cappello *et al.* [484] who could not detect MUC2 mucin in tumoral or nontumoral prostatic tissues. Zhang *et al.* [178], on the other hand, found that *MUC2* is expressed by normal prostate gland as well as by 73% of primary prostate cancers and by 90-100% of metastatic prostate adenocarcinoma. The results reported by Osunkoya *et al.* [179] are somewhere in the middle between these investigations: they found that *MUC2* is expressed in 100% of mucinous adenocarcinoma of the prostate, and with significantly lower frequency (24%) in nonmucinous adenocarcinoma of the prostate. Importantly, Osunkoya *et al.* [179], also found that 100% of the benign prostate glands were MUC2-negative.

In a comprehensive study, Legrier *et al.* [485] analyzed expression of mucin genes including *MUC2* in hormone-dependent and hormone-independent prostate cancer. Using a hormone-dependent poorly differentiated prostate adenocarcinoma xenograft, PAC120, and several hormone-independent variants of this tumor, the authors showed that, although *MUC2* expression was observed in both types of tumors, the loss of hormone dependence was marked by irreversible mucinous or neuro-endocrine histological alterations, associated with a constant increase in the expression of three gel-forming mucin genes - *MUC2*, *MUC5B* and *MUC6*. These data point to mucinous differentiation as an important step in the acquisition of hormone independence in prostate cancer. Although Legrier *et al.* [485] highlighted the role of *MUC2* in transition from hormone-dependent prostate cancer to hormone-independent state, a better understanding of the role of MUC2 mucin in the physiology of the prostate gland and the pathogenesis of prostate cancer requires further investigations.

Testis: No comprehensive studies of *MUC2* gene expression in testis have been performed, and what is known comes from the rare nonsystematic investigations based on a limited number of specimens. According to Zhang *et al.* [178], the *MUC2* gene is not expressed in normal testis, but this conclusion is based on a small number of specimens tested. In the pathologically changed testicular tissues, *MUC2* was expressed in goblet cells of mucinous cystadenoma of testis [486] – a finding that correlates with the results reported by Naito *et al.* [487]. These limited studies provide some information about *MUC2* activity in testicular tissues, but do not permit drawing conclusions about the biological functions of the MUC2 mucin in the testis. Further studies with large numbers of samples are needed to receive statistically reliable results regarding MUC2 mucin expression in normal and disease-affected testis.

4.7.6. Expression of *MUC2* Gene in the Female Reproductive Tract

Gel-forming mucins play an important role in the physiology and pathology of the female reproductive organs [183, 185, 488]. Importantly, expression of a particular mucin is organ- and tissue-specific and hormone-dependent. Aberrant expression of MUC2 and other secretory mucins are often observed in pathology.

Vagina: Vagina is almost *terra incognita* with regard to *MUC2* gene expression. We could find only one study, in which only one specimen of vaginal mucosa was analyzed [183]; it was MUC2-negative. Does vaginal epithelium really not express MUC2 mucin? The answer to this question awaits more studies.

Uterine cervix: The uterine cervix has been studied much more than other female genital organs with regard to MUC2 mucin expression. In the studies of Gipson *et al.* [183] and Audie *et al.* [488], the normal endocervical epithelium was found to express several mucin genes with relatively weak focal expression of *MUC2*. In other studies, the expression of *MUC2* was not detected in normal cervix [185, 186, 489], and found to be increased during transformation of the normal cervical tissue into malignant lesions. According to Reithdorf *et al.* [185], *MUC2* expression was observed in 100% of immature squamous cervical metaplasia, but was not detected in mature squamous and tubular metaplasias and in microglandular hyperplasia. These authors note that "the expression of MUC2 in cells of immature metaplasias is a surprising phenomenon that could be explained by changes in the expression of mucins during the metaplastic process" [185]. In contrast to expression in benign lesions, *MUC2* expression is observed relatively frequently in cervical adenocarcinoma [186]. Zhao *et al.* [489] detected expression of *MUC2* in 72% of endocervical adenocarcinoma, and speculated that neo-expression of *MUC2* could be associated with intestinal metaplasia of the endocervical epithelium. Importantly, intestinal type mucinous adenocarcinomas of different origin usually contain large amounts of MUC2-positive goblet and non-goblet cells, while the endocervical type of mucinous adenocarcinoma contains a smaller number of MUC2-positive cells, located mainly in clusters. Baker *et al.* [186] reported expression of MUC2 mucin in 25% of cervical adenocarcinoma *in situ*, in 43% of endocervical invasive lesions, in 50% of cervical adenosquamous carcinoma and in 40% of adenocarcinomas, confirming the earlier observation that progression from benign to malignant lesions may be associated with increased *MUC2* expression. In agreement with the aforementioned studies, Shintaku *et al.* [490] observed substantial cytoplasmic expression of MUC2 and nuclear expression of CDX2 in colloidal carcinoma of the intestinal type in the uterine cervix. Interestingly, the tumor cells appeared to be MUC5AC and MUC6-negative. This observation is of great importance, as the mucin profile in the studied colloidal carcinoma appeared to be opposite to that observed in normal endocervical glands, which are usually MUC5AC and MUC6-positive, but MUC2-negative.

MUC2 and *MUC6* expression in the gastrointestinal immunophenotype of cervical adenocarcinoma was extensively examined by Mikami *et al.* [491]. They found that lobular endocervical glandular hyperplasia has a high rate of *MUC6* expression (75%) and a relatively low rate (15%) of *MUC2* expression. The *MUC2* expression was most frequently observed in the intestinal type of adenocarcinomas, with a rate of 85%. Only 14% of endocervical-type adenocarcinomas were positive for MUC2 mucin. Importantly, both invasive adenocarcinomas and adenocarcinomas *in situ* expressed *MUC2* at approximately the same rates. Thus, the literature documents the relatively high level of *MUC2* gene activity in the uterine cervix during carcinogenesis.

Endometrium: During the adult life of a woman, the endometrial epithelium is constantly influenced by steroid hormones, which regulate the proliferative and secretory phases of the menstrual cycle. They also control expression of mucin genes in endometrium. The relationship between the phases of the menstrual cycle and expression of the MUC2 mucin has been studied. According to Alameda *et al.* [184], MUC2 is hardly detected in endometrium during the proliferative phase, whereas its expression is relatively high in the secretory phase. MUC2-positive reaction was observed in 37.5% of specimens obtained at the secretory phase of the cycle. MUC2 mucin was expressed in about 9% of simple hyperplasia samples and could not be detected at all in complex hyperplasia, although the differences between the two types of hyperplasias were not statistically significant. Low level of *MUC2* expression was found also in this study in all samples of endometrial adenocarcinoma, showing that expression of *MUC2* gene decreases in endometrial neoplastic transformation. These results correlate well with those reported by Hebbar *et al.* [187], who observed extremely low to negligible levels of MUC2 in all endometrial adenocarcinoma samples studied, findings confirmed by Baker *et al.* [186]. In one of the most comprehensive studies of mucin gene expression in endometrial carcinomas, Morrison *et al.* [492] found only two MUC2-positive tumors (0.6%) in 310 cases. In contrast, Zhao *et al.* [489] reported that 93% endometrial adenocarcinomas tested (14 in a series 15 tumors) were MUC2-positive with amount of MUC2-positive cells varying in different samples from 9% to 49%.

In conclusion, endometrial tumors appear to represent a non-homogeneous group with regard to *MUC2* gene expression: a subgroup of neoplasms with low expression level of *MUC2*, and a subgroup with a relatively high level of *MUC2* expression. Further studies of a large number of endometrial neoplasms classified according to histological and gene-specific markers, including the *MUC2* gene, are required.

Ovary: The recent review of Aubert *et al.* [493] and other studies [14, 494, 495] indicate that the superficial epithelium of the normal ovaries does not express mucin genes, while malignant transformation of ovarian epithelial cells is accompanied by differential expression of these genes. Ovarian mucinous tumors are classified into 3 types: mucinous adenoma (MA), mucinous borderline tumor (MB), and mucinous adenocarcinoma (MC) [496]. It has been shown that 71% of the benign MA and MB demonstrate intestinal phenotype with high level of *MUC2* expression. Definite levels of *MUC2* expression has been detected in ovarian clear cell and endometrioid adenocarcinomas as well as in mucinous and serous cystadenocarcinomas [497].

Hirabayashi *et al.* [498] found that expression of MUC2 glycoprotein is progressively increased from MA to MB and further to MC. These findings correlate well with those of previous studies that demonstrated an increase in *MUC2* expression concomitantly with transition from MA to MC [499, 500]. MC with high level of *MUC2* expression was reported associated with poor prognosis [498]. However, according to Dong *et al.* [499], the presence of MUC2 was not associated with altered survival but was inversely associated with high tumour grade. Hirabayashi *et al.* [498] described four different phenotypes of ovarian mucinous tumors: 1) intestinal pattern - MUC2(+)/MUC5AC(-)/MUC6(-)/CD10(+); 2) gastrointestinal pattern - MUC2(+)/MUC5AC(+)/MUC6(+)/CD10(-); 3) gastric pattern - MUC2(-)/MUC5AC(+/-)/MUC6(-/+)/CD10(-) and 4) unclassified pattern - MUC2(-)/MUC5AC(-)/MUC6(-)/CD10(-). The authors concluded that a close association between carcinogenesis and intestinal metaplasia is observed in major ovarian mucinous neoplasms. Feng *et al.* [500] also point to the role of MUC2 mucin in pathogenesis of ovarian tumours. Thus, although mucin genes are completely suppressed in normal ovarian epithelium, ovarian carcinogenesis is strongly associated with a high level of mucin gene expression, in general, and with *MUC2* expression, in particular. The presence of MUC2 mucin was inversely associated with high tumour grade but was not associated with altered survival.

4.7.7. Expression of *MUC2* Gene in the Visual System

Mucins have been ascribed several functions in the eye, including stabilizing the tear film integrity, providing a smooth and refractive surface of high optical quality over the cornea, lubricating the corneal and conjunctival epithelial surfaces, and preventing desiccation of the ocular surface [188, 501]. However, the information concerning *MUC2* expression in the ocular tissues is limited and controversial. McKenzie *et al.* [190] showed that although conjunctiva expresses MUC2 mRNA, the level of expression is very low, with the number of MUC2 transcripts detected in conjunctiva 5900 times lower than those of MUC5AC transcripts. Nevertheless, the MUC2 mRNA is translated into a detectable amount of MUC2 protein. Paulsen *et al.* [502] found expression of MUC2 as well as other gel-forming mucins (MUC5AC and MUC5B) in goblet cells, and in the intraepithelial mucous glands of the lacrimal sac and the nasolacrimal duct. Spurr-Michaud *et al.* [503] consistently detected low level of the MUC2 mucin in human tear fluid. In contrast, Schafer *et al.* [504] could not find MUC2 mRNA in human lacrimal glands, but easily detected transcripts of other gel-forming mucin genes. The absence of MUC2 expression in normal human lacrimal glands as well as in patients with dry eyes was reported also by Paulsen *et al.* [505].

Alterations in the mucin gene expression associated with eye diseases have been reported in several studies [505-507]. Patients with atopic keratoconjunctivitis, an allergic disease characterized by changes in expression of the gel-forming mucins [506, 507], exhibited a decreased level of the MUC5AC mucin and an increased level of MUC2. Such reciprocal alterations in expression of *MUC2* and *MUC5AC* genes may represent a defense mechanism that compensates the loss of protection brought about by the MUC5AC mucin deficiency [507]. A reduced level of *MUC2* expression was described in primary acquired dacryo-stenosis. *MUC2* expression was detected in 13% of samples obtained from patients with nasolacrimal duct stenosis, but was absent in all samples of lacrimal sac stenosis [502].

The studies cited above cover nearly all available investigations in which expression of *MUC2* in ocular tissues was examined in both normal and pathologic conditions. Further study of the subject is required.

4.7.8. Expression of *MUC2* Gene in the Audio System

The expression of *MUC2* by epithelial cells of organs related to the audio system have been studied, including the nasal and paranasal sinus mucosa and epithelium of the middle ear [173, 174, 508-515]. Aust *et al.* [173] detected MUC2 mRNA expression in normal nasal mucosa, as did Kim *et al.* [174]. Both groups observed up-regulation of *MUC2* gene expression in nasal polyps compared with control nasal mucosa. Ali and Pearson [508] reported expression of several mucin genes, including *MUC2*, in the paranasal sinuses of patients with chronic sinusitis: *MUC2* expression was found in 89% of the specimens tested, but the mean level of *MUC2* expression was lower than that of *MUC5AC* and *MUC5B*. Importantly, chronic sinusitis shows a reciprocal relationship between *MUC2* and *MUC5AC* expression: high levels of *MUC2* gene expression correlated with low levels of *MUC5AC*, and *vice versa*. This type of inverse relationship was also significant in nasal polyps [510].

Studies of *MUC2* gene expression in patients with cystic fibrosis (CF) revealed conflicting results. Some studies reported *MUC2* up-regulation in nasal mucosa of CF patients while others reported equal levels of *MUC2* expression in normal and CF nasal mucosa. The down-regulation of *MUC2* expression in nasal mucosa of CF patients was also reported [510, 511].

Ali and Pearson [508] found that the development of laryngeal carcinoma is associated with up-regulation of *MUC2* gene expression. The level of the expression was not significantly related to survival.

Mucins are considered to play an important role in the physiology of the middle ear and the pathogenesis of middle ear diseases. *MUC2* expression was detected in the human and animal normal middle ear epithelium (MEE) both *in vivo* and *in vitro* [192, 512]. MUC2 mucin was implicated in the development of acute and chronic otitis media (OM) in human and experimental animals, but findings of the studies are not uniform [192, 508, 513-515]. According to Ubell *et al.* [193], the levels of *MUC2* expression in human MEE are significantly increased in patients with OM compared with controls: six times higher in the OM samples than in controls; five times higher in recurrent OM samples; and eight times higher in OM samples with effusion (OME). Importantly, *MUC2* up-regulation was observed both in acute otitis events and in chronic OM. These data correlate well with the results obtained by Kirschner [192], but not with the findings of Ali and Pearson [508] who emphasized no *MUC2* expression in MEE which did display a substantial level of *MUC5B* and *MUC4* activity. Hutton *et al.* [514] also found no *MUC2* expression in human MEE, and Lin *et al.* [515] could not detect *MUC2* expression in mucoid OM.

Thus, while MUC2 mucin plays a definite role in the pathology of the upper airways and middle ear, the precise functions of the MUC2 mucin in these processes are not well understood. The available information raises more questions than answers. Hence, further studies are required to clarify the role of the MUC2 glycoprotein in the pathogenesis of diseases associated with the audio system.

4.7.9. Expression of *MUC2* Gene in the Mammary Gland

While *MUC2* expression is entirely absent in the normal breast [516-518], pathological changes in the mammary gland are often associated with aberrant expression of the gene. *MUC2* is expressed in malignant breast lesions such as ductal carcinoma *in situ* (DCIS), and further up-regulation of the gene is associated with transition of DCIS to lobular carcinoma *in situ* (LCIS), a precursor of highly aggressive invasive lobular carcinoma [518, 519]. Rates of *MUC2* expression in breast cancer vary in different studies. Zhang *et al.* [517] observed expression of *MUC2* in 43% of breast carcinomas tested. Berois *et al.* [520] detected only one MUC2-positive sample in a series of 11 breast carcinomas, which indicated a rather low rate (9%) of *MUC2* expression in breast cancer. Almost the same low rate (8.3%) was detected in breast carcinomas by Rakha *et al.* [521]. Expression of *MUC2* was observed in 19% [518] and 25 % [522] of invasive ductal carcinomas and in 11% of breast carcinomas *in situ* [518]. The expression rate of *MUC2* in mucinous carcinomas was significantly higher than in invasive ductal carcinomas [523]. For instance, Matsukita *et al.* [439] reported that 94% of breast mucinous carcinomas expressed MUC2 mucin, whereas only 15% of invasive ductal carcinomas samples were MUC2-positive. Chu and Chang [524] detected MUC2 in 100% of mucinous carcinomas of the breast, but in only 11% of invasive ductal carcinomas. According to Hanski *et al.*

[165], ectopic over-expression of *MUC2* is the common feature of mucinous carcinomas of many organs, including colon, pancreas, ovary and breast. The high expression rate of *MUC2* in mucinous carcinomas suggests that high production of MUC2 mucin may act as a barrier mechanism to cancerous extension. An inverse trend was noted between MUC2 expression, on the one hand, and lymph node stage and vascular invasion status, on the other hand [521]. In line with these findings, Adsay *et al.* [366] found that 100% of colloid (pure mucinous) breast carcinomas express MUC2 mucin, whereas only 6% of non-mucinous ductal carcinomas were MUC2-positive. MUC2 glycoprotein, a potential tumor suppressor [145], has been shown to accumulate in the stroma surrounding the colloid carcinoma cells where it "acts as a containing factor, slackening the spread of the cells" [366]. It appears that MUC2 mucin can indeed restrain tumor cells from expansion, although a recent study of mucin expression in lobular carcinoma of the mammary gland with histocytoid feature (HLC) raised doubts about the barrier defense ability attributed to MUC2 mucin [525]. These authors showed that 75% of HLCs express MUC2, whereas 100% of classical lobular carcinomas of the breast are MUC2-negative. Importantly, MUC2-positive HLC showed significantly shorter disease-free interval and survival time than lobular carcinomas without MUC2 expression. The authors concluded that "the expression of "non-mammary mucins" such as MUC2 and MUC5AC in the HLC is characteristic and indicates the more malignant transformation of tumor cells and poorer prognosis". Of note, the different behaviors of mucinous breast carcinomas and HLCs may be explained by the difference in intratumoral distribution of the MUC2 mucin: it is present in the cell cytoplasm in HLC and abundant in the extracellular mucosal lake in mucinous tumors.

The studies detailed above show that MUC2 mucin plays a definite role in breast carcinogenesis, fulfilling different functions in different clinical forms of the disease. More work is needed to determine the association of MUC2 expression with the biological properties, histological differentiation and clinical outcome of breast carcinomas. As a parameter for predicting prognosis and survival of patients with breast cancer, expression of MUC2 requires further study, which could yield a system of classification of breast tumors that would include mucin gene expression, particularly expression of the *MUC2* gene.

This chapter has demonstrated that MUC2 mucin is a complex glycoprotein consisting of numerous domains with the potential to carry out various functions under both physiological and pathological conditions. The multiple functions performed by MUC2 mucin will be discussed in detail in Chapters 12 and 13.

REFERENCES

[1] Allen A, Hutton DA, Pearson JP. The MUC2 gene product: a human intestinal mucin. Int J Biochem Cell Biol 1998;30:797-801.

[2] Gum JR, Jr., Hicks JW, Toribara NW, *et al.* The human MUC2 intestinal mucin has cysteine-rich subdomains located both upstream and downstream of its central repetitive region. J Biol Chem 1992;267:21375-83.

[3] Gum JR, Jr., Hicks JW, Toribara NW, Siddiki B, Kim YS. Molecular cloning of human intestinal mucin (MUC2) cDNA. Identification of the amino terminus and overall sequence similarity to prepro-von Willebrand factor. J Biol Chem 1994;269:2440-6.

[4] Pigny P, Guyonnet-Duperat V, Hill AS, *et al.* Human mucin genes assigned to 11p15.5: identification and organization of a cluster of genes. Genomics 1996;38:340-52.

[5] Byrd JC, Bresalier RS. Mucins and mucin binding proteins in colorectal cancer. Cancer Metastasis Rev 2004;23:77-99.

[6] Gum JR, Byrd JC, Hicks JW, *et al.* Molecular cloning of human intestinal mucin cDNAs. Sequence analysis and evidence for genetic polymorphism. J Biol Chem 1989;264:6480-7.

[7] Gum JR, Jr. Human mucin glycoproteins: varied structures predict diverse properties and specific functions. Biochem Soc Trans 1995;23:795-9.

[8] Gum JR, Hicks JW, Kim YS. Identification and characterization of the MUC2 (human intestinal mucin) gene 5'-flanking region: promoter activity in cultured cells. Biochem J 1997;325 (Pt 1):259-67.

[9] Baeckstrom D, Karlsson N, Hansson GC. Purification and characterization of sialyl-Le(a)-carrying mucins of human bile; evidence for the presence of MUC1 and MUC3 apoproteins. J Biol Chem 1994;269:14430-7.

[10] Audie JP, Janin A, Porchet N, *et al.* Expression of human mucin genes in respiratory, digestive, and reproductive tracts ascertained by *in situ* hybridization. J Histochem Cytochem 1993;41:1479-85.

[11] Jany BH, Gallup MW, Yan PS, *et al.* Human bronchus and intestine express the same mucin gene. J Clin Invest 1991;87:77-82.

[12] Gerard C, Eddy RL, Jr., Shows TB. The core polypeptide of cystic fibrosis tracheal mucin contains a tandem repeat structure. Evidence for a common mucin in airway and gastrointestinal tissue. J Clin Invest 1990;86:1921-7.

[13] Ogata S, Uehara H, Chen A, Itzkowitz SH. Mucin gene expression in colonic tissues and cell lines. Cancer Res 1992;52:5971-8.

[14] Ho SB, Niehans GA, Lyftogt C, *et al.* Heterogeneity of mucin gene expression in normal and neoplastic tissues. Cancer Res 1993;53:641-51.

[15] Andrianifahanana M, Moniaux N, Batra SK. Regulation of mucin expression: mechanistic aspects and implications for cancer and inflammatory diseases. Biochim Biophys Acta 2006;1765:189-222.

[16] Velcich A, Palumbo L, Selleri L, Evans G, Augenlicht L. Organization and regulatory aspects of the human intestinal mucin gene (MUC2) locus. J Biol Chem 1997;272:7968-76.

[17] Aslam F, Palumbo L, Augenlicht LH, Velcich A. The Sp family of transcription factors in the regulation of the human and mouse MUC2 gene promoters. Cancer Res 2001;61:570-6.

[18] Nogami H, Ohmori H, Li JD, *et al.* Sp1 protein contributes to airway-specific rat MUC 2 mucin gene transcription. Gene 1997;198:191-201.

[19] van Klinken BJ, Einerhand AW, Duits LA, *et al.* Gastrointestinal expression and partial cDNA cloning of murine Muc2. Am J Physiol 1999;276:G115-24.

[20] Davies DE, Polosa R, Puddicombe SM, Richter A, Holgate ST. The epidermal growth factor receptor and its ligand family: their potential role in repair and remodeling in asthma. Allergy 1999;54:771-83.

[21] Perrais M, Pigny P, Copin MC, Aubert JP, Van Seuningen I. Induction of MUC2 and MUC5AC mucins by factors of the epidermal growth factor (EGF) family is mediated by EGF receptor/Ras/Raf/extracellular signal-regulated kinase cascade and Sp1. J Biol Chem 2002;277:32258-67.

[22] McNamara N, Gallup M, Khong A, *et al.* Adenosine up-regulation of the mucin gene, MUC2, in asthma. Faseb J 2004;18:1770-2.

[23] Siedow A, Szyf M, Gratchev A, *et al. De novo* expression of the Muc2 gene in pancreas carcinoma cells is triggered by promoter demethylation. Tumour Biol 2002;23:54-60.

[24] Vincent A, Perrais M, Desseyn JL, *et al.* Epigenetic regulation (DNA methylation, histone modifications) of the 11p15 mucin genes (MUC2, MUC5AC, MUC5B, MUC6) in epithelial cancer cells. Oncogene 2007;26:6566-76.

[25] Traber PG, Silberg DG. Intestine-specific gene transcription. Annu Rev Physiol 1996;58:275-97.

[26] Taylor JK, Boll W, Levy T, *et al.* Comparison of intestinal phospholipase A/lysophospholipase and sucrase-isomaltase genes suggest a common structure for enterocyte-specific promoters. DNA Cell Biol 1997;16:1419-28.

[27] Sakaguchi T, Gu X, Golden HM, *et al.* Cloning of the human claudin-2 5'-flanking region revealed a TATA-less promoter with conserved binding sites in mouse and human for caudal-related homeodomain proteins and hepatocyte nuclear factor-1alpha. J Biol Chem 2002;277:21361-70.

[28] Yamamoto H, Bai YQ, Yuasa Y. Homeodomain protein CDX2 regulates goblet-specific MUC2 gene expression. Biochem Biophys Res Commun 2003;300:813-8.

[29] Mesquita P, Jonckheere N, Almeida R, *et al.* Human MUC2 mucin gene is transcriptionally regulated by Cdx homeodomain proteins in gastrointestinal carcinoma cell lines. J Biol Chem 2003;278:51549-56.

[30] Guo RJ, Suh ER, Lynch JP. The role of Cdx proteins in intestinal development and cancer. Cancer Biol Ther 2004;3:593-601.

[31] Almeida R, Silva E, Santos-Silva F, *et al.* Expression of intestine-specific transcription factors, CDX1 and CDX2, in intestinal metaplasia and gastric carcinomas. J Pathol 2003;199:36-40.

[32] Martin MG, Wang J, Solorzano-Vargas RS, *et al.* Regulation of the human Na(+)-glucose cotransporter gene, SGLT1, by HNF-1 and Sp1. Am J Physiol Gastrointest Liver Physiol 2000;278:G591-603.

[33] Yussa M, Lohr U, Su K, Pick L. The nuclear receptor Ftz-F1 and homeodomain protein Ftz interact through evolutionarily conserved protein domains. Mech Dev 2001;107:39-53.

[34] Molkentin JD. The zinc finger-containing transcription factors GATA-4, -5, and -6. Ubiquitously expressed regulators of tissue-specific gene expression. J Biol Chem 2000;275:38949-52.
[35] Divine JK, McCaul SP, Simon TC. HNF-1alpha and endodermal transcription factors cooperatively activate Fabpl: MODY3 mutations abrogate cooperativity. Am J Physiol Gastrointest Liver Physiol 2003;285:G62-72.
[36] Krasinski SD, Van Wering HM, Tannemaat MR, Grand RJ. Differential activation of intestinal gene promoters: functional interactions between GATA-5 and HNF-1 alpha. Am J Physiol Gastrointest Liver Physiol 2001;281:G69-84.
[37] Boudreau F, Rings EH, van Wering HM, *et al.* Hepatocyte nuclear factor-1 alpha, GATA-4, and caudal related homeodomain protein Cdx2 interact functionally to modulate intestinal gene transcription. Implication for the developmental regulation of the sucrase-isomaltase gene. J Biol Chem 2002;277:31909-17.
[38] van der Sluis M, Melis MH, Jonckheere N, *et al.* The murine Muc2 mucin gene is transcriptionally regulated by the zinc-finger GATA-4 transcription factor in intestinal cells. Biochem Biophys Res Commun 2004;325:952-60.
[39] van der Sluis M, Vincent A, Bouma J, *et al.* Forkhead box transcription factors Foxa1 and Foxa2 are important regulators of Muc2 mucin expression in intestinal epithelial cells. Biochem Biophys Res Commun 2008;369:1108-13.
[40] Buisine MP, Devisme L, Savidge TC, *et al.* Mucin gene expression in human embryonic and fetal intestine. Gut 1998;43:519-24.
[41] Corfield AP, Carroll D, Myerscough N, Probert CS. Mucins in the gastrointestinal tract in health and disease. Front Biosci 2001;6:D1321-57.
[42] Morrisey EE, Ip HS, Lu MM, Parmacek MS. GATA-6: a zinc finger transcription factor that is expressed in multiple cell lineages derived from lateral mesoderm. Dev Biol 1996;177:309-22.
[43] Morrisey EE, Ip HS, Tang Z, Lu MM, Parmacek MS. GATA-5: a transcriptional activator expressed in a novel temporally and spatially-restricted pattern during embryonic development. Dev Biol 1997;183:21-36.
[44] Gao X, Sedgwick T, Shi YB, Evans T. Distinct functions are implicated for the GATA-4, -5, and -6 transcription factors in the regulation of intestine epithelial cell differentiation. Mol Cell Biol 1998;18:2901-11.
[45] Fang R, Olds LC, Santiago NA, Sibley E. GATA family transcription factors activate lactase gene promoter in intestinal Caco-2 cells. Am J Physiol Gastrointest Liver Physiol 2001;280:G58-67.
[46] Zaret K. Developmental competence of the gut endoderm: genetic potentiation by GATA and HNF3/fork head proteins. Dev Biol 1999;209:1-10.
[47] Tamura S, Wang XH, Maeda M, Futai M. Gastric DNA-binding proteins recognize upstream sequence motifs of parietal cell-specific genes. Proc Natl Acad Sci U S A 1993;90:10876-80.
[48] Yuasa Y. Control of gut differentiation and intestinal-type gastric carcinogenesis. Nat Rev Cancer 2003;3:592-600.
[49] Ookawa K, Kudo T, Aizawa S, Saito H, Tsuchida S. Transcriptional activation of the MUC2 gene by p53. J Biol Chem 2002;277:48270-5.
[50] Li JD, Feng W, Gallup M, *et al.* Activation of NF-kappaB *via* a Src-dependent Ras-MAPK-pp90rsk pathway is required for Pseudomonas aeruginosa-induced mucin overproduction in epithelial cells. Proc Natl Acad Sci U S A 1998;95:5718-23.

[51] Weinstein SL, Sanghera JS, Lemke K, DeFranco AL, Pelech SL. Bacterial lipopolysaccharide induces tyrosine phosphorylation and activation of mitogen-activated protein kinases in macrophages. J Biol Chem 1992;267:14955-62.

[52] Li JD, Dohrman AF, Gallup M, *et al.* Transcriptional activation of mucin by Pseudomonas aeruginosa lipopolysaccharide in the pathogenesis of cystic fibrosis lung disease. Proc Natl Acad Sci U S A 1997;94:967-72.

[53] Jono H, Shuto T, Xu H, *et al.* Transforming growth factor-beta -Smad signaling pathway cooperates with NF-kappa B to mediate nontypeable Haemophilus influenzae-induced MUC2 mucin transcription. J Biol Chem 2002;277:45547-57.

[54] Suzuki S, Takeuchi K, Ishinaga H, Basbaum C, Majima Y. Leukotriene D4 upregulates MUC2 gene transcription in human epithelial cells. Pharmacology 2008;81:221-8.

[55] Dahlen SE, Hedqvist P, Hammarstrom S, Samuelsson B. Leukotrienes are potent constrictors of human bronchi. Nature 1980;288:484-6.

[56] Wu J, Gong J, Geng J, Song Y. Deoxycholic acid induces the overexpression of intestinal mucin, MUC2, *via NF-κB* signaling pathway in human esophageal adenocarcinoma cells. BMC Cancer 2008;8:333.

[57] Smythies LE, Waites KB, Lindsey JR, *et al.* Helicobacter pylori-induced mucosal inflammation is Th1 mediated and exacerbated in IL-4, but not IFN-gamma, gene-deficient mice. J Immunol 2000;165:1022-9.

[58] Karttunen R, Karttunen T, Ekre HP, MacDonald TT. Interferon gamma and interleukin 4 secreting cells in the gastric antrum in Helicobacter pylori positive and negative gastritis. Gut 1995;36:341-5.

[59] Takahashi K, Yamanaka S. Induction of pluripotent stem cells from mouse embryonic and adult fibroblast cultures by defined factors. Cell 2006;126:663-76.

[60] Que J, Okubo T, Goldenring JR, *et al.* Multiple dose-dependent roles for Sox2 in the patterning and differentiation of anterior foregut endoderm. Development 2007;134:2521-31.

[61] Hokari R, Lee H, Crawley SC, *et al.* Vasoactive intestinal peptide upregulates MUC2 intestinal mucin *via* CREB/ATF1. Am J Physiol Gastrointest Liver Physiol. 2005;289:G949-59.

[62] Kim SW, Hong JS, Ryu SH, *et al.* Regulation of mucin gene expression by CREB *via* a nonclassical retinoic acid signaling pathway. Mol Cell Biol 2007;27:6933-47.

[63] Dawson TM, Ginty DD. CREB family transcription factors inhibit neuronal suicide. Nat Med 2002;8:450-1.

[64] Reusch JE, Colton LA, Klemm DJ. CREB activation induces adipogenesis in 3T3-L1 cells. Mol Cell Biol 2000;20:1008-20.

[65] Song S, Byrd JC, Mazurek N, *et al.* Galectin-3 modulates MUC2 mucin expression in human colon cancer cells at the level of transcription *via* AP-1 activation. Gastroenterology 2005;129:1581-91.

[66] Yoshii T, Fukumori T, Honjo Y, *et al.* Galectin-3 phosphorylation is required for its anti-apoptotic function and cell cycle arrest. J Biol Chem 2002;277:6852-7.

[67] Perillo NL, Marcus ME, Baum LG. Galectins: versatile modulators of cell adhesion, cell proliferation, and cell death. J Mol Med (Berl) 1998;76:402-12.

[68] Cortegano I, Pozo V, Cardaba B, *et al.* Interaction between galectin-3 and FcgammaRII induces down-regulation of IL-5 gene: implication of the promoter sequence IL-5REIII. Glycobiology 2000;10:237-42.

[69] Bresalier RS, Byrd JC, Wang L, Raz A. Colon cancer mucin: a new ligand for the beta-galactoside-binding protein galectin-3. Cancer Res 1996;56:4354-7.

[70] Burger-van Paassen N, Vincent A, Puiman PJ, *et al.* The regulation of intestinal mucin MUC2 expression by short-chain fatty acids: implications for epithelial protection. Biochem J 2009;420:211-9.

[71] Alvaro A, Sola R, Rosales R, *et al.* Gene expression analysis of a human enterocyte cell line reveals downregulation of cholesterol biosynthesis in response to short-chain fatty acids. IUBMB Life 2008;60:757-64.

[72] Sauer J, Richter KK, Pool-Zobel BL. Physiological concentrations of butyrate favorably modulate genes of oxidative and metabolic stress in primary human colon cells. J Nutr Biochem 2007;18:736-45.

[73] Daly K, Shirazi-Beechey SP. Microarray analysis of butyrate regulated genes in colonic epithelial cells. DNA Cell Biol 2006;25:49-62.

[74] Davie JR. Inhibition of histone deacetylase activity by butyrate. J Nutr 2003;133:2485S-93S.

[75] Vincent A, Ducourouble MP, Van Seuningen I. Epigenetic regulation of the human mucin gene MUC4 in epithelial cancer cell lines involves both DNA methylation and histone modifications mediated by DNA methyltransferases and histone deacetylases. Faseb J 2008;22:3035-45.

[76] Jaenisch R, Bird A. Epigenetic regulation of gene expression: how the genome integrates intrinsic and environmental signals. Nat Genet 2003;33 Suppl:245-54.

[77] Turker MS. Gene silencing in mammalian cells and the spread of DNA methylation. Oncogene 2002;21:5388-93.

[78] Jenuwein T, Allis CD. Translating the histone code. Science 2001;293:1074-80.

[79] Hanski C, Riede E, Gratchev A, *et al.* MUC2 gene suppression in human colorectal carcinomas and their metastases: *in vitro* evidence of the modulatory role of DNA methylation. Lab Invest 1997;77:685-95.

[80] Gratchev A, Bohm C, Riede E, *et al.* Regulation of mucin MUC2 gene expression during colon carcinogenesis. Ann N Y Acad Sci 1998;859:180-3.

[81] Ho JJ, Han SW, Pan PL, *et al.* Methylation status of promoters and expression of MUC2 and MUC5AC mucins in pancreatic cancer cells. Int J Oncol 2003;22:273-9.

[82] Gratchev A, Siedow A, Bumke-Vogt C, *et al.* Regulation of the intestinal mucin MUC2 gene expression *in vivo*: evidence for the role of promoter methylation. Cancer Lett 2001;168:71-80.

[83] Hamada T, Goto M, Tsutsumida H, *et al.* Mapping of the methylation pattern of the MUC2 promoter in pancreatic cancer cell lines, using bisulfite genomic sequencing. Cancer Lett 2005;227:175-84.

[84] Mesquita P, Peixoto AJ, Seruca R, *et al.* Role of site-specific promoter hypomethylation in aberrant MUC2 mucin expression in mucinous gastric carcinomas. Cancer Lett 2003;189:129-36.

[85] Sylvester PA, Myerscough N, Warren BF, *et al.* Differential expression of the chromosome 11 mucin genes in colorectal cancer. J Pathol 2001;195:327-35.

[86] Yamada N, Hamada T, Goto M, *et al.* MUC2 expression is regulated by histone H3 modification and DNA methylation in pancreatic cancer. Int J Cancer 2006;119:1850-7.

[87] de Bustros A, Nelkin BD, Silverman A, *et al.* The short arm of chromosome 11 is a "hot spot" for hypermethylation in human neoplasia. Proc Natl Acad Sci U S A 1988;85:5693-7.

[88] Bartel DP. MicroRNAs: genomics, biogenesis, mechanism, and function. Cell 2004;116:281-97.

[89] He L, Hannon GJ. MicroRNAs: small RNAs with a big role in gene regulation. Nat Rev Genet 2004;5:522-31.

[90] Ji Q, Karnak D, Hao P, Wang R, Xu L. No small matter: microRNAs - key regulators of cancer stem cells. Int J Clin Exp Med 3:84-7.

[91] Biemer-Huttmann AE, Walsh MD, McGuckin MA, *et al.* Mucin core protein expression in colorectal cancers with high levels of microsatellite instability indicates a novel pathway of morphogenesis. Clin Cancer Res 2000;6:1909-16.

[92] Manne U, Weiss HL, Grizzle WE. Racial differences in the prognostic usefulness of MUC1 and MUC2 in colorectal adenocarcinomas. Clin Cancer Res 2000;6:4017-25.

[93] Matsuda K, Masaki T, Watanabe T, *et al.* Clinical significance of MUC1 and MUC2 mucin and p53 protein expression in colorectal carcinoma. Jpn J Clin Oncol 2000;30:89-94.

[94] Sternberg LR, Byrd JC, Hansson GC, Liu KF, Bresalier RS. Alternative splicing of the human MUC2 gene. Arch Biochem Biophys 2004;421:21-33.

[95] Toribara NW, Gum JR, Jr., Culhane PJ, *et al.* MUC-2 human small intestinal mucin gene structure. Repeated arrays and polymorphism. J Clin Invest 1991;88:1005-13.

[96] Jeong YH, Kim MC, Ahn EK, *et al.* Rare exonic minisatellite alleles in MUC2 influence susceptibility to gastric carcinoma. PLoS One 2007;2:e1163.

[97] Kralovicova J, Gaunt TR, Rodriguez S, *et al.* Variants in the human insulin gene that affect pre-mRNA splicing: is -23HphI a functional single nucleotide polymorphism at IDDM2? Diabetes 2006;55:260-4.

[98] Yeo G, Hoon S, Venkatesh B, Burge CB. Variation in sequence and organization of splicing regulatory elements in vertebrate genes. Proc Natl Acad Sci U S A 2004;101:15700-5.

[99] Dirksen WP, Hampson RK, Sun Q, Rottman FM. A purine-rich exon sequence enhances alternative splicing of bovine growth hormone pre-mRNA. J Biol Chem 1994;269:6431-6.

[100] Pret AM, Balvay L, Fiszman MY. Regulated splicing of an alternative exon of beta-tropomyosin pre-mRNAs in myogenic cells depends on the strength of pyrimidine-rich intronic enhancer elements. DNA Cell Biol 1999;18:671-83.

[101] Gabellini N. A polymorphic GT repeat from the human cardiac Na+Ca2+ exchanger intron 2 activates splicing. Eur J Biochem 2001;268:1076-83.

[102] McGuckin MA, Devine PL, Ward BG. Early steps in the biosynthesis of MUC2 epithelial mucin in colon cancer cells. Biochem Cell Biol 1996;74:87-93.

[103] Rose MC, Voynow JA. Respiratory tract mucin genes and mucin glycoproteins in health and disease. Physiol Rev 2006;86:245-78.

[104] Rogers DF. Physiology of airway mucus secretion and pathophysiology of hypersecretion. Respir Care 2007;52:1134-46; discussion 46-9.

[105] Asker N, Axelsson MA, Olofsson SO, Hansson GC. Dimerization of the human MUC2 mucin in the endoplasmic reticulum is followed by a N-glycosylation-dependent transfer of the mono- and dimers to the Golgi apparatus. J Biol Chem 1998;273:18857-63.

[106] Bell SL, Xu G, Khatri IA, *et al.* N-linked oligosaccharides play a role in disulphide-dependent dimerization of intestinal mucin Muc2. Biochem J 2003;373:893-900.

[107] Dekker J, Strous GJ. Covalent oligomerization of rat gastric mucin occurs in the rough endoplasmic reticulum, is N-glycosylation-dependent, and precedes initial O-glycosylation. J Biol Chem 1990;265:18116-22.

[108] McCool DJ, Forstner JF, Forstner GG. Synthesis and secretion of mucin by the human colonic tumour cell line LS180. Biochem J 1994;302 (Pt 1):111-8.
[109] van Klinken BJ, Einerhand AW, Buller HA, Dekker J. The oligomerization of a family of four genetically clustered human gastrointestinal mucins. Glycobiology 1998;8:67-75.
[110] McCool DJ, Okada Y, Forstner JF, Forstner GG. Roles of calreticulin and calnexin during mucin synthesis in LS180 and HT29/A1 human colonic adenocarcinoma cells. Biochem J 1999;341 (Pt 3):593-600.
[111] Bell SL, Xu G, Forstner JF. Role of the cystine-knot motif at the C-terminus of rat mucin protein Muc2 in dimer formation and secretion. Biochem J 2001;357:203-9.
[112] Groux-Degroote S, Harduin-Lepers A, Delannoy P. Biosynthesis of mucin O-glycan chains in normal and pathological states. In: Van Seuningen I, editor. The epithelial mucins:structure/function Roles in cancer and inflammatory diseases. Kerala: Research Singpost; 2008. p. 39-54.
[113] Miyoshi E, Moriwaki K, Nakagawa T. Biological function of fucosylation in cancer biology. J Biochem 2008;143:725-9.
[114] Coughtrie MW. Sulfation through the looking glass--recent advances in sulfotransferase research for the curious. Pharmacogenomics J 2002;2:297-308.
[115] Ten Hagen KG, Fritz TA, Tabak LA. All in the family: the UDP-GalNAc:polypeptide N-acetylgalactosaminyltransferases. Glycobiology 2003;13:1R-16R.
[116] Ju T, Brewer K, D'Souza A, Cummings RD, Canfield WM. Cloning and expression of human core 1 beta1,3-galactosyltransferase. J Biol Chem 2002;277:178-86.
[117] Iwai T, Inaba N, Naundorf A, *et al.* Molecular cloning and characterization of a novel UDP-GlcNAc:GalNAc-peptide beta1,3-N-acetylglucosaminyltransferase (beta 3Gn-T6), an enzyme synthesizing the core 3 structure of O-glycans. J Biol Chem 2002;277:12802-9.
[118] Sasaki K, Kurata-Miura K, Ujita M, *et al.* Expression cloning of cDNA encoding a human beta-1,3-N-acetylglucosaminyltransferase that is essential for poly-N-acetyllactosamine synthesis. Proc Natl Acad Sci U S A 1997;94:14294-9.
[119] Harduin-Lepers A, Mollicone R, Delannoy P, Oriol R. The animal sialyltransferases and sialyltransferase-related genes: a phylogenetic approach. Glycobiology 2005;15:805-17.
[120] Tytgat KM, Bovelander FJ, Opdam FJ, *et al.* Biosynthesis of rat MUC2 in colon and its analogy with human MUC2. Biochem J 1995;309 (Pt 1):221-9.
[121] Renes IB, Boshuizen JA, Van Nispen DJ, *et al.* Alterations in Muc2 biosynthesis and secretion during dextran sulfate sodium-induced colitis. Am J Physiol Gastrointest Liver Physiol 2002;282:G382-9.
[122] Raouf AH, Tsai HH, Parker N, *et al.* Sulphation of colonic and rectal mucin in inflammatory bowel disease: reduced sulphation of rectal mucus in ulcerative colitis. Clin Sci (Lond) 1992;83:623-6.
[123] Tytgat KM, Buller HA, Opdam FJ, *et al.* Biosynthesis of human colonic mucin: Muc2 is the prominent secretory mucin. Gastroenterology 1994;107:1352-63.
[124] Brockhausen I, Dowler T, Paulsen H. Site directed processing: role of amino acid sequences and glycosylation of acceptor glycopeptides in the assembly of extended mucin type O-glycan core 2. Biochim Biophys Acta 2009;1790:1244-57.
[125] Robbe-Masselot C, Michalski JC, Capon C. Tumour-associated antigens of mucin O-glycans. Clinical relevance of glycobiology in cancer and inflammatory diseases of the epthelium. In: Van Seuningen I, editor. The epithelial mucins:structure/function Roles in cancer and inflammatory diseases. Kerala: Research Signpost; 2008. p. 55-73.

[126] Axelsson MA, Asker N, Hansson GC. O-glycosylated MUC2 monomer and dimer from LS 174T cells are water-soluble, whereas larger MUC2 species formed early during biosynthesis are insoluble and contain nonreducible intermolecular bonds. J Biol Chem 1998;273:18864-70.

[127] Egea G, Franci C, Gambus G, *et al.* cis-Golgi resident proteins and O-glycans are abnormally compartmentalized in the RER of colon cancer cells. J Cell Sci 1993;105 (Pt 3):819-30.

[128] Khatri IA, Forstner GG, Forstner JF. Susceptibility of the cysteine-rich N-terminal and C-terminal ends of rat intestinal mucin muc 2 to proteolytic cleavage. Biochem J 1998;331 (Pt 1):323-30.

[129] Bowen DJ, Collins PW. Insights into von Willebrand factor proteolysis: clinical implications. Br J Haematol 2006;133:457-67.

[130] Verweij CL, Diergaarde PJ, Hart M, Pannekoek H. Full-length von Willebrand factor (vWF) cDNA encodes a highly repetitive protein considerably larger than the mature vWF subunit. Embo J 1986;5:1839-47.

[131] Masson PL, Ferin J. [Protein constituents of the cervix mucus]. Gynecol Obstet (Paris) 1969;68:419-32.

[132] Creeth JM, Bhaskar KR, Horton JR, *et al.* The separation and characterization of bronchial glycoproteins by density-gradient methods. Biochem J 1977;167:557-69.

[133] Xu G, Huan LJ, Khatri IA, *et al.* cDNA for the carboxyl-terminal region of a rat intestinal mucin-like peptide. J Biol Chem 1992;267:5401-7.

[134] Lidell ME, Johansson ME, Hansson GC. An autocatalytic cleavage in the C terminus of the human MUC2 mucin occurs at the low pH of the late secretory pathway. J Biol Chem 2003;278:13944-51.

[135] Chan P, Risler JL, Raguenez G, Salier JP. The three heavy-chain precursors for the inter-alpha-inhibitor family in mouse: new members of the multicopper oxidase protein group with differential transcription in liver and brain. Biochem J 1995;306 (Pt 2):505-12.

[136] Goto A, Kumagai T, Kumagai C, *et al.* A Drosophila haemocyte-specific protein, hemolectin, similar to human von Willebrand factor. Biochem J 2001;359:99-108.

[137] Sheng ZQ, Hull SR, Carraway KL. Biosynthesis of the cell surface sialomucin complex of ascites 13762 rat mammary adenocarcinoma cells from a high molecular weight precursor. J Biol Chem 1990;265:8505-10.

[138] Sheng Z, Wu K, Carraway KL, Fregien N. Molecular cloning of the transmembrane component of the 13762 mammary adenocarcinoma sialomucin complex. A new member of the epidermal growth factor superfamily. J Biol Chem 1992;267:16341-6.

[139] Xu G, Bell SL, McCool D, Forstner JF. The cationic C-terminus of rat Muc2 facilitates dimer formation post translationally and is subsequently removed by furin. Eur J Biochem 2000;267:2998-3004.

[140] Hosaka M, Nagahama M, Kim WS, *et al.* Arg-X-Lys/Arg-Arg motif as a signal for precursor cleavage catalyzed by furin within the constitutive secretory pathway. J Biol Chem 1991;266:12127-30.

[141] Perez-Vilar J, Eckhardt AE, DeLuca A, Hill RL. Porcine submaxillary mucin forms disulfide-linked multimers through its amino-terminal D-domains. J Biol Chem 1998;273:14442-9.

[142] Troxler RF, Offner GD, Zhang F, Iontcheva I, Oppenheim FG. Molecular cloning of a novel high molecular weight mucin (MG1) from human sublingual gland. Biochem Biophys Res Commun 1995;217:1112-9.

[143] Perez-Vilar J, Hill RL. Identification of the half-cystine residues in porcine submaxillary mucin critical for multimerization through the D-domains. Roles of the CGLCG motif in the D1- and D3-domains. J Biol Chem 1998;273:34527-34.
[144] Buisine MP, Porchet N, Van Seuningen I. Mucin expression and regulation during development and cell differentiation. In: Van Seuningen I, editor. The epithelial mucins: structure/function Roles in cancer nd inflammatory diseases. Kerala: Research Signpost; 2008. p. 75-94.
[145] Velcich A, Yang W, Heyer J, *et al.* Colorectal cancer in mice genetically deficient in the mucin Muc2. Science 2002;295:1726-9.
[146] Ho SB, Shekels LL, Toribara NW, *et al.* Mucin gene expression in normal, preneoplastic, and neoplastic human gastric epithelium. Cancer Res 1995;55:2681-90.
[147] Filipe MI, Linehan JM, Durrant LG, *et al.* Expression of a peptide epitope of the colonic mucin MUC2 in precursor lesions to gastric carcinoma. Eur J Cancer Prev 1996;5:287-95.
[148] Lee HW, Yang DH, Kim HK, *et al.* Expression of MUC2 in gastric carcinomas and background mucosae. J Gastroenterol Hepatol 2007;22:1336-43.
[149] Buisine MP, Devisme L, Degand P, *et al.* Developmental mucin gene expression in the gastroduodenal tract and accessory digestive glands. II. Duodenum and liver, gallbladder, and pancreas. J Histochem Cytochem 2000;48:1667-76.
[150] Buisine MP, Devisme L, Maunoury V, *et al.* Developmental mucin gene expression in the gastroduodenal tract and accessory digestive glands. I. Stomach. A relationship to gastric carcinoma. J Histochem Cytochem 2000;48:1657-66.
[151] Chambers JA, Hollingsworth MA, Trezise AE, Harris A. Developmental expression of mucin genes MUC1 and MUC2. J Cell Sci 1994;107 (Pt 2):413-24.
[152] Buisine MP, Devisme L, Copin MC, *et al.* Developmental mucin gene expression in the human respiratory tract. Am J Respir Cell Mol Biol 1999;20:209-18.
[153] Jeffery PK, Gaillard D, Moret S. Human airway secretory cells during development and in mature airway epithelium. Eur Respir J 1992;5:93-104.
[154] Reid CJ, Harris A. Developmental expression of mucin genes in the human gastrointestinal system. Gut 1998;42:220-6.
[155] Leroy X, Devisme L, Buisine MP, *et al.* Expression of human mucin genes during normal and abnormal renal development. Am J Clin Pathol 2003;120:544-50.
[156] Lopez-Ferrer A, Barranco C, de Bolos C. Apomucin expression and association with Lewis antigens during gastric development. Appl Immunohistochem Mol Morphol 2001;9:42-8.
[157] Deren JS. Development of structure and function in the fetal and newborn stomach. Am J Clin Nutr 1971;24:144-59.
[158] Deplancke B, Gaskins HR. Microbial modulation of innate defense: goblet cells and the intestinal mucus layer. Am J Clin Nutr 2001;73:1131S-41S.
[159] Meslin JC, Fontaine N, Andrieux C. Variation of mucin distribution in the rat intestine, caecum and colon: effect of the bacterial flora. Comp Biochem Physiol A Mol Integr Physiol 1999;123:235-9.
[160] Chadee K, Meerovitch E. Entamoeba histolytica: early progressive pathology in the cecum of the gerbil (Meriones unguiculatus). Am J Trop Med Hyg 1985;34:283-91.
[161] Mack DR, Michail S, Wei S, McDougall L, Hollingsworth MA. Probiotics inhibit enteropathogenic E. coli adherence *in vitro* by inducing intestinal mucin gene expression. Am J Physiol 1999;276:G941-50.
[162] Atuma C, Strugala V, Allen A, Holm L. The adherent gastrointestinal mucus gel layer: thickness and physical state *in vivo*. Am J Physiol Gastrointest Liver Physiol 2001;280:G922-9.

[163] Hasnain SZ, Wang H, Ghia JE, *et al.* Mucin gene deficiency in mice impairs host resistance to an enteric parasitic infection. Gastroenterology138:1763-71.

[164] Forstner JP, Forstner GG. Gastrointestinal mucins. In: Johnson LR, editor. Physiology of the gastrointestinal tract; 1994. p. 1255-83.

[165] Hanski C, Hofmeier M, Schmitt-Graff A, *et al.* Overexpression or ectopic expression of MUC2 is the common property of mucinous carcinomas of the colon, pancreas, breast, and ovary. J Pathol 1997;182:385-91.

[166] Jass JR, Roberton AM. Colorectal mucin histochemistry in health and disease: a critical review. Pathol Int 1994;44:487-504.

[167] Sprong RC, Schonewille AJ, van der Meer R. Dietary cheese whey protein protects rats against mild dextran sulfate sodium-induced colitis: role of mucin and microbiota. J Dairy Sci 2010;93:1364-71.

[168] Alos L, Lujan B, Castillo M, *et al.* Expression of membrane-bound mucins (MUC1 and MUC4) and secreted mucins (MUC2, MUC5AC, MUC5B, MUC6 and MUC7) in mucoepidermoid carcinomas of salivary glands. Am J Surg Pathol 2005;29:806-13.

[169] Handra-Luca A, Lamas G, Bertrand JC, Fouret P. MUC1, MUC2, MUC4, and MUC5AC expression in salivary gland mucoepidermoid carcinoma: diagnostic and prognostic implications. Am J Surg Pathol 2005;29:881-9.

[170] Andrianifahanana M, Moniaux N, Schmied BM, *et al.* Mucin (MUC) gene expression in human pancreatic adenocarcinoma and chronic pancreatitis: a potential role of MUC4 as a tumor marker of diagnostic significance. Clin Cancer Res 2001;7:4033-40.

[171] Copin MC, Buisine MP, Devisme L, *et al.* Normal respiratory mucosa, precursor lesions and lung carcinomas: differential expression of human mucin genes. Front Biosci 2001;6:D1264-75.

[172] Copin MC, Devisme L, Buisine MP, *et al.* From normal respiratory mucosa to epidermoid carcinoma: expression of human mucin genes. Int J Cancer 2000;86:162-8.

[173] Aust MR, Madsen CS, Jennings A, Kasperbauer JL, Gendler SJ. Mucin mRNA expression in normal and vasomotor inferior turbinates. Am J Rhinol 1997;11:293-302.

[174] Kim SS, Kim KS, Lee JG, *et al.* Levels of intracellular protein and messenger RNA of mucin and lysozyme in normal human nasal and polyp epithelium. Laryngoscope 2000;110:276-80.

[175] Hovenberg HW, Davies JR, Herrmann A, Linden CJ, Carlstedt I. MUC5AC, but not MUC2, is a prominent mucin in respiratory secretions. Glycoconj J 1996;13:839-47.

[176] Durrant LG, Jacobs E, Price MR. Production of monoclonal antibodies recognising the peptide core of MUC2 intestinal mucin. Eur J Cancer 1994;30A:355-63.

[177] Lagow E, DeSouza MM, Carson DD. Mammalian reproductive tract mucins. Hum Reprod Update 1999;5:280-92.

[178] Zhang S, Zhang HS, Reuter VE, *et al.* Expression of potential target antigens for immunotherapy on primary and metastatic prostate cancers. Clin Cancer Res 1998;4:295-302.

[179] Osunkoya AO, Adsay NV, Cohen C, Epstein JI, Smith SL. MUC2 expression in primary mucinous and nonmucinous adenocarcinoma of the prostate: an analysis of 50 cases on radical prostatectomy. Mod Pathol 2008;21:789-94.

[180] Seo JT, Lee JS, Jun JH, Yang MH. Expression of mucin genes in the human testis and its relationship to spermatogenesis. Yonsei Med J 2005;46:667-72.

[181] Retz M, Lehmann J, Roder C, *et al.* Differential mucin MUC7 gene expression in invasive bladder carcinoma in contrast to uniform MUC1 and MUC2 gene expression in both normal urothelium and bladder carcinoma. Cancer Res 1998;58:5662-6.
[182] Walsh MD, Hohn BG, Thong W, *et al.* Mucin expression by transitional cell carcinomas of the bladder. Br J Urol 1994;73:256-62.
[183] Gipson IK, Ho SB, Spurr-Michaud SJ, *et al.* Mucin genes expressed by human female reproductive tract epithelia. Biol Reprod 1997;56:999-1011.
[184] Alameda F, Mejias-Luque R, Garrido M, de Bolos C. Mucin genes (MUC2, MUC4, MUC5AC, and MUC6) detection in normal and pathological endometrial tissues. Int J Gynecol Pathol 2007;26:61-5.
[185] Riethdorf L, O'Connell JT, Riethdorf S, Cviko A, Crum CP. Differential expression of MUC2 and MUC5AC in benign and malignant glandular lesions of the cervix uteri. Virchows Arch 2000;437:365-71.
[186] Baker AC, Eltoum I, Curry RO, *et al.* Mucinous expression in benign and neoplastic glandular lesions of the uterine cervix. Arch Pathol Lab Med 2006;130:1510-5.
[187] Hebbar V, Damera G, Sachdev GP. Differential expression of MUC genes in endometrial and cervical tissues and tumors. BMC Cancer 2005;5:124.
[188] Gipson IK, Inatomi T. Cellular origin of mucins of the ocular surface tear film. Adv Exp Med Biol 1998;438:221-7.
[189] Inatomi T, Spurr-Michaud S, Tisdale AS, *et al.* Expression of secretory mucin genes by human conjunctival epithelia. Invest Ophthalmol Vis Sci 1996;37:1684-92.
[190] McKenzie RW, Jumblatt JE, Jumblatt MM. Quantification of MUC2 and MUC5AC transcripts in human conjunctiva. Invest Ophthalmol Vis Sci 2000;41:703-8.
[191] Berry M, Ellingham RB, Corfield AP. Human preocular mucins reflect changes in surface physiology. Br J Ophthalmol 2004;88:377-83.
[192] Kerschner JE. Mucin gene expression in human middle ear epithelium. Laryngoscope 2007;117:1666-76.
[193] Ubell ML, Kerschner JE, Wackym PA, Burrows A. MUC2 expression in human middle ear epithelium of patients with otitis media. Arch Otolaryngol Head Neck Surg 2008;134:39-44.
[194] Lin J, Ho S, Shekels L, Paparella MM, Kim Y. Mucin gene expression in the rat middle ear: an improved method for RNA harvest. Ann Otol Rhinol Laryngol 1999;108:762-8.
[195] Bai CH, Song SY, Kim YD. The inhibitory effect of the leukotriene receptor antagonist on leukotriene D4-induced MUC2/5AC gene expression and mucin secretion in human airway epithelial cells. Auris Nasus Larynx 2007;34:203-6.
[196] Basbaum C, Gallup M, Gum J, Kim Y, Jany B. Modification of mucin gene expression in the airways of rats exposed to sulfur dioxide. Biorheology 1990;27:485-9.
[197] Li D, Wang D, Majumdar S, *et al.* Localization and up-regulation of mucin (MUC2) gene expression in human nasal biopsies of patients with cystic fibrosis. J Pathol 1997;181:305-10.
[198] Plantier L, Crestani B, Wert SE, *et al.* Ectopic respiratory epithelial cell differentiation in bronchiolised distal airspaces in idiopathic pulmonary fibrosis. Thorax 66:651-7.
[199] Rogers DF. Airway mucus hypersecretion in asthma: an undervalued pathology? Curr Opin Pharmacol 2004;4:241-50.
[200] Davies JR, Herrmann A, Russell W, *et al.* Respiratory tract mucins: structure and expression patterns. Novartis Found Symp 2002;248:76-88; discussion -93, 277-82.

[201] Kirkham S, Sheehan JK, Knight D, Richardson PS, Thornton DJ. Heterogeneity of airways mucus: variations in the amounts and glycoforms of the major oligomeric mucins MUC5AC and MUC5B. Biochem J 2002;361:537-46.
[202] Ordonez CL, Khashayar R, Wong HH, *et al.* Mild and moderate asthma is associated with airway goblet cell hyperplasia and abnormalities in mucin gene expression. Am J Respir Crit Care Med 2001;163:517-23.
[203] Caramori G, Di Gregorio C, Carlstedt I, *et al.* Mucin expression in peripheral airways of patients with chronic obstructive pulmonary disease. Histopathology 2004;45:477-84.
[204] Ho J, Kim Y. Do mucins promote tumor-cell metastasis. Int J Oncol 1995;7:913-26.
[205] Koga T, Hashimoto S, Sugio K, *et al.* Lung adenocarcinoma with bronchioloalveolar carcinoma component is frequently associated with foci of high-grade atypical adenomatous hyperplasia. Am J Clin Pathol 2002;117:464-70.
[206] Mori M, Rao SK, Popper HH, Cagle PT, Fraire AE. Atypical adenomatous hyperplasia of the lung: a probable forerunner in the development of adenocarcinoma of the lung. Mod Pathol 2001;14:72-84.
[207] Awaya H, Takeshima Y, Yamasaki M, Inai K. Expression of MUC1, MUC2, MUC5AC, and MUC6 in atypical adenomatous hyperplasia, bronchioloalveolar carcinoma, adenocarcinoma with mixed subtypes, and mucinous bronchioloalveolar carcinoma of the lung. Am J Clin Pathol 2004;121:644-53.
[208] Hayashi H, Ito T, Yazawa T, *et al.* Reduced expression of p27/Kip1 is associated with the development of pulmonary adenocarcinoma. J Pathol 2000;192:26-31.
[209] Seki N, Takasu T, Mandai K, *et al.* Expression of eukaryotic initiation factor 4E in atypical adenomatous hyperplasia and adenocarcinoma of the human peripheral lung. Clin Cancer Res 2002;8:3046-53.
[210] Travis WD, Colby B, Corrin B. Histological Typing of Lung and Pleural Tumours. 3rd edition ed. Berlin: Springer; 1999.
[211] Barsky SH, Cameron R, Osann KE, Tomita D, Holmes EC. Rising incidence of bronchioloalveolar lung carcinoma and its unique clinicopathologic features. Cancer 1994;73:1163-70.
[212] Nishiumi N, Abe Y, Inoue Y, *et al.* Use of 11p15 mucins as prognostic factors in small adenocarcinoma of the lung. Clin Cancer Res 2003;9:5616-9.
[213] Osako M, Yonezawa S, Siddiki B, *et al.* Immunohistochemical study of mucin carbohydrates and core proteins in human pancreatic tumors. Cancer 1993;71:2191-9.
[214] Kitamura H, Yonezawa S, Tanaka S, Kim YS, Sato E. Expression of mucin carbohydrates and core proteins in carcinomas of the ampulla of Vater: their relationship to prognosis. Jpn J Cancer Res 1996;87:631-40.
[215] Higashi M, Yonezawa S, Ho JJ, *et al.* Expression of MUC1 and MUC2 mucin antigens in intrahepatic bile duct tumors: its relationship with a new morphological classification of cholangiocarcinoma. Hepatology 1999;30:1347-55.
[216] Stacher E, Ullmann R, Halbwedl I, *et al.* Atypical goblet cell hyperplasia in congenital cystic adenomatoid malformation as a possible preneoplasia for pulmonary adenocarcinoma in childhood: A genetic analysis. Hum Pathol 2004;35:565-70.
[217] Inamura K, Satoh Y, Okumura S, *et al.* Pulmonary adenocarcinomas with enteric differentiation: histologic and immunohistochemical characteristics compared with metastatic colorectal cancers and usual pulmonary adenocarcinomas. Am J Surg Pathol 2005;29:660-5.

[218] Tsao MS, Fraser RS. Primary pulmonary adenocarcinoma with enteric differentiation. Cancer 1991;68:1754-7.

[219] Weidner N. Pulmonary adenocarcinoma with intestinal-type differentiation. Ultrastruct Pathol 1992;16:7-10.

[220] Hatanaka K, Tsuta K, Watanabe K, Sugino K, Uekusa T. Primary pulmonary adenocarcinoma with enteric differentiation resembling metastatic colorectal carcinoma: a report of the second case negative for cytokeratin 7. Pathol Res Pract 2011;207:188-91.

[221] Yousem SA. Pulmonary intestinal-type adenocarcinoma does not show enteric differentiation by immunohistochemical study. Mod Pathol 2005;18:816-21.

[222] Rossi G, Murer B, Cavazza A, *et al.* Primary mucinous (so-called colloid) carcinomas of the lung: a clinicopathologic and immunohistochemical study with special reference to CDX-2 homeobox gene and MUC2 expression. Am J Surg Pathol 2004;28:442-52.

[223] Mesquita P, Almeida R, Van Seuningen I, David L. Coordinated expression of MUC2 and CDX-2 in mucinous carcinomas of the lung can be explained by the role of CDX-2 as transcriptional regulator of MUC2. Am J Surg Pathol 2004;28:1254-5.

[224] Seregni E, Botti C, Lombardo C, *et al.* Pattern of mucin gene expression in normal and neoplastic lung tissues. Anticancer Res 1996;16:2209-13.

[225] Arul GS, Moorghen M, Myerscough N, *et al.* Mucin gene expression in Barrett's oesophagus: an *in situ* hybridisation and immunohistochemical study. Gut 2000;47:753-61.

[226] Endo T, Tamaki K, Arimura Y, *et al.* Expression of sulfated carbohydrate chain and core peptides of mucin detected by monoclonal antibodies in Barrett's esophagus and esophageal adenocarcinoma. J Gastroenterol 1998;33:811-5.

[227] Glickman JN, Blount PL, Sanchez CA, *et al.* Mucin core polypeptide expression in the progression of neoplasia in Barrett's esophagus. Hum Pathol 2006;37:1304-15.

[228] Baldus SE, Zirbes TK, Engel S, *et al.* Correlation of the immunohistochemical reactivity of mucin peptide cores MUC1 and MUC2 with the histopathological subtype and prognosis of gastric carcinomas. Int J Cancer 1998;79:133-8.

[229] Utsunomiya T, Yonezawa S, Sakamoto H, *et al.* Expression of MUC1 and MUC2 mucins in gastric carcinomas: its relationship with the prognosis of the patients. Clin Cancer Res 1998;4:2605-14.

[230] Paulsen FP, Varoga D, Paulsen AR, Corfield A, Tsokos M. Prognostic value of mucins in the classification of ampullary carcinomas. Hum Pathol 2006;37:160-7.

[231] Nagata K, Horinouchi M, Saitou M, *et al.* Mucin expression profile in pancreatic cancer and the precursor lesions. J Hepatobiliary Pancreat Surg 2007;14:243-54.

[232] Chang SK, Dohrman AF, Basbaum CB, *et al.* Localization of mucin (MUC2 and MUC3) messenger RNA and peptide expression in human normal intestine and colon cancer. Gastroenterology 1994;107:28-36.

[233] Weiss AA, Babyatsky MW, Ogata S, Chen A, Itzkowitz SH. Expression of MUC2 and MUC3 mRNA in human normal, malignant, and inflammatory intestinal tissues. J Histochem Cytochem 1996;44:1161-6.

[234] Blank M, Klussmann E, Kruger-Krasagakes S, *et al.* Expression of MUC2-mucin in colorectal adenomas and carcinomas of different histological types. Int J Cancer 1994;59:301-6.

[235] Bax DA, Haringsma J, Einerhand AW, *et al.* MUC4 is increased in high grade intraepithelial neoplasia in Barrett's oesophagus and is associated with a proapoptotic Bax to Bcl-2 ratio. J Clin Pathol 2004;57:1267-72.

[236] Guillem P, Billeret V, Buisine MP, *et al.* Mucin gene expression and cell differentiation in human normal, premalignant and malignant esophagus. Int J Cancer 2000;88:856-61.

[237] Spechler SJ. Clinical practice. Barrett's Esophagus. N Engl J Med 2002;346:836-42.

[238] Hu Y, Jones C, Gellersen O, *et al.* Pathogenesis of Barrett esophagus: deoxycholic acid up-regulates goblet-specific gene MUC2 in concert with CDX2 in human esophageal cells. Arch Surg 2007;142:540-4; discussion 4-5.

[239] Endo H, Iijima K, Asanuma K, *et al.* Exogenous luminal nitric oxide exposure accelerates columnar transformation of rat esophagus. Int J Cancer 127:2009-19.

[240] Burjonrappa SC, Reddimasu S, Nawaz Z, *et al.* Mucin expression profile in Barrett's, dysplasia, adenocarcinoma sequence in the esophagus. Indian J Cancer 2007;44:1-5.

[241] Warson C, Van De Bovenkamp JH, Korteland-Van Male AM, *et al.* Barrett's esophagus is characterized by expression of gastric-type mucins (MUC5AC, MUC6) and TFF peptides (TFF1 and TFF2), but the risk of carcinoma development may be indicated by the intestinal-type mucin, MUC2. Hum Pathol 2002;33:660-8.

[242] Szachnowicz S, Cecconello I, Ribeiro U, *et al.* Mucin pattern reflects the origin of the adenocarcinoma in Barrett's esophagus: a retrospective clinical and laboratorial study. World J Surg Oncol 2009;7:27.

[243] Brown IS, Whiteman DC, Lauwers GY. Foveolar type dysplasia in Barrett esophagus. Mod Pathol 2010;23:834-43.

[244] Barrett NR. Chronic peptic ulcer of the oesophagus and'oesophagitis'. Br J Surg 1950;38:175-82.

[245] Allison PR, Johnstone AS. The oesophagus lined with gastric mucous membrane. Thorax 1953;8:87-101.

[246] Hayward J. The lower end of the oesophagus. Thorax 1961;16:36-41.

[247] Bremner CG, Lynch VP, Ellis FH, Jr. Barrett's esophagus: congenital or acquired? An experimental study of esophageal mucosal regeneration in the dog. Surgery 1970;68:209-16.

[248] Peters JH, Avisar N. The molecular pathogenesis of Barrett's esophagus: common signaling pathways in embryogenesis metaplasia and neoplasia. J Gastrointest Surg 14 Suppl 1:S81-7.

[249] Seery JP. Stem cells of the oesophageal epithelium. J Cell Sci 2002;115:1783-9.

[250] Barbera M, Fitzgerald RC. Cellular origin of Barrett's metaplasia and oesophageal stem cells. Biochem Soc Trans 2010;38:370-3.

[251] Anisfeld AM, Kast-Woelbern HR, Lee H, *et al.* Activation of the nuclear receptor FXR induces fibrinogen expression: a new role for bile acid signaling. J Lipid Res 2005;46:458-68.

[252] Souza RF, Krishnan K, Spechler SJ. Acid, bile, and CDX: the ABCs of making Barrett's metaplasia. Am J Physiol Gastrointest Liver Physiol 2008;295:G211-8.

[253] Kazumori H, Ishihara S, Kinoshita Y. Roles of caudal-related homeobox gene Cdx1 in oesophageal epithelial cells in Barrett's epithelium development. Gut 2009;58:620-8.

[254] Lorentz O, Duluc I, Arcangelis AD, *et al.* Key role of the Cdx2 homeobox gene in extracellular matrix-mediated intestinal cell differentiation. J Cell Biol 1997;139:1553-65.

[255] Steininger H, Pfofe DA, Muller H, Haag-Sunjic G, Fratianu V. Expression of CDX2 and MUC2 in Barrett's mucosa. Pathol Res Pract 2005;201:573-7.

[256] Groisman GM, Amar M, Meir A. Expression of the intestinal marker Cdx2 in the columnar-lined esophagus with and without intestinal (Barrett's) metaplasia. Mod Pathol 2004;17:1282-8.

[257] Phillips RW, Frierson HF, Jr., Moskaluk CA. Cdx2 as a marker of epithelial intestinal differentiation in the esophagus. Am J Surg Pathol 2003;27:1442-7.

[258] Moons LM, Bax DA, Kuipers EJ, *et al.* The homeodomain protein CDX2 is an early marker of Barrett's oesophagus. J Clin Pathol 2004;57:1063-8.

[259] Avissar NE, Toia L, Hu Y, *et al.* Bile acid alone, or in combination with acid, induces CDX2 expression through activation of the epidermal growth factor receptor (EGFR). J Gastrointest Surg 2009;13:212-22.

[260] Vaninetti N, Williams L, Geldenhuys L, *et al.* Regulation of CDX2 expression in esophageal adenocarcinoma. Mol Carcinog 2009;48:965-74.

[261] Akyurek N, Akyol G, Dursun A, Yamac D, Gunel N. Expression of MUC1 and MUC2 mucins in gastric carcinomas: their relationship with clinicopathologic parameters and prognosis. Pathol Res Pract 2002;198:665-74.

[262] Lee HS, Lee HK, Kim HS, *et al.* MUC1, MUC2, MUC5AC, and MUC6 expressions in gastric carcinomas: their roles as prognostic indicators. Cancer 2001;92:1427-34.

[263] Mitsuuchi M, Hinoda Y, Itoh F, *et al.* Expression of MUC2 gene in gastric regenerative, metaplastic, and neoplastic epithelia. J Clin Lab Anal 1999;13:259-65.

[264] Lauren P. The Two Histological Main Types of Gastric Carcinoma: Diffuse and So-Called Intestinal-Type Carcinoma. an Attempt at a Histo-Clinical Classification. Acta Pathol Microbiol Scand 1965;64:31-49.

[265] Nakamura K, Sugano H, Takagi K. Carcinoma of the stomach in incipient phase: its histogenesis and histological appearances. Gann 1968;59:251-8.

[266] Mulligan RM. Histogenesis and biologic behavior of gastric carcinoma. Pathol Annu 1972;7:349-415.

[267] Hamilton SR, Aaltonen LA. Pathology and genetics of tumours of the digestive system. World health organization classification of tumours. Lyon: IARC Press; 2000. p. 37-67.

[268] Goseki N, Takizawa T, Koike M. Differences in the mode of the extension of gastric cancer classified by histological type: new histological classification of gastric carcinoma. Gut 1992;33:606-12.

[269] Inada K, Nakanishi H, Fujimitsu Y, *et al.* Gastric and intestinal mixed and solely intestinal types of intestinal metaplasia in the human stomach. Pathol Int 1997;47:831-41.

[270] Tajima Y, Nakanishi Y, Yoshino T, *et al.* Clinicopathological study of early adenocarcinoma of the gastric cardia: comparison with early adenocarcinoma of the distal stomach and esophagus. Oncology 2001;61:1-9.

[271] Tajima Y, Shimoda T, Nakanishi Y, *et al.* Gastric and intestinal phenotypic marker expression in gastric carcinomas and its prognostic significance: immunohistochemical analysis of 136 lesions. Oncology 2001;61:212-20.

[272] Kabashima A, Yao T, Sugimachi K, Tsuneyoshi M. Relationship between biologic behavior and phenotypic expression in intramucosal gastric carcinomas. Hum Pathol 2002;33:80-6.

[273] Lee OJ, Kim HJ, Kim JR, Watanabe H. The prognostic significance of the mucin phenotype of gastric adenocarcinoma and its relationship with histologic classifications. Oncol Rep 2009;21:387-93.

[274] Yamazaki K, Tajima Y, Makino R, *et al.* Tumor differentiation phenotype in gastric differentiated-type tumors and its relation to tumor invasion and genetic alterations. World J Gastroenterol 2006;12:3803-9.

[275] Yang HY, Lieska N, Shao D, Kriho V, Pappas GD. Proteins of the intermediate filament cytoskeleton as markers for astrocytes and human astrocytomas. Mol Chem Neuropathol 1994;21:155-76.

[276] Caligaris-Cappio F, Ghia P. The normal counterpart to the chronic lymphocytic leukemia B cell. Best Pract Res Clin Haematol 2007;20:385-97.

[277] Leteurtre E, Zerimech F, Piessen G, *et al.* Relationships between mucinous gastric carcinoma, MUC2 expression and survival. World J Gastroenterol 2006;12:3324-31.

[278] Reis CA, David L, Correa P, *et al.* Intestinal metaplasia of human stomach displays distinct patterns of mucin (MUC1, MUC2, MUC5AC, and MUC6) expression. Cancer Res 1999;59:1003-7.

[279] Conze T, Carvalho AS, Landegren U, *et al.* MUC2 mucin is a major carrier of the cancer-associated sialyl-Tn antigen in intestinal metaplasia and gastric carcinomas. Glycobiology 2010;20:199-206.

[280] Hammerschmidt M, Brook A, McMahon AP. The world according to hedgehog. Trends Genet 1997;13:14-21.

[281] Li XL, Eishi Y, Bai YQ, *et al.* Expression of the SRY-related HMG box protein SOX2 in human gastric carcinoma. Int J Oncol 2004;24:257-63.

[282] Shiotani A, Kamada T, Yamanaka Y, *et al.* Sonic hedgehog and CDX2 expression in the stomach. J Gastroenterol Hepatol 2008;23 Suppl 2:S161-6.

[283] Asonuma S, Imatani A, Asano N, *et al.* Helicobacter pylori induces gastric mucosal intestinal metaplasia through the inhibition of interleukin-4-mediated HMG box protein Sox2 expression. Am J Physiol Gastrointest Liver Physiol 2009;297:G312-22.

[284] Mizoshita T, Inada K, Tsukamoto T, *et al.* Expression of Cdx1 and Cdx2 mRNAs and relevance of this expression to differentiation in human gastrointestinal mucosa--with special emphasis on participation in intestinal metaplasia of the human stomach. Gastric Cancer 2001;4:185-91.

[285] Liu Q, Teh M, Ito K, *et al.* CDX2 expression is progressively decreased in human gastric intestinal metaplasia, dysplasia and cancer. Mod Pathol 2007;20:1286-97.

[286] Silberg DG, Sullivan J, Kang E, *et al.* Cdx2 ectopic expression induces gastric intestinal metaplasia in transgenic mice. Gastroenterology 2002;122:689-96.

[287] Bai YQ, Yamamoto H, Akiyama Y, *et al.* Ectopic expression of homeodomain protein CDX2 in intestinal metaplasia and carcinomas of the stomach. Cancer Lett 2002;176:47-55.

[288] Roessler K, Monig SP, Schneider PM, *et al.* Co-expression of CDX2 and MUC2 in gastric carcinomas: correlations with clinico-pathological parameters and prognosis. World J Gastroenterol 2005;11:3182-8.

[289] Sobala GM, O'Connor HJ, Dewar EP, *et al.* Bile reflux and intestinal metaplasia in gastric mucosa. J Clin Pathol 1993;46:235-40.

[290] Bernstein H, Bernstein C, Payne CM, Dvorakova K, Garewal H. Bile acids as carcinogens in human gastrointestinal cancers. Mutat Res 2005;589:47-65.

[291] Xu Y, Watanabe T, Tanigawa T, *et al.* Bile acids induce cdx2 expression through the farnesoid x receptor in gastric epithelial cells. J Clin Biochem Nutr 2010;46:81-6.

[292] Konturek PC, Konturek SJ, Brzozowski T. Helicobacter pylori infection in gastric cancerogenesis. J Physiol Pharmacol 2009;60:3-21.

[293] Correa P, Houghton J. Carcinogenesis of Helicobacter pylori. Gastroenterology 2007;133:659-72.

[294] Correa P, Piazuelo MB, Camargo MC. Etiopathogenesis of gastric cancer. Scand J Surg 2006;95:218-24.

[295] Correa P, Piazuelo MB. Natural history of Helicobacter pylori infection. Dig Liver Dis 2008;40:490-6.

[296] Matsuda K, Yamauchi K, Matsumoto T, *et al.* Quantitative analysis of the effect of Helicobacter pylori on the expressions of SOX2, CDX2, MUC2, MUC5AC, MUC6, TFF1, TFF2, and TFF3 mRNAs in human gastric carcinoma cells. Scand J Gastroenterol 2008;43:25-33.

[297] Satoh K, Mutoh H, Eda A, *et al.* Aberrant expression of CDX2 in the gastric mucosa with and without intestinal metaplasia: effect of eradication of Helicobacter pylori. Helicobacter 2002;7:192-8.

[298] Kaneko T, Ota H, Hayama M, Akamatsu T, Katsuyama T. Helicobacter pylori infection produces expression of a secretory component in gastric mucous cells. Virchows Arch 2000;437:514-20.

[299] Semino-Mora C, Doi SQ, Marty A, *et al.* Intracellular and interstitial expression of Helicobacter pylori virulence genes in gastric precancerous intestinal metaplasia and adenocarcinoma. J Infect Dis 2003;187:1165-77.

[300] Wang RQ, Fang DC. Effects of Helicobacter pylori infection on mucin expression in gastric carcinoma and pericancerous tissues. J Gastroenterol Hepatol 2006;21:425-31.

[301] Mejias-Luque R, Linden SK, Garrido M, *et al.* Inflammation modulates the expression of the intestinal mucins MUC2 and MUC4 in gastric tumors. Oncogene 2010;29:1753-62.

[302] Lee HW, Ahn DH, Crawley SC, *et al.* Phorbol 12-myristate 13-acetate up-regulates the transcription of MUC2 intestinal mucin *via* Ras, ERK, and NF-kappa B. J Biol Chem 2002;277:32624-31.

[303] Bleuming SA, Kodach LL, Garcia Leon MJ, *et al.* Altered bone morphogenetic protein signaling in the Helicobacter pylori-infected stomach. J Pathol 2006;209:190-7.

[304] Barros R, Pereira B, Duluc I, *et al.* Key elements of the BMP/SMAD pathway co-localize with CDX2 in intestinal metaplasia and regulate CDX2 expression in human gastric cell lines. J Pathol 2008;215:411-20.

[305] Varas A, Hager-Theodorides AL, Sacedon R, *et al.* The role of morphogens in T-cell development. Trends Immunol 2003;24:197-206.

[306] Lowrey JA, Stewart GA, Lindey S, *et al.* Sonic hedgehog promotes cell cycle progression in activated peripheral CD4(+) T lymphocytes. J Immunol 2002;169:1869-75.

[307] Shibata D, Weiss LM. Epstein-Barr virus-associated gastric adenocarcinoma. Am J Pathol 1992;140:769-74.

[308] Osato T, Imai S. Epstein-Barr virus and gastric carcinoma. Semin Cancer Biol 1996;7:175-82.

[309] Lee HS, Chang MS, Yang HK, Lee BL, Kim WH. Epstein-barr virus-positive gastric carcinoma has a distinct protein expression profile in comparison with epstein-barr virus-negative carcinoma. Clin Cancer Res 2004;10:1698-705.

[310] Hirano N, Tsukamoto T, Mizoshita T, *et al.* Down regulation of gastric and intestinal phenotypic expression in Epstein-Barr virus-associated stomach cancers. Histol Histopathol 2007;22:641-9.

[311] Uozaki H, Chong JM, Fujimoto E, *et al.* Soft and hard keratin expression in Epstein-Barr-virus-associated gastric carcinoma. Anticancer Res 2005;25:3183-90.

[312] Barua RR, Uozaki H, Chong JM, *et al.* Phenotype analysis by MUC2, MUC5AC, MUC6, and CD10 expression in Epstein-Barr virus-associated gastric carcinoma. J Gastroenterol 2006;41:733-9.

[313] Fukayama M, Hino R, Uozaki H. Epstein-Barr virus and gastric carcinoma: virus-host interactions leading to carcinoma. Cancer Sci 2008;99:1726-33.
[314] Au WY, Pang A, Chan EC, *et al.* Epstein-barr virus-related gastric adenocarcinoma: an early secondary cancer post hemopoietic stem cell transplantation. Gastroenterology 2005;129:2058-63.
[315] Duelli D, Lazebnik Y. Cell-to-cell fusion as a link between viruses and cancer. Nat Rev Cancer 2007;7:968-76.
[316] Korbling M, Katz RL, Khanna A, *et al.* Hepatocytes and epithelial cells of donor origin in recipients of peripheral-blood stem cells. N Engl J Med 2002;346:738-46.
[317] Tatematsu M, Tsukamoto T, Inada K. Stem cells and gastric cancer: role of gastric and intestinal mixed intestinal metaplasia. Cancer Sci 2003;94:135-41.
[318] Takaishi S, Okumura T, Wang TC. Gastric cancer stem cells. J Clin Oncol 2008;26:2876-82.
[319] Clarke MF, Dick JE, Dirks PB, *et al.* Cancer stem cells--perspectives on current status and future directions: AACR Workshop on cancer stem cells. Cancer Res 2006;66:9339-44.
[320] Prince ME, Sivanandan R, Kaczorowski A, *et al.* Identification of a subpopulation of cells with cancer stem cell properties in head and neck squamous cell carcinoma. Proc Natl Acad Sci U S A 2007;104:973-8.
[321] Hermann PC, Huber SL, Herrler T, *et al.* Distinct populations of cancer stem cells determine tumor growth and metastatic activity in human pancreatic cancer. Cell Stem Cell 2007;1:313-23.
[322] Dalerba P, Dylla SJ, Park IK, *et al.* Phenotypic characterization of human colorectal cancer stem cells. Proc Natl Acad Sci U S A 2007;104:10158-63.
[323] Saikawa Y, Fukuda K, Takahashi T, *et al.* Gastric carcinogenesis and the cancer stem cell hypothesis. Gastric Cancer 2010;13:11-24.
[324] Houghton J, Stoicov C, Nomura S, *et al.* Gastric cancer originating from bone marrow-derived cells. Science 2004;306:1568-71.
[325] Krause DS, Theise ND, Collector MI, *et al.* Multi-organ, multi-lineage engraftment by a single bone marrow-derived stem cell. Cell 2001;105:369-77.
[326] Maitra A, Hruban RH. Pancreatic cancer. Annu Rev Pathol 2008;3:157-88.
[327] Ranganathan P, Harsha HC, Pandey A. Molecular alterations in exocrine neoplasms of the pancreas. Arch Pathol Lab Med 2009;133:405-12.
[328] Reya T, Morrison SJ, Clarke MF, Weissman IL. Stem cells, cancer, and cancer stem cells. Nature 2001;414:105-11.
[329] Takaori K. Current understanding of precursors to pancreatic cancer. J Hepatobiliary Pancreat Surg 2007;14:217-23.
[330] Gudjonsson T, Magnusson MK. Stem cell biology and the cellular pathways of carcinogenesis. Apmis 2005;113:922-9.
[331] Thayer SP, di Magliano MP, Heiser PW, *et al.* Hedgehog is an early and late mediator of pancreatic cancer tumorigenesis. Nature 2003;425:851-6.
[332] Miyamoto Y, Maitra A, Ghosh B, *et al.* Notch mediates TGF alpha-induced changes in epithelial differentiation during pancreatic tumorigenesis. Cancer Cell 2003;3:565-76.
[333] Leach SD. Epithelial differentiation in pancreatic development and neoplasia: new niches for nestin and Notch. J Clin Gastroenterol 2005;39:S78-82.
[334] Jensen JN, Cameron E, Garay MV, *et al.* Recapitulation of elements of embryonic development in adult mouse pancreatic regeneration. Gastroenterology 2005;128:728-41.

[335] Hruban RH, Adsay NV, Albores-Saavedra J, *et al.* Pathology of genetically engineered mouse models of pancreatic exocrine cancer: consensus report and recommendations. Cancer Res 2006;66:95-106.

[336] Li C, Heidt DG, Dalerba P, *et al.* Identification of pancreatic cancer stem cells. Cancer Res 2007;67:1030-7.

[337] Lee CJ, Dosch J, Simeone DM. Pancreatic cancer stem cells. J Clin Oncol 2008;26:2806-12.

[338] Olempska M, Eisenach PA, Ammerpohl O, *et al.* Detection of tumor stem cell markers in pancreatic carcinoma cell lines. Hepatobiliary Pancreat Dis Int 2007;6:92-7.

[339] Sergeant G, Vankelecom H, Gremeaux L, Topal B. Role of cancer stem cells in pancreatic ductal adenocarcinoma. Nat Rev Clin Oncol 2009;6:580-6.

[340] Immervoll H, Hoem D, Sakariassen PO, Steffensen OJ, Molven A. Expression of the "stem cell marker" CD133 in pancreas and pancreatic ductal adenocarcinomas. BMC Cancer 2008;8:48.

[341] Hruban RH, Adsay NV, Albores-Saavedra J, *et al.* Pancreatic intraepithelial neoplasia: a new nomenclature and classification system for pancreatic duct lesions. Am J Surg Pathol 2001;25:579-86.

[342] Hruban RH, Takaori K, Klimstra DS, *et al.* An illustrated consensus on the classification of pancreatic intraepithelial neoplasia and intraductal papillary mucinous neoplasms. Am J Surg Pathol 2004;28:977-87.

[343] Maitra A, Fukushima N, Takaori K, Hruban RH. Precursors to invasive pancreatic cancer. Adv Anat Pathol 2005;12:81-91.

[344] Brat DJ, Lillemoe KD, Yeo CJ, Warfield PB, Hruban RH. Progression of pancreatic intraductal neoplasias to infiltrating adenocarcinoma of the pancreas. Am J Surg Pathol 1998;22:163-9.

[345] Albores-Saavedra J, Weimersheimer-Sandoval M, Chable-Montero F, *et al.* The foamy variant of pancreatic intraepithelial neoplasia. Ann Diagn Pathol 2008;12:252-9.

[346] van Heek NT, Meeker AK, Kern SE, *et al.* Telomere shortening is nearly universal in pancreatic intraepithelial neoplasia. Am J Pathol 2002;161:1541-7.

[347] Sohn TA, Yeo CJ, Cameron JL, *et al.* Intraductal papillary mucinous neoplasms of the pancreas: an updated experience. Ann Surg 2004;239:788-97; discussion 97-9.

[348] Iacobuzio-Donahue CA, Klimstra DS, Adsay NV, *et al.* Dpc-4 protein is expressed in virtually all human intraductal papillary mucinous neoplasms of the pancreas: comparison with conventional ductal adenocarcinomas. Am J Pathol 2000;157:755-61.

[349] Sahin F, Maitra A, Argani P, *et al.* Loss of Stk11/Lkb1 expression in pancreatic and biliary neoplasms. Mod Pathol 2003;16:686-91.

[350] Furukawa T, Kloppel G, Volkan Adsay N, *et al.* Classification of types of intraductal papillary-mucinous neoplasm of the pancreas: a consensus study. Virchows Arch 2005;447:794-9.

[351] Belyaev O, Seelig MH, Muller CA, *et al.* Intraductal papillary mucinous neoplasms of the pancreas. J Clin Gastroenterol 2008;42:284-94.

[352] Ringel J, Lohr M. The MUC gene family: their role in diagnosis and early detection of pancreatic cancer. Mol Cancer 2003;2:9.

[353] Klimstra DS, Pitman MB, Hruban RH. An algorithmic approach to the diagnosis of pancreatic neoplasms. Arch Pathol Lab Med 2009;133:454-64.

[354] Sato N, Goggins M. The role of epigenetic alterations in pancreatic cancer. J Hepatobiliary Pancreat Surg 2006;13:286-95.

[355] Pasca di Magliano M, Sekine S, Ermilov A, *et al.* Hedgehog/Ras interactions regulate early stages of pancreatic cancer. Genes Dev 2006;20:3161-73.

[356] Wang Z, Zhang Y, Banerjee S, Li Y, Sarkar FH. Notch-1 down-regulation by curcumin is associated with the inhibition of cell growth and the induction of apoptosis in pancreatic cancer cells. Cancer 2006;106:2503-13.

[357] Adsay NV, Merati K, Andea A, *et al.* The dichotomy in the preinvasive neoplasia to invasive carcinoma sequence in the pancreas: differential expression of MUC1 and MUC2 supports the existence of two separate pathways of carcinogenesis. Mod Pathol 2002;15:1087-95.

[358] Luttges J, Feyerabend B, Buchelt T, Pacena M, Kloppel G. The mucin profile of noninvasive and invasive mucinous cystic neoplasms of the pancreas. Am J Surg Pathol 2002;26:466-71.

[359] Yonezawa S, Sueyoshi K, Nomoto M, *et al.* MUC2 gene expression is found in noninvasive tumors but not in invasive tumors of the pancreas and liver: its close relationship with prognosis of the patients. Hum Pathol 1997;28:344-52.

[360] Nakamura A, Horinouchi M, Goto M, *et al.* New classification of pancreatic intraductal papillary-mucinous tumour by mucin expression: its relationship with potential for malignancy. J Pathol 2002;197:201-10.

[361] Terris B, Dubois S, Buisine MP, *et al.* Mucin gene expression in intraductal papillary-mucinous pancreatic tumours and related lesions. J Pathol 2002;197:632-7.

[362] Adsay NV, Merati K, Basturk O, *et al.* Pathologically and biologically distinct types of epithelium in intraductal papillary mucinous neoplasms: delineation of an "intestinal" pathway of carcinogenesis in the pancreas. Am J Surg Pathol 2004;28:839-48.

[363] Bonhomme C, Duluc I, Martin E, *et al.* The Cdx2 homeobox gene has a tumour suppressor function in the distal colon in addition to a homeotic role during gut development. Gut 2003;52:1465-71.

[364] Eda A, Osawa H, Satoh K, *et al.* Aberrant expression of CDX2 in Barrett's epithelium and inflammatory esophageal mucosa. J Gastroenterol 2003;38:14-22.

[365] Seno H, Oshima M, Taniguchi MA, *et al.* CDX2 expression in the stomach with intestinal metaplasia and intestinal-type cancer: Prognostic implications. Int J Oncol 2002;21:769-74.

[366] Adsay NV, Merati K, Nassar H, *et al.* Pathogenesis of colloid (pure mucinous) carcinoma of exocrine organs: Coupling of gel-forming mucin (MUC2) production with altered cell polarity and abnormal cell-stroma interaction may be the key factor in the morphogenesis and indolent behavior of colloid carcinoma in the breast and pancreas. Am J Surg Pathol 2003;27:571-8.

[367] Liszka L, Zielinska-Pajak E, Pajak J, Golka D. Colloid carcinoma of the pancreas: review of selected pathological and clinical aspects. Pathology 2008;40:655-63.

[368] Luttges J, Zamboni G, Longnecker D, Kloppel G. The immunohistochemical mucin expression pattern distinguishes different types of intraductal papillary mucinous neoplasms of the pancreas and determines their relationship to mucinous noncystic carcinoma and ductal adenocarcinoma. Am J Surg Pathol 2001;25:942-8.

[369] Sanada Y, Yoshida K. A case of benign intraductal papillary mucinous neoplasm of the pancreas containing two major subtypes. J Gastrointestin Liver Dis 2008;17:457-60.

[370] Yonezawa S, Sato E. Expression of mucin antigens in human cancers and its relationship with malignancy potential. Pathol Int 1997;47:813-30.
[371] Yonezawa S, Higashi M, Yamada N, Yokoyama S, Goto M. Significance of mucin expression in pancreatobiliary neoplasms. J Hepatobiliary Pancreat Sci 2010;17:108-24.
[372] Pantano F, Baldi A, Santini D, *et al.* MUC2 but not MUC5 expression correlates with prognosis in radically resected pancreatic cancer patients. Cancer Biol Ther 2009;8:996-9.
[373] Oren M. Decision making by p53: life, death and cancer. Cell Death Differ 2003;10:431-42.
[374] Levi E, Klimstra DS, Andea A, Basturk O, Adsay NV. MUC1 and MUC2 in pancreatic neoplasia. J Clin Pathol 2004;57:456-62.
[375] Ji Y, Lou WH, Jin DY, *et al.* A series of 64 cases of pancreatic cystic neoplasia from an institutional study of China. World J Gastroenterol 2006;12:7380-7.
[376] Yoon WJ, Lee JK, Lee KH, *et al.* Cystic neoplasms of the exocrine pancreas: an update of a nationwide survey in Korea. Pancreas 2008;37:254-8.
[377] Thompson LD, Becker RC, Przygodzki RM, Adair CF, Heffess CS. Mucinous cystic neoplasm (mucinous cystadenocarcinoma of low-grade malignant potential) of the pancreas: a clinicopathologic study of 130 cases. Am J Surg Pathol 1999;23:1-16.
[378] Zamboni G, Scarpa A, Bogina G, *et al.* Mucinous cystic tumors of the pancreas: clinicopathological features, prognosis, and relationship to other mucinous cystic tumors. Am J Surg Pathol 1999;23:410-22.
[379] Adsay NV. Cystic neoplasia of the pancreas: pathology and biology. J Gastrointest Surg 2008;12:401-4.
[380] Fischer HP, Zhou H. Pathogenesis of carcinoma of the papilla of Vater. J Hepatobiliary Pancreat Surg 2004;11:301-9.
[381] Hutchins GF, Draganov PV. Cystic neoplasms of the pancreas: a diagnostic challenge. World J Gastroenterol 2009;15:48-54.
[382] Chu PG, Schwarz RE, Lau SK, Yen Y, Weiss LM. Immunohistochemical staining in the diagnosis of pancreatobiliary and ampulla of Vater adenocarcinoma: application of CDX2, CK17, MUC1, and MUC2. Am J Surg Pathol 2005;29:359-67.
[383] Heinrich S, Clavien PA. Ampullary cancer. Curr Opin Gastroenterol 2010;26:280-5.
[384] Kimura W, Futakawa N, Yamagata S, *et al.* Different clinicopathologic findings in two histologic types of carcinoma of papilla of Vater. Jpn J Cancer Res 1994;85:161-6.
[385] Sessa F, Furlan D, Zampatti C, *et al.* Prognostic factors for ampullary adenocarcinomas: tumor stage, tumor histology, tumor location, immunohistochemistry and microsatellite instability. Virchows Arch 2007;451:649-57.
[386] Zhou H, Schaefer N, Wolff M, Fischer HP. Carcinoma of the ampulla of Vater: comparative histologic/immunohistochemical classification and follow-up. Am J Surg Pathol 2004;28:875-82.
[387] Nakanuma Y, Sasaki M, Ikeda H, *et al.* Pathology of peripheral intrahepatic cholangiocarcinoma with reference to tumorigenesis. Hepatol Res 2008;38:325-34.
[388] Wittekind C, Fischer HP, Ponchon T. Combined hepatocellular and cholangiocarcinoma. In: Hamilton SR, Aaltonen LA, editors. Pathology and Genetics Tumours of the Digestive System World Health Organization of Tumours. Lion: IARC Press; 2000. p. 181.
[389] Libbrecht L. Hepatic progenitor cells in human liver tumor development. World J Gastroenterol 2006;12:6261-5.
[390] Fausto N. Mouse liver tumorigenesis: models, mechanisms, and relevance to human disease. Semin Liver Dis 1999;19:243-52.

[391] Dumble ML, Croager EJ, Yeoh GC, Quail EA. Generation and characterization of p53 null transformed hepatic progenitor cells: oval cells give rise to hepatocellular carcinoma. Carcinogenesis 2002;23:435-45.

[392] Theise ND, Saxena R, Portmann BC, *et al.* The canals of Hering and hepatic stem cells in humans. Hepatology 1999;30:1425-33.

[393] Theise ND, Yao JL, Harada K, *et al.* Hepatic'stem cell' malignancies in adults: four cases. Histopathology 2003;43:263-71.

[394] Roskams T. Liver stem cells and their implication in hepatocellular and cholangiocarcinoma. Oncogene 2006;25:3818-22.

[395] Yovchev MI, Grozdanov PN, Zhou H, *et al.* Identification of adult hepatic progenitor cells capable of repopulating injured rat liver. Hepatology 2008;47:636-47.

[396] Gulluoglu MG, Ozden I, Poyanli A, Cevikbas U, Ariogul O. Intraductal growth-type mucin-producing peripheral cholangiocarcinoma associated with biliary papillomatosis. Ann Diagn Pathol 2007;11:34-8.

[397] Sasaki M, Nakanuma Y, Kim YS. Characterization of apomucin expression in intrahepatic cholangiocarcinomas and their precursor lesions: an immunohistochemical study. Hepatology 1996;24:1074-8.

[398] Park SY, Roh SJ, Kim YN, *et al.* Expression of MUC1, MUC2, MUC5AC and MUC6 in cholangiocarcinoma: prognostic impact. Oncol Rep 2009;22:649-57.

[399] Kloppel G, Kosmahl M. Is the intraductal papillary mucinous neoplasia of the biliary tract a counterpart of pancreatic papillary mucinous neoplasm? J Hepatol 2006;44:249-50.

[400] Zen Y, Sasaki M, Fujii T, *et al.* Different expression patterns of mucin core proteins and cytokeratins during intrahepatic cholangiocarcinogenesis from biliary intraepithelial neoplasia and intraductal papillary neoplasm of the bile duct--an immunohistochemical study of 110 cases of hepatolithiasis. J Hepatol 2006;44:350-8.

[401] Amaya S, Sasaki M, Watanabe Y, *et al.* Expression of MUC1 and MUC2 and carbohydrate antigen Tn change during malignant transformation of biliary papillomatosis. Histopathology 2001;38:550-60.

[402] Sasaki M, Nakanuma Y. Expression of mucin core protein of mammary type in primary liver cancer. Hepatology 1994;20:1192-7.

[403] Sasaki M, Nakanuma Y. Abnormal expression of MUC1 apomucin and mature MUC1 mucin in biliary epithelial cells in various cystic liver diseases. Hepatology 1996;24:539-43.

[404] Sasaki M, Ikeda H, Nakanuma Y. Expression profiles of MUC mucins and trefoil factor family (TFF) peptides in the intrahepatic biliary system: physiological distribution and pathological significance. Prog Histochem Cytochem 2007;42:61-110.

[405] Nakanuma Y, Xu J, Harada K, *et al.* Pathological spectrum of intrahepatic cholangiocarcinoma arising in non-biliary chronic advanced liver diseases. Pathol Int 2011;61:298-305.

[406] Sasaki M, Nakanuma Y, Ho SB, Kim YS. Cholangiocarcinomas arising in cirrhosis and combined hepatocellular-cholangiocellular carcinomas share apomucin profiles. Am J Clin Pathol 1998;109:302-8.

[407] Yamashita K, Yonezawa S, Tanaka S, *et al.* Immunohistochemical study of mucin carbohydrates and core proteins in hepatolithiasis and cholangiocarcinoma. Int J Cancer 1993;55:82-91.

[408] Aishima S, Kuroda Y, Nishihara Y, *et al.* Gastric mucin phenotype defines tumour progression and prognosis of intrahepatic cholangiocarcinoma: gastric foveolar type is associated with aggressive tumour behaviour. Histopathology 2006;49:35-44.
[409] Goto M, Shibahara H, Tamada S, *et al.* Aberrant expression of pyloric gland-type mucin in mucin-producing bile duct carcinomas: a clear difference between the core peptide and the carbohydrate moiety. Pathol Int 2005;55:464-70.
[410] Cao Y, Blohm D, Ghadimi BM, *et al.* Mucins (MUC1 and MUC3) of gastrointestinal and breast epithelia reveal different and heterogeneous tumor-associated aberrations in glycosylation. J Histochem Cytochem 1997;45:1547-57.
[411] Cao Y, Karsten U, Otto G, Bannasch P. Expression of MUC1, Thomsen-Friedenreich antigen, Tn, sialosyl-Tn, and alpha2,6-linked sialic acid in hepatocellular carcinomas and preneoplastic hepatocellular lesions. Virchows Arch 1999;434:503-9.
[412] Yuan SF, Li KZ, Wang L, *et al.* Expression of MUC1 and its significance in hepatocellular and cholangiocarcinoma tissue. World J Gastroenterol 2005;11:4661-6.
[413] Inagaki Y, Xu H, Nakata M, *et al.* Clinicopathology of sialomucin: MUC1, particularly KL-6 mucin, in gastrointestinal, hepatic and pancreatic cancers. Biosci Trends 2009;3:220-32.
[414] Xu HL, Inagaki Y, Seyama Y, *et al.* Expression of KL-6 mucin, a human MUC1 mucin, in intrahepatic cholangiocarcinoma and its potential involvement in tumor cell adhesion and invasion. Life Sci 2009;85:395-400.
[415] Shoda J, Oda K, Suzuki H, *et al.* Etiologic significance of defects in cholesterol, phospholipid, and bile acid metabolism in the liver of patients with intrahepatic calculi. Hepatology 2001;33:1194-205.
[416] Nakanuma Y, Yamaguchi K, Ohta G, Terada T. Pathologic features of hepatolithiasis in Japan. Hum Pathol 1988;19:1181-6.
[417] Shoda J, Tanaka N, Osuga T. Hepatolithiasis--epidemiology and pathogenesis update. Front Biosci 2003;8:e398-409.
[418] Sasaki M, Nakanuma Y, Ho SB, Kim YS. Increased MUC6 apomucin expression is a characteristic of reactive biliary epithelium in chronic viral hepatitis. J Pathol 1998;185:191-8.
[419] Wang HH, Afdhal NH, Gendler SJ, Wang DQ. Evidence that gallbladder epithelial mucin enhances cholesterol cholelithogenesis in MUC1 transgenic mice. Gastroenterology 2006;131:210-22.
[420] Lee SP. Pathogenesis of biliary sludge. Hepatology 1990;12:200S-3S; discussion 3S-5S.
[421] Smith BF. Human gallbladder mucin binds biliary lipids and promotes cholesterol crystal nucleation in model bile. J Lipid Res 1987;28:1088-97.
[422] Smith BF. Gallbladder mucin as a pronucleating agent for cholesterol monohydrate crystals in bile. Hepatology 1990;12:183S-6S; discussion 6S-8S.
[423] Lee SP, Carey MC, LaMont JT. Aspirin prevention of cholesterol gallstone formation in prairie dogs. Science 1981;211:1429-31.
[424] Hulten O. Formation of gallstones. II. Acta Chir Scand. 1968;134:557-61.
[425] Pearson JP, Foster SN. Mucus glycoprotein content of human cholesterol gallstones. Digestion 1987;36:132-40.
[426] Shiffman ML, Shamburek RD, Schwartz CC, *et al.* Gallbladder mucin, arachidonic acid, and bile lipids in patients who develop gallstones during weight reduction. Gastroenterology 1993;105:1200-8.

[427] Ishikawa A, Sasaki M, Ohira S, *et al.* Aberrant expression of CDX2 is closely related to the intestinal metaplasia and MUC2 expression in intraductal papillary neoplasm of the liver in hepatolithiasis. Lab Invest 2004;84:629-38.
[428] Lee KT, Liu TS. Mucin gene expression in gallbladder epithelium. J Formos Med Assoc 2002;101:762-8.
[429] Sasaki M, Tsuneyama K, Nakanuma Y. Aberrant expression of trefoil factor family 1 in biliary epithelium in hepatolithiasis and cholangiocarcinoma. Lab Invest 2003;83:1403-13.
[430] Sasaki M, Ikeda H, Ohira S, Ishikawa A, Nakanuma Y. Expression of trefoil factor family 1, 2, and 3 peptide is augmented in hepatolithiasis. Peptides 2004;25:763-70.
[431] Tomasetto C, Masson R, Linares JL, *et al.* pS2/TFF1 interacts directly with the VWFC cysteine-rich domains of mucins. Gastroenterology 2000;118:70-80.
[432] Sasaki M, Huang SF, Chen MF, *et al.* Expression of deleted in malignant brain tumor-1 (DMBT1) molecule in biliary epithelium is augmented in hepatolithiasis: possible participation in lithogenesis. Dig Dis Sci 2003;48:1234-40.
[433] Ho SB, Shekels LL, Toribara NW, *et al.* Altered mucin core peptide expression in acute and chronic cholecystitis. Dig Dis Sci 2000;45:1061-71.
[434] Fijneman RJ, Peham JR, van de Wiel MA, *et al.* Expression of Pla2g2a prevents carcinogenesis in Muc2-deficient mice. Cancer Sci 2008;99:2113-9.
[435] Van der Sluis M, De Koning BA, De Bruijn AC, *et al.* Muc2-deficient mice spontaneously develop colitis, indicating that MUC2 is critical for colonic protection. Gastroenterology 2006;131:117-29.
[436] Yang K, Popova NV, Yang WC, *et al.* Interaction of Muc2 and Apc on Wnt signaling and in intestinal tumorigenesis: potential role of chronic inflammation. Cancer Res 2008;68:7313-22.
[437] Blache P, van de Wetering M, Duluc I, *et al.* SOX9 is an intestine crypt transcription factor, is regulated by the Wnt pathway, and represses the CDX2 and MUC2 genes. J Cell Biol 2004;166:37-47.
[438] Yonezawa S, Goto M, Yamada N, Higashi M, Nomoto M. Expression profiles of MUC1, MUC2, and MUC4 mucins in human neoplasms and their relationship with biological behavior. Proteomics 2008;8:3329-41.
[439] Matsukita S, Nomoto M, Kitajima S, *et al.* Expression of mucins (MUC1, MUC2, MUC5AC and MUC6) in mucinous carcinoma of the breast: comparison with invasive ductal carcinoma. Histopathology 2003;42:26-36.
[440] Ilhan O, Han U, Onal B, Celik SY. Prognostic significance of MUC1, MUC2 and MUC5AC expressions in gastric carcinoma. Turk J Gastroenterol 2010;21:345-52.
[441] Podolsky DK. Inflammatory bowel disease (1). N Engl J Med 1991;325:928-37.
[442] Tytgat KM, Opdam FJ, Einerhand AW, Buller HA, Dekker J. MUC2 is the prominent colonic mucin expressed in ulcerative colitis. Gut 1996;38:554-63.
[443] Kim YS, Ho SB. Intestinal goblet cells and mucins in health and disease: recent insights and progress. Curr Gastroenterol Rep 2010;12:319-30.
[444] Van Klinken BJ, Van der Wal JW, Einerhand AW, Buller HA, Dekker J. Sulphation and secretion of the predominant secretory human colonic mucin MUC2 in ulcerative colitis. Gut 1999;44:387-93.
[445] Lu P, Burger-van Paassen N, van der Sluis M, *et al.* Colonic gene expression patterns of mucin muc2 knockout mice reveal various phases in colitis development. Inflamm Bowel Dis 2011;17:2047-57.

[446] Strugala V, Dettmar PW, Pearson JP. Thickness and continuity of the adherent colonic mucus barrier in active and quiescent ulcerative colitis and Crohn's disease. Int J Clin Pract 2008;62:762-9.

[447] Myllarniemi H, Nickels J. Scanning electron microscopy of Crohn's disease and ulcerative colitis of the colon. Virchows Arch A Pathol Anat Histol 1980;385:343-50.

[448] McCormick DA, Horton LW, Mee AS. Mucin depletion in inflammatory bowel disease. J Clin Pathol 1990;43:143-6.

[449] Hanski C, Born M, Foss HD, *et al.* Defective post-transcriptional processing of MUC2 mucin in ulcerative colitis and in Crohn's disease increases detectability of the MUC2 protein core. J Pathol 1999;188:304-11.

[450] Tytgat KM, Buller HA, Einerhand AW, van Deventer SJ, Dekker J. [Role of mucins in inflammatory intestinal diseases]. Ned Tijdschr Geneeskd 1994;138:2386-90.

[451] Tytgat KM, van der Wal JW, Einerhand AW, Buller HA, Dekker J. Quantitative analysis of MUC2 synthesis in ulcerative colitis. Biochem Biophys Res Commun 1996;224:397-405.

[452] Tytgat KM, Lygidakis NJ. Crohn's disease. Hepatogastroenterology 1990;37:1-5.

[453] Nieuw Amerongen AV, Bolscher JG, Bloemena E, Veerman EC. Sulfomucins in the human body. Biol Chem 1998;379:1-18.

[454] Shaoul R, Okada Y, Cutz E, Marcon MA. Colonic expression of MUC2, MUC5AC, and TFF1 in inflammatory bowel disease in children. J Pediatr Gastroenterol Nutr 2004;38:488-93.

[455] Larsson JM, Karlsson H, Crespo JG, *et al.* Altered O-glycosylation profile of MUC2 mucin occurs in active ulcerative colitis and is associated with increased inflammation. Inflamm Bowel Dis 2011;17:2299-307.

[456] Hinoda Y, Akashi H, Suwa T, *et al.* Immunohistochemical detection of MUC2 mucin core protein in ulcerative colitis. J Clin Lab Anal 1998;12:150-3.

[457] Longman RJ, Poulsom R, Corfield AP, *et al.* Alterations in the composition of the supramucosal defense barrier in relation to disease severity of ulcerative colitis. J Histochem Cytochem 2006;54:1335-48.

[458] Heazlewood CK, Cook MC, Eri R, *et al.* Aberrant mucin assembly in mice causes endoplasmic reticulum stress and spontaneous inflammation resembling ulcerative colitis. PLoS Med 2008;5:e54.

[459] Satsangi J, Jewell DP, Rosenberg WM, Bell JI. Genetics of inflammatory bowel disease. Gut 1994;35:696-700.

[460] Mirza MM, Lee J, Teare D, *et al.* Evidence of linkage of the inflammatory bowel disease susceptibility locus on chromosome 16 (IBD1) to ulcerative colitis. J Med Genet 1998;35:218-21.

[461] Hugot JP, Laurent-Puig P, Gower-Rousseau C, *et al.* Mapping of a susceptibility locus for Crohn's disease on chromosome 16. Nature 1996;379:821-3.

[462] Vinall LE, Hill AS, Pigny P, *et al.* Variable number tandem repeat polymorphism of the mucin genes located in the complex on 11p15.5. Hum Genet 1998;102:357-66.

[463] Swallow DM, Vinall LE, Gum JR, *et al.* Ulcerative colitis is not associated with differences in MUC2 mucin allele length. J Med Genet 1999;36:859-60.

[464] Moehle C, Ackermann N, Langmann T, *et al.* Aberrant intestinal expression and allelic variants of mucin genes associated with inflammatory bowel disease. J Mol Med (Berl) 2006;84:1055-66.

[465] Corfield AP, Myerscough N, Bradfield N, *et al.* Colonic mucins in ulcerative colitis: evidence for loss of sulfation. Glycoconj J 1996;13:809-22.

[466] Dvorak AM, Osage JE, Monahan RA, Dickersin GR. Crohn's disease: transmission electron microscopic studies. III. Target tissues. Proliferation of and injury to smooth muscle and the autonomic nervous system. Hum Pathol 1980;11:620-34.
[467] Trabucchi E, Mukenge S, Baratti C, *et al.* Differential diagnosis of Crohn's disease of the colon from ulcerative colitis: ultrastructure study with the scanning electron microscope. Int J Tissue React 1986;8:79-84.
[468] Gersemann M, Becker S, Kubler I, *et al.* Differences in goblet cell differentiation between Crohn's disease and ulcerative colitis. Differentiation 2009;77:84-94.
[469] Buisine MP, Desreumaux P, Leteurtre E, *et al.* Mucin gene expression in intestinal epithelial cells in Crohn's disease. Gut 2001;49:544-51.
[470] Kaneko Y, Nakamura T, Hayama M, *et al.* Altered expression of CDX-2, PDX-1 and mucin core proteins in "Ulcer-associated cell lineage (UACL)" in Crohn's disease. J Mol Histol 2008;39:161-8.
[471] Davies JA. The Kidney Development Database. Dev Genet 1999;24:194-8.
[472] Glassberg KI. Normal and abnormal development of the kidney: a clinician's interpretation of current knowledge. J Urol 2002;167:2339-50; discussion 50-1.
[473] Leroy X, Copin MC, Devisme L, *et al.* Expression of human mucin genes in normal kidney and renal cell carcinoma. Histopathology 2002;40:450-7.
[474] Lau SK, Weiss LM, Chu PG. Differential expression of MUC1, MUC2, and MUC5AC in carcinomas of various sites: an immunohistochemical study. Am J Clin Pathol 2004;122:61-9.
[475] Hayashi T, Matsubara A, Ohara S, *et al.* Immunohistochemical analysis of Reg IV in urogenital organs: Frequent expression of Reg IV in prostate cancer and potential utility as serum tumor marker. Oncol Rep 2009;21:95-100.
[476] Chu PG, Lau SK, Weiss LM, Jiang Z. Intestinal type of mucinous borderline tumor arising from mixed epithelial and stromal tumor of kidney. Virchows Arch 2009;455:389-94.
[477] Jankovic Velickovic L, Katic V, Hattori T, *et al.* Differences in the expression of mucins in various forms of cystitis glandularis. Pathol Res Pract 2007;203:653-8.
[478] Ohtsuki Y, Fukumoto T, Okada Y, *et al.* Immunohistochemical and ultrastructural characterization of the signet-ring cell carcinoma component in a case of urothelial carcinoma of the urinary bladder. Med Mol Morphol 2010;43:96-101.
[479] Marques ML, D'Alessandro GS, Chade DC, *et al.* Primary mucinous adenocarcinoma of the bladder with signet-ring cells: case report. Sao Paulo Med J 2007;125:297-9.
[480] Cordon-Cardo C, Wartinger DD, Melamed MR, Fair W, Fradet Y. Immunopathologic analysis of human urinary bladder cancer. Characterization of two new antigens associated with low-grade superficial bladder tumors. Am J Pathol 1992;140:375-85.
[481] Fradet Y, LaRue H, Parent-Vaugeois C, *et al.* Monoclonal antibody against a tumor-associated sialoglycoprotein of superficial papillary bladder tumors and cervical condylomas. Int J Cancer 1990;46:990-7.
[482] Tavora F, Epstein JI. Urothelial carcinoma with abundant myxoid stroma. Hum Pathol 2009;40:1391-8.
[483] Cozzi PJ, Wang J, Delprado W, *et al.* MUC1, MUC2, MUC4, MUC5AC and MUC6 expression in the progression of prostate cancer. Clin Exp Metastasis 2005;22:565-73.
[484] Cappello F, Rappa F, David S, Anzalone R, Zummo G. Immunohistochemical evaluation of PCNA, p53, HSP60, HSP10 and MUC-2 presence and expression in prostate carcinogenesis. Anticancer Res 2003;23:1325-31.

[485] Legrier ME, de Pinieux G, Boye K, *et al.* Mucinous differentiation features associated with hormonal escape in a human prostate cancer xenograft. Br J Cancer 2004;90:720-7.
[486] Nokubi M, Kawai T, Mitsu S, Ishikawa S, Morinaga S. Mucinous cystadenoma of the testis. Pathol Int 2002;52:648-52.
[487] Naito S, Yamazumi K, Yakata Y, *et al.* Immunohistochemical examination of mucinous cystadenoma of the testis. Pathol Int 2004;54:355-9.
[488] Audie JP, Tetaert D, Pigny P, *et al.* Mucin gene expression in the human endocervix. Hum Reprod 1995;10:98-102.
[489] Zhao S, Hayasaka T, Osakabe M, *et al.* Mucin expression in nonneoplastic and neoplastic glandular epithelia of the uterine cervix. Int J Gynecol Pathol 2003;22:393-7.
[490] Shintaku M, Kushima R, Abiko K. Colloid carcinoma of the intestinal type in the uterine cervix: mucin immunohistochemistry. Pathol Int 2010;60:119-24.
[491] Mikami Y, Kiyokawa T, Hata S, *et al.* Gastrointestinal immunophenotype in adenocarcinomas of the uterine cervix and related glandular lesions: a possible link between lobular endocervical glandular hyperplasia/pyloric gland metaplasia and'adenoma malignum'. Mod Pathol 2004;17:962-72.
[492] Morrison C, Merati K, Marsh WL, Jr., *et al.* The mucin expression profile of endometrial carcinoma and correlation with clinical-pathologic parameters. Appl Immunohistochem Mol Morphol 2007;15:426-31.
[493] Aubert S, Van Seuningen I, Leroy X. Mucin in the uro-genital tract. Potential for therapeutic approaches using mucins. In: Van Seuningen I, editor. The epithelial mucins: structure/function Roles in cancer and inflammatory diseases. Kerala: Research Signpost; 2008. p. 249-72.
[494] Drapkin R, Crum CP, Hecht JL. Expression of candidate tumor markers in ovarian carcinoma and benign ovary: evidence for a link between epithelial phenotype and neoplasia. Hum Pathol 2004;35:1014-21.
[495] Pryse-Davies J, Dewhurst CJ. The development of the ovary and uterus in the foetus, newborn and infant: a morphological and enzyme histochemical study. J Pathol 1971;103:5-25.
[496] Jaffe ES, Harris NL, Stein H, Vardiman JW, editors. WHO classification of tumor pathology and genetics of tumors of the breast and female genital organs. Lyon: IARC 2004.
[497] Giuntoli RL, 2nd, Rodriguez GC, Whitaker RS, Dodge R, Voynow JA. Mucin gene expression in ovarian cancers. Cancer Res 1998;58:5546-50.
[498] Hirabayashi K, Yasuda M, Kajiwara H, *et al.* Alterations in mucin expression in ovarian mucinous tumors: immunohistochemical analysis of MUC2, MUC5AC, MUC6, and CD10 expression. Acta Histochem Cytochem 2008;41:15-21.
[499] Dong Y, Walsh MD, Cummings MC, *et al.* Expression of MUC1 and MUC2 mucins in epithelial ovarian tumours. J Pathol 1997;183:311-7.
[500] Feng H, Ghazizadeh M, Konishi H, Araki T. Expression of MUC1 and MUC2 mucin gene products in human ovarian carcinomas. Jpn J Clin Oncol 2002;32:525-9.
[501] Argueso P, Balaram M, Spurr-Michaud S, *et al.* Decreased levels of the goblet cell mucin MUC5AC in tears of patients with Sjogren syndrome. Invest Ophthalmol Vis Sci 2002;43:1004-11.
[502] Paulsen FP, Corfield AP, Hinz M, *et al.* Characterization of mucins in human lacrimal sac and nasolacrimal duct. Invest Ophthalmol Vis Sci 2003;44:1807-13.

[503] Spurr-Michaud S, Argueso P, Gipson I. Assay of mucins in human tear fluid. Exp Eye Res 2007;84:939-50.

[504] Schafer G, Hoffmann W, Berry M, Paulsen F. [Lacrimal gland-associated mucins. Age related production and their role in the pathophysiology of dry eye]. Ophthalmologe 2005;102:175-83.

[505] Paulsen F, Langer G, Hoffmann W, Berry M. Human lacrimal gland mucins. Cell Tissue Res 2004;316:167-77.

[506] Dogru M, Okada N, Asano-Kato N, *et al.* Alterations of the ocular surface epithelial mucins 1, 2, 4 and the tear functions in patients with atopic keratoconjunctivitis. Clin Exp Allergy 2006;36:1556-65.

[507] Dogru M, Okada N, Asano-Kato N, *et al.* Atopic ocular surface disease: implications on tear function and ocular surface mucins. Cornea 2005;24:S18-S23.

[508] Ali MS, Pearson JP. Upper airway mucin gene expression: a review. Laryngoscope 2007;117:932-8.

[509] Ali MS, Wilson JA, Bennett M, Pearson JP. Mucin gene expression in hypertrophic adenoids. Acta Otolaryngol 2007;127:1080-5.

[510] Ali MS, Wilson JA, Bennett M, Pearson JP. Mucin gene expression in nasal polyps. Acta Otolaryngol 2005;125:618-24.

[511] Voynow JA, Selby DM, Rose MC. Mucin gene expression (MUC1, MUC2, and MUC5/5AC) in nasal epithelial cells of cystic fibrosis, allergic rhinitis, and normal individuals. Lung 1998;176:345-54.

[512] Kerschner JE, Khampang P, Samuels T. Extending the chinchilla middle ear epithelial model for mucin gene investigation. Int J Pediatr Otorhinolaryngol 2010;74:980-5.

[513] Kubba H, Pearson JP, Birchall JP. The aetiology of otitis media with effusion: a review. Clin Otolaryngol Allied Sci 2000;25:181-94.

[514] Hutton DA, Fogg FJ, Kubba H, Birchall JP, Pearson JP. Heterogeneity in the protein cores of mucins isolated from human middle ear effusions: evidence for expression of different mucin gene products. Glycoconj J 1998;15:283-91.

[515] Lin J, Tsuboi Y, Rimell F, *et al.* Expression of mucins in mucoid otitis media. J Assoc Res Otolaryngol 2003;4:384-93.

[516] Mukhopadhyay P, Chakraborty S, Ponnusamy MP, *et al.* Mucins in the pathogenesis of breast cancer: implications in diagnosis, prognosis and therapy. Biochim Biophys Acta 2011;1815:224-40.

[517] Zhang S, Zhang HS, Cordon-Cardo C, Ragupathi G, Livingston PO. Selection of tumor antigens as targets for immune attack using immunohistochemistry: protein antigens. Clin Cancer Res 1998;4:2669-76.

[518] Walsh MD, McGuckin MA, Devine PL, Hohn BG, Wright RG. Expression of MUC2 epithelial mucin in breast carcinoma. J Clin Pathol 1993;46:922-5.

[519] Pereira MB, Dias AJ, Reis CA, Schmitt FC. Immunohistochemical study of the expression of MUC5AC and MUC6 in breast carcinomas and adjacent breast tissues. J Clin Pathol 2001;54:210-3.

[520] Berois N, Varangot M, Sonora C, *et al.* Detection of bone marrow-disseminated breast cancer cells using an RT-PCR assay of MUC5B mRNA. Int J Cancer 2003;103:550-5.

[521] Rakha EA, Boyce RW, Abd El-Rehim D, *et al.* Expression of mucins (MUC1, MUC2, MUC3, MUC4, MUC5AC and MUC6) and their prognostic significance in human breast cancer. Mod Pathol 2005;18:1295-304.

[522] Diaz LK, Wiley EL, Morrow M. Expression of epithelial mucins Muc1, Muc2, and Muc3 in ductal carcinoma *in situ* of the breast. Breast J 2001;7:40-5.
[523] O'Connell JT, Shao ZM, Drori E, Basbaum CB, Barsky SH. Altered mucin expression is a field change that accompanies mucinous (colloid) breast carcinoma histogenesis. Hum Pathol 1998;29:1517-23.
[524] Chu JS, Chang KJ. Mucin expression in mucinous carcinoma and other invasive carcinomas of the breast. Cancer Lett 1999;142:121-7.
[525] Kasashima S, Kawashima A, Zen Y, *et al.* Expression of aberrant mucins in lobular carcinoma with histiocytoid feature of the breast. Virchows Arch 2007;450:397-403.

Mucins – Potential Regulators of Cell, 2013, 145-245

CHAPTER 5

Gel-Forming Mucin MUC5AC

Abstract: The MUC5AC mucin plays an essential role in homeostasis of respiratory, gastrointestinal and urogenital tracts under physiological conditions and fulfills various important functions in human pathology. This chapter describes the molecular structure of the *MUC5AC* gene and the mechanisms that regulate its activity. It analyzes the biochemical and biophysical properties of the MUC5AC glycoprotein, its biosynthesis and posttranslational modifications, as well as its expression in the respiratory, gastrointestinal and male and female urogenital organs under physiological and pathological conditions. The roles of the MUC5AC mucin in eye physiology and pathology and in breast cancer are also discussed.

Keywords: MUC5AC, domain, biosynthesis, promoter, regulation, expression.

5.1. STRUCTURAL CHARACTERISTICS OF *MUC5AC* GENE AND MUC5AC MUCIN

The gel-forming mucin MUC5AC is encoded be the *MUC5AC* gene. Initially, three MUC5 cDNAs were cloned from tracheo-bronchial tissues and designated MUC5A, MUC5B and MUC5C [1]. The partial MUC5A and MUC5C cDNAs were subsequently shown to be derived from the same gene, which was therefore designated the *MUC5AC* gene. The MUC5B cDNA represents a separate gene, *MUC5B* [2, 3]. Cloning of the mucin genes is extremely difficult due to the highly repetitive structure and extremely large size of the mucin mRNAs. Several attempts to isolate and sequence the *MUC5AC* gene resulted in identification of different parts of the gene [3-6], until the combined efforts of several laboratories succeeded in sequencing its full length [3, 4, 7-12]. The gene consists of 48 exons and 47 intrones of different lengths [12-14] (Fig. **1**). The N- and C-terminal parts of the MUC5AC protein consist of 1858 and 1167 amino acids, respectively, and the central tandem repeat (TR)-containing region consists of 2500 amino acids [9]. Thus, the whole MUC5AC mucin is 5525 amino acids in length. Theoretically, such a large protein must be encoded by an mRNA of 16.6 kb, which is in line with the experimentally estimated size of the MUC5AC mRNA of about 17 kb [8, 13]. While investigators generally agree about the structure of the MUC5AC protein, there is disagreement about the length of different regions of the mucin. For instance, Van de Bovenkamp *et al.* [9] detected 1858 amino acids

Joseph Z. Zaretsky and Daniel H. Wreschner

comprising the MUC5AC N-terminus in a human gastric cDNA library, while Li *et al.* [11] found this region cloned from the human lung epithelial carcinoma cell line to be much shorter, consisting of only 1100 amino acids. According to Escande *et al.* [12], the N-terminal part of *MUC5AC* isolated from human trachea encodes 1336 amino acids. Clearly, further studies are needed to clarify this issue and the dependence of the *MUC5AC* gene characteristics on the cell type used for its cloning and expression.

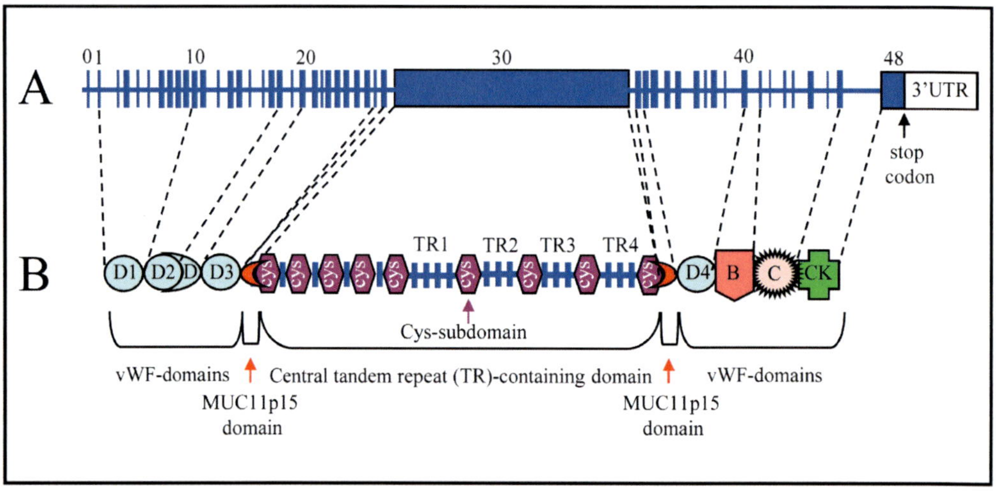

Figure 1: MUC5AC exon/domain structural relationship. A – exon structure of the MUC5AC gene, B – domain structure of the MUC5AC protein (based on the data reported in [12-14]).

The N-terminal region of the MUC5AC protein contains a signal peptide, four cysteine-rich domains similar to the D-domains of the human pro-von Willebrand factor (vWF), and a short domain corresponding to the MUC11p15-type domain described by Desseyn *et al.* [14] (Fig. **1**). Importantly, the positions of the cysteine residues in the MUC5AC D-domains correspond to those at the N-termini of human MUC2 and MUC5B mucins [14-16]. In addition, the N-terminal region of the MUC5AC contains seven potential N-glycosylation consensus sites (Asn-X-Ser/Thr) [12]. Although there are some discrepancies in the size of the *MUC5AC* gene, all investigators emphasized the high level of nucleotide homology and structural similarity between this gene and *MUC2, MUC5B* and *vWF* [6, 10, 12, 14, 15, 17]. However, in contrast to vWF and gel-forming mucins MUC2, MUC5B and MUC6, the D1-domain of the MUC5AC mucin contains a leucine "zipper" sequence between the residues 273 and 300. Its function is not well established. According to van de Bovenkamp *et al.* [9], the leucine "zipper" is apparently involved in non-covalent homo-oligomerization *via* the N-termini of MUC5AC dimers. It may also participate in stabilization of non-covalent interactions between two D1- domains of different MUC5AC molecules, and thereby contribute to the process of MUC5AC multimerization. This MUC5AC-specific component, which appears to be important for MUC5AC biochemistry, needs further investigation.

The central region of the MUC5AC mucin is encoded by one central exon 30 (~10.5 kb) consisting of several domains: nine cysteine-rich domains interspersed by two PTS-domains, containing irregular tandem repeats, and four tandem repeat domains (TR1-TR4) composed of a variable number of regular tandem repeats (Fig. **1**). In addition, there are two short MUC11p15 type domains at the flanks of the central region [12]. Each cysteine-rich domain (Cys-domain) contains 10 cysteine residues per 110 amino acid residues. The tandem repeat domains are not similar, containing different numbers of tandem repeat units, each composed of 8 amino acids (TTSTTSAP or GTTPSPVP). TR1 has 124 repeat units, TR2 has 17 units, TR3 has 34 units, and TR4 has 66 units. TRs are highly O-glycasylated, and because O-glycosylation influences the size, shape and mass of the MUC5AC mucin, the TRs contribute to the biophysical and biochemical properties of the mucin [18-21]. Besides multiple O-glycan binding sites, each TR domain contains at least one potential N-glycosylation site. According to Escande *et al.* [12], "each TR domain is preceded and followed by short unique sequences of 21aa and 30aa, respectively, which are almost identical to each other".

The C-terminal region of *MUC5AC* located 3'-downstream to the central exon has been well characterized [4, 5, 10]. This region contains 18 exons that are translated into several domains, including the D4-, B-, C- and CK-domains [10, 22, 23] (Fig. **1**). Analysis of the amino acid sequences of the C-terminal region revealed 12 potential N-glycosylation sites [5, 10]. Importantly, the positions of some N-glycan-specific sites as well as cysteine residues of the MUC5AC C-terminal region correspond to those in MUC2 and MUC5B mucins, indicating the common evolutionary history of these genes [12].

5.2. REGULATION OF *MUC5AC* GENE EXPRESSION

There is convincing evidence that epithelial cells respond to pathogens and environmental stress agents by altering gene expression, particularly mucin gene transcription. Multiple stimuli including Gram-positive and Gram-negative bacterial pathogens, viruses, cytokines, growth factors and others have been implicated in the activation of *MUC5AC* and other gel-forming mucin genes [18, 23-33]. Most of the effectors induced by these stimuli are targeted to the *MUC5AC* promoter, where the specific binding of different transcription factors to their cognate *cis*-elements takes place.

5.2.1. *MUC5AC* Promoter

The promoter region of the *MUC5AC* gene has been cloned by several laboratories. Bioinformatics and molecular biology techniques allowed identification of numerous *cis*-elements in this region specific for binding of multiple transcription factors (Fig. **2**). Currently, the TATA box and putative binding sites for various nuclear factors have been identified in the *MUC5AC* genomic DNA located 5'-upstream to the transcription start site: these include Sp1, GRE, AP-1, AP-2, PEA3, SMAD3 and SMAD4, HIF1, FOXA2, BKLF, EGFR and TGFα, LEF, T3Rα, CRE, c-Ets, YY1, Cdx, NKX2, HNF4 and NFκB [34-39]. A high degree of homology is observed between human *MUC5AC* promoter sequence and its mouse and rat orthologs, demonstrating evolutionary conservation of the main regulatory elements that control transcription of the human *MUC5AC* and other mammalian *Muc5ac* genes [40].

MUC5AC transcription is mediated mainly through the c-Src and MAPK kinase pathways, much like *MUC2*, but, unlike *MUC2*, *MUC5AC* may also be transcriptionally activated by protein kinase C-mediated events, and by other signaling pathways [41, 42]. Importantly, numerous stimuli may activate one and the same signaling pathway, while a single stimulus may induce different pathways; these activities depend on cell type and/or conditions. The following are several examples illustrating the roles of different transcription factors and signal transduction pathways in *MUC5AC* transcriptional regulation.

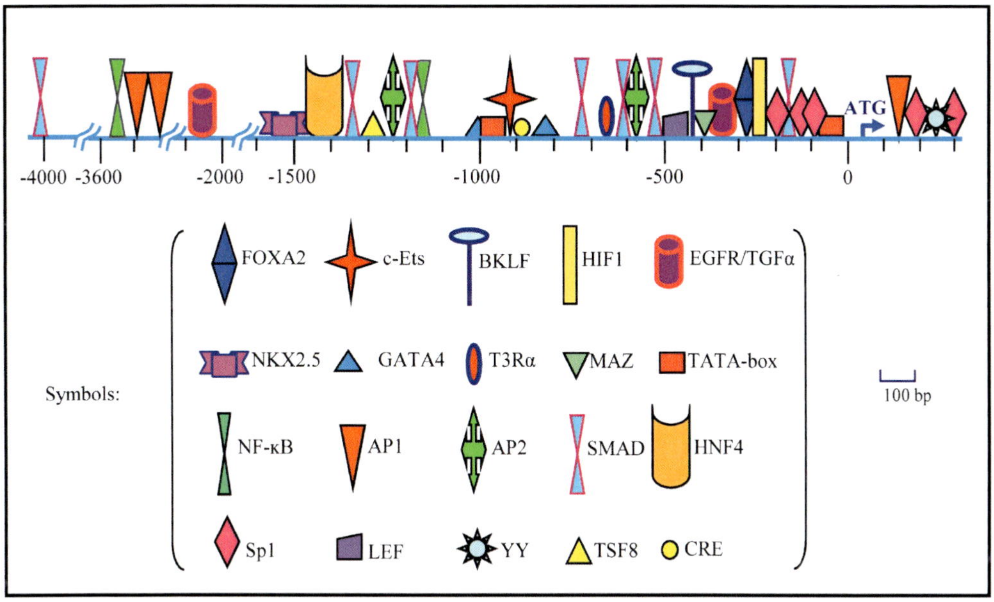

Figure 2: Transcription factor *cis*-element map of the *MUC5AC* promoter (based on the data reported in [34-39]).

5.2.1.1. IL-13-Mediated Regulation of MUC5AC Gene Expression

Mucin hyper-secretion is an evolutionarily ancient innate defense mechanism retained in mammalian epithelial cells to protect the host against pathogens and other noxious factors. The normal respiratory epithelium synthesizes and secretes a fixed amount of mucins, whose production in case of "emergency" can be up-regulated [35]. In a healthy individual, there are a few goblet cells located distal to the trachea that produce mucins, including MUC5AC [43]. In asthma and other chronic inflammatory diseases of airways, goblet cell hyperplasia or metaplasia associated with substantially elevated numbers of goblet cells producing excessive amounts of mucins is constantly observed [33]. Among the many cytokines (TNFα, IL-1β, IL-4, IL-5, IL-6, IL-9, IL-10 and IL-13) and growth factors (EGF, TGFα and TGFβ) implicated in the regulation of *MUC5AC* transcription in chronic airway diseases, IL-13 and EGFR were shown to play the central roles in the pathogenesis of allergic asthma [18, 33, 40, 44, 45].

Two important signaling pathways associated with EGFR and the IL-13/IL-4 Rα complex are implicated in the up-regulation of *Muc5ac* in allergic inflamed airways in mice [46-50]. Mucin hyper-production can be induced in the inflamed airway by several products of Th2 lymphocytes, including the cytokines IL-4, IL-5, IL-9, IL-10 and IL-13, albeit with different roles in the activation process. In the absence of IL-13, the Th2 cells can not stimulate mucus production even in the presence of IL-4 and IL-5 [48, 51]. Moreover, mucus hyper-secretion in transgenic mice with excessive amounts of IL-4, IL-5 and IL-9 was found to be IL-13 dependent [48, 52, 53]. On the other hand, when these three interleukines were ablated, IL-13 alone could promote mucous metaplasia and *Muc5ac* over-expression [48, 54]. The crucial role of IL-13 in goblet cell metaplasia and *Muc5ac* over-expression was seen in an *in vivo* mouse model of asthma [55] and in an HBEC culture *in vitro* [27]. Untreated cultures exhibited only a few goblet cells, while treatment of the cultures with IL-13 resulted in up to a 10-fold increase in number of goblet cells, leading to an increased level of MUC5AC mucin compared with control cultures. Thus, IL-13 appears to be an absolutely indispensable and powerful activator of *MUC5AC* transcription associated with goblet cell metaplasia. However, IL-13 cannot directly regulate transcription of *MUC5AC*, as the binding sites specific for STAT6, a principal signaling molecule activated by IL-13, are absent from the human *MUC5AC* promoter and the promoters of the mammalian orthologs of the *MUC5AC* gene [39, 40]. These *cis*-STAT6 specific elements are also indispensable for IL-13-mediated signaling [46, 56].

IL-13 was found to activate *MUC5AC* transcription in an airway epithelial cell culture *via* the TGFβ_2-mediated pathway [31]. Jonckheere *et al.* [39] described transcriptional activation of the murine *Muc5ac* mucin gene in epithelial rectal cancer cells through the TGFβ/Smad4 signaling mechanism potentiated by Sp1. In other words, the "direct" effects of IL-13 on *MUC5AC* expression can be mediated *via* the "indirect" activation of TGFβ_2/SMAD signaling through a STAT6-dependent pathway. Recent studies show that the human *MUC5AC* promoter contains several SMAD *cis*-elements [39, 40] (Fig. **2**). The findings of Jonckheere *et al.* [39] point to the participation of TGFβ_2/SMAD signaling in *Muc5ac* and *MUC5AC* transcriptional regulation, and may explain, at least partially, the mechanism by which IL-13 activates *MUC5AC* transcription.

Another indirect mechanism involved in IL-13-activated *Muc5ac* super-expression is STAT6-dependent down-regulation of the gene coding for the Foxa2 transcription factor, a critical negative regulator of *Muc5ac* [45, 57, 58]. Careful inspection of the transcription factor binding sites located close to the TATA-box and transcription start site in the *MUC5AC* and *Muc5ac* promoters shows specific and meaningful combination of several *cis*-elements, including those specific for Sp1, SMAD4, HIF-1 and FOXA2 transcription factors. This combination is an important instrument of negative/positive regulation of the *MUC5AC* gene transcription [45]. FOXA2 is a powerful negative regulator of the *MUC5AC* gene and a key factor in the regulation of genetic programs controlling the Th2 cell-conducted reactions [45, 58]. Inhibition of *FOXA2* gene expression activates the transcription of several cytokine genes (IL-13, IL-4, IL-5, IL-17) that are expressed upon Th2 cell-mediated pulmonary inflammation. The expression of *Foxa2* is inhibited during allergen- or IL-13-induced goblet cell metaplasia and Th2-mediated inflammation [49, 57]. Under inflammation, IL-13 interacts with its cognate receptor on the surface of the goblet cells, after which the STAT6 transcription factor is activated, leading in turn to inhibition of FOXA2 production and activation of HIF-1α expression. These two transcription factors act reciprocally in regard with the *MUC5AC* gene activity: FOXA2 inhibits *MUC5AC* gene expression while HIF-1α activates transcription of this gene through binding to its promoter [45]. Notably, TGFβ stabilizes HIF-1α protein, thereby increasing its stimulating effect on *MUC5AC* transcription [59]. Interestingly, HIF-1α is activated by the EGFR signaling pathway (EGFR/Ras/Raf/MAPK), one of the main signaling pathways required for mucous metaplasia and *MUC5AC* expression in a variety of *in vivo* and *in vitro* systems [49, 50, 60, 61]. In addition, IL-13 increases expression of the MUC5AC glycoprotein by stimulating production of epidermal growth factor receptor (EGFR) [62].

5.2.1.2. Function of EGFR in Regulation of MUC5AC Gene Expression

EGFR has critical functions in airway epithelial cell physiology [50, 63]. Its expression is up-regulated in the airway epithelium of patients with asthma, cystic fibrosis (CF), and chronic pulmonary inflammatory diseases [64-68]. Activation of EGFR is an important step in up-regulation of *MUC5AC* expression and in goblet cell metaplasia. The available data suggest that the EGFR-activated pathway is involved in regulation of mucin production by several stress-related allergens, viruses, growth factors such as EGF and TGFα, bacterial LPS, cigarette smoke, neutrophil elastase, phorbol 12-myristate 13-acetate (PMA) and other agents [60-62, 69, 70]. Several comprehensive studies have been carried out on the role of EGFR signaling in *MUC5AC* expression. Nadel and co-workers [30, 64, 71] showed that, in lung, binding of ligands to the EGFR activates MAPK cascade and induces increase in MUC5AC mRNA expression. Van Seuningen's group [70] established that EGF- and TGFα-mediated up-regulation of *MUC5AC* occurs through binding of EGFR to the *MUC5AC* promoter and activation of the Ras/Raf/ERK-signaling pathway. The transcription factor(s) activated by EGF or TGFα cooperate with the Sp1 ubiquitous transcription factor to form a complex that binds to the *MUC5AC* promoter at the -202/-1 region. Evan's group [40, 45] considerably advanced the understanding of EGFR–mediated *MUC5AC* expression by analysis of the intracellular mechanisms associated with activation of EGFR and consequent signal transduction leading to up-regulation of *MUC5AC* expression. Activation of the *HIF1α* gene by EGF through signaling cascade EGFR/Ras/Raf/MAPK has a marked effect on the activity of the *MUC5AC* gene, as the *MUC5AC* gene contains HIF1α-specific binding sites in close proximity to the SMAD4/Sp1 site in the *MUC5AC* promoter. The binding of HIF1α to the *MUC5AC* promoter may explain the modes of interaction between signaling pathways activated by IL-13, TGFβ and EGF that lead to activation of *MUC5AC* transcription. HIF1α- and TGFβ-directed signalings are connected to each other *via* protein-protein interaction between HIF1α and SMAD3 [72], or *via* autocrine activation of the *TGFβ2* promoter through HIF1α and SMAD3 binding [73]. According to Young *et al.* [40], the close proximity of Smad4 and Hif1α *cis*-elements to each other in the *Muc5ac* promoter suggests that these cellular stress response signals may interact to mediate *Muc5ac* production in response to inflammatory and tissue damage *in vivo*. Moreover, in contrast to *MUC5AC* and *TGFβ* promoters, which do not contain STAT6 consensus motif and therefore their expression cannot be affected directly by STAT6 signaling, promoter regions of all mammalian *HIF1α* genes contain canonical STAT6 binding motifs. Taken together, these data suggest a model in which IL-13 and EGF regulate the differential expression of *HIF1α* in the airway epithelium and thereby mediate the efficiency of SMAD-mediated transcriptional activation of genes involved in Th2-induced inflammatory reaction. This model is strongly supported by two observations: mutations in the HIF1α-specific *cis*-element in the *Muc5ac* promoter dramatically reduce transcriptional activity of the *Muc5ac* gene, and binding of HIF1α to the *Muc5ac* promoter is induced by IL-13 and/or EGF stimulation [40].

A new player recently emerged in goblet cell metaplasia with the discovery that transgenic over-expression of SPDEF (Sam Pointed Domain-containing ETS transcription Factor) in the murine airway epithelium caused spontaneous mucous metaplasia [74, 75]. IL-13 exposure caused STAT6-dependent over-expression of SPDEF in epithelial cells, followed by mucosal metaplasia. Moreover, over-production of SPDEF induced down-regulation of *Foxa2* expression, placing it upstream to *Foxa2* in goblet cell differentiation. SPDEF was shown to be both necessary and sufficient for induction of a transcriptional program controlling goblet cell differentiation [44, 75]. All these findings demonstrate the important role SPDEF plays in normal differentiation and pathological metaplasia of goblet cells.

Based on the above data, Curran and Cohn [44] suggested an attractive hypothesis that may explain the mechanism of goblet cell differentiation. According to this hypothesis, goblet cell differentiation requires two signals: the first one activates EGFR on the ciliated cell through the effects of EGF, TGFα or O_2; the second one involves binding of IL-13 to its receptor. The first signal leads to inhibition of epithelial cell apoptosis, and the cells that survived stress induced by stress-specific effectors may differentiate into goblet cells if provided with IL-13. Upon IL-13 and STAT6 activation, up-regulation of SPDEF and GABA(A)R (γ–aminobutyric acid receptor) occurs, followed by down-regulation of *FOXA2*. These effects lead to activation of the transcription of *MUC5AC* and other mucin genes specific to the goblet cell phenotype.

Thus, one of the conclusions emerged from the above data is the importance of the EGFR signaling to the regulation of *MUC5AC* transcription – a conclusion supported by other studies. Hewson *et al.* [76] showed that EGFR activation is a critical event for *MUC5AC* expression induced by phorbol 12-myristate 13-acetate (PMA) in human bronchial cells. The authors showed that upstream to EGFR phosphorylation, PMA activates novel protein kinase C isoforms, PKCδ and PKCθ, which activate metalloproteinase TACE necessary for processing of TGFα pro-ligand into mature ligand, which in turn activates EGFR. The authors also established that EGFR downstream signaling occurs exclusively *via* the Ras/Raf/MEK/ERK pathway and leads to binding of the Sp1 protein to the specific *cis*-elements located at the -192/-63 region of the *MUC5AC* promoter.

5.2.1.3. Role of Bacteria and Bacterial Toxins in Expression of the MUC5AC Gene

Bacteria and bacterial toxins play a specific role in airway physiology and pathology by inducing alterations in expression of several mucin genes. *Streptococcus pneumoniae (S. pneumoniae)*, the most common bacterial pathogen isolated from patients with otitis media and chronic obstructive pulmonary disease (COPD), has a strong effect on *MUC5AC* expression. Pneumolysin, a cytolytic toxin produced by *S. pneumoniae,* acts as a key virulence factor in respiratory infections and induces alterations in *MUC5AC* transcription [77-80]. Li and co-workers found that this toxin regulates *MUC5AC* gene expression *via* positive ERK1/2 MAPK and negative JNK1/2MAPK signaling pathways [77, 79]. The two regulatory mechanisms were found to have common steps, including interaction of pneumolysin with TLR4 receptor and consequent activation of the MyD88/IRAK/TRAF6 complex. From this point the signaling pathways separate: the positive signaling pathway comprises the Ras-Raf1-MEK1/2-ERK1/2-ATF2-AP-1 chain, while the negative regulation links to the MEKK3-JNKK1/2-JNK1/2-c-Jun-AP-1 cascade. Importantly, transcription factor AP-1 participates in both pathways and serves the opposite functions of activator and inhibitor of *MUC5AC* transcription. This was demonstrated by mutation analysis of two AP1-binding sites in the *MUC5AC* promoter: the distal AP-1 site located at the -3567/-3570 bp sequence mediates the positive regulation of *S. pneumoniae*-induced *MUC5AC* over-expression *via* ERK1/2 MAPK signaling pathways; the proximal AP-1 *cis*-element located at the -3535/-3529 bp sequence mediated the negative regulation of the *MUC5AC* gene *via* the JNK1/2 MAPK signaling pathway [77, 79].

Pseudomonas aeruginosa (*P. aeruginosa*) is another pathogen that induces up-regulation of mucin gene expression. Several components of *P. aeruginosa*, including a variety of proteases, exotoxin A, exoenzyme S and endotoxin LPS, may induce over-expression of the *MUC5AC* and *MUC2* genes [81-84]. The best studied among them in this context is LPS. Dohrman *et al.* [34] reported that both *MUC5AC* and *MUC2* responded to *P. aeruginosa* infection and to purified LPS in different *in vivo* and *in vitro* models in a similar manner. In the case of the *MUC2* gene, LPS activates signaling cascade LPS-LBP-Ras-Raf-1-MEK1/2-ERK1/2-pp90rsk-NF-κB(p50/p65), resulting in binding of NF-κB transcription factor to its cognate *cis*-element located between -1458 and -1430 bp in the *MUC2* promoter. Activation of this cascade up-regulates *MUC2* transcription in airway epithelial cells [37, 38]. Like the *MUC2* promoter, the *MUC5AC* promoter also contains two NF-κB specific binding sites, although which of them participates in LPS-activated signaling is not known and awaits clarification. From the earlier studies it appeared that *MUC5AC* over-expression induced by *P. aeruginosa* and/or LPS was controlled, like in *MUC2*, by the LPS-LBP-Ras-Raf-1-MEK1/2-ERK1/2-pp90rsk-NF-κB(p50/p65) cascade. However, subsequent studies showed that regulation of *MUC5AC* in *P. aeruginosa*-infected cells is more complex. As noted by Yan *et al.* [85], "in contrast to the relatively well-known mechanism by which MUC2 mucin is up-regulated by PA-LPS, the signaling mechanism underlying PA-LPS-induced MUC5AC mucin over-expression is still elusive". Some progress in these studies was made by later analysis that demonstrated the involvement of EGFR in this mechanism [86]. Kohri *et al.* [86] found that both LPS and *P. aeruginosa* supernatant (PA-sup) induced over-expression of *MUC5AC* in parallel with *de novo* EGFR expression and phosphorylation of downstream p44/42 MAPK. Research from this same laboratory showed that activation of TGFα, a ligand of the epidermal growth factor receptor important for activation of EGFR by phosphorylation, precedes LPS- and PA-sup-induced *MUC5AC* over-expression associated with EGFR-MAPK signaling [60]. The authors also documented the critical role that metalloprotease TACE plays in processing of the TGFα-precursor (pro-TGFα), thereby mediating PA-sup- and LPS-induced *MUC5AC* expression by means of TGFα-dependent EGFR phosphorylation and downstream signaling.

Additional progress in understanding the mechanism by which *P. aeruginosa* infection up-regulates *MUC5AC* expression was made by Yan *et al.* [85]. Their study showed that in human epithelial cells NCI- H292, *P. aeruginosa* utilizes reactive oxygen species (ROS) for up-regulation of *MUC5AC* mucin expression. PA-LPS induces production of ROS through PKC-NADPH oxidase signaling pathway. PA-PLS-generated ROS then releases TGFα, which in turn leads to up-regulation of *MUC5AC* expression through the EGFR-MAPK pathway. According to a recent study by Song *et al.* [87], PA-LPS may induce *MUC5AC* gene expression in both an ATP-dependent and ATP-independent manner. It appears that ATP-independent *MUC5AC* up-regulation is associated with the signaling pathways described above. On the contrary, the ATP-dependent *MUC5AC* up-regulation is realized through LPS interaction with TLR4 receptor, followed by extracellular secretion of ATP, which further interacts with epithelial P2Y2 purinergic receptor, leading to *MUC5AC* over-expression. Both ATP and its metabolite adenosine are believed to promote Ca^{2+}-dependent mucin secretion *via* activation of P2Y2 receptor [88]. Taken together, the data suggest that LPS triggers *MUC5AC* gene expression through several pathways, and that ATP-dependent and ATP-independent pathways are equally important for LPS-induced *MUC5AC* expression. In line with the concept that several mechanisms may transmit LPS signaling into *MUC5AC* transcriptional activity, Hauber *et al.* [89] showed that product of the human calcium-activated chloride channel 1 gene, *hCLCA1*, may play a role in LPS-induced *MUC5AC* expression in human airway mucosa.

Human neutrophil elastase (HNE) may also function as an activator of *MUC5AC* gene expression. HNE is present in high concentration in airway secretion of patients with chronic inflammatory airway diseases and is one of the factors that induce goblet cell metaplasia and up-regulation of the mucin genes [90-92]. The mechanism by which neutrophil elastase up-regulates *MUC5AC* expression appears to be similar to that operating in PA-sup and LPS stimulations of the *MUC5AC* gene reported by Yan *et al.* [85]. The similarity is not surprising as neutrophil elastase is a component of PA-sup and may activate the PA-sup-directed pathway in airway epithelial cells. The signaling pathway described by Yan *et al.* [85] consists of the PA-sup-ROS-TACE-TGFα-EGFR-MAPK cascade. Shao and Nadel [93] found that the first reaction in the HNE-activated process is the activation of protein kinase C (PKC), which in turn generates ROS, thereby activating the downstream elements of the signaling cascade. A recent finding that transcription factor Nrf2 down-regulates *MUC5AC* transcription by inhibition of HNE-mediated ROS production [94] indicates that HNE and Nrf2 may function as the reciprocal regulators of *MUC5AC* expression in epithelial cells under oxidative stress conditions.

5.2.1.4. Effect of ROS, Mechanical Stress and NRG1β1 on MUC5AC Gene Expression

One of the phenomena discovered by the above studies is the increased production of MUC5AC mucin in the airway epithelia under ROS-mediated oxidative stress conditions. Jang *et al.* [95] recently identified an important participant in the interplay between mucin genes and ROS which is the tyrosine phosphotase, SHP-1, involved in bronchial mucin production during oxidative stress. SHP-1 is known to be a negative regulator of intercellular signaling [96], and *SHP-1* knockout homozygous mice spontaneously develop a Th2-dominant immune response associated with enhanced production of airway mucins [97]. These findings are consistent with the report that suppression of the SHP-1 function promotes increase in the intracellular ROS concentration, which in turn up-regulates mucin expression [98]. On the other hand, increased intracellular ROS content leads to inhibition of *SHP-1* activity [99, 100]. It is also well known that SHP-1 is involved in regulation of development and survival of different cell types by activation of several signaling pathways, including MAPK, Jak2, EGFR, P13K, and NF-κB [96, 101]. In bronchial cell culture exposed to H_2O_2-mediated oxidative stress [95], the p38 MAPK signaling pathway is involved in MUC5AC mucin hyper-secretion associated with suppression of *SHP-1*. Thus, in the respiratory system, SHP-1 phosphotase plays an important role in mucin production *via* regulation of oxidative stress [95].

Transcriptional up-regulation of some mucin genes is a typical reaction of the airway epithelium not only to chemically active stress stimuli such as LPS, PMA, various toxins and enzymes, but also to mechanical stress. Tschumperlin *et al.* [102] studied the influence of mechanical stress on gene regulation and found that in both mouse airways and human cell culture the compressive mechanical stresses trigger activation of the epidermal growth factor receptor. In a subsequent study, Park and Tschumperlin [103] showed that chronic intermittent mechanical stress induces *MUC5AC* over-expression through activation of EGFR and $TGF\beta_2$ signaling. Interestingly, up-regulation of *MUC5AC* transcription was associated with down-regulation of *FOXA2* expression. This phenomenon is consistent with the EGFR-dependent alterations in expression of *MUC5AC* and *FOXA2* observed in NHBE cells treated with IL-13 [49] and demonstrates the universality of the regulatory mechanisms activated in airway epithelium in response to different stimuli.

Growth factors are the active participants of signaling pathways activated in epithelial cells in the context of mucin over-expression. Recently, a new mediator of *MUC5AC* and *MUC5B* transcription, growth factor neuregulin 1β1 (NRG1β1), was identified [104]. NRG1β1 was shown earlier to operate through Erb receptors that are homologs of EGFR [66]. Kettle *et al.* [104] demonstrated that NRG1β1 regulates the expression of *MUC5AC* and *MUC5B* in HBEC cell culture through interaction with Erb2 and Erb3 receptors, followed by activation of the p38MAPK, ERK1/2 and P13K pathways.

5.2.1.5. Cytokines as Regulators of MUC5AC Gene Activity

Interleukine-1β (IL-1β) is one of the cytokines whose activation is associated with inflammosomes, the molecular platforms activated upon cellular infection or stress. These events trigger the maturation of pro-inflammatory cytokines to engage the innate immune defense mechanism by coordinated inflammatory reactions [105]. IL-1β has been implicated in the pathogenesis of many inflammatory diseases, including asthma [106, 107]. One of the innate immune reactions activated by IL-1β in response to infection or stress is up-regulation of the *MUC5AC* gene transcriptional activity [108]. Several studies elucidate the mechanism by which IL-1β regulates *MUC5AC* expression. Using specific inhibitors, Kim *et al.* [109] established that IL-1β activates MAPK (ERK or p38) pathways, which induce cyclooxygenase-2 (COX-2) production followed by prostaglandin E_2 (PGE_2) synthesis that results in *MUC5AC* over-expression in airway epithelial cells. Further progress in deciphering the regulatory mechanism of *MUC5AC* expression in airway by IL-1β was made by Fujisawa *et al.* [110] who found that the activating effect of both IL-1β and IL-17A cytokines on *MUC5AC* transcription is carried out by utilization of the NF-κB pathway. The importance of the NF-κB transcription factor in airway inflammation, particularly in asthma and COPD, has been well documented [111, 112]. The involvement of NF-κB in *MUC5AC* up-regulation by various stimuli including bacterial lipoproteins and cytokine TNFα has also been shown [113-115]. The significance of Fujisawa *et al.'s* study [110] is in identification of molecular events associated with NFκB participation in direct induction of *MUC5AC* expressions by IL-1β and IL-17A cytokines, potent inducers of MUC5AC mucin over-production. The authors showed that NFκB plays a crucial role in both IL-1β- and IL-17A-induced *MUC5AC* expression in well-differentiated primary NHBE and HBE1 cultured cells. The indispensability of NFκB in this process was evidenced by suppression of *MUC5AC* expression in cells treated with the NFκB-specific inhibitors.

Moreover, specific binding of NFκB subunits to the NFκB *cis*-element located at -3594/-3582 bp in the *MUC5AC* promoter was demonstrated by using ChIP analysis of IL-1β- or IL-17A-treated cells. Importantly, although the *MUC5AC* promoter contains the proximal and distal NFκB binding sites, only the distal one located at -3594/-3582 bp is involved in cytokine-directed *MUC5AC* up-regulation. The authors pointed out that the observed participation of NFκB signaling in IL-1β- and IL-17A-induced activation of the *MUC5AC* gene is consistent with the involvement of the NFκB transcription factor in IL-1β- and IL17A-directed activation of other genes [116-119]. In summary, this study [110] highlights an important function of NFκB as a regulator of the cytokine-induced *MUC5AC* gene expression.

5.2.1.6. CREB: Role in Regulation of MUC5AC Gene Expression

As clearly seen from the above sections, different regions of the *MUC5AC* promoter and different transcription factor binding sites play important and specific roles in *MUC5AC* transcriptional activaty. In addition to the *cis*-elements described above, the binding site specific for the CREB transcription factor, *cis*-CRE, was also found in the *MUC5AC* promoter [120, 121]. This *cis*-element is also of great importance to *MUC5AC* expression. Up-regulation of *MUC5AC* gene transcriptional activity by IL-4 in airway epithelial cells occurred through interaction of the CREB protein with the cognate binding site located at -878/-870 bp in the *MUC5AC* promoter [120]. CREB is known as a cAMP responsive *cis*-CRE binding protein, a powerful effecter of growth factors, hormones, retinoids, cytokines and prostaglandins [121]. Multiple signaling pathways are involved in CREB activation through various up-stream kinases, including PKA, PKC, MAPK-activated protein-2, Akt, p90RSK and mitogen and stress-activated kinase MSK [122-127]. Kim *et al.* [120] found that ERK1/2, but not p38 and JNK signaling, is essential for IL-4-induced *MUC5AC* gene expression. One ofthe downstream participants in this process is RSK1 which activates CREB. The latter directly binds to the *cis*-CRE element in the *MUC5AC* promoter, thereby activating transcription of the *MUC5AC* gene. However, as emphasized by the authors, CREB may also interact with the *MUC5AC* promoter indirectly through formation of a complex with other transcription factors, for instance, with NFκB, as was observed in various experimental systems [128-130]. Such interactions facilitate formation of various transcription-competent complexes, which in turn involve diverse signaling pathways in regulation of *MUC5AC* gene expression [120, 127]. For example, CREB interacts with c-Ets1 protein, resulting in ATP-dependent *MUC5AC* gene expression [131]. ATP was found to increase *MUC5AC* gene transcription in the primary nasal epithelium and pulmonary adenocarcinoma cell line NCI-H292 *via* PLCβ3-mediated activation of the Akt-ERK1/2-p38-RSK1-CREB pathway. Importantly, in these models the transcription factor CREB does not directly bind to the specific CRE-binding site in the *MUC5AC* promoter, but instead forms a complex with the c-Ets1 protein, which, through binding to the Est-specific *cis*-element located at -938 to -930 bp in the *MUC5AC* promoter, activates transcription of the *MUC5AC* gene. The CREB interaction with the c-Ets1 protein results in phosphorylation of the latter, enabling it to participate in ATP-induced *MUC5AC* expression.

Retinoic acid (RA) plays an essential role in the development and normal physiology of airway epithelium [132-135]. While the classical mechanism of RA-mediated signal transduction involves interaction with retinoic acid receptors (RAR) [136, 137], Aggarwal *et al.* [138] found that RA may use also a nonclassical mechanism to regulate gene expression through activation of the CREB protein. It was shown previously that classical RARα has an important function in induction of the expression of the *MUC2, MUC5AC* and *MUC5B* mucin genes [134]. However, Kim *et al.* [139] demonstrated transcriptional up-regulation of these genes *via* a nonclassical retinoic acid signaling pathway by activation of the CREB protein. This pathway includes activation of PKCα, which transmits the activating signal to CREB *via* the Raf/Mek/ERK/RSK cascade. Importantly, PKCα is absolutely indispensable in this signaling pathway as its inhibition completely blocks the RA-induced *MUC5AC* transcriptional up-regulation. The downstream signaling includes translocation of the activated RSK from the cytoplasm to the nucleus, where it phosphorylates CREB, which then binds to a *cis*-acting CRE sequence located at the -878/-871 bp sequence of the *MUC5AC* promoter. Mutation of this CRE motif sequence abolishes the responsiveness of the *MUC5AC* promoter-directed reporter plasmid to RA. Notably, two different regulatory mechanisms, the nonclassical retinoic acid pathway and the IL-4-mediated signaling cascade, activate *MUC5AC* transcription through utilization of the same -878/-871 *cis*-CRE element in the *MUC5AC* promoter. Taken together, these data indicate that signaling pathways activated by different stimuli may compete for the same binding site in the *MUC5AC* promoter. It appears, therefore, that the indicated CRE-site is at a junction of different signaling pathways and therefore plays a critical role in regulating the *MUC5AC* gene expression in response to different agents. Additional evidence of the importance of this CRE *cis*-element for *MUC5AC* transcriptional regulation comes from a study showing that the over-expression of *MUC5AC* induced by prostaglandin $F_{2\alpha}$ is also mediated through binding of the transcription factor CREB to the -878/-871 *cis*-CRE site [140].

5.2.1.7. Tobacco Smoke: Influence on MUC5AC Gene Expression

Tobacco smoke is a mixture of various irritant molecules including acetaldehyde, hydroquinone, formaldehyde, benzo[a]pyrene, cresol, nicotine, catechol, acrolein, coumarin, anthracene, nitrogen oxides, and heavy metals [141, 142]. Each of the components of tobacco smoke has the potential to induce alterations in gene expression, including expression of mucin genes. Some ingredients of tobacco smoke activate expression of a given gene while others inhibit transcription of the same gene. Thus, an alteration in gene expression induced by tobacco smoke is always the sum total of interactions between different pathways activated by different tobacco smoke components. The final outcome may depend on composition and relative concentrations of different smoke components as well as on cell types used for analysis. These factors differ from study to study, making it difficult to compare results.

Tobacco smoke usually up-regulates expression of various transcription factors which, in turn, may activate or suppress the target genes including mucin genes [143]. In experiments conducted by Baginski *et al.* [144], cigarette smoke increased mucin gene expression and mucin production by activation of the same signal transduction pathways as those activated by LPS, TNFα, EGFR and amphiregulin in human airway epithelial cells. Tobacco smoke was also shown to up-regulate *MUC5AC via* TACE, P13K and peroxysome proliferater-activated receptor-γ (PPAR-γ) [29, 145, 146]. However, TGFβ-mediated signaling activated by cigarette smoke suppressed mucus hyper-secretion in the study performed by Baraldo *et al.* [147]. Some reports implicated the EGFR- and Sp1-mediated signaling cascades operating *via* the distal Sp1 binding sites located at the -200/-1 region of the *MUC5AC* promoter in up-regulation of *MUC5AC* gene expression by cigarette smoke [69, 70]. In contrast to these reports, Basbaum's group [142, 148] suggested that expression of *MUC5AC* is controlled mostly by the mechanism dependent on two AP-1-sites located at the -3577/-3571 and -3565/-3553 bp of *MUC5AC* promoter. As established in these studies, stimulation of NADPH oxidase expression is the earliest event induced in smoke-treated cells leading to generation of ROS that in turn activates at least two downstream signaling cascades: one that involves activation of TACE (TNFα-converting enzyme), amphiregulin and EGFR [148], and another one that activates Src and JNK [142]. The induction of MUC5AC mucin expression depends on the cooperation of the two pathways that results in activation of the JunD and Fra-2 transcription factors, which in turn bind to the AP-1 sites and activate transcription of the *MUC5AC* gene.

Mishra *et al.* [149] showed that nicotine, the major constituent of cigarette smoke, significantly increases expression of the Muc5ac mRNA. In parallel, it dramatically decreases expression of Th2 cytokines including IL-4, IL-13 and other chemokines known to be powerful inducers of *MUC5AC* expression [27, 48, 52-55]. It is likely, therefore, that the effects of nicotine on *Muc5ac* are independent of its effects on IL-13/IL-4 expression. The authors consider that the strong interactions between nicotine and GABA receptors observed in neuronal cells [150] implicate GABA receptors in *MUC5AC* up-regulation by nicotine in airway epithelial cells as well. This suggestion is supported by Xiang *et al.'s* finding [151] that a GABAergic system in airway epithelium is essential for mucus overproduction in asthma.

Another component of tobacco smoke, acrolien, also increases expression of rat, mouse and human MUC5AC mRNAs and proteins [42, 152-154]. The proposed mechanism includes MMP-9- and probably also ADAM17-dependent activation of EGFR [155, 156].

Thus, one may conclude that tobacco smoke has multiple activating effects on *MUC5AC* transcription. Further studies are needed to clarify how components of tobacco smoke, alone and in combination, activate *MUC5AC* expression.

5.2.1.8. Role of Leuкotrienes, Aquaporins and Glucocorticoids in MUC5AC Gene Expression

In addition to the factors and stimuli described above, many other agents up-regulate *MUC5AC* expression, including leukotrienes, aquaporins and lipid metabolites [157-161]. Their contributions to *MUC5AC* up-regulation have been partially investigated. Leukotriene D4 was reported to activate *MUC5AC* expression through interaction with leukotriene receptor. It has been shown that the leukotriene receptor antagonist pranlukast hydrate inhibits leukotriene D4-induced *MUC5AC* gene expression [157]. Residual oil fly ash (ROFA), a particulate air pollutant, increases MUC5AC steady state mRNA *via* a phosphotyrosine-dependent pathway [159], while NO up-regulates MUC5AC mRNA *via* PKCα and PKCδ-ERK pathways [160]. An unexpected discovery was recently made by Chen *et al.* [158] who found that aquaporins are involved in regulation of *MUC5AC* expression. The molecular mechanisms of this phenomenon have not been identified, although the connection between *MUC5AC* expression and activity of the aquaporin 5 encoding gene has been established.

In contrast to the agents that up-regulate *MUC5AC* production, the glucocorticoids inhibit *MUC5AC* expression [162-164]. The mechanism is complex and depends on the ligand-activated glucocorticoid receptor (GR) [165]. It is suggested that the GR activated by glucocorticoid is translocated to the nucleus, where it is homodimerized and binds to the GR-responsive elements (*cis*-GRE) in the promoter region of a target gene, thereby affecting gene expression [165, 166]. However, in addition to the *cis*-GR-mediated mechanism of gene regulation, there is also a *trans*-GR-dependent mechanism that is utilized by genes that lack *cis*-GRE sites in their promoters. In these cases, the activated GR regulates a target gene expression through *trans*-interaction with the NF-κB or AP-1 transcription factors [165-167]. Chen *et al.* [164] showed that repression of *MUC5AC* transcriptional activity by dexamethasone is mediated through binding of the GR to the *cis*-GRE sites located at the -930/-912 and -369/-351 bp sequences in the *MUC5AC* promoter. Interestingly, of the five *cis*-GRE sites present in the *MUC5AC* promoter, only two are involved in *MUC5AC* repression in A549 cell line. Komatsu *et al.* [163] established that dexamethasone inhibits non-typeable *Haemophilus influenzae*-induced *MUC5AC* expression *via* the MAPK phospatase-1 (MKP-1)-dependent suppression of p38 MAPK. Dexamethasone up-regulates MKP-1 expression, which in turn leads to p38 dephosphorylation and subsequent inhibition of bacteria-induced *MUC5AC* expression.

In agreement with the findings of Chen *et al.* [164], Martinez-Anton *et al.* [168] observed down-regulation of gel-forming mucin expression, including *MUC5AC*, by prednisone and budesonide in nasal polyps. Interestingly, the same treatment increased expression of the membrane-bound MUC1 and MUC4 mucins.

5.2.2. Epigenetic Regulation of *MUC5AC* Gene Expression

It is becoming increasingly obvious that epigenetic factors play substantial roles in regulating mucin gene expression. DNA methylation and histone H3 lysine-9 modification are involved in the regulation of *MUC1* expression in cancer cells [169]. Methylation of promoter CpG islands in cancer cells contributes also to regulation of another membrane-bound mucin gene, *MUC4* [170]. The gel-forming mucin genes were also shown to be regulated by epigenetic mechanisms. Yamada *et al.* [171] reported that expression of the *MUC2* in pancreatic cancer cells is regulated by histone H3 modification and DNA methylation. There is little information on the involvement of epigenetic mechanisms in the regulation of *MUC5AC*. Nevertheless, methylation of the CpG sequences in the distal region of the *MUC5AC* promoter was detected and implicated in regulation of *MUC5AC* transcription [172]. More studies are needed on the role epigenetic mechanisms play in the regulation of *MUC5AC* expression.

In summary, multiple signal transduction pathways are involved in the regulation of *MUC5AC* expression, which may be affected by various agents including bacteria, toxins, pollutants, enzymes, metabolites and chemicals – and the list is constantly growing. Future studies will delineate the mechanisms regulating *MUC5AC* expression induced by different agents and open the way to the development of new therapeutic approaches aimed at repairing impaired *MUC5AC* gene expression.

5.3. MUC5 mRNA: STABILITY AND ALTERNATIVE SPLICING

The biological molecules regulating stability of the MUC5AC mRNA are of great importance to the expression of *MUC5AC*. Neutrophil elastase (NE), cytokine

TNFα and interleukine IL-8 [92, 154, 173] are some of the agents that upregulate *MUC5AC* expression by increasing the stability of its mRNA. Their biological effects on the stability of MUC5AC mRNA are significant. RNA stability assays demonstrate an NE-induced increase in MUC5AC mRNA half-life from 4.5 hours in resting cells to 14.75 hours in the NE-treated cells [92]. Importantly, a proteolytically active NE is required for induction of MUC5AC mRNA expression. TNFα also induces a 2.5-fold increase in MUC5AC mRNA half-life [154]. The mechanisms involved in the NE- and TNFα-mediated increase of MUC5AC mRNA stability are not known. In contrast, interleukine IL-8 stabilizes the MUC5AC mRNA post-transcriptionally by the classical mechanism that involves binding of the RNA-binding proteins to *cis*-AU-rich sequences present in the 3'-untranslated region (3'-UTR) of the mRNA stabilized [173-175]. Apparently the same mechanism could be utilized by NE and TNFα, as three short potential AU-rich sequences were identified in the MUC5AC 3'-UTR [4, 5]. However, there is no evidence that this mechanism is activated in the NE- and TNFα-treated cells. Although interaction of the RNA-binding proteins with AU-rich sequences of the 3'-UTR is considered as the most studied and best understood mechanism that determines mRNA stability. Other mechanisms, such as "formation of stem-loop structure within the 3'-UTR that serves as protein-binding sites or protein binding within the coding region of mRNAs", have also been described [154]. The presented data show that various complex mechanisms have the potential to ensure and maintain MUC5AC mRNA stability. Future studies are likely to reveal regulatory mechanisms that contribute to MUC5AC mRNA stability and regulation of *MUC5AC* transcriptional activity.

The MUC5AC genomic DNA is located between 5'-UTR and 3'-UTR and comprises 48 exons and 47 introns [176]. The ratio of the exon-coding sequences to intron sequences is 1:4. This region contains multiple classical splice donor and splice acceptor sites that participate in generation of the only known form of the MUC5AC mRNA. There are, however, numerous cryptic donor and acceptor splice sites in the MUC5AC genomic sequence (J. Zaretsky, unpublished results), which, together with the aforementioned classical splice sites, represent enormous potentials for alternative splicing and production of multiple mRNA and protein isoforms. Meanwhile, no additional MUC5AC isoforms have been described. The only gel-forming mucin transcript known to undergo alternative splicing is the

MUC2 primary mRNA [177]. From the standpoint of adaptive evolution, such a situation appears to be biologically unjustified, as it narrows the adaptability of mucin gene products and increases their functional vulnerability. As noted by the Russian writer A. Chekhov, if in the beginning of a play a rifle is hanging on the wall, it should fire a shot at the end of the play. Back to gel-forming mucins, we believe that if evolution has conserved multiple cryptic and noncryptis splice sites in the gel-forming mucin genes, they should be utilized by the splice machinery in the appropriate cells under the appropriate conditions and at the appropriate times. More studies are needed to determine those cells, times and conditions, and to explore the possible role of the alternative splicing mechanism in expressing the potentials embedded in the gel-forming mucin genes, including *MUC5AC*.

5.4. MUC5AC GLYCOPROTEIN: BIOSYNTHESIS AND PROCESSING

5.4.1. MUC5AC Apomucin: Synthesis and Dimerization

Like other gel-forming mucins, MUC5AC apomucin is synthesized in the rough endoplasmic reticulum (ER) of the specialized epithelial goblet cells [178, 179]. It takes only 2 hours for a native MUC5AC mucin glycoprotein to be synthesized, post-translationally modified, packed into secretion vesicles, and, finally, secreted into extracellular space [178] (Fig. **3**). The mucin polypeptide is synthesized within 15-20 minutes. N-glycosylation, C-mannosylation, folding, auto-proteolysis at the C-terminus, and disulfide-linked dimerization occur in the endoplasmic reticulum within the first hour. Then, the partially modified MUC5AC precursor is transferred to the Golgi for final extensive O-glycosylation [180].

Using subcellular fractionation, Asker *et al.* [181] established that MUC5AC apomucin forms dimers in the endoplasmic reticulum in an N-glycosylation-dependent manner similar to that of MUC2 apomucin; and, like with MUC2, inhibition of N-glycosylation prevents transfer of MUC5AC monomers and dimers to the Golgi. The observed similarity in biosynthesis of MUC5AC and MUC2 is not surprising since both mucins demonstrate a high level of sequence homology, especially in the number and positions of the cysteine residues in the "cysteine knot" domain responsible for dimer formation. Despite these similarities, MUC5AC and MUC2 mucins do not form heterodimers when synthesized in the same cell [17, 181]. This suggests that the sequences outside

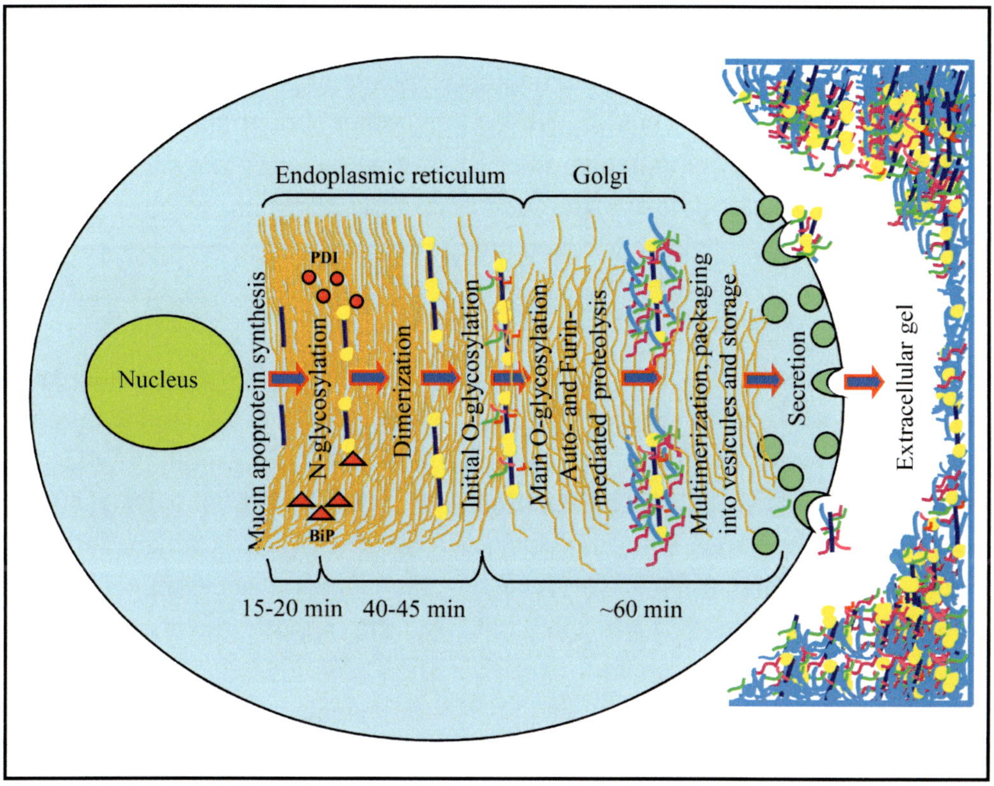

Figure 3: Biosynthesis and processing of the MUC5AC mucin (based on the data reported in [178, 179, 188, 191, 200]).

the "cysteine knot" may be involved in dimerization and may determine specific interactions during dimer formation [181]. Although biosynthesis of MUC5AC and MUC2 have many features in common, it would be wrong to assume that all stages of biosynthesis and maturation of these glycoproteins are identical or similar. As shown by McCool *et al.* [182], at the early stage of biosynthesis (within 5 minutes) MUC2 interacts with the chaperones calreticulin (CRT) and calnexin (CLN), components of the quality-control apparatus that regulate intracellular protein transport. CTR and CLN bind transiently to MUC2 mucin precursors, but once the precursors are completely oligomerized and moved from the endoplasmic reticulum to the Golgi, the chaperons are no longer associated with the mucin. Importantly, interaction of MUC2 with chaperones is an N-glycosylation-dependent process. Dimerization of the MUC5AC mucin is also dependent on N-glycosylation, although no interaction between MUC5AC and either CRT or CLN has been detected. Thus, the immature unfolded molecule of MUC2 likely contains the unique structural elements important for the binding of chaperones, which are absent or shielded by other structural elements in the MUC5AC molecule. Alternatively, other ER-resident chaperones such as BiP and PDI may participate in the maturation and transportation of the MUC5AC precursors from the endoplasmic reticulum to the Golgi.

5.4.2. O-Glycosylation of MUC5AC Precursor

The molecular mass of a MUC5AC polypeptide dimer present in endoplasmic reticulum before O-glycosylation is about 1MDa. Addition of O-glycans to the MUC5AC precursor increases the MUC5AC dimer mass up to 5 MDa, and further polymerization develops gigantic mucin molecules of 50 MDa [180, 183, 184]. Despite the large size and specific structure of the dimers, all sugar acceptor sites on the MUC5AC precursor molecule are accessible to various glycosyltransferases when a nascent MUC5AC polypeptide traverses the Golgi. O-glycosylation is initiated in the Golgi by N-acetyl-galactosaminyl peptidyltransferase-mediated transfer of N-acetylgalactosamine (GaNAc) to serine or threonine residues on the apomucin molecule. GaNAc is the only structural element shared by hundreds of extended and branched O-glycan chains found in mucins. Each GaNAc is then elongated by the stepwise addition of hexoses or sialic acid. More than 30 glycosyltransferases have been identified; at least a dozen of them participate in O-glycosylation of mucin precursors [185]. These glycosyl-transferases attach the multiple O-glycan chains to the polypeptide backbone in a highly specific manner. There is growing evidence that the apomucin amino acid sequences influence the first glycosylation reaction associated with GaNAc-attachment [186-188]. Notably, activity of glycosyltransferases is affected not only by the mucin amino acid sequence, but also by distribution of the neighboring glycan residues [189, 190]. The amino acid motif $X^{2}\underline{TPXP}^{6}$ is thought to serve as a signal for mucin-type O-glycosylation [186, 187]. This motif is present in each MUC5AC tandem repeat unit (GTTPS**P**VP). The P->A mutation in this sequence (GTTPS***A***VP) creates some steric hindrances to the GaNTases' attempt to reach essential Thr/Ser residues in this motif, resulting in a dramatic decrease in the O-glycosylation rate [188]. The proline residues present in the O-glycosylation target site determine molecular conformation of the site, which appears to be of great importance for recognition by GaNTases of the MUC5AC glycosylation sites [191]. Moreover, the ability of different GaNTases to recognize a substrate depends specifically on the GalNac residue(s) already attached to the peptide backbone. For instance, if the GaNTase-T9 requires only one GalNAc residue for recognition and further step-by-step addition of GalNAc residues to the mucine backbone, the GaNTase-T7 is active only when two GalNAc residues are already present on the MUC5AC polypeptide [188].

Mucins expressed in different organs or tissues display distinct patterns of O-linked glycans due to cell- and tissue-specific expression of different glycosyltransferases [192, 193]. Moreover, even in the same cell, the O-glycan profile of a mucin molecule may change in response to different stimuli. Wu *et al.* [194] showed that treatment of human pancreatic MIAPaCa cells with inflammation-specific cytokine TNFα significantly affected glycosylation of MUC5AC mucin. TNFα induced synthesis of mucin molecules with elongated polylactosamine chains associated with the loss of terminal β1,3-linked galactose (Gal) or N-acetylgalactosamine. Interestingly, the inflammation-induced changes in glycosylation are more prominent than the alterations in properties of MUC5AC core protein [194].

Initiation of MUC5AC core protein O-glycosylation takes place in the *cis*-Golgi or in the intermediate compartment located between the endoplasmic reticulum and the Golgi apparatus. O-glycosylation is continued in the *trans*-Golgi-network by elongation of the backbone sugar chain, which is ultimately modified by the termination reactions [195, 196]. Sheehan *et al.* [178] found that the first major O-glycosylation step, addition of N-acetyl-galactosamine to mucin backbone, lasts approximately 1-2 hours, although about 20% of dimers are still non-glycosylated even after 8 hours. Nevertheless, the average time for completion of MUC5AC glycosylation and the secretion of mature molecules into medium is 2-4 hours [178].

Five types of monosaccharides (N-acetylgalactosamine, N-acetylglucosamine, galactose, fucose and sialic acid) are commonly found in the O-glycan chains of human mucins that include MUC5AC. By varying the composition of these carbohydrate residues, the O-glycosylation biosynthetic process yields a wide spectrum of oligosaccharide structures on the mucin backbone [197]. The impact of

O-glycans on physico-chemical properties and functions of mucins is difficult to overestimate. Glycosylation extends and stiffens a mucin molecule [198]. It results in the large volume of mucin in solution, which is important for development of the defense gel [180, 199]. One of the consequences of the conformational changes associated with glycosylation is the ability of a huge polymerized mucin to be packaged into relatively small secretory vesicles [200-202]. The O-glycans on mucin molecules are specific ligands that can interact with the carbohydrate adhesins on bacterial fimbrae, resulting in the trapping of various microorganisms in the mucus layer [203]. Specific glycan structures on MUC5AC corresponding to the blood-specific antigens Le^b and sialyl-Le^x act as ligands for the bacterium *Helicobacter pylori* [204, 205]. The MUC5AC-specific O-glycans undergo alterations and play specific roles in diseases such as cancer and inflammation [206].

In addition to N- and O-glycosylation, the third type of protein glycosylation, C-mannosylation, is also associated with the MUC5AC glycoprotein maturation process. C-mannose residues are attached to the C-mannosylation acceptor WXXW motifs [207] present on the N-terminal side of all nine Cys-subdomains. C-mannosylation of MUC5AC mucin occurs within 5 minutes of the mucin biosynthesis and takes place in the endoplasmic reticulum. Nonmannosylated mucin was shown to be retained in ER and not secreted from the cell. C-mannosylation seems to be required during the early stages of mucin biosynthesis, either for the folding of the domains or for the transportation of the precursor molecule from the endoplasmic reticulum [202].

5.4.3. Secretion of MUC5AC Glycoprotein

As noted above, the mature, fully glycosylated and polymerized mucin is a large anionic polymer of 50 MDa in mass and up to 10 μm in length [178]. This huge polymeric structure has to be packed into the secretory vesicle measuring 1μm in diameter. The mechanism of packing is not well understood although many efforts have been made to study the process [200, 201, 208-211]. In a study of MUC5AC packing [200], the mucin vesicle lumen was found to be divided into two phases: a mobile phase and an immobile matrix. In the mobile phase, mucin polymers are able to diffuse, albeit very slowly. In the immobile matrix, mucin molecules are retained by a pH-dependent mechanism. The incorporation of mucin from a mobile into an immobile, osmotically inert phase permits packing of a large amount of polymer molecules. The proposed mechanism suggests the presence in the vesicle compartments not only of mucin molecules, but of other proteins, lipids and ion molecules as well, which provide secretory vesicles with several additional functions important for intravesicle formation of disulfide bonds between D- domains [212, 213], proteolytic processing [214], and even terminal glycosylation [200]. Another function critical for regulation of vesicle exocytosis is associated with intra-luminal Ca^{2+}, which is present in the mobile phase [215]. Recent studies [45, 216] showed that mucin secretion is regulated separately from mucin production, principally by extracellular triphosphate nucleotides (NTP) such as ATP and UTP. Binding of NTPs to purino-receptors P2Y on the luminal surface of airway secretory cells generates intracellular second messengers, which in turn activate the exocytic proteins Munc13-2 and synaptotagmin-2 [45, 216-221]. Two pathways specific for exocytic mucin secretion from goblet cells have been described: one a baseline mucin secretory pathway regulated by Munc13-2 protein [217], and the other a synaptotagmin-2-dependent pathway responsible for the regulated secretion [216, 222].

5.4.4. Auto-Cleavage of MUC5AC Mucin Glycoprotein

Proteolytic cleavage is an important step in the transition of a precursor mucin polypeptide molecule to a mature protein [214, 223-226]. Proteolytic processing has a role in mucin packing into secretory vesicles, as it affects macromolecular structure of mucin polymers, thereby determining the physical properties of mucus gel [180]. Davis *et al.* [227] found that the MUC5AC mucin in airway secretion from CF patients is present in two states, soluble and insoluble, and suggested that the soluble fraction represents a product of proteolytical modification of the precursor mucin molecules, whereas the insoluble fraction consists of nonmodified molecules. The ability of MUC5AC to undergo proteolytic processing was later confirmed by Lidell and Hansson [226] who showed that MUC5AC is auto-cleaved at GDPH sequence located in the D4-domain of the human MUC5AC mucin molecule. The cleavage disrupts the bond between Asp and Pro in the GDPH (Gly-Asp-Pro-His) sequence. Earlier, these authors observed auto-cleavage of the human MUC2 mucin at the same GDPH site [214]. MUC2 cleavage is a relatively slow nonenzymatic process triggered by a pH below 6, which takes place in the late regions of the secretory pathway. In contrast, cleavage of the MUC5AC molecule is initiated in the neutral endoplasmic reticulum. However, this process is accelerated at a pH below 6.5, typical of the late secretary pathway compartments. Pulse-chase experiments established that cleavage occurs 30 minutes after the beginning of apomucin synthesis and remains at the same level through the chase time. It is notable that the amount of a cleaved protein seems to be relatively higher in the medium than in the intracellular compartments, suggesting that further cleavage might occur after the protein has passed the ER, and even in the secreted vesicles [214]. The cleavage generates a reactive group at the new C-terminus that could link the modified MUC5AC polypeptide to a primary amine with the potential to form cross-links with itself and/or other molecules [214, 226]. The cleavage may be important for maturation of a mucin molecule both in physiological conditions, as well as in pathological conditions characterized by changes in the pH within cells or at the epithelial surfaces that may result in dramatic aberrations of the mucus gel [180].

5.5. EXPRESSION OF *MUC5AC* GENE UNDER PHYSIOLOGICAL CONDITIONS

Gel-forming mucins play an important role in different stages of cell and organism life, including embryogenesis, fetal development and adulthood. What is known about the expression of the *MUC5AC* gene in different organs under physiological conditions – a subject that has not been thoroughly studied – is described in this section.

5.5.1. Expression of *MUC5AC* Gene in the Respiratory System

Little information is available about the relationship between the expression of *MUC5AC* and the level of cytodifferentiation of human airway epithelial cells. Nevertheless, a study performed with the rat analog of the human *MUC5AC* gene suggests that expression of this gene correlates with mucous differentiation [228]. Reid *et al.* [229] examined expression of several mucin genes, including *MUC5AC*, in human fetal tissues during respiratory tract development, and demonstrated that MUC5AC, MUC4 and MUC5B mucins are the major components of the airway mucus in the developing human lung. The pattern of *MUC5AC* expression resembled that of *MUC2*, being restricted to individual goblet cells in the airway epithelium. According to Reid *et al.* [229], MUC5AC mRNA was not detected at the 13th week of gestation, but by weeks 17 and 18 it was expressed in tracheal goblet cells. Notably, although the goblet cells continued to express MUC5AC at week 23 of gestation, its expression was lower than at week 18. MUC5AC mRNA was not detected in the small bronchioles or in distal parts of the lung.

These data differ to some extent from those of Buisine *et al.* [230], who also studied mucin gene expression in the embryonic, fetal, and adult tissues of the human respiratory tract. The first expression of the MUC5AC mRNA was detected in their study in the surface epithelium of lobular bronchi at 13 weeks of gestation. At 18 weeks the MUC5AC mRNA expression was seen in cells of the main bronchi, but not in the trachea, lobular bronchi, bronchioles or terminal sacs. The expression of *MUC5AC* was first detected in trachea and lobular bronchi at 23 weeks of gestation. At 26 weeks the amount of MUC5AC mRNA was decreased to a barely detectable level and returned to the substantial expression at 27 weeks. In adult lungs, high level of *MUC5AC* expression was observed in goblet cells of trachea, main and lobular bronchi. In bronchioles and alveoles, the MUC5AC mRNA was not detected at any stage of embryonic and fetal development, and not in adult lungs. It should be pointed out that although the patterns of *MUC5AC* expression generally resembled those of *MUC2*, the expression of the latter was detectable as early as 9.5 weeks of gestation and could be constantly measured during the rest of the embryonic, fetal and adult stages of lung development in practically all lung compartments except terminal sacs and alveoles. As stressed by Buisine *et al.* [230], during embryogenesis, *MUC5AC* expression was restricted to the epithelial folds and gland ducts, and was detected much more rarely in goblet cells. However, in adults, MUC5AC mRNA was found in all goblet cells of the surface epithelium as well as in gland ducts. Notably, *MUC5AC* was not expressed in submucosal glands at any stage of development or adulthood. The general conclusion derived from these laboratory studies is that MUC5AC mucin is expressed in the adult respiratory tract mainly by goblet cells. Together with MUC5B, it represents the main component of airway mucin secretion [179, 229-232].

5.5.2. Expression of *MUC5AC* Gene in the Gastrointestinal System

Comprehensive studies of the gel-forming mucin expression in human embryonic and fetal gastrointestinal tract, including accessory glands such as pancreas, liver and gallbladder, were carried out by Buisine and colleagues [233-235]. They found that a moderate level of MUC5AC mRNA expression can be detected already at 8 weeks of gestation in the primitive gut, including embryonic stomach region destined to develop into the antrum. However, after 8 weeks, MUC5AC mRNAs are inconsistently found only in clusters of epithelial cells, both in villi and crypts. At 12 weeks, the *MUC5AC* gene is expressed in the ileum but not in the colon. Importantly, after 13 weeks of gestation, expression of the *MUC5AC* gene is repressed and MUC5AC mRNA is not detected in any region of the intestine, although it is expressed in stomach. Interestingly, MUC5AC mRNA expression in stomach is observed at 10.5 weeks of gestation in epithelium of both the antrum and fundus, with much stronger intensity of expression in the antrum. After 10.5 weeks of gestation, *MUC5AC* expression in gastric epithelium is homogenous and of moderate intensity, irrespective of the region. From week 18, the MUC5AC mRNA could be constantly detected in the surface epithelium of developing and adult stomach, although it could not be detected in stomach glands. Taking into account that the rudiments of gastric pits develop in the fundus by 9-10 weeks of gestation, and in the antrum and cardia by 12 weeks, and mucus glands appear at 13-14 weeks of gestation [236, 237], it seems that expression of *MUC5AC* in the gastrointestinal tract is induced before the occurrence of epithelial cytodifferentiation. These data demonstrate the dependence of *MUC5AC* expression on the stage of stomach development. According to Buisine *et al.* [235], the MUC5AC mucin may play a role in gastroduodenal mucus gland formation.

In duodenum, expression of the MUC5AC mRNA is restricted to crypts and is not detected before 18 and after 26 weeks of gestation [235]. There is no information about expression of the *MUC5AC* gene between these two time points, but it is known that it is not expressed in developing duodenum at any other gestational age, or in the adult organ.

In embryonic liver, *MUC5AC* expression is not detected in hepatoblasts or primordial hepatocytes at any gestational age. Epithelial cells of intrahepatic bile ducts also do not express any mucin genes. The *MUC5AC* gene is generally not transcribed in the adult liver, but some specimens exhibit a low expression of *MUC5AC* in epithelial cells of intrahepatic bile ducts [235].

Like in liver, the expression of *MUC5AC* in pancreas appears to be strongly repressed [235]. The gallbladder is the only organ of the pancreato-hepato-biliary system in which expression of the *MUC5AC* gene is observed. The MUC5AC mRNA has been detected in the surface epithelium and epithelial folds of gallbladder since the 18th week of gestation. It is also expressed in the adult organ [235].

5.5.3. Expression of *MUC5AC* Gene in the Male Urogenital Tract

Discussion of the expression of mucin genes in the urogenital tract has to be prefaced by a caveat: while mucins of the female reproductive tract have been relatively well studied [238-241], little is known about the mucin expression profiles in the male reproductive organs and in those regions of urine-producing and urine-excreting systems (kidney, ureters and bladder) that are common to both genders [242].

Kidney: Renal development is a complex process that includes expression of more than 300 genes [243-245]. Some mucin genes, including *MUC1, MUC3* and *MUC6*, are expressed in normal fetal and/or adult kidneys, while *MUC5AC* expression has never been detected in these organs [243, 244, 246].

Ureters, bladder and urethra: The data concerning mucin expression in upper and lower urinary tract are controversial. According to the study performed by N'Dow *et al.* [242], epithelium of urethra and urinary bladder does not express *MUC5AC*. Jankovic Velickovic and colleagues [247] and Kunze and collaborators [248] confirm these data. In contrast to these authors, Russo *et al.* [249] demonstrated strong expression of a battery of mucin genes, including *MUC5AC*, in bladder urothelium. These investigators emphasized that *MUC5AC* was expressed in apical cells of the bladder urothelium. These cells were distributed in clusters and were "filled with the mucin". In urethral epithelium, expression of *MUC5AC* was localized to the glands of Littre present in the *lamina propria* of the urethra. The authors stressed that *MUC5AC* was the only gel-forming mucin actively expressed in bladder and urethra. It is difficult to compare the cited conflicting results as they were obtained in nonhomogenous populations of patients and by different methods and antibodies.

Foreskin: The expression of MUC5AC mRNA in the foreskin was documented by Russo *et al.* [249] in the only research found by us in the literature on expression of *MUC5AC* in the foreskin. The results were obtained on a limited number of species and require confirmation.

Testis and vas deferens: In addition to the analysis of mucin gene expression in the urinary tract, Rosso *et al.* [249] also studied the activity of mucin genes in the male reproductive organs. They found that all "testicular tissues", as well as the cells lining the *vas deferens*, expressed both membrane-bound and gel-forming mucins, including MUC5AC mucin. However, the authors did not specify the cells of the "testicular tissues" – which normally contain Sertoli and Leydig cells, germ cells, macrophages, mastocytes, blood vessel cells and nervous cells – expressing MUC5AC. It should be noted, that the data on mucin expression in *vas deferens* were based on only one sample tested. More statistically validated data are needed for objective evaluation of mucin gene expression in the testis.

Epididymis and seminal vesicle: Expression of the *MUC5AC* gene in the epididymis, seminal vesicle and seminal plasma has not been detected to date, although other mucin genes that were expressed concomitantly with *MUC5AC* in "testicular tissues" and *vas deferens* demonstrated high levels of expression in these organs [249].

Prostate: Russo *et al.* [249] reported that MUC5AC is the only gel-forming mucin expressed in prostate. In contrast, Cozzi *et al.* [250] and Zhang *et al.* [251] did not find evidence of the MUC5AC mucin expression in normal prostate tissues. Cozzi *et al.* [250] also could not detect *MUC5AC* expression in benign and malignant prostate tumors, whereas other investigators reported expression of the MUC5AC antigen in 9%-11% of primary and metastatic prostate adenocarcinomas [251, 252].

In summary, there is baseline information on mucin gene expression profiles in the male urogenital tract epithelia, but more systematic and statistically reliable studies are needed to clarify the expression of mucin genes, including *MUC5AC*, in embryonic, fetal and adult male urogenital tissues.

5.5.4. Expression of *MUC5AC* Gene in the Female Reproductive System

Mucus plays an important role in reproductive function and defense of the female reproductive tract [239, 253]. Qualitative and quantitative characteristics of the expressed mucins influence the properties of the mucin gel that provides a protective covering for the female reproductive ways and regulates sperm penetration into the uterus. Alterations in the mucus content are related to the physiological changes associated with steroid hormone status during the menstrual cycle, as well as to pathological conditions such as infections and malignant processes [254-256]. Different regions of the female reproductive tract respond to steroid hormones in a specific manner, resulting in a specific profile of mucin expression in each region. Surprisingly, there are no publications (at least in the PubMed Database) on the expression of the *MUC5AC* gene in the embryonic and fetal female reproductive organs, in contrast to numerous reports on the expression of this gene in adult reproductive organs [253, 254, 256-274]. The gene is highly expressed in some organs and tissues of the female reproductive tract [253, 257, 258, 268, 269], and suppressed in others [253].

Vagina and ectocervix: The mucus defense of the normal human vagina and the ectocervical part of the uterus is carried out mainly by two membrane-bound mucins, MUC1 and MUC4 [253, 257, 258]. Expression of gel-forming mucins in vaginal epithelium has not been reported.

Uterine endocervix: A number of reports on expression of mucin genes in uterus have been published [253, 259, 260, 263, 264, 275]. Cervical mucins play a critical role in reproductive physiology. They determine the rheological properties of the cervical mucus, thereby regulating interaction between spermatozoa and ova [259, 260, 263, 264, 275]. Interestingly, different regions of this organ exhibit different sets of the expressed mucins. While the ectocervix does not express any gel-forming mucins, the endocervical part of the uterus expresses at least four secreted mucins: three gel-forming mucins, MUC5AC, MUC5B, MUC6, and one soluble mucin, MUC8 [239, 253]. Importantly, while the expression of *MUC5AC* and *MUC6* is detectable mainly by Northern blot analysis and *in situ* hybridization, the expression of *MUC4* and *MUC5B* genes could be observed at protein level [256, 266]. In addition to the indicated gel-forming mucins, Audie *et al.* [264] found that human endocervix also expresses MUC2 mucin. Thus, practically all gel-forming mucins are expressed in this region of the female reproductive tract. Interestingly, Muc5ac mucin is not expressed in mouse reproductive organs [267]. In contrast, not only does normal human endocervival epithelium express the MUC5AC mucin, but all kinds of dysplasia and adenocarcinomas that develop in this location are MUC5AC-positive [257]. However, as noted by Riethdorf *et al.* [268], the expression of *MUC5AC* is diminished in most neoplastic glandular lesions, a tendency that was also observed by Zhao *et al.* [258]. The findings of different studies with regard to *MUC5AC* expression in endocervix are not uniform. According to Zhao *et al.* [258], only surface epithelium of the endocervix expresses MUC5AC, while the crypt epithelium is mostly MUC5AC-negative. In contrast, Audie *et al.* [269] and Riethdorf *et al.* [268] found equal expression of MUC5AC both in the surface and glandular epithelia of the endocervix.

Although mucin expression in the female reproductive epithelium is known to be sensitive to estrogen/progesteron ratio alterations, this phenomenon was not observed in studies performed by Audie *et al.* [264] and Gipson *et al.* [253]: expression of *MUC5AC*, *MUC5B* and *MUC2* in endocervix was not influenced by hormonal status during the ovarian cycle. The insensitivity of *MUC5AC* expression to alterations of estrogen and progesterone concentrations was observed also in the ocular surface epithelia [267].

<u>*Endometrium*</u>: The expression of MUC5AC in menstrual, secretory and proliferative endometrium, examined by Gipson *et al.* [253], was not detected by either Northern blot analysis, *in situ* hybridisation or immunohistochemistry, whereas Alameda *et al.* [241] did detect focal expression of the MUC5AC mucin in about 13% of samples of post-menstrual hyperplastic endometrium by immunohistochemistry. Interestingly, MUC5AC mucin expression was detected by Baker *et al.* [257] in endometrial carcinomas in 44% of studied samples. Almost the same frequency of *MUC5AC* expression in endometrial carcinomas was observed by Zhao *et al.* [258] and Riethdorf *et al.* [268], while the rate reported by Alameda *et al.* [241] was higher (61.7%). The data suggest that MUC5AC mucin may carry out different functions in normal and malignant tissues of endocervix and endometrium.

<u>*Fallopian tubes*</u>: The expression of a number of mucin genes, including *MUC5AC*, in normal fallopian tubes has been analyzed, with no finding of *MUC5AC* expression [253, 268]. It appears that among secreted mucins, MUC9 is the only one expressed in normal fallopian tubes [270, 271].

<u>*Ovary*</u>: Normal ovaries do not contain goblet or glandular cells, and, therefore, as noted by Giuntoli *et al.* [272], normal ovarian tissue would not be expected to express secreted mucin genes. In agreement with this assumption, Ho *et al.* [273] and Giuntoli *et al.* [272] could not detect expression of the *MUC5AC*, *MUC5B* or

MUC2 genes in these tissues. Surprisingly, spontaneously transformed, but nonmalignant, ovarian epithelial cell lines derived from normal ovarian epithelium did express two mucin genes, *MUC1* and *MUC5AC*, whereas *MUC2, MUC3, MUC4* and *MUC5B* were not expressed [272]. Moreover, benign and borderline ovarian tumors, which are histologically closer to normal ovarian tissue than to malignant tumors, exhibited intense expression of *MUC5AC* in 86% of samples tested [274].

In summary, the endocervix is the only part of the female reproductive tract that expresses MUC5AC under normal conditions, whereas the other regions are MUC5AC-negative. Apparently, malignant transformation of the endocervical epithelium correlates with a decrease in expression of the *MUC5AC* gene, while the same processes in other regions of the female reproductive tract are associated with *de novo* synthesis of MUC5AC mucin. The role of different cell types in the regulation of *MUC5AC* gene activity, and the bivalent role of MUC5AC mucin in cell malignization will be discussed in Chapter 12.

5.6. EXPRESSION OF *MUC5AC* GENE IN PATHOLOGY

The development of pathological processes in epithelial cells and tissues is often associated with dramatic alterations in mucin gene expression. This phenomenon enables the use of the mucin gene products (mRNAs and/or antigenic determinants) as diagnostic markers for specific types of pathology. The expression of the *MUC5AC* gene in different organs, tissues and cells under pathological conditions and the role of MUC5AC mucin in the development of various diseases are discussed in the following sections.

5.6.1. Expression of *MUC5AC* Gene in Pathology of the Respiratory System

The MUC5AC mucin is expressed in normal respiratory epithelium exclusively by mucus-secreting goblet cells [179, 229, 230, 232, 276-278]. Multiple studies have shown that under various pathological conditions, such as malignant tumors or inflammatory diseases of the respiratory system, changes in expression level and/or distribution profile of the MUC5AC mucin can be exploited for diagnostic purposes [42, 279-281].

Malignant tumors: Early on, *in vitro* studies showed that many airway carcinoma cell lines express MUC5AC mRNA, at different levels in different cell lines [282, 283]. Some carcinoma cell lines (Calu-3, NCI-H292 and A-549) produce substantial amounts of MUC5AC transcripts while others (Calu-6, RPMI2650 and A-427) are MUC5AC-negative [284]. The levels of *MUC5AC* expression also differ *in vivo* in different types of airway tumors [276, 285-290]. An interesting dynamics of *MUC5AC* expression was observed by Lopez-Ferrer *et al.* [289] in bronchial epithelial cells differently affected by malignant process: 100% of bronchial carcinoma cells expressed MUC5AC mucin; 94% of cells located distal to bronchial carcinoma were MUC5AC-positive (Fig. **4**); and 66% of cells in the peritumoral epithelium were positive in regard with MUC5AC expression.

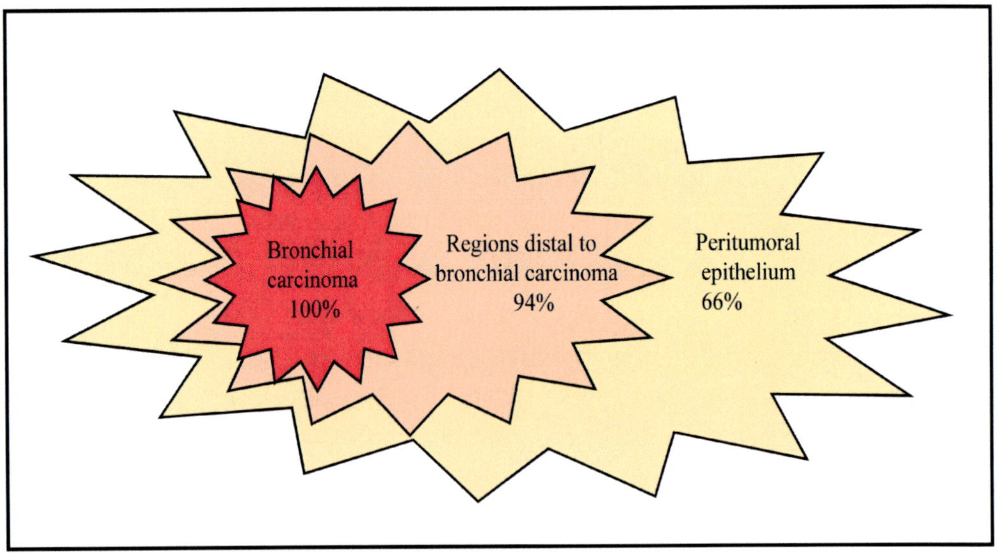

Figure 4: Expression of the *MUC5AC* gene in malignant bronchial tissues (the data extracted from [289]).

Progressively smaller percentages of MUC5AC-expressing cells were detected by the authors [289] in squamous metaplastic lesions (50%), squamous cell carcinoma (29%), adenocarcinoma (15%), and small cell lung carcinoma (9%) (Fig. **5**). Higher expression levels of *MUC5AC* in adenocarcinoma compared with squamous cell carcinoma were reported by Copin *et al.* [276] and Brambilla *et al.* [291]. Yu *et al.* [285] found a high level of *MUC5AC* expression in non-small cell lung carcinoma, and a correlation between the level of the *MUC5AC* gene expression and occurrence of early post-operative metastasis.

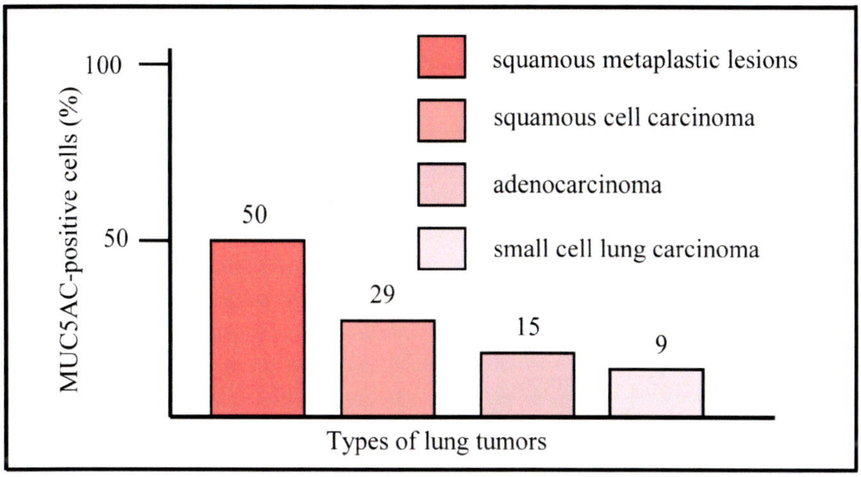

Figure 5: MUC5AC expression in lung tumors (the data extracted from [289]).

The different levels of *MUC5AC* expression observed in different types of lung cancer are not surprising in view of the great diversity of morphological patterns in primary lung adenocarcinomas, both in the different subtypes (acinar, papillary, bronchoalveolar, solid with mucin formation, and mixed types) and within the same tumor [288, 291]. Bronchoalveolar carcinoma (BAC) is a particular subset of adenocarcinomas associated with different histological subtypes: mucinous, non-mucinous and mixed. Mucinous BAC (m-BAC) is of special interest. As reported by Copin *et al.* [288], the level of expression of the *MUC5AC* gene in all m-BAC cases was high and the pattern diffuse while no MUC5AC mRNAs were detected in any of the non-mucinous BAC (n-BAC) cases. The constant high level of *MUC5AC* expression in m-BAC is associated with a well-differentiated phenotype similar to that of the goblet cell in normal airway epithelium. This type of tumor has a non-invasive growth pattern and better prognosis than n-BAC [287]. In contrast, poorly differentiated adenocarcinomas such as n-BAC, epidermoid metaplasia and epidermoid carcinomas do not express *MUC5AC* [288].

To better understand the role of mucins, including MUC5AC, in malignant transformation of airway epithelial cells, Awaya *et al.* [290] examined the expression of *MUC1*, *MUC2*, *MUC5AC* and *MUC6* genes in atypical adenomatous hyperplasia (AAH), m-BAC, n-BAC, and mixed adenocarcinoma (MX). AAH is considered a premalignant precursor of BAC [288, 291-295]. These authors showed that the levels of *MUC5AC*, *MUC2* and *MUC6* expression increase significantly during progression from AAH through BAC to MX, while expression of *MUC1* decreases. Alterations in the expression of these mucins were associated with de-differentiation of bronchial epithelium. Comparison of these data with those reported by Copin *et al.* [287, 288] discloses a major discrepancy between the studies. The reason is not obvious. It may be related to the use of different methods and different antibodies in different laboratories, or different sources of experimental materials. More studies are needed to better understand the mechanisms of lung carcinogenesis and the role MUC5AC and other mucins play in this process, particularly in the development of m-BAC.

Gel-forming mucins are also expressed by several other types of lung carcinoma. Mucin-producing carcinomas of the lung can be subdivided into m-BAC, solid adenocarcinoma (SA) with mucin production, mucinous ("colloid") adenocarcinoma, mucinous cystadenocarcinoma, signet-ring cell carcinoma (SRCC), and mucoepidermoid carcinoma [296, 297]. Three of these tumors –SRCC, SA and m-BAC – show increased amounts of mucins in their cytoplasm. At the same time they display different biological and clinical properties, associated, in particular, with differences in *MUC5AC* expression [296]. SRCC and SA exhibit low rates of *MUC5AC* expression (25.5% and 21.1%, respectively), while m-BAC has very high frequency of this mucin expression (94.7%). Immunohistochemical scoring and hierarchical clustering demonstrate the similarity of the immunophenotypes of SRCC and SA and their difference from the m-BAC phenotype. SRCC and SA have the alveolar lining cell phenotype, whereas m-BAC is clustered with gastric foveolar cells and bronchial goblet cells. Together, these data indicate that the histogenesis of m-BAC differs from that of SRCC and SA. Castillo *et al.* [298] recently reported that SRCCs of sinonasal tract are highly positive for MUC5AC mucin, a finding that contradicts the results of Tsuta *et al.* [296]. Comparison of these two studies is complicated by the fact that Castillo *et al.* [298] examined only 5 cases of SRCC while Tsuta *et al.* [296] studied 40 SRCC tumor samples. Moreover, the locations of the tumors studied by the two groups were different which may also contribute to difference in gene expression. Studies based on larger numbers of tumors are required to explain the observed discrepancies.

Inflammatory diseases: Numerous studies [44, 180, 299] have highlighted the critical functions airway mucins perform in defending the respiratory tract against pathogens and environmental challenges. Under normal physiological conditions, secreted mucins, in particular the MUC5AC and MUC5B mucins, define the rheological properties of mucus gel. Over-production of mucins under pathological conditions is an important factor in the morbidity and mortality of such chronic airway diseases as asthma, CF and COPD. There have been numerous studies linking changes in mucin gene expression, including *MUC5AC*, with respiratory diseases [18, 300-303]. While there are few mucus-producing goblet cells distal to trachea in healthy individuals [43], the numbers of cells producing excessive amounts of mucus is dramatically increased in asthma and COPD [33]. Less than 5% of airway epithelial cells are goblet cells in healthy individuals, and the percentage rises to 25% in patients with fatal asthma [304]. Post-mortem analysis of lung sections from patients with fatal asthma showed a 20-30 fold increase in amount of goblet cells compared with those from control subjects without asthma [305]. This rapid and dramatic increase in amount of goblet cells is termed "mucous metaplasia" [44, 45, 306]. Using mouse allergic model, Young *et al.* [40] reported selective up-regulation of *MUC5AC* expression in metaplastic airways of antigen-challenged mice, demonstrating a central role of the gene in mucous metaplasia. Expression of *MUC5AC* is tightly regulated by a complex interaction of activator and repressor sites in the *MUC5AC* promoter. Activation and suppression of the *MUC5AC* transcription in secretory goblet cells in the bronchial airways in response to tissue stress/damage signals is specifically regulated by cytokines [40, 45].

MUC5AC was shown to be a principal gel-forming mucin up-regulated in airway inflammation, in particular in asthma [45, 107, 301, 302]. Voynow [299] reported that the same mucins found in normal airway secretions (MUC5AC and MUC5B) are present in the asthmatic airway discharge in substantially increased concentrations. Although both *MUC5AC* and *MUC5B* can be up-regulated by inflammatory mediators, patients with asthma tend to produce MUC5AC rich mucus. As suggested by Lai and Rogers [107], this phenomenon occurs "because the mediators involved activate signal pathways that favor *MUC5AC* gene expression". It has been reported that even mild asthmatics have goblet cell hyperplasia associated with increased expression of MUC5AC mRNA and protein [302]. The over-expression of *MUC5AC* and mucus hyper-secretion related to goblet cell hyperplasia appears to be linked to disease severity in asthma, leading to death by mucus plugging of the airway lumen.

COPD is currently defined as a disease of airflow limitation associated with an abnormal inflammatory response of the lungs to noxious factors or gases, including cigarette smoke and its noxious components, which are considered important triggers of COPD pathogenesis [307, 308]. Although different factors cause asthma and COPD, both diseases show the increase in goblet cell population and mucus production associated with MUC5AC mucin over-expression [33]. Indeed, the progression of COPD and chronic bronchitis has been found to be strongly associated with accumulation of mucus in the lumen of the small bronchioles [309]. The selective increase in *MUC5AC* expression was reported as the first specific change in mucin expression in the peripheral airways of COPD patients [307]. It is noteworthy that in addition to *MUC5AC*, *MUC5B* is also over-expressed in COPD [310-312], and some authors consider MUC5B to be the predominant form of mucins over-expressed in COPD [33, 311]. Agents that activate expression of the *MUC5AC* gene in COPD patients include, among others, cigarette smoke, neutrophil elastase, IL-13, EGF and TNFα, as well as components of Gram-positive and Gram-negative bacteria, all known to be involved in the pathogenesis of chronic lung inflammation [32, 34, 68, 92, 154, 312, 313].

The respiratory form of CF is associated with initial mucus obstruction of small airways, with progression to obstruction of large airways [299, 314, 315]. CF is characterized by mucus colonization with pathogenic bacteria and altered posttranslational glycosylation of mucins. Interestingly, the lower respiratory tract is colonized very early in the course of the disease, while the upper regions represented by nasal epithelium remain free of bacteria and are not involved in the pathological process for a long period of time [299]. The role of mucins, particularly MUC5AC, in the pathogenesis of CF is unclear. Comparison of mucin expression in non-affected nasal epithelium and lower airways from CF patients performed by Voynow *et al.* [316] showed an increased expression of MUC5AC compared with MUC1 and MUC2 mucins. Martinez-Anton *et al.* [317], on the other hand, could not detect any difference in *MUC5AC* expression between normal nasal epithelium and nasal epithelium affected by CF. As noted by Copin *et al.* [315] it has not been ever evidenced that CF results in mucin hyper-secretion although goblet cell hyperplasia was demonstrated in CF airways. Still other studies show that the amount of MUC5AC and MUC5B mucins is significantly reduced in CF airways sputum compared with normal airway mucus, most dramatically for MUC5AC [227, 303, 318]. However, Dohrman *et al.* [34] observed up-regulation of MUC5AC mRNA in both human nasal polyp and bronchial tissues of CF patients. Clearly, the role of MUC5AC mucin in CF pathogenesis has not been unequivocally established. Taking into account that airway CF is one of the high death rate pathologies, more studies are needed to clarify the role of MUC5AC mucin in development of this disease.

5.6.2. Expression of *MUC5AC* Gene in Pathology of the Digestive System

Pathological processes in the digestive system may lead to malignant diseases such as cancer of esophagus, stomach, colon, pancreas, gallbladder and liver. Another group of serious chronic diseases of the gastrointestinal tract includes gastritis and stomach ulcer, gallbladder stone disease, chlecystitis, ulcerative collitis and Chron's disease. Development of these types of pathology involves alterations in expression of mucin genes. In some diseases, such alterations serve defensive functions with regard to epithelial cells, whereas in other types they facilitate pathogenic processes. In the following sections, the role of the *MUC5AC* gene and its products in the pathogenesis of various diseases of the gastrointestinal tract is discussed.

<u>*Esophagus*</u>: The esophgeal mucosa is protected by mucus gel consisting of several substances, the main ones of which are mucins produced by the salivary and submucosal glands [319]. Normal esophageal epithelium produces two membrane-bound mucins, MUC1 and MUC4, and one gel-forming mucin, MUC5B [320]. Other gel-forming mucins, namely MUC2, MUC5AC and MUC6, are not expressed by normal esophageal mucosa [269, 320, 321]. The changes in mucin expression profile are specific to different pathological conditions; of these,

Barrett's metaplasia and esophageal adenocarcinoma are the most extensively investigated, however, the results obtained are controversial.

Barrett's esophagus (BE) is a premalignant condition predisposing to esophageal adenocarcinoma. It is defined as a change arising from normal multilayered esophageal epithelium that begins as gastric-type metaplasia and progresses to intestinal-type metaplastic lesion [322-327]. In comprehensive studies of mucin and trefoil peptide (TFF) expression in Barrett's esophagus Dekker's group [322, 323] found that in the first stage of BE pathogenesis a single layer of columnar epithelium replaces the normal stratified epithelium. While normal epithelium does not express secretory mucins and TFF-peptides, the metaplastic epithelium strongly expresses MUC5AC co-localized with TFF1. MUC6 expression was shown to be associated with expression of TFF3, and TFF2 is synthesized in cells that produce both MUC5AC and MUC6 mucins. The phenotype of esophageal epithelium at this stage of Barrett's metaplasia mimics the gastric-type pattern of mucin and TFF expression [328-330]. These data are in agreement with those reported by Van de Bovenkamp *et al.* [322], Guillem *et al.* [320] and Arul *et al.* [321], all of whom also observed *de novo MUC5AC* and *MUC6* expression in gastric-type metaplasia of Barrett's disease.

At the next stage of Barrett's disease, the gastric-type metaplasia is transformed into intestinal metaplasia, which is accompained by activation of *MUC2* expression. Concomitantly, the coordinated expressions of other mucins, including MUC5AC, and TFFs undergo dramatic changes: expression of MUC5AC is significantly increased while expression of MUC5B and MUC6 are decreased [322, 323]. These findings were confirmed by Chaves *et al.* [324] who observed strong expression of MUC5AC in all cases of Barrett's metaplasia studied. These authors showed that in the areas of metaplasia, MUC5AC apomucin was present in both columnar non-goblet and goblet cells, mainly at the surface epithelium with focal expression in the deep glandular structures. They also detected MUC2expression in the metaplastic areas with intestinal component, whereas regions lacking intestinal metaplastic features were MUC2-negative.

In partial agreement with these studies, Glickman *et al.* [331] and Burjonrappa *et al.* [332] observed high expression of MUC5AC in both gastric and intestinal metaplasias, while Guillem *et al.* [320] detected high level of MUC5AC expression only in gastric-type metaplastic lesions. Down-regulation of the *MUC5AC* gene was observed by these authors in the intestinal metaplasia samples. As noted by Burjonrappa *et al.* [332], "metaplastic epithelium may reflect an adoptive response to the changes occurred in the normal esophageal mucosa with MUC5AC offering protection from gastric acid and MUC2 associated with protection from bile". The trefoil peptides (TFF1 and TFF2) associated with mucosal repair act synergistically with mucins (in particular with MUC5AC) to protect epithelial tissues [333].

Interestingly, although most studies investigated both gastric- and intestinal-type metaplasias, it is unclear whether these metaplasias develop simultaneously or successively. According to Burjonrappa *et al.* [332], all intestinal metaplasia samples contain regions with gastric-type metaplasia, suggesting a common origin for both types of metaplasia. The findings of other laboratories [320, 334] suggest activation of multipotent esophageal stem cells that give rise to gastric and intestinal metaplasias following destruction of normal squamous epithelium.

In addition to metaplastic changes in Barrett's esophagus, dysplastic abnormalities of different grades may occur during progression of the disease to adenocarcinoma. At this stage, esophageal epithelium strongly expresses MUC5AC mucin [320, 322-324, 331, 332]. In Burjonrappa *et al.'s* study [332], 100% of dysplastic lesions, including epithelium adjacent to dysplasia and dysplastic regions of indefinite or low grade, were strongly MUC5AC-positive, while in high grade dysplasia lesions lower percentages of cells (84-86%) expressed MUC5AC mucin. The decrease in MUC5AC expression associated with increase in grade of dysplasia correlates well with findings that progress from dysplasia to Barrett's adenocarcinoma is accompanied by a sharp decrease in percentage of MUC5AC-positive cells (32-33%) [331, 332]. Chaves *et al.* [324], on the other hand, observed no reduction in expression of MUC5AC in Barrett's cancer in a series of 46 patients. They did not see any changes in dynamics of MUC5AC expression associated with transition from one stage of the disease to another: 100% of lesions expressed MUC5AC at the metaplastic, dysplastic and neoplastic stages.

Brown *et al.* [335] recently brought evidence of a nonintestinal pathway of neoplastic development in Barrett's esophagus consisting of gastric metaplasia-foveolar dysplasia-adenocarcinoma sequence. Foveolar-type dysplasia (nonadenomatous type) occurred in about 50% of dysplastic lesions in Barrett's esophagus, all of them expressing MUC5AC but none expressing the typical markers of intestinal differentiation (MUC2, CDX2 and villin). In contrast, intestinal-type dysplasia (adematous type) usually expresses markers of intestinal differentiation but not MUC5AC. This study indicates that the previously established gastric metaplasia-intestinal metaplasia-dysplasia-adenocarcinoma sequence is not the only pathway of malignant development in esophagus – that another pathway that excludes intestinal differentiation is observed equally often.

Several rare pathologies associated with Barrett's esophagus have been described. Kushima *et al.* [336] reported a case of pyloric gland adenoma arising in Barrett's esophagus. Immunohistochemically, almost all tumor cells were strongly MUC5AC- and MUC6-positive, but did not express MUC2 or CD10. Another rare cancer associated with Barrett's esophagus, esophageal polypoid dysplasia of gastric foveolar phenotype with focal intramucosal carcinoma, was reported by Asthana *et al.* [337]. Histologically, this lesion corresponds to an exuberant polypoid gastric epithelium with areas of low-grade dysplasia, high-grade dysplasia, and focal intramucosal extremely well-differentiated adenocarcinoma. This lesion exhibited a pattern of apomucin expression consistent with a gastric foveolar phenotype: MUC5AC - diffusely positive, MUC6 and MUC1 - focally positive, and MUC2 - negative.

In summary, numerous studies document the strong expression of *MUC5AC* gene in various pathological conditions of esophagus, including Barrett's esophagus, Barrett's adenocarcinoma, esophageal dysplasia, and BE-associated pyloric gland adenoma. However, the precise functions of MUC5AC glycoprotein in the pathogenesis of these diseases remain to be discovered.

<u>Stomach</u>: Changes in expression of *MUC5AC* gene in stomach are associated mainly with two pathological conditions: *Helicobacter pilory* chronic infection and malignant transformation of gastric mucosa which often demonstrate causal links.

a) Helicobacter pilory infection: *Helicobacter pilory* (*H. pilory*) is known to reside in the gastric mucus layer of about 50% of the world's population [205,

338, 339], and is recognized as a cause of gastric diseases, including chronic gastritis, gastric and duodenal ulcers, atrophic gastritis, gastric carcinoma, and gastric mucosa-associated lymphoid tissue lymphoma [340-343]. There is a definite association between *H. pilory* infection and gastric mucin expression: MUC5AC mucin was reported to function as a putative *H. pilory* receptor [344-350], and changes in expression of gastric mucins, including MUC5AC, were reported in numerous studies of *H. pilory* infection [338, 348-350]. However, it must be noted that the the data reported by these studies are often controversial, precluding drawing definite conclusions about the influence of *H. pilory* infection on the *MUC5AC* gene expression. According to Kang *et al.* [340], the contradictions may be explained by differing methodologies.

MUC5AC mucin is highly expressed by foveolar epithelium of both stomach antrum and body in healthy gastric mucosa [234, 273, 348, 349, 351-354]. Byrd *et al.* [348, 355], Beil *et al.* [356] and Slomiany and Slomiany [357] all reported that *H. pilory* infection inhibits synthesis of the MUC5AC mucin. Partial inhibition or under-expression of *MUC5AC* gene in *H. pilory*-infected gastric mucosa was described by Van den Brink *et al.* [346], Byrd *et al.* [348], Babu *et al.* [350], and Reis *et al.* [352, 353]. Babu *et al.* [350] studied the expression of gel-forming mucins, including MUC5AC, in normal and *H. pilory*-infected pre-neoplastic and neoplastic human gastric epithelium and showed progressive decrease in *MUC5AC* expression from normal epithelium through atrophic gastritis, intestinal metaplasia and dysplasia to carcinoma. Wang and Fang [338] showed that only 23.8% of *H. pilory*-positive gastric carcinomas expressed MUC5AC, whereas up to 71.4% of *H. pilory*-negative carcinomas synthesized and secreted this mucin. Correlation between *H. pilory* infection and down-regulation of *MUC5AC* expression in gastric biopsy specimens was also noted by Kocer *et al.* [347]. Direct documentation of the relationship between *H. pilory* infection and decrease in *MUC5AC* expression was obtained by Morgenstern *et al.* [358], who showed that *H. pilory* infection induced down-regulation of *MUC5AC* expression, and eradication of the bacteria restored *MUC5AC* expression to the pre-infection level.

In contrast to the above studies, Kaneko *et al.* [359] observed induction of *MUC5AC* expression by *H. pilory* infection. High level of *MUC5AC* expression in patients with *H. pilory*-associated gastritis was detected also by Marques *et al.*

[349]. These authors noted that there were no significant differences in *MUC5AC* expression between *H. pilory*-positive and *H. pilory*-negative groups. Kang *et al.* [340] conducted the most comprehensive *in vivo* study of *MUC5AC* expression, based on analysis of 136 *H. pilory*-positive and 88 *H. pilory*-negative patients, and found no differences in *MUC5AC* expression between the groups. They did, however, find down-regulation of *MUC5AC* and *MUC6* expression in the dysplasia plus cancer group. Similarly, the *H. pilory*-positive patients with atrophic gastritis and intestinal metaplasia demonstrated lower *MUC5AC* and *MUC6* expression than *H. pilory*-positive patients in the control group without signs of gastritis and metaplasia. The authors draw attention to the fact that expression of the MUC5AC and MUC6 mucins tended to be lower also in the group of patients with the same pathology but not infected with *H. pilory*.

Thus, it is not clear how *H. pilory* infection influences expression of the *MUC5AC* gene in stomach. Further studies are needed to evaluate the role of this microorganism and its products in regulation of *MUC5AC* gene transcription, MUC5AC mRNA translation, and processing of the MUC5AC apomucin.

b) Gastric cancer: The pathogenesis of gastric cancer is a multifactorial and multistep process [360, 361]. One of the risk factors leading to development of gastric cancer is *H. pilory*-associated progressive gastritis [362]. Intestinal metaplasia (IM) is one of the early lesions identified in the cascade of events preceding the development of gastric carcinoma [363, 364]. IM is a manifestation of atrophic mucosa occurring as a result of chronic *H. pilory* gastritis [365, 366]. Intestinal-type gastric cancers are thought to evolve through a process that begins with superficial gastritis and progresses through atrophy followed by the development of dysplasia and finally carcinoma [362]. Mucin expression is altered in the precancerous stages of gastric carcinoma [352, 353, 367-369]. The membrane-bound MUC1 mucin, the gel-forming mucins MUC5AC and MUC6, and *de novo* synthesized MUC2 mucin are co-expressed in incomplete metaplasia [368, 369], and, with the exception of MUC2, their expression is inhibited or down-regulated in complete metaplasia and carcinoma [368-370]. Gastric cancer is characterized by major substitution of MUC5AC and MUC6 mucins by MUC3 and MUC4 [354, 371-373]. According to Shiotani *et al.* [374], in gastric mucosa without IM, MUC5AC is expressed by the foveolar and mucous neck cells of both the antrum and corpus, while in gastric mucosa with complete IM (type I), MUC5AC is detected only in the goblet cells of the antrum and corpus in 6% and 13% of cases, respectively. In incomplete metaplasia (type II), MUC5AC is expressed in 83% of patients in the antrum and in 57% in the corpus, whereas in the mixed variant with prevalence of incomplete IM its expression is observed in 94% and 90% of analyzed samples of antrum and corpus, respectively. Thus, expression of MUC5AC is detected most often in type II IM and in the mixed metaplastic lesions.

This conclusion is in agreement with the data reported previously by other groups [375, 376]. Ho *et al.* [354] stressed the striking differences in both the quality and quantity of mucin gene expression in gastric cancer compared with normal stomach. The amounts of MUC5AC and MUC6 mRNAs and corresponding mucin glycoproteins were markedly lower in most carcinomas compared with normal gastric specimens. In this study, the process of neoplastic transformation in stomach was associated with a decrease in expression of MUC1, MUC5AC and MUC6 mucins, and with aberrant expression of MUC2, MUC3 and MUC4. These observations are in line with reports in which changes in expression of MUC5AC and MUC2 were used to predict prognosis in patients with gastric carcinoma [377, 378].

It has been established that IM of gastric epithelium is followed by four stages in gastric carcinogenesis: gastric reactive hyperplasia (RH), indefinite dysplasia (IDys), dysplasia (Dys) of low or high grade, and intestinal gastric carcinoma (IGC) [379]. The rate of MUC5AC expression decreases gradually during gastric carcinogenesis, from 86% in RH to 77% in IDys, to 29% in Dys, and to 7% in IGC [379]. As underlined by Dong *et al.* [379], RH preserves most of the typical features of normal gastric mucosa, including expression of mucins, and the lesion is therefore considered inflammatory and not neoplastic. According to expression of *MUC5AC, MUC6* and *APC* genes and proliferation and differentiation parameters, IDys is defined as a hyperplastic lesion. The abnormalities observed at this stage are closer to those of Dys, but distinct from those of RH. The rates of *MUC5AC, MUC6* and *APC* expression in Dys are significantly lower than those observed at the RH and IDys stages, but not significantly different from those of the IGC stage. Thus, Dys was categorized as a non-invasive neoplasia or an intraepithelial neoplasia [379]. Dong *et al.* [379] analyzed seven parameters –

MUC5AC, MUC6, APC, p53, EGFR, K-67 and PCNA – and showed that expression of MUC5AC in combination with K-67 and PCNA could best differentiate the four stages. Thus, on the basis of the discussed data, MUC5AC may be considered a gastric carcinoma marker.

However, data from other studies do not corroborate this conclusion. Wongham *et al.* [380], for instance, could not detect MUC5AC antigen in serum of patients with stomach cancer. And, importantly, these and other investigators did find the MUC5AC mucin in serum of patients with cholangiocarcinoma and showed that this glycoprotein is a reliable tumor marker [380-382]. Xu *et al.* [206] attributed the failure to detect MUC5AC in the above-described studies to use of an mAb not suitable for the MUC5AC epitope specifically expressed in gastric cancer. It is well known that the glycosylation patterns (antigenic epitopes) and expression levels of mucins in normal gastric mucosa are different from those in gastric carcinomas [354, 371, 383-385]. The clear lack of appropriate diagnostic tools specific for gastric cancer led some investigators to suggest the alternatively glycosylated MUC5AC mucin epitopes M1-a, M1-b, M1-g and M1-f as diagnostic targets [386], and the corresponding monoclonal antibodies Muc5A01, Muc5A02, Muc5A03 and 21M1 as the diagnostic tools [206].

Most of the studies on the role of MUC5AC mucin in gastric carcinogenesis have yielded conflicting results. Some examples will suffice to illustrate this statement. Dong *et al.* [379] observed gradual decrease in MUC5AC expression during malignant transformation of gastric epithelium from normal mucosa to carcinoma, while Xu *et al.* [206] detected elevated levels of the MUC5AC mRNA in gastric carcinoma cells and high concentrations of the MUC5AC mucin antigens in serum of gastric cancer patients. Ho *et al.* [354] reported a decrease in MUC5AC expression in gastric cancer compared with normal gastric epithelium, while Pinto-de-Sousa *et al.* [387, 388] described a high rate of MUC5AC expression in diffuse-type gastric carcinoma. These last authors confirmed previous findings that diffuse gastric carcinomas usually express MUC5AC mucin and exhibit patterns of differentiation characteristic of infiltrative gastric carcinomas [389, 390]. Toki *et al.* [391] also reported a high rate of MUC5AC expression associated with advanced gastric adenocarcinomas. Choi *et al.* [392] noted that non-mucinous gastric carcinomas have a higher rate of MUC5AC expression compared with mucinous type carcinomas, which expressed decreased levels MUC5AC or did not express this mucin at all. Interestingly, no association of MUC5AC expression with a particular subtype of gastric carcinoma was observed in a comprehensive study by Leteurtre *et al.* [393], although other studies [394-396] detected a high rate of the subtype-specific MUC5AC expression that reached 65%.

The discrepancies in the data on *MUC5AC* gene expression in gastric cancer indicate the need for more studies. A recently published paper on the role of genetic variations of the *MUC5AC* gene in the development of stomach cancer [397] emphasized the potential of the MUC5AC mucin in the oncogenic process.

Pancreas: Pancreas is one of the central organs in a mammalian organism, including humans, as it performs several dispatcher functions important for the whole body, and for the digestive system in particular. Pathology associated with the pancreas often involves aberrations in mucin gene expression, including the *MUC5AC* gene. The alterations relevant to pathogenesis of pancreatic diseases are discussed in the following section.

a) Pancreatic intraepithelial, intraductal papillary and mucinous cystic neoplasias: Pancreatic cancer (PC) is one of the most widespread malignancies in the western world [398]. PC can be of exocrine or endocrine type, with prevalence of the former [399]. It has been hypothesized that PC originates from pancreatic stem cells as a result of genetic abnormalities that accumulate in genes critical for self-renewal pathways [399-404]. Among the potential candidates for pancreatic cancer stem cells, acinar and centroacinar cells are the most important [405-413]. Three histologically distinct precursors of ductal adenocarcinomas have been described: pancreatic intraepithelial neoplasias (PanIN), intraductal papillary mucinous neoplasias (IPMN) and mucinous cystic neoplasias (MCN) [414-416]. According to degree of cell and tissue atypia, PanIN can be further sub-classified into PanIN-1A, PanIN-1B, PanIN-2 and PanIN-3. The last one as well as IPMNs and MCNs have the potential to progress to invasive adenocarcinoma [399, 416, 417].

Gene expression analysis revealed abnormalities in expression of some genes, including mucin genes, specific for the development of pancreatic cancer from benign adenoma to invasive carcinoma (Fig. 6). Expression of MUC5AC mucin in normal and malignant pancreas has been thoroughly studied [418-421], however with conflicting results.

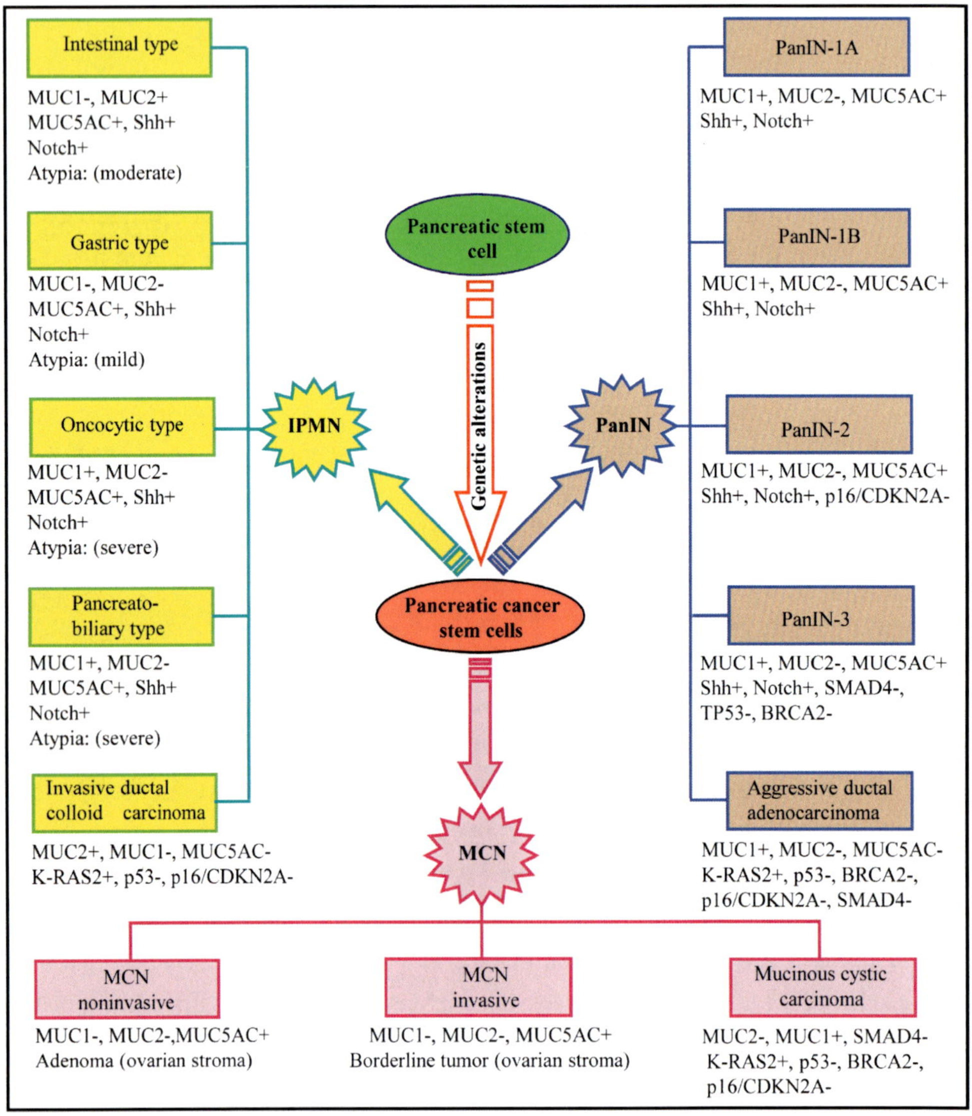

Figure 6: Gene expression in different types of pancreatic neoplasms – PanIN, IPMN and MCN (based on the data reported in [399-417]).

For instance, Andrianfahanana *et al.* [419] "consistently detected MUC1, MUC5AC, MUC5B and MUC6 in normal pancreatic samples", while Nagata *et al.* [420] contended that "MUC2 and MUC5AC are never expressed in normal pancreatic tissue". According to Ji *et al.* [422], the normal pancreatic duct epithelium expresses a moderate level of MUC5AC mucin. The discrepancies in the data describing *MUC5AC* expression in normal pancreas have been noted by many investigators [419, 421-428]. These discrepancies have been attributed to the use of methods with different specificities and sensitivities. They have also been associated with the artificial effects of proteases present in pancreatic fluids or tissues that degrade antigens or antibodies, thereby affecting MUC5AC immunodetection [429]. Further studies with uniformly sensitive methods are needed to resolve the discrepancies. Knowledge of *MUC5AC* expression in normal pancreas is of great importance as it is a starting point for analysis of the behavior of the *MUC5AC* gene in pancreas pathology. It is also important to establish the pattern of MUC5AC glycosylation in normal pancreas, since its alteration is germane to pathological conditions [194, 420, 430, 431]. In some cases, changes in apomucin expression are negligible while alterations in quantity and composition of glycans are much larger and may be helpful in diagnosis. For example, elevated MUC5AC apomucin expression was observed in only 35% of pancreatic cancer patients, but highly prevalent and distinct glycan alteration was detected in 65% of cases [431].

While it is not clear whether *MUC5AC* is expressed in normal pancreatic tissues, expression of *MUC5AC* in all types of premalignant and highly neoplastic tissues of pancreas is well documented [420, 423, 427, 432]. Some authors observed stepwise elevation of *MUC5AC* expression during progression of pancreatic malignant process through PanIN-1A, PanIN-1B-PanIN-2-PanIN-3-DAC sequence [433, 434], while others consider neoplastic transformation of normal pancreatic epithelium to be associated with *de novo* high level expression of *MUC5AC* [420, 435, 436]. Sanada *et al.* [434] showed that *MUC5AC* is not expressed in normal pancreas and is very rarely expressed (only 12.5%) in PanIN-1A stage. However, during progression of the PanIN-1A stage to the PanIN-1B, PanIN-2 and PanIN-3 stages, the frequency and level of *MUC5AC* expression gradually but significantly increase from 35% in PanIN-1B to 72.1% in PanIN-2 and further to 87.1% in PanIN-3 (Fig. 7). Unexpectedly, invasive carcinomas, especially poorly differentiated adenocarcinomas, show a significant decrease in

MUC5AC expression. These data are in agreement with the results of Kigure *et al.* [433] who also observed slow elevation of *MUC5AC* expression in PanINs depending on the lesion grade. In addition, Kigure *et al.* [433] demonstrated the association between progression of PanIN and expression of MUC1, MUC2, MUC5AC and E-cadherin.

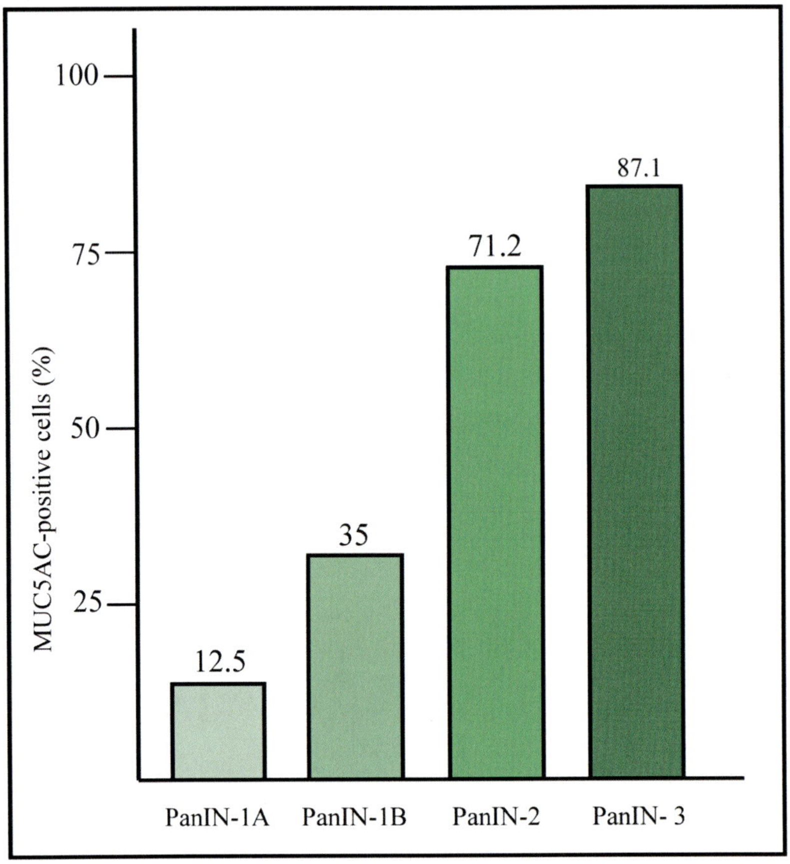

Figure 7: Expression of MUC5AC in PanINs (the data extracted from [434]).

The functional meaning of this association was recently established by Inaguma *et al.* [437], who showed that transcription factor GLI1 facilitates migration and invasion of pancreatic cancer cells through the MUC5AC-mediated attenuation of E-cadherin. The GLTI1-up-regulated MUC5AC glycoprotein was expressed in the intercellular junction between pancreatic adenocarcinoma cells and interfered with the membrane localization of E-cadherin. This interference attenuated E-cadherin-mediated intercellular adhesion and facilitated migration and invasion of pancreatic adenocarcinoma cells. The impact of MUC5AC on the adhesive and invasive abilities of pancreatic cancer cells was further clarified by Yamazoe *et al.* [438] and Hoshi *et al.* [428]. These investigators demonstrated that si-RNA-mediated knockdown of *MUC5AC*, usually over-expressed in pancreatic ductal carcinoma, reduced the adhesive, invasive and metastatic potentials of pancreatic cancer cell lines. These effects resulted from down-regulation of several genes encoding integrins, metalloprotease-3 (MMP3) and vascular endothelial growth factor (VEGF), which, in turn, are associated with suppression of the *MUC5AC* gene. In addition, *MUC5AC* down-regulation induced attenuation of Erk1/2 expression. Thus, MUC5AC mucin may contribute to the invasive and metastatic abilities of pancreatic cancer cells by enhancing expression of integrins, MMP3 and VEGF, and by activating the Erk pathway.

Interestingly, the *in vivo* and *in vitro* study by Hoshi *et al.* [428] showed that *MUC5AC* knockdown induced by short interfering RNA (si-RNA) significantly reduced the tumorigenesity and suppressed tumor growth of human pancreatic cancer cell lines SW1990 and BxPC3 *in vivo*. In addition, antibodies against cancer cells were detected in the sera from the mice bearing MUC5AC-knockdown tumors but not MUC5AC-expressing tumors. These results indicate to the ability of MUC5AC mucin glycoprotein, expressed on the surface of pancreatic cancer cells, to support escape of these cells from immunosurveillance. Hence, a new potential function of this mucin as immuno-suppresive agent emerges from this study and highlights its important role in pancreatic cancer progression. However, as follows from the study of Luka *et al.* [439], MUC5AC mucin may also function as an immuno-stimulating agent. These authors showed that the tumor-associated MUC5AC-related antigen, NPC-IC, secreted by pancreatic and colonic tumor cells into blood, has the potential to induce production of specific anti-MUC5AC monoclonal antibody. The NPC-IC antigen is expressed on tumor associated MUC5AC, but does not occur on MUC5AC from normal tissue. Based on these data, this antigen might be considered as a specific target to aid in the diagnosis of pancreatic and colonic cancers and the immunological monitoring of cancer treatment regimens.

The significance of MUC5AC mucin expression in the development of pancreatobiliary neoplasms established in the above studies was recently confirmed and extended by other investigators [427, 440]. Yonezawa *et al.* [427] reported that high *de novo* expression of *MUC5AC* was observed in most types of pancreatobiliary neoplasms, including all grades of PanINs, all types of IPMNs and mucin-producing bile duct tumors, biliary intraepithelial neoplasia, pancreatic ductal adenocarcinomas, and intrahepatic cholangiocarcinomas. Yamada *et al.* [440] established that such *de novo* expression of *MUC5AC* is regulated by DNA methylation and histone H3-K9 modification of the *MUC5AC* promoter. Altogether, these and numerous other studies indicate that high *de novo* expression of *MUC5AC* is an early and important event in the pathogenesis of pancreatic cancer. However, many questions await clarification. What signals are responsible for activation of the *MUC5AC* gene transcription? Is the expression of MUC5AC mucin sufficient for the beginning of the transformation process? What factors can inhibit *MUC5AC* activation? More studies are needed to clarify these issues.

IPMNs are intraductal pancreatic mucinous neoplasias arising within the main pancreatic duct or its branches [441]. Four histologically distinct subtype of IPMN have been identified [442, 443]– gastric, intestinal, pancreatobiliary and oncocytic – which may progress in a stepwise manner to intraductal carcinoma and further to invasive colloid carcinoma [423]. With regard to mucin expression, each IPMN subtype is characterized by a specific expression profile of the *MUC1*, *MUC2* and *MUC5AC* genes. Importantly, *MUC5AC* is expressed in all subtypes of IPMN, while *MUC2* is expressed only in the intestinal one. High level of *MUC1* expression is usually observed in oncocytic and pancreatobiliary subtypes; gastric and intestinal subtypes are commonly *MUC1*-negative [444-458]. In the study of Adsay *et al.* [459], 54% of IPMNs expressed *MUC2* and only 20% were MUC1-positive. Interestingly, transition of IPMNs to colloid carcinoma was associated with increased frequency of *MUC2* expression that reached 100% of MUC2-positive tumors, and decreased frequency of MUC1-positive lesions from 20% to 0%. Thus, it appears that a specific profile of *MUC1*, *MUC2* and *MUC5AC* expression may be indicative in differential diagnostics and follow-up of tumor progression. However, a recent study by Lee *et al.* [460] cast doubt on this supposition when they found no significant differences in expression of *MUC1*, *MUC2* and *MUC5AC* between benign and malignant IPMNs. The observed discrepancies in the data demands further investigation of this issue.

Mucinous cystic neoplsms (MCNs), the third histologically distinct group of pancreatic tumors, which usually appear in women and contain ovarian-type stroma, are located in the pancreatic tail, and are separated from the pancreatic ducts [461]. MCNs are less malignant than IPMNs: the rate of malignant potential of IPMNs ranges from 6% to 92%, while that of neoplastic transformation of MCNs ranges from 6-36% [462]. There are two types of MCNs: infiltrating and non-infiltrating. Both types express *MUC5AC* [422]. Expression of *MUC2* is not typical for MCNs. *MUC1* is expressed only in infiltrating MCNs, while non-infiltrating lesions are usually MUC1-negative [423, 432].

Besides malignant mucinous neoplasms of pancreas, there is a group of benign pancreatic cystic lesions (serous cystadenomas + pseudocysts). While all the benign lesions express MUC5AC mucin, the glycan components of their MUC5AC glycoprotein molecules differ from those in malignant tumors – a difference reflected in different antigenic landscapes of these molecules. These properties of MUC5AC may be useful in differential diagnosis [430]. Interestingly, cancer-specific alteration of mucin glycosylation, including glycosylation of MUC5AC, was shown to be stimulated by pro-inflamatory signaling in pancreatic cancer cells [194].

b) Ampulary cancer: Among pancreatic cancers, carcinoma of the papilla Vater (also known as ampulary cancer) is of special interest. It arises from the unique location of Vater's valve at the border between two different types of mucosa: epithelium of the distal bile ducts within the pancreatic head, and epithelial cells characteristic of the duodenum [463-465]. The exact location of ampulary carcinomas determines their histological characteristics, relating them to either intestinal or pancreatobiliary subtypes [466]. The former subtype usually expresses MUC2 in combination with intestinal transcription factor CDX2 and cytokeratin 20 (CK20); tumors of the pancreatobiliary subtype are mostly negative for these proteins [464-469]. A significant part of carcinomas of the papilla Vater associated mainly with pancreatic component expresses MUC5AC mucin, and exhibits metaplastic gastric and foveolar transformation [463]. Matsubayashi *et al.* [468] noted that in their study all papilla tumors of incomplete intestinal subtype and all lesions of metaplastic pancreatobiliary subtype were

MUC5AC-positive, while the tumors of complete intestinal subtype and ordinary pancreatobiliary subtype were negative for the mucin.

<u>*Hepatobiliary organs*: </u>Like the pancreas, the liver and billiary organs play extremely important roles in human physiology. Pathology of the hepatobiliary system is associated mainly with inflammation in the bile ducts, gallbladder stone disease and cholecystitis, as well as with malignant tumors of the liver and hepatobiliary tree including gallbladder. Analysis of *MUC5AC* expression in hepatobiliary pathology is of both basic science and clinical importance.

a) Hepatobiliary cancer: The main malignant lesion of the hepatobiliary system is cholangiocarcinoma, accounting for about 3% of all gastrointestinal cancers [470, 471]. Depending on location, it is classified as an intrahepatic (ICC) or extrahepatic (ECC) tumor [470, 472]. ICCs are malignant neoplasms arising from intrahepatic biliary epithelial cells distributed at all segments or levels of the intrahepatic biliary tree. Most cases develop in an otherwise normal liver, although a portion of tumors arise in livers with pathological conditions such as hepatolithiasis, primary sclerosing cholangitis or liver chirrhosis [473-479]. ECCs are also highly malignant tumors arising from the ductal epithelium of the extrahepatic bile duct. Most of cholangiocarcinomas are well-to-moderately differentiated adenocarcinomas [479]. Risk factors for ECC include cholangitis, biliary cirrhosis, choleolithiasis and choledochal cysts [480, 481].

Classification based on the anatomical parameters indicates that ICCs arise within the liver parenchyma, whereas ECCs occur in the biliary duct tree within the hepatoduodenal ligament and gallbladder. ECCs can be further divided into hilar and distal tumors [482]. ICCs are divided into mass-forming, infiltrating periductal- and intraductal-growing types. *MUC5AC* expression has been studied in most ICC and ECC subtypes. One of the relatively well-studied lesions is the mucin-producing intraductal-growing type ICC, also known as intrahepatic biliary intraductal papillary mucinous neoplasia (b-IPMN) [483-485]. This type of ICC is strikingly similar to pancreatic IPMN (p-IPMN) in histopathological features and production of a large amount of mucins, including MUC5AC [486-488]. Two distinct neoplastic lesions precede invasive ICC: biliary intraepithelial neoplasia (BilIN) and intraductal papillary neoplasia (IPN-B) [489-491]. Multi-step carcinogenesis has been suggested in ICCs arising from BilIN and IPN-B, with an important role of mucins in this process [489, 491]. Zen *et al.* [492] examined the phenotype changes of mucin core proteins, including MUC5AC, during cholangiocarcinogenesis from BilIN and IPN-B to ICC and found a high level of MUC5AC expression in 89% of BilINs and 96% of IPN-Bs. Combined lesions of ICC and IPN-B expressed MUC5AC in 100% of the examined tumors; combined lesions of ICC and BilIN expressed MUC5AC in only 83% of neoplasms.

On the basis of histology and mucin expression patterns, four subtypes of b-IPMN are currently defined: gastric (MUC1-, MUC2-, MUC5AC+), intestinal (MUC1-, MUC2+, MUC5AC+), pancreatobiliary (MUC1+, MUC2-, MUC5AC+) and oncocytic (MUC1+, MUC2-, MUC5AC+) [442, 453]. Expression of MUC5AC is characteristic of all subtypes of b-IPMNs [453, 476]. While Bamrungphon *et al.* [476] consider MUC5AC to be the most reliable marker for detection of cholangiocarcinomas, with sensitivity and specificity of 71% and 90%, respectively, Jan *et al.* [493] detected only focal and weak expression of MUC5AC in the mass-forming peripheral cholangiocarcinomas. Park *et al.* [470] also described low rate of MUC5AC expression (27%) in mass-forming ICC, in contrast to higher rates of expression in tumors with intraductal growth (50%) and lesions with periductal infiltrating growth (72%). Mall *et al.* [494] observed substantial expression of MUC5AC in 46% of intrahepatic central and hilar tumors, mainly in moderate differentiated carcinomas, while 54% of cholangiocarcinomas remained MUC5AC-negative. The same frequency of MUC5AC expression (45%) in liver cholangiocarcinomas was observed by Lau *et al.* [495].

A study of MUC5AC expression in ICCs and ECCs recently published by Park *et al.* [470] showed that only 47.1% of ICC tumors express MUC5AC compared to 70.6% of ECCs. Lee *et al.* [496] also reported more frequently expression of MUC5AC in ECCs than in ICCs. Both research groups found the difference to be statistically significant. Moreover, there is a significant correlation between MUC5AC expression and a higher tumor grade both for ICC and ECC. There is no statistical proof, however, of an association between MUC5AC expression and poor survival of patients [470]. At the same time, MUC5AC-expressing gastric foveolar type of ICC is more often associated with aggressive tumor development compared with the less aggressive behavior of the pyloric gland type [497, 498]. Interestingly, cholangiocarcinomas associated with cirrhosis or with hepatocellular carcinoma do not express MUC5AC [499], or express it at a very low level [470, 500]. Several studies reported data indicating that the serum content of MUC5AC point to poor prognosis [380, 390, 497, 498].

A number of pathological conditions accompanying malignant transformation in the hepato-biliary system have been found to influence mucin expression in the neoplastic lesions. Among these conditions are the inflammatory diseases - cholecystitis, hepatitis and hepatolithiatis.

b) Inflammatory diseases (cholecystitis and hepatolithiasis): The literature is sparse on the effect of inflammation on the expression of mucin genes in human gallbladder epithelium and billary ducts. It has been established that several mucin genes, namely *MUC1, MUC2, MUC3, MUC5AC, MUC5B* and *MUC6*, are expressed in cultured human gallbladder epithelial cells as well as in carcinomas arising in billary ducts [500, 501]. Expression of the same set of genes has also been observed in tissues of freshly isolated inflamed gallbladder. Expression of MUC5AC glycoprotein in biliary epithelial cells was shown to be weak and focal, although almost every one of the studied samples had positive MUC5AC-specific RNA-labeling detected by *in situ* hybridization (83%) or Northern blot analysis (100%) [502].

Inflammation of the gallbladder (cholecystitis) is usually associated with stone formation. Among 104 cases of acute and chronic cholecystitis analyzed by Ho *et al.* [503], only 11 were free of gallstones. Mucins have been shown to be structural components of gallstones, and mucin hyper-secretion in the gallbladder plays an important role in stone formation and pathogenesis of gallstone disease and cholecystitis [504-508]. Normal gallbladder epithelium is characterized by the uniform expression of MUC3, MUC5B and MUC5AC mucins, which decreases with increasing degree of inflammation. Only 52% of specimens with inflammation expressed MUC5AC compared to 100% of normal gallbladder samples. On the other hand, MUC5AC was expressed in the normal large bile ducts in only 4% of the analyzed specimens whereas 89% of the hepatolithiasis specimens tested were MUC5AC-positive [509].

Vilkin *et al.* [510] reported expression of MUC5AC mucin in 95% of gallbladder bile samples and in 85.7% of gallbladder mucosal specimens. A high level of *MUC5AC* gene expression in the inflamed gallbladder was associated with pigmented brown stones, a report confirmed by Kim *et al.* [511] who detected up to 1558-fold increase in the expression levels of MUC5AC mRNA in gallstone patients compared with control subjects. According to Sasaki *et al.* [509], the distribution profiles of MUC5AC and MUC6 apomucins in large bile ducts in hepatolithiasis resemble those of gastric mucosa [328, 512, 513], indicating that in hepatolithiasis the intrahepatic biliary epithelium undergoes transformation to gastric type epithelium. Interestingly, Nagata *et al.* [514] observed co-expression of gastric and biliary phenotype in pyloric-gland type adenoma of the gallbladder, with MUC5AC expression in 38% of tumors. Importantly, Sasaki *et al.* [499] observed transition to gastric type mucosa with specific MUC5AC expression in nondysplastic epithelia of gallbladder, dysplasias and carcinomas of this organ, as well as in cholecystitis and cholecystolithiasis. The authors concluded that the phenomenon of "gastric mucosa transformation" in gallbladder characterized by high expression of MUC5AC mucin may be related to chronic inflammation and lithogenesis, which, according to epidemiological data, precede development of gallbladder carcinoma [515].

In contrast to biliary epithelial cells, which express MUC5AC in both normal and pathological conditions, hepatocytes and hepatocellular carcinomas are usually MUC5AC-negative [495, 500].

<u>*Intestinal tract*:</u> Normal epithelium of adult intestinal tract docs not express MUC5AC mucin, although its expression is constantly detected in human embryonic intestine up to 8 weeks of gestation. Later in development, MUC5AC expression is significantly reduced, and after the 18th week of gestation MUC5AC mRNA is not detectable [233, 235, 516, 517]. In pathology (inflammatory conditions and malignant diseases), aberrant expression of several mucin genes, including *MUC5AC*, has been described [518-529].

a) Intestinal inflammatory diseases: In inflammatory bowel disease (IBD), the quantity and quality of gel-forming mucins are affected by several factors: reduction in the number of goblet cells, decrease in the content of some mucins, and/or changes in their glycosylation [530-535]. MUC5AC and TFF1 are *de nove* expressed in scattered goblet cells in IBD and Crohn's disease (CD) [518]. Both proteins are expressed in abundance in goblet cells of the inflamed duodenum [536, 537], but not in normal small bowel and colon [538-540]. Shaoul *et al.* [518, 536, 537] showed that MUC5AC mucin is expressed in IBD and in small bowel gastric type metaplasia, while Tytgat *et al.* [531] could not detect MUC5AC mRNA in colonic biopsies from a patient with mild ulcerative colitis (UC). Longman *et al.* [519] also found no expression of MUC5AC in UC. On the other hand, Buisine *et al.* [517] reported activation of *MUC5AC* expression in specific areas of affected mucosa in CD; the healthy ilial mucosa of these patients exhibited a pattern of mucins identical to normal controls, with no expression of MUC5AC, MUC5B, MUC6 and MUC7, moderate expression of MUC1 and MUC4, and high expression of MUC2 and MUC3 mucins.

The specific expression pattern of mucins characterized by the appearance of MUC5AC, MUC5B and MUC6 mRNAs and corresponding mucin polypeptides, on the one hand, and disappearance of MUC2, on the other hand, was observed in the regions of ilial mucosa adjacent to the ulceration area in the so-called ulceration-associated cell lineage (UACL). The UACL was described by Wright *et al.* [541] as a specific anatomical structure appearing in close proximity to the ulcerated areas and containing three main components: the acinar portion; the duct, which arises from acinus; and the surface cells, which migrate through the duct and replace the indigenous lineages. MUC5AC mucin was shown to be expressed in epithelial cells of the surface and the upper part of the ducts, while MUC6 was expressed in epithelial cells of acinar glands and the deeper part of the ducts [517]. Kaneko *et al.* [520] also reported expression of MUC5AC in mucous cells with a foveolar structure, and MUC6 in cells with a glandular structure. The UACL expresses a number of peptides implicated in the repair of damaged gastrointestinal mucosa, notably, members of the trefoil peptide family (TFF) [542]. Longman *et al.* [521] demonstrated that MUC5AC and TFF1 are co-localized in distal ductular and surface elements of the UACL, while MUC6 and TFF2 are co-localized to acinar and proximal structures of the lineage. According to Buisine *et al.* [517], MUC5AC and MUC6 may be involved also in epithelial wound healing after mucosal injury in IBD, UC and CD. The coordinated localization of trefoil peptides and mucins in UACL suggests they may assist each other in protection and repair of gastrointestinal mucosa.

b) Intestinal malignant diseases: It has been noted that IBD and UC predispose to development of colorectal carcinoma [543], although UC-related cancers appear to have a histogenic pathway distinct from that of sporadic colorectal cancers [544]. Mucin-coding genes are among the genes that demonstrate different expression in sporadic colonic and UC-associated neoplasms (UCAN). The proportion of MUC5AC-positive tumors in a group of UCAN is higher (80%) than among sporadic adenocarcinomas (43%) [544]. Low grade and high grade UC-associated dysplasias are MUC5AC-positive in 89% and 90%, respectively. The percentage of neoplastic cells expressing MUC5AC in UCAN tends to decrease with the severity of dysplasia, diminishing to 73% in UC-related adenocarcinomas. According to Tatsumi *et al.* [544], these data suggest that MUC5AC plays an important role in UC-associated tumorigenesis, especially in the initial steps of the carcinogenic process.

Evidence is accumulating that a substantial portion of sporadic colorectal cancers develop *via* the recently recognized "serrated polyp-neoplasia pathway", different from the conventional adenoma-carcinoma sequence [523]. The group of serrated polyps of the large intestine is composed of hyperplastic polyps (HP), sessile serrated adenomas (SSA), traditional serrated adenomas (TSA), and mixed polyps (MP) [545]. Fujita *et al.* [523] examined mucin core protein expression in serrated polyps of the large intestine and found that MUC5AC is expressed throughout the entire length of the crypts in HPs and SSAs. Interestingly, MUC5AC expression tends to be decreased at the basal portion of the crypts in TAS even when other parts of the crypts are MUC5AC-immunoreactive. The frequencies of MUC5AC expression in HPs and SSAs are similar and relatively high (75% and 80%, respectively), while in TAS the frequency reaches only 43%. The same level of MUC5AC expression in HPs was detected by Hirono *et al.* [546]. Mochizuka *et al.* [522] observed up-regulation of MUC5AC expression in all types of serrated polyps of colon and rectum. Thus, it appears that the development of serrated polyps in large intestine is associated with high activity of the *MUC5AC* gene.

Although the serrated polyp-neoplasia pathway is observed in a substantial portion of colorectal malignancies, the conventional adenoma-dysplasia-adenocarcinoma sequence is the main vector in development of cancer in large intestine. Most sporadic colon cancers are thought to develop from adenomas

[547, 548]. Adenomatous polyps are benign premalignant lesions with a malignant potential that progress from tubular adenoma through tubulovillous to villous adenomas. The sequentially developed dysplasia ultimately leads to invasive carcinoma [548]. Changes in mucin expression in adenomas of the large bowel have been reported, which indicate *de novo* synthesis of two gastric mucins, MUC5AC and MUC6 [549, 550] – alterations that correlate with villous histology, adenoma size and degree of dysplasia [529, 550].

Notably, data on the expression of mucins in different tissues obtained by mucin-specifc antibodies may vary depending on whether anti-VNTR or non-VNTR antibodies are used [525]. Myerscough *et al.* [525] noted that VNTR-specific antibodies can detect precursor forms of a mucin but not the mature, fully glycosylated molecules. In contrast, the non-VNTR antibodies, which are not subject to interference by glycosylation, can react with mucin molecules at all stages of biosynthesis and therefore can detect both non-modified and modified mucins expressed in different cell compartments. Using anti-VNTR antibodies, several groups found high levels of *de novo* MUC5AC expression in adenomas (71%), reflecting a significant decrease in mucin production as the degree of tissue dysplasia rose [529, 551]. As shown by Bara *et al.* [551], only 12%-29% of tumors expressed MUC5AC when dysplastic lesions progressed to the adenocarcinoma stage. According to these findings, it appears that more aggressive colon adenocarcinomas express less MUC5AC mucin. This assumption was confirmed by Kocer *et al.* [527] in their study on the relationship between MUC5AC expression and prognosis of colorectal carcinomas. They found that overall disease-free survival of patients with MUC5AC-negative tumors was lower than that of patients with MUC5AC-positive tumors. Thus, the expression rate of MUC5AC detected by anti-VNTR antibodies is high in tumors with a better prognosis, although it declines with increasing malignant potential of the tumors.

Interestingly, Myerscough *et al.* [525] observed up-regulation of MUC5AC expression in all types of adenomas and transitional mucosa analyzed by non-VNTR antibodies.They did not, however, find a correlation between the level of MUC5AC expression and size of lesion, grade of dysplasia or subcellular localization. In contrast to MUC5AC, the use of non-VNTR MUC2-specific antibodies enabled detection of a substantial change in the subcellular staining of MUC2 in all three groups of adenomas (tubular, tubulovillous and villous) compared with the control colon. These results strongly suggest that alterations in the cellular processing of MUC2 are specific to the adenoma-carcinoma sequence, while induction of MUC5AC expression appears to be a common early event in the formation of adenoma lesions. Further studies are needed to determine what changes, if any, take place in MUC5AC subcellular localization during colon carcinogenesis.

An interesting observation of association of *MUC5AC* expression in mucinous carcinomas with microsatellite instability (MSI) was reported [552, 553]. Biemer-Huttmann *et al.* [554] described expression of MUC5AC in 77% of the high microsatellite-unstable colon cancers compared with 28% of microsatellite-stable cancers. These data show that colorectal carcinomas of high MSI status are more often positive for MUC5AC than microsatellite-stable tumors. Messerini *et al.* [555] consider that MSI may directly influence mucus production by altering genes involved in mucin biosynthesis or degradation.

In summary, it appears that the MUC5AC mucin plays an essential role in pathology of the large bowel both in inflammatory diseases and cancer. Each of the two types of pathological conditions is associated with specific dynamics in MUC5AC expression, indicating important functions of this glycoprotein in the processes associated with inflammation and malignant transformation of the intestinal epithelium.

5.6.3. Expression of *MUC5AC* Gene in Pathology of the Female Reproductive Tract

Mucins play an important role in reproductive function and defense of the female reproductive tract [253]. The expression of an individual mucin is associated with a specific region of the tract, and depends on the stage of the menstrual cycle. Under physiological conditions, expression of *MUC5AC* gene was observed mainly in endocervix, to a lesser extent in endometrium [253], and not in ectocervix, vagina or Fallopian tube [264, 556, 557]. Audie *et al.* [264] found that *MUC5AC* is mildly expressed in surface epithelium of the endocervix and endometrium at the follicular phase of ovulatory cycle, not expressed at the ovulatory phase, and highly expressed in both the surface and glandular epithelium at the luteal phase. In contrast to these results, Alameda *et al.* [241] detected no *MUC5AC* expression in normal endometrium. This discrepancy in results may be accounted for by different methods for MUC5AC detection in the studies. Another reason may be detection of different products of the MUC5AC gene in the two studies: the MUC5AC mRNA in one [264] and the MUC5AC protein in the other [241].

a) Endometrial tumors: Transition of normal reproductive epithelium to benign and malignant tumors is associated with changes in the activity of mucin genes, including alterations in *MUC5AC* expression. Reithdorf *et al.* [268] noted that cells synthesizing MUC5AC were observed in 100% of immature squamous metaplasia samples, but in none of the mature squamous and tubal metaplasia specimens. In contrast to these data, Baker *et al.* [257] reported that all benign lesions of endocervix, including tubal metaplasias and grandular hyperplasias, strongly expressed MUC5AC, although only a small portion of the benign lesions in endometrium expressed the MUC5AC mucin. Alameda *et al.* [241] also reported that only a small portion (~13%) of endometrial samples with simple hyperplasia expresses MUC5AC, and does so at a low level. Interestingly, MUC5AC was not detected in the specimens characterized by complex type hyperplasia of endometrium in this study. The expression of both gastric mucins, MUC5AC and MUC6, was observed, however, in lobular endocervical glandular hyperplasya – a distinct benign glandular lesion of gastric type [557].

Studies of *MUC5AC* gene expression in endometrial adenocarcinomas gave extremely controversial results. Alameda *et al.* [241] observed expression of the MUC5AC mucin in about 62% of 34 adenocarcinoma samples studied. Baker *et al.* [257] found MUC5AC expression in only 22% of endometrial carcinomas. Morrison *et al.* [558] could detect expression of the mucin in only 1 out of 310 samples examined (0.3%). Hebbar *et al.* [266] observed extremely low level of MUC5AC expression in endometrial malignancies. Such dramatic differences in MUC5AC expression reported by different investigators attest to the existence of some unknown factors that determine *MUC5AC* gene expression in endometrial carcinoma. More studies are needed to clarify this issue.

b) Cervical malignancies: The results from studies on the relationship between *MUC5AC* gene activity and cervical malignancies are also ambiguous. Baker *et al.*

[257] found that the *MUC5AC* gene is differently expressed in different forms of cervical cancer: MUC5AC-positiveness was observed in 38% of adenocarcinoma *in situ*, in 67% of simple adenocarcinomas, in 77% of invasive adenocarcinomas, and in 100% of adenosquamous carcinoma. Reithdorf *et al.* [268] also detected high level of MUC5AC mucin expression in cervical adenocarcinoma *in situ* and in invasive forms of cervical carcinoma, and diminished expression in most neoplastic glandular lesions. Zhao *et al.* [258] reported a marked decrease in MUC5AC expression in endocervical adenocarcinoma compared with normal endocervical epithelium. And finally, Shintaku *et al.* [559] reported that the colloid carcinoma of the intestinal type that develops in uterine cervix was MUC5AC-negative. Taken together, the data suggest that MUC5AC mucin may have specific functions in carcinogenesis of some types of cervical malignancies, with further studies needed to evaluate these functions.

c) Ovarian tumors: MUC5AC mucin has been also implicated in progression of ovarian cancer [274]. Mucinous tumors of the ovary have different phenotypes depending on mucin expression profile. Mall *et al.* [560] described expression of MUC5AC mucin in ovarian teratoma with cysts lined by colonic and respiratory mucosa. The respiratory component of the teratoma expressed MUC1 mucin, while the colonic component was positive for MUC2 and partially for MUC6 glycoproteins. Interestingly, MUC5AC mucin was expressed in both components of the tumor. Earlier, Bara *et al.* [561] showed that ovarian mucinous cysts, but not ovarian cysts of other histological types, expressed oncofetal M1antigen, encoded by the *MUC5AC* gene [386].

Boman *et al.* [274] analyzed the mucin gene transcripts in a series of benign adenomas and borderline mucinous tumors of the ovary. Intense expression of the *MUC5AC* gene, characteristic of gastric phenotype of tumor surface cells, was found in 86% of the examined specimens. These results are in accord with ultrastructural, histochemical and immunohistochemical data showing that in ovarian mucinous tumors, mucin-secreting cells often contain features of gastric cells [562-564]. According to Boman *et al.* [274], the gastric differentiation is an "early and almost constant event in ovarian mucinous tumorigenesis". This conclusion is in line with the data reported by Hirabayashi *et al.* [565] who studied mucin expression profiles in three main groups of ovarian mucinous tumors – mucinous adenomas, mucinous borderline lesions, and mucinous adenocarcinomas [566]. Based on the mucin expression profile and expression of cell specific marker CD10, the examined tumors were classified into four phenotypes: intestinal, gastrointestinal, gastric and unclassified. It has been established that gastrointestinal pattern and the expression of MUC2 and CD10 are increased from adenomas to carcinomas. In contrast, the expression of gastric pattern and activity of *MUC5AC* gene are decreased from adenoma to carcinoma. Hence, intestinal metaplasia, associated with more aggressive malignant lesions in the ovary, may arise from the gastric-like epithelium developed in mucinous adenomas. On the whole, there is close association between carcinogenesis and intestinal metaplasia in major ovarian mucinous tumors [565].

The ovarian epithelium differs from the uterine epithelium. As noted by Giuntoli *et al.* [272], "normal ovaries do not have goblet cells and glands histologically, and, therefore, the normal tissue would not be expected to express secretory mucin genes". However, non-malignant ovarian epithelial cell lines do express MUC5AC mRNA, although other secretory mucin genes are not expressed in these cells [272]. Malignant ovarian tissues of different histological types (endometrioid, mucinous, serous and mixed mesodermal tumors) exhibit mRNAs transcribed by *MUC2, MUC5B* and *MUC6* genes in addition to mRNA of the *MUC5AC* gene. The change from negative to positive expression of these mucin genes in malignant ovarian cells suggests that new onset expression of secretory mucins is associated with the carcinogenic process in the ovarian epithelium [274].

In summary, the analysis of *MUC5AC* expression in female reproductive organs shows that this mucin is an active and important participant in the physiological and pathological processes in reproductive epithelium. The dynamics of the *MUC5AC* gene expression in normal, metaplastic, borderline and highly malignant epithelial lesions may reflect the processes associated with carcinogenic transformation in female reproductive tract.

5.6.4. Expression of *MUC5AC* Gene in Pathology of the Male Urogenital Tract

Little is known about the mucin gene repertoire of the male urogenital tract. It has been much less studied in male tissues than in female organs, hence the data presented below are limited and in some cases not statistically reliable.

Kidney: According to available information, the *MUC5AC* gene is expressed in human kidney. Leroy *et al.* [243] could not detect its expression during either normal or abnormal renal development. No *MUC5AC* expression was observed in normal adult kidneys and renal cell carcinomas [246]. Clearly more studies are needed to evaluate the role, if any, of MUC5AC mucin in normal kidney physiology and in pathogenesis of renal diseases.

Urinary bladder: Like kidney epithelium, normal urothelium of the bladder shows MUC5AC-negativity in immunohistochemical study [248] and in *in situ* hybridization analysis [242]. However, aberrant neo-expression of MUC5AC mucin was observed in primary adenocarcinomas, signet ring cell carcinomas and transitional cell carcinoma of the bladder [248, 567]. MUC5AC was also detected in cystitis glandularis, a metaplastic lesion of bladder occurring in the presence of chronic inflammation [247].

Prostate: With regard to mucin expression, prostate has been studied more intensively than other male reproductive organs. High expression of MUC5AC mucin (M1 antigen) in the superficial collumnal cells of the *veru montanum* area of prostate were described by Daher *et al.* [568]. The majority of MUC5AC-positive cells were found in the posterior part of the urethra and mainly in the periutricular area. Moreover, the presence of MUC5AC-positive cells in fetal and adult urethral prostatic epithelium as well as in some tumors arising in this area was also evidenced [568]. Russo *et al.* [249] confirmed these data by detection of MUC5AC mRNA and protein in 67% of normal human prostates studied. MUC5AC mucin expression was observed also by Zhang *et al.* [252] in 16% of prostate glandular epithelia specimens, in 9% of primary prostate carcinoma, and in 11% of metastatic prostate cancer. However, Cozzi *et al.* [250] disproved the earlier reported data by showing that both normal and cancer prostate tissues are MUC5AC-negative. In agreement with their data [250] are results obtained by Legrier *et al.* [569] on mucin expression in human hormone-dependent and hormone-independent prostate tumors. Anti-MUC5AC antibodies stained the cytoplasm of rare cells in hormone-dependent adenocarcinoma, but did not stain a single cell in the hormone-independent variant of the tumor. These conflicting results preclude drawing conclusions about the relationship between MUC5AC expression and the function of the prostate in both normal physiological conditions and in pathology, showing the need for further study.

Testis, epididymis, seminal vesicle, urethra and foreskin: Very few studies have examined expression of mucin genes in testis, epididymis, seminal vesicle, urethra and foreskin. Russo *et al.* [249] used a sensitive PCR method and found expression of *MUC5AC* gene in 1 of 3 specimens of testis, in 1 of 1 specimen of *vas deferens*, in 3 of 7 samples of normal urethra, and in 2 of 4 foreskin samples. In another study, the expression of MUC5AC at the protein level was detected in the mucinous cystadenoma of testis [570], while in still another study the the same type of tumor was MUC5AC-negative [571].

In summary, the role of MUC5AC mucin in the male urogenital organs has been studied insufficiently [249]. The inconsistent results obtained from only a few samples indicate the necessity for further systematic study using up-to-date methods and larger numbers of clinical and experimental specimens.

5.6.5. Expression of *MUC5AC* Gene in Pathology of the Mammary Gland

Normal breast epithelium does not express MUC5AC mucin, while the pathologically changed epithelium of the mammary gland does express this glycoprotein [251, 572-574]. Several types of breast adenocarcinomas actively express the MUC5AC mucin [251, 572-574]. Pereira *et al.* [572] detected MUC5AC glycoprotein in 5 of 68 cases (7%) of invasive carcinomas, including one specimen of pure colloid carcinoma; it is noteworthy that none of the ductal hyperplasia specimens without atypia was positive for MUC5AC. Zhang *et al.* [251] found 5 of 6 breast carcinomas (83%) to be MUC5AC-positive, while Matsukita *et al.* [573] found 12% of mucinous adenocarcinomas and 4% of intraductal carcinomas to express the MUC5AC mucin. Rakha *et al.* [574] reported that 37% of 1447 invasive breast carcinomas expressed the MUC5AC mucin; interestingly, its expression was not associated with any of the clinico-pathological variables including prognostic factors.

Thus, according to the literature, *MUC5AC* is expressed in breast cancer in the range from 7% [572] to 83% [251] of cases, depending on histological type and malignant potential of the tumor. Further studies are needed to clarify the biological role of the non-breast MUC5AC mucin in pathogenesis of different types of breast tumors.

5.6.6. Expression of *MUC5AC* Gene in Pathology of the Eye

All kinds of mucins – membrane-bound, gel-forming and small soluble ones – are expressed by the ocular surface epithelia, including conjunctival goblet cells, and by the lacrimal duct and gland [575-581]. MUC5AC mucin is a hydrophilic glycoprotein involved in maintaining water on the surface of the eye. As the main component of the tear film [582-584], MUC5AC glycoprotein is believed to provide tear stability, prevent ocular surface drying out [585, 586], and protect the ocular epithelium against infection, mechanical and chemical injuries [575].

The role of MUC5AC mucin in pathogenesis of ocular diseases has been also investigated. Kurnet *et al.* [587] demonstated that repetitive application of allergens to mouse conjunctiva results in reduction in the number of conjunctival goblet cells and decreases Muc5ac mRNA expression. Dogru *et al.* [588] found MUC5AC mRNA expression significantly decreased in eyes of patients with corneal ulcer compared with control subjects. The authors showed that ocular surface inflammation, tear film instability and decreased expression of *MUC5AC* are important factors in the pathogenesis of allergic non-infectious corneal shield ulcers that occur in patients with atopic keratoconjunctivitis (AKC).

In contrast to AKC, patients with another allergic disease of eyes, vernal keratoconjunctivitis (VKC), have increased numbers of conjunctival goblet cells and increased levels of *MUC5AC* expression [589, 590]. The over-production of the MUC5AC mucin suggests a protective mechanism aimed at clearing allergens from the ocular surface [575].

Dry eyes is a common feature of several pathological conditions occurring *via* different pathogenetic pathways. Dry eyes is a symptom of complete androgen insensitivity syndrome (CAIS). A significant reduction in the expression of the *MUC5AC* gene associated with dryness and reduced tear film stability in CAIS has been reported [591]. Dry eyes is one of the main features of autoimmune diseases including Sjogren's syndrome. This systemic autoimmune disease is characterized by dryness of epithelial surfaces of eyes, mouth and other mucosae [575]. The ocular surface of patients affected by Sjogren's dry eyes displays decreased levels of MUC5AC mucin expression in conjunctival epithelium and in tears [592, 593]. The reduction in MUC5AC mucin content observed in patients with this syndrome results from a depletion in the number of MUC5AC-producing goblet cells [594-597].

In addition to diseases, therapeutic procedures may also affect MUC5AC mucin expression. Liu *et al.* [585] recently reported a change in the MUC5AC content of tear fluid of glaucoma patients after short-term treatment with anti-glaucoma eye drops and phacotrabeculectomy. The level of MUC5AC mucin after the eye drops was two times lower than in healthy control individuals, and after surgery was also significantly decreased, returning to pre-surgery levels only after six months. It has been established that the preservatives in eye drops can decrease the density of goblet cells in the conjunctival epithelium, which, in turn, decreases the MUC5AC content in tears [598-600].

In conclusion, the MUC5AC mucin plays an essential role in homeostasis of the ocular wet-surfaced epithelium. Its expression is important for protection of mucosal surface, lubrication, and clearance of allergens, pathogens and debris. Alterations of the *MUC5AC* gene expression at the ocular surface are associated with a number of diseases. Further understanding of the role MUC5AC mucin plays in maintaining the integrity of ocular mucosa may contribute to the development of new therapeutic modalities.

REFERENCES

[1] Nguyen VC, Aubert JP, Gross MS, *et al.* Assignment of human tracheobronchial mucin gene(s) to 11p15 and a tracheobronchial mucin-related sequence to chromosome 13. Hum Genet 1990;86:167-72.

[2] Dufosse J, Porchet N, Audie JP, *et al.* Degenerate 87-base-pair tandem repeats create hydrophilic/hydrophobic alternating domains in human mucin peptides mapped to 11p15. Biochem J 1993;293 (Pt 2):329-37.

[3] Guyonnet Duperat V, Audie JP, Debailleul V, *et al.* Characterization of the human mucin gene MUC5AC: a consensus cysteine-rich domain for 11p15 mucin genes? Biochem J 1995;305 (Pt 1):211-9.

[4] Meezaman D, Charles P, Daskal E, *et al.* Cloning and analysis of cDNA encoding a major airway glycoprotein, human tracheobronchial mucin (MUC5). J Biol Chem 1994;269:12932-9.

[5] Lesuffleur T, Roche F, Hill AS, *et al.* Characterization of a mucin cDNA clone isolated from HT-29 mucus-secreting cells. The 3' end of MUC5AC? J Biol Chem 1995;270:13665-73.

[6] Van Klinken BJ, Dekker J, Buller HA, Einerhand AW. Mucin gene structure and expression: protection *vs.* adhesion. Am J Physiol 1995;269:G613-27.

[7] Klomp LW, Van Rens L, Strous GJ. Cloning and analysis of human gastric mucin cDNA reveals two types of conserved cysteine-rich domains. Biochem J 1995;308 (Pt 3):831-8.
[8] Debailleul V, Laine A, Huet G, *et al.* Human mucin genes MUC2, MUC3, MUC4, MUC5AC, MUC5B, and MUC6 express stable and extremely large mRNAs and exhibit a variable length polymorphism. An improved method to analyze large mRNAs. J Biol Chem 1998;273:881-90.
[9] van de Bovenkamp JH, Hau CM, Strous GJ, *et al.* Molecular cloning of human gastric mucin MUC5AC reveals conserved cysteine-rich D-domains and a putative leucine zipper motif. Biochem Biophys Res Commun 1998;245:853-9.
[10] Buisine MP, Desseyn JL, Porchet N, *et al.* Genomic organization of the 3'-region of the human MUC5AC mucin gene: additional evidence for a common ancestral gene for the 11p15.5 mucin gene family. Biochem J 1998;332 (Pt 3):729-38.
[11] Li D, Gallup M, Fan N, Szymkowski DE, Basbaum CB. Cloning of the amino-terminal and 5'-flanking region of the human MUC5AC mucin gene and transcriptional up-regulation by bacterial exoproducts. J Biol Chem 1998;273:6812-20.
[12] Escande F, Aubert JP, Porchet N, Buisine MP. Human mucin gene MUC5AC: organization of its 5'-region and central repetitive region. Biochem J 2001;358:763-72.
[13] Moniaux N, Escande F, Porchet N, Aubert JP, Batra SK. Structural organization and classification of the human mucin genes. Front Biosci 2001;6:D1192-206.
[14] Desseyn JL, Buisine MP, Porchet N, Aubert JP, Laine A. Genomic organization of the human mucin gene MUC5B. cDNA and genomic sequences upstream of the large central exon. J Biol Chem 1998;273:30157-64.
[15] Gum JR, Jr., Hicks JW, Toribara NW, *et al.* The human MUC2 intestinal mucin has cysteine-rich subdomains located both upstream and downstream of its central repetitive region. J Biol Chem 1992;267:21375-83.
[16] Gum JR, Jr., Hicks JW, Toribara NW, Siddiki B, Kim YS. Molecular cloning of human intestinal mucin (MUC2) cDNA. Identification of the amino terminus and overall sequence similarity to prepro-von Willebrand factor. J Biol Chem 1994;269:2440-6.
[17] van Klinken BJ, Einerhand AW, Buller HA, Dekker J. The oligomerization of a family of four genetically clustered human gastrointestinal mucins. Glycobiology 1998;8:67-75.
[18] Rose MC, Voynow JA. Respiratory tract mucin genes and mucin glycoproteins in health and disease. Physiol Rev 2006;86:245-78.
[19] Rose MC, Voter WA, Brown CF, Kaufman B. Structural features of human tracheobronchial mucus glycoprotein. Biochem J 1984;222:371-7.
[20] Rose MC, Voter WA, Sage H, Brown CF, Kaufman B. Effects of deglycosylation on the architecture of ovine submaxillary mucin glycoprotein. J Biol Chem 1984;259:3167-72.
[21] Cheng PW, Boucher RC, Yankaskas JR, Boat TF. In: Mastella G, Quinton PM, editors. Cellular and molecular basis of cystic fibrosis. San Francisco: San Francisco Press; 1988. p. 233-8.
[22] Desseyn JL, Aubert JP, Van Seuningen I, Porchet N, Laine A. Genomic organization of the 3' region of the human mucin gene MUC5B. J Biol Chem 1997;272:16873-83.
[23] Nollet S, Forgue-Lafitte ME, Kirkham P, Bara J. Mapping of two new epitopes on the apomucin encoded by MUC5AC gene: expression in normal GI tract and colon tumors. Int J Cancer 2002;99:336-43.
[24] Song KS, Lee WJ, Chung KC, *et al.* Interleukin-1 β and tumor necrosis factor-α induce MUC5AC overexpression through a mechanism involving ERK/p38 mitogen-activated protein kinases-MSK1-CREB activation in human airway epithelial cells. J Biol Chem 2003;278:23243-50.

[25] Gray T, Coakley R, Hirsh A, *et al.* Regulation of MUC5AC mucin secretion and airway surface liquid metabolism by IL-1β in human bronchial epithelia. Am J Physiol Lung Cell Mol Physiol 2004;286:L320-30.

[26] Vermeer PD, Harson R, Einwalter LA, Moninger T, Zabner J. Interleukin-9 induces goblet cell hyperplasia during repair of human airway epithelia. Am J Respir Cell Mol Biol 2003;28:286-95.

[27] Atherton HC, Jones G, Danahay H. IL-13-induced changes in the goblet cell density of human bronchial epithelial cell cultures: MAP kinase and phosphatidylinositol 3-kinase regulation. Am J Physiol Lung Cell Mol Physiol 2003;285:L730-9.

[28] Chen Y, Thai P, Zhao YH, *et al.* Stimulation of airway mucin gene expression by interleukin (IL)-17 through IL-6 paracrine/autocrine loop. J Biol Chem 2003;278:17036-43.

[29] Shao MX, Nakanaga T, Nadel JA. Cigarette smoke induces MUC5AC mucin overproduction *via* tumor necrosis factor-α-converting enzyme in human airway epithelial (NCI-H292) cells. Am J Physiol Lung Cell Mol Physiol 2004;287:L420-7.

[30] Takeyama K, Dabbagh K, Lee HM, *et al.* Epidermal growth factor system regulates mucin production in airways. Proc Natl Acad Sci USA 1999;96:3081-6.

[31] Chu HW, Balzar S, Seedorf GJ, *et al.* Transforming growth factor-β2 induces bronchial epithelial mucin expression in asthma. Am J Pathol 2004;165:1097-106.

[32] Fischer BM, Voynow JA. Neutrophil elastase induces MUC5AC gene expression in airway epithelium *via* a pathway involving reactive oxygen species. Am J Respir Cell Mol Biol 2002;26:447-52.

[33] Turner J, Jones CE. Regulation of mucin expression in respiratory diseases. Biochem Soc Trans 2009;37:877-81.

[34] Dohrman A, Miyata S, Gallup M, *et al.* Mucin gene (MUC 2 and MUC 5AC) upregulation by Gram-positive and Gram-negative bacteria. Biochim Biophys Acta 1998;1406:251-9.

[35] Basbaum C, Lemjabbar H, Longphre M, *et al.* Control of mucin transcription by diverse injury-induced signaling pathways. Am J Respir Crit Care Med. 1999;160:S44-8.

[36] Pigny P, Pratt WS, Laine A, *et al.* The MUC5AC gene: RFLP analysis with the Jer58 probe. Hum Genet 1995;96:367-8.

[37] Li JD, Dohrman AF, Gallup M, *et al.* Transcriptional activation of mucin by Pseudomonas aeruginosa lipopolysaccharide in the pathogenesis of cystic fibrosis lung disease. Proc Natl Acad Sci USA 1997;94:967-72.

[38] Li JD, Feng W, Gallup M, *et al.* Activation of NF-κB *via* a Src-dependent Ras-MAPK-pp90rsk pathway is required for Pseudomonas aeruginosa-induced mucin overproduction in epithelial cells. Proc Natl Acad Sci USA 1998;95:5718-23.

[39] Jonckheere N, Van Der Sluis M, Velghe A, *et al.* Transcriptional activation of the murine Muc5ac mucin gene in epithelial cancer cells by TGF-β/Smad4 signaling pathway is potentiated by Sp1. Biochem J 2004;377:797-808.

[40] Young HW, Williams OW, Chandra D, *et al.* Central role of Muc5ac expression in mucous metaplasia and its regulation by conserved 5' elements. Am J Respir Cell Mol Biol 2007;37:273-90.

[41] Longphre M, Li D, Gallup M, *et al.* Allergen-induced IL-9 directly stimulates mucin transcription in respiratory epithelial cells. J Clin Invest 1999;104:1375-82.

[42] Leikauf GD, Borchers MT, Prows DR, Simpson LG. Mucin apoprotein expression in COPD. Chest 2002;121:166S-82S.

[43] Williams OW, Sharafkhaneh A, Kim V, Dickey BF, Evans CM. Airway mucus: From production to secretion. Am J Respir Cell Mol Biol 2006;34:527-36.

[44] Curran DR, Cohn L. Advances in mucous cell metaplasia: a plug for mucus as a therapeutic focus in chronic airway disease. Am J Respir Cell Mol Biol 2010;42:268-75.
[45] Evans CM, Kim K, Tuvim MJ, Dickey BF. Mucus hypersecretion in asthma: causes and effects. Curr Opin Pulm Med 2009;15:4-11.
[46] Kuperman DA, Huang X, Koth LL, *et al.* Direct effects of interleukin-13 on epithelial cells cause airway hyperreactivity and mucus overproduction in asthma. Nat Med 2002;8:885-9.
[47] Kuperman DA, Huang X, Nguyenvu L, *et al.* IL-4 receptor signaling in Clara cells is required for allergen-induced mucus production. J Immunol 2005;175:3746-52.
[48] Whittaker L, Niu N, Temann UA, *et al.* Interleukin-13 mediates a fundamental pathway for airway epithelial mucus induced by CD4 T cells and interleukin-9. Am J Respir Cell Mol Biol 2002;27:593-602.
[49] Zhen G, Park SW, Nguyenvu LT, *et al.* IL-13 and epidermal growth factor receptor have critical but distinct roles in epithelial cell mucin production. Am J Respir Cell Mol Biol 2007;36:244-53.
[50] Burgel PR, Nadel JA. Roles of epidermal growth factor receptor activation in epithelial cell repair and mucin production in airway epithelium. Thorax 2004;59:992-6.
[51] Walter DM, McIntire JJ, Berry G, *et al.* Critical role for IL-13 in the development of allergen-induced airway hyperreactivity. J Immunol 2001;167:4668-75.
[52] Justice JP, Crosby J, Borchers MT, *et al.* CD4(+) T cell-dependent airway mucus production occurs in response to IL-5 expression in lung. Am J Physiol Lung Cell Mol Physiol 2002;282:L1066-74.
[53] Temann UA, Ray P, Flavell RA. Pulmonary overexpression of IL-9 induces Th2 cytokine expression, leading to immune pathology. J Clin Invest 2002;109:29-39.
[54] Temann UA, Laouar Y, Eynon EE, Homer R, Flavell RA. IL9 leads to airway inflammation by inducing IL13 expression in airway epithelial cells. Int Immunol 2007;19:1-10.
[55] Zuhdi Alimam M, Piazza FM, Selby DM, *et al.* Muc-5/5ac mucin messenger RNA and protein expression is a marker of goblet cell metaplasia in murine airways. Am J Respir Cell Mol Biol 2000;22:253-60.
[56] Kuperman D, Schofield B, Wills-Karp M, Grusby MJ. Signal transducer and activator of transcription factor 6 (Stat6)-deficient mice are protected from antigen-induced airway hyperresponsiveness and mucus production. J Exp Med 1998;187:939-48.
[57] Wan H, Kaestner KH, Ang SL, *et al.* Foxa2 regulates alveolarization and goblet cell hyperplasia. Development 2004;131:953-64.
[58] Chen G, Wan H, Luo F, *et al.* Foxa2 programs Th2 cell-mediated innate immunity in the developing lung. J Immunol 2010;184:6133-41.
[59] McMahon S, Charbonneau M, Grandmont S, Richard DE, Dubois CM. Transforming growth factor β1 induces hypoxia-inducible factor-1 stabilization through selective inhibition of PHD2 expression. J Biol Chem 2006;281:24171-81.
[60] Shao MX, Ueki IF, Nadel JA. Tumor necrosis factor α-converting enzyme mediates MUC5AC mucin expression in cultured human airway epithelial cells. Proc Natl Acad Sci USA 2003;100:11618-23.
[61] Basbaum C, Li D, Gensch E, Gallup M, Lemjabbar H. Mechanisms by which gram-positive bacteria and tobacco smoke stimulate mucin induction through the epidermal growth factor receptor (EGFR). Novartis Found Symp 2002;248:171-6; discussion 6-80, 277-82.
[62] Shim JJ, Dabbagh K, Ueki IF, *et al.* IL-13 induces mucin production by stimulating epidermal growth factor receptors and by activating neutrophils. Am J Physiol Lung Cell Mol Physiol 2001;280:L134-40.
[63] Hackel PO, Zwick E, Prenzel N, Ullrich A. Epidermal growth factor receptors: critical mediators of multiple receptor pathways. Curr Opin Cell Biol 1999;11:184-9.

[64] Takeyama K, Fahy JV, Nadel JA. Relationship of epidermal growth factor receptors to goblet cell production in human bronchi. Am J Respir Crit Care Med 2001;163:511-6.
[65] Amishima M, Munakata M, Nasuhara Y, *et al.* Expression of epidermal growth factor and epidermal growth factor receptor immunoreactivity in the asthmatic human airway. Am J Respir Crit Care Med 1998;157:1907-12.
[66] Polosa R, Puddicombe SM, Krishna MT, *et al.* Expression of c-erbB receptors and ligands in the bronchial epithelium of asthmatic subjects. J Allergy Clin Immunol 2002;109:75-81.
[67] Voynow JA, Fischer BM, Roberts BC, Proia AD. Basal-like cells constitute the proliferating cell population in cystic fibrosis airways. Am J Respir Crit Care Med 2005;172:1013-8.
[68] de Boer WI, Hau CM, van Schadewijk A, *et al.* Expression of epidermal growth factors and their receptors in the bronchial epithelium of subjects with chronic obstructive pulmonary disease. Am J Clin Pathol 2006;125:184-92.
[69] Takeyama K, Jung B, Shim JJ, *et al.* Activation of epidermal growth factor receptors is responsible for mucin synthesis induced by cigarette smoke. Am J Physiol Lung Cell Mol Physiol 2001;280:L165-72.
[70] Perrais M, Pigny P, Copin MC, Aubert JP, Van Seuningen I. Induction of MUC2 and MUC5AC mucins by factors of the epidermal growth factor (EGF) family is mediated by EGF receptor/Ras/Raf/extracellular signal-regulated kinase cascade and Sp1. J Biol Chem 2002;277:32258-67.
[71] Takeyama K, Dabbagh K, Jeong Shim J, *et al.* Oxidative stress causes mucin synthesis *via* transactivation of epidermal growth factor receptor: role of neutrophils. J Immunol 2000;164:1546-52.
[72] Sanchez-Elsner T, Botella LM, Velasco B, *et al.* Synergistic cooperation between hypoxia and transforming growth factor-β pathways on human vascular endothelial growth factor gene expression. J Biol Chem 2001;276:38527-35.
[73] Zhang H, Akman HO, Smith EL, *et al.* Cellular response to hypoxia involves signaling *via* Smad proteins. Blood 2003;101:2253-60.
[74] Park KS, Korfhagen TR, Bruno MD, *et al.* SPDEF regulates goblet cell hyperplasia in the airway epithelium. J Clin Invest 2007;117:978-88.
[75] Chen G, Korfhagen TR, Xu Y, *et al.* SPDEF is required for mouse pulmonary goblet cell differentiation and regulates a network of genes associated with mucus production. J Clin Invest 2009;119:2914-24.
[76] Hewson CA, Edbrooke MR, Johnston SL. PMA induces the MUC5AC respiratory mucin in human bronchial epithelial cells, *via* PKC, EGF/TGF-α, Ras/Raf, MEK, ERK and Sp1-dependent mechanisms. J Mol Biol 2004;344:683-95.
[77] Ha U, Lim JH, Jono H, *et al.* A novel role for IκB kinase (IKK) α and IKKβ in ERK-dependent up-regulation of MUC5AC mucin transcription by Streptococcus pneumoniae. J Immunol 2007;178:1736-47.
[78] Lim JH, Ha U, Sakai A, *et al.* Streptococcus pneumoniae synergizes with nontypeable Haemophilus influenzae to induce inflammation *via* upregulating TLR2. BMC Immunol 2008;9:40.
[79] Ha UH, Lim JH, Kim HJ, *et al.* MKP1 regulates the induction of MUC5AC mucin by Streptococcus pneumoniae pneumolysin by inhibiting the PAK4-JNK signaling pathway. J Biol Chem 2008;283:30624-31.
[80] Lim JH, Stirling B, Derry J, *et al.* Tumor suppressor CYLD regulates acute lung injury in lethal Streptococcus pneumoniae infections. Immunity 2007;27:349-60.

[81] Oppenheimer EH, Esterly JR. Pathology of cystic fibrosis review of the literature and comparison with 146 autopsied cases. Perspect Pediatr Pathol 1975;2:241-78.
[82] Klinger JD, Tandler B, Liedtke CM, Boat TF. Proteinases of Pseudomonas aeruginosa evoke mucin release by tracheal epithelium. J Clin Invest 1984;74:1669-78.
[83] Doring G, Obernesser HJ, Botzenhart K, *et al.* Proteases of Pseudomonas aeruginosa in patients with cystic fibrosis. J Infect Dis 1983;147:744-50.
[84] Liu PV. Extracellular toxins of Pseudomonas aeruginosa. J Infect Dis 1974;130 Suppl:S94-9.
[85] Yan F, Li W, Jono H, *et al.* Reactive oxygen species regulate Pseudomonas aeruginosa lipopolysaccharide-induced MUC5AC mucin expression *via* PKC-NADPH oxidase-ROS-TGF-α signaling pathways in human airway epithelial cells. Biochem Biophys Res Commun 2008;366:513-9.
[86] Kohri K, Ueki IF, Shim JJ, *et al.* Pseudomonas aeruginosa induces MUC5AC production *via* epidermal growth factor receptor. Eur Respir J 2002;20:1263-70.
[87] Song KS, Kim HJ, Kim K, Lee JG, Yoon JH. Regulator of G-protein signaling 4 suppresses LPS-induced MUC5AC overproduction in the airway. Am J Respir Cell Mol Biol 2009;41:40-9.
[88] Tarran R, Button B, Boucher RC. Regulation of normal and cystic fibrosis airway surface liquid volume by phasic shear stress. Annu Rev Physiol 2006;68:543-61.
[89] Hauber HP, Goldmann T, Vollmer E, *et al.* LPS-induced mucin expression in human sinus mucosa can be attenuated by hCLCA inhibitors. J Endotoxin Res 2007;13:109-16.
[90] Breuer R, Christensen TG, Lucey EC, Stone PJ, Snider GL. An ultrastructural morphometric analysis of elastase-treated hamster bronchi shows discharge followed by progressive accumulation of secretory granules. Am Rev Respir Dis 1987;136:698-703.
[91] Breuer R, Christensen TG, Niles RM, Stone PJ, Snider GL. Human neutrophil elastase causes glycoconjugate release from the epithelial cell surface of hamster trachea in organ culture. Am Rev Respir Dis 1989;139:779-82.
[92] Voynow JA, Young LR, Wang Y, *et al.* Neutrophil elastase increases MUC5AC mRNA and protein expression in respiratory epithelial cells. Am J Physiol 1999;276:L835-43.
[93] Shao MX, Nadel JA. Neutrophil elastase induces MUC5AC mucin production in human airway epithelial cells *via* a cascade involving protein kinase C, reactive oxygen species, and TNF-α-converting enzyme. J Immunol 2005;175:4009-16.
[94] Qi L, Xiangdong Z, Hongmei Y, Xiaohong N, Xiaoyan X. Regulation of neutrophil elastase-induced MUC5AC expression by nuclear factor erythroid-2 related factor 2 in human airway epithelial cells. J Investig Med 2010;58:730-6.
[95] Jang MK, Kim SH, Lee KY, *et al.* The tyrosine phosphatase, SHP-1, is involved in bronchial mucin production during oxidative stress. Biochem Biophys Res Commun 2010;393:137-43.
[96] Chong ZZ, Maiese K. The Src homology 2 domain tyrosine phosphatases SHP-1 and SHP-2: diversified control of cell growth, inflammation, and injury. Histol Histopathol 2007;22:1251-67.
[97] Oh SY, Zheng T, Kim YK, *et al.* A critical role of SHP-1 in regulation of type 2 inflammation in the lung. Am J Respir Cell Mol Biol 2009;40:568-74.
[98] Krotz F, Engelbrecht B, Buerkle MA, *et al.* The tyrosine phosphatase, SHP-1, is a negative regulator of endothelial superoxide formation. J Am Coll Cardiol 2005;45:1700-6.
[99] Meng TC, Fukada T, Tonks NK. Reversible oxidation and inactivation of protein tyrosine phosphatases *in vivo*. Mol Cell 2002;9:387-99.
[100] Alonso A, Sasin J, Bottini N, *et al.* Protein tyrosine phosphatases in the human genome. Cell 2004;117:699-711.

[101] Kohri K, Ueki IF, Nadel JA. Neutrophil elastase induces mucin production by ligand-dependent epidermal growth factor receptor activation. Am J Physiol Lung Cell Mol Physiol 2002;283:L531-40.

[102] Tschumperlin DJ, Dai G, Maly IV, *et al.* Mechanotransduction through growth-factor shedding into the extracellular space. Nature 2004;429:83-6.

[103] Park JA, Tschumperlin DJ. Chronic intermittent mechanical stress increases MUC5AC protein expression. Am J Respir Cell Mol Biol 2009;41:459-66.

[104] Kettle R, Simmons J, Schindler F, *et al.* Regulation of neuregulin 1β1-induced MUC5AC and MUC5B expression in human airway epithelium. Am J Respir Cell Mol Biol 2010;42:472-81.

[105] Schroder K, Tschopp J. The inflammasomes. Cell 2010;140:821-32.

[106] Barnes PJ. Cytokines as mediators of chronic asthma. Am J Respir Crit Care Med 1994;150:S42-9.

[107] Lai HY, Rogers DF. Mucus hypersecretion in asthma: intracellular signaling pathways as targets for pharmacotherapy. Curr Opin Allergy Clin Immunol 2010;10:67-76.

[108] Thai P, Loukoianov A, Wachi S, Wu R. Regulation of airway mucin gene expression. Annu Rev Physiol 2008;70:405-29.

[109] Kim YD, Kwon EJ, Park DW, *et al.* Interleukin-1β induces MUC2 and MUC5AC synthesis through cyclooxygenase-2 in NCI-H292 cells. Mol Pharmacol 2002;62:1112-8.

[110] Fujisawa T, Velichko S, Thai P, *et al.* Regulation of airway MUC5AC expression by IL-1β and IL-17A; the NF-κB paradigm. J Immunol 2009;183:6236-43.

[111] Hart LA, Krishnan VL, Adcock IM, Barnes PJ, Chung KF. Activation and localization of transcription factor, nuclear factor-κB, in asthma. Am J Respir Crit Care Med 1998;158:1585-92.

[112] Di Stefano A, Caramori G, Oates T, *et al.* Increased expression of nuclear factor-κB in bronchial biopsies from smokers and patients with COPD. Eur Respir J 2002;20:556-63.

[113] Chen R, Lim JH, Jono H, *et al.* Nontypeable Haemophilus influenzae lipoprotein P6 induces MUC5AC mucin transcription *via* TLR2-TAK1-dependent p38 MAPK-AP1 and IKKβ-IκBα-NF-κB signaling pathways. Biochem Biophys Res Commun 2004;324:1087-94.

[114] Kraft M, Adler KB, Ingram JL, *et al.* Mycoplasma pneumoniae induces airway epithelial cell expression of MUC5AC in asthma. Eur Respir J 2008;31:43-6.

[115] Lora JM, Zhang DM, Liao SM, *et al.* Tumor necrosis factor-α triggers mucus production in airway epithelium through an IκB kinase β-dependent mechanism. J Biol Chem 2005;280:36510-7.

[116] Issa R, Xie S, Lee KY, *et al.* GRO-α regulation in airway smooth muscle by IL-1β and TNF-α: role of NF-κB and MAP kinases. Am J Physiol Lung Cell Mol Physiol 2006;291:L66-74.

[117] Dinarello CA. Biologic basis for interleukin-1 in disease. Blood 1996;87:2095-147.

[118] Huang F, Kao CY, Wachi S, *et al.* Requirement for both JAK-mediated PI3K signaling and ACT1/TRAF6/TAK1-dependent NF-κB activation by IL-17A in enhancing cytokine expression in human airway epithelial cells. J Immunol 2007;179:6504-13.

[119] Kao CY, Kim C, Huang F, Wu R. Requirements for two proximal NF-κB binding sites and IκB-zeta in IL-17A-induced human β-defensin 2 expression by conducting airway epithelium. J Biol Chem 2008;283:15309-18.

[120] Kim CH, Kim KE, Yoon JH, Song KS. Upregulation of MUC5AC gene expression by IL-4 through CREB in human airway epithelial cells. J Cell Biochem 2009;108:974-81.

[121] Aggarwal S, Kim SW, Ryu SH, Chung WC, Koo JS. Growth suppression of lung cancer cells by targeting cyclic AMP response element-binding protein. Cancer Res 2008;68:981-8.

[122] Gonzalez GA, Yamamoto KK, Fischer WH, *et al.* A cluster of phosphorylation sites on the cyclic AMP-regulated nuclear factor CREB predicted by its sequence. Nature 1989;337:749-52.
[123] Colbran JL, Roach PJ, Fiol CJ, *et al.* cAMP-dependent protein kinase, but not the cGMP-dependent enzyme, rapidly phosphorylates delta-CREB, and a synthetic delta-CREB peptide. Biochem Cell Biol 1992;70:1277-82.
[124] Yamamoto KK, Gonzalez GA, Biggs WH, 3rd, Montminy MR. Phosphorylation-induced binding and transcriptional efficacy of nuclear factor CREB. Nature 1988;334:494-8.
[125] Tan Y, Rouse J, Zhang A, *et al.* FGF and stress regulate CREB and ATF-1 *via* a pathway involving p38 MAP kinase and MAPKAP kinase-2. Embo J 1996;15:4629-42.
[126] Du K, Montminy M. CREB is a regulatory target for the protein kinase Akt/PKB. J Biol Chem 1998;273:32377-9.
[127] Song KS, Seong JK, Chung KC, *et al.* Induction of MUC8 gene expression by interleukin-1 β is mediated by a sequential ERK MAPK/RSK1/CREB cascade pathway in human airway epithelial cells. J Biol Chem 2003;278:34890-6.
[128] Parry GC, Mackman N. Role of cyclic AMP response element-binding protein in cyclic AMP inhibition of NF-κB-mediated transcription. J Immunol 1997;159:5450-6.
[129] Shenkar R, Yum HK, Arcaroli J, Kupfner J, Abraham E. Interactions between CBP, NF-κB, and CREB in the lungs after hemorrhage and endotoxemia. Am J Physiol Lung Cell Mol Physiol 2001;281:L418-26.
[130] Yalcin A, Koulich E, Mohamed S, Liu L, D'Mello SR. Apoptosis in cerebellar granule neurons is associated with reduced interaction between CREB-binding protein and NF-κB. J Neurochem 2003;84:397-408.
[131] Song KS, Lee TJ, Kim K, Chung KC, Yoon JH. cAMP-responding element-binding protein and c-Ets1 interact in the regulation of ATP-dependent MUC5AC gene expression. J Biol Chem 2008;283:26869-78.
[132] Gray TE, Guzman K, Davis CW, Abdullah LH, Nettesheim P. Mucociliary differentiation of serially passaged normal human tracheobronchial epithelial cells. Am J Respir Cell Mol Biol 1996;14:104-12.
[133] Gudas LJ. Molecular mechanisms of retinoid action. Am J Respir Cell Mol Biol 1990;2:319-20.
[134] Koo JS, Jetten AM, Belloni P, *et al.* Role of retinoid receptors in the regulation of mucin gene expression by retinoic acid in human tracheobronchial epithelial cells. Biochem J 1999;338 (Pt 2):351-7.
[135] Koo JS, Yoon JH, Gray T, *et al.* Restoration of the mucous phenotype by retinoic acid in retinoid-deficient human bronchial cell cultures: changes in mucin gene expression. Am J Respir Cell Mol Biol 1999;20:43-52.
[136] Chambon P. A decade of molecular biology of retinoic acid receptors. Faseb J 1996;10:940-54.
[137] Giguere V. Retinoic acid receptors and cellular retinoid binding proteins: complex interplay in retinoid signaling. Endocr Rev 1994;15:61-79.
[138] Aggarwal S, Kim SW, Cheon K, *et al.* Nonclassical action of retinoic acid on the activation of the cAMP response element-binding protein in normal human bronchial epithelial cells. Mol Biol Cell 2006;17:566-75.
[139] Kim SW, Hong JS, Ryu SH, *et al.* Regulation of mucin gene expression by CREB *via* a nonclassical retinoic acid signaling pathway. Mol Cell Biol 2007;27:6933-47.
[140] Chung WC, Ryu SH, Sun H, Zeldin DC, Koo JS. CREB mediates prostaglandin F2α-induced MUC5AC overexpression. J Immunol 2009;182:2349-56.
[141] Smith CJ, Hansch C. The relative toxicity of compounds in mainstream cigarette smoke condensate. Food Chem Toxicol 2000;38:637-46.

[142] Gensch E, Gallup M, Sucher A, *et al.* Tobacco smoke control of mucin production in lung cells requires oxygen radicals AP-1 and JNK. J Biol Chem 2004;279:39085-93.
[143] Zhao J, Harper R, Barchowsky A, Di YP. Identification of multiple MAPK-mediated transcription factors regulated by tobacco smoke in airway epithelial cells. Am J Physiol Lung Cell Mol Physiol 2007;293:L480-90.
[144] Baginski TK, Dabbagh K, Satjawatcharaphong C, Swinney DC. Cigarette smoke synergistically enhances respiratory mucin induction by proinflammatory stimuli. Am J Respir Cell Mol Biol 2006;35:165-74.
[145] Lee SY, Kang EJ, Hur GY, *et al.* Peroxisome proliferator-activated receptor-gamma inhibits cigarette smoke solution-induced mucin production in human airway epithelial (NCI-H292) cells. Am J Physiol Lung Cell Mol Physiol 2006;291:L84-90.
[146] Lee SY, Kang EJ, Hur GY, *et al.* The inhibitory effects of rebamipide on cigarette smoke-induced airway mucin production. Respir Med 2006;100:503-11.
[147] Baraldo S, Bazzan E, Turato G, *et al.* Decreased expression of TGF-β type II receptor in bronchial glands of smokers with COPD. Thorax 2005;60:998-1002.
[148] Lemjabbar H, Li D, Gallup M, *et al.* Tobacco smoke-induced lung cell proliferation mediated by tumor necrosis factor α-converting enzyme and amphiregulin. J Biol Chem 2003;278:26202-7.
[149] Mishra NC, Rir-Sima-Ah J, Langley RJ, *et al.* Nicotine primarily suppresses lung Th2 but not goblet cell and muscle cell responses to allergens. J Immunol 2008;180:7655-63.
[150] Fujii S, Jia Y, Yang A, Sumikawa K. Nicotine reverses GABAergic inhibition of long-term potentiation induction in the hippocampal CA1 region. Brain Res 2000;863:259-65.
[151] Xiang YY, Wang S, Liu M, *et al.* A GABAergic system in airway epithelium is essential for mucus overproduction in asthma. Nat Med 2007;13:862-7.
[152] Borchers MT, Wert SE, Leikauf GD. Acrolein-induced MUC5ac expression in rat airways. Am J Physiol 1998;274:L573-81.
[153] Borchers MT, Wesselkamper S, Wert SE, Shapiro SD, Leikauf GD. Monocyte inflammation augments acrolein-induced Muc5ac expression in mouse lung. Am J Physiol 1999;277:L489-97.
[154] Borchers MT, Carty MP, Leikauf GD. Regulation of human airway mucins by acrolein and inflammatory mediators. Am J Physiol 1999;276:L549-55.
[155] Deshmukh HS, Case LM, Wesselkamper SC, *et al.* Metalloproteinases mediate mucin 5AC expression by epidermal growth factor receptor activation. Am J Respir Crit Care Med 2005;171:305-14.
[156] Deshmukh HS, Shaver C, Case LM, *et al.* Acrolein-activated matrix metalloproteinase 9 contributes to persistent mucin production. Am J Respir Cell Mol Biol 2008;38:446-54.
[157] Bai CH, Song SY, Kim YD. The inhibitory effect of the leukotriene receptor antagonist on leukotriene D4-induced MUC2/5AC gene expression and mucin secretion in human airway epithelial cells. Auris Nasus Larynx 2007;34:203-6.
[158] Chen Z, Wang X, Gao L, *et al.* Regulation of MUC5AC mucin secretion by depletion of AQP5 in SPC-A1 cells. Biochem Biophys Res Commun 2006;342:775-81.
[159] Longphre M, Li D, Li J, *et al.* Lung mucin production is stimulated by the air pollutant residual oil fly ash. Toxicol Appl Pharmacol 2000;162:86-92.
[160] Song JS, Kang CM, Yoo MB, *et al.* Nitric oxide induces MUC5AC mucin in respiratory epithelial cells through PKC and ERK dependent pathways. Respir Res 2007;8:28.
[161] Adler KB, Akley NJ, Glasgow WC. Platelet-activating factor provokes release of mucin-like glycoproteins from guinea pig respiratory epithelial cells *via* a lipoxygenase-dependent mechanism. Am J Respir Cell Mol Biol 1992;6:550-6.

[162] Liu J, Zhang Z, Xu Y, Xing L, Zhang H. Effects of glucocorticoid on IL-13-induced Muc5ac expression in airways of mice. J Huazhong Univ Sci Technolog Med Sci 2004;24:575-7.
[163] Komatsu K, Jono H, Lim JH, *et al.* Glucocorticoids inhibit nontypeable Haemophilus influenzae-induced MUC5AC mucin expression *via* MAPK phosphatase-1-dependent inhibition of p38 MAPK. Biochem Biophys Res Commun 2008;377:763-8.
[164] Chen Y, Nickola TJ, DiFronzo NL, Colberg-Poley AM, Rose MC. Dexamethasone-mediated repression of MUC5AC gene expression in human lung epithelial cells. Am J Respir Cell Mol Biol 2006;34:338-47.
[165] Schoneveld OJ, Gaemers IC, Lamers WH. Mechanisms of glucocorticoid signaling. Biochim Biophys Acta 2004;1680:114-28.
[166] Dostert A, Heinzel T. Negative glucocorticoid receptor response elements and their role in glucocorticoid action. Curr Pharm Des 2004;10:2807-16.
[167] Webster JC, Cidlowski JA. Mechanisms of Glucocorticoid-receptor-mediated Repression of Gene Expression. Trends Endocrinol Metab 1999;10:396-402.
[168] Martinez-Anton A, de Bolos C, Alobid I, *et al.* Corticosteroid therapy increases membrane-tethered while decreases secreted mucin expression in nasal polyps. Allergy 2008;63:1368-76.
[169] Yamada N, Nishida Y, Tsutsumida H, *et al.* MUC1 expression is regulated by DNA methylation and histone H3 lysine 9 modification in cancer cells. Cancer Res 2008;68:2708-16.
[170] Yamada N, Nishida Y, Tsutsumida H, *et al.* Promoter CpG methylation in cancer cells contributes to the regulation of MUC4. Br J Cancer 2009;100:344-51.
[171] Yamada N, Hamada T, Goto M, *et al.* MUC2 expression is regulated by histone H3 modification and DNA methylation in pancreatic cancer. Int J Cancer 2006;119:1850-7.
[172] Yamada N, Yokoyama S, Nishida Y. 68th Annual Meeting of the Japanese Cancer Society; 2009; Yokohama; 2009.
[173] Bautista MV, Chen Y, Ivanova VS, *et al.* IL-8 regulates mucin gene expression at the posttranscriptional level in lung epithelial cells. J Immunol 2009;183:2159-66.
[174] de Moor CH, Meijer H, Lissenden S. Mechanisms of translational control by the 3' UTR in development and differentiation. Semin Cell Dev Biol 2005;16:49-58.
[175] Moore MJ. From birth to death: the complex lives of eukaryotic mRNAs. Science 2005;309:1514-8.
[176] Desseyn JL, Gouyer V, Tetaert D. Architecture of the gel-forming mucins. In: van Seuningen I, editor. The Epithelial mucins: Structure/Function Roles in Cancer and Inflammatory Disease. Kerala, India,: Research Signpost; 2008. p. 1-16.
[177] Sternberg LR, Byrd JC, Hansson GC, Liu KF, Bresalier RS. Alternative splicing of the human MUC2 gene. Arch Biochem Biophys 2004;421:21-33.
[178] Sheehan JK, Kirkham S, Howard M, *et al.* Identification of molecular intermediates in the assembly pathway of the MUC5AC mucin. J Biol Chem 2004;279:15698-705.
[179] Hovenberg HW, Davies JR, Carlstedt I. Different mucins are produced by the surface epithelium and the submucosa in human trachea: identification of MUC5AC as a major mucin from the goblet cells. Biochem J 1996;318 (Pt 1):319-24.
[180] Thornton DJ, Rousseau K, McGuckin MA. Structure and function of the polymeric mucins in airways mucus. Annu Rev Physiol 2008;70:459-86.
[181] Asker N, Axelsson MA, Olofsson SO, Hansson GC. Human MUC5AC mucin dimerizes in the rough endoplasmic reticulum, similarly to the MUC2 mucin. Biochem J 1998;335 (Pt 2):381-7.
[182] McCool DJ, Okada Y, Forstner JF, Forstner GG. Roles of calreticulin and calnexin during mucin synthesis in LS180 and HT29/A1 human colonic adenocarcinoma cells. Biochem J 1999;341 (Pt 3):593-600.

[183] Thornton DJ, Davies JR, Kraayenbrink M, *et al.* Mucus glycoproteins from'normal' human tracheobronchial secretion. Biochem J 1990;265:179-86.

[184] Sheehan JK, Brazeau C, Kutay S, *et al.* Physical characterization of the MUC5AC mucin: a highly oligomeric glycoprotein whether isolated from cell culture or *in vivo* from respiratory mucous secretions. Biochem J 2000;347 Pt 1:37-44.

[185] Brockhausen I, Schachter H. In: Gaius HJ, Gabius S, editors. Glycosciences: Status and Perspectives. New York: Chapman & Hall; 1997. p. 79-113.

[186] Yoshida A, Suzuki M, Ikenaga H, Takeuchi M. Discovery of the shortest sequence motif for high level mucin-type O-glycosylation. J Biol Chem 1997;272:16884-8.

[187] Asada M, Orikasa N, Yoneda A, *et al.* The AATPAP sequence is a very efficient signal for O-glycosylation in CHO cells. Glycoconj J 1999;16:321-6.

[188] Tetaert D, Ten Hagen KG, Richet C, *et al.* Glycopeptide N-acetylgalactosaminyltransferase specificities for O-glycosylated sites on MUC5AC mucin motif peptides. Biochem J 2001;357:313-20.

[189] Gerken TA, Raman J, Fritz TA, Jamison O. Identification of common and unique peptide substrate preferences for the UDP-GalNAc:polypeptide α-N-acetylgalactosaminyltransferases T1 and T2 derived from oriented random peptide substrates. J Biol Chem 2006;281:32403-16.

[190] Gerken TA, Tep C, Rarick J. Role of peptide sequence and neighboring residue glycosylation on the substrate specificity of the uridine 5'-diphosphate-α-N-acetylgalactosamine:polypeptide N-acetylgalactosaminyl transferases T1 and T2: kinetic modeling of the porcine and canine submaxillary gland mucin tandem repeats. Biochemistry 2004;43:9888-900.

[191] Hennebicq S, Tetaert D, Soudan B, *et al.* Influence of the amino acid sequence on the MUC5AC motif peptide O-glycosylation by human gastric UDP-GalNAc:polypeptide N-acetylgalactosaminyltransferase(s). Glycoconj J 1998;15:275-82.

[192] Ten Hagen KG, Fritz TA, Tabak LA. All in the family: the UDP-GalNAc:polypeptide N-acetylgalactosaminyltransferases. Glycobiology 2003;13:1R-16R.

[193] Young WW, Jr., Holcomb DR, Ten Hagen KG, Tabak LA. Expression of UDP-GalNAc:polypeptide N-acetylgalactosaminyltransferase isoforms in murine tissues determined by real-time PCR: a new view of a large family. Glycobiology 2003;13:549-57.

[194] Wu YM, Nowack DD, Omenn GS, Haab BB. Mucin glycosylation is altered by pro-inflammatory signaling in pancreatic-cancer cells. J Proteome Res 2009;8:1876-86.

[195] Roth J, Wang Y, Eckhardt AE, Hill RL. Subcellular localization of the UDP-N-acetyl-D-galactosamine: polypeptide N-acetylgalactosaminyltransferase-mediated O-glycosylation reaction in the submaxillary gland. Proc Natl Acad Sci USA 1994;91:8935-9.

[196] Schweizer A, Clausen H, van Meer G, Hauri HP. Localization of O-glycan initiation, sphingomyelin synthesis, and glucosylceramide synthesis in Vero cells with respect to the endoplasmic reticulum-Golgi intermediate compartment. J Biol Chem 1994;269:4035-41.

[197] Lamblin G, Degroote S, Perini JM, *et al.* Human airway mucin glycosylation: a combinatory of carbohydrate determinants which vary in cystic fibrosis. Glycoconj J 2001;18:661-84.

[198] Jentoft N. Why are proteins O-glycosylated? Trends Biochem Sci. 1990;15:291-4.

[199] Forstner JF, Forstner GG. In: Johnston LR, editor. Physiology of the gastrointestinal tract. New-York: Raven Press; 1994. p. 1245-83.

[200] Perez-Vilar J, Olsen JC, Chua M, Boucher RC. pH-dependent intraluminal organization of mucin granules in live human mucous/goblet cells. J Biol Chem 2005;280:16868-81.

[201] Perez-Vilar J, Mabolo R, McVaugh CT, Bertozzi CR, Boucher RC. Mucin granule intraluminal organization in living mucous/goblet cells. Roles of protein post-translational modifications and secretion. J Biol Chem 2006;281:4844-55.

[202] Perez-Vilar J, Randell SH, Boucher RC. C-Mannosylation of MUC5AC and MUC5B Cys subdomains. Glycobiology 2004;14:325-37.

[203] Robbe-Masselot C, Michalski JC, Capon C. Tumour-associated antigens of mucin O-glycans. Clinical relevance of glycobiology in cancer and inflammatory diseases of the epithelium. In: van Seuningen I, editor. The epithelial mucins: structure/function Roles in cancer and inflammatory diseases Kerala, India: Research Signpost; 2008. p. 55-73.

[204] Linden S, Nordman H, Hedenbro J, *et al.* Strain- and blood group-dependent binding of Helicobacter pylori to human gastric MUC5AC glycoforms. Gastroenterology 2002;123:1923-30.

[205] Van de Bovenkamp JH, Mahdavi J, Korteland-Van Male AM, *et al.* The MUC5AC glycoprotein is the primary receptor for Helicobacter pylori in the human stomach. Helicobacter 2003;8:521-32.

[206] Xu Y, Zhang L, Hu G. Potential application of alternatively glycosylated serum MUC1 and MUC5AC in gastric cancer diagnosis. Biologicals 2009;37:18-25.

[207] Vliegenthart JF, Casset F. Novel forms of protein glycosylation. Curr Opin Struct Biol 1998;8:565-71.

[208] Verdugo P. Goblet cells secretion and mucogenesis. Annu Rev Physiol 1990;52:157-76.

[209] Verdugo P, Deyrup-Olsen I, Aitken M, Villalon M, Johnson D. Molecular mechanism of mucin secretion: I. The role of intragranular charge shielding. J Dent Res 1987;66:506-8.

[210] Verdugo P, Aitken M, Langley L, Villalon MJ. Molecular mechanism of product storage and release in mucin secretion. II. The role of extracellular Ca^{++}. Biorheology 1987;24:625-33.

[211] Perez-Vilar J. Mucin granule intraluminal organization. Am J Respir Cell Mol Biol 2007;36:183-90.

[212] Perez-Vilar J, Eckhardt AE, DeLuca A, Hill RL. Porcine submaxillary mucin forms disulfide-linked multimers through its amino-terminal D-domains. J Biol Chem 1998;273:14442-9.

[213] Godl K, Johansson ME, Lidell ME, *et al.* The N terminus of the MUC2 mucin forms trimers that are held together within a trypsin-resistant core fragment. J Biol Chem 2002;277:47248-56.

[214] Lidell ME, Johansson ME, Hansson GC. An autocatalytic cleavage in the C terminus of the human MUC2 mucin occurs at the low pH of the late secretory pathway. J Biol Chem 2003;278:13944-51.

[215] Quesada I, Chin WC, Verdugo P. ATP-independent luminal oscillations and release of Ca^{2+} and H^+ from mast cell secretory granules: implications for signal transduction. Biophys J 2003;85:963-70.

[216] Davis CW, Dickey BF. Regulated airway goblet cell mucin secretion. Annu Rev Physiol 2008;70:487-512.

[217] Zhu Y, Ehre C, Abdullah LH, *et al.* Munc13-2-/- baseline secretion defect reveals source of oligomeric mucins in mouse airways. J Physiol 2008;586:1977-92.

[218] Pang ZP, Melicoff E, Padgett D, *et al.* Synaptotagmin-2 is essential for survival and contributes to Ca^{2+} triggering of neurotransmitter release in central and neuromuscular synapses. J Neurosci 2006;26:13493-504.

[219] Bahra P, Mesher J, Li S, Poll CT, Danahay H. P2Y2-receptor-mediated activation of a contralateral, lanthanide-sensitive calcium entry pathway in the human airway epithelium. Br J Pharmacol 2004;143:91-8.

[220] Kemp PA, Sugar RA, Jackson AD. Nucleotide-mediated mucin secretion from differentiated human bronchial epithelial cells. Am J Respir Cell Mol Biol 2004;31:446-55.

[221] Ehre C, Zhu Y, Abdullah LH, *et al.* nPKCepsilon, a P2Y2-R downstream effector in regulated mucin secretion from airway goblet cells. Am J Physiol Cell Physiol 2007;293:C1445-54.
[222] Tuvim MJ, Mospan AR, Burns KA, *et al.* Synaptotagmin 2 couples mucin granule exocytosis to Ca^{2+} signaling from endoplasmic reticulum. J Biol Chem 2009;284:9781-7.
[223] Berry M, Ellingham RB, Corfield AP. Human preocular mucins reflect changes in surface physiology. Br J Ophthalmol 2004;88:377-83.
[224] Thornton DJ, Gray T, Nettesheim P, *et al.* Characterization of mucins from cultured normal human tracheobronchial epithelial cells. Am J Physiol Lung Cell Mol Physiol 2000;278:L1118-28.
[225] Wickstrom C, Carlstedt I. N-terminal cleavage of the salivary MUC5B mucin. Analogy with the Van Willebrand propolypeptide? J Biol Chem 2001;276:47116-21.
[226] Lidell ME, Hansson GC. Cleavage in the GDPH sequence of the C-terminal cysteine-rich part of the human MUC5AC mucin. Biochem J 2006;399:121-9.
[227] Davies JR, Svitacheva N, Lannefors L, Kornfalt R, Carlstedt I. Identification of MUC5B, MUC5AC and small amounts of MUC2 mucins in cystic fibrosis airway secretions. Biochem J 1999;344 Pt 2:321-30.
[228] Guzman K, Bader T, Nettesheim P. Regulation of MUC5 and MUC1 gene expression: correlation with airway mucous differentiation. Am J Physiol 1996;270:L846-53.
[229] Reid CJ, Gould S, Harris A. Developmental expression of mucin genes in the human respiratory tract. Am J Respir Cell Mol Biol 1997;17:592-8.
[230] Buisine MP, Devisme L, Copin MC, *et al.* Developmental mucin gene expression in the human respiratory tract. Am J Respir Cell Mol Biol 1999;20:209-18.
[231] Seregni E, Botti C, Massaron S, *et al.* Structure, function and gene expression of epithelial mucins. Tumori 1997;83:625-32.
[232] Groneberg DA, Eynott PR, Oates T, *et al.* Expression of MUC5AC and MUC5B mucins in normal and cystic fibrosis lung. Respir Med 2002;96:81-6.
[233] Buisine MP, Devisme L, Savidge TC, *et al.* Mucin gene expression in human embryonic and fetal intestine. Gut 1998;43:519-24.
[234] Buisine MP, Devisme L, Maunoury V, *et al.* Developmental mucin gene expression in the gastroduodenal tract and accessory digestive glands. I. Stomach. A relationship to gastric carcinoma. J Histochem Cytochem 2000;48:1657-66.
[235] Buisine MP, Devisme L, Degand P, *et al.* Developmental mucin gene expression in the gastroduodenal tract and accessory digestive glands. II. Duodenum and liver, gallbladder, and pancreas. J Histochem Cytochem 2000;48:1667-76.
[236] Deren JS. Development of structure and function in the fetal and newborn stomach. Am J Clin Nutr 1971;24:144-59.
[237] Grand RJ, Watkins JB, Torti FM. Development of the human gastrointestinal tract. A review. Gastroenterology 1976;70:790-810.
[238] Carson DD, DeSouza MM, Kardon R, *et al.* Mucin expression and function in the female reproductive tract. Hum Reprod Update 1998;4:459-64.
[239] Lagow E, DeSouza MM, Carson DD. Mammalian reproductive tract mucins. Hum Reprod Update 1999;5:280-92.
[240] Gipson IK. Mucins of the human endocervix. Front Biosci 2001;6:D1245-55.
[241] Alameda F, Mejias-Luque R, Garrido M, de Bolos C. Mucin genes (MUC2, MUC4, MUC5AC, and MUC6) detection in normal and pathological endometrial tissues. Int J Gynecol Pathol 2007;26:61-5.
[242] N'Dow J, Pearson JP, Bennett MK, Neal DE, Robson CN. Mucin gene expression in human urothelium and in intestinal segments transposed into the urinary tract. J Urol 2000;164:1398-404.

[243] Leroy X, Devisme L, Buisine MP, *et al.* Expression of human mucin genes during normal and abnormal renal development. Am J Clin Pathol 2003;120:544-50.
[244] Davies JA. The Kidney Development Database. Dev Genet 1999;24:194-8.
[245] Glassberg KI. Normal and abnormal development of the kidney: a clinician's interpretation of current knowledge. J Urol 2002;167:2339-50; discussion 50-1.
[246] Leroy X, Copin MC, Devisme L, *et al.* Expression of human mucin genes in normal kidney and renal cell carcinoma. Histopathology 2002;40:450-7.
[247] Jankovic Velickovic L, Katic V, Hattori T, *et al.* Differences in the expression of mucins in various forms of cystitis glandularis. Pathol Res Pract 2007;203:653-8.
[248] Kunze E, Krassenkova I, Fayyazi A. Tumor-associated neoexpression of the pS2 peptide and MUC5AC mucin in primary adenocarcinomas and signet ring cell carcinomas of the urinary bladder. Histol Histopathol 2008;23:539-48.
[249] Russo CL, Spurr-Michaud S, Tisdale A, *et al.* Mucin gene expression in human male urogenital tract epithelia. Hum Reprod 2006;21:2783-93.
[250] Cozzi PJ, Wang J, Delprado W, *et al.* MUC1, MUC2, MUC4, MUC5AC and MUC6 expression in the progression of prostate cancer. Clin Exp Metastasis 2005;22:565-73.
[251] Zhang S, Zhang HS, Cordon-Cardo C, Ragupathi G, Livingston PO. Selection of tumor antigens as targets for immune attack using immunohistochemistry: protein antigens. Clin Cancer Res 1998;4:2669-76.
[252] Zhang S, Zhang HS, Reuter VE, *et al.* Expression of potential target antigens for immunotherapy on primary and metastatic prostate cancers. Clin Cancer Res 1998;4:295-302.
[253] Gipson IK, Ho SB, Spurr-Michaud SJ, *et al.* Mucin genes expressed by human female reproductive tract epithelia. Biol Reprod 1997;56:999-1011.
[254] Carson DD, Farrar JD, Laidlaw J, Wright DA. Selective activation of the N-glycosylation apparatus in uteri by estrogen. J Biol Chem 1990;265:2947-55.
[255] Surveyor GA, Gendler SJ, Pemberton L, *et al.* Expression and steroid hormonal control of Muc-1 in the mouse uterus. Endocrinology 1995;136:3639-47.
[256] Hoffman LH, Olson GE, Carson DD, Chilton BS. Progesterone and implanting blastocysts regulate Muc1 expression in rabbit uterine epithelium. Endocrinology 1998;139:266-71.
[257] Baker AC, Eltoum I, Curry RO, *et al.* Mucinous expression in benign and neoplastic glandular lesions of the uterine cervix. Arch Pathol Lab Med 2006;130:1510-5.
[258] Zhao S, Hayasaka T, Osakabe M, *et al.* Mucin expression in nonneoplastic and neoplastic glandular epithelia of the uterine cervix. Int J Gynecol Pathol 2003;22:393-7.
[259] Wolf DP, Blasco L, Khan MA, Litt M. Human cervical mucus. I. Rheologic characteristics. Fertil Steril 1977;28:41-6.
[260] Wolf DP, Blasco L, Khan MA, Litt M. Human cervical mucus. II. Changes in viscoelasticity during the ovulatory menstrual cycle. Fertil Steril 1977;28:47-52.
[261] Wolf DP, Blasco L, Khan MA, Litt M. Human cervical mucus. III Isolation and characterization of rheologically active mucin. Fertil Steril 1977;28:53-8.
[262] Wolf DP, Blasco L, Khan MA, Litt M. Human cervical mucus. IV Viscoelastisity and sperm penetrability during the ovulatory menstrual cycle. Fertil Steril 1978;30:163-9.
[263] Yudin AI, Hanson FW, Katz DF. Human cervical mucus and its interaction with sperm: a fine-structural view. Biol Reprod 1989;40:661-71.
[264] Audie JP, Tetaert D, Pigny P, *et al.* Mucin gene expression in the human endocervix. Hum Reprod 1995;10:98-102.
[265] Gipson IK, Spurr-Michaud S, Moccia R, *et al.* MUC4 and MUC5B transcripts are the prevalent mucin messenger ribonucleic acids of the human endocervix. Biol Reprod 1999;60:58-64.

[266] Hebbar V, Damera G, Sachdev GP. Differential expression of MUC genes in endometrial and cervical tissues and tumors. BMC Cancer 2005;5:124.
[267] Lange C, Fernandez J, Shim D, *et al.* Mucin gene expression is not regulated by estrogen and/or progesterone in the ocular surface epithelia of mice. Exp Eye Res 2003;77:59-68.
[268] Riethdorf L, O'Connell JT, Riethdorf S, Cviko A, Crum CP. Differential expression of MUC2 and MUC5AC in benign and malignant glandular lesions of the cervix uteri. Virchows Arch 2000;437:365-71.
[269] Audie JP, Janin A, Porchet N, *et al.* Expression of human mucin genes in respiratory, digestive, and reproductive tracts ascertained by *in situ* hybridization. J Histochem Cytochem 1993;41:1479-85.
[270] Lapensee L, Paquette Y, Bleau G. Allelic polymorphism and chromosomal localization of the human oviductin gene (MUC9). Fertil Steril 1997;68:702-8.
[271] Malette B, Paquette Y, Merlen Y, Bleau G. Oviductins possess chitinase- and mucin-like domains: a lead in the search for the biological function of these oviduct-specific ZP-associating glycoproteins. Mol Reprod Dev 1995;41:384-97.
[272] Giuntoli RL, 2nd, Rodriguez GC, Whitaker RS, Dodge R, Voynow JA. Mucin gene expression in ovarian cancers. Cancer Res 1998;58:5546-50.
[273] Ho SB, Niehans GA, Lyftogt C, *et al.* Heterogeneity of mucin gene expression in normal and neoplastic tissues. Cancer Res 1993;53:641-51.
[274] Boman F, Buisine MP, Wacrenier A, *et al.* Mucin gene transcripts in benign and borderline mucinous tumours of the ovary: an *in situ* hybridization study. J Pathol 2001;193:339-44.
[275] Wolf DP BL, Khan MA and Litt M. Human cervical mucus. Fertil Steril 1978;30:163.
[276] Copin MC, Devisme L, Buisine MP, *et al.* From normal respiratory mucosa to epidermoid carcinoma: expression of human mucin genes. Int J Cancer 2000;86:162-8.
[277] Andrianifahanana M, Moniaux N, Batra SK. Regulation of mucin expression: mechanistic aspects and implications for cancer and inflammatory diseases. Biochim Biophys Acta 2006;1765:189-222.
[278] Groneberg DA, Peiser C, Dinh QT, *et al.* Distribution of respiratory mucin proteins in human nasal mucosa. Laryngoscope 2003;113:520-4.
[279] Fahy JV. Goblet cell and mucin gene abnormalities in asthma. Chest 2002;122:320S-6S.
[280] Jeffery PK, Li D. Airway mucosa: secretory cells, mucus and mucin genes. Eur Respir J 1997;10:1655-62.
[281] Rogers DF. Mucus pathophysiology in COPD: differences to asthma, and pharmacotherapy. Monaldi Arch Chest Dis 2000;55:324-32.
[282] Finkbeiner WE, Carrier SD, Teresi CE. Reverse transcription-polymerase chain reaction (RT-PCR) phenotypic analysis of cell cultures of human tracheal epithelium, tracheobronchial glands, and lung carcinomas. Am J Respir Cell Mol Biol 1993;9:547-56.
[283] Yu CJ, Yang PC, Shew JY, *et al.* Mucin mRNA expression in lung adenocarcinoma cell lines and tissues. Oncology 1996;53:118-26.
[284] Berger JT, Voynow JA, Peters KW, Rose MC. Respiratory carcinoma cell lines. MUC genes and glycoconjugates. Am J Respir Cell Mol Biol 1999;20:500-10.
[285] Yu CJ, Yang PC, Shun CT, *et al.* Overexpression of MUC5 genes is associated with early post-operative metastasis in non-small-cell lung cancer. Int J Cancer 1996;69:457-65.
[286] Yu CJ, Shew JY, Shun CT, *et al.* Quantitative analysis of mRNA encoding MUC1, MUC2, and MUC5AC genes: a correlation between specific mucin gene expression and sialomucin expression in non-small cell lung cancer. Am J Respir Cell Mol Biol 1998;18:643-52.
[287] Copin MC, Buisine MP, Devisme L, *et al.* Normal respiratory mucosa, precursor lesions and lung carcinomas: differential expression of human mucin genes. Front Biosci 2001;6:D1264-75.

[288] Copin MC, Buisine MP, Leteurtre E, *et al.* Mucinous bronchioloalveolar carcinomas display a specific pattern of mucin gene expression among primary lung adenocarcinomas. Hum Pathol 2001;32:274-81.
[289] Lopez-Ferrer A, Curull V, Barranco C, *et al.* Mucins as differentiation markers in bronchial epithelium. Squamous cell carcinoma and adenocarcinoma display similar expression patterns. Am J Respir Cell Mol Biol 2001;24:22-9.
[290] Awaya H, Takeshima Y, Yamasaki M, Inai K. Expression of MUC1, MUC2, MUC5AC, and MUC6 in atypical adenomatous hyperplasia, bronchioloalveolar carcinoma, adenocarcinoma with mixed subtypes, and mucinous bronchioloalveolar carcinoma of the lung. Am J Clin Pathol 2004;121:644-53.
[291] Brambilla E, Travis WD, Colby TV, Corrin B, Shimosato Y. The new World Health Organization classification of lung tumours. Eur Respir J 2001;18:1059-68.
[292] Koga T, Hashimoto S, Sugio K, *et al.* Lung adenocarcinoma with bronchioloalveolar carcinoma component is frequently associated with foci of high-grade atypical adenomatous hyperplasia. Am J Clin Pathol 2002;117:464-70.
[293] Niho S, Yokose T, Suzuki K, *et al.* Monoclonality of atypical adenomatous hyperplasia of the lung. Am J Pathol 1999;154:249-54.
[294] Mori M, Rao SK, Popper HH, Cagle PT, Fraire AE. Atypical adenomatous hyperplasia of the lung: a probable forerunner in the development of adenocarcinoma of the lung. Mod Pathol 2001;14:72-84.
[295] Yamasaki M, Takeshima Y, Fujii S, *et al.* Correlation between genetic alterations and histopathological subtypes in bronchiolo-alveolar carcinoma and atypical adenomatous hyperplasia of the lung. Pathol Int 2000;50:778-85.
[296] Tsuta K, Ishii G, Nitadori J, *et al.* Comparison of the immunophenotypes of signet-ring cell carcinoma, solid adenocarcinoma with mucin production, and mucinous bronchioloalveolar carcinoma of the lung characterized by the presence of cytoplasmic mucin. J Pathol 2006;209:78-87.
[297] Nagarjun Rao R, Moran CA, Suster S. Histiocytic disorders of the lung. Adv Anat Pathol 2010;17:12-22.
[298] Castillo C, Buob D, Mortuaire G, *et al.* Signet-ring cell adenocarcinoma of sinonasal tract: an immunohistochemical study of the mucins profile. Arch Pathol Lab Med 2007;131:961-4.
[299] Voynow JA. What does mucin have to do with lung disease? Paediatr Respir Rev 2002;3:98-103.
[300] Rogers DF. Physiology of airway mucus secretion and pathophysiology of hypersecretion. Respir Care 2007;52:1134-46; discussion 46-9.
[301] Kuyper LM, Pare PD, Hogg JC, *et al.* Characterization of airway plugging in fatal asthma. Am J Med 2003;115:6-11.
[302] Ordonez CL, Khashayar R, Wong HH, *et al.* Mild and moderate asthma is associated with airway goblet cell hyperplasia and abnormalities in mucin gene expression. Am J Respir Crit Care Med 2001;163:517-23.
[303] Hogg JC, Chu FS, Tan WC, *et al.* Survival after lung volume reduction in chronic obstructive pulmonary disease: insights from small airway pathology. Am J Respir Crit Care Med 2007;176:454-9.
[304] Shimura S, Andoh Y, Haraguchi M, Shirato K. Continuity of airway goblet cells and intraluminal mucus in the airways of patients with bronchial asthma. Eur Respir J 1996;9:1395-401.
[305] Aikawa T, Shimura S, Sasaki H, Ebina M, Takishima T. Marked goblet cell hyperplasia with mucus accumulation in the airways of patients who died of severe acute asthma attack. Chest 1992;101:916-21.

[306] Evans CM, Williams OW, Tuvim MJ, *et al.* Mucin is produced by clara cells in the proximal airways of antigen-challenged mice. Am J Respir Cell Mol Biol 2004;31:382-94.

[307] Caramori G, Di Gregorio C, Carlstedt I, *et al.* Mucin expression in peripheral airways of patients with chronic obstructive pulmonary disease. Histopathology 2004;45:477-84.

[308] Global Initiative for Chronic Obstructive Lung Disease (COPD) Global Strategy for the Diagnosis, Management and Prevention of Chronic Obstructive Pulmonary Disease; 2001; Bethesda, MD: NIH; 2001. p. NIH Publication No 2701A.

[309] Saetta M, Turato G, Baraldo S, *et al.* Goblet cell hyperplasia and epithelial inflammation in peripheral airways of smokers with both symptoms of chronic bronchitis and chronic airflow limitation. Am J Respir Crit Care Med 2000;161:1016-21.

[310] Kirkham S, Sheehan JK, Knight D, Richardson PS, Thornton DJ. Heterogeneity of airways mucus: variations in the amounts and glycoforms of the major oligomeric mucins MUC5AC and MUC5B. Biochem J 2002;361:537-46.

[311] Kirkham S, Kolsum U, Rousseau K, *et al.* MUC5B is the major mucin in the gel phase of sputum in chronic obstructive pulmonary disease. Am J Respir Crit Care Med 2008;178:1033-9.

[312] Kim EY, Battaile JT, Patel AC, *et al.* Persistent activation of an innate immune response translates respiratory viral infection into chronic lung disease. Nat Med 2008;14:633-40.

[313] Foster WM, Adler KB, Crews AL, *et al.* MARCKS-related peptide modulates *in vivo* the secretion of airway Muc5ac. Am J Physiol Lung Cell Mol Physiol 2010;299:L345-52.

[314] Drumm ML, Ziady AG, Davis PB. Genetic Variation and Clinical Heterogeneity in Cystic Fibrosis. Annu Rev Pathol 2012;7:267-82.

[315] Copin MC, Perrais M, van Seuningen I. Mucins in the pathophysiological lung. Expression and roles in immuno-and gene-based therapies. In: van Seuningen I, editor. The Epithelial Mucins: Structure/Function Roles in Cancer and Inflammatory Diseases. Kerala, India: Research Signpost; 2008. p. 109-23.

[316] Voynow JA, Selby DM, Rose MC. Mucin gene expression (MUC1, MUC2, and MUC5/5AC) in nasal epithelial cells of cystic fibrosis, allergic rhinitis, and normal individuals. Lung 1998;176:345-54.

[317] Martinez-Anton A, Debolos C, Garrido M, *et al.* Mucin genes have different expression patterns in healthy and diseased upper airway mucosa. Clin Exp Allergy 2006;36:448-57.

[318] Henke MO, Renner A, Huber RM, Seeds MC, Rubin BK. MUC5AC and MUC5B Mucins Are Decreased in Cystic Fibrosis Airway Secretions. Am J Respir Cell Mol Biol 2004;31:86-91.

[319] Hopwood D, Coghill G, Sanders DS. Human oesophageal submucosal glands. Their detection mucin, enzyme and secretory protein content. Histochemistry 1986;86:107-12.

[320] Guillem P, Billeret V, Buisine MP, *et al.* Mucin gene expression and cell differentiation in human normal, premalignant and malignant esophagus. Int J Cancer 2000;88:856-61.

[321] Arul GS, Moorghen M, Myerscough N, *et al.* Mucin gene expression in Barrett's oesophagus: an *in situ* hybridisation and immunohistochemical study. Gut 2000;47:753-61.

[322] Van De Bovenkamp JH, Korteland-Van Male AM, Warson C, *et al.* Gastric-type mucin and TFF-peptide expression in Barrett's oesophagus is disturbed during increased expression of MUC2. Histopathology 2003;42:555-65.

[323] Warson C, Van De Bovenkamp JH, Korteland-Van Male AM, *et al.* Barrett's esophagus is characterized by expression of gastric-type mucins (MUC5AC, MUC6) and TFF peptides (TFF1 and TFF2), but the risk of carcinoma development may be indicated by the intestinal-type mucin, MUC2. Hum Pathol 2002;33:660-8.

[324] Chaves P, Cruz C, Dias Pereira A, *et al.* Gastric and intestinal differentiation in Barrett's metaplasia and associated adenocarcinoma. Dis Esophagus 2005;18:383-7.

[325] Haggitt RC. Barrett's esophagus, dysplasia, and adenocarcinoma. Hum Pathol 1994;25:982-93.
[326] Hamilton SR, Smith RR. The relationship between columnar epithelial dysplasia and invasive adenocarcinoma arising in Barrett's esophagus. Am J Clin Pathol 1987;87:301-12.
[327] Mendes de Almeida JC, Chaves P, Pereira AD, Altorki NK. Is Barrett's esophagus the precursor of most adenocarcinomas of the esophagus and cardia? A biochemical study. Ann Surg 1997;226:725-33; discussion 33-5.
[328] Ho SB, Roberton AM, Shekels LL, *et al.* Expression cloning of gastric mucin complementary DNA and localization of mucin gene expression. Gastroenterology 1995;109:735-47.
[329] Hanby AM, Poulsom R, Singh S, *et al.* Spasmolytic polypeptide is a major antral peptide: distribution of the trefoil peptides human spasmolytic polypeptide and pS2 in the stomach. Gastroenterology 1993;105:1110-6.
[330] Taupin D, Pedersen J, Familari M, *et al.* Augmented intestinal trefoil factor (TFF3) and loss of pS2 (TFF1) expression precedes metaplastic differentiation of gastric epithelium. Lab Invest 2001;81:397-408.
[331] Glickman JN, Blount PL, Sanchez CA, *et al.* Mucin core polypeptide expression in the progression of neoplasia in Barrett's esophagus. Hum Pathol 2006;37:1304-15.
[332] Burjonrappa SC, Reddimasu S, Nawaz Z, *et al.* Mucin expression profile in Barrett's, dysplasia, adenocarcinoma sequence in the esophagus. Indian J Cancer2007;44:1-5.
[333] Labouvie C, Machado JC, Carneiro F, *et al.* Differential expression of mucins and trefoil peptides in native epithelium, Barrett's metaplasia and squamous cell carcinoma of the oesophagus. J Cancer Res Clin Oncol 1999;125:71-6.
[334] Shields HM, Sawhney RA, Zwas F, *et al.* Scanning electron microscopy of the human esophagus: application to Barrett's esophagus, a precancerous lesion. Microsc Res Tech 1995;31:248-56.
[335] Brown IS, Whiteman DC, Lauwers GY. Foveolar type dysplasia in Barrett esophagus. Mod Pathol 2010;23:834-43.
[336] Kushima R, Vieth M, Mukaisho K, *et al.* Pyloric gland adenoma arising in Barrett's esophagus with mucin immunohistochemical and molecular cytogenetic evaluation. Virchows Arch 2005;446:537-41.
[337] Asthana N, Mandich D, Ligato S. Esophageal polypoid dysplasia of gastric foveolar phenotype with focal intramucosal carcinoma associated with Barrett's esophagus. Am J Surg Pathol 2008;32:1581-5.
[338] Wang RQ, Fang DC. Effects of Helicobacter pylori infection on mucin expression in gastric carcinoma and pericancerous tissues. J Gastroenterol Hepatol 2006;21:425-31.
[339] Linden S, Semino-Mora C, Liu H, Rick J, Dubois A. Role of mucin Lewis status in resistance to Helicobacter pylori infection in pediatric patients. Helicobacter 2010;15:251-8.
[340] Kang HM, Kim N, Park YS, *et al.* Effects of Helicobacter pylori Infection on gastric mucin expression. J Clin Gastroenterol 2008;42:29-35.
[341] Peterson WL. Helicobacter pylori and peptic ulcer disease. N Engl J Med 1991;324:1043-8.
[342] Parsonnet J, Friedman GD, Vandersteen DP, *et al.* Helicobacter pylori infection and the risk of gastric carcinoma. N Engl J Med 1991;325:1127-31.
[343] Zucca E, Bertoni F, Roggero E, *et al.* Molecular analysis of the progression from Helicobacter pylori-associated chronic gastritis to mucosa-associated lymphoid-tissue lymphoma of the stomach. N Engl J Med 1998;338:804-10.
[344] Magalhaes A, Reis CA. Helicobacter pylori adhesion to gastric epithelial cells is mediated by glycan receptors. Braz J Med Biol Res 2010;43:611-8.

[345] Takahashi T, Matsumoto T, Nakamura M, *et al.* A novel *in vitro* infection model of Helicobacter pylori using mucin-producing murine gastric surface mucous cells. Helicobacter 2004;9:302-12.
[346] Van den Brink GR, Tytgat KM, Van der Hulst RW, *et al.* H pylori colocalises with MUC5AC in the human stomach. Gut 2000;46:601-7.
[347] Kocer B, Ulas M, Ustundag Y, *et al.* A confirmatory report for the close interaction of Helicobacter pylori with gastric epithelial MUC5AC expression. J Clin Gastroenterol 2004;38:496-502.
[348] Byrd JC, Yan P, Sternberg L, *et al.* Aberrant expression of gland-type gastric mucin in the surface epithelium of Helicobacter pylori-infected patients. Gastroenterology 1997;113:455-64.
[349] Marques T, David L, Reis C, Nogueira A. Topographic expression of MUC5AC and MUC6 in the gastric mucosa infected by Helicobacter pylori and in associated diseases. Pathol Res Pract 2005;201:665-72.
[350] Babu SD, Jayanthi V, Devaraj N, Reis CA, Devaraj H. Expression profile of mucins (MUC2, MUC5AC and MUC6) in Helicobacter pylori infected pre-neoplastic and neoplastic human gastric epithelium. Mol Cancer 2006;5:10.
[351] Cooke CL, An HJ, Kim J, *et al.* Modification of gastric mucin oligosaccharide expression in rhesus macaques after infection with Helicobacter pylori. Gastroenterology 2009;137:1061-71, 71 e1-8.
[352] Reis CA, David L, Nielsen PA, *et al.* Immunohistochemical study of MUC5AC expression in human gastric carcinomas using a novel monoclonal antibody. Int J Cancer 1997;74:112-21.
[353] Reis CA, David L, Correa P, *et al.* Intestinal metaplasia of human stomach displays distinct patterns of mucin (MUC1, MUC2, MUC5AC, and MUC6) expression. Cancer Res 1999;59:1003-7.
[354] Ho SB, Shekels LL, Toribara NW, *et al.* Mucin gene expression in normal, preneoplastic, and neoplastic human gastric epithelium. Cancer Res 1995;55:2681-90.
[355] Byrd JC, Yunker CK, Xu QS, Sternberg LR, Bresalier RS. Inhibition of gastric mucin synthesis by Helicobacter pylori. Gastroenterology 2000;118:1072-9.
[356] Beil W, Enss ML, Muller S, *et al.* Role of vacA and cagA in Helicobacter pylori inhibition of mucin synthesis in gastric mucous cells. J Clin Microbiol 2000;38:2215-8.
[357] Slomiany BL, Slomiany A. Disruption in gastric mucin synthesis by Helicobacter pylori lipopolysaccharide involves ERK and p38 mitogen-activated protein kinase participation. Biochem Biophys Res Commun 2002;294:220-4.
[358] Morgenstern S, Koren R, Moss SF, *et al.* Does Helicobacter pylori affect gastric mucin expression? Relationship between gastric antral mucin expression and H. pylori colonization. Eur J Gastroenterol Hepatol 2001;13:19-23.
[359] Kaneko T, Ota H, Hayama M, Akamatsu T, Katsuyama T. Helicobacter pylori infection produces expression of a secretory component in gastric mucous cells. Virchows Arch 2000;437:514-20.
[360] Dixon MF. Prospects for intervention in gastric carcinogenesis: reversibility of gastric atrophy and intestinal metaplasia. Gut 2001;49:2-4.
[361] Meining A, Morgner A, Miehlke S, Bayerdorffer E, Stolte M. Atrophy-metaplasia-dysplasia-carcinoma sequence in the stomach: a reality or merely an hypothesis? Best Pract Res Clin Gastroenterol 2001;15:983-98.
[362] Correa P. Helicobacter pylori and gastric carcinogenesis. Am J Surg Pathol 1995;19 Suppl 1:S37-43.
[363] Iida F, Kusama J. Gastric carcinoma and intestinal metaplasia. Significance of types of intestinal metaplasia upon development of gastric carcinoma. Cancer 1982;50:2854-8.

[364] Khattab AZ, Nasif WA, Lotfy M. MUC2 and MUC6 apomucins expression in human gastric neoplasm: an immunohistochemical analysis. Med Oncol 2011;28 Suppl 1:S207-11
[365] El-Zimaity HM, Ramchatesingh J, Saeed MA, Graham DY. Gastric intestinal metaplasia: subtypes and natural history. J Clin Pathol 2001;54:679-83.
[366] El-Zimaity HM, Ota H, Graham DY, Akamatsu T, Katsuyama T. Patterns of gastric atrophy in intestinal type gastric carcinoma. Cancer 2002;94:1428-36.
[367] Nogueira AM, Machado JC, Carneiro F, *et al.* Patterns of expression of trefoil peptides and mucins in gastric polyps with and without malignant transformation. J Pathol 1999;187:541-8.
[368] Craanen ME, Blok P, Dekker W, Ferwerda J, Tytgat GN. Prevalence of subtypes of intestinal metaplasia in gastric antral mucosa. Dig Dis Sci 1991;36:1529-36.
[369] Rokkas T, Filipe MI, Sladen GE. Detection of an increased incidence of early gastric cancer in patients with intestinal metaplasia type III who are closely followed up. Gut 1991;32:1110-3.
[370] Reis CA, David L, Carvalho F, *et al.* Immunohistochemical study of the expression of MUC6 mucin and co-expression of other secreted mucins (MUC5AC and MUC2) in human gastric carcinomas. J Histochem Cytochem 2000;48:377-88.
[371] Lee HS, Lee HK, Kim HS, *et al.* MUC1, MUC2, MUC5AC, and MUC6 expressions in gastric carcinomas: their roles as prognostic indicators. Cancer 2001;92:1427-34.
[372] Carrato C, Balague C, de Bolos C, *et al.* Differential apomucin expression in normal and neoplastic human gastrointestinal tissues. Gastroenterology 1994;107:160-72.
[373] Jass JR, Walsh MD. Altered mucin expression in the gastrointestinal tract: a review. J Cell Mol Med 2001;5:327-51.
[374] Shiotani A, Haruma K, Uedo N, *et al.* Histological risk markers for non-cardia early gastric cancer. Pattern of mucin expression and gastric cancer. Virchows Arch 2006;449:652-9.
[375] Silva E, Teixeira A, David L, *et al.* Mucins as key molecules for the classification of intestinal metaplasia of the stomach. Virchows Arch 2002;440:311-7.
[376] Tanaka H, Tsukamoto T, Mizoshita T, *et al.* Expression of small intestinal and colonic phenotypes in complete intestinal metaplasia of the human stomach. Virchows Arch 2005;447:806-15.
[377] Mizoshita T, Tsukamoto T, Nakanishi H, *et al.* Expression of Cdx2 and the phenotype of advanced gastric cancers: relationship with prognosis. J Cancer Res Clin Oncol 2003;129:727-34.
[378] Tajima Y, Shimoda T, Nakanishi Y, *et al.* Gastric and intestinal phenotypic marker expression in gastric carcinomas and its prognostic significance: immunohistochemical analysis of 136 lesions. Oncology 2001;61:212-20.
[379] Dong B, Xie YQ, Chen K, *et al.* Differences in biological features of gastric dysplasia, indefinite dysplasia, reactive hyperplasia and discriminant analysis of these lesions. World J Gastroenterol 2005;11:3595-600.
[380] Wongkham S, Sheehan JK, Boonla C, *et al.* Serum MUC5AC mucin as a potential marker for cholangiocarcinoma. Cancer Lett 2003;195:93-9.
[381] Boonla C, Wongkham S, Sheehan JK, *et al.* Prognostic value of serum MUC5AC mucin in patients with cholangiocarcinoma. Cancer 2003;98:1438-43.
[382] Wittel UA, Goel A, Varshney GC, Batra SK. Mucin antibodies - new tools in diagnosis and therapy of cancer. Front Biosci 2001;6:D1296-310.
[383] Sakamoto H, Yonezawa S, Utsunomiya T, *et al.* Mucin antigen expression in gastric carcinomas of young and old adults. Hum Pathol 1997;28:1056-65.

[384] Carneiro F, Amado M, David L, Clausen H, Sobrinho-Simoes M. Glycosylation features of gastric carcinoma initiation and progression. A review with emphasis on simple mucin-type carbohydrates and histo-blood group antigens of the Lewis system. Eur J Cancer Prev 1994;3 Suppl 2:39-46.
[385] David L, Nesland JM, Clausen H, Carneiro F, Sobrinho-Simoes M. Simple mucin-type carbohydrate antigens (Tn, sialosyl-Tn and T) in gastric mucosa, carcinomas and metastases. APMIS Suppl. 1992;27:162-72.
[386] Bara J, Chastre E, Mahiou J, *et al.* Gastric M1 mucin, an early oncofetal marker of colon carcinogenesis, is encoded by the MUC5AC gene. Int J Cancer 1998;75:767-73.
[387] Pinto-de-Sousa J, David L, Reis CA, *et al.* Mucins MUC1, MUC2, MUC5AC and MUC6 expression in the evaluation of differentiation and clinico-biological behaviour of gastric carcinoma. Virchows Arch 2002;440:304-10.
[388] Pinto-de-Sousa J, Reis CA, David L, Pimenta A, Cardoso-de-Oliveira M. MUC5B expression in gastric carcinoma: relationship with clinico-pathological parameters and with expression of mucins MUC1, MUC2, MUC5AC and MUC6. Virchows Arch 2004;444:224-30.
[389] Fiocca R, Villani L, Tenti P, *et al.* Characterization of four main cell types in gastric cancer: foveolar, mucopeptic, intestinal columnar and goblet cells. An histopathologic, histochemical and ultrastructural study of "early" and "advanced" tumours. Pathol Res Pract 1987;182:308-25.
[390] Ming SC. Gastric carcinoma. A pathobiological classification. Cancer 1977;39:2475-85.
[391] Toki F, Takahashi A, Aihara R, *et al.* Relationship between clinicopathological features and mucin phenotypes of advanced gastric adenocarcinoma. World J Gastroenterol 2010;16:2764-70.
[392] Choi JS, Kim MA, Lee HE, Lee HS, Kim WH. Mucinous gastric carcinomas: clinicopathologic and molecular analyses. Cancer 2009;115:3581-90.
[393] Leteurtre E, Zerimech F, Piessen G, *et al.* Relationships between mucinous gastric carcinoma, MUC2 expression and survival. World J Gastroenterol 2006;12:3324-31.
[394] Hamilton SR, Aaltonen LA. WHO Classification of Tumours. Lyon: IARC Press; 2000. p. 37-67.
[395] Mulligan RM. Histogenesis and biologic behavior of gastric carcinoma. Pathol Annu 1972;7:349-415.
[396] Goseki N, Takizawa T, Koike M. Differences in the mode of the extension of gastric cancer classified by histological type: new histological classification of gastric carcinoma. Gut 1992;33:606-12.
[397] Jia Y, Persson C, Hou L, *et al.* A comprehensive analysis of common genetic variation in MUC1, MUC5AC, MUC6 genes and risk of stomach cancer. Cancer Causes Control 2010;21:313-21.
[398] Ranganathan P, Harsha HC, Pandey A. Molecular alterations in exocrine neoplasms of the pancreas. Arch Pathol Lab Med 2009;133:405-12.
[399] Maitra A, Hruban RH. Pancreatic cancer. Annu Rev Pathol 2008;3:157-88.
[400] Reya T, Morrison SJ, Clarke MF, Weissman IL. Stem cells, cancer, and cancer stem cells. Nature 2001;414:105-11.
[401] Takaori K. Current understanding of precursors to pancreatic cancer. J Hepatobiliary Pancreat Surg 2007;14:217-23.
[402] Thayer SP, di Magliano MP, Heiser PW, *et al.* Hedgehog is an early and late mediator of pancreatic cancer tumorigenesis. Nature 2003;425:851-6.
[403] Miyamoto Y, Maitra A, Ghosh B, *et al.* Notch mediates TGF α-induced changes in epithelial differentiation during pancreatic tumorigenesis. Cancer Cell 2003;3:565-76.

[404] Prasad NB, Biankin AV, Fukushima N, *et al.* Gene expression profiles in pancreatic intraepithelial neoplasia reflect the effects of Hedgehog signaling on pancreatic ductal epithelial cells. Cancer Res 2005;65:1619-26.

[405] Leach SD. Epithelial differentiation in pancreatic development and neoplasia: new niches for nestin and Notch. J Clin Gastroenterol 2005;39:S78-82.

[406] Jensen JN, Cameron E, Garay MV, *et al.* Recapitulation of elements of embryonic development in adult mouse pancreatic regeneration. Gastroenterology 2005;128:728-41.

[407] Hruban RH, Adsay NV, Albores-Saavedra J, *et al.* Pathology of genetically engineered mouse models of pancreatic exocrine cancer: consensus report and recommendations. Cancer Res 2006;66:95-106.

[408] Li C, Heidt DG, Dalerba P, *et al.* Identification of pancreatic cancer stem cells. Cancer Res 2007;67:1030-7.

[409] Lee CJ, Dosch J, Simeone DM. Pancreatic cancer stem cells. J Clin Oncol 2008;26:2806-12.

[410] Hermann PC, Huber SL, Herrler T, *et al.* Distinct populations of cancer stem cells determine tumor growth and metastatic activity in human pancreatic cancer. Cell Stem Cell 2007;1:313-23.

[411] Olempska M, Eisenach PA, Ammerpohl O, *et al.* Detection of tumor stem cell markers in pancreatic carcinoma cell lines. Hepatobiliary Pancreat Dis Int 2007;6:92-7.

[412] Sergeant G, Vankelecom H, Gremeaux L, Topal B. Role of cancer stem cells in pancreatic ductal adenocarcinoma. Nat Rev Clin Oncol 2009;6:580-6.

[413] Immervoll H, Hoem D, Sakariassen PO, Steffensen OJ, Molven A. Expression of the "stem cell marker" CD133 in pancreas and pancreatic ductal adenocarcinomas. BMC Cancer 2008;8:48.

[414] Hruban RH, Adsay NV, Albores-Saavedra J, *et al.* Pancreatic intraepithelial neoplasia: a new nomenclature and classification system for pancreatic duct lesions. Am J Surg Pathol 2001;25:579-86.

[415] Hruban RH, Takaori K, Klimstra DS, *et al.* An illustrated consensus on the classification of pancreatic intraepithelial neoplasia and intraductal papillary mucinous neoplasms. Am J Surg Pathol 2004;28:977-87.

[416] Maitra A, Fukushima N, Takaori K, Hruban RH. Precursors to invasive pancreatic cancer. Adv Anat Pathol 2005;12:81-91.

[417] Brat DJ, Lillemoe KD, Yeo CJ, Warfield PB, Hruban RH. Progression of pancreatic intraductal neoplasias to infiltrating adenocarcinoma of the pancreas. Am J Surg Pathol 1998;22:163-9.

[418] Balague C, Audie JP, Porchet N, Real FX. *In situ* hybridization shows distinct patterns of mucin gene expression in normal, benign, and malignant pancreas tissues. Gastroenterology 1995;109:953-64.

[419] Andrianifahanana M, Moniaux N, Schmied BM, *et al.* Mucin (MUC) gene expression in human pancreatic adenocarcinoma and chronic pancreatitis: a potential role of MUC4 as a tumor marker of diagnostic significance. Clin Cancer Res 2001;7:4033-40.

[420] Nagata K, Horinouchi M, Saitou M, *et al.* Mucin expression profile in pancreatic cancer and the precursor lesions. J Hepatobiliary Pancreat Surg 2007;14:243-54.

[421] Cao W, Adley BP, Liao J, *et al.* Mucinous nonneoplastic cyst of the pancreas: apomucin phenotype distinguishes this entity from intraductal papillary mucinous neoplasm. Hum Pathol 2010;41:513-21.

[422] Ji Y, Xu JF, Kuang TT, *et al.* Mucin profile of the pancreatic mucinous cystic neoplasms. Chin Med J (Engl) 2006;119:328-30.

[423] Ringel J, Lohr M. The MUC gene family: their role in diagnosis and early detection of pancreatic cancer. Mol Cancer 2003;2:9.
[424] Kim GE, Bae HI, Park HU, *et al.* Aberrant expression of MUC5AC and MUC6 gastric mucins and sialyl Tn antigen in intraepithelial neoplasms of the pancreas. Gastroenterology 2002;123:1052-60.
[425] Balague C, Gambus G, Carrato C, *et al.* Altered expression of MUC2, MUC4, and MUC5 mucin genes in pancreas tissues and cancer cell lines. Gastroenterology 1994;106:1054-61.
[426] Yamada N, Nishida Y, Yokoyama S, *et al.* Expression of MUC5AC, an early marker of pancreatobiliary cancer, is regulated by DNA methylation in the distal promoter region in cancer cells. J Hepatobiliary Pancreat Sci 2010;17:844-54.
[427] Yonezawa S, Higashi M, Yamada N, Yokoyama S, Goto M. Significance of mucin expression in pancreatobiliary neoplasms. J Hepatobiliary Pancreat Sci 2010;17:108-24.
[428] Hoshi H, Sawada T, Uchida M, *et al.* Tumor-associated MUC5AC stimulates *in vivo* tumorigenicity of human pancreatic cancer. Int J Oncol 2011;38:619-27.
[429] Forgue-Lafitte ME, Arambam R, Bara J. Proteases present in some pancreatic cyst fluids may affect mucin immunoassay by degrading antibodies and antigens. Pancreas 2010;39:1070-6.
[430] Haab BB, Porter A, Yue T, *et al.* Glycosylation variants of mucins and CEACAMs as candidate biomarkers for the diagnosis of pancreatic cystic neoplasms. Ann Surg 2010;251:937-45.
[431] Yue T, Goldstein IJ, Hollingsworth MA, *et al.* The prevalence and nature of glycan alterations on specific proteins in pancreatic cancer patients revealed using antibody-lectin sandwich arrays. Mol Cell Proteomics 2009;8:1697-707.
[432] Luttges J, Feyerabend B, Buchelt T, Pacena M, Kloppel G. The mucin profile of noninvasive and invasive mucinous cystic neoplasms of the pancreas. Am J Surg Pathol 2002;26:466-71.
[433] Kigure S. Immunohistochemical study of the association between the progression of pancreatic ductal lesions and the expression of MUC1, MUC2, MUC5AC, and E-cadherin. Rinsho Byori 2006;54:447-52.
[434] Sanada Y, Yoshida K, Ohara M, *et al.* Histopathologic evaluation of stepwise progression of pancreatic carcinoma with immunohistochemical analysis of gastric epithelial transcription factor SOX2: comparison of expression patterns between invasive components and cancerous or nonneoplastic intraductal components. Pancreas 2006;32:164-70.
[435] Yonezawa S, Horinouchi M, Osako M, *et al.* Gene expression of gastric type mucin (MUC5AC) in pancreatic tumors: its relationship with the biological behavior of the tumor. Pathol Int 1999;49:45-54.
[436] Takaori K, Kobashi Y, Matsusue S, Matsui K, Yamamoto T. Clinicopathological features of pancreatic intraepithelial neoplasias and their relationship to intraductal papillary-mucinous tumors. J Hepatobiliary Pancreat Surg 2003;10:125-36.
[437] Inaguma S, Kasai K, Ikeda H. GLI1 facilitates the migration and invasion of pancreatic cancer cells through MUC5AC-mediated attenuation of E-cadherin. Oncogene 2011;30:714-23.
[438] Yamazoe S, Tanaka H, Sawada T, *et al.* RNA interference suppression of mucin 5AC (MUC5AC) reduces the adhesive and invasive capacity of human pancreatic cancer cells. J Exp Clin Cancer Res 2010;29:53.
[439] Luka J, Arlen PM, Bristol A. Development of a serum biomarker assay that differentiates tumor-associated MUC5AC (NPC-1C ANTIGEN) from normal MUC5AC. J Biomed Biotechnol 2011;2011:ID 934757.

[440] Yamada N, Nishida Y, Yokoyama S, *et al.* Expression of MUC5AC, an early marker of pancreatobiliary cancer, is regulated by DNA methylation in the distal promoter region in cancer cells. J Hepatobiliary Pancreat Sci 2010;17:844-54.

[441] Sohn TA, Yeo CJ, Cameron JL, *et al.* Intraductal papillary mucinous neoplasms of the pancreas: an updated experience. Ann Surg 2004;239:788-97; discussion 97-9.

[442] Furukawa T, Kloppel G, Volkan Adsay N, *et al.* Classification of types of intraductal papillary-mucinous neoplasm of the pancreas: a consensus study. Virchows Arch 2005;447:794-9.

[443] Belyaev O, Seelig MH, Muller CA, *et al.* Intraductal papillary mucinous neoplasms of the pancreas. J Clin Gastroenterol 2008;42:284-94.

[444] Ban S, Satoh H, Satoh M, *et al.* Invasive ductal carcinoma of the pancreas tail with noninvasive growth through the nondilated main pancreatic duct and macroscopically cystic invasive carcinomatous glands. Ann Diagn Pathol 2010;15:476-80.

[445] Lee SY, Choi DW, Jang KT, *et al.* High expression of intestinal-type mucin (MUC2) in intraductal papillary mucinous neoplasms coexisting with extrapancreatic gastrointestinal cancers. Pancreas 2006;32:186-9.

[446] Yonezawa S, Nakamura A, Horinouchi M, Sato E. The expression of several types of mucin is related to the biological behavior of pancreatic neoplasms. J Hepatobiliary Pancreat Surg 2002;9:328-41.

[447] Ban S, Naitoh Y, Ogawa F, *et al.* Intraductal papillary mucinous neoplasm (IPMN) of the gastric-type with focal nodular growth of the arborizing papillae: a case of high-grade transformation of the gastric-type IPMN. Virchows Arch 2006;449:112-6.

[448] Ban S, Naitoh Y, Mino-Kenudson M, *et al.* Intraductal papillary mucinous neoplasm (IPMN) of the pancreas: its histopathologic difference between 2 major types. Am J Surg Pathol 2006;30:1561-9.

[449] Rouzbahman M, Serra S, Adsay NV, *et al.* Oncocytic papillary neoplasms of the biliary tract: a clinicopathological, mucin core and Wnt pathway protein analysis of four cases. Pathology 2007;39:413-8.

[450] Siveke JT, Einwachter H, Sipos B, *et al.* Concomitant pancreatic activation of Kras(G12D) and Tgfa results in cystic papillary neoplasms reminiscent of human IPMN. Cancer Cell 2007;12:266-79.

[451] Yonezawa S, Higashi M, Yamada N, Goto M. Precursor lesions of pancreatic cancer. Gut Liver 2008;2:137-54.

[452] Nishigami T, Onodera M, Torii I, *et al.* Comparison between mucinous cystic neoplasm and intraductal papillary mucinous neoplasm of the branch duct type of the pancreas with respect to expression of CD10 and cytokeratin 20. Pancreas 2009;38:558-64.

[453] Yaman B, Nart D, Yilmaz F, *et al.* Biliary intraductal papillary mucinous neoplasia: three case reports. Virchows Arch 2009;454:589-94.

[454] Maker AV, Katabi N, Gonen M, *et al.* Pancreatic cyst fluid and serum mucin levels predict dysplasia in intraductal papillary mucinous neoplasms of the pancreas. Ann Surg Oncol 2011;18:199-206.

[455] Terris B, Dubois S, Buisine MP, *et al.* Mucin gene expression in intraductal papillary-mucinous pancreatic tumours and related lesions. J Pathol 2002;197:632-7.

[456] Adsay NV, Merati K, Basturk O, *et al.* Pathologically and biologically distinct types of epithelium in intraductal papillary mucinous neoplasms: delineation of an "intestinal" pathway of carcinogenesis in the pancreas. Am J Surg Pathol 2004;28:839-48.

[457] Nakamura A, Horinouchi M, Goto M, *et al.* New classification of pancreatic intraductal papillary-mucinous tumour by mucin expression: its relationship with potential for malignancy. J Pathol 2002;197:201-10.

[458] Pantano F, Baldi A, Santini D, *et al.* MUC2 but not MUC5 expression correlates with prognosis in radically resected pancreatic cancer patients. Cancer Biol Ther 2009;8:996-9.
[459] Adsay NV, Merati K, Andea A, *et al.* The dichotomy in the preinvasive neoplasia to invasive carcinoma sequence in the pancreas: differential expression of MUC1 and MUC2 supports the existence of two separate pathways of carcinogenesis. Mod Pathol 2002;15:1087-95.
[460] Lee JH, Lee KT, Park J, *et al.* Predictive factors associated with malignancy of intraductal papillary mucinous pancreatic neoplasms. World J Gastroenterol 2010;16:5353-8.
[461] Klimstra DS, Pitman MB, Hruban RH. Tumors of pancreas. Arch Pathol Lab Med 2009;133:454-64
[462] Hutchins GF, Draganov PV. Cystic neoplasms of the pancreas: a diagnostic challenge. World J Gastroenterol 2009;15:48-54.
[463] Fischer HP, Zhou H. Pathogenesis of carcinoma of the papilla of Vater. J Hepatobiliary Pancreat Surg 2004;11:301-9.
[464] Chu PG, Schwarz RE, Lau SK, Yen Y, Weiss LM. Immunohistochemical staining in the diagnosis of pancreatobiliary and ampulla of Vater adenocarcinoma: application of CDX2, CK17, MUC1, and MUC2. Am J Surg Pathol 2005;29:359-67.
[465] Heinrich S, Clavien PA. Ampullary cancer. Curr Opin Gastroenterol 2010;26:280-5.
[466] Kimura W, Futakawa N, Yamagata S, *et al.* Different clinicopathologic findings in two histologic types of carcinoma of papilla of Vater. Jpn J Cancer Res 1994;85:161-6.
[467] Sessa F, Furlan D, Zampatti C, *et al.* Prognostic factors for ampullary adenocarcinomas: tumor stage, tumor histology, tumor location, immunohistochemistry and microsatellite instability. Virchows Arch 2007;451:649-57.
[468] Matsubayashi H, Watanabe H, Yamaguchi T, *et al.* Differences in mucus and K-ras mutation in relation to phenotypes of tumors of the papilla of vater. Cancer 1999;86:596-607.
[469] Sanada Y, Yoshida K, Konishi K, *et al.* Expression of gastric mucin MUC5AC and gastric transcription factor SOX2 in ampulla of vater adenocarcinoma: comparison between expression patterns and histologic subtypes. Oncol Rep 2006;15:1157-61.
[470] Park SY, Roh SJ, Kim YN, *et al.* Expression of MUC1, MUC2, MUC5AC and MUC6 in cholangiocarcinoma: prognostic impact. Oncol Rep 2009;22:649-57.
[471] Khan SA, Davidson BR, Goldin R, *et al.* Guidelines for the diagnosis and treatment of cholangiocarcinoma: consensus document. Gut 2002;51 Suppl 6:VI1-9.
[472] Shaib Y, El-Serag HB. The epidemiology of cholangiocarcinoma. Semin Liver Dis 2004;24:115-25.
[473] Nakanuma Y, Terada T, Tanaka Y, Ohta G. Are hepatolithiasis and cholangiocarcinoma aetiologically related? A morphological study of 12 cases of hepatolithiasis associated with cholangiocarcinoma. Virchows Arch A Pathol Anat Histopathol 1985;406:45-58.
[474] Nakanuma Y, Sasaki M, Terada T, Harada K. Intrahepatic peribiliary glands of humans. II. Pathological spectrum. J Gastroenterol Hepatol 1994;9:80-6.
[475] Wee A, Ludwig J, Coffey RJ, Jr., LaRusso NF, Wiesner RH. Hepatobiliary carcinoma associated with primary sclerosing cholangitis and chronic ulcerative colitis. Hum Pathol 1985;16:719-26.
[476] Bamrungphon W, Prempracha N, Bunchu N, *et al.* A new mucin antibody/enzyme-linked lectin-sandwich assay of serum MUC5AC mucin for the diagnosis of cholangiocarcinoma. Cancer Lett 2007;247:301-8.
[477] Honda N, Cobb C, Lechago J. Bile duct carcinoma associated with multiple von Meyenburg complexes in the liver. Hum Pathol 1986;17:1287-90.
[478] Terada T, Kida T, Nakanuma Y, *et al.* Intrahepatic cholangiocarcinomas associated with nonbiliary cirrhosis. A clinicopathologic study. J Clin Gastroenterol 1994;18:335-42.

[479] Lim JH, Park CK. Pathology of cholangiocarcinoma. Abdom Imaging 2004;29:540-7.
[480] Welzel TM, Graubard BI, El-Serag HB, *et al.* Risk factors for intrahepatic and extrahepatic cholangiocarcinoma in the United States: a population-based case-control study. Clin Gastroenterol Hepatol 2007;5:1221-8.
[481] Shaib YH, El-Serag HB, Nooka AK, *et al.* Risk factors for intrahepatic and extrahepatic cholangiocarcinoma: a hospital-based case-control study. Am J Gastroenterol 2007;102:1016-21.
[482] Aljiffry M, Walsh MJ, Molinari M. Advances in diagnosis, treatment and palliation of cholangiocarcinoma: 1990-2009. World J Gastroenterol 2009;15:4240-62.
[483] Chen TC, Nakanuma Y, Zen Y, *et al.* Intraductal papillary neoplasia of the liver associated with hepatolithiasis. Hepatology 2001;34:651-8.
[484] Chen MF, Jan YY, Chen TC. Clinical studies of mucin-producing cholangiocellular carcinoma: a study of 22 histopathology-proven cases. Ann Surg 1998;227:63-9.
[485] Kim YI, Yu ES, Kim ST. Intraductal variant of peripheral cholangiocarcinoma of the liver with Clonorchis sinensis infection. Cancer 1989;63:1562-6.
[486] Ji Y, Fan J, Zhou J, *et al.* Intraductal papillary neoplasms of bile duct. A distinct entity like its counterpart in pancreas. Histol Histopathol 2008;23:41-50.
[487] Kloppel G, Kosmahl M. Is the intraductal papillary mucinous neoplasia of the biliary tract a counterpart of pancreatic papillary mucinous neoplasm? J Hepatol 2006;44:249-50.
[488] Zen Y, Fujii T, Itatsu K, *et al.* Biliary papillary tumors share pathological features with intraductal papillary mucinous neoplasm of the pancreas. Hepatology 2006;44:1333-43.
[489] Shimonishi T, Sasaki M, Nakanuma Y. Precancerous lesions of intrahepatic cholangiocarcinoma. J Hepatobiliary Pancreat Surg 2000;7:542-50.
[490] Nakanuma Y, Harada K, Ishikawa A, Zen Y, Sasaki M. Anatomic and molecular pathology of intrahepatic cholangiocarcinoma. J Hepatobiliary Pancreat Surg 2003;10:265-81.
[491] Nakanuma Y, Sasaki M, Ishikawa A, *et al.* Biliary papillary neoplasm of the liver. Histol Histopathol 2002;17:851-61.
[492] Zen Y, Sasaki M, Fujii T, *et al.* Different expression patterns of mucin core proteins and cytokeratins during intrahepatic cholangiocarcinogenesis from biliary intraepithelial neoplasia and intraductal papillary neoplasm of the bile duct--an immunohistochemical study of 110 cases of hepatolithiasis. J Hepatol 2006;44:350-8.
[493] Jan YY, Yeh TS, Yeh JN, Yang HR, Chen MF. Expression of epidermal growth factor receptor, apomucins, matrix metalloproteinases, and p53 in rat and human cholangiocarcinoma: appraisal of an animal model of cholangiocarcinoma. Ann Surg 2004;240:89-94.
[494] Mall AS, Tyler MG, Ho SB, *et al.* The expression of MUC mucin in cholangiocarcinoma. Pathol Res Pract 2010;206:805-9.
[495] Lau SK, Weiss LM, Chu PG. Differential expression of MUC1, MUC2, and MUC5AC in carcinomas of various sites: an immunohistochemical study. Am J Clin Pathol 2004;122:61-9.
[496] Lee MJ, Lee HS, Kim WH, Choi Y, Yang M. Expression of mucins and cytokeratins in primary carcinomas of the digestive system. Mod Pathol 2003;16:403-10.
[497] Sasaki M, Nakanuma Y, Kim YS. Characterization of apomucin expression in intrahepatic cholangiocarcinomas and their precursor lesions: an immunohistochemical study. Hepatology 1996;24:1074-8.
[498] Aishima S, Kuroda Y, Nishihara Y, *et al.* Gastric mucin phenotype defines tumour progression and prognosis of intrahepatic cholangiocarcinoma: gastric foveolar type is associated with aggressive tumour behaviour. Histopathology 2006;49:35-44.

[499] Sasaki M, Yamato T, Nakanuma Y. Expression of sialyl-Tn, Tn and T antigens in primary liver cancer. Pathol Int 1999;49:325-31.

[500] Sasaki M, Nakanuma Y, Ho SB, Kim YS. Cholangiocarcinomas arising in cirrhosis and combined hepatocellular-cholangiocellular carcinomas share apomucin profiles. Am J Clin Pathol 1998;109:302-8.

[501] Campion JP, Porchet N, Aubert JP, L'Helgoualc'h A, Clement B. UW-preservation of cultured human gallbladder epithelial cells: phenotypic alterations and differential mucin gene expression in the presence of bile. Hepatology 1995;21:223-31.

[502] Vandenhaute B, Buisine MP, Debailleul V, *et al.* Mucin gene expression in biliary epithelial cells. J Hepatol 1997;27:1057-66.

[503] Ho SB, Shekels LL, Toribara NW, *et al.* Altered mucin core peptide expression in acute and chronic cholecystitis. Dig Dis Sci 2000;45:1061-71.

[504] Lee SP. Pathogenesis of biliary sludge. Hepatology 1990;12:200S-3S; discussion 3S-5S.

[505] Smith BF. Gallbladder mucin as a pronucleating agent for cholesterol monohydrate crystals in bile. Hepatology 1990;12:183S-6S; discussion 6S-8S.

[506] Carlei F, Lygidakis NJ, Barsotti P, *et al.* Mucin-producing cells and endocrine cells of gallbladder epithelium in patients with uncomplicated cholelithiasis. Hepatogastroenterology 1991;38:149-53.

[507] Bar Dayan Y, Vilkin A, Niv Y. Gallbladder mucin plays a role in gallstone formation. Eur J Intern Med 2004;15:411-4.

[508] Wang HH, Afdhal NH, Gendler SJ, Wang DQ. Targeted disruption of the murine mucin gene 1 decreases susceptibility to cholesterol gallstone formation. J Lipid Res 2004;45:438-47.

[509] Sasaki M, Nakanuma Y, Kim YS. Expression of apomucins in the intrahepatic biliary tree in hepatolithiasis differs from that in normal liver and extrahepatic biliary obstruction. Hepatology 1998;27:54-61.

[510] Vilkin A, Nudelman I, Morgenstern S, *et al.* Gallbladder inflammation is associated with increase in mucin expression and pigmented stone formation. Dig Dis Sci 2007;52:1613-20.

[511] Kim HJ, Kim SH, Chae GB, Lee SJ, Kang CD. Increased expression of mucin 5AC mRNA and decreased expression of epidermal growth-factor receptor mRNA in gallstone patients. Tohoku J Exp Med 2008;214:139-44.

[512] De Bolos C, Garrido M, Real FX. MUC6 apomucin shows a distinct normal tissue distribution that correlates with Lewis antigen expression in the human stomach. Gastroenterology 1995;109:723-34.

[513] Kim YS, Gum JR, Jr. Diversity of mucin genes, structure, function, and expression. Gastroenterology 1995;109:999-1001.

[514] Nagata S, Ajioka Y, Nishikura K, *et al.* Co-expression of gastric and biliary phenotype in pyloric-gland type adenoma of the gallbladder: immunohistochemical analysis of mucin profile and CD10. Oncol Rep 2007;17:721-9.

[515] Rosai J. In: Rosai J, editor. Ackerman's Surgical Pathology. 8th edn. St. Louis,: Mosby; 1996. p. 943-68

[516] Reid CJ, Harris A. Developmental expression of mucin genes in the human gastrointestinal system. Gut 1998;42:220-6.

[517] Buisine MP, Desreumaux P, Leteurtre E, *et al.* Mucin gene expression in intestinal epithelial cells in Crohn's disease. Gut 2001;49:544-51.

[518] Shaoul R, Okada Y, Cutz E, Marcon MA. Colonic expression of MUC2, MUC5AC, and TFF1 in inflammatory bowel disease in children. J Pediatr Gastroenterol Nutr 2004;38:488-93.

[519] Longman RJ, Poulsom R, Corfield AP, *et al.* Alterations in the composition of the supramucosal defense barrier in relation to disease severity of ulcerative colitis. J Histochem Cytochem 2006;54:1335-48.

[520] Kaneko Y, Nakamura T, Hayama M, *et al.* Altered expression of CDX-2, PDX-1 and mucin core proteins in "Ulcer-associated cell lineage (UACL)" in Crohn's disease. J Mol Histol 2008;39:161-8.

[521] Longman RJ, Douthwaite J, Sylvester PA, *et al.* Coordinated localisation of mucins and trefoil peptides in the ulcer associated cell lineage and the gastrointestinal mucosa. Gut 2000;47:792-800.

[522] Mochizuka A, Uehara T, Nakamura T, Kobayashi Y, Ota H. Hyperplastic polyps and sessile serrated'adenomas' of the colon and rectum display gastric pyloric differentiation. Histochem Cell Biol 2007;128:445-55.

[523] Fujita K, Hirahashi M, Yamamoto H, *et al.* Mucin core protein expression in serrated polyps of the large intestine. Virchows Arch 2010;457:443-9.

[524] Owens SR, Chiosea SI, Kuan SF. Selective expression of gastric mucin MUC6 in colonic sessile serrated adenoma but not in hyperplastic polyp aids in morphological diagnosis of serrated polyps. Mod Pathol 2008;21:660-9.

[525] Myerscough N, Sylvester PA, Warren BF, *et al.* Abnormal subcellular distribution of mature MUC2 and *de novo* MUC5AC mucins in adenomas of the rectum: immunohistochemical detection using non-VNTR antibodies to MUC2 and MUC5AC peptide. Glycoconj J 2001;18:907-14.

[526] Sylvester PA, Myerscough N, Warren BF, *et al.* Differential expression of the chromosome 11 mucin genes in colorectal cancer. J Pathol 2001;195:327-35.

[527] Kocer B, Soran A, Erdogan S, *et al.* Expression of MUC5AC in colorectal carcinoma and relationship with prognosis. Pathol Int 2002;52:470-7.

[528] Shi C, Scudiere JR, Cornish TC, *et al.* Clear cell change in colonic tubular adenoma and corresponding colonic clear cell adenocarcinoma is associated with an altered mucin core protein profile. Am J Surg Pathol 2010;34:1344-50.

[529] Bartman AE, Sanderson SJ, Ewing SL, *et al.* Aberrant expression of MUC5AC and MUC6 gastric mucin genes in colorectal polyps. Int J Cancer 1999;80:210-8.

[530] Pullan RD, Thomas GA, Rhodes M, *et al.* Thickness of adherent mucus gel on colonic mucosa in humans and its relevance to colitis. Gut 1994;35:353-9.

[531] Tytgat KM, Opdam FJ, Einerhand AW, Buller HA, Dekker J. MUC2 is the prominent colonic mucin expressed in ulcerative colitis. Gut 1996;38:554-63.

[532] Podolsky DK, Isselbacher KJ. Composition of human colonic mucin. Selective alteration in inflammatory bowel disease. J Clin Invest 1983;72:142-53.

[533] Raouf AH, Tsai HH, Parker N, *et al.* Sulphation of colonic and rectal mucin in inflammatory bowel disease: reduced sulphation of rectal mucus in ulcerative colitis. Clin Sci (Lond) 1992;83:623-6.

[534] Culling CF, Reid PE, Dunn WL. A histochemical comparison of the O-acylated sialic acids of the epithelial mucins in ulcerative colitis, Crohn's disease, and normal controls. J Clin Pathol 1979;32:1272-7.

[535] Smithson JE, Campbell A, Andrews JM, *et al.* Altered expression of mucins throughout the colon in ulcerative colitis. Gut 1997;40:234-40.

[536] Shaoul R, Marcon MA, Okada Y, Cutz E, Forstner G. Gastric metaplasia: a frequently overlooked feature of duodenal biopsy specimens in untreated celiac disease. J Pediatr Gastroenterol Nutr 2000;30:397-403.

[537] Shaoul R, Marcon P, Okada Y, Cutz E, Forstner G. The pathogenesis of duodenal gastric metaplasia: the role of local goblet cell transformation. Gut 2000;46:632-8.
[538] Van Klinken BJ, Dekker J, Buller HA, de Bolos C, Einerhand AW. Biosynthesis of mucins (MUC2-6) along the longitudinal axis of the human gastrointestinal tract. Am J Physiol 1997;273:G296-302.
[539] Sotozono M, Okada Y, Sasagawa T, *et al.* Novel monoclonal antibody, SO-MU1, against human gastric MUC5AC apomucin. J Immunol Methods 1996;192:87-96.
[540] Rio MC, Chenard MP, Wolf C, *et al.* Induction of pS2 and hSP genes as markers of mucosal ulceration of the digestive tract. Gastroenterology 1991;100:375-9.
[541] Wright NA, Pike C, Elia G. Induction of a novel epidermal growth factor-secreting cell lineage by mucosal ulceration in human gastrointestinal stem cells. Nature 1990;343:82-5.
[542] Wright NA, Poulsom R, Stamp G, *et al.* Trefoil peptide gene expression in gastrointestinal epithelial cells in inflammatory bowel disease. Gastroenterology 1993;104:12-20.
[543] Harpaz N, Talbot IC. Colorectal cancer in idiopathic inflammatory bowel disease. Semin Diagn Pathol 1996;13:339-57.
[544] Tatsumi N, Kushima R, Vieth M, *et al.* Cytokeratin 7/20 and mucin core protein expression in ulcerative colitis-associated colorectal neoplasms. Virchows Arch 2006;448:756-62.
[545] Snover DC, Jass JR, Fenoglio-Preiser C, Batts KP. Serrated polyps of the large intestine: a morphologic and molecular review of an evolving concept. Am J Clin Pathol 2005;124:380-91.
[546] Hirono H, Ajioka Y, Watanabe H, *et al.* Bidirectional gastric differentiation in cellular mucin phenotype (foveolar and pyloric) in serrated adenoma and hyperplastic polyp of the colorectum. Pathol Int 2004;54:401-7.
[547] Jass JR, Stewart SM, Stewart J, Lane MR. Hereditary non-polyposis colorectal cancer--morphologies, genes and mutations. Mutat Res 1994;310:125-33.
[548] Vogelstein B, Kinzler KW. The multistep nature of cancer. Trends Genet 1993;9:138-41.
[549] Buisine MP, Janin A, Maunoury V, *et al.* Aberrant expression of a human mucin gene (MUC5AC) in rectosigmoid villous adenoma. Gastroenterology 1996;110:84-91.
[550] Ho SB, Ewing SL, Montgomery CK, Kim YS. Altered mucin core peptide immunoreactivity in the colon polyp-carcinoma sequence. Oncol Res 1996;8:53-61.
[551] Bara J, Loisillier F, Burtin P. Antigens of gastric and intestinal mucous cells in human colonic tumours. Br J Cancer 1980;41:209-21.
[552] Ward R, Meagher A, Tomlinson I, *et al.* Microsatellite instability and the clinicopathological features of sporadic colorectal cancer. Gut 2001;48:821-9.
[553] Alexander J, Watanabe T, Wu TT, *et al.* Histopathological identification of colon cancer with microsatellite instability. Am J Pathol 2001;158:527-35.
[554] Biemer-Huttmann AE, Walsh MD, McGuckin MA, *et al.* Mucin core protein expression in colorectal cancers with high levels of microsatellite instability indicates a novel pathway of morphogenesis. Clin Cancer Res 2000;6:1909-16.
[555] Messerini L, Vitelli F, De Vitis LR, *et al.* Microsatellite instability in sporadic mucinous colorectal carcinomas: relationship to clinico-pathological variables. J Pathol 1997;182:380-4.
[556] Andersch-Bjorkman Y, Thomsson KA, Holmen Larsson JM, Ekerhovd E, Hansson GC. Large scale identification of proteins, mucins, and their O-glycosylation in the endocervical mucus during the menstrual cycle. Mol Cell Proteomics 2007;6:708-16.
[557] Ota H, Harada O, Uehara T, Hayama M, Ishii K. Aberrant expression of TFF1, TFF2, and PDX1 and their diagnostic value in lobular endocervical glandular hyperplasia. Am J Clin Pathol 2011;135:253-61.

[558] Morrison C, Merati K, Marsh WL, Jr., *et al.* The mucin expression profile of endometrial carcinoma and correlation with clinical-pathologic parameters. Appl Immunohistochem Mol Morphol 2007;15:426-31.

[559] Shintaku M, Kushima R, Abiko K. Colloid carcinoma of the intestinal type in the uterine cervix: mucin immunohistochemistry. Pathol Int 2010;60:119-24.

[560] Mall AS, Tyler M, Lotz Z, *et al.* The characterisation of mucin in a mature ovarian teratoma occurring in an eight year old patient. Int J Med Sci 2007;4:115-23.

[561] Bara J, Gautier R, Mouradian P, Decaens C, Daher N. Oncofetal mucin M1 epitope family: characterization and expression during colonic carcinogenesis. Int J Cancer 1991;47:304-10.

[562] Langley FA, Cummins PA, Fox H. An ultrastructural study of mucin secreting epithelia in ovarian neoplasms. Acta Pathol Microbiol Scand Suppl 1972;233:76-86.

[563] Tenti P, Aguzzi A, Riva C, *et al.* Ovarian mucinous tumors frequently express markers of gastric, intestinal, and pancreatobiliary epithelial cells. Cancer 1992;69:2131-42.

[564] Shiozawa T, Tsukahara Y, Ishii K, *et al.* Histochemical demonstration of gastrointestinal mucins in ovarian mucinous cystadenoma. Acta Pathol Jpn 1992;42:104-10.

[565] Hirabayashi K, Yasuda M, Kajiwara H, *et al.* Alterations in mucin expression in ovarian mucinous tumors: immunohistochemical analysis of MUC2, MUC5AC, MUC6, and CD10 expression. Acta Histochem Cytochem 2008;41:15-21.

[566] Jaffe ES, Harris NL, Stein H, Vardiman JV, (eds). WHO Classification of Pathology and Genetics of Tumours of the Breast and Female Genital Organs. Lyon: IARC Press; 2004.

[567] Kunze E, Francksen B, Schulz H. Expression of MUC5AC apomucin in transitional cell carcinomas of the urinary bladder and its possible role in the development of mucus-secreting adenocarcinomas. Virchows Arch 2001;439:609-15.

[568] Daher N, Gonzales J, Gautier R, Bara J. Evidence of mucin M1 antigens in seminal plasma and normal cells of human prostatic urethra in relation to embryonic development and tumors. Prostate 1990;16:57-69.

[569] Legrier ME, de Pinieux G, Boye K, *et al.* Mucinous differentiation features associated with hormonal escape in a human prostate cancer xenograft. Br J Cancer 2004;90:720-7.

[570] Naito S, Yamazumi K, Yakata Y, *et al.* Immunohistochemical examination of mucinous cystadenoma of the testis. Pathol Int 2004;54:355-9.

[571] Nokubi M, Kawai T, Mitsu S, Ishikawa S, Morinaga S. Mucinous cystadenoma of the testis. Pathol Int 2002;52:648-52.

[572] Pereira MB, Dias AJ, Reis CA, Schmitt FC. Immunohistochemical study of the expression of MUC5AC and MUC6 in breast carcinomas and adjacent breast tissues. J Clin Pathol 2001;54:210-3.

[573] Matsukita S, Nomoto M, Kitajima S, *et al.* Expression of mucins (MUC1, MUC2, MUC5AC and MUC6) in mucinous carcinoma of the breast: comparison with invasive ductal carcinoma. Histopathology 2003;42:26-36.

[574] Rakha EA, Boyce RW, Abd El-Rehim D, *et al.* Expression of mucins (MUC1, MUC2, MUC3, MUC4, MUC5AC and MUC6) and their prognostic significance in human breast cancer. Mod Pathol 2005;18:1295-304.

[575] Mantelli F, Argueso P. Functions of ocular surface mucins in health and disease. Curr Opin Allergy Clin Immunol 2008;8:477-83.

[576] Gipson IK, Argueso P. Role of mucins in the function of the corneal and conjunctival epithelia. Int Rev Cytol 2003;231:1-49.

[577] Inatomi T, Spurr-Michaud S, Tisdale AS, Gipson IK. Human corneal and conjunctival epithelia express MUC1 mucin. Invest Ophthalmol Vis Sci 1995;36:1818-27.

[578] Inatomi T, Spurr Michaud S, Tisdale AS, *et al.* Expression of secretory mucin genes by human conjunctival epithelia. Invest Ophthalmol Vis Sci 1996;37:1684-92.

[579] Argueso P, Spurr-Michaud S, Russo CL, Tisdale A, Gipson IK. MUC16 mucin is expressed by the human ocular surface epithelia and carries the H185 carbohydrate epitope. Invest Ophthalmol Vis Sci 2003;44:2487-95.
[580] Jumblatt MM, McKenzie RW, Steele PS, Emberts CG, Jumblatt JE. MUC7 expression in the human lacrimal gland and conjunctiva. Cornea 2003;22:41-5.
[581] Paulsen F, Langer G, Hoffmann W, Berry M. Human lacrimal gland mucins. Cell Tissue Res 2004;316:167-77.
[582] Jumblatt JE, Cunningham LT, Li Y, Jumblatt MM. Characterization of human ocular mucin secretion mediated by 15(S)-HETE. Cornea 2002;21:818-24.
[583] Jumblatt MM, McKenzie RW, Jumblatt JE. MUC5AC mucin is a component of the human precorneal tear film. Invest Ophthalmol Vis Sci 1999;40:43-9.
[584] Gipson IK. Distribution of mucins at the ocular surface. Exp Eye Res 2004;78:379-88.
[585] Liu W, Li H, Lu D, *et al.* The tear fluid mucin 5AC change of primary angle-closure glaucoma patients after short-term medications and phacotrabeculectomy. Mol Vis 2010;16:2342-6.
[586] Ohashi Y, Dogru M, Tsubota K. Laboratory findings in tear fluid analysis. Clin Chim Acta 2006;369:17-28.
[587] Kunert KS, Keane-Myers AM, Spurr-Michaud S, Tisdale AS, Gipson IK. Alteration in goblet cell numbers and mucin gene expression in a mouse model of allergic conjunctivitis. Invest Ophthalmol Vis Sci 2001;42:2483-9.
[588] Dogru M, Okada N, Asano-Kato N, *et al.* Atopic ocular surface disease: implications on tear function and ocular surface mucins. Cornea 2005;24:S18-S23.
[589] Aragona P, Romeo GF, Puzzolo D, Micali A, Ferreri G. Impression cytology of the conjunctival epithelium in patients with vernal conjunctivitis. Eye (Lond) 1996;10 (Pt 1):82-5.
[590] Bonini S, Coassin M, Aronni S, Lambiase A. Vernal keratoconjunctivitis. Eye (Lond) 2004;18:345-51.
[591] Mantelli F, Moretti C, Micera A, Bonini S. Conjunctival mucin deficiency in complete androgen insensitivity syndrome (CAIS). Graefes Arch Clin Exp Ophthalmol 2007;245:899-902.
[592] Argueso P, Balaram M, Spurr-Michaud S, *et al.* Decreased levels of the goblet cell mucin MUC5AC in tears of patients with Sjogren syndrome. Invest Ophthalmol Vis Sci 2002;43:1004-11.
[593] Gipson IK, Hori Y, Argueso P. Character of ocular surface mucins and their alteration in dry eye disease. Ocul Surf 2004;2:131-48.
[594] Tei M, Spurr-Michaud SJ, Tisdale AS, Gipson IK. Vitamin A deficiency alters the expression of mucin genes by the rat ocular surface epithelium. Invest Ophthalmol Vis Sci 2000;41:82-8.
[595] Pisella PJ, Brignole F, Debbasch C, *et al.* Flow cytometric analysis of conjunctival epithelium in ocular rosacea and keratoconjunctivitis sicca. Ophthalmology 2000;107:1841-9.
[596] Nelson JD, Havener VR, Cameron JD. Cellulose acetate impressions of the ocular surface. Dry eye states. Arch Ophthalmol 1983;101:1869-72.
[597] Rivas L, Oroza MA, Perez-Esteban A, Murube-del-Castillo J. Morphological changes in ocular surface in dry eyes and other disorders by impression cytology. Graefes Arch Clin Exp Ophthalmol 1992;230:329-34.
[598] Pisella PJ, Pouliquen P, Baudouin C. Prevalence of ocular symptoms and signs with preserved and preservative free glaucoma medication. Br J Ophthalmol 2002;86:418-23.

[599] Pauly A, Brignole-Baudouin F, Labbe A, *et al.* New tools for the evaluation of toxic ocular surface changes in the rat. Invest Ophthalmol Vis Sci 2007;48:5473-83.
[600] Albietz JM, Bruce AS. The conjunctival epithelium in dry eye subtypes: effect of preserved and non-preserved topical treatments. Curr Eye Res 2001;22:8-18.

CHAPTER 6

Gel-Forming Mucin MUC5B

Abstract: The MUC5B mucin plays an important role in the homeostasis of respiratory, gastrointestinal and female reproductive tracts under physiological conditions and fulfills various important functions in human pathology. In this chapter, the molecular structure of the *MUC5B* gene and the mechanisms that regulate its activity are considered. Biochemical and biophysical properties of the MUC5B glycoprotein, its biosynthesis and posttranslational modifications as well as expression in the respiratory, gastrointestinal and female reproductive organs under physiological and pathological conditions are analyzed. The impact of MUC5B mucin in physiology and pathology of respiratory, gastrointestinal and female reproductive tracts and of such organs as eye, middle ear, breast and thyroid is discussed.

Keywords: MUC5B, evolution, domain, biosynthesis, promoter, regulation, expression.

6.1. GENERAL CHARACTERISTICS OF *MUC5B* GENE

The *MUC5B* gene encoding the gel-forming mucin glycoprotein MUC5B has been identified at the 3'-end of the cluster comprised of four genes – *MUC6, MUC2, MUC5AC* and *MUC5B* – and located between *HRAS* and *IGF2* on the short arm of chromosome 11 at locus 15.5 [1]. It evolved from the ancestor gene common to other gel-forming mucin genes, including *MUC2, MUC5AC, MUC6* and *MUC19*, and to the *vWF*gene encoding the von Willibrand Factor (vWF). According to Desseyn *et al.* [2, 3], a common ancestral gene was duplicated and diverged into two progenitor genes, one - the progenitor of *MUC6-MUC2* genes, and the other - the progenitor of *MUC5ACB*. Duplication of the *MUC5ACB* progenitor gene gave rise to two genes, *MUC5AC* and *MUC5B* (Fig. **1**).

MUC5B has been cloned in several laboratories, and the complete genomic DNA (acc # AJ012453.1, NM_002458.2, Y0988.2) and cDNA sequences (acc # Z72496, U78531.1) are now available [4-10]. The general structure of MUC5B mucin corresponds to that of MUC2 and MUC5AC: it consists of a central part encoded by a single unusually large exon of 10713 bp, and the N- and C-terminal flanking regions. The *MUC5B* genomic sequence (39 kb) covers 49 exons and 48 introns [2]. This structure was established by comparing the *MUC5B* genomic

Joseph Z. Zaretsky and Daniel H. Wreschner

DNA sequences with the corresponding cDNA sequences [7-9]. Desseyn and collaborators [7-9] reported 30 introns upstream and 18 introns downstream to the central large exon (exon 30). Most of the *MUC5B* introns are in the same position, and are related to the same class, as the corresponding introns in *vWF*. These data indicate the common evolution of these genes.

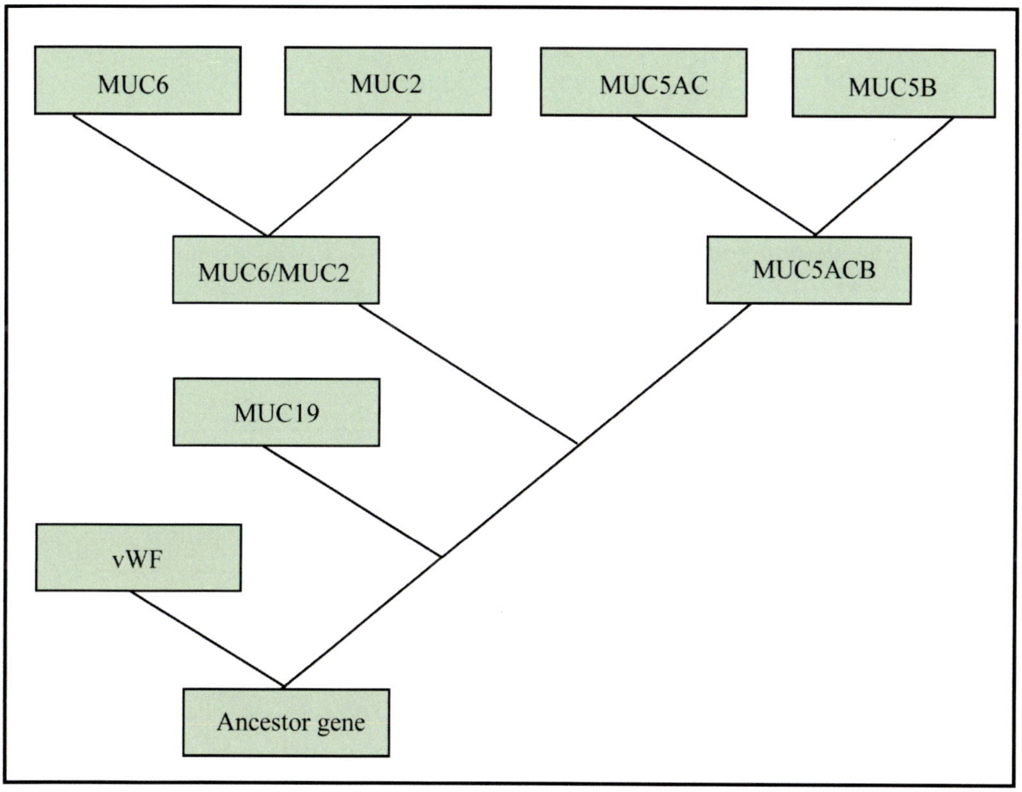

Figure 1: Evolutionary history of the *MUC5B* gene (based on the data reported in [2, 3]).

6.2. DOMAIN STRUCTURE OF MUC5B MUCIN

The central exon of *MUC5B* encodes a 3570-aa tandem repeat-containing mucin domain (PST-domain) unique for each gel-forming mucin (Fig. **2**). The N-terminal region of MUC5B mRNA encoding the D1-D2-D'-D3-domains is assembled by the splicing of 29 exons located upstream to the central (30th) exon; the nucleotide sequences encoding the C-terminally located D4-B-C-CK-domains are fused by the splicing of 18 exons located downstream to the central exon.

Both the N- and C-terminal domains are homologous to the corresponding domains of vWF and gel-forming mucins MUC2 and MUC5AC [7, 8, 11-16].

Importantly, three gel-forming mucin genes comprising the 11p15.5 cluster, namely *MUC2*, *MUC5AC* and *MUC6*, are characterized by genetic restriction fragment length polymorphism due to variations in the number of tandem repeats (TR) [17-19]. In contrast to other genes of the family, *MUC5B* shows no allele length variation in TRs of the central domain [20]. However, genetically determined polymorphism was detected in other exons and introns of the gene [21]. For example, intron 36 of *MUC5B* is made up almost entirely of perfect 59 bp-containing direct repeats. The number of repeats varies in individuals from three to eight, demonstrating allelic polymorphism [22]. Importantly, each 59 bp repeat contains one site able to bind a 42-kD nuclear factor (NF1-MUC5B) produced by mucus-secreting cells [23]. This factor may play a role in splicing and/or in pre-mRNA stability [24].

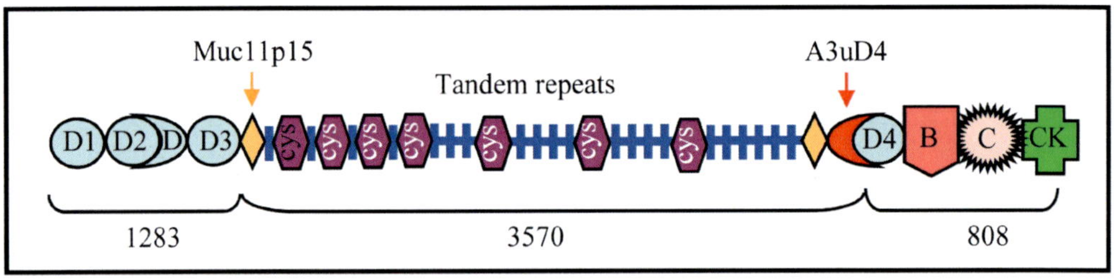

Figure 2: Domain structure of the MUC5B gel-forming mucin (based on the data from [4-10]).

In addition to the mucin-specific central domain, the *MUC5B* gene contains several cysteine-rich domains characteristic of other gel-forming mucins and the vWF glycoprotein. The C-terminally-located cysteine-rich CK-domain of the MUC5B mucin contains 85 amino acids, including 11 cysteine residues. This domain is similar to the corresponding domains of other gel-forming mucins and vWF both structurally and functionally, being responsible for initial disulfide-linked dimer formation [2, 25-28].

The central mucin-specific domain contains seven cysteine-rich subdomains (Cys-1-7 subdomains), which are distributed unevenly within the central domain (Fig. **2**). Each Cys-subdomain comprises a 108-aa polypeptide containing 10 highly conserved cysteine residues [7]. Cys-1-subdomain is located at the very N-terminus of the central domain. Each Cys-subdomain is followed by R-sequences, some of which (R1-RV) contain TRs while others (R01, R02 and R03) do not. The R01 (64 aa) and R02 (174 aa) sequences are located at the C-terminal ends of Cys-1 and Cys-2, respectively. The R03 (75 aa) subdomain occupies the C-terminus of the central domain. Its amino acid sequence is different from the sequence having the same location in the MUC5AC [4, 14] and MUC2 [12] proteins. The RI-RV sequences contain different numbers of a 29 aa-containing repeat unit: the subdomains RI, RII and RIV contain 11 imperfectly conserved repeats; RIII contains 17 repeat units; and RV is composed of 22 irregular units. Four of these subdomains (RI-RIV) are followed by unique 111 aa-containing sequences defined as R-end subdomains [7]. Interestingly, close examination of the MUC5B amino acid sequence [7] disclosed a super-repeat unit, four copies of which are present within the central exon of *MUC5B*. These super-repeats, designated UpA, UpB, UpC and UpD, consist of 11 imperfect repeats of 29 amino acid residues each followed by a 111 aa-containing R-end subdomain and a 108 aa-containing Cys-subdomain. Importantly, each R-subdomain contains one potential N-glycosylation site and numerous O-glycosylation sites. The central mucin-specific domain of MUC5B glycoprotein is flanked by two similar MUC11p15 domains, each encoded by one separate exon of 198 bp [8].

The N-terminal region of *MUC5B* encompasses 15143 bp of genomic DNA containing 29 exons and 29 introns. This genomic DNA fragment does not incorporate all the *MUC5B* promoter sequences, but does contain a short 36 bp fragment of the *MUC5B* 5'-UTR as part of exon 1. The 5'-end of exon 1 contains transcription start site (TSS). The ATG translation start codone is located 36 bp downstream to the TSS. The 5'- end of exon 1 is represented by 10 nucleotides of the 3850 bp-containing open reading frame sequence encoding a polypeptide of 1283 aa, comprising four vWF-like D-domains (D1- D2 - D'- D3) and the MUC11p15 domain [8]. This peptide is rich in cysteine and proline residues, but contains much less serine and threonine amino acids. There are 10 N-glycosylation target sites in the N-terminal sequence of MUC5B. All these structures are highly conserved among gel-forming mucins of various species [13, 15, 16, 29-31].

The C-terminal region of the *MUC5B* gene encompasses 10690 bp located downstream to the central exon. This region is composed of 18 exons and 18 introns. The 2423 nucleotide-containing open reading frame encodes a peptide of 808 aa rich in cysteine and proline, but free of serine and threonine [9]. This part of the MUC5B mucin is similar to the N-terminal region and different from the mucin-like central region. Several N-glycosylation-specific sites have been identified in the C-terminal part of the MUC5B mucin and most are preserved also in MUC2 and MUC5AC [9].

The 3'-region of the MUC5B glycoprotein consists of six domains: 5'-MUC11p15-A3uD4-D4-B-C-CK-3'. Interestingly, the first three domains are conserved in vWF, MUC2 and MUC5AC and some mucin-like proteins [7, 9, 32-34], whereas the B- and C-like domains are found only in MUC2 and MUC5AC mucins and not in the other aforementioned proteins. Importantly, while the vWF glycoprotein contains three copies of B-domain and two copies of C-domain, MUC5B mucin contains only single copies of these domains, reflecting differences in the evolution of *vWF* and *MUC5B*. On the other hand, the CK-domain is strictly conserved in all gel-forming mucin glycoproteins and in TGFβ [35]. Desseyn *et al.* [9] suggest combining the human gel-forming mucins MUC5B, MUC5AC, MUC2, their animal homologues and TGFβ in a new CK subfamily.

The mouse *Muc5b* gene and its human counterpart share an identical gene organization: their exons and introns are all in the same positions in the gene sequences [10]. There are, however, differences between them. The mucins have different numbers of N-glycosylation sites and Cys-subdomains [10]. There is no significant similarity in their intronic sequences. *Muc5b* contains no tandemly repeated GA-rich or GC-rich intronic sequences specific for transcription factors Sp1 and NF1 that are present in human *MUC5B* genomic sequences and contribute to regulation of *MUC5B* gene expression [23, 36]. According to Escande *et al.* [10], this indicates different regulation of *MUC5B* and *Muc5b* expression.

6.3. REGULATION OF THE *MUC5B* GENE EXPRESSION

Although more than thirteen years had passed since the *MUC5B* gene was cloned, the regulation of its expression at the transcriptional level is still poorly understood. Nevertheless, the available studies, although short in numbers, give some basic ideas regarding the mechanisms involved in regulation of *MUC5B* gene expression, including those operating at the promoter and epigenetic levels.

6.3.1. *MUC5B* promoter

Thanks in large part to the efforts of Van Seuningen and collaborators [36, 37], the nucleotide sequence of the 2 kb *MUC5B* promoter region has been established and analyzed. The authors found that the *MUC5B* promoter contains two TATA-boxes and numerous putative binding sites for numerous transcription factors, including Sp1, AP-1, NF-κB, TTF1, TGT3, HNF3β, N-Myc/Max, GRE, CREB, ATF, STAT, RORA1, GKLF/NRF2, and others (Fig. **3**).

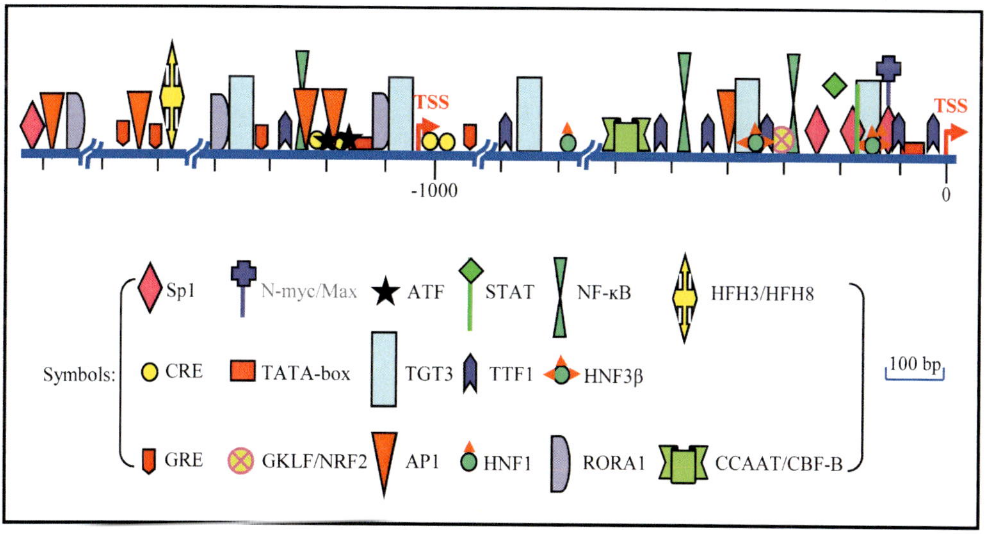

Figure 3: Transcription factor *cis*-element map of the *MUC5B* promoter (based on the data reported in [36, 37, 41-43]).

The proximal TATA-box is located at the bases -32/-26 and the distal one at -1144/-1134 bp from the transcription start site (TSS). Importantly, both TATA boxes appear to be functionally active and cell-specific: the distal promoter is 10 times more active than the proximal one in gastric cancer cell line KATO-III, and the two promoters show the same activity in AGS cells [37]. The functional activity of the distal promoter was confirmed by primer extension experiments, which identified the second transcription start site in the *MUC5B* promoter located 23 bp downstream to the distal TATA box. Binding assay demonstrated that transcription factor ATF1 binds to a corresponding *cis*-element in the distal promoter and activates transcription of the *MUC5B* gene in KATO-III cells. On the other hand, protein Sp1, which binds to both promoters, activates only the proximal promoter. Interestingly, Sp3, another member of the Sp family known to compete with Sp1 for the same binding site, has no effect on the proximal promoter but strongly inhibits the distal promoter in both gastric cell lines studied. In colon cell line, the proximal but not the distal promoter drives transcription of the *MUC5B* gene. Importantly, in addition to *cis*-elements that activate transcription, the 5'-UTR region of the *MUC5B* gene also contains elements that suppress *MUC5B* promoter activity in transfection assays [36].

There are several transcription factor binding sites in the *MUC5B* promoter but only a few have been tested experimentally in the context of their involvement in *MUC5B* transcriptional regulation.

Sp1-mediated transcription of MUC5B: Transcription factor Sp1 has been implicated in regulation of the *MUC5B* gene activity. Chang's group [38, 39] used promoter-reporter gene expression assays, site-directed mutagenesis and immunoprecipitation to show that Sp1 transcription factor participates in regulation of *MUC5B* expression *via* binding to *cis*-Sp1 sites in the *MUC5B* promoter. Interestingly, among three Sp1-binding sites tested in these experiments, only two of them, Sp1-1 and Sp1-2, located at -122/-114 bp and -196/-184 bp, respectively, were involved in *MUC5B* transcription in NHBE and HBE1 cells. Sp1-1 was important for basal promoter activity supporting constitutive *MUC5B* expression and Sp1-2 directed Phorbol-12-myristate-13-acetate (PMA)-induced *MUC5B* expression.

NFκB-specific activation of MUC5B gene: Transcription factor NFκB also participated in regulation of *MUC5B* transcription. It was found that cigarette smoke induced *Muc5b* expression in mouse middle ear epithelium by activation of NFκB transcription factor, followed by its binding to the NFκB-specific *cis*-elements located in the *Muc5b* promoter at -102/-92 bp and -516/-503 bp sites [40].

CREB-conducted MUC5B expression: Several studies showed that retinoids activate transcription of mucin genes including *MUC5B* [41-43]. As noted in the previous chapter, one of the pathways that activate *MUC5AC* transcription involves retinoid receptors on the cell membrane. These receptors transmit extracellular signals to the cell nucleus through the PKCα-Raf-MEK-ERK-RSK-CREB cascade. The phosphorylated CREB protein binds to specific *cis*-CRE sites in the *MUC5AC* promoter, thereby activating transcription of the *MUC5AC* gene [44]. Like *MUC5AC*, *MUC5B* also contains several *cis*-CRE sites (-1252 bp, -1165 bp, -1032 bp, and -1011 bp) in its promoter [37, 44]. The direct and indirect data indicate the possible involvement of these *cis*-elements in activation of *MUC5B* by a mechanism analogous to the one operating in *MUC5AC* [41, 45]. Koo *et al.* [42, 45] showed in a cell culture differentiation model that retinoids induce sequential activation of three gel-forming genes, *MUC2, MUC5AC* and *MUC5B. MUC2* was activated first, constituting a marker of early stage mucous differentiation, then followed *MUC5AC* and finally *MUC5B*, constituting a marker of advanced stage differentiation. Direct evidence of retinoic acid (RA) involvement in regulation of *MUC5B* transcription was brought by Chen *et al.* [43], who showed that the *MUC5B* promoter responded to RA by a 5-fold increase in expression of a reporter gene in transfection assay. Although the authors identified the several *cis*-CRE elements in the *MUC5B* promoter, they did not specify the CRE-sites that participated in activation of the *MUC5B* promoter/reporter gene construct in response to RA. More studies are needed to clarify this issue. Importantly, RA may activate expression of *MUC2, MUC5AC* and *MUC5B* through the retinoic receptor (classical pathway) and through non-classical pathways [44]. Notably, the activating effect of RA can be modified by various biologically active agents, such as thyroid hormone T_3, resulting in down-regulation of mucin genes [46]. The above studies indicate that RA regulation of mucin genes, including *MUC5B*, is a gene- and cell-specific process that may involve several distinct signaling cascades.

Glucocorticoid-mediated suppression of MUC5B activity: In addition to the already mentioned *cis*-elements, the *MUC5B* promoter contains several glucocorticoid receptor binding elements, *cis*-GRE sites [36, 37]. A number of studies showed that glucocorticoids substantially decrease expression of *MUC5B* [47, 48]. In this regard, the *MUC5B* gene is regulated in the same manner as *MUC5AC,* which also contains several *cis*-GRE sites and is also down-regulated by dexamethasone [49-51]. However, in contrast to the *MUC5AC* gene, in which the effect of dexamethasone is mediated through binding of the glucocorticoid receptor to two identified *cis*-GRE elements [51], it is unknown which of three GRE binding sites in the *MUC5B* promoter are involved in *MUC5B* transcriptional regulation. It is possible that glucocorticoid receptor interacts with other transcription factors, such as NFκB or AP-1, which in turn interact with *cis*-elements specific to these transcription factors [52, 53].

NRG1β1-specific up-regulation of MUC5B expression: The similar response of *MUC5B* and *MUC5AC* to various stimuli is not surprising. Many structural elements of the genes demonstrate strong evolutionary conservation. One of the stimuli to which both genes respond in a similar manner is growth factor neuregulin 1β1 (NRG1β1). Recently, Kettle *et al.* [54] demonstrated that NRG1β1 up-regulates transcription of both the *MUC5AC* and *MUC5B* genes in an HBEC cell culture. NRG1β1 has at least 15 alternatively spliced isoforms [55] and is a member of the NRG growth factor family comprised of several signaling proteins (NRG1, NRG2, NRG3 and NRG4) that mediate multiple cell-cell interactions *via* the receptor tyrosine kinase of the ErbB family [56]. Of the 18 growth factors tested by Kettle *et al.* [54], only NRG1β1 activated both *MUC5AC* and *MUC5B* genes. NRG1β1 up-regulates expression of these genes through interactions with Erb2 and Erb3 receptors that result in activation of the p38MAPK, ERK1/2 and P13K pathways [54]. Importantly, *MUC5B* could also be up-regulated by NRG1α isoform, while *MUC5AC* could not. Two isoforms of NRG1 – NRG1α and NRG1β – differ in the C-terminal region of the EGF-like domain [57, 58], a difference that appears to be critical for the activation of *MUC5B* by NRG1α.

Cytokine-regulated expression of MUC5B gene: The difference between *MUC5AC* and *MUC5B* genes is not restricted to their responsiveness to NRG1α, but is reflected also in their sensitivity to IL-13 cytokine. IL-13 strongly activates *MUC5AC*, but does not affect expression of *MUC5B*. The same difference was found in their reaction to IL-1β and TNFα cytokines. IL-1β up-regulated *MUC5AC* in a time- and dose-dependent manner in human pulmonary carcinoma cell lines [59-61], but had no effect on expression of the gene in any of the cell systems examined in other studies [61, 62]. TNFα elicited no response in *MUC5B* [63, 64] while *MUC5AC* was sensitive to the cytokine [64].

In contrast to the IL-1β and IL-13 cytokines, IL-6 and IL-17 have a strong activating effect on *MUC5B*. Taken together, the above studies show the differential effect of various cytokines on the expression of the *MUC5B* gene. It is important to note that there is no consensus on the role inflammatory cytokines play in mucin gene expression. The conflicting results obtained in the studies on the influence of IL-4 and IL-13 on expression of the *MUC2*, *MUC5AC* and *MUC5B* genes [64-70] confirm the above notion.

The role of different cytokines in transcriptional regulation of *MUC5B* has been studied much less than in *MUC2* and *MUC5AC*. The research of Chen *et al.* [64] is especially important in this context. This group studied the effect of a panel of 19 cytokines on expression of *MUC5B* in tracheobronchial epithelial cells and found that only IL-6 and IL-17 strongly induced expression of the gene. Stimulation of *MUC5B* expression by IL-6 in goblet HT29-MTX cells was observed earlier by Smirnova *et al.* [71]. Interestingly, in this cell system, IL-6 stimulated expression of the *MUC5B* gene but not secretion of its product, MUC5B mucin, while both processes responded positively to IL-6 treatment in the *MUC5AC* gene. Some details of the mechanisms by which IL-6 and IL-17 stimulate *MUC5B* expression were elucidated by Chen *et al.* [64], who established that the effect of IL-17 is mediated partially through IL-6 by a JAK2-dependent autocrine/paracrine loop. Further study is needed to identify the *cis*-elements in the *MUC5B* promoter involved in IL-6 and IL-17 activation of this gene. Expression of the *MUC5AC* gene in NHBE and HBE1 airway cell lines is activated by IL-17A through the ERK pathway and by utilization of the *cis*-NFκB site [60]. By analogy, NFκB binding site(s) present in *MUC5B* promoter may also participate in IL-17-mediated *MUC5B* expression. This suggestion is consistent with the many observations of NFκB involevement in IL17A-directed activation of numerous other genes [72-75]. Returning to the study of Chen and colleagues [64], it must be noted that although their results strongly support the potential of IL-6 and IL-17 to activate *MUC5B* in cell culture, these effects could not be fully reproduced *in vivo* in transgenic mice [76].

<u>*Bacterial LPS – stimulator of MUC5B transcription:*</u> Bacterial lipopolysaccharide (LPS), the major surface membrane component of almost all Gram-negative bacteria, is one of the strongest stimulators of mucin gene expression both *in vitro*

[61, 77, 78] and *in vivo* [79-82]. *In vitro*, LPS up-regulated expression of the *MUC5B* gene in a time- and dose-dependent manner [83], leading Smirnova and collaborators [83] to suggest that LPS-stimulatory effect on *MUC5B* expression is mediated through IL-8-activated pathways. IL-8 indeed induced abundance of MUC5B mRNAs in a time- and dose-dependent manner in another study (albeit not by up-regulating its mRNA production but by increasing its stability) [84].

Thus, it appears that the transcriptional up-regulation of *MUC5B* expression by LPS is based on the following mechanisms. The significant structural and evolutionary similarities between MUC5AC and MUC5B mucins [1, 3, 11, 36] suggest that the LPS-LBP-Ras-Raf-1-MEK1/2-ERK1/2-pp90rsk-NF-κB(p50/p65) cascade involved in LPS-induced up-regulation of *MUC5AC* expression [85] participates also in LPS-stimulated expression of *MUC5B*. Accordingly, NFκB *cis*-elements present in the *MUC5B* promoter may be recruited by this cascade for LPS-induced *MUC5B* expression. However, this is not the only cascade that activates *MUC5AC* in response to LPS. Another pathway consists of LPS-PKC-NADPH-ROS-TGFα-EGFR-MAPK [86-88]. Several studies point to a high probability that thise pathway is also utilized for regulation of *MUC5B* expression. Recently, it has been shown that EGFR participates in oxidative stress-induced over-expression of the *MUC5B* gene. It is known that *MUC5B* is expressed in airway epithelium, particularly in diseases associated with oxidative stress [89, 90]. Cigarette smoke, one of the etiological agents of chronic obstructive pulmonary disease (COPD), induces inflammation in airway epithelium and increases intracellular ROS concentration, which, in turn, increases hyaluronan concentration followed by its depolymerization [91-93] and consequent CD44-EGFR-$^{44/42}$MAPK activation [94-96]. All these events lead to over-expression of *MUC5B* in response to cigarette smoke [97], although the specific component of the smoke that triggers the cascade is unknown [63, 98].

In addition to the aforementioned mechanisms utilized by bacterial LPS, it may also activates expression of *MUC5B* and *MUC5AC* genes through an ATP-dependent signaling mechanism. The nucleotide triphosphates ATP and UTP are potent regulators of *MUC5AC* over-expression and goblet cell-mediated mucin release [99, 100]. Accumulating evidence suggests the involvement of this mechanism also in *MUC5B* activation. Chen *et al.* [101] analyzed the roles of extracellular nucleotide triphosphates, including UTP and ATP, in regulation of *MUC5AC* and *MUC5B* expression and secretion. They found that ATP and its analogs had no stimulatory effect on transcriptional activity of these genes in TBE cells, but they could affect secretion of MUC5AC and MUC5B mucins. These results are consistent with previous findings on the role of ATP in release of MUC5AC glycoprotein from human airway cells *in vitro* [102]. In contrast to ATP, UTP demonstrated a unique dual role in up-regulation of *MUC5AC* and *MUC5B* and stimulation of MUC5AC and MUC5B mucin secretion. Both processes were stimulated in a time- and dose-dependent manner *in vitro* and *in vivo*. Importantly, the two processes, expression and secretion, are regulated by different signaling pathways. These pathways are apparently initiated at the receptor-G-protein complex (P2Y receptors) [103, 104], inasmuch as both processes can be abrogated by a specific inhibitor. However, after initial interaction with P2Y receptors, the signaling pathways that regulate the two processes diverge: UTP- and ATP- stimulated secretion is controlled by a PLC/PKC-dependent pathway, whereas regulation of *MUC5AC* and *MUC5B* expression by UTP is independent of the PLC/PKC pathway but controlled by the MAPK cascade. The use of specific inhibitors enables differentiating between two signaling systems [101]. The presented data highlight the important and specific role of NTP-dependent mechanisms in regulation of mucin gene expression and secretion, in general, and *MUC5B,* in particular. However, more studies are needed to evaluate all links in the signaling chains leading to *MUC5B* over-expression and secretion. Factors upstream to ATP may also operate in regulation of the downstream cascade. It was recently shown that LPS-induced ATP-dependent mechanism of *MUC5AC* expression is mediated through LPS interaction with TLR4 receptor followed by extracellular secretion of ATP, which further interacts with epithelial $P2Y_2$ purinergic receptor to induce *MUC5AC* over-expression [100, 105]. Although precise mechanisms by which LPS activates *MUC5B* gene have not been identified, direct and indirect data suggest that all mechanisms participating in *MUC5AC* activation may also be involved in activation of *MUC5B* expression.

<u>*Leptin-dependent expression of the MUC5B gene:*</u> A new player recently emerged on the mucin scene: leptin, a hormone synthesized primarily by adipocytes but

produced by many other cells as well. In addition to regulation of genes involved in metabolic processes, leptin regulates mucin gene activity [106-110]. It up-regulates *MUC5B* and to a lesser degree *MUC5AC* in human airway epithelial cells [111]. Up-regulation of mucin genes by leptin is dependent on activation of leptin receptor, which transduces an activating signal from cell membrane to cell nucleus through the ERK1/2 or p38 MAPK pathways and further to the *MUC5B* promoter [111].

<u>Role of *MUC5B* promoter nucleotide polymorphism in allele-specific *MUC5B* expression:</u> Recently, an interesting observation has been made by Kamio *et al.* [112] who analyzed the role of nucleotide polymorphism in the promoter regions of several mucin genes in expression of these genes. No polymorphism was found in the promoter region of *MUC1* while two to four polymorphisms were detected in the promoter regions of *MUC2, MUC4, MUC5AC* and *MUC7* genes. In contrast, the *MUC5B* promoter contained 10 single nucleotide polymorphisms and a dinucleotide insertion/deletion polymorphism in almost every 180 nucleotides in the 2-kb promoter region [112, 113]. At least three of these polymorphisms had a significant effect on expression profile of the reported gene in transient transfection assays, and on *MUC5B* expression in patients with diffuse panbronchiolitis [112]. As mentioned above, the *MUC5B* gene shows little, if any, variation in numbers of tandem repeat [20]. Instead, relatively large numbers of nucleotide polymorphisms are present in the *MUC5B* promoter region, implying individual differences in *MUC5B* expression levels [112].

It is known that, in addition to transcriptional regulation of gene expression associated with promoter regions, the first intron of some genes also serves regulatory functions [114]. Intron I of *MUC5B* contains several *cis*-elements specific for AP-2α and GATA-1 transcription factors [36]. The first intron of *MUC5B* contains a DNA segment of 259 bp comprising eight clustered, tandemly repeated GA boxes and a CACCC box that bind Sp1 factor. The role of these elements in transcriptional regulation of *MUC5B* is unclear.

Still another region of the *MUC5B* gene that may provide regulatory functions lies between the central tandem repeat domain encoding sequence and the 3'-end of the gene [23]. This 18 bp region contains a GGGCGG-box (a potential Sp1-

binding site), which, however, does not bind a classical Sp1 transcription factor. Instead, it interacts with a 42 kDa nuclear factor NF1-MUC5B detected in HT-29 MTX cells in which it activates transcription of *MUC5B*. The properties of NF1-MUC5B protein are unknown; it may turn out to be a cell-specific transcription factor. Interestingly, parental HT-29 cells do not express MUC5B glycoprotein, however, treatment of these cells with methotrexat (MTX) stimulates them to differentiate into MUC5B mucin-producing cells [14, 115].

6.3.2. Epigenetic Regulation of *MUC5B* Gene Activity

The epigenetic tools, DNA methylation and histone modifications, regulate expression of genes important for various physiological processes (*e.g.* epithelial cell differentiation, innate immunity, mucus defense) and aberrant processes under pathological conditions, (*e.g.* malignant diseases) [116-121]. Epigenetic regulation of the genes encoding gel-forming mucins is poorly studied. To the best of our knowledge, only one paper has been published on epigenetic regulation of the 11p15 mucin genes (*MUC2, MUC5AC, MUC5B, MUC6*), that of Van Seuningen's group [116]. Their results highlight important aspects of epigenetic mechanisms that control gene expression. They showed that: 1) gel- forming mucin genes comprising the 11p15 cluster are differently regulated by epigenetic mechanisms - the *MUC2* and *MUC5B* promoters are under strong epigenetic regulation, *MUC5AC* is rarely affected by DNA methylation and histone modification, and *MUC6* is practically insensitive to epigenetic effects; 2) *MUC5B* silencing is the result of extensive methylation of its distal promoter and site-specific methylation of the proximal promoter region; 3) *MUC5B* methylation is cell specific and controlled by DNA methyltransferase 1; 4) methylation of the *MUC5B* promoter is associated with stages of cell proliferation and differentiation; 5) histone modification of *MUC5B* is a cell-specific process: in one cell type *MUC5B* repression is associated with histone H3 deacetylation and K9H3 and K27H3 trimethylation, while in the other cells its expression is observed when histones H3 and H4 are hyperacetylated; 6) the impact of methylation and histone deacetylation on *MUC5B* repression is different in various cells.

Comparing *MUC5B* epigenetic status in normal and cancer cells at different stages of differentiation, Vincent *et al.* [116] came to the conclusion that epigenetic silencing of both *MUC5B* and *MUC2* is a specific mechanism used by cancer cells to maintain undifferentiated cells. The mechanism also operates in normal cells but not in a differentiation process. The authors emphasize the importance of these findings for cancer diagnosis and prognosis. Their results are in agreement with previous findings of a correlation between expression of the 11p15 mucin gene cluster in normally differentiating cells [45, 122] and differentiated tumors [123, 124].

6.4. PROPERTIES OF MUC5B mRNA

The human *MUC5B* gene comprised of 49 exons and 48 introns spans 40.1 kb of genomic DNA and encodes mRNA of 17.5 kb [7-9, 36, 37]. This mRNA is the single *MUC5B*-specific transcript detected to date in cells producing MUC5B apomucin; that is, no alternatively spliced MUC5B mRNA isoforms have been recognized. An important question is whether the MUC5B primary transcript has the potential to undergo alternative splicing. Analysis of the MUC5B intronic sequences suggests that, in some yet unknown conditions, alternatively spliced variants of MUC5B primary transcript may occur. In addition to the central mucin-specific exon, at least six introns and three exons contain unique tandemly repeated sequences of different lengths. Intron 36 is of special interest [9, 21]: it is made up almost entirely of perfect direct repeats of 59 bp that vary in number between individuals from three to eight. Each repeat contains the binding sites for NF1-MUC5B nuclear factor [23] and Hand1/E47 transcription factor, both of which may play a role in splicing [24] – a possibility strengthened by the fact that each of 59 bp-repeats contains at least one consensus-splicing donor sequence. An earlier report describes several alternatively spliced isoforms obtained by deletion and/or insertion mechanisms at six alternative splicing sites in the tandem repeat region in the *hTERT VNTR2-2* gene [125]. These data suggest that *MUC5B* gene with its intronic repeats does indeed have the potential for alternative splicing. As noted by Desseyn and colleagues [22], "it remains to be clarified whether the machinery of splicing recognizes only the first repeat as the donor site", or other splicing competent sites may also be utilized.

Although the mouse *Muc5b* and human *MUC5B* genes share an identical gene organization (49 exons and 48 introns), their intronic sequences are completely different [10]. Moreover, transcription of *Muc5b* results in two mRNAs of 14932 bp and 14860 bp in length: the first one is the full-length mRNA containing all 49 exons and the second mRNA is an alternatively spliced variant of the first one that does not include exon 1. The full-length mRNA is a predominant form of the *Muc5b* transcript, although the alternative mRNA is also present and detectable in laryngo-tracheal cells. *Muc5b* mRNAs were transcribed also in stomach and duodenum, but not in salivary glands and hepatobiliary tract, the main locations of human *MUC5B* expression [10]. Thus, despite substantial structural similarity, mouse *Muc5b* and human *MUC5B* genes possess differences that may account for their different potentials for alternative splicing. More studies are needed to evaluate the capacity of the human *MUC5B* primary transcript to undergo alternative splicing in different cells and tissues under various conditions.

6.5. BIOSYNTHESIS OF MUC5B GLYCOPROTEIN

6.5.1. General Characteristics

The MUC5B mucin possesses many structural and functional properties common to other members of the gel-forming mucin family and to vWF [126], including the biosynthesis of these glycoproteins [127]. Generally, biosynthesis of the MUC5B mucin follows the same stages as that of MUC2 and MUC5AC. The MUC5B apomocin is synthesized on polyribosomes bound to endoplasmic reticulum (ER), followed by N-glycosylation and dimerization through interchain disulfide bond formation between the C-terminal domains of the precursor monomers. MUC5B dimers are transported from ER to the Golgi where they are initially O-glycosylated in the *cis*-Golgi compartments. Upon reaching the *trans*-Golgi compartments, oligomerization takes place by interchain disulfide bond formation between N-terminal D-domains. Next, the oligomeric forms are combined into multimeric complexes, which are stored in dehydrated form within secretory vesicles, and released by constitutively and/or regulated secretion in response to environmental stimuli followed by gel formation [127-129]. These mechanisms allow immediate response to external challenges with no requirement for *de novo* mucin synthesis [130]. While the described stages are common to all gel-forming mucins, there are certain peculiarities in the synthesis, maturation and secretion of the MUC5B mucin. These peculiarities result from differences in amino acid sequences in the corresponding domains of MUC5B and other gel-forming mucins, and from the fact that different mucins are produced by different cells [89, 131, 132]. One such peculiarity is associated with the existence of two different glycoforms of MUC5B mucin, defined as low- and high-charged glycoforms [133], which are synthesized by different cells within the same submucosal gland [134]. Other specific features of MUC5B biosynthesis are linked to C- and N-terminal cleavages of the immature molecules [134-136]. Different rates of the early steps in biosynthesis of different gel-forming mucins have also been observed [126].

6.5.2. N-Glycosylation and Polymerization of MUC5B Apomucin

The monomeric precursors of the MUC2, MUC5AC and MUC6 mucins, with molecular weights of 600, 500 and 400 kDa, respectively, were detected soon after the beginning of polypeptide synthesis, and in only 0.5-1 hour the corresponding dimers appeared [126]. This dimerization of the MUC5B precursors (470 kDa) was much slower than that of other gel-forming mucins. It occurred only 4 hours after the beginning of biosynthesis, although a small quantity of MUC5B dimers were detected earlier [126]. Dimerization of secreted mucins (*e.g.* MUC2) was facilitated by specific N-glycosylation and occurred together with internalization of the nascent peptides into rough endoplasmic reticulum; this process preceded interaction with the specific ER-chaperones, important for appropriate folding and oligomerization in the ER [137, 138]. The lack of N-glycans on MUC2 precursor molecules was shown to slow down the process of dimerization, while their presence at specific sites on the MUC2 backbone appeared to be important for efficient oligomerization in ER [126]. The same effect was observed in porcine submaxillary mucin (PSM), a porcine ortholog of human MUC5B mucin; although N-glycosylation was not absolutely required for PSM dimer formation and multimerization, it was very important for secretion. The underglycosylated PSM species were poorly secreted and rapidly degraded after secretion into extracellular medium [128]. On the other hand, N-glycosylation was not required for the secretion of rat submandibular gland mucin glycoprotein (RSMG), a rat ortholog of human MUC5B mucin [139].

Thus, the role of N-glycosylation in biosynthesis of gel-forming mucins in general, and MUC5B in particular, is not well understood. On the one hand, N-glycosylation occurs at the very beginning of the multistep biosynthesis of gel-forming mucins, while on the other hand, its functions appear to be necessary at the very end of this process, at the secretory stage. The N-glycan functions in mucin glycoprotein biosynthesis become even more confusing when one considers that lack of N-glycans on the mucin molecule (*e.g.* MUC2) affects efficiency of mucin dimerization, suggesting a possible role of N-glycans in mucin glycoprotein maturation at early stages. It is generally accepted that MUC5B probably undergoes N-glycosylation and oligomerization by mechanisms common to other gel-forming mucins, even though to the best of our knowledge the role of N-glycosylation in MUC5B di- and oligomerization has not been specifically studied.

Multimerization of mucin dimers occurs in the *trans*-Golgi compartments. The molecular mechanisms of this compartmentalization are not clear. Studies on the N-terminal D-domains of vWF showed that the CGLCG motif detected in the D1-, D2- and D3-domains seems to play a critical role in Golgi targeting [140]. Among the gel-forming mucins structurally related to vWF, this motif was found in the D1-, D2-, and D3-domains of MUC5AC and MUC5B mucins, but only in the D1- and D3-domains of MUC2 [127]. The existence of similar motifs in the active sites of disulfide isomerase raises the question of whether this motif is directly involved in disulfide bond formation during multimerization of mucin molecules conducted in acidic medium of the *trans*-Golgi compartment, or participates only in targeting of mucin dimers to the *trans*-Golgi.

6.5.3. O-Glycosylation of MUC5B Precursor

Gel-forming mucin O-glycosylation is a multistep process resulting in covalent binding of different types of multi-branched sugar chains to an apomucin molecule. Despite variable carbohydrates attached to the polypeptide backbone, MUC2 and MUC5AC glycoproteins can be separated more or less homogeneously by electrophoresis, isoelectrofocusing or density centrifugation. In contrast to MUC2 and MUC5AC mucins, the population of MG1/MUC5B, a salivary mucin encoded by the *MUC5B* gene [134, 141, 142], is highly heterogeneous [133, 143], being composed almost entirely of differently glycosylated forms of the apomucin molecules [142]. Different salivary glands produce distinct populations of MG1/MUC5B glycoforms varying in buoyant density and degree of sulfation and sialylation of their glycan chains [134, 142, 144, 145]. Recently, Veerman *et al.* [146] showed that individual cells within one gland produce MUC5B molecules carrying a unique carbohydrate signature while other cells in the same gland synthesize different MUC5B glycoforms. According to the authors, this finding suggests that diversity in mucin molecules may be much greater than assumed.

Characterization of the MUC5B carbohydrates by Hansson's group [147] revealed an enormously heterogeneous set of neutral, sialylated and sulfated oligosaccharides of varying lengths. Neutral oligosaccharides, which comprise 56% of all carbohydrates attached to MG1/MUC5B, have an average of 13 sugar residues, the sialylated oligosaccharide chains (26%) generally contain 17 sugar residues, and the sulfated fraction (19%) contains 41 sugar residues [147]. There are several fucose-containing blood group epitopes (H, Le^a, Le^x, Le^b and Le^y) on the MUC5B molecule [147]. Human oligomeric mucin MG1/MUC5B contains 292 O-linked oligosaccharide chains *per* monomer unit, of which 118 have sialic acids [148, 149].

According to most studies, O-glycosylation is initiated in the *cis*-Golgi compartments by addition of GalNAc to serine and threonine. Interestingly, attachment of GalNAc to serine and threonine usually occurs simultaneously. In contrast to other O-glycosylated proteins, attachment of GalNAc to serine and threonine residues of rat submandibular gland (RSMG) mucin does not occur simultaneously [139]. These results may be explained by the nature of the amino acid sequence of the polypeptide target in primary O-glycosylation: specific threonine residues are the preferred sites for addition of GalNAC; proline in the +3 position of threonine residue enhances primary glycosylation; an inverse relationship between the size of the adjacent glycans and the rate of GalNAc addition exists; and GalNac-transferases can distinguish between substrates having different amino acid sequences possessing different O-glycosylation sites. Based on these data [150], the nonsimultaneous addition of GalNAc to a polypeptide backbone seems to be a rule that applies not only to the RSMG mucin, a rat ortholog of human MUC5B mucin.

In addition to the role amino acid sequences play in O-glycosylation of mucin molecules, the participation of TRs in this process has been also studied [151, 152]. In a study of chimeric mucin peptides carrying increasing numbers of MUC5B TR units [151], the number of carbohydrate epitopes was proportional to the number of TRs; the occurrence of novel carbohydrate structures was not affected by the number of repeat units. Thus, the specificity of mucin O-glycosylation does not depend on the number of TRs present in the mucin molecule.

It has been known for a long time that differently charged MUC5B molecules have different content of fucose and sulfate [133, 142]: the lower charged isoform has higher fucose content and the higher charged isoform has higher sulfate content. No differences in sialic acid content were observed [142]. These data are consistent with an interesting observation made by Shori *et al.* [153]: while the core glycosylation of MG1/MUC5B glycoprotein based on composition of neutral sugars does not vary significantly between individuals, terminal decoration over the underlying core carbohydrate scaffolding varies greatly between individuals. This variability results, in particular, from different sulfation, sialylation and fucosylation of individual MG1/MUC5B mucins. The authors emphasize that different levels of glycosylation and differences in terminal decoration by highly charged sugars lead to extensive heterogeneity of multimeric mucin isoforms. Individual variability in properties of mucin molecules may have advantages in clearance of bacterial infections [154] and disadvantages in pathogenesis of some diseases [155]. In consistence with the aforementioned observations there are results reported by Thomsson *et al.* [156], who also found high individual variability of O-glycans on human salivary MUC5B mucin, however, in contrast to Shori *et al.'* interpretation [153], attributed the variation to the blood group antigens and expression of secretor fucosyltransferase.

High variability in carbohydrate content is characteristic of the MUC5B salivary mucin produced by various organs including cervix. Cervical MUC5B mucin contains shorter sugar chains and a different ratio of neutral to acidic sugars compared with salivary mucin. Most of the carbohydrate chains on cervical mucin are neutral, sialylated, and sulfated oligosaccharides that vary from two to nine sugar residues in length [157, 158]. Importantly, the composition and ratio of the indicated sugars and the quantity of MUC5B glycoprotein are strongly dependent on concentration of estrogen and progesterone. Varying levels of these hormones are responsible for the cyclic changes in the amount of mucus and its rheological (*i.e.* viscoelasticity) and biological (*i.e.* sperm penetrability) properties in the endocervix [159-162]. The presence of negatively charged terminal carbohydrates on the MUC5B polypeptide appears to affect its rheological parameters, as they contribute to the distension and flexibility of the mucin protein coil. There is a reciprocal relationship between neutral (fucose) and negatively (sialyc acid) charged terminal sugars in human cervical mucus during the normal menstrual cycle: neutral carbohydrates in terminal positions are abundant during the ovulatory phase, whereas negatively charged residues reach maximal level during the preovulatory phase [159, 163-165]. Additional evidence of steroid hormone influence on amount and composition of sugars on a MUC5B molecule throughout the menstrual cycle comes from the central role these variations play in cervical mucus hydration and viscoelasticity [166]. In particular, the amounts of Tn-antigen (Galβ1-3GalNAcα1) and N-acetyllactosamine (Gal1β1-4GlcNAcβ1-3) oligosaccharides on MUC5B, and MUC5B backbone molecules are increased during the first half of the cycle, reach maximal level at midcycle, and dramatically drop at the end of the cycle. Taken together, these data indicate that cervical mucus contains more mucin and its carbohydrate content is higher and less acidic at midcycle than during the periovulatory phase, resulting in more hydrated and less viscous mucus gel.

This conclusion is in agreement with the recent findings of Brunelli *et al.* [167] that structural changes of MUC5B mucin during the menstrual cycle depend on pH alterations in the cervical canal, which, in turn, might be connected to changes in the content of neutral and negatively charged carbohydrates on MUC5B mucin. It has been established that pH is an important determinant of sperm-mucus interaction: mucus permeability to sperm peaks at midcycle, when cervical pH is relatively high and mucin sugars attached to MUC5B are less acidic, and decreases outside the ovulatory period, when pH is low and mucin-bound carbohydrate are more acidic [166, 168]. Brunelli *et al.* [167] showed that in the preovulatory sperm-impermeable phase, mucus, composed mostly of MUC5B mucin, is arranged in compact fiber-like structures, while ovulatory sperm-permeable mucus is composed of floating globules of mucin aggregates. The authors succeeded in demonstrating that the switch from globular ovulatory to fibrous preovulatory mucus depends largely on a pH-driven mechanism (Fig. **4**).

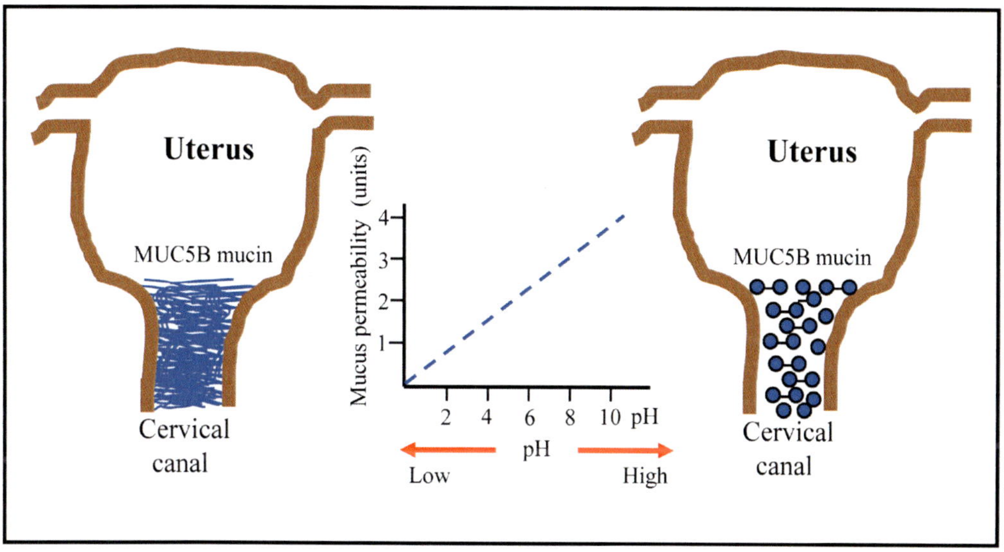

Figure 4: pH-dependent fiber-globule transition of the MUC5B mucin (based on the data reported in [167]).

Analysis of the MUC5B primary structure by using of the Miyazawa-Jernigan hydrophobicity scale shows that pH-sensitive domains in the MUC5B mucin molecule are associated with flexible regions prone to drive aggregation [167]. Thus, the biophysical properties of the MUC5B mucin molecule, determined by the primary structure of its polypeptide backbone and dynamic alterations in carbohydrate content, permit the fiber-to-globule switch in cervical mucus during the menstrual cycle.

We have discussed changes in the MUC5B-associated carbohydrates in different organs under physiological conditions. Different types of pathology are also associated with alterations in MUC5B content and with the arrangement of the carbohydrates attached to its polypeptide skeleton. Conflicting data has been reported by most of the laboratories studying this subject. Studies of MUC5B carbohydrate changes associated with cystic fibrosis (CF) are a case in point. Human bronchial epithelial cell cultures from healthy controls and CF patients disclosed the same highly variable glycan structure, suggesting that there were no differences in either the specific mucins or their O-glycans between the CF phenotype under noninfected/noninflammatory conditions of cell culture *in vitro* and CF patients *in vivo* [169]. In another study, no significant differences were found in glycosylation of MUC5B or MUC5AC mucins between CF patients, non-CF subjects with severe lung disease, and control subjects with non-diseased lung [170]. On the other hand, this same group found reduced sulfation, increased sialylation and reduced fucosylation in CF patients compared with controls [171]. Other studies report more sialic acid, less hexose [172], higher sulfate uptake and more sulfated high-molecular mass glycans [173] in CF. Many of these studies [169-173] suggest that changes, if any, in mucin-attached carbohydrates in CF result from infection/inflammatory-dependent glycosylation induced by accompanying bacterial infection.

6.5.4. C-Mannosylation of MUC5B Core Polypeptide

In addition to N- and O-glycosylation, MUC5B undergoes a third type of glycosylation, C-glycosylation, also called C-mannosylation as it involves covalent attachment of a single α-mannose residue to the C_2 carbon atom of the first Trp in the WXXW peptide motif [174]. The WXXW sequence is the major acceptor site for the glycosyltransferase involved in this kind of glycosylation [175-178]. The C-mannosylation specific motif is present at the N-terminal region of the Cys-subdomains interspersed among O-glycosylated regions of MUC5B and MUC5AC mucins. Recombinant Cys-subdomains of MUC5B and MUC5AC mucins underwent C-mannosylation in an *in vitro* study [179]. According to pulse-chase experiments, C-mannosylation occurs very early in biosynthesis, most likely in the endoplasmic reticulum, and is required for appropriate folding of the cys-subdomain and/or for ER export. Perez-Vilar *et al.*'s study [179] is the single investigation of MUC5B mucin C-mannosylation. More studies are required on the mechanisms of this type of glycosylation, its cell-, tissue-, and organ-specificity, and its functions in both normal and pathological conditions.

6.5.5. Proteolytic Modification of MUC5B Mucin by C- and N-Terminal Cleavages

It has been well established that proteolytic processing at the N- and C- termini of polymeric gel-forming mucins is part of their posttranslational modifications

[130], and that an auto-catalytic cleavage at GDPH site in the C-terminus of human MUC2 and MUC5AC mucins occurs at the low pH in the late secretory pathway [135, 180]. This cleavage may have functional consequences as it generates a reactive Asp at the C-terminal end that could interact with other molecules by covalent bond formation. Such interaction was observed between chondroitin 4-sulfate and human blood protein pre-alpha-inhibitor [181] A linkage of this type in mucins could be physiologically important. Different physiological and pathological conditions that have a lowered pH (CF for instance) may induce an auto-cleavage reaction at the GDPH site of MUC2 or MUC5AC mucins, with the consequent interaction of the modified mucin with different proteins or carbohydrates. Although the GDPH site has not been found in the MUC5B amino acid sequence [135], C-terminal cleavage of the MUC5B mucin isolated from salivary gland, and from mucosa of the respiratory tract and endocervix, have been detected [134] (Fig. **5**). Whether this cleavage is auto-proteolytic or conducted by a protease is unknown. Nevertheless, if the resulting fragments contain reactive amino acid residues, further interactions associated with formation of new protein complexes are possible.

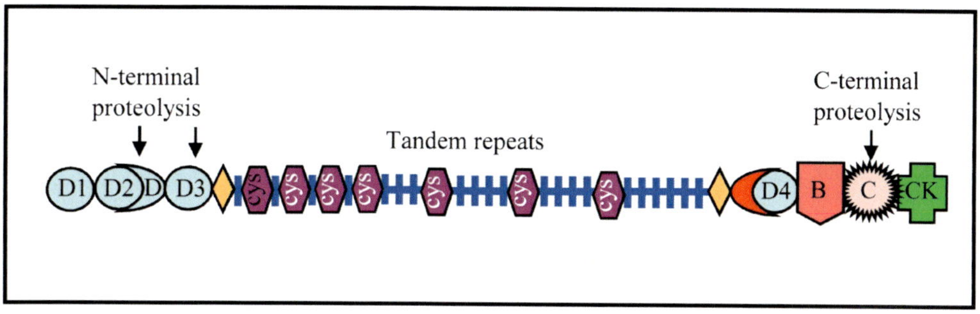

Figure 5: Proteolytic modifications of the MUC5B mucin (based on the data reported in [134,136, 182]).

In addition to C-terminal cleavage, MUC5B also undergoes N-terminal proteolysis [136, 182]. In contrast to C-terminal cleavage, whose mechanism is unknown, that of N-terminal cleavage(s) appears to be more or less defined and resembles the relevant cleavage(s) of the vWF glycoprotein [136]. During biosynthesis and processing of the vWF protein, its CK domain is involved in disulfide-bond-mediated dimerization between the C-terminal regions of monomers, while D1- and D2-domains participate in oligomerization *via* the N-terminal ends. The D1- and D2-domains of pre-propolypeptide vWF do not form disulfide bond; instead, they provide alignment of the D3- and A- domains, which then form disulfide bonds. When D1- and D2-domains have carried out their anchor tasks and disulfide bonds have been formed, D1- and D2-domains are removed by cleavage that takes place in the D'-domain [183, 184] It has been shown that MUC5B undergoes analogous proteolysis resulting in removal of D1- and D2-domains during MUC5B oligomerization [136, 182]. In addition to this cleavage, one more proteolytic event takes place in the D3-domain of MUC5B close to the N-terminus of the central TR-containing domain [136]. Thus, it appears that the major part of the D1-D3 region serves auxiliary functions and its removal does not affect the main rheological and defense functions of the MUC5B mucin. However, it must be noted that the functional significance of proteolytic modification of the gel-forming mucin, in general, and MUC5B, in particular, is not fully understood. According to Thornton *et al.* [130], proteolytic processing may operate both inside and outside the cell: inside the cell, it may participate in mucin packaging within secretory vesicles, while outside the cell, proteolytic modification may affect macromolecular properties of mucin molecules important for the quality of mucin gel. The fact that secreted MUC5B mucin exists in two forms, soluble and insoluble, indicates the importance of the extracellular role of mucin proteolysis [185]; at the same time, removal of some parts of the mucin molecule inside the cell may be crucial for incorporation of gigantic polymers into vesicles with restricted capacity [127, 186]. A better understanding of how different modifications contribute to mucin functioning must take into account that each modification of a precursor molecule operates in concert with other modifications and/or auxiliary factors in order to create the optimal structure with the properties necessary for specific functions.

6.5.6. Packaging, Secretion and Unpackaging of MUC5B Mucin

It is not known for certain whether maturation of mucin precursors is completed before incorporation of mucin molecules within secretory vesicles, or takes place also within vesicles. It is also not clear how the gigantic mucin molecules are transported through the endoplasmic reticulum to Golgi compartments and get packaged into a vesicle whose capacity is much smaller than the volume of mucin polymer molecules [187, 188]. Studies with de-glycosylated bovine salivary mucins (an ortholog of human MUC5B) showed that in ER, the largely uncharged and nonglycosylated mucin monomers and dimers form highly flexible random coils, which may develop more compact structures that facilitate transportation of mucin precursors from ER to the Golgi [189, 190]. In the proximal Golgi compartment, O-glycans, added to the mucin Ser/Thr-rich central domain, reduce flexibility of mucin chains and lead to formation of entanglements [189]. Important changes occur in the *trans*- Golgi compartment associated with intraluminal increase in H^{+} and Ca^{2+} ion concentration, inter-dimeric disulfide bond formation, and extensive sulfation and sialylation of the mucin-bound oligosaccharide chains [127, 191, 192], resulting in formation of polyanionic mucin macromolecules that subsequently develop gel structure [193]. Under certain conditions of, among others, temperature, pH, salts, and polymer characteristics, the polyanionic gel surrounded by fluid is able to transform from a swollen to a condensed phase, facilitating compact packaging of the mucin conglomerates as granules within vesicles [194]. It appears that Ca^{2+} cations are absolutely indispensible elements of the packaging mechanism. By shielding the negative charges carried by sulfate and sialic acid groups, Ca^{2+} may induce condensation of mucin molecules as part of a mechanism orchestrating the efficient packaging of mucin molecules inside a vesicle [186, 195-197]. Using MUC5B as a model, Raynal *et al.* [198] showed that reversible, calcium-dependent protein cross-link modulates mucin supermolecular organization.

Still other factors may smooth the process of mucin packaging. C-mannosylation of MUC5B and MUC5AC mucins may play a definite role in the folding of the extended mucin molecules that eases their packaging into vesicles [179]. Factors that increase mucin flexibility – such as interaction of MUC5B mucin with histatin or statherin [199] – may also assist mucin incorporation into the vesicle. Whatever the mechanism, the intravesicle mucins are a partially condensed, pH-dependent network surrounded by a mobile fluid [130, 200]. Because volume transitions of a gel are reversible [194], mucin intravesicle packaging (gel condensation) and vesicle exocytosis (expansion of the intravesicle mucin matrix into extracellular space) [186, 201] may utilize a single mechanism [193].

Kesimer *et al.* [129] recently carried out a detailed analysis of the unpackaging of the gel-forming MUC5B mucin and its organization after granular release. It is known from the studies of Verdugo and coworkers [195, 196] that upon release in exocytosis, the mucin molecules unfold very rapidly, within seconds, and the packaged protein expands its entire granule structure all at once rather than in a stepwise manner starting from the granule surface edge. This process strongly depends on the exchange of calcium with sodium. It has been suggested that efficient and rapid unpackaging requires that mucins in the granule matrix be folded in a nonrandom conformation [195]. Later on, this suggestion was supported by microscopy evidence that the mucin granule matrix is indeed organized in large supramolecular nematic crystalline structures [202]. In agreement with these data, Kesimer *et al.* [129] found that recently secreted MUC5B mucin molecules, many of which were only partially unfolded and still in the process of expansion from their granular form, maintained a circular structure of flexible chains of 100-200 nm in length attached to protein-rich nodes of about 10-20 nm in diameter. Each node was surrounded by an average of 4-8 of these flexible chains, most the length of a single subunit but some as long as two subunits. There are 10 or more such nodes in the structure expanded from one granule.

The mechanism controlling unpackaging of the granule mucin matrix is based on Ca^{2+}/Na^{+} exchange, as suggested by Kesimer *et al.* [129]. This ion exchange, which is the driving force of mucin granule expansion [186, 203, 204], triggers the mucus gel phase transition from condensed to solvated phase and drives the post-exocytotic swelling of mucus granules [186, 205]. Inside the cell, the granules are stored at low pH medium with Ca^{2+} as a dominant ion. Once the secretory pore is formed, exchange of divalent Ca^{2+} by monovalent Na^{+} begins, causing the gel to swell by osmosis. In contrast, the opposite process (gel-condensation, or granule formation) is associated with the movement of Na^{+} ions outside and Ca^{2+} inside the loose mucin gel, which induces condensation of mucin matrix structures that results in formation of the dense mucin-containing granules.

Fully processed mucins are stored in cytoplasm of mucin-producing epithelial cells in large secretory vesicles known as mucin granules [206]. As described in the previous chapters, secretory granules and small mucin-containing vesicles are slowly but constantly secreted from the cell, thereby mediating constitutive mucin secretion [206, 207]. In contrast, immediate robust discharge of mucin granules can be effected by Ca^{2+} -regulated mucin secretion, which causes a rapid reaction triggered by various physiological and pathological stimuli including cytokines, nucleotides, proteases and chemicals, and bacterial and viral infection [208]. These two main types of secretion are common to all gel-forming mucins studied, although there are cell-specific (goblet and glandular) differences and differences between various animal models. More research is needed to delineate the mechanisms that regulate mucin secretion in different cells and animal models under physiological conditions and in diseases.

6.6. EXPRESSION OF *MUC5B* GENE UNDER PHYSIOLOGICAL CONDITIONS

Under physiological conditions, *MUC5B* is expressed mainly in the upper and lower airways, in some organs of the digestive system (salivary glands, esophagus, pancreas and gallbladder), and in the female reproductive tract [209, 210]. The patterns of *MUC5B* expression in these organs during normal embryo-fetal development and in adult tissues are analyzed in the following sections.

6.6.1. Respiratory Tract

The first expression of the *MUC5B* gene in the respiratory tract is detected at 13 weeks of gestation in surface epithelium of the main and lobular bronchi [211-213]. From the 18th week, it is also expressed in trachea, but not in terminal sacs and alveoles, which remain MUC5B-negative also in adult lungs [211]. The pattern of MUC5B expression during fetal development follows the differentiation pathway of submucosal glands: during early development (13 weeks gestation), *MUC5B* mRNA is expressed in both surface tracheal epithelium and in developing submucosal glands, and by week 23 its expression is more prominent in the glands [212-214].

MUC5B is the major gel-forming mucin expressed in the adult respiratory tract [130, 132, 215], where it is expressed mainly in mucous tubules of the submucosal glands [121, 216]. However, MUC5B mucin may be synthesized also in the surface epithelium [134]. Thornton *et al.* [217] showed that cultured normal human tracheobronchial epithelial cells (NHTBE) which do not form glandular structures do produce MUC5B mucin. In normal adult airways, MUC5B is synthesized mainly in mucous cells of the submucosal glands [89, 132, 134]. Like in salivary glands, MUC5B mucin occurs in airway submucosal glands in two different glycoforms generated by different cells within the same gland [134].

6.6.2. Digestive System

The developmental patterns of the *MUC5B* gene expression in the digestive system have been studied by several laboratories. The earliest expression of *MUC5B* in human development was detected by *in situ* hybridization at 8 weeks of gestation in the embryonic stomach [218]. At 27 weeks, it was observed in stomach fundus and antrum. In fundus, MUC5B-positive epithelial cells were located at the surface and pits, whereas in antrum they were restricted to epithelial folds and mucous glands. Importantly, although *MUC5B* was expressed in embryonic and fetal human stomach during the period of glandular cytodifferentiation (up to 27 weeks gestation), its expression was never detected in normal adult stomach [218].

Expression of *MUC5B* in submucosal glands of esophagus was identified by *in situ* hybridization [219] and by immunohistochemistry [220-222]. In embryonic, fetal or adult intestine, it was not detected [223, 224]. In the fetal duodenum, MUC5B mRNA was detected in crypts at 18 weeks of gestation, but, importantly, not at any other gestational period [225]. Buisine *et al.* [225] could not detect its expression in adult duodenum. However, in contrast to the results obtained by *in situ* hybridization [225], RT-PCR and immunohistochemical analysis [226] revealed expression of *MUC5B* both in Brunner glands and in mucosa of the adult duodenum. In general, *MUC5B* is expressed during gastric and duodenal ontogenesis. According to Buisine *et al.* [225], the transient expression of *MUC5B* in fetal stomach and duodenum may indicate that "epithelial cells in fetal stomach and duodenum respond to similar stimuli and possess similar mechanisms of turning on and off MUC5B mRNA expression".

Among the accessory glands of the digestive system, salivary glands demonstrate the highest expression of MUC5B mucin [141, 185, 210, 227]. The MUC5B glycoprotein is secreted by mucus acinar cells in all salivary glands [228]. Analysis of MUC5B expression in different types of salivary glands performed by Veerman *et al.* [146] showed that two distinct glycoforms of MUC5B (sulfo-Lewis^{a}+ and sulfo-Lewis^{a}-) are produced by different cells of the same glandular unit. No information is available on expression of *MUC5B* gene in human salivary glands during embryogenesis and fetal development. On the other hand, its expression during the embryonic and fetal periods in other accessory glands of the digestive system, including pancreas, liver and the biliary tract, has been thoroughly studied. In pancreas, the MUC5B mRNA was first detected at 26 weeks of gestation, mainly in the epithelial cells of interlobular ducts. Parenchymal cells of Langerhans islets appear to be MUC5B-negative [225]. Light expression of *MUC5B* is usually observed in interlobular ducts in the normal adult pancreas [225, 229]. Studies of fetal liver showed that neither hepatoblasts and hepatocystes nor cells of intrahepatic bile ducts express MUC5B at any gestational stage [225], whereas both the surface epithelium and the epithelial folds of fetal gallbladder do express MUC5B during fetal development. The earliest expression of MUC5B mRNA in this organ found at 18 weeks of gestation [225]; the *MUC5B* expression throughout the rest of gestation in fetal gallbladder and in adult normal gallblader has been also detected.

6.6.3. Female Reproductive Tract

The *MUC5B* gene is known to be expressed in the female reproductive organs, mainly in the endocervix [230-232], where both endocervical epithelium and the submucosal glands express MUC5B mucin [209]. The extent of the expression of *MUC5B* is dependent on the stage of the menstrual cycle. The maximum level of the MUC5B mRNA in endocervical epithelium was detected just before midcycle, when the amount of MUC5B protein per unit of total protein in cervical mucus also reached its peak [161, 233]. During this midcycle period, changes take place in the mucus pH that alter MUC5B mucin structure and permeability of the mucus gel, thereby influencing sperm transition to the uterus [167].

6.6.4. Other Organs

MUC5B mucin was also detected, albeit to lesser extent, in normal human lacrimal glands and nasolacrimal duct [234, 235], in human and mouse normal middle ear epithelium [236, 237], and in normal thyroid tissues [238]. Further studies are needed for evaluation of the *MUC5B* gene activity in various organs, tissues and cells under physiological conditions and estimation of functions the MUC5B glycoprotein fulfills in these organs.

6.7. EXPRESSION OF *MUC5B* GENE IN PATHOLOGY

6.7.1. Respiratory Tract

Changes in the expression and/or distribution of MUC5B mucin accompany various inflammatory diseases of the airways as well as lung cancer [43, 123, 239-241].

Inflammatory and allergic diseases: MUC5B and MUC5AC are the major mucin glycoproteins found in sputum of patients with asthma [133, 242], CF [243] and COPD [89, 244, 245]. MUC5B is highly expressed in asthma and was found to be a main constituent of mucus plugs from the lungs of patients who died in status asthmaticus [133, 246, 247]. Moreover, mucin fraction of an asthmatic respiratory mucous plug is dominated by the low-charge glycoform (96%) of the MUC5B glycoprotein [247]. Studies on mucus hyper-secretion in an experimental murine asthma model show Muc5b to be both over-expressed and over-secreted [248-250]. Thus, the amount of intracellular Muc5b in epithelial cells of asthmatic airways reflects the balance between the rates of mucin production and secretion [248].

In chronic bronchitis and COPD, the MUC5B mucin is over-expressed and secreted into the lumen of the small airways [130, 251, 252], a finding in line with Caramori *et al.* [89] who showed that "COPD is specifically associated with increased expression of *MUC5B* in the bronchiolar lumen". Kettle *et al.* [54] recently found that over-expression of *MUC5B* in COPD and other inflammatory lung diseases is induced by neuregulin 1β1, a powerful activator of MUC5B expression. In chronic inflammatory airway diseases, including chronic bronchitis, chronic rhinosinusitis, nasal polyposis and CF, submucosal glands become enlarged and goblet cells appear in increasing numbers in the distal airways where they are normally not present [253, 254]. These changes lead to the pattern of airway mucin expression specific to chronic respiratory pathologies. Martinez-Anton *et al.* [253] found different MUC5B expression levels in healthy and diseased upper airway mucosa: higher in nasal polyps (NP) than in normal mucosa, and higher in NP patients with CF compared with patients with simple bilateral NP and normal mucosa. While only 42% of normal mucosa samples expressed MUC5B, 78% of NP and 100% of CF specimens were MUC5B-positive. Increased expression of MUC5B in trachea, bronchi and peripheral lung from CF patients was observed also by Groneberg *et al.* [132]. Thus, according to these studies, hyper-secretion of MUC5B is a characteristic feature of CF. Nevertheless, Rubin [255] noted that "it has never been proven that there is mucin hyper-secretion in the CF airways". According to Henke *et al.* [256], concentrations of MUC5B and MUC5AC mucins in sputum of CF patients normalized per volume, weight and total protein content were lower than those in normal individuals. Although these results suggest reduced expression and/or secretion of mucins by CF cells, possible degradation of MUC5B and MUC5AC proteins by proteases present in sputum should be taken into account. Interestingly, when levels of MUC5B and MUC5AC mucins in secretion from differentiated NHBE cells established from lung tissues of control and CF patients were compared, there were no differences in the their mucin content [169]. Several studies reported altered glycosylation of CF mucins, including MUC5B [154, 171], although Rose and Voynow consider that "this remains to be conclusively demonstrated" [257].

While there is some disagreement about the MUC5B expression level in CF airways, there is consensus regarding cell patterns that express MUC5B mucin in CF. Groneberg *et al.* [132] observed over-production of MUC5B not only in submucosal glands of CF patients, but also in epithelial goblet cells. The ability of goblet cells to produce MUC5B in CF patients and in cell culture *in vitro* was demonstrated in several studies [250, 253, 258, 259]. It appears that expression of MUC5B in goblet cells is characteristic not only for CF patients but also for patients with other inflammatory airway diseases. Expression of MUC5B both in glands and goblet cells was described by Chen and co-workers [43] in patients with emphysema and usual interstitial pneumonitis. Immunohistochemical study of the lung tissues from patients with diffuse panbronchitis [112] and chronic obstructive pulmonary disease [260] revealed abundant expression of MUC5B in bronchial glands and in the increased number of goblet cells on the bronchial surface, where *MUC5B* expression in the normal lung is generally negligible. Rose *et al.* [261] noted that the pattern of MUC5B expression described by Chen *et al.* [43] and Kamio *et al.* [112] mimics that of MUC5B in early fetal development. These observations suggest that expression of MUC5B reflects "either trans-differentiation of ciliated cells to a secretory phenotype or differentiation of progenitor cells" [261].

Summing up, the MUC5B mucin is over-expressed in different inflammatory conditions in the respiratory tract, indicating the active role MUC5B mucin plays in the pathogenesis of airway inflammatory diseases.

<u>Lung cancer</u>: The World Health Organization classifies lung cancer [262] into several subtypes (Fig. **6**), not all of which have been equally studied with regard to mucin expression. Among the various subtypes, there are MUC5B-positive and MUC5B-negative forms. Copin *et al.* [240] compared the expression of human mucin genes in normal respiratory mucosa and in lesions with typical epithelial hyperplasia, squamous metaplasia, dysplasia and carcinoma histology. They found that *MUC5B* was lightly expressed in the surface epithelium, where it is confined to some goblet cells, and decreased in expression from trachea to intrapulmonary bronchi. It was strongly expressed in collecting duct cells and in mucous cell of submucosal glands, but was not detectable in bronchioles and alveolar epithelial cells. The premalignant lesions characterized by basal cell and mucous cell hyperplasias were found to be MUC5B-positive, but lesions with squamous cell metaplasia and dysplasia as well as epidermoid carcinomas *in situ* and invasive carcinomas were all MUC5B-negative. The absence of *MUC5B* expression in epidermoid carcinoma [240] correlates with a practically undetectable amount of MUC5B mRNA in mucoepidermoid cell line NCI-H292 [264].

Expression of MUC5B and MUC5AC mucins was shown to be related to mucus formation in lung adenocarcinomas [123]. In Copin and co-workers' comprehensive study [263] of mucin expression in bronchoalveolar carcinomas (BAC) and non-bronchoalveolar lung carcinomas (non-BAC), 89% of mucinous BACs (m-BAC) express MUC5B mRNA (Fig. **6**), while among non-mucinous BAC (nm-BAC) tumors only 12% synthesize MUC5B specific transcripts. A higher expression of MUC5B mRNA (23%) was observed in non-BAC solid tumors that secreted mucus. Importantly, MUC5B protein was detected in 100% of m-BACs. In the group of nm-BAC, only 42% were MUC5B-positive when tested by specific immunochemical assay. Almost the same rate of MUC5B-positive tumors (55%) was observed among solid mucus-producing non-BAC.

Interestingly, lung tumors of smokers had higher MUC5B and MUC5AC mRNAs expression ratios than those of nonsmokers [265]. Tumors with increased expression of mucin genes tended to be associated with post-operative relapse, especially when *MUC5B* and *MUC5AC* genes were over-expressed.

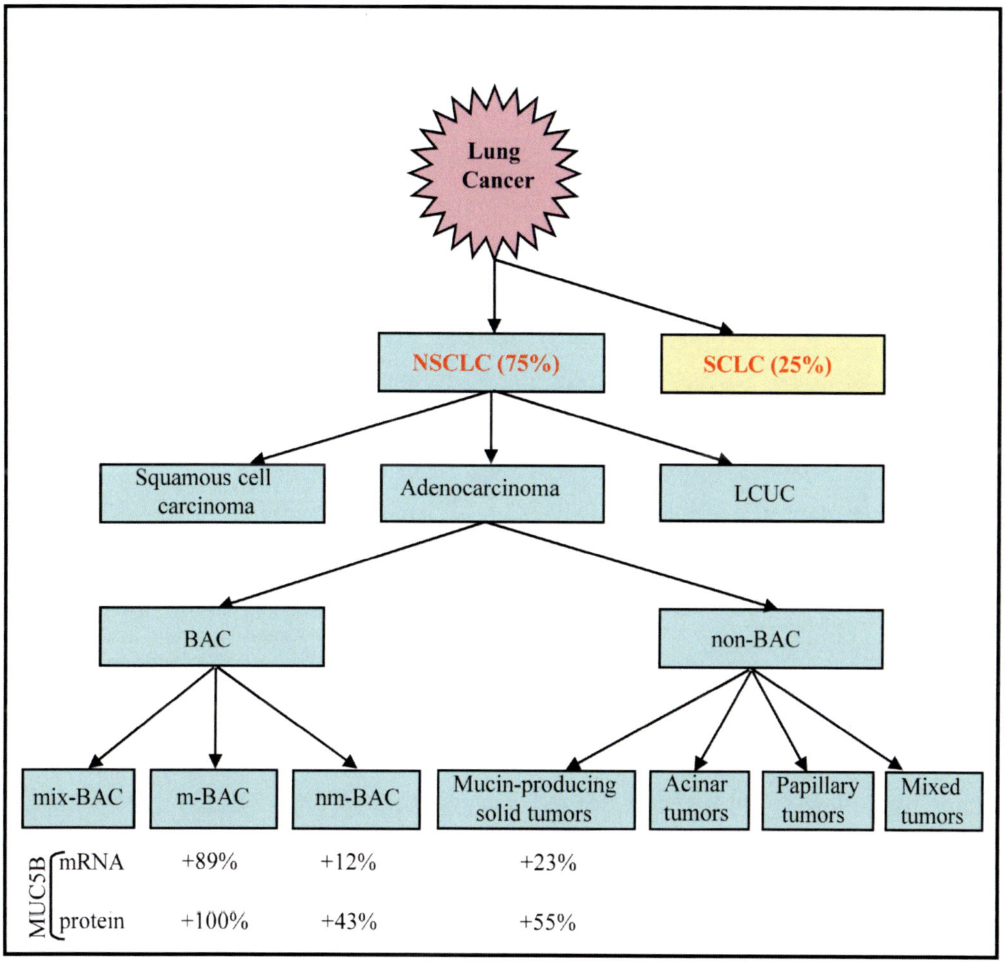

Figure 6: Expression of the *MUC5B* gene in lung tumors (based on the data reported in [262, 263]).

<u>*Adenocarcinomas of upper airways*</u>: Less information is available on mucin expression in adenocarcinomas of upper airways (sinonasal tract) compared with lung tumors. This type of tumor is classified as salivary type, intestinal type, non-intestinal type, or metastatic. In a study by Castilo *et al.* [241] of mucin expression in signet-ring cell adenocarcinoma (SRCA), a rare variant of sinonasal intestine-type adenocarcinoma, all neoplasms expressed MUC2 and MUC5B

mucins at very high levels, as measured by immunohistochemistry. Lower expression was typical of MUC5AC glycoprotein, whereas MUC6 mucin was not expressed in this type of tumors. The authors noted that expression of mucin genes in SRCA is very similar to that of gastrointestinal adenocarcinomas.

In summary, it is clear that data on *MUC5B* expression in airway neoplasms are sparse, and that many aspects of lung cancer pathogenesis associated with expression of the gene await further investigations.

6.7.2. Digestive System

Malignant tumors of salivary glands: The MUC5B mucin is constantly expressed in mucous acini of salivary glands [228, 266-270], in differing amounts at different ages. Ruhl *et al.* [271] observed increasing amounts of MUC5B in whole saliva during the first year of life. According to Denny *et al.* [270], the amount of mucins in whole saliva seems to decline with age. However, no significant age-dependent differences in the amount of MUC5B were found between 3-, 14- and 20-25-year-old individuals by Sonesson *et al.* [272]. According to the authors [272], these observations reflect changes in mucin expression associated with significant changes in oral environment, shifts in nutrition from fluids to solids, and variations in exposure to oral diseases but do not dependent on age. Individuals with high incidence of caries exhibit a significant decrease or even absence of salivary mucins MUC5B (MG1) and MUC7 (MG2) [273].

Among the malignant tumors originating in major and minor salivary glands, mucoepidermoid carcinoma (MEC) is the most frequent [274]. Alos *et al.* [267] found that *MUC5B* gene was expressed in 82% of MEC tumors, although each tumor had less than 50% of the MUC5B-positive cells. The MUC5B mucin expression and staining pattern was found to be similar to that of the MUC5AC protein; it was present in both the cytoplasm of the mucous and the columnar cells. Both mucins are expressed in most MEC, but mainly in low-grade tumors. Differential diagnosis of MEC includes mucin-rich variant of salivary duct carcinoma (mr-SDC), a relatively uncommon aggressive neoplasm [275]. Simpson *et al.* [276] analyzed mucin expression in several cases of mr-SDCs and found a consistent expression of MUC5B mucin combined with expression of

MUC2 and MUC6 glycoproteins. In the opinion of the authors, this profile of mucin expression might be useful for diagnostic purposes.

<u>*Sjogren syndrome (dry tongue disease)*</u>: Decreased amount of MUC5B mucin is found on the surface of anterior tongue in dry mouth patients compared with control subjects, in whom the amount of MUC5B mucin on the anterior tongue is higher than on other surfaces [277]. Pramanik *et al.* [277] noted that the oral cavity in dry mouth individuals can retain MUC5B and other salivary proteins, but the distribution of these components of saliva is changed and functional integrity is uncertain. In contrast, Alliende *et al.* [155] found no significant differences in MUC5B mRNA and protein levels between controls and dry mouth patients with Sjogren syndrome, but did observe changes in glycosylation of the MUC5B mucin. The levels of sulfo-Lewis^a antigen ($SO_3Gal\beta$-3GlcNAc) were notably decreased in gland extracts from patients compared with controls, and the number of sulfo-Lewis^a - positive mucous acini was reduced in patients. The authors concluded that disorganization of the basal lamina observed in patients with Sjogren syndrome may lead to de-differentiation of acinar mucous cells, consequently altering sulfation of the MUC5B mucin. These changes may explain xerostomia in these patients.

Further studies of different salivary gland and tongue diseases are needed for better understanding of MUC5B mucin's role in oral cavity pathophysiology.

<u>*Esophagus (Barrett's disease)*</u>: Expression of the gel-forming mucins in esophagus has been studied mostly in premalignant and neoplastic esophageal tissues. The scant information on *MUC5B* expression in inflammatory conditions of esophagus comes from a comparison of inflamed and noninflamed specimens of Barrett's eosophagus (BE) by Warson *et al.* [222]. The extent of *MUC5B* expression differs dramatically among the clinical groups of BE patients, with higher levels in the inflamed group than in the ulcerated group. However, the gene is most widely expressed in the noninflamed and dysplasia groups, with the same level of expression in the two groups [222].

Barrett's esophagus is a precancerous lesion that is thought to progress to adenocarcinoma *via* the metaplasia-dysplasia-adenocarcinoma sequence [278].

Guillem and co-workers [220] showed that *MUC5B* is the only mucin gene expressed in normal submucosal glands of esophagus, a finding confirmed by Arul *et al.* [219] at both the mRNA and protein levels. Unexpectedly, this gene was not expressed in any of the tested premalignant and malignant lesions of esophagus, including gastric and intestinal metaplasias, low and high grade dysplasias, and Barrett's adenocarcinomas [219, 220]. In contrast, Dekker's group [221, 222] detected MUC5B in the deeper glands of Barrett's esophagus, albeit to a relatively limited extent. Importantly, MUC5B expression was co-localized with MUC6 and TFF3 in nearly all Barrett's patients at the gastric-metaplasia stage. This type of protein co-expression changed at the next stage of Barrett's esophagus progression, the intestinal stage, when MUC5B is co-localized with MUC5AC. When mucin expression was examined in four clinical groups (BE without inflammation, BE with inflammation, ulcerating BE and BE with dysplasia), the *MUC5B* gene expression was the lowest in the ulcerated BE group, higher in the inflamed lesions, and most widely and strongly expressed in the noninflamed and dysplasia groups. The authors concluded that specific gel-forming mucins and TFF proteins are reliable markers for metaplastic BE in the preneoplastic progression to adenocarcinoma of esophagus [221, 222].

<u>Stomach</u>: The role of MUC5B expression in pathological conditions of stomach has been studies insufficiently. Although the available information is scant, it contains data related to Helicobacter infection and gastric cancer.

a) Helicobacter pylori infection: Since its discovery, *Helicobacter pylori* (*H. pylori*) has been identified as the major cause of gastritis, gastric ulcer, gastric atrophy, and gastric carcinoma [279-282]. The role of this bacterium in expression of *MUC5B* gene and functions of MUC5B mucin in *H. pylori*-induced pathology has not been well studied.

Because MUC5B is a mucin secreted into the oral cavity, and the mouth is the portal to the gastrointestinal system, the possible role of this mucin in *H. pylori* infection has become a focus of interest [283-287]. It has been shown that salivary MUC5B mucin functions as a receptor for binding of *H. pylori* bacterium *via* the carbohydrate sulfo-Lewis^{a}-epitope on the MUC5B glycoprotein [288]. As reported by Silva *et al.* [286], individuals with *H. pylori*-associated gastric diseases were found to have higher levels of salivary MUC5B expression than individuals without gastric diseases. This finding suggests that individuals with higher levels of salivary MUC5B in the oral cavity could be at higher risk for *H. pylori* infection in the stomach and consequent gastric diseases.

Indeed, *H. pylori* infection induces increased MUC5B expression not only in oral cavity, but also in stomach. Schmitz *et al.* [289] showed that *Helicobacter felis* infection in experimental mouse mimics the changes seen in the human stomach infected with *H. pylori*. The observed changes in mouse included increased expression of Muc5b and Muc4 and loss of Muc5ac. These results are in agreement with previous data showing that *H. pylori* infection affects, in a specific manner, all major secretory cell populations in the human antrum [290]. *H. pylori* infection led to a coordinated increase in cells expressing MUC5B and MUC6 at the expense of MUC5AC-producing cells. These changes in cell populations correlate with an increase in MUC5B and MUC6 mucin expression and a decrease in expression of MUC5AC glycoprotein and TFF1 and TFF2 peptides [290]. However, MUC5B expression is hardly detected in gastric metaplasia of duodenum, which is frequently associated with *H. pylori* infection and characterized by replacement of intestinal epithelium with gastric mucosa cells [291].

The importance of up-regulation of *MUC5B* expression in *H. pylori* infection was highlighted recently by the discovery that MUC5B mucin participates in all modes of adhesion utilized by *H. pylori* during the binding to human mucins in the oral and gastric niches [292, 293].

In summary, the question of how *H. pylori* infection influences *MUC5B* gene expression in gastric mucosa remains to be answered. Further studies are needed to evaluate the contribution of *H. pylori* infection to regulation of *MUC5B* transcription, synthesis of MUC5B mucin and its secretion in gastric epithelium and submucosal glands.

b) Gastric carcinoma: Development of gastric carcinoma is a multistep process. Numerous histopathological classifications have been proposed for the morphologically varied gastric tumors, including Lauren's, Mulligand's, Goseki's and the WHO classification [294-297]. All are based at least in part on the quantity of mucus and mucins in an individual tumor. According to Leteurte *et al.* [294], "mucin gene expression patterns in gastric cancer may reflect a precise state of differentiation at the cell level not recognized in used morphologic classification system". These authors consider that, whatever the classification, MUC5B expression is not associated with a particular subtype of gastric carcinoma. For example, MUC5B was expressed in 57%, 67%, 40% and 54% of gastric carcinomas classified as belonging to groups I, II, III and IV of Goseki's system. Pinto-de-Sousa *et al.* [298] consider that generally mucin expression can be used to evaluate differentiation patterns of gastric carcinoma. They showed that MUC5AC expression is associated with diffuse gastric type of early gastric carcinomas, while MUC2 expression is typical of the mucinous intestinal type. They also found a significant association between expression of MUC5B and MUC5AC in gastric carcinoma, but, like Leteurte *et al.* [294], could not relate it to the specific histological type of tumor. The authors suggest that MUC5B expression may be characteristic of the "unclassified histological type of gastric carcinoma according to Lauren's classification". On the one hand, MUC5B expression correlated with the absence of venous invasion, while on the other hand its expression was not detected in any so-called "complete intestinal subtype of gastric carcinoma" [298].

Despite the multiple findings of *MUC5B* expression in gastric cancer cells *in vitro* [37, 299] and in gastric carcinomas *in vivo* [218, 298], it is uncertain how MUC5B mucin contributes to the histological type and biological behavior of gastric carcinoma [37, 218, 298-300]. Nevertheless, based on the fact that *MUC5B* is not expressed by normal gastric mucosa and only briefly expressed during fetal development, Pinto-de-Sousa *et al.* [298] concluded that aberrant or *de novo* expression of MUC5B mucin in gastric carcinoma might be an oncofetal marker in adults and constitute a valuable diagnostic tool in gastric cancer.

Summing up, although there is evidence that expression of *MUC5B* gene might be specific for gastric cancer (at least for some subtypes), more studies are needed on its role in gastric carcinogenesis and the possible usefulness of the MUC5B-specific probes in diagnosis and follow-up of gastric cancer patients.

Hepatobiliary tract: Epithelial cells lining the bile ducts and gallbladder have several similar functions, including production of mucins. These mucins, MUC5B in particular, protect the biliary tree and gallbladder from the noxious effects of bile produced by hepatocytes and transported to the gallbladder through a complex system of canals, ductules and ducts [301]. MUC5B and other mucins are also involved in the pathogenesis of hepatobiliary diseases, including acute and chronic cholecystitis, hepatolithiasis, gallstone disease and cholangiocarcinoma [302-305].

a) Inflammatory diseases: Little is known about the synthesis and secretion of MUC5B in patients with hepatolithiasis, choledocholithiasis, cholangitis and cholecystitis [306]. There is, however, evidence of the participation of this mucin in the pathogenesis of hepatobiliary diseases. It serves a dual function in the hepatobiliary tract. On the one hand the MUC5B glycoprotein is expressed in normal human gallbladder epithelial cells [307, 308] and in the normal bile [306], while on the other hand its hypersecretion accelerates gallstone formation *in vitro* and *in vivo* [303, 306, 309, 310]. Vandenhaute *et al.* [304] analyzed expression of the gel-forming and membrane-bound mucins in gallbladder, intrahepatic bile ducts and liver and found that hepatocytes do not express any mucins except MUC3. In the biliary epithelial cells of liver, MUC5B was expressed only in the intrahepatic large bile ducts, while the small bile ducts expressed no mucin genes. Moderate expression of the *MUC5B* gene was observed in surface epithelial cells of gallbladder and in the cells of deep folds. MUC5B mRNA-positive cells were detected by Northern blot analysis or *in situ* hybridization in 100% of gallbladders collected by these authors from patients with cholecystitis [304]. Vilkin *et al.* [301] observed MUC5B expression in both the superficial epithelium and the deep folds of the gallbladder at the same levels as those reported by Vandenhaute *et al.* [304]. Ho *et al.* [303] found significant expression of MUC5B in surface epithelial cells similar to its expression in the deep folds. These authors [303] reported decreasing expression of MUC5B from 100% in normal gallbladder to 84.2% in mild chronic cholecystitis, 82.9% in chronic cholecystitis, and 72.7% in cholecystitis with features of acute and chronic disease. However, Vilkin *et al.* [301] did not observe such a trend when they compared MUC5B expression in cholecystitic and normal gallbladders: both gallbladders expressed MUC5B in 100% of the specimens.

In conclusion, in view of the possible involvement of MUC5B in pathology of gallbladder, more studies are needed to determine the role that MUC5B mucin plays in the pathogenesis of gallstone disease associated with different types of stones and different degrees of inflammation.

b) Malignant tumors: Chronic proliferative cholangitis and hepatothiliasis have been suggested to undergo a progressive change to atypical epithelial hyperplasia-dysplasia-cholangiocarcinoma [311]. Clinically, hepatolithiasis has been regarded as a risk factor for cholangiocarcinoma [302, 312]. In a study of the expression of *MUC5B* in normal gallbladder specimens and specimens from patients with hepatolithiasis and cholangiocarcinoma, only 33% of the cholangiocarcinoma samples were MUC5B-positive, while in the rest of cholangiocarcinomas tested (67%), expression of the MUC5B mRNA in tumor cells was decreased to undetectable level compared with the surrounding non-neoplastic biliary epithelium [302]. A PubMed search turned up only a few reports describing expression of MUC5B mucin in cholangiocarcinoma [302, 313-316], pointing to the need for additional investigations in this area.

Topographically, cholangiocarcinomas may be of two types: intrahepatic and extrahepatic [312, 317, 318]. Intrahepatic cholangiocarcinoma is further subdivided into mass-forming, periductal-infiltrating and intraductal-growing types [319]. The intraductal-growing type has been described as mucin-producing intrahepatic cholangio-carcinoma or intrahepatic (biliary) intraductal papillary mucinous neoplasia (b-IPMN) [320, 321]. On the basis of histology and mucin gene expression, b-IPMN is subdivided further into four subtypes: pancreatobiliary, gastric, intestinal and oncocytic carcinomas [313-315]. These subtypes have been identified only in the last decade, hence little is known at present about the association of these tumors with *MUC5B* expression. Rouzbahman *et al.* [316] found diffuse cytoplasmic expression of MUC5B apomucin in 3 tumors out of 4 oncocytic papillary neoplasms of the biliary tract. Expression of *MUC5B* was associated with different levels (0%-70%) of expression of other mucins. To the best of our knowledge, this is the only publication describing expression of MUC5B in oncocytic b-IPMN. Many questions remain unclear concerning the expression of *MUC5B* in hepatobiliary tumors.

Pancreas: Although the dynamics and topography of MUC5B mucin expression in embryonic and normal adult pancreas have been thoroughly studied [210, 225, 229, 322, 323], there are only a few reports on expression of MUC5B in pathological pancreas [229, 323, 324]. Nevertheless, this limited information contains important data on the subject. According to Balague *et al.* [323], *MUC5B* is homogeneously expressed in normal pancreatic ducts; the normal pattern of MUC5B mRNA expression in pancreas is seen also in obstructive chronic pancreatitis; only 50% of pancreatic adenocarcinomas express MUC5B transcripts; and no MUC5B mRNA is observed in papillary ductal hyperplasia.

These findings are only partially in agreement with those of Andrianifahanana *et al.* [229] who found that only 60% of pancreatic tumor cell lines were MUC5B mRNA-positive, while in an *in vivo* study there was no down-regulation of the *MUC5B* gene in diseased pancreas. In the latter study, 100% of the pancreatic adenocarcinoma specimens and 90% of the chronic pancreatitis specimens were MUC5B-positive. Interestingly, in contrast to Andrianifahanana *et al.* [229] but in agreement with Balague *et al.* [323], Zhang *et al.* [324] found MUC5B mucin expressed in only 20% of pancreatic cancers.

In summary, the available information from studies on the expression of *MUC5B* in pathological pancreas is not only limited but is also controversial. The discrepancies call for further investigation.

Colorectal intestine: Various mucins play different roles in the physiology and pathology of the intestine [325, 326]. The predominant mucin genes expressed in the normal colorectum are *MUC1-MUC4* [327-330], whereas *MUC5B* and *MUC6* are expressed at low levels [300, 331]. The expression and functions of MUC5B mucin in small and large intestine have been studied less than these activities of other gel-forming mucins including MUC2 or MUC5AC. Nevertheless, it is known that MUC5B is expressed in the adult [124, 300, 307] but not in the embryonic colon [223, 332]; is not detected in normal ileal mucosa [333]; and is expressed in specific patterns in the inflamed and malignant intestine [124, 300, 307].

a) Inflammatory bowel diseases: A number of studies have been performed on expression of *MUC5B* gene in inflammatory bowel diseases (IBD), including

Crohn's disease (CD) and ulcerative colitis (UC). Buisine *et al.* [333] analyzed the expression of several membrane-bound and gel-forming mucins, including MUC5B, in ileal mucosa of patients with CD, and found that mucin genes display an abnormal expression pattern in the mucosa adjacent to the ulcerations. This region of the affected mucosa, known as UACL (ulceration associated cell lineage), displays a unique glandular structure that develops *de novo* at the site of chronic ulceration in Crohn's and gastroduodenal ulcers [334, 335]. *De novo* expression of MUC5B mRNA and protein was detected throughout the UACL [333]. Longman *et al.* [334] also noted that expression of MUC5B, MUC5AC and MUC6 in small bowel affected by active CD is limited to the UACL, and that the MUC5B glycoprotein is localized in the secretory vesicles and cytoplasm of cells in both ductular and acinar elements of the UACL. Interestingly, unlike in CD, *de novo* ectopic expression of MUC5B mucin was not detected in UC even at the very severe stage of disease.

b) Colorectal cancer: Expression of *MUC5B* in colorectal cancer has been studied extensively *in vitro* in human colon adenocarcinoma cell lines [36, 336]. MUC5B mRNA was found in all 19 human colon adenocarcinoma cell lines tested in one study [336]. Moreover, *MUC5B* expression in colon cancer was cell specific: transcriptional activity of *MUC5B* promoter was high in the colon cancer differentiated mucus-secreting LS174T cell line, while transcription of the recombinant plasmid directed by *MUC5B* promoter was blocked in the colon cancer cell line Caco-2, which has an enterocyte-specific pattern of differentiation. Differential expression of mucins in human colon cancer cells was also observed with regard to their specific resistance to anticancer drugs [337]. MUC1 mucin is the major mucin expressed in the clone SF7 of enterocyte-like HT29 cells (HT29-5F7), resistant to 5-fluoruracil (5-FU) and methotrexate (MTX); MUC5B is the main expressed mucin in another clone of these cells, HT29-5M12, demonstrating the same drug resistance as HT29-5F7 cells. The mucus-secreting cells HT29-5F12, which are resistant to 5-FU but sensitive to MTX, express mainly the MUC2 mucin, while mucus-secreting clone HT29-5M21, which is resistant to MTX but sensitive to 5-FU, expresses both MUC5B and MUC5AC [337]. Clearly, mucins differ in chemoresistance, but the precise nature of this phenomenon remains to be delineated. As noted by Leteurtre *et al.*

[337], "it is tempting to speculate that mucins play a role in the ability of tumor cells to escape chemotherapy".

In addition to *in vitro* studies, *MUC5B* expression has also been analyzed in specimens from patients with colon malignancies. Myerscough *et al.* [338] found minor MUC5B mRNA expression only in the tubulovillous adenomas, and no expression in normal rectum mucosa, tubular and villous adenomas. In another study from this group [124], no MUC5B protein was detected in the diseased tissues of patients with colorectal cancer when examined by immunohistochemistry using non-VNTR MUC5B specific antibodies. Thus, solid adenocarcinomas of colon origin seem to be MUC5B-negative. In contrast, pseudomyxoma peritonei, an extracellular mucin-containing ascite associated with mucinous tumors of gastrointestinal tract and ovaries [339-341], appears to express MUC5B mucin. Biochemical and immunohistochemical analysis revealed the presence of three gel-forming mucins (MUC2, MUC5AC and MUC5B) in the mucus of a patient with pseudomyxoma peritonei that originated in the primary appendiceal adenocarcinoma [340]. Although MUC2 and MUC5AC glycoproteins were the main components of the ascite and were expressed in 75% of tumor cells, MUC5B mucin was present in myxoma and was expressed in about 25% of carcinoma cells of appendix.

Taken together, the data show that MUC5B mucin plays a role in pathological processes in the intestine, although the mechanisms are unknown. Further study is needed to understand the role of MUC5B glycoprotein in pathogenesis of bowel diseases.

6.8. EXPRESSION OF *MUC5B* GENE IN PATHOLOGY OF THE MALE UROGENITAL TRACT

The *MUC5B* gene is poorly expressed in the normal male urogenital tract. Its expression was not detected in male urogenital tissues and organs during embryonic and fetal development [342], and not in normal adult kidney, bladder, urethra, prostate, foreskin, epididymis, *vas deferens* and seminal vesicles [343-345]. The only exceptions are testis [346], bulbourethral glands [347] and seminal plasma [345].

Very little information is available on *MUC5B* expression in urogenital organs under pathological conditions. No expression of the gene was seen in renal dysplasia, autosomal recessive polycystic kidney disease, clear renal cell carcinoma, or papillary renal cell carcinoma [342, 343]. Low but detectable levels of *MUC5B* expression were observed in primary and metastatic prostate cancer [346]. Importantly, *MUC5B* is preferentially expressed in a hormone-independent variant of prostate adenocarcinoma. According to Legrier *et al.* [348], the loss of hormone dependence in prostate cancer is associated with irreversible histological alterations of mucinous or neuroendocrine character, marked by activation of gel-forming mucin expression, including MUC5B. These data point to mucinous differentiation as an important step in the acquisition of hormone independence in prostate cancer [348].

The shortage of information on MUC5B mucin expression in the male urogenital tract under physiological and pathological conditions points to the need for further investigation.

6.9. EXPRESSION OF *MUC5B* GENE IN PATHOLOGY OF THE FEMALE REPRODUCTIVE TRACT

More data are available on the activity of the *MUC5B* gene and the functions of the MUC5B glycoprotein in normal and diseased organs of the female reproductive system. The MUC5B glycoprotein is the main mucin of the cervical canal, where it undergoes dramatic physical, chemical, rheological and hydrodynamic changes during different stages of the menstrual cycle [349]. These changes affect the protective barrier of the genital tract against bacterial and chemical damages, and influence permeability of mucus gel in the cervical canal, thus contributing to reproductive success [160-162, 166, 167, 231, 233]. Importantly, these changes are detectable both at the transcriptional and translational levels as well as at the level of posttranslational modification [166, 232].

The involvement of the MUC5B mucin in various pathological processes in the female genital tract has been studied less than the participation of the MUC5B glycoprotein in reproductive physiology. While it is known that normal ovaries do not express secretory mucins, as this organ does not have histologically defined goblet cells and submucosal glands, Boman *et al.* [350] found expression of MUC5B mRNA in 64% of benign and 30% of borderline ovarian tumors. These findings conflict with those of Guintole *et al.* [351] who found MUC5B mRNA rarely expressed in epithelial ovarian cancers, including clear cell adenocarcinomas, mucinous and serous cystadenocarcinomas, and mixed mesodermal tumors of the ovary. Interestingly, Ju *et al.* [352] compared the expression of MUC5B mRNA in primary ovarian cancer tissues from chemosensitive and chemoresistant tumors and found that an increase in chemoresistance was associated with a decrease in expression of the *MUC5B* gene.

Mall *et al.* [353] reported expression of MUC5B mucin in a mature ovarian teratoma containing colonic and respiratory mucosa. Expression of *MUC5B* in endometrial and cervical cancers has been also studied, but in a very limited number of tumors. Hebbar *et al.* [354] found an increase in *MUC5B* transcriptional activity in endometrial and cervical carcinomas.

On the whole, the available data do not allow a reliable evaluation of the role MUC5B mucin plays in the pathogenesis of female reproductive tract diseases. There is no data about the participation of this mucin in inflammatory diseases of female reproductive organs. More studies are needed to fill in this gap.

6.10. EXPRESSION OF *MUC5B* GENE IN PATHOLOGY OF THE MAMMARY AND THYROID GLANDS

Only a few studies have been undertaken to evaluate participation of the MUC5B mucin in pathological processes in the mammary and thyroid glands.

Mammary gland: Normal mammary gland expresses almost no MUC5B mucin [355]. Sonora *et al.* [355] analyzed MUC5B apomucin expression in breast cancer and non-malignant breast tissues and found that expression of MUC5B has cytoplasmic and perinuclear distribution in ductal carcinoma and predominantly apical distribution in colloid carcinoma. Both localization and intensity of the expression varied according to the type of tumor. The MUC5B apomucin was detected in 81% of breast cancer tissues and, importantly, it was found in only 42% of normal-appearing breast mucosa samples adjacent to cancer lesions, while control normal breast samples were absolutely MUC5B-negative (Fig. 7). There was a high correlation between MUC5B expression and aggressiveness of the malignant process: whereas only 31% of ductal carcinoma *in situ* samples were MUC5B-positive, 100% of invasive ductal carcinomas expressed MUC5B.

MUC5B mRNA was detected in the human breast cancer cell line MCF7 *in vitro* and in the disseminated breast cancer cells *in vivo*, while 100% negative results were obtained when *MUC5B* expression was examined in peripheral blood mononuclear cells from normal individuals [356, 357]. These results suggest that *MUC5B* expression could be an early event during human breast carcinogenesis. Importantly, there was a correlation between MUC5B protein expression in primary tumors and MUC5B mRNA in bone marrow samples from the tumor-carrying patients [355]. Altogether, these results indicate that MUC5B mRNA might serve a specific molecular marker for the monitoring of breast cancer cell dissemination [356, 357].

Regarding nonmalignant mammary gland diseases, expression of MUC5B mucin was detected in fibroadenomas, in fibrocystic disease, and in sclerosing papilomas [355, 358].

In summary, MUC5B mucin appears to be an active participant in breast carcinogenesis and might become a target for molecular diagnostics and therapy of breast cancer.

<u>*Thyroid gland*</u>: In contrast to the extensively studied mucin expression in digestive, respiratory and urogenital tracts, much less information is available on expression of these glycoproteins, including MUC5B, in normal and neoplastic thyroid tissues. The only publication describing expression of MUC5B in thyroid gland comes from Magro *et al.* [238], who examined the expression pattern of seven mucins in normal, hyperplastic, benign neoplastic, and papillary-type carcinoma (PTC) tissues of thyroid. Of all mucin genes studied (*MUC1-MUC6*), *MUC5B* demonstrated the highest level of expression in all samples investigated. Despite this, the biological role of this mucin in thyroid is not clear. Interestingly, thyroid transcription factor-1 (TTF-1) completely inhibits expression of mouse *Muc5b*. Further investigations are needed to better understand the functions of *MUC5B* in the physiology and pathology of the thyroid.

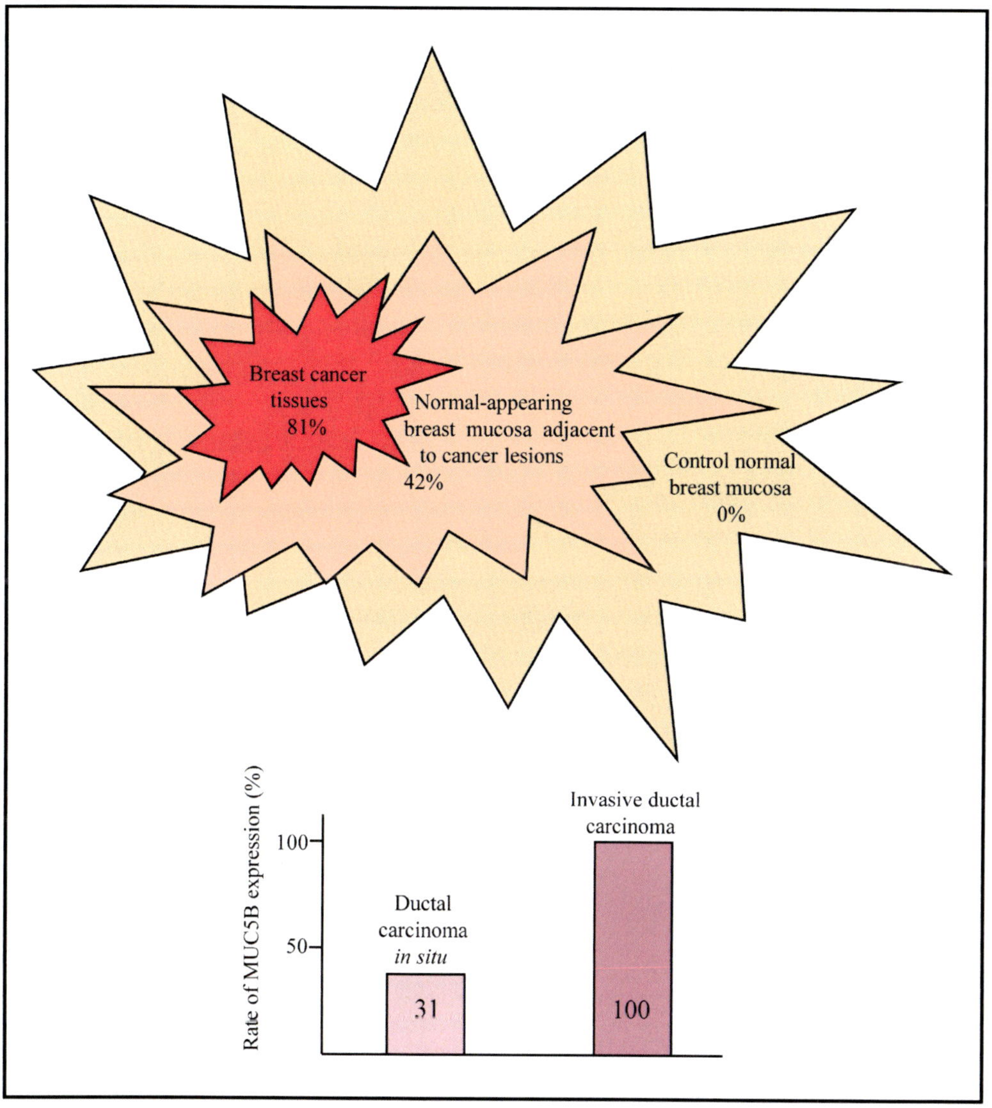

Figure 7: MUC5B expression in breast cancer (based on the data reported in [355]).

6.11. EXPRESSION OF *MUC5B* GENE IN PATHOLOGY OF THE EYE

Mucins have a great impact on eye functions. Goblet cells and intraepithelial mucous glands of the lacrimal sac and the nasolacrimal duct, as well as columnar cells of the efferent tear duct system, express several mucins, including MUC5B [235]. These mucins are important for integrity of mucous epithelia. They influence the rheological and antimicrobial properties of the tear film, suggesting an important role in normal physiology and in pathological processes.

Lacrimal glands and nasolacrimal ducts synthesize and secrete MUC5B protein [234, 359, 360]. Interestingly, human tears contain a number of mucins (MUC1, MUC2, MUC4, MUC5AC and MUC16), but do not have MUC5B and MUC7 glycoproteins [161, 361, 362]. This may be explained by the suggestion of Spurr-Michaud *et al.* [361] that MUC5B and MUC7 mucins are retained in the lacrimal gland, where they are produced and perhaps undergo local degradation.

The role of MUC5B mucin in the pathogenesis of eye diseases has been studied extensively. It has been shown that the amount of MUC5B is substantially increased in the lacrimal glands of patients who received treatment for dry eye syndrome [359]. A new flavonoid derivative, DA-6034, a candidate for treatment of dry eye disease, was found to increase production of several mucins, including MUC5B, thereby maintaining occular surface integrity [363]. Taking into account the role of mucins in ocular integrity and homeostasis, Paulsen *et al.* [235] compared mucin expression in normal human lacrimal sac and nasolacrimal duct, and in primary acquired nasolacrimal duct obstruction (PANDO). Reduced levels of MUC5B, MUC5AC and MUC2 mRNAs were observed in PANDO compared with normal mucosa. Since PANDO is associated with epiphora, these results support the view that secretory gel-forming mucins ease the tear flow through the efferent tear ducts. Thus, as concluded by these and other authors [231], disorders in mucin balance play an important role in the pathogenesis of dacryostenosis, dacryocystitis and dacryolithiasis.

6.12. EXPRESSION OF *MUC5B* GENE IN PATHOLOGY OF THE MIDDLE EAR

Variations in mucin production and secretion are known to be important in the pathophysiology of otitis media [364, 365]. Most of the 21 mucin genes identified so far, including *MUC5B*, are expressed, at least at mRNA level, in the normal middle ear epithelium in human, mouse and chinchilla [236, 237]. Expression of these genes was detected both *in vivo* and *in vitro*. Some conflicting data have been reported: Kershner *et al.* [236, 237] found a high level of mRNA expression of most, if not all, of the tested mucin genes, including *MUC5B*; others [366-369] who analyzed expression of mucin genes at the protein level, found only MUC5B glycoprotein expressed and secreted by the middle ear epithelium. An explanation for this discrepancy can be found in Jansen *et al.*'s report [370] who found surprisingly low correlation between protein levels and corresponding mRNA levels within various tissue samples. Based on this analysis, Preciado *et al.* [371] concluded that "studies looking solely at mRNA levels in middle ear tissue specimens cannot conclusively extrapolate presence or absence of proteins, especially within secretions". They also attributed the dearth of studies in which expression of mucin proteins in middle ear effusions have been comprehensively examined to the lack of reliable molecular tools for mucin glycoprotein detection in middle ear secretions.

Nevertheless, some investigators have examined mucin expression in otitis media both in the middle ear epithelium (MEE) and in the effusion. Lin *et al.* [368] showed that normal MEE and eustachian tube express distinct mucin profiles: while MEE expresses only MUC5B mucin, normal human eustachian tube epithelium synthesizes four mucin glycoproteins: MUC5B, MUC1, MUC4 and MUC5AC. In this study, production of MUC5B and MUC4 was up-regulated 4.2- and 6-fold, respectively, both in middle ear chronic otitis and in mucoid otitis media. This up-regulation was accompanied by extensive proliferation of MUC5B- and MUC4-producing cells. Still another study suggested dependence of mucin hyper-production on cytokines produced by $CD4^+$ and $CD8^+$ T-cells and/or $CD68^+$ monocyte macrophages [369]. This suggestion is in line with the observations of the stimulating role of different cytokines in expression of mucin genes reported by several laboratories [371-375]. The results reported by Lin *et al.* [368] were confirmed by others [366, 367, 371, 375] who showed that MUC5B is the predominant mucin glycoprotein in otitis media.

Summing up, there is ample evidence of the participation of MUC5B mucin in middle ear inflammatory processes, even though the precise role of this mucin in otitis media and other inflammatory conditions is not clear. It is also not clear whether MUC5B participates in malignant ear diseases as no information is available on this issue, at least not in the PubMed Database.

REFERENCES

[1] Pigny P, Guyonnet-Duperat V, Hill AS, *et al.* Human mucin genes assigned to 11p15.5: identification and organization of a cluster of genes. Genomics 1996;38:340-52.

[2] Desseyn JL, Aubert JP, Porchet N, Laine A. Evolution of the large secreted gel-forming mucins. Mol Biol Evol 2000;17:1175-84.

[3] Desseyn JL, Buisine MP, Porchet N, *et al.* Evolutionary history of the 11p15 human mucin gene family. J Mol Evol 1998;46:102-6.

[4] Meezaman D, Charles P, Daskal E, *et al.* Cloning and analysis of cDNA encoding a major airway glycoprotein, human tracheobronchial mucin (MUC5). J Biol Chem 1994;269:12932-9.

[5] Keates AC, Nunes DP, Afdhal NH, Troxler RF, Offner GD. Molecular cloning of a major human gall bladder mucin: complete C-terminal sequence and genomic organization of MUC5B. Biochem J 1997;324 (Pt 1):295-303.

[6] Offner GD, Nunes DP, Keates AC, Afdhal NH, Troxler RF. The amino-terminal sequence of MUC5B contains conserved multifunctional D domains: implications for tissue-specific mucin functions. Biochem Biophys Res Commun 1998;251:350-5.

[7] Desseyn JL, Guyonnet-Duperat V, Porchet N, Aubert JP, Laine A. Human mucin gene MUC5B, the 10.7-kb large central exon encodes various alternate subdomains resulting in a super-repeat. Structural evidence for a 11p15.5 gene family. J Biol Chem 1997;272:3168-78.

[8] Desseyn JL, Buisine MP, Porchet N, Aubert JP, Laine A. Genomic organization of the human mucin gene MUC5B. cDNA and genomic sequences upstream of the large central exon. J Biol Chem 1998;273:30157-64.

[9] Desseyn JL, Aubert JP, Van Seuningen I, Porchet N, Laine A. Genomic organization of the 3' region of the human mucin gene MUC5B. J Biol Chem 1997;272:16873-83.

[10] Escande F, Porchet N, Aubert JP, Buisine MP. The mouse Muc5b mucin gene: cDNA and genomic structures, chromosomal localization and expression. Biochem J 2002;363:589-98.

[11] Buisine MP, Desseyn JL, Porchet N, *et al.* Genomic organization of the 3'-region of the human MUC5AC mucin gene: additional evidence for a common ancestral gene for the 11p15.5 mucin gene family. Biochem J 1998;332 (Pt 3):729-38.

[12] Gum JR, Jr., Hicks JW, Toribara NW, *et al.* The human MUC2 intestinal mucin has cysteine-rich subdomains located both upstream and downstream of its central repetitive region. J Biol Chem 1992;267:21375-83.

[13] Gum JR, Jr., Hicks JW, Toribara NW, Siddiki B, Kim YS. Molecular cloning of human intestinal mucin (MUC2) cDNA. Identification of the amino terminus and overall sequence similarity to prepro-von Willebrand factor. J Biol Chem 1994;269:2440-6.

[14] Lesuffleur T, Roche F, Hill AS, *et al.* Characterization of a mucin cDNA clone isolated from HT-29 mucus-secreting cells. The 3' end of MUC5AC? J Biol Chem 1995;270:13665-73.

[15] Li D, Gallup M, Fan N, Szymkowski DE, Basbaum CB. Cloning of the amino-terminal and 5'-flanking region of the human MUC5AC mucin gene and transcriptional up-regulation by bacterial exoproducts. J Biol Chem 1998;273:6812-20.

[16] Mancuso DJ, Tuley EA, Westfield LA, *et al.* Structure of the gene for human von Willebrand factor. J Biol Chem 1989;264:19514-27.

[17] Debailleul V, Laine A, Huet G, *et al.* Human mucin genes MUC2, MUC3, MUC4, MUC5AC, MUC5B, and MUC6 express stable and extremely large mRNAs and exhibit a variable length polymorphism. An improved method to analyze large mRNAs. J Biol Chem 1998;273:881-90.
[18] Gum JR, Byrd JC, Hicks JW, *et al.* Molecular cloning of human intestinal mucin cDNAs. Sequence analysis and evidence for genetic polymorphism. J Biol Chem 1989;264:6480-7.
[19] Pigny P, Pratt WS, Laine A, *et al.* The MUC5AC gene: RFLP analysis with the Jer58 probe. Hum Genet 1995;96:367-8.
[20] Vinall LE, Hill AS, Pigny P, *et al.* Variable number tandem repeat polymorphism of the mucin genes located in the complex on 11p15.5. Hum Genet 1998;102:357-66.
[21] Ahn EK, Kim WJ, Kwon JA, *et al.* Variants of MUC5B minisatellites and the susceptibility of bladder cancer. DNA Cell Biol 2009;28:169-76.
[22] Desseyn JL, Rousseau K, Laine A. Fifty-nine bp repeat polymorphism in the uncommon intron 36 of the human mucin gene MUC5B. Electrophoresis 1999;20:493-6.
[23] Pigny P, Van Seuningen I, Desseyn JL, *et al.* Identification of a 42-kDa nuclear factor (NF1-MUC5B) from HT-29 MTX cells that binds to the 3' region of human mucin gene MUC5B. Biochem Biophys Res Commun 1996;220:186-91.
[24] Nakamura Y, Koyama K, Matsushima M. VNTR (variable number of tandem repeat) sequences as transcriptional, translational, or functional regulators. J Hum Genet 1998;43:149-52.
[25] Voorberg J, Fontijn R, Calafat J, *et al.* Assembly and routing of von Willebrand factor variants: the requirements for disulfide-linked dimerization reside within the carboxy-terminal 151 amino acids. J Cell Biol 1991;113:195-205.
[26] Perez-Vilar J, Eckhardt AE, Hill RL. Porcine submaxillary mucin forms disulfide-bonded dimers between its carboxyl-terminal domains. J Biol Chem 1996;271:9845-50.
[27] Perez-Vilar J, Hill RL. The carboxyl-terminal 90 residues of porcine submaxillary mucin are sufficient for forming disulfide-bonded dimers. J Biol Chem 1998;273:6982-8.
[28] Bell SL, Khatri IA, Xu G, Forstner JF. Evidence that a peptide corresponding to the rat Muc2 C-terminus undergoes disulphide-mediated dimerization. Eur J Biochem 1998;253:123-31.
[29] Klomp LW, Van Rens L, Strous GJ. Cloning and analysis of human gastric mucin cDNA reveals two types of conserved cysteine-rich domains. Biochem J 1995;308 (Pt 3):831-8.
[30] Joba W, Hoffmann W. Similarities of integumentary mucin B.1 from Xenopus laevis and prepro-von Willebrand factor at their amino-terminal regions. J Biol Chem 1997;272:1805-10.
[31] Eckhardt AE, Timpte CS, DeLuca AW, Hill RL. The complete cDNA sequence and structural polymorphism of the polypeptide chain of porcine submaxillary mucin. J Biol Chem 1997;272:33204-10.
[32] Hardy DM, Garbers DL. A sperm membrane protein that binds in a species-specific manner to the egg extracellular matrix is homologous to von Willebrand factor. J Biol Chem 1995;270:26025-8.
[33] Baker ME. Invertebrate vitellogenin is homologous to human von Willebrand factor. Biochem J 1988;256:1059-61.
[34] Mayadas TN, Wagner DD. Vicinal cysteines in the prosequence play a role in von Willebrand factor multimer assembly. Proc Natl Acad Sci U S A 1992;89:3531-5.

[35] Meitinger T, Meindl A, Bork P, *et al.* Molecular modeling of the Norrie disease protein predicts a cystine knot growth factor tertiary structure. Nat Genet 1993;5:376-80.
[36] Van Seuningen I, Perrais M, Pigny P, Porchet N, Aubert JP. Sequence of the 5'-flanking region and promoter activity of the human mucin gene MUC5B in different phenotypes of colon cancer cells. Biochem J 2000;348 Pt 3:675-86.
[37] Perrais M, Pigny P, Buisine MP, *et al.* Aberrant expression of human mucin gene MUC5B in gastric carcinoma and cancer cells. Identification and regulation of a distal promoter. J Biol Chem 2001;276:15386-96.
[38] Wu DY, Wu R, Chen Y, Tarasova N, Chang MM. PMA stimulates MUC5B gene expression through an Sp1-based mechanism in airway epithelial cells. Am J Respir Cell Mol Biol 2007;37:589-97.
[39] Yuan-Chen Wu D, Wu R, Reddy SP, Lee YC, Chang MM. Distinctive epidermal growth factor receptor/extracellular regulated kinase-independent and -dependent signaling pathways in the induction of airway mucin 5B and mucin 5AC expression by phorbol 12-myristate 13-acetate. Am J Pathol 2007;170:20-32.
[40] Preciado D, Lin J, Wuertz B, Rose M. Cigarette smoke activates NF kappa B and induces Muc5b expression in mouse middle ear cells. Laryngoscope 2008;118:464-71.
[41] Choi JY, Cho KN, Yoon JH. All-trans retinoic acid induces mucociliary differentiation in a human cholesteatoma epithelial cell culture. Acta Otolaryngol 2004;124:30-5.
[42] Koo JS, Jetten AM, Belloni P, *et al.* Role of retinoid receptors in the regulation of mucin gene expression by retinoic acid in human tracheobronchial epithelial cells. Biochem J 1999;338 (Pt 2):351-7.
[43] Chen Y, Zhao YH, Di YP, Wu R. Characterization of human mucin 5B gene expression in airway epithelium and the genomic clone of the amino-terminal and 5'-flanking region. Am J Respir Cell Mol Biol 2001;25:542-53.
[44] Kim SW, Hong JS, Ryu SH, *et al.* Regulation of mucin gene expression by CREB *via* a nonclassical retinoic acid signaling pathway. Mol Cell Biol 2007;27:6933-47.
[45] Koo JS, Yoon JH, Gray T, *et al.* Restoration of the mucous phenotype by retinoic acid in retinoid-deficient human bronchial cell cultures: changes in mucin gene expression. Am J Respir Cell Mol Biol 1999;20:43-52.
[46] Gray T, Koo JS, Nettesheim P. Regulation of mucous differentiation and mucin gene expression in the tracheobronchial epithelium. Toxicology 2001;160:35-46.
[47] Ishinaga H, Takeuchi K, Kishioka C, Yagawa M, Majima Y. Effects of dexamethasone on mucin gene expression in cultured human nasal epithelial cells. Laryngoscope 2002;112:1436-40.
[48] Martinez-Anton A, de Bolos C, Alobid I, *et al.* Corticosteroid therapy increases membrane-tethered while decreases secreted mucin expression in nasal polyps. Allergy 2008;63:1368-76.
[49] Liu J, Zhang Z, Xu Y, Xing L, Zhang H. Effects of glucocorticoid on IL-13-induced Muc5ac expression in airways of mice. J Huazhong Univ Sci Technolog Med Sci 2004;24:575-7.
[50] Komatsu K, Jono H, Lim JH, *et al.* Glucocorticoids inhibit nontypeable Haemophilus influenzae-induced MUC5AC mucin expression *via* MAPK phosphatase-1-dependent inhibition of p38 MAPK. Biochem Biophys Res Commun 2008;377:763-8.

[51] Chen Y, Nickola TJ, DiFronzo NL, Colberg-Poley AM, Rose MC. Dexamethasone-mediated repression of MUC5AC gene expression in human lung epithelial cells. Am J Respir Cell Mol Biol 2006;34:338-47.

[52] Schoneveld OJ, Gaemers IC, Lamers WH. Mechanisms of glucocorticoid signaling. Biochim Biophys Acta 2004;1680:114-28.

[53] Dostert A, Heinzel T. Negative glucocorticoid receptor response elements and their role in glucocorticoid action. Curr Pharm Des 2004;10:2807-16.

[54] Kettle R, Simmons J, Schindler F, *et al.* Regulation of neuregulin 1beta1-induced MUC5AC and MUC5B expression in human airway epithelium. Am J Respir Cell Mol Biol 2010;42:472-81.

[55] Falls DL. Neuregulins: functions, forms, and signaling strategies. Exp Cell Res 2003;284:14-30.

[56] Breuleux M. Role of heregulin in human cancer. Cell Mol Life Sci 2007;64:2358-77.

[57] Holmes WE, Sliwkowski MX, Akita RW, *et al.* Identification of heregulin, a specific activator of p185erbB2. Science 1992;256:1205-10.

[58] Jaradat M, Stapleton C, Tilley SL, *et al.* Modulatory role for retinoid-related orphan receptor alpha in allergen-induced lung inflammation. Am J Respir Crit Care Med 2006;174:1299-309.

[59] Kim YD, Kwon EJ, Park DW, *et al.* Interleukin-1beta induces MUC2 and MUC5AC synthesis through cyclooxygenase-2 in NCI-H292 cells. Mol Pharmacol 2002;62:1112-8.

[60] Fujisawa T, Velichko S, Thai P, *et al.* Regulation of airway MUC5AC expression by IL-1beta and IL-17A; the NF-kappaB paradigm. J Immunol 2009;183:6236-43.

[61] Koo JS, Kim YD, Jetten AM, Belloni P, Nettesheim P. Overexpression of mucin genes induced by interleukin-1 beta, tumor necrosis factor-alpha, lipopolysaccharide, and neutrophil elastase is inhibited by a retinoic acid receptor alpha antagonist. Exp Lung Res 2002;28:315-32.

[62] Enss ML, Cornberg M, Wagner S, *et al.* Proinflammatory cytokines trigger MUC gene expression and mucin release in the intestinal cancer cell line LS180. Inflamm Res 2000;49:162-9.

[63] Borchers MT, Carty MP, Leikauf GD. Regulation of human airway mucins by acrolein and inflammatory mediators. Am J Physiol 1999;276:L549-55.

[64] Chen Y, Thai P, Zhao YH, *et al.* Stimulation of airway mucin gene expression by interleukin (IL)-17 through IL-6 paracrine/autocrine loop. J Biol Chem 2003;278:17036-43.

[65] Jayawickreme SP, Gray T, Nettesheim P, Eling T. Regulation of 15-lipoxygenase expression and mucus secretion by IL-4 in human bronchial epithelial cells. Am J Physiol 1999;276:L596-603.

[66] Rose MC, Piazza FM, Chen YA, *et al.* Model systems for investigating mucin gene expression in airway diseases. J Aerosol Med 2000;13:245-61.

[67] Dabbagh K, Takeyama K, Lee HM, *et al.* IL-4 induces mucin gene expression and goblet cell metaplasia *in vitro* and *in vivo*. J Immunol 1999;162:6233-7.

[68] Zuhdi Alimam M, Piazza FM, Selby DM, *et al.* Muc-5/5ac mucin messenger RNA and protein expression is a marker of goblet cell metaplasia in murine airways. Am J Respir Cell Mol Biol 2000;22:253-60.

[69] Singer M, Lefort J, Vargaftig BB. Granulocyte depletion and dexamethasone differentially modulate airways hyperreactivity, inflammation, mucus accumulation, and secretion induced by rmIL-13 or antigen. Am J Respir Cell Mol Biol 2002;26:74-84.

[70] Kim CH, Song KS, Koo JS, *et al.* IL-13 suppresses MUC5AC gene expression and mucin secretion in nasal epithelial cells. Acta Otolaryngol 2002;122:638-43.
[71] Smirnova MG, Kiselev SL, Birchall JP, Pearson JP. Up-regulation of mucin secretion in HT29-MTX cells by the pro-inflammatory cytokines tumor necrosis factor-alpha and interleukin-6. Eur Cytokine Netw 2001;12:119-25.
[72] Issa R, Xie S, Lee KY, *et al.* GRO-alpha regulation in airway smooth muscle by IL-1beta and TNF-alpha: role of NF-kappaB and MAP kinases. Am J Physiol Lung Cell Mol Physiol 2006;291:L66-74.
[73] Dinarello CA. Biologic basis for interleukin-1 in disease. Blood 1996;87:2095-147.
[74] Huang F, Kao CY, Wachi S, *et al.* Requirement for both JAK-mediated PI3K signaling and ACT1/TRAF6/TAK1-dependent NF-kappaB activation by IL-17A in enhancing cytokine expression in human airway epithelial cells. J Immunol 2007;179:6504-13.
[75] Kao CY, Kim C, Huang F, Wu R. Requirements for two proximal NF-kappaB binding sites and IkappaB-zeta in IL-17A-induced human beta-defensin 2 expression by conducting airway epithelium. J Biol Chem 2008;283:15309-18.
[76] DiCosmo BF, Geba GP, Picarella D, *et al.* Airway epithelial cell expression of interleukin-6 in transgenic mice. Uncoupling of airway inflammation and bronchial hyperreactivity. J Clin Invest 1994;94:2028-35.
[77] Choi J, Klinkspoor JH, Yoshida T, Lee SP. Lipopolysaccharide from Escherichia coli stimulates mucin secretion by cultured dog gallbladder epithelial cells. Hepatology 1999;29:1352-7.
[78] Li JD, Dohrman AF, Gallup M, *et al.* Transcriptional activation of mucin by Pseudomonas aeruginosa lipopolysaccharide in the pathogenesis of cystic fibrosis lung disease. Proc Natl Acad Sci U S A 1997;94:967-72.
[79] Enss ML, Muller H, Schmidt-Wittig U, *et al.* Effects of perorally applied endotoxin on colonic mucins of germfree rats. Scand J Gastroenterol 1996;31:868-74.
[80] Hunter SE, Singla AK, Prazma J, *et al.* Mucin production in the middle ear in response to lipopolysaccharides. Otolaryngol Head Neck Surg 1999;120:884-8.
[81] Shimizu T, Takahashi Y, Takeuchi K, Majima Y, Sakakura Y. Role of neutrophil elastase in endotoxin-induced mucus hypersecretion in rat nasal epithelium. Ann Otol Rhinol Laryngol 2000;109:1049-54.
[82] Kim YT, Jung HH, Ko TO, Kim SJ. Up-regulation of MUC5AC mRNA expression in endotoxin-induced otitis media. Acta Otolaryngol 2001;121:364-70.
[83] Smirnova MG, Guo L, Birchall JP, Pearson JP. LPS up-regulates mucin and cytokine mRNA expression and stimulates mucin and cytokine secretion in goblet cells. Cell Immunol 2003;221:42-9.
[84] Bautista MV, Chen Y, Ivanova VS, *et al.* IL-8 regulates mucin gene expression at the posttranscriptional level in lung epithelial cells. J Immunol 2009;183:2159-66.
[85] Dohrman A, Miyata S, Gallup M, *et al.* Mucin gene (MUC 2 and MUC 5AC) upregulation by Gram-positive and Gram-negative bacteria. Biochim Biophys Acta 1998;1406:251-9.
[86] Yan F, Li W, Jono H, *et al.* Reactive oxygen species regulate Pseudomonas aeruginosa lipopolysaccharide-induced MUC5AC mucin expression *via* PKC-NADPH oxidase-ROS-TGF-alpha signaling pathways in human airway epithelial cells. Biochem Biophys Res Commun 2008;366:513-9.
[87] Kohri K, Ueki IF, Shim JJ, *et al.* Pseudomonas aeruginosa induces MUC5AC production *via* epidermal growth factor receptor. Eur Respir J 2002;20:1263-70.

[88] Shao MX, Ueki IF, Nadel JA. Tumor necrosis factor alpha-converting enzyme mediates MUC5AC mucin expression in cultured human airway epithelial cells. Proc Natl Acad Sci U S A 2003;100:11618-23.

[89] Caramori G, Di Gregorio C, Carlstedt I, *et al.* Mucin expression in peripheral airways of patients with chronic obstructive pulmonary disease. Histopathology 2004;45:477-84.

[90] Groneberg DA, Eynott PR, Lim S, *et al.* Expression of respiratory mucins in fatal status asthmaticus and mild asthma. Histopathology 2002;40:367-73.

[91] Casalino-Matsuda SM, Monzon ME, Forteza RM. Epidermal growth factor receptor activation by epidermal growth factor mediates oxidant-induced goblet cell metaplasia in human airway epithelium. Am J Respir Cell Mol Biol 2006;34:581-91.

[92] Forteza R, Conner GE, Salathe M. Hylaluronan in the airways. In: Garg HG, Hales CA, editors. Chemistry and Biology of Hyaluronan. Kidlington, UK: Elsevier Ltd; 2004. p. 323-38.

[93] Manzanares D, Monzon ME, Savani RC, Salathe M. Apical oxidative hyaluronan degradation stimulates airway ciliary beating *via* RHAMM and RON. Am J Respir Cell Mol Biol 2007;37:160-8.

[94] Kobayashi H, Suzuki M, Kanayama N, *et al.* CD44 stimulation by fragmented hyaluronic acid induces upregulation of urokinase-type plasminogen activator and its receptor and subsequently facilitates invasion of human chondrosarcoma cells. Int J Cancer 2002;102:379-89.

[95] Tsatas D, Kanagasundaram V, Kaye A, Novak U. EGF receptor modifies cellular responses to hyaluronan in glioblastoma cell lines. J Clin Neurosci 2002;9:282-8.

[96] Wakahara K, Kobayashi H, Yagyu T, *et al.* Bikunin down-regulates heterodimerization between CD44 and growth factor receptors and subsequently suppresses agonist-mediated signaling. J Cell Biochem 2005;94:995-1009.

[97] Casalino-Matsuda SM, Monzon ME, Day AJ, Forteza RM. Hyaluronan fragments/CD44 mediate oxidative stress-induced MUC5B up-regulation in airway epithelium. Am J Respir Cell Mol Biol 2009;40:277-85.

[98] Leikauf GD, Borchers MT, Prows DR, Simpson LG. Mucin apoprotein expression in COPD. Chest 2002;121:166S-82S.

[99] Lethem MI, Dowell ML, Van Scott M, *et al.* Nucleotide regulation of goblet cells in human airway epithelial explants: normal exocytosis in cystic fibrosis. Am J Respir Cell Mol Biol 1993;9:315-22.

[100] Song KS, Kim HJ, Kim K, Lee JG, Yoon JH. Regulator of G-protein signaling 4 suppresses LPS-induced MUC5AC overproduction in the airway. Am J Respir Cell Mol Biol 2009;41:40-9.

[101] Chen Y, Zhao YH, Wu R. Differential regulation of airway mucin gene expression and mucin secretion by extracellular nucleotide triphosphates. Am J Respir Cell Mol Biol 2001;25:409-17.

[102] Roger P, Gascard JP, Bara J, *et al.* ATP induced MUC5AC release from human airways *in vitro*. Mediators Inflamm 2000;9:277-84.

[103] Choi JY, Namkung W, Shin JH, Yoon JH. Uridine-5'-triphosphate and adenosine triphosphate gammaS induce mucin secretion *via* Ca2+-dependent pathways in human nasal epithelial cells. Acta Otolaryngol 2003;123:1080-6.

[104] Wong AM, Chow AW, Au SC, Wong CC, Ko WH. Apical *vs.* basolateral P2Y(6) receptor-mediated Cl(-) secretion in immortalized bronchial epithelia. Am J Respir Cell Mol Biol 2009;40:733-45.

[105] Tarran R, Button B, Boucher RC. Regulation of normal and cystic fibrosis airway surface liquid volume by phasic shear stress. Annu Rev Physiol 2006;68:543-61.
[106] Loffreda S, Yang SQ, Lin HZ, *et al.* Leptin regulates proinflammatory immune responses. Faseb J 1998;12:57-65.
[107] Myers MG, Jr. Leptin receptor signaling and the regulation of mammalian physiology. Recent Prog Horm Res 2004;59:287-304.
[108] Fantuzzi G, Faggioni R. Leptin in the regulation of immunity, inflammation, and hematopoiesis. J Leukoc Biol 2000;68:437-46.
[109] Bruno A, Chanez P, Chiappara G, *et al.* Does leptin play a cytokine-like role within the airways of COPD patients? Eur Respir J 2005;26:398-405.
[110] Tsuchiya T, Shimizu H, Horie T, Mori M. Expression of leptin receptor in lung: leptin as a growth factor. Eur J Pharmacol 1999;365:273-9.
[111] Woo HJ, Yoo WJ, Bae CH, *et al.* Leptin up-regulates MUC5B expression in human airway epithelial cells *via* mitogen-activated protein kinase pathway. Exp Lung Res 2010;36:262-9.
[112] Kamio K, Matsushita I, Hijikata M, *et al.* Promoter analysis and aberrant expression of the MUC5B gene in diffuse panbronchiolitis. Am J Respir Crit Care Med 2005;171:949-57.
[113] Kamio K, Matsushita I, Tanaka G, *et al.* Direct determination of MUC5B promoter haplotypes based on the method of single-strand conformation polymorphism and their statistical estimation. Genomics 2004;84:613-22.
[114] Fedorova L, Fedorov A. Introns in gene evolution. Genetica. 2003;118:123-31.
[115] Dahiya R, Lesuffleur T, Kwak KS, *et al.* Expression and characterization of mucins associated with the resistance to methotrexate of human colonic adenocarcinoma cell line HT29. Cancer Res 1992;52:4655-62.
[116] Vincent A, Perrais M, Desseyn JL, *et al.* Epigenetic regulation (DNA methylation, histone modifications) of the 11p15 mucin genes (MUC2, MUC5AC, MUC5B, MUC6) in epithelial cancer cells. Oncogene 2007;26:6566-76.
[117] Vincent A, Van Seuningen I. Epigenetics, stem cells and epithelial cell fate. Differentiation 2009;78:99-107.
[118] Munshi A, Shafi G, Aliya N, Jyothy A. Histone modifications dictate specific biological readouts. J Genet Genomics 2009;36:75-88.
[119] Herceg Z. Epigenetic information in chromatin and cancer. Eur J Cancer 2009;45 Suppl 1:442-4.
[120] Parra M. Epigenetic events during B lymphocyte development. Epigenetics 2009;4:462-8.
[121] Herman JG, Baylin SB. Gene silencing in cancer in association with promoter hypermethylation. N Engl J Med 2003;349:2042-54.
[122] Blache P, van de Wetering M, Duluc I, *et al.* SOX9 is an intestine crypt transcription factor, is regulated by the Wnt pathway, and represses the CDX2 and MUC2 genes. J Cell Biol 2004;166:37-47.
[123] Copin MC, Buisine MP, Devisme L, *et al.* Normal respiratory mucosa, precursor lesions and lung carcinomas: differential expression of human mucin genes. Front Biosci 2001;6:D1264-75.
[124] Sylvester PA, Myerscough N, Warren BF, *et al.* Differential expression of the chromosome 11 mucin genes in colorectal cancer. J Pathol 2001;195:327-35.
[125] Yi X, White DM, Aisner DL, *et al.* An alternate splicing variant of the human telomerase catalytic subunit inhibits telomerase activity. Neoplasia 2000;2:433-40.

[126] van Klinken BJ, Einerhand AW, Buller HA, Dekker J. The oligomerization of a family of four genetically clustered human gastrointestinal mucins. Glycobiology 1998;8:67-75.
[127] Perez-Vilar J, Hill RL. The structure and assembly of secreted mucins. J Biol Chem 1999;274:31751-4.
[128] Perez-Vilar J, Eckhardt AE, DeLuca A, Hill RL. Porcine submaxillary mucin forms disulfide-linked multimers through its amino-terminal D-domains. J Biol Chem 1998;273:14442-9.
[129] Kesimer M, Makhov AM, Griffith JD, Verdugo P, Sheehan JK. Unpacking a gel-forming mucin: a view of MUC5B organization after granular release. Am J Physiol Lung Cell Mol Physiol 2010;298:L15-22.
[130] Thornton DJ, Rousseau K, McGuckin MA. Structure and function of the polymeric mucins in airways mucus. Annu Rev Physiol 2008;70:459-86.
[131] Hovenberg HW, Davies JR, Carlstedt I. Different mucins are produced by the surface epithelium and the submucosa in human trachea: identification of MUC5AC as a major mucin from the goblet cells. Biochem J 1996;318 (Pt 1):319-24.
[132] Groneberg DA, Eynott PR, Oates T, *et al.* Expression of MUC5AC and MUC5B mucins in normal and cystic fibrosis lung. Respir Med 2002;96:81-6.
[133] Thornton DJ, Howard M, Khan N, Sheehan JK. Identification of two glycoforms of the MUC5B mucin in human respiratory mucus. Evidence for a cysteine-rich sequence repeated within the molecule. J Biol Chem 1997;272:9561-6.
[134] Wickstrom C, Davies JR, Eriksen GV, Veerman EC, Carlstedt I. MUC5B is a major gel-forming, oligomeric mucin from human salivary gland, respiratory tract and endocervix: identification of glycoforms and C-terminal cleavage. Biochem J 1998;334 (Pt 3):685-93.
[135] Lidell ME, Johansson ME, Hansson GC. An autocatalytic cleavage in the C terminus of the human MUC2 mucin occurs at the low pH of the late secretory pathway. J Biol Chem 2003;278:13944-51.
[136] Wickstrom C, Carlstedt I. N-terminal cleavage of the salivary MUC5B mucin. Analogy with the Van Willebrand propolypeptide? J Biol Chem 2001;276:47116-21.
[137] Bell SL, Xu G, Khatri IA, *et al.* N-linked oligosaccharides play a role in disulphide-dependent dimerization of intestinal mucin Muc2. Biochem J 2003;373:893-900.
[138] McCool DJ, Okada Y, Forstner JF, Forstner GG. Roles of calreticulin and calnexin during mucin synthesis in LS180 and HT29/A1 human colonic adenocarcinoma cells. Biochem J 1999;341 (Pt 3):593-600.
[139] Nehrke K, Tabak LA. Biosynthesis of a low-molecular-mass rat submandibular gland mucin glycoprotein in COS7 cells. Biochem J 1997;323 (Pt 2):497-502.
[140] Sadler JE. Biochemistry and genetics of von Willebrand factor. Annu Rev Biochem 1998;67:395-424.
[141] Nielsen PA, Bennett EP, Wandall HH, *et al.* Identification of a major human high molecular weight salivary mucin (MG1) as tracheobronchial mucin MUC5B. Glycobiology 1997;7:413-9.
[142] Thornton DJ, Khan N, Mehrotra R, *et al.* Salivary mucin MG1 is comprised almost entirely of different glycosylated forms of the MUC5B gene product. Glycobiology 1999;9:293-302.
[143] Thornton DJ, Howard M, Devine PL, Sheehan JK. Methods for separation and deglycosylation of mucin subunits. Anal Biochem 1995;227:162-7.
[144] Veerman EC, van den Keybus PA, Valentijn-Benz M, Nieuw Amerongen AV. Isolation of different high-Mr mucin species from human whole saliva. Biochem J 1992;283 (Pt 3):807-11.

[145] Bolscher J, Veerman E, Van Nieuw Amerongen A, Tulp A, Verwoerd D. Distinct populations of high-M(r) mucins secreted by different human salivary glands discriminated by density-gradient electrophoresis. Biochem J 1995;309 (Pt 3):801-6.
[146] Veerman EC, van den Keijbus PA, Nazmi K, *et al.* Distinct localization of MUC5B glycoforms in the human salivary glands. Glycobiology 2003;13:363-6.
[147] Thomsson KA, Prakobphol A, Leffler H, *et al.* The salivary mucin MG1 (MUC5B) carries a repertoire of unique oligosaccharides that is large and diverse. Glycobiology 2002;12:1-14.
[148] Levine MJ, Reddy MS, Tabak LA, *et al.* Structural aspects of salivary glycoproteins. J Dent Res 1987;66:436-41.
[149] Zalewska A, Zwierz K, Zolkowski K, Gindzienski A. Structure and biosynthesis of human salivary mucins. Acta Biochim Pol 2000;47:1067-79.
[150] Brockhausen I, Dowler T, Paulsen H. Site directed processing: role of amino acid sequences and glycosylation of acceptor glycopeptides in the assembly of extended mucin type O-glycan core 2. Biochim Biophys Acta 2009;1790:1244-57.
[151] Silverman HS, Sutton-Smith M, McDermott K, *et al.* The contribution of tandem repeat number to the O-glycosylation of mucins. Glycobiology 2003;13:265-77.
[152] Silverman HS, Parry S, Sutton-Smith M, *et al.* *In vivo* glycosylation of mucin tandem repeats. Glycobiology 2001;11:459-71.
[153] Shori DK, Kariyawasam HH, Knight RA, *et al.* Sulphation of the salivary mucin MG1 (MUC-5B) is not correlated to the degree of its sialylation and is unaffected by cystic fibrosis. Pflugers Arch 2001;443 Suppl 1:S50-4.
[154] Shori DK, Genter T, Hansen J, *et al.* Altered sialyl- and fucosyl-linkage on mucins in cystic fibrosis patients promotes formation of the sialyl-Lewis X determinant on salivary MUC-5B and MUC-7. Pflugers Arch 2001;443 Suppl 1:S55-61.
[155] Alliende C, Kwon YJ, Brito M, *et al.* Reduced sulfation of muc5b is linked to xerostomia in patients with Sjogren syndrome. Ann Rheum Dis 2008;67:1480-7.
[156] Thomsson KA, Schulz BL, Packer NH, Karlsson NG. MUC5B glycosylation in human saliva reflects blood group and secretor status. Glycobiology 2005;15:791-804.
[157] Carlstedt I, Lindgren H, Sheehan JK. The macromolecular structure of human cervical-mucus glycoproteins. Studies on fragments obtained after reduction of disulphide bridges and after subsequent trypsin digestion. Biochem J 1983;213:427-35.
[158] Yurewicz EC, Matsuura F, Moghissi KS. Structural studies of sialylated oligosaccharides of human midcycle cervical mucin. J Biol Chem 1987;262:4733-9.
[159] Moghissi KS, Syner FN. Studies on human cervical mucus: mucoids and their relation to sperm penetration. Fertil Steril 1970;21:234-9.
[160] Morales P, Roco M, Vigil P. Human cervical mucus: relationship between biochemical characteristics and ability to allow migration of spermatozoa. Hum Reprod 1993;8:78-83.
[161] Gipson IK, Moccia R, Spurr-Michaud S, *et al.* The Amount of MUC5B mucin in cervical mucus peaks at midcycle. J Clin Endocrinol Metab 2001;86:594-600.
[162] Gipson IK, Spurr-Michaud S, Moccia R, *et al.* MUC4 and MUC5B transcripts are the prevalent mucin messenger ribonucleic acids of the human endocervix. Biol Reprod 1999;60:58-64.
[163] Iacobelli S, Garcea N, Angeloni C. Biochemistry of cervical mucus: a comparative analysis of the secretion from preovulatory, postovulatory, and pregnancy periods. Fertil Steril 1971;22:727-34.

[164] Chantler E, Debruyne E. Factors regulating the changes in cervical mucus in different hormonal states. Adv Exp Med Biol 1977;89:131-41.
[165] Gilks CB, Reid PE, Clement PB, Owen DA. Histochemical changes in cervical mucus-secreting epithelium during the normal menstrual cycle. Fertil Steril 1989;51:286-91.
[166] Argueso P, Spurr-Michaud S, Tisdale A, Gipson IK. Variation in the amount of T antigen and N-acetyllactosamine oligosaccharides in human cervical mucus secretions with the menstrual cycle. J Clin Endocrinol Metab 2002;87:5641-8.
[167] Brunelli R, Papi M, Arcovito G, *et al.* Globular structure of human ovulatory cervical mucus. Faseb J 2007;21:3872-6.
[168] Eggert-Kruse W, Kohler A, Rohr G, Runnebaum B. The pH as an important determinant of sperm-mucus interaction. Fertil Steril 1993;59:617-28.
[169] Holmen JM, Karlsson NG, Abdullah LH, *et al.* Mucins and their O-Glycans from human bronchial epithelial cell cultures. Am J Physiol Lung Cell Mol Physiol 2004;287:L824-34.
[170] Schulz BL, Sloane AJ, Robinson LJ, *et al.* Mucin glycosylation changes in cystic fibrosis lung disease are not manifest in submucosal gland secretions. Biochem J 2005;387:911-9.
[171] Schulz BL, Sloane AJ, Robinson LJ, *et al.* Glycosylation of sputum mucins is altered in cystic fibrosis patients. Glycobiology 2007;17:698-712.
[172] Chace KV, Leahy DS, Martin R, *et al.* Respiratory mucous secretions in patients with cystic fibrosis: relationship between levels of highly sulfated mucin component and severity of the disease. Clin Chim Acta 1983;132:143-55.
[173] Mendicino J, Sangadala S. Synthesis of sulfated oligosaccharides by cystic fibrosis trachea epithelial cells. Mol Cell Biochem 1999;201:141-9.
[174] Hofsteenge J, Muller DR, de Beer T, *et al.* New type of linkage between a carbohydrate and a protein: C-glycosylation of a specific tryptophan residue in human RNase Us. Biochemistry 1994;33:13524-30.
[175] Krieg J, Hartmann S, Vicentini A, *et al.* Recognition signal for C-mannosylation of Trp-7 in RNase 2 consists of sequence Trp-x-x-Trp. Mol Biol Cell 1998;9:301-9.
[176] Doucey MA, Hess D, Blommers MJ, Hofsteenge J. Recombinant human interleukin-12 is the second example of a C-mannosylated protein. Glycobiology 1999;9:435-41.
[177] Loffler A, Doucey MA, Jansson AM, *et al.* Spectroscopic and protein chemical analyses demonstrate the presence of C-mannosylated tryptophan in intact human RNase 2 and its isoforms. Biochemistry 1996;35:12005-14.
[178] Vliegenthart JF, Casset F. Novel forms of protein glycosylation. Curr Opin Struct Biol 1998;8:565-71.
[179] Perez-Vilar J, Randell SH, Boucher RC. C-Mannosylation of MUC5AC and MUC5B Cys subdomains. Glycobiology 2004;14:325-37.
[180] Lidell ME, Hansson GC. Cleavage in the GDPH sequence of the C-terminal cysteine-rich part of the human MUC5AC mucin. Biochem J 2006;399:121-9.
[181] Enghild JJ, Salvesen G, Hefta SA, *et al.* Chondroitin 4-sulfate covalently cross-links the chains of the human blood protein pre-alpha-inhibitor. J Biol Chem 1991;266:747-51.
[182] Davies JR, Herrmann A, Russell W, *et al.* Respiratory tract mucins: structure and expression patterns. Novartis Found Symp 2002;248:76-88; discussion -93, 277-82.
[183] Bonthron DT, Handin RI, Kaufman RJ, *et al.* Structure of pre-pro-von Willebrand factor and its expression in heterologous cells. Nature 1986;324:270-3.
[184] Verweij CL, Hart M, Pannekoek H. Expression of variant von Willebrand factor (vWF) cDNA in heterologous cells: requirement of the pro-polypeptide in vWF multimer formation. Embo J 1987;6:2885-90.

[185] Wickstrom C, Christersson C, Davies JR, Carlstedt I. Macromolecular organization of saliva: identification of'insoluble' MUC5B assemblies and non-mucin proteins in the gel phase. Biochem J 2000;351 Pt 2:421-8.
[186] Verdugo P. Goblet cells secretion and mucogenesis. Annu Rev Physiol 1990;52:157-76.
[187] Thornton DJ, Davies JR, Kraayenbrink M, *et al.* Mucus glycoproteins from'normal' human tracheobronchial secretion. Biochem J 1990;265:179-86.
[188] Sheehan JK, Howard M, Richardson PS, Longwill T, Thornton DJ. Physical characterization of a low-charge glycoform of the MUC5B mucin comprising the gel-phase of an asthmatic respiratory mucous plug. Biochem J 1999;338 (Pt 2):507-13.
[189] Shogren R, Gerken TA, Jentoft N. Role of glycosylation on the conformation and chain dimensions of O-linked glycoproteins: light-scattering studies of ovine submaxillary mucin. Biochemistry 1989;28:5525-36.
[190] Gerken TA. Biophysical approaches to salivary mucin structure, conformation and dynamics. Crit Rev Oral Biol Med 1993;4:261-70.
[191] Thiele C, Huttner WB. Protein and lipid sorting from the trans-Golgi network to secretory granules-recent developments. Semin Cell Dev Biol 1998;9:511-6.
[192] Burgoyne RD, Morgan A. Secretory granule exocytosis. Physiol Rev 2003;83:581-632.
[193] Perez-Vilar J. Mucin granule intraluminal organization. Am J Respir Cell Mol Biol 2007;36:183-90.
[194] Li Y, Tanaka T. Phase transitions of gels. Annu Rev Mater Sci 1992;22:243-77.
[195] Verdugo P, Aitken M, Langley L, Villalon MJ. Molecular mechanism of product storage and release in mucin secretion. II. The role of extracellular Ca++. Biorheology 1987;24:625-33.
[196] Verdugo P, Deyrup-Olsen I, Aitken M, Villalon M, Johnson D. Molecular mechanism of mucin secretion: I. The role of intragranular charge shielding. J Dent Res 1987;66:506-8.
[197] Baconnais S, Delavoie F, Zahm JM, *et al.* Abnormal ion content, hydration and granule expansion of the secretory granules from cystic fibrosis airway glandular cells. Exp Cell Res. 2005;309:296-304.
[198] Raynal BD, Hardingham TE, Sheehan JK, Thornton DJ. Calcium-dependent protein interactions in MUC5B provide reversible cross-links in salivary mucus. J Biol Chem 2003;278:28703-10.
[199] Iontcheva I, Oppenheim FG, Offner GD, Troxler RF. Molecular mapping of statherin- and histatin-binding domains in human salivary mucin MG1 (MUC5B) by the yeast two-hybrid system. J Dent Res 2000;79:732-9.
[200] Perez-Vilar J, Olsen JC, Chua M, Boucher RC. pH-dependent intraluminal organization of mucin granules in live human mucous/goblet cells. J Biol Chem 2005;280:16868-81.
[201] Bansil R, Turner BS. Mucin structure, aggregation, physiological functions and biomedical applications. Curr Opin Colloid Interf Sci 2006;11:164-70.
[202] Viney C, Huber AE, Verdugo P. Liquid crystalline order in mucus. Macromolecules 1993;26:852-5.
[203] Fernandez JM, Villalon M, Verdugo P. Reversible condensation of mast cell secretory products *in vitro*. Biophys J 1991;59:1022-7.
[204] Tam PY, Verdugo P. Control of mucus hydration as a Donnan equilibrium process. Nature 1981;292:340-2.
[205] Verdugo PDobnPB, 2007. Dynamics of biopolymer networks. Polymer Bull 2007;58:139-43.
[206] Rogers DF. The airway goblet cell. Int J Biochem Cell Biol 2003;35:1-6.

[207] McCool DJ, Forstner JF, Forstner GG. Regulated and unregulated pathways for MUC2 mucin secretion in human colonic LS180 adenocarcinoma cells are distinct. Biochem J 1995;312 (Pt 1):125-33.

[208] Jackson AD. Airway goblet-cell mucus secretion. Trends Pharmacol Sci 2001;22:39-45.

[209] Audie JP, Janin A, Porchet N, *et al.* Expression of human mucin genes in respiratory, digestive, and reproductive tracts ascertained by *in situ* hybridization. J Histochem Cytochem 1993;41:1479-85.

[210] Andrianifahanana M, Moniaux N, Batra SK. Regulation of mucin expression: mechanistic aspects and implications for cancer and inflammatory diseases. Biochim Biophys Acta 2006;1765:189-222.

[211] Buisine MP, Devisme L, Copin MC, *et al.* Developmental mucin gene expression in the human respiratory tract. Am J Respir Cell Mol Biol 1999;20:209-18.

[212] Reid CJ, Gould S, Harris A. Developmental expression of mucin genes in the human respiratory tract. Am J Respir Cell Mol Biol 1997;17:592-8.

[213] Engelhardt JF, Schlossberg H, Yankaskas JR, Dudus L. Progenitor cells of the adult human airway involved in submucosal gland development. Development 1995;121:2031-46.

[214] Sharma P, Dudus L, Nielsen PA, *et al.* MUC5B and MUC7 are differentially expressed in mucous and serous cells of submucosal glands in human bronchial airways. Am J Respir Cell Mol Biol 1998;19:30-7.

[215] Kirkham S, Sheehan JK, Knight D, Richardson PS, Thornton DJ. Heterogeneity of airways mucus: variations in the amounts and glycoforms of the major oligomeric mucins MUC5AC and MUC5B. Biochem J 2002;361:537-46.

[216] Bernacki SH, Nelson AL, Abdullah L, *et al.* Mucin gene expression during differentiation of human airway epithelia *in vitro*. Muc4 and muc5b are strongly induced. Am J Respir Cell Mol Biol 1999;20:595-604.

[217] Thornton DJ, Gray T, Nettesheim P, *et al.* Characterization of mucins from cultured normal human tracheobronchial epithelial cells. Am J Physiol Lung Cell Mol Physiol 2000;278:L1118-28.

[218] Buisine MP, Devisme L, Maunoury V, *et al.* Developmental mucin gene expression in the gastroduodenal tract and accessory digestive glands. I. Stomach. A relationship to gastric carcinoma. J Histochem Cytochem 2000;48:1657-66.

[219] Arul GS, Moorghen M, Myerscough N, *et al.* Mucin gene expression in Barrett's oesophagus: an *in situ* hybridisation and immunohistochemical study. Gut 2000;47:753-61.

[220] Guillem P, Billeret V, Buisine MP, *et al.* Mucin gene expression and cell differentiation in human normal, premalignant and malignant esophagus. Int J Cancer 2000;88:856-61.

[221] Van De Bovenkamp JH, Korteland-Van Male AM, Warson C, *et al.* Gastric-type mucin and TFF-peptide expression in Barrett's oesophagus is disturbed during increased expression of MUC2. Histopathology 2003;42:555-65.

[222] Warson C, Van De Bovenkamp JH, Korteland-Van Male AM, *et al.* Barrett's esophagus is characterized by expression of gastric-type mucins (MUC5AC, MUC6) and TFF peptides (TFF1 and TFF2), but the risk of carcinoma development may be indicated by the intestinal-type mucin, MUC2. Hum Pathol 2002;33:660-8.

[223] Buisine MP, Devisme L, Savidge TC, *et al.* Mucin gene expression in human embryonic and fetal intestine. Gut 1998;43:519-24.

[224] Buisine MP, Aubert JP, Walker WA, Savidge TC. Developmental patterns of mucin gene expression in human fetal small intestinal xenografts maintained in severe-combined immunodeficient mice. Pediatr Res 2003;53:898-904.

[225] Buisine MP, Devisme L, Degand P, *et al.* Developmental mucin gene expression in the gastroduodenal tract and accessory digestive glands. II. Duodenum and liver, gallbladder, and pancreas. J Histochem Cytochem 2000;48:1667-76.
[226] Paulsen FP, Varoga D, Paulsen AR, Corfield A, Tsokos M. Prognostic value of mucins in the classification of ampullary carcinomas. Hum Pathol 2006;37:160-7.
[227] Troxler RF, Offner GD, Zhang F, Iontcheva I, Oppenheim FG. Molecular cloning of a novel high molecular weight mucin (MG1) from human sublingual gland. Biochem Biophys Res Commun 1995;217:1112-9.
[228] Piludu M, Rayment SA, Liu B, *et al.* Electron microscopic immunogold localization of salivary mucins MG1 and MG2 in human submandibular and sublingual glands. J Histochem Cytochem 2003;51:69-79.
[229] Andrianifahanana M, Moniaux N, Schmied BM, *et al.* Mucin (MUC) gene expression in human pancreatic adenocarcinoma and chronic pancreatitis: a potential role of MUC4 as a tumor marker of diagnostic significance. Clin Cancer Res 2001;7:4033-40.
[230] Audie JP, Tetaert D, Pigny P, *et al.* Mucin gene expression in the human endocervix. Hum Reprod 1995;10:98-102.
[231] Gipson IK, Ho SB, Spurr-Michaud SJ, *et al.* Mucin genes expressed by human female reproductive tract epithelia. Biol Reprod 1997;56:999-1011.
[232] Andersch-Bjorkman Y, Thomsson KA, Holmen Larsson JM, Ekerhovd E, Hansson GC. Large scale identification of proteins, mucins, and their O-glycosylation in the endocervical mucus during the menstrual cycle. Mol Cell Proteomics 2007;6:708-16.
[233] Gipson IK. Mucins of the human endocervix. Front Biosci 2001;6:D1245-55.
[234] Paulsen F, Langer G, Hoffmann W, Berry M. Human lacrimal gland mucins. Cell Tissue Res 2004;316:167-77.
[235] Paulsen FP, Corfield AP, Hinz M, *et al.* Characterization of mucins in human lacrimal sac and nasolacrimal duct. Invest Ophthalmol Vis Sci 2003;44:1807-13.
[236] Kerschner JE. Mucin gene expression in human middle ear epithelium. Laryngoscope 2007;117:1666-76.
[237] Kerschner JE, Li J, Tsushiya K, Khampang P. Mucin gene expression and mouse middle ear epithelium. Int J Pediatr Otorhinolaryngol 2010;74:864-8.
[238] Magro G, Schiappacassi M, Perissinotto D, *et al.* Differential expression of mucins 1-6 in papillary thyroid carcinoma: evidence for transformation-dependent post-translational modifications of MUC1 *in situ*. J Pathol 2003;200:357-69.
[239] Rogers DF. Mucus pathophysiology in COPD: differences to asthma, and pharmacotherapy. Monaldi Arch Chest Dis 2000;55:324-32.
[240] Copin MC, Devisme L, Buisine MP, *et al.* From normal respiratory mucosa to epidermoid carcinoma: expression of human mucin genes. Int J Cancer 2000;86:162-8.
[241] Castillo C, Buob D, Mortuaire G, *et al.* Signet-ring cell adenocarcinoma of sinonasal tract: an immunohistochemical study of the mucins profile. Arch Pathol Lab Med 2007;131:961-4.
[242] Rose MC, Kaufman B, Martin BM. Proteolytic fragmentation and peptide mapping of human carboxyamidomethylated tracheobronchial mucin. J Biol Chem 1989;264:8193-9.
[243] Davies JR, Svitacheva N, Lannefors L, Kornfalt R, Carlstedt I. Identification of MUC5B, MUC5AC and small amounts of MUC2 mucins in cystic fibrosis airway secretions. Biochem J 1999;344 Pt 2:321-30.
[244] Voynow JA, Gendler SJ, Rose MC. Regulation of mucin genes in chronic inflammatory airway diseases. Am J Respir Cell Mol Biol 2006;34:661-5.

[245] Voynow JA. What does mucin have to do with lung disease? Paediatr Respir Rev 2002;3:98-103.
[246] Rogers DF. Physiology of airway mucus secretion and pathophysiology of hypersecretion. Respir Care 2007;52:1134-46; discussion 46-9.
[247] Sheehan JK, Richardson PS, Fung DC, Howard M, Thornton DJ. Analysis of respiratory mucus glycoproteins in asthma: a detailed study from a patient who died in status asthmaticus. Am J Respir Cell Mol Biol 1995;13:748-56.
[248] Evans CM, Kim K, Tuvim MJ, Dickey BF. Mucus hypersecretion in asthma: causes and effects. Curr Opin Pulm Med 2009;15:4-11.
[249] Nguyen LP, Omoluabi O, Parra S, *et al.* Chronic exposure to beta-blockers attenuates inflammation and mucin content in a murine asthma model. Am J Respir Cell Mol Biol 2008;38:256-62.
[250] Zhu Y, Ehre C, Abdullah LH, *et al.* Munc13-2-/- baseline secretion defect reveals source of oligomeric mucins in mouse airways. J Physiol 2008;586:1977-92.
[251] Prescott E, Lange P, Vestbo J. Chronic mucus hypersecretion in COPD and death from pulmonary infection. Eur Respir J 1995;8:1333-8.
[252] Hogg JC, Chu F, Utokaparch S, *et al.* The nature of small-airway obstruction in chronic obstructive pulmonary disease. N Engl J Med 2004;350:2645-53.
[253] Martinez-Anton A, Debolos C, Garrido M, *et al.* Mucin genes have different expression patterns in healthy and diseased upper airway mucosa. Clin Exp Allergy 2006;36:448-57.
[254] Shimura S, Andoh Y, Haraguchi M, Shirato K. Continuity of airway goblet cells and intraluminal mucus in the airways of patients with bronchial asthma. Eur Respir J 1996;9:1395-401.
[255] Rubin BK. Mucus structure and properties in cystic fibrosis. Paediatr Respir Rev 2007;8:4-7.
[256] Henke MO, Renner A, Huber RM, Seeds MC, Rubin BK. MUC5AC and MUC5B Mucins Are Decreased in Cystic Fibrosis Airway Secretions. Am J Respir Cell Mol Biol 2004;31:86-91.
[257] Rose MC, Voynow JA. Respiratory tract mucin genes and mucin glycoproteins in health and disease. Physiol Rev 2006;86:245-78.
[258] Groneberg DA, Peiser C, Dinh QT, *et al.* Distribution of respiratory mucin proteins in human nasal mucosa. Laryngoscope 2003;113:520-4.
[259] Hajj R, Lesimple P, Nawrocki-Raby B, *et al.* Human airway surface epithelial regeneration is delayed and abnormal in cystic fibrosis. J Pathol 2007;211:340-50.
[260] Kirkham S, Kolsum U, Rousseau K, *et al.* MUC5B is the major mucin in the gel phase of sputum in chronic obstructive pulmonary disease. Am J Respir Crit Care Med 2008;178:1033-9.
[261] Rose MC, Nickola TJ, Voynow JA. Airway mucus obstruction: mucin glycoproteins, MUC gene regulation and goblet cell hyperplasia. Am J Respir Cell Mol Biol 2001;25:533-7.
[262] Travis WD, Colbey TV, Corrin B. Histological Typing of Lung and Pleural Tumors. International Histological Classification of Tumours. 3 ed. Geneva, Switzerland: WHO; 1999.
[263] Copin MC, Buisine MP, Leteurtre E, *et al.* Mucinous bronchioloalveolar carcinomas display a specific pattern of mucin gene expression among primary lung adenocarcinomas. Hum Pathol 2001;32:274-81.
[264] Berger JT, Voynow JA, Peters KW, Rose MC. Respiratory carcinoma cell lines. MUC genes and glycoconjugates. Am J Respir Cell Mol Biol 1999;20:500-10.

[265] Yu CJ, Yang PC, Shun CT, *et al.* Overexpression of MUC5 genes is associated with early post-operative metastasis in non-small-cell lung cancer. Int J Cancer 1996;69:457-65.
[266] Nielsen PA, Mandel U, Therkildsen MH, Clausen H. Differential expression of human high-molecular-weight salivary mucin (MG1) and low-molecular-weight salivary mucin (MG2). J Dent Res 1996;75:1820-6.
[267] Alos L, Lujan B, Castillo M, *et al.* Expression of membrane-bound mucins (MUC1 and MUC4) and secreted mucins (MUC2, MUC5AC, MUC5B, MUC6 and MUC7) in mucoepidermoid carcinomas of salivary glands. Am J Surg Pathol 2005;29:806-13.
[268] Piras M, Hand AR, Tore G, Ledda GP, Piludu M. Ultrastructural localization of salivary mucins MUC5B and MUC7 in human labial glands. Eur J Oral Sci 2010;118:14-8.
[269] Kouznetsova I, Gerlach KL, Zahl C, Hoffmann W. Expression analysis of human salivary glands by laser microdissection: differences between submandibular and labial glands. Cell Physiol Biochem 2010;26:375-82.
[270] Denny PC, Denny PA, Klauser DK, *et al.* Age-related changes in mucins from human whole saliva. J Dent Res 1991;70:1320-7.
[271] Ruhl S, Rayment SA, Schmalz G, Hiller KA, Troxler RF. Proteins in whole saliva during the first year of infancy. J Dent Res 2005;84:29-34.
[272] Sonesson M, Wickstrom C, Kinnby B, Ericson D, Matsson L. Mucins MUC5B and MUC7 in minor salivary gland secretion of children and adults. Arch Oral Biol 2008;53:523-7.
[273] Banderas-Tarabay JA, Zacarias-D'Oleire IG, Garduno-Estrada R, Aceves-Luna E, Gonzalez-Begne M. Electrophoretic analysis of whole saliva and prevalence of dental caries. A study in Mexican dental students. Arch Med Res 2002;33:499-505.
[274] Ellis GL, Auclair PL. Tumors of the Salivary Glands Atlas of Tumor Pathology
Third Series ed. Washington, D.C.: Armed Forces Institute of Pathology; 1996. p. 155-72.
[275] Ellis GL, Auclair PL. Tumors of the Salivary Glands Atlas of Tumor Pathology Third Series ed. Washington, D.C: Armed Forces Institute of Pathology; 1996. p. 324-33.
[276] Simpson RH, Prasad AR, Lewis JE, Skalova A, David L. Mucin-rich variant of salivary duct carcinoma: a clinicopathologic and immunohistochemical study of four cases. Am J Surg Pathol 2003;27:1070-9.
[277] Pramanik R, Osailan SM, Challacombe SJ, Urquhart D, Proctor GB. Protein and mucin retention on oral mucosal surfaces in dry mouth patients. Eur J Oral Sci 2010;118:245-53.
[278] Jankowski JA, Wright NA, Meltzer SJ, *et al.* Molecular evolution of the metaplasia-dysplasia-adenocarcinoma sequence in the esophagus. Am J Pathol 1999;154:965-73.
[279] Van de Bovenkamp JH, Mahdavi J, Korteland-Van Male AM, *et al.* The MUC5AC glycoprotein is the primary receptor for Helicobacter pylori in the human stomach. Helicobacter 2003;8:521-32.
[280] Marshall BJ, McGechie DB, Rogers PA, Glancy RJ. Pyloric Campylobacter infection and gastroduodenal disease. Med J Aust 1985;142:439-44.
[281] Marshall BJ, Warren JR. Unidentified curved bacilli in the stomach of patients with gastritis and peptic ulceration. Lancet 1984;1:1311-5.
[282] Parsonnet J, Friedman GD, Vandersteen DP, *et al.* Helicobacter pylori infection and the risk of gastric carcinoma. N Engl J Med 1991;325:1127-31.
[283] Cave DR. How is Helicobacter pylori transmitted? Gastroenterology 1997;113:S9-14.
[284] Thomas E, Jiang C, Chi DS, Li C, Ferguson DA, Jr. The role of the oral cavity in Helicobacter pylori infection. Am J Gastroenterol 1997;92:2148-54.

[285] Nieuw Amerongen AV, Bolscher JG, Bloemena E, Veerman EC. Sulfomucins in the human body. Biol Chem 1998;379:1-18.

[286] Silva DG, Stevens RH, Macedo JM, *et al.* Higher levels of salivary MUC5B and MUC7 in individuals with gastric diseases who harbor Helicobacter pylori. Arch Oral Biol 2009;54:86-90.

[287] Veerman EC, Bank CM, Namavar F, *et al.* Sulfated glycans on oral mucin as receptors for Helicobacter pylori. Glycobiology 1997;7:737-43.

[288] Bosch JA, de Geus EJ, Ligtenberg TJ, *et al.* Salivary MUC5B-mediated adherence (*ex vivo*) of Helicobacter pylori during acute stress. Psychosom Med 2000;62:40-9.

[289] Schmitz JM, Durham CG, Ho SB, Lorenz RG. Gastric mucus alterations associated with murine Helicobacter infection. J Histochem Cytochem 2009;57:457-67.

[290] Van De Bovenkamp JH, Korteland-Van Male AM, Buller HA, Einerhand AW, Dekker J. Infection with Helicobacter pylori affects all major secretory cell populations in the human antrum. Dig Dis Sci 2005;50:1078-86.

[291] Wyatt JI, Rathbone BJ, Sobala GM, *et al.* Gastric epithelium in the duodenum: its association with Helicobacter pylori and inflammation. J Clin Pathol 1990;43:981-6.

[292] Linden SK, Wickstrom C, Lindell G, Gilshenan K, Carlstedt I. Four modes of adhesion are used during Helicobacter pylori binding to human mucins in the oral and gastric niches. Helicobacter 2008;13:81-93.

[293] Walz A, Odenbreit S, Mahdavi J, Boren T, Ruhl S. Identification and characterization of binding properties of Helicobacter pylori by glycoconjugate arrays. Glycobiology 2005;15:700-8.

[294] Leteurtre E, Zerimech F, Piessen G, *et al.* Relationships between mucinous gastric carcinoma, MUC2 expression and survival. World J Gastroenterol 2006;12:3324-31.

[295] Mulligan RM. Histogenesis and biologic behavior of gastric carcinoma. Pathol Annu 1972;7:349-415.

[296] Goseki N, Takizawa T, Koike M. Differences in the mode of the extension of gastric cancer classified by histological type: new histological classification of gastric carcinoma. Gut 1992;33:606-12.

[297] Hamilton SR, Aaltonen LA. Pathology and Genetics of Tumors of Digestive System. World Health Organization Classification of Tumors. Lyon: IARC Press; 2000. p. 37-67.

[298] Pinto-de-Sousa J, Reis CA, David L, Pimenta A, Cardoso-de-Oliveira M. MUC5B expression in gastric carcinoma: relationship with clinico-pathological parameters and with expression of mucins MUC1, MUC2, MUC5AC and MUC6. Virchows Arch 2004;444:224-30.

[299] Carvalho F, David L, Aubert JP, *et al.* Mucins and mucin-associated carbohydrate antigens expression in gastric carcinoma cell lines. Virchows Arch 1999;435:479-85.

[300] Carrato C, Balague C, de Bolos C, *et al.* Differential apomucin expression in normal and neoplastic human gastrointestinal tissues. Gastroenterology 1994;107:160-72.

[301] Vilkin A, Nudelman I, Morgenstern S, *et al.* Gallbladder inflammation is associated with increase in mucin expression and pigmented stone formation. Dig Dis Sci 2007;52:1613-20.

[302] Lee KT, Liu TS. Altered mucin gene expression in stone-containing intrahepatic bile ducts and cholangiocarcinomas. Dig Dis Sci 2001;46:2166-72.

[303] Ho SB, Shekels LL, Toribara NW, *et al.* Altered mucin core peptide expression in acute and chronic cholecystitis. Dig Dis Sci 2000;45:1061-71.

[304] Vandenhaute B, Buisine MP, Debailleul V, *et al.* Mucin gene expression in biliary epithelial cells. J Hepatol 1997;27:1057-66.

[305] Kim HJ, Kim SH, Chae GB, Lee SJ, Kang CD. Increased expression of mucin 5AC mRNA and decreased expression of epidermal growth-factor receptor mRNA in gallstone patients. Tohoku J Exp Med 2008;214:139-44.

[306] Vilkin A, Geller A, Levi Z, Niv Y. Mucin gene expression in bile of patients with and without gallstone disease, collected by endoscopic retrograde cholangiography. World J Gastroenterol 2009;15:2367-71.

[307] van Klinken BJ, Dekker J, van Gool SA, *et al.* MUC5B is the prominent mucin in human gallbladder and is also expressed in a subset of colonic goblet cells. Am J Physiol 1998;274:G871-8.

[308] Campion JP, Porchet N, Aubert JP, L'Helgoualc'h A, Clement B. UW-preservation of cultured human gallbladder epithelial cells: phenotypic alterations and differential mucin gene expression in the presence of bile. Hepatology 1995;21:223-31.

[309] Smith BF. Gallbladder mucin as a pronucleating agent for cholesterol monohydrate crystals in bile. Hepatology 1990;12:183S-6S; discussion 6S-8S.

[310] Gallinger S, Taylor RD, Harvey PR, Petrunka CN, Strasberg SM. Effect of mucous glycoprotein on nucleation time of human bile. Gastroenterology 1985;89:648-58.

[311] Nakanuma Y, Terada T, Tanaka Y, Ohta G. Are hepatolithiasis and cholangiocarcinoma aetiologically related? A morphological study of 12 cases of hepatolithiasis associated with cholangiocarcinoma. Virchows Arch A Pathol Anat Histopathol 1985;406:45-58.

[312] Su CH, Shyr YM, Lui WY, P'Eng FK. Hepatolithiasis associated with cholangiocarcinoma. Br J Surg 1997;84:969-73.

[313] Ji Y, Fan J, Zhou J, *et al.* Intraductal papillary neoplasms of bile duct. A distinct entity like its counterpart in pancreas. Histol Histopathol 2008;23:41-50.

[314] Kloppel G, Kosmahl M. Is the intraductal papillary mucinous neoplasia of the biliary tract a counterpart of pancreatic papillary mucinous neoplasm? J Hepatol 2006;44:249-50.

[315] Shibahara H, Tamada S, Goto M, *et al.* Pathologic features of mucin-producing bile duct tumors: two histopathologic categories as counterparts of pancreatic intraductal papillary-mucinous neoplasms. Am J Surg Pathol 2004;28:327-38.

[316] Rouzbahman M, Serra S, Adsay NV, *et al.* Oncocytic papillary neoplasms of the biliary tract: a clinicopathological, mucin core and Wnt pathway protein analysis of four cases. Pathology 2007;39:413-8.

[317] Shaib Y, El-Serag HB. The epidemiology of cholangiocarcinoma. Semin Liver Dis 2004;24:115-25.

[318] Park SY, Roh SJ, Kim YN, *et al.* Expression of MUC1, MUC2, MUC5AC and MUC6 in cholangiocarcinoma: prognostic impact. Oncol Rep 2009;22:649-57.

[319] Yaman B, Nart D, Yilmaz F, *et al.* Biliary intraductal papillary mucinous neoplasia: three case reports. Virchows Arch 2009;454:589-94.

[320] Chen TC, Nakanuma Y, Zen Y, *et al.* Intraductal papillary neoplasia of the liver associated with hepatolithiasis. Hepatology 2001;34:651-8.

[321] Chen MF, Jan YY, Chen TC. Clinical studies of mucin-producing cholangiocellular carcinoma: a study of 22 histopathology-proven cases. Ann Surg 1998;227:63-9.

[322] Balague C, Gambus G, Carrato C, *et al.* Altered expression of MUC2, MUC4, and MUC5 mucin genes in pancreas tissues and cancer cell lines. Gastroenterology 1994;106:1054-61.

[323] Balague C, Audie JP, Porchet N, Real FX. *In situ* hybridization shows distinct patterns of mucin gene expression in normal, benign, and malignant pancreas tissues. Gastroenterology 1995;109:953-64.

[324] Zhang S, Zhang HS, Cordon-Cardo C, Ragupathi G, Livingston PO. Selection of tumor antigens as targets for immune attack using immunohistochemistry: protein antigens. Clin Cancer Res 1998;4:2669-76.
[325] Van Seuningen I, Leteurtre E, Pigny P. Mucins in pancreas and hepato-biliary tract. Expression, regulation, biomarkers and therapy. In: Van Seuningen I, editor. The Epithelial mucins: structure/function Roles in Cancer and Inflamatory diseases. Kerala, India: Research Signpost 2008. p. 233-48.
[326] Byrd JC, Bresalier RS. Mucins and mucin binding proteins in colorectal cancer. Cancer Metastasis Rev 2004;23:77-99.
[327] Williams SJ, McGuckin MA, Gotley DC, *et al.* Two novel mucin genes down-regulated in colorectal cancer identified by differential display. Cancer Res 1999;59:4083-9.
[328] Chang SK, Dohrman AF, Basbaum CB, *et al.* Localization of mucin (MUC2 and MUC3) messenger RNA and peptide expression in human normal intestine and colon cancer. Gastroenterology 1994;107:28-36.
[329] Tytgat KM, Buller HA, Opdam FJ, *et al.* Biosynthesis of human colonic mucin: Muc2 is the prominent secretory mucin. Gastroenterology 1994;107:1352-63.
[330] Ogata S, Uehara H, Chen A, Itzkowitz SH. Mucin gene expression in colonic tissues and cell lines. Cancer Res 1992;52:5971-8.
[331] Bartman AE, Sanderson SJ, Ewing SL, *et al.* Aberrant expression of MUC5AC and MUC6 gastric mucin genes in colorectal polyps. Int J Cancer 1999;80:210-8.
[332] Reid CJ, Harris A. Developmental expression of mucin genes in the human gastrointestinal system. Gut 1998;42:220-6.
[333] Buisine MP, Desreumaux P, Leteurtre E, *et al.* Mucin gene expression in intestinal epithelial cells in Crohn's disease. Gut 2001;49:544-51.
[334] Longman RJ, Douthwaite J, Sylvester PA, *et al.* Coordinated localisation of mucins and trefoil peptides in the ulcer associated cell lineage and the gastrointestinal mucosa. Gut 2000;47:792-800.
[335] Wright NA, Pike C, Elia G. Induction of a novel epidermal growth factor-secreting cell lineage by mucosal ulceration in human gastrointestinal stem cells. Nature 1990;343:82-5.
[336] Iida S, Tsuiji H, Nemoto Y, *et al.* Expression of mucin genes and carbohydrate epitopes in 19 human colon carcinoma cell lines. Oncol Res 1998;10:407-14.
[337] Leteurtre E, Gouyer V, Rousseau K, *et al.* Differential mucin expression in colon carcinoma HT-29 clones with variable resistance to 5-fluorouracil and methotrexate. Biol Cell 2004;96:145-51.
[338] Myerscough N, Sylvester PA, Warren BF, *et al.* Abnormal subcellular distribution of mature MUC2 and *de novo* MUC5AC mucins in adenomas of the rectum: immunohistochemical detection using non-VNTR antibodies to MUC2 and MUC5AC peptide. Glycoconj J 2001;18:907-14.
[339] Hoorens PR, Rinaldi M, Li RW, *et al.* Genome wide analysis of the bovine mucin genes and their gastrointestinal transcription profile. BMC Genomics 2011;12:140.
[340] Mall AS, Chirwa N, Govender D, *et al.* MUC2, MUC5AC and MUC5B in the mucus of a patient with pseudomyxoma peritonei: biochemical and immunohistochemical study. Pathol Int 2007;57:537-47.
[341] Ronnett BM, Kurman RJ, Zahn CM, *et al.* Pseudomyxoma peritonei in women: a clinicopathologic analysis of 30 cases with emphasis on site of origin, prognosis, and

relationship to ovarian mucinous tumors of low malignant potential. Hum Pathol 1995;26:509-24.

[342] Leroy X, Devisme L, Buisine MP, *et al.* Expression of human mucin genes during normal and abnormal renal development. Am J Clin Pathol 2003;120:544-50.

[343] Leroy X, Copin MC, Devisme L, *et al.* Expression of human mucin genes in normal kidney and renal cell carcinoma. Histopathology 2002;40:450-7.

[344] N'Dow J, Pearson JP, Bennett MK, Neal DE, Robson CN. Mucin gene expression in human urothelium and in intestinal segments transposed into the urinary tract. J Urol 2000;164:1398-404.

[345] Russo CL, Spurr-Michaud S, Tisdale A, *et al.* Mucin gene expression in human male urogenital tract epithelia. Hum Reprod 2006;21:2783-93.

[346] Zhang S, Zhang HS, Reuter VE, *et al.* Expression of potential target antigens for immunotherapy on primary and metastatic prostate cancers. Clin Cancer Res 1998;4:295-302.

[347] Piludu M, Hand AR, Cossu M, Piras M. Immunocytochemical localization of MG1 mucin in human bulbourethral glands. J Anat 2009;214:179-82.

[348] Legrier ME, de Pinieux G, Boye K, *et al.* Mucinous differentiation features associated with hormonal escape in a human prostate cancer xenograft. Br J Cancer 2004;90:720-7.

[349] Wolf DP, Blasco L, Khan MA, Litt M. Human cervical mucus. I. Rheologic characteristics. Fertil Steril 1977;28:41-6.

[350] Boman F, Buisine MP, Wacrenier A, *et al.* Mucin gene transcripts in benign and borderline mucinous tumours of the ovary: an *in situ* hybridization study. J Pathol 2001;193:339-44.

[351] Giuntoli RL, 2nd, Rodriguez GC, Whitaker RS, Dodge R, Voynow JA. Mucin gene expression in ovarian cancers. Cancer Res 1998;58:5546-50.

[352] Ju W, Yoo BC, Kim IJ, *et al.* Identification of genes with differential expression in chemoresistant epithelial ovarian cancer using high-density oligonucleotide microarrays. Oncol Res 2009;18:47-56.

[353] Mall AS, Tyler M, Lotz Z, *et al.* The characterisation of mucin in a mature ovarian teratoma occurring in an eight year old patient. Int J Med Sci 2007;4:115-23.

[354] Hebbar V, Damera G, Sachdev GP. Differential expression of MUC genes in endometrial and cervical tissues and tumors. BMC Cancer 2005;5:124.

[355] Sonora C, Mazal D, Berois N, *et al.* Immunohistochemical analysis of MUC5B apomucin expression in breast cancer and non-malignant breast tissues. J Histochem Cytochem 2006;54:289-99.

[356] Berois N, Varangot M, Sonora C, *et al.* Detection of bone marrow-disseminated breast cancer cells using an RT-PCR assay of MUC5B mRNA. Int J Cancer 2003;103:550-5.

[357] Varangot M, Barrios E, Sonora C, *et al.* Clinical evaluation of a panel of mRNA markers in the detection of disseminated tumor cells in patients with operable breast cancer. Oncol Rep 2005;14:537-45.

[358] Mukhopadhyay P, Chakraborty S, Ponnusamy MP, *et al.* Mucins in the pathogenesis of breast cancer: implications in diagnosis, prognosis and therapy. Biochim Biophys Acta 2011;1815:224-40.

[359] Schafer G, Hoffmann W, Berry M, Paulsen F. [Lacrimal gland-associated mucins. Age related production and their role in the pathophysiology of dry eye]. Ophthalmologe 2005;102:175-83.

[360] Paulsen F, Corfield A, Hinz M, *et al.* [Tear outflow. Impact of mucins and TFF-peptides]. Ophthalmologe 2004;101:19-24.
[361] Spurr-Michaud S, Argueso P, Gipson I. Assay of mucins in human tear fluid. Exp Eye Res 2007;84:939-50.
[362] Jumblatt MM, McKenzie RW, Steele PS, Emberts CG, Jumblatt JE. MUC7 expression in the human lacrimal gland and conjunctiva. Cornea 2003;22:41-5.
[363] Choi SM, Seo MJ, Lee YG, *et al.* Effects of DA-6034, a flavonoid derivative, on mucin-like glycoprotein and ocular surface integrity in a rabbit model. Arzneimittelforschung 2009;59:498-503.
[364] Carrie S, Hutton DA, Birchall JP, Green GG, Pearson JP. Otitis media with effusion: components which contribute to the viscous properties. Acta Otolaryngol 1992;112:504-11.
[365] Kubba H, Pearson JP, Birchall JP. The aetiology of otitis media with effusion: a review. Clin Otolaryngol Allied Sci 2000;25:181-94.
[366] Ali MS, Pearson JP. Upper airway mucin gene expression: a review. Laryngoscope 2007;117:932-8.
[367] Elsheikh MN, Mahfouz ME. Up-regulation of MUC5AC and MUC5B mucin genes in nasopharyngeal respiratory mucosa and selective up-regulation of MUC5B in middle ear in pediatric otitis media with effusion. Laryngoscope 2006;116:365-9.
[368] Lin J, Tsuprun V, Kawano H, *et al.* Characterization of mucins in human middle ear and Eustachian tube. Am J Physiol Lung Cell Mol Physiol 2001;280:L1157-67.
[369] Lin J, Tsuboi Y, Rimell F, *et al.* Expression of mucins in mucoid otitis media. J Assoc Res Otolaryngol 2003;4:384-93.
[370] Jansen RC, Nap JP, Mlynarova L. Errors in genomics and proteomics. Nat Biotechnol 2002;20:19.
[371] Preciado D, Goyal S, Rahimi M, *et al.* MUC5B Is the predominant mucin glycoprotein in chronic otitis media fluid. Pediatr Res 2010;68:231-6.
[372] Smirnova MG, Birchall JP, Pearson JP. *In vitro* study of IL-8 and goblet cells: possible role of IL-8 in the aetiology of otitis media with effusion. Acta Otolaryngol 2002;122:146-52.
[373] Kawano H, Haruta A, Tsuboi Y, *et al.* Induction of mucous cell metaplasia by tumor necrosis factor alpha in rat middle ear: the pathological basis for mucin hyperproduction in mucoid otitis media. Ann Otol Rhinol Laryngol 2002;111:415-22.
[374] Kerschner JE, Meyer TK, Wohlfeill E. Middle ear epithelial mucin production in response to interleukin 1beta exposure *in vitro*. Otolaryngol Head Neck Surg 2003;129:128-35.
[375] Schousboe LP, Rasmussen LM, Ovesen T. Induction of mucin and adhesion molecules in middle ear mucosa. Acta Otolaryngol 2001;121:596-601.

Mucins – Potential Regulators of Cell Functions, 2013, 316-386

CHAPTER 7

Gel-Forming Mucin MUC6

Abstract: The MUC6 mucin plays an essential role mainly in homeostasis of the gastrointestinal tract, and in pathogenesis of many diseases associated with the gastrointestinal, respiratory, and male and female urogenital systems. The structure of the *MUC6* gene and the mechanisms that regulate its activity are analyzed in this chapter. The biochemical and biophysical properties of the MUC6 glycoprotein, its biosynthesis, posttranslational modifications, and expression in various cells and tissues are discussed.

Keywords: MUC6, glycoprotein, mucin, structure, biosynthesis, expression.

7.1. *MUC6* GENE: CHROMOSOMAL LOCALIZATION AND EVOLUTIONARY HISTORY

The *MUC6* gene, encoding the main human gastric gel-forming mucin, belongs to a cluster of four gel-forming mucin genes – *MUC6*, *MUC2*, *MUC5AC* and *MUC5B* – located on human chromosome 11p15.5 [1]. *MUC6* occupies the very 5'-position at the telomeric end of a 400 kb cluster that is located 38.5 kb upstream to the *MUC2* gene. *MUC6* and *MUC2* have a head-to-head orientation, indicating their opposite-directed transcription [2, 3]. Such orientation of two genes suggests that the upstream 5'-UTR comprising the regulatory sequences of *MUC6* and *MUC2* may contain common regulatory elements that determine the reciprocal pattern of expression [3].

MUC6 and *MUC2* together with the two other gel-forming genes (*MUC5AC* and *MUC5B*) are clustered in a highly recombination active region [1]. The recombination rate in the 11p15.5 MUC gene region is fifty times greater than the genome average for corresponding distance [4]. Such a rate reflects the recombination hot spots in this gene complex and may also point to mechanisms operating in the evolution of 11p15.5 genes. All four genes show sequence similarity in their nontandem repeat domains, and are thought to have arisen from a common ancestor by gene duplication [5] and intragenic sequence duplication [6, 7]. Interestingly, although *MUC6* is adjucent to *MUC2* on the chromosome and demonstrates a high level of sequence similarity to it, the two genes belong, according to Desseyn *et al.* [8], to different subfamilies. The phylogenic analysis

Joseph Z. Zaretsky and Daniel H. Wreschner

performed by Desseyn *et al.* [8] revealed three subfamilies of mucin genes: one of *MUC6* alone; one of animal mucins *FIM-B.1*, *PSM* and *BSM*; and one of *MUC2*, *MUC5AC* and *MUC5B*. In contrast to the other three human gel-forming genes, *MUC6* does not contain a 108 aa Cys-subdomain and the D4-B-C-domains found in *MUC2*, *MUC5AC* and *MUC5B* (Fig. **1**). C-terminal CK-domain is present, however, in all four genes that make up the 11p15.5 chromosome cluster [8, 9].

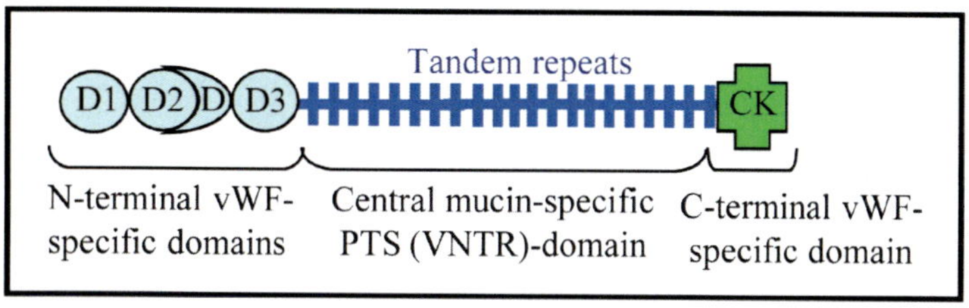

Figure 1: Domain structure of the MUC6 gel-forming mucin (based on the data from [3, 9, 12]).

Comparison of the nucleotide sequences and genomic organization of the human gel-forming mucin genes suggests that, during evolution, *MUC6* and *MUC19* lost the sequences encoding three C-terminal domains, namely D4-, B- and C-domains, present in other gel-forming mucin genes between the central TR-containing domain and the 3'-end located CK-domain. Hence, while all human gel-forming genes and the gene encoding von Willebrand factor (vWF) have a common ancestor, the evolutionary history of *MUC6* apparently is differed in some steps from that of the other gel-forming mucin genes. According to the hypothesis suggested by Desseyn *et al.* [8], a common ancestor underwent duplication that resulted in two progenitor genes: the progenitor of the vWF gene and the progenitor of all 11p15.5 cluster genes (and probably of *MUC19* gene). During evolution, a part of the cluster's progenitor molecules developed into *MUC2, MUC5AC and MUC5B*, while another part by loosing of Cys and D4-B-C-domains had been transformed into *MUC6*. Apparently, retention of the C-domain in progenior led to occurrence of *MUC19* (Fig. **2**).

Despite some structural differences, all four genes, *MUC6, MUC2, MUC5AC* and *MUC5B*, encode secreted gel-forming mucins [10, 11] with specific individual patterns of expression under physiological conditions and in pathology [12-14]. *MUC6* is highly expressed in the pyloric glands of stomach, pancreas and

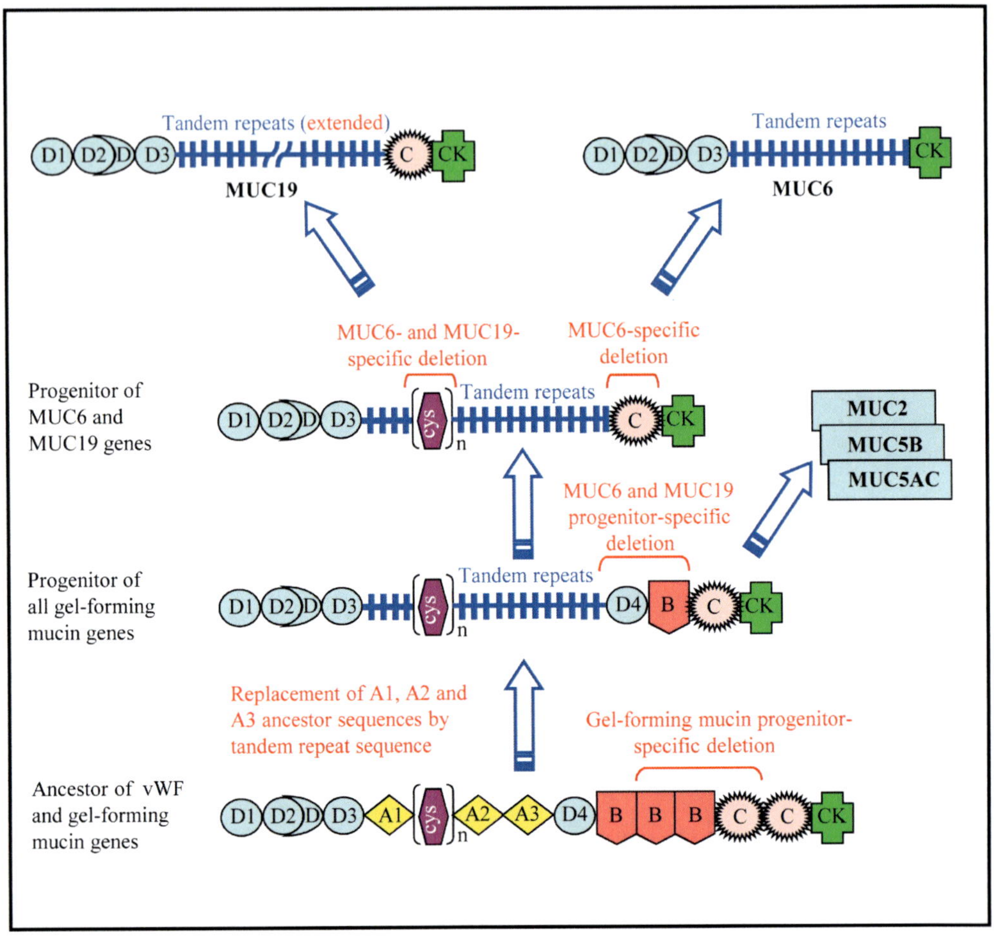

Figure 2: Evolutionary pathway of the MUC6 and other gel-forming mucin glycoproteins (based on the data reported in [8]).

gallbladder. *MUC2* is the main mucin expressed in the intestine. The *MUC5AC* gene is highly expressed in the bronchus and gastric surface epithelium, and *MUC5B* is mainly expressed in the submaxillary and bronchus glands [15, 16]. As noted by Pigny *et al.* [1], the order of four gel-forming mucin genes on the 11p15.5 chromosomal locus corresponds to the order (in terms of the anterior-posterior axis) of epithelial organs where these genes are preferentially expressed during early stages of development [17]. The gastric MUC6 mucin is thought to play a unique role in the protection of the gastrointestinal tract against acid pH and proteases. In addition to the general need of mucosal surfaces to be protected from microorganisms, desiccation, and mechanical damage, the gastric epithelium needs specific cytoprotection from low pH and proteolytic enzyme digestion [12, 18, 19]. This protection is provided by the gastric mucin MUC6.

7.2. STRUCTURE OF *MUC6* GENE AND DOMAIN COMPOSITION OF MUC6 MUCIN

In 2004, Rousseau *et al.* [3] wrote: "The full sequences of the mucins and their gene structures are not all publicly available despite the fact that the first cDNAs were isolated some 15 years ago... The mucin genes were for a long time located in holes in the gene map and have been among the last to be included in the human genome sequences... Even with the availability of complete genomic sequences from the Human Genome Mapping Project and sophisticated gene prediction programs, it is not trivial to extract protein domain and exon structure information for the mucin gene".

The gap in our knowledge of the genomic organization of the *MUC6* gene is significant to this day. Part of the *MUC6* gene was cloned in 1993 [12], but its full sequence remains elusive. Nevertheless, the several fragments that have been sequenced (accession nombers: AY312160, U977698, AC083984, AC139749, GI29135707 and GI30794567) [3, 9, 12] show that *MUC6* is composed of an N-terminal 30 exons encoding D1-, D2-, D'- and D3-domains, one large central exon 31 that encodes a mucin-specific domain containing variable numbers of tandem repeats (VNTRs), and two exons encoding the C-terminal CK-domain. The TR-containing domain of the MUC6 mucin protein is less well characterized than that of the MUC2 glycoprotein. Conflicting numbers of TRs have been reported: in one study (clones AC139749 and AC083984), 4-5 poorly conserved repeats were identified [12], while in another study the shortest of the alleles examined contained at least 15 repeat units [20]. The *MUC6* gene appears to be unusual with respect to the length of an individual repeat unit: 507 bp, which is several times longer than any other mucin repeat units cloned to date and described in the literature [12, 18].

The uniqueness of the *MUC6* gene is also evidenced by its VNTR-polymorphism. As shown by Vinall *et al.* [20], the four genes of the mucin gene 11p15.5 cluster can be divided into two groups: the *MUC6* and *MUC2* genes at the telomeric side of the cluster, which show considerable variation in length due to VNTR; and the *MUC5AC* and *MUC5B* genes at the centromeric side of the cluster, characterized by small (*MUC5AC*) or no (*MUC5B*) allelic variation in number of TR. According to these authors [20], the interruption of TRs of MUC5AC, and especially of MUC5B, by cysteine-rich domains limits VNTR mutations and stabilizes the structure of the gene, while the long stretches of simple repeats in MUC6 and MUC2 are conducive to VNTR mutation. *MUC6* polymorphism has been associated with different diseases, including gastric cancer and *Helicobacter pylori* (*H. pylori*) infection [20-24].

The N-terminal region of the MUC6 mucin located upstream to the TR-containing domain shows 36.2% amino acid identity with the corresponding region of MUC2 [3], and contains the same vWF domains (D1, D2, D' and D3) as other gel-forming mucins [25, 26]. In contrast, the C-terminal region of MUC6 mucin differs significantly from MUC2, MUC5AC and MUC5B. Importantly, while by C-terminal region *MUC6* is differed from other 11p15.5 cluster genes, it is very similar to *MUC19* located on chromosome 12. The close relationship between these two genes was confirmed by phylogenetic analysis [27, 28]. The genomic organization of human *MUC6* is highly conserved through evolution [3, 29, 30]. Its mouse homolog, *Muc6*, clustered at mouse chromosome 7 band F5 together with the other gel-forming mucin genes, has the same genomic structure and orientation as its human counterpart *MUC6* [31]. *Muc6* consists of 33 exons spread over 28.8 kb from the putative ATG codon to the poly(A) signal sequence including the large (11.9 kb) TR-containing exon 31. Although N- and C-terminal regions of mouse *Muc6* and human *MUC6* genes exhibit high levels of identity and similarity, the TR-containing domain sequences in the two species are quite different. As noted by Desseyn and Laine [30], the nonconservation of TR sequences may reflect species-dependent functions. The absence in human MUC6 and mouse Muc6 of cysteine-rich subdomains interrupting the TR-containing domain in MUC5AC and MUC5B might be due to the necessity to maximize resistance of the mucin core peptide to proteolysis by minimizing naked parts of the core peptide [9]. Phylogenetic analysis of *Muc6* confirmed the biochemical and molecular biology data and demonstrated the evolutionary relationship between the mouse *Muc6* and human *MUC6* genes [30, 31].

7.3. MUC6 MUCIN: BIOSYNTHESIS AND POSTTRANSLATIONAL MODIFICATIONS

The biogenesis of the MUC6 mucin has not been sufficiently studied, although some aspects of the biosynthesis, glycosylation and structural transformation of MUC6 apomucin during processing have been relatively well investigated. Taking into account that all gel-foming mucins of human and animal origin demonstrate high levels of sequence and structure conservation, one may assume that biogenesis of all gel-forming mucin glycoproteins, including MUC6 mucin, display similar spacio-temporal parameters and follow the same stages of maturation [32-37]. Based on this assumption, the following information on biosynthesis of MUC6 mucin will be supplemented with data from studies on other gel-forming mucins.

7.3.1. Biosynthesis and Structural Transformations of MUC6 Precursor

The pulse-chase approach has shown that the first molecules of the human MUC6 apomucin (400 kDa-precursor) occur in cells 30 min after pulse labeling [38]. The size of the MUC6 precursor correlates well with that of the MUC6 mRNA, estimated to be 9.5 kb [12]. The dynamics of MUC6 precursor biosynthesis corresponds to that of human MUC2 and MUC5AC mucins [32] and rat gastric mucin, a rat homolog of human MUC6 mucin [39]. Dekker and Strous [34] described the steps of rat gastric mucin precursor biosynthesis, which is initiated in the rough endoplasmic reticulum (RER) by translation of mucin mRNA and is completed in the Golgi complex by posttranslational modifications. The authors consider that the process of the biosynthesis of this particular mucin can be assumed to hold for other members of the gel-forming mucin group. They showed that synthesis of the mucin's precursor in the RER is followed by oligomerization, and that oligomerization of precursor monomers is generally completed by the development of di- and trimers within 1 hour of chase in pulse-chase experiments. Importantly, under reducing conditions, all oligomeric forms undergo depolymerization, resulting in accumulation of monomeric mucin molecules.

Co-translational N-glycosylation appears to be the key event determining the efficiency of precursor oligomerization. The requirement of N-glycans for mucin oligomerization was demonstrated by experiments in which N-glycosylation was blocked by specific inhibitors [34]. Under physiological conditions, transition of monomers into di- and trimers occurred within 30 minutes, while it took more than 6 hours when N-glycosylation was inhibited [34]. Interestingly, the monomeric nonglycosylated precursor molecules were retained in the endoplasmic reticulum, while N-glycosylated oligomerized precursors were transported to the Golgi complex for further modifications and maturation. The efficiency of transportation does not depend on the type of oligomers (di-, tri- or tetramers). It appears that an absolute requirement for efficient RER-to-Golgi transport is the presence of N-linked glycans. However, it is unlikely that the N-glycans *per se* contribute to this process, as the configuration of N-linked oligosaccharides was shown to be irrelevant for the transport of mucin molecules. The N-linked glycans most likely participate in protein folding to facilitate the intra- and intermolecular disulfide bond formation important for oligomerization, which in turn regulates the transportation mechanism [34, 40-44].

Studies have been conducted to elucidate the biophysical and biochemical properties of the MUC6 mono- and oligomers. The isopycnic density of human MUC6 mucin, as determined by CsCl/guanidinium chloride density gradient centrifugation, is 1.45g/ml [45]. Interestingly, MUC6 glycoproteins isolated from different parts of the same stomach (cardia, fundus and antrum) have slightly different densities, indicating a different degree of glycosylation of MUC6 in different stomach glands. MUC6 also appears to be nonhomogeneous based on charge density. MUC6 mucin molecules can be separated into neutral, moderately and highly charged populations [45] that differ in degree of sulfation and sialyc acid content. Analysis of charge density of heavily glycosylated porcine stomach mucin, a porcine analog of human MUC6, allowed Yakubov *et al.* [46] to show that "mucin molecule possesses both positively and negatively charged domains, so they must exist in a state of balanced charge regulation, which may be realized through multiple mesoscale dipole moments within the mucin domains".

Several important structural features of the MUC6 mucin were established by physical methods. Electron-microscopy revealed that both monomeric and oligomeric forms of rat homolog of human MUC6 appear to be filamentous molecules that vary in length from 300 to 3000 nm. The 300 nm molecule represents the smallest unit of the mucin oligomer, as the length of the oligomer is a multiple of the monomer length [47]. Interestingly, electron-microscope images visualized oligomers as small globular domains separated from each other by a linear monomer unit. As Dekker *et al.* [47] pointed out, these globular structures represent attachment sites of two monomeric subunits that are linked together end-to-end in the oligomers. The occurrence of oligomers longer than dimers indicates that both the N- and C-termini of the monomer molecules are able to form intermolecular disulfide bonds.

The molecular structure of porcine gastric mucin Orthana (genetically close to human MUC6 type [48]) was recently studied by a number of complementary methods including atomic force microscope, transmission electron microscope, dynamic light scattering, static light scattering, zeta potential rheology, microrheology, small-angle neutron scattering, small-angle X-ray scattering, pulsed-gradient spin-echo nuclear magnetic resonance, optical waveguide lightmode spectroscopy, and ball-on-disk tribometry [46, 48-55]. Taken together, these studies showed that not only do oligomers show globular structures at the point of subunit attachment, but hydrophobic globules are present also at the N- and C-termini of monomer molecules. As shown by Yakubov and colalborators [48, 50], the structure of porcin gastric mucin in bulk solutions corresponds to a daisy-chain random coil, with the majority of molecules adopting a dumbbell-like configuration, which supposes the presence of two hydrophobic globules per monomer chain separated by a heavily glycosylated spacer. In other words, this mucin adopts a double-globular comb structure. As pointed out by the authors, the globular structure is determined most probably by electrostatic and hydrophobic interactions inside the nonglycosylated subunits of the mucin molecule, and occurs when the energies of these interactions are comparable to the surface-free energy of the globules – a state that occurs mainly due to the presence of extensively hydrated carbohydrate comb segments [46, 48]. Thus, the discrepancy between recent findings and those reported by Dekker *et al.* [47] almost twenty years ago can be explained by the sensitivity of the methods used in the studies.

It is well known that at low pH and high concentration, mammalian gastric mucins undergo aggregation and form a gel that is essential for protection of the stomach from auto-digestion [49, 52]. Study of this process by atomic force microscope provided direct visual evidence of gastric mucin aggregation at pH 4 and gel formation at pH 2. At pH 5-7, mucin molecules have extended filamentous configuration, whereas at pH 4 and below they tend to cluster together. While the sugar side-chains are important in maintaining the solubility and extended structure of mucin molecules, they are not directly involved in the aggregation process, as removal of sugar side chains did not affect aggregation and gel formation. According to Hong *et al.* [49] and Cao *et al.* [56], the process of aggregation and gel formation is based on the hydrophobic and electrostatic interactions between less glycosylated or nonglycosylated portions of the mucin molecule. Thus, the physiological demand to keep stomach pH at the low values necessary for food digestion correlates well with the ability of gastric mucins to form a protective gel at these values.

Defense of gastric epithelia by mucin gel is associated with adsorption of the gel on the epithelial surface – a process much less studied than gel-formation itself. McColl *et al.* [51] recently analyzed the dependence of mucin adsorption on temperature. Using porcine gastric analog of human MUC6 mucin, they measured the kinetics of mucin adsorption and desorption on a silica-like hydrophilic surface across a temperature range of 25-60°C. The analysis showed that the area occupied per molecule diminishes with increasing temperature both in the bulk and adsorbed states. The mucin contracted parallel with temperature increase, placing it in the class of "natively open" proteins. It retains the contracted conformation upon adsorption. Due to the conformational rearrangement, the specific interaction energy governing desorption greatly increases with temperature, resulting in a regulated surface coating. The conclusion from McColl *et al.'s* study [51] is that the initial mucin-epithelial surface interaction creates "strain" within the mucin molecule, which is used to drive internal conformational changes that facilitate desorption.

7.3.2. O-Glycosylation of MUC6 Polypeptide Precursor

O-glycosylation of the large central Ser/Thr-rich domain is one of the important steps in processing and maturation of a mucin molecule, in general, and the MUC6 protein, in particular. It helps the apomucin polypeptide to acquire the properties of a functionally competent mucin molecule. Studies of mucins and other glycoproteins showed that O-glycosylation – the addition of O-linked

GalNAc residues to Ser and/or Thr residues of the polypeptide chain – occurs mainly in the *cis*-Golgi, although initial O-glycosylation can occur co-translationally in the RER [57-60]. As shown by Dekker and Strous [34], the initial O-glycosylation of rat homolog of MUC6 is a continuous process that begins on nascent polypeptides in the RER, whereas the major addition of GalNAc takes place in a late RER compartment and the *cis*-Golgi. These data unambiguously indicate that oligomerization of monomers precedes the O-glycosylation stage of the MUC6 mucin maturation. This is in agreement with findings that MUC2, MUC5AC, MUC5B and MUC6 apoproteins all form dimers prior to major O-glycosylation [32, 61].

The first step of the mucin-type O-glycosylation pathway under physiological conditions is, as mentioned above, the linking of a GalNAc residue to serine and/or threonine residues by one of the polypeptide N-acetyl-galactosaminyltransferases, which are characterized by both substrate and tissue specificity [62-64]. The GalNAc residue added to polypeptide substrate creates a Tn-antigen (GalNAc-O-Ser/Thr), which is the innermost O-linked structure usually masked by sugar residues added at the following steps of the physiological process of mucin O-glycosylation. However, under pathological conditions such as malignant transformation, mucin O-glycosylation often stops at the stage of Tn-antigen biosynthesis, resulting in abnormal expression of this antigen by cancer cells. Of note, MUC6 mucin overexpressed in breast carcinoma epithelium, tissue that does not normally express this mucin, may function as a substrate for over-production of Tn-antigen [62]. The synthesis of Tn-antigen most probably occurs in the RER as initial O-glycosylation of MUC6 mucin. Further processing of the oligosaccharide chains linked to apomucin scafolding polypeptide includes biosynthesis of O-glycan core structures. Addition of β1,3-linked galactose and β1,3-linked N-acetylglucosamine results in the formation of Core1 and Core3, respectively. Core2 and Core4 structures are further modified by galactosylation, sialylation, fucosylation, sulfation or elongation reactions [65-67]. At the last stage of normal mucin O-glycosylation process, terminal sugar is added to the core structures. There are two antigens at the terminal epitopes that cap the oligosaccharide chains of MUC6 mucin, Lewis^x and Lewis^y, which belong to the type 2 structures [68, 69]. Interestingly, the type 1 antigens, Lewis^a and Lewis^b, are associated with another gastric mucin, MUC5AC glycoprotein [16, 69, 70]. All the intermediate and terminal reactions mentioned above take place in the Golgi compartments. The transition from naked apomucin to the fully glycosylated mature form is a ralatively slow process that requires not less than 3 hours [39]. In the rat, the first mature analog of the human MUC6 mucin is detected 4 hours from the beginning of mucin native polypeptide synthesis.

Thus, the biogenesis of the MUC6 mucin, although not fully studied, emerges as a multi-step process that is initiated at ribosomes of RER, continues in the Golgi compartments, and is completed like in other gel-forming mucins in the secretory vesicles secreted from the cell. Neither precise mechanisms of MUC6 incorporation into a vesicle nor the processes occurring inside a vesicle have been investigated and require further study.

7.4. REGULATION OF *MUC6* GENE EXPRESSION

Generally, two groups of mechanisms regulate expression of a gene: one group acts at the genomic DNA structures (promoter, introns) through interaction of transcription factors with specific *cis*-elements; another group operates at the epigenomic level *via* interplay of modified histones and/or methylation of specific sites in the genomic DNA. Although *MUC6* was cloned 18 years ago [12], the mechanisms regulating its expression have not been thoroughly studied. The promoter region of the *MUC6* gene was cloned and sequenced only in 2005 [71], and since then only a few papers have appeared on the regulation of *MUC6* expression. Many questions concerning *MUC6* transcription and its regulation remain to be answered.

7.4.1. MUC6 mRNA: General Characteristics

One of the issues needing clarification is the nature of MUC6 mRNA and the mechanisms of its processing. It is known that the mRNAs of all the gel-forming mucin genes are unusually large and have a high degree of polydispersity [72-77]. This polydispersity may be related to rapid turnover or alternative splicing or expression of several related genes encoding transcripts of comparable lengths; or, it may result from specific properties of the mucin transcripts due to a particular conformation of a large TR-containing domain that perturbs the properties of the molecules during electrophoresis or centrifugation. Debailleul *et al.* [78] studied this issue and showed that all gel-forming genes, including *MUC6*, express mRNAs that are stable and not degradable, but are unusually large (14-24 kb), comparable in size to the biggest mRNAs ever found in eukaryotes, including titin mRNA (23 kb) and human nebulin mRNA (20.7 kb) [79, 80]. Depending on the source (different individuals), the size of MUC6 mRNA varies between 16.5 and 18 kb, indicating allelic variation of these transcripts related to the VNTR polymorphisms. Interestingly, MUC6 mRNAs isolated from different regions of the stomach (fundus and antrum) of one individual, were 17.5 kb and 16.5 kb, respectively. The authors [79] attributed these results to expression of the two different alleles that could encode two main precursors of different sizes, like rat gastric and colon apomucins [39, 81]. Further study is needed, however, to rule out or confirm the alternative splicing of the primary transcript common to both locations. In the previous chapters we noted that the alternative splicing mechanisms are used much more rarely by secreted mucin transcripts compared to membrane-bound mucins. However, it is difficult to imagine that these mechanisms are not exploited for realization of the multiple structural and functional potentials embedded in the nucleotide sequences of the gel-forming mucin genes. Future studies may uncover splice variants of mRNAs transcribed from the gel-forming mucin genes, including *MUC6*; such studies will require a very wide spectrum of cells and tissues and different biological states and conditions.

7.4.2. *MUC6* Promoter: Role in Regulation of *MUC6* Gene Expression

Analysis of 1307 bp of the 5'-flanking region of the *MUC6* gene (Fig. **3**) showed that the *MUC6* promoter contains *cis*-elements specific for TATA-box (-35 to -29 bp) and several transcription factors such as NFκB (-173 to -164 bp) and Sp1/3 (-530 to -521 and -847 to -838 bp) [71]. Luciferase assays coupled with gradual deletion of these *cis*-elements showed that NFκB as well as Sp1 and Sp3 transcription factors actively up-regulate *MUC6* transcription. The transcription start site (TSS) is located 28 bp downstream to the TATA-box. Involvement of the Sp1 and Sp3 *cis*-elements in transcriptional regulation was demonstrated by electrophoretic mobility shift assay. The active role of NFκB transcription factor was demonstrated by the fact that TNFα, one of the cytokines that activates NFκB, induced strong up-regulation of *MUC6* transcription. Still more evidence of the positive role of NFκB in regulation of *MUC6* expression came from Inoue *et al.* [82], who showed that *MUC6* expression in human tracheal surface epithelial cells and submucosal glands was up-regulated by infection with rhinovirus (RV), a known activator of NFκB [83]. It is possible that, analogous to RV-induced *MUC5AC* transcription, RV-triggered *MUC6* expression is also mediated through the RV-initiated MEK-MAPK/ERK/NFκB pathway *via* activation of the NFκB transcription factor.

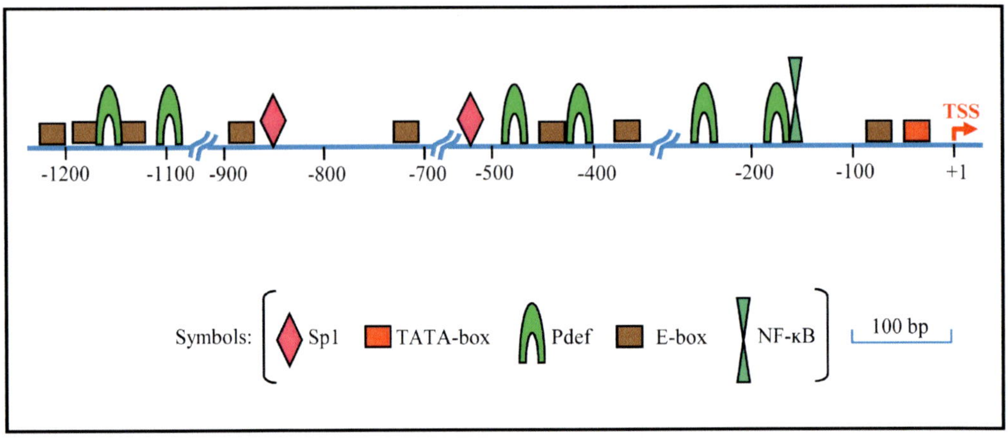

Figure 3: Transcription factor *cis*-element map of the *MUC6* promoter (based on the data reported in [71]).

Although Sakai *et al.* [71] found in the *MUC6* promoter only three transcription factor-specific *cis*-elements (NFκB, Sp1 and Sp3), the length of the promoter region and small number of already known *cis*-elements suggest that the *MUC6* promoter contains other not yet experimentally identified *cis*-elements that may participate in *MUC6* transcriptional regulation. For example, the *cis*-elements specific for estrogen receptors (ER) must also be present in the *MUC6* promoter, as steroid hormone was shown to up-regulate expresssion of *MUC6* in an ER-positive cell line but not in an ER-negative one [84]. Apparently, *cis*-STAT1 element(s), a potential binding site for STAT1 transcription factor activated by interferon γ, is also present in the *MUC6* promoter, an assumption based on Kang *et al.*'s finding [85] that interferon γ activates both the STAT1 protein and MUC6 mucin expression in NCI-N87 cells.

Other interesting data apparently relevant to regulation of *MUC6* expression were recently reported by Sekine *et al.* [86]. The authors found that Hath1, a member of the HLH transcription factor family, up-regulated transcription of *MUC6* and *MUC5AC* genes in gastric cells. It is known that Hath1 transcription factor acts through binding to the E-box (CANNTG) in the promoter region of the target gene [87]. An examination of the 1307 bp *MUC6* promoter sequence [71] performed by us (J. Zaretsky, unpublished data) showed that the *MUC6* promoter contains 9 perfectly conserved E-boxes: -64/-59 bp, -373/-368 bp, -433/-428 bp, -710/-705 bp, -885/-880 bp, -1162/-1156 bp, -1187/-1181 bp, -1236/-1231 bp, and -1266/-1261 bp. These findings and Sekine *et al.*'s observations [86] point to the possible role of Hath transcription factor in transcriptional regulation of the *MUC6* gene. Sekine *et al.* [86] found that knockdown of the Hath1 gene by means of siRNA in human gastric cell line TGBC11TKB significantly decreased the expression of both *MUC6* and *MUC5AC*. In contrast, overexpression of Math1, a mouse homolog of Hath1, in Hath1-negative cells enhanced *MUC6* and *MUC5AC* expression. Collectively, these data show that Hath1 is indeed a powerful regulator of *MUC6* and *MUC5AC* activities. Hath1 acts as a tumor suppressor in stomach. The direct connection of the *MUC6* expression with Hath1 puts *MUC6* in the same tumor suppression pathway in which Hath1 is an active participant.

In contrast to the Hath1 transcription factor, the Sox2 transcription regulator, a member of the transcription factor family containing a Sry-like high-mobility group (HMG) box, has the potential to suppress transcription of the *MUC6* gene [88]. Interestingly, whereas Hath1 uniformly up-regulates both *MUC6* and *MUC5AC*, Sox2 acts in a differential manner: it down-regulates *MUC6*, but up-regulates *MUC5AC*. Apparently, such differential response of two gastric genes to Sox2 results from the properties of Sox2 as a key regulator of gastric differentiation in mammals [89, 90]. It stimulates expression of gastric differentiation markers including *MUC5AC*, simultaneously repressing expression of other gastric genes such as *MUC6* and that encoding pepsinogen [91].

Recently, one more repressor of the *MUC6* gene was discovered by Xiao *et al.* [92] who reported that Hedghog signaling pathway regulates expression of many genes, including *MUC6*. Sonic Hedgehog (Shh), one of the Hedgehog signaling pathway effectors, is known to be a powerful regulator of epithelial cell functions and differentiation in the adult stomach [93-97], and its loss or down-regulation triggers a number of molecular events associated with epithelial-to-mesenchymal transition of gastric epithelial cells [97]. The up-regulation of *MUC6* expression, a mucous neck cell lineage marker [92], is one of these events, indicating the reciprocal relationship between *Shh* and *MUC6*.

Among different families of transcription factors involved in regulation of *MUC6* expression, the Ets protein family plays a specific role. This became obvious after discovery of the Spdef (Pdef) transcription factor, a member of the Ets protein family, whose activity is restricted to specific epithelial cell subsets and differentiation stages [98-101]. Spdef (Pdef) is unique among Ets transcription factors for its DNA-binding specificity and restricted expression in hormone-regulated cells of prostate and other organs [99]. The consensus sequence of the Pdef-binding site is NG/A/CCA/C<u>GGA**T**</u>G/C/A, with core sequence GGAT, which differs from the GGA**A**-sequence canonical for Ets transcription factors [99]. Horst *et al.* [102] recently demonstrated the requirement of Spdef for mucus gland cell function in the gastric antrum. They showed that Spedf is highly expressed in mouse stomach and regulates 713 genes in mouse antrum, including *Muc6* and *Tff2*, the two specific markers of antral mucus gland cells [103, 104]. Loss of Spdef dramatically reduced expression of *Muc6* in the mouse antrum [102].

When the promoter sequence of the human *MUC6* gene was analyzed for the presence of Spdef-specific binding sites with the core sequence GGAT, six Spdef-specific binding sites were found (J. Zaretsky, unpublished data). One of the *cis*-Spdef sites located at -172/-169 bp overlaps the *cis*-NFκB binding site, suggesting competition of two transcription factors for the same promoter position as a possible instrument in regulation of *MUC6* transcription. Thus, although only three transcription factor binding sites specific for NFκB and Sp1/Sp3 have been experimentally established in the *MUC6* promoter [71], additional *cis*-elements that might interact with transcription factors – such as Sox2, Shh, Hath1 and Spdef – are apparently also present in the 5'-flanking sequence of the *MUC6* gene and their function may be identified in future studies.

7.4.3. Epigenetic Regulation of *MUC6* Gene Activity

The importance of epigenetic mechanisms in transcriptional regulation of many genes has been established. However, their role in regulation of *MUC6* expression is not clear. Only one comprehensive study [105] that of Van Seuningen's group has been carried out on epigenetic mechanisms in transcriptional regulation of the clustered gel-forming mucin genes, including *MUC6*. This study revealed strong epigenetic regulation of *MUC2* and *MUC5B*, and rare and weak epigenetic regulation of *MUC5AC*. *MUC2* was regulated by site-specific DNA methylation associated with establishment of a repressive histone code, while *MUC5B* gene silencing was kept under control by hypermethylation of the promoter region. There was no causal connection between methylation of *MUC6* promoter and *MUC6* gene activity. Despite the high number of CpG sites throughout the *MUC6* promoter, and identification of several key methylated sites at -806, -651, -282, -271 and -72 bp, *MUC6* could not be reactivated by demethylation agents in any of the cancer cell lines studied. There was a specific pattern of cytosine methylation in LS174 and HT-29 STD cells correlated with a specific level of *MUC6* expression in these cells, but their treatment with 5-aza did not change the expression profile of the gene [105]. Thus, *MUC6* appears to be the exception among the genes clustered at 11p15. At present, there is no explanation for the peculiar behavior of *MUC6*; it may be associated with the peculiarities in its evolutionary history which is differed from that of other three genes in the cluster [8]. More studies are neeeded to determine whether and how epigenetic mechanisms regulate transcription of the *MUC6* gene.

7.5. EXPRESSION OF *MUC6* GENE UNDER PHYSIOLOGICAL CONDITIONS

Expression of the *MUC6* gene in different organs and tissues under physiological conditions has been extensively investigated. Its role in the physiology of epithelial adult tissues is determined by the extremely high resistance of the MUC6 mucin glycoprotein to a number of noxious factors, such as acid pH, proteases, bile and chemical agents [18]. Organs and cells that must function in contact with noxious factors have developed defense mechanisms that include the fast and robust biosynthesis and secretion of the MUC6 mucin in response to stress induced by the indicated factors. Among the affected organs are stomach, gallbladder, pancreas, and some regions of the female and male reproductive tracts. The airway epithelium also comes into contact with noxious agents, but the MUC6 glycoprotein appears to serve a minor function in defense of the respiratory tract. In the following sections, the available information regarding expression of *MUC6* gene in embryonal, fetal and adult tissues under normal physiological conditions is presented.

7.5.1. Expression of *MUC6* Gene in the Respiratory Tract

Analyses of *MUC6* expression in human embryonal airway tissues performed by Reid *et al.* [106] and Buisine *et al.* [107] turned up no MUC6 mRNA expression in lung or trachea epithelium at any gestational age. The adult airway tissues also appeared to be MUC6-negative, as reported by Reid *et al.* [106]. However, de Bolos *et al.* [69] did observe *MUC6* expression in adult tracheal and larynx surface epithelium, albeit at a low level, and Bernacki *et al.* [108] found its expression up-regulated in bronchial and nasal cells upon differentiation *in vitro*. Although MUC6 is not considered an important secreted mucin in airway epithelial cells, the data reported by Bernacki *et al.* [108] suggests involvement of this mucin in differentiation of respiratory epithelial cells. Taken together, the results indicate the site-specific and differentiation-dependent manner of *MUC6* expression in the respiratory tract. It is of note that in all the cited studies except that of Bernacki *et al.* [108], *MUC6* expression was studied by *in situ* hybridization or immunohistochemical methods. These methods are less sensitive than RT-PCR, leading one to speculate that analysis of *MUC6* expression by RT-PCR might give a more accurate picture of its expression in the respiratory tract.

7.5.2. Expression of *MUC6* Gene in the Gastrointestinal Tract

Numerous studies of mucin gene expression in the gastrointestinal tract led investigators to the conclusion that these genes are expressed in the organs of the digestive system in a strongly selective manner. An important observation, from the physiological and developmental points of view, was made by Van Klinken *et al.* [109]. In a study of the expression of the gel-forming mucins along the longitudinal axis of the human gastrointestinal tract, *MUC6* was expressed in the stomach but not in the small or large intestine, demonstrating region-specific expression of the gene along the tract. Audie *et al.* [13] also noted that mucin genes, including *MUC6*, were expressed throughout the gastrointestinal tract in a site-specific manner. All these data suggest selectivity in expression of *MUC6* in the gastrointestinal tract. As noted by Cornfeld [110], "each region of the gastrointestinal tract has characteristic functional requirements and the properties of the mucus produced at each site are adopted to cope with these functions".

<u>*Salivary glands*:</u> Salivary glands are important components of the digestive system. The role of the MUC6 mucin in the physiology and pathology of salivary

glands is poorly understood. According to Alos *et al.* [111], *MUC6* is never expressed in normal salivary glands, while its expression is often observed in salivary neoplasias. However, Kusafuka *et al.* [112] described a rare benign epithelial tumor of salivary gland, cystadenoma, histologically close to normal salivary tissues, that expressed MUC6 mucin in an amount detectable by immunohistochemistry. The role of this mucin in the physiology and pathology of the salivary glands needs further study.

Esophagus: Like in salivary glands, the expression of MUC6 mucin in esophagus has also been poorly studied. The few studies available give conflicting results. Warson *et al.* [113] reported that none of the secretory mucins is expressed in the normal stratified epithelium of the esophagus, while Arul *et al.* [114] detected MUC6 mRNA and protein in deep glands and crypts of normal esophagus. Additional studies are required to determine the mode of *MUC6* expression in normal esophageal epithelium, and the role of this mucin, if any, in the physiology of this region of the gastrointestinal tract.

Stomach: According to data from a number of laboratories, the MUC6 mucin glycoprotein is a well-documented marker of stomach epithelial cells [12, 103, 115, 116]. Although this statement is generally correct, it is restricted mainly to the post-natal normal adult stomach. During embryogenesis and fetal development of the gastrointstinal tract, other types of cells also express MUC6 mRNA and protein, albeit transiently. Thus, *MUC6* is expressed in small intestine and pancreas from 13 weeks of gestation. Interestingly, the expression of *MUC6* in pancreatic cells occurs early in development and even precedes its expression in stomach, where it appears only at 23 weeks of gestation. In embryonic and adult stomach, MUC6 mucin is expressed in the neck mucous cells but not in the surface mucous cells, which express MUC5AC [117].

Intestine: Conflicting results on *MUC6* expression in the embryonic and fetal intestine were obtained by different groups of investigators. Reid and Harris [117] found MUC6 mRNA in the small intestine from 13 weeks through 23 weeks of gestation, albeit at very low levels. Buisine *et al.* [118] could not detect MUC6 mRNA in primitive gut and embryonic and fetal intestine from 6.5 to 27 weeks of gestation. No expression of *MUC6* was found in normal human ileal mucosa [118], whereas it was observed in one-week-old pigs in conventional jejunal and ileal mucous epithelial cells [119]. MUC6 mRNA was not detected in colon at any gestational age.

Gallbladder: The MUC6 mucin is the most consistently expressed mucin glycoprotein in the gallbladder [120]. According to Ho *et al.* [121], 86% of the analyzed normal gallbladder specimens from humans expressed MUC6 in an amount equal to 28% of that expressed in normal stomach. Interestingly, MUC6 displays a gradient in the level of expression in the gallbladder, but whether the content of the MUC6 mucin in bile is different in the upper and deeper regions of the gallbladder folds is not known [122].

Pancreas: MUC6 mRNA occurs in the pancreas early in gestation (at 13 weeks), but only in the epithelium of small ducts and in developing acini. In the later stages, the distribution of the MUC6 transcripts follows the developing pancreatic ducts with expression in the epithelium of small ducts, centroacinar cells and acini [117]. MUC6 is one of the major high-molecular weight secreted mucins expressed in the post-natal and adult pancreas [123-125], with a consistently strong expression in pancreatic centroacinar cells [124]. Pancreatic ducts, on the other hand, express MUC6 heterogeneously, with the most consistent expression in the interlobular ducts and less in the intralobular and intercalated ducts. The main pancreatic duct expresses a moderate level of MUC6 mucin while pancreatic islets are MUC6-negative as follows from the numerous studies [125, 126]. Nagata *et al.* [127] found that periductal glands in normal pancreas also express MUC6. Thus, according to the cited studies, all epithelial surfaces of the normal pancreas that have constant physical contact with bile synthesize and secrete the MUC6 mucin.

7.5.3. Expression of *MUC6* Gene in the Female Reproductive Tract

Mucins play an important role in reproductive function and defense of the female reproductive tract [128-131], where they demonstrate organ- and region-specific expression. For example, MUC9 glycoprotein is expressed mainly in the oviduct [132, 133], whereas MUC6 is dominant in the uterus, specifically in the endometrium and endocervix [128, 129]. As defined by Andersch-Bjorkman *et al.* [129], cervix is an "open entry into the female endometrial and abdominal cavities". The strategic location of this "entry" dictates the composition and expression of a set of mucins that have to protect the uterus from pathogens and noxious agents and allow the sperm to enter at the ovulation stage. The set includes, in addition to other mucins, three gel-forming mucins – MUC6, MUC5AC and MUC5B. These glycoproteins possess similar characteristics and are expressed at the same locations – a redundancy that ensures protection of the reproductive process. However, each of the mucins also fulfills a specific function in the protection reactions. MUC6 defends vaginal vault and endocervix from acid vaginal secretion, similar to its functions in the stomach and seminal vesicles where it protects their inner surfaces from luminal low pH-fluid rich in proteases [103].

Under physiological conditions, expression of the *MUC6* gene in the female reproductive tract has been detected in endocervix and endometrium during the proliferative stage of the menstrual cycle, but not in the secretory period. Also in post-menopausal women, only epithelial cells of endocervix and endometrium produce MUC6 glycoprotein. No *MUC6* expression was detected in vagina, ectocervix or fallopian tubes [128]. Apparently, MUC6 mucin produced by endocervical epithelium is secreted to ectocervix and vagina to provide them with a defensive mucin layer.

7.5.4. Expression of the *MUC6* Gene in the Male Urogenital Tract

The *MUC6* gene is expressed also in the male urogenital organs [134-140]. Expression of *MUC6* in kidney, bladder, prostate, epididymis and seminal vesicles has been better studied than in other organs of the tract but still insufficiently [136-140]. Nevertheless, some data have been obtained that provide basic information about *MUC6* expression in the male urogenital tract during embryonic and fetal development and in normal adults.

<u>Kidney</u>: The role of MUC6 mucin as a protecter of epithelial cells from various harmful agents is well known. It may also be an important participant in epithelial organogenesis. Reid and Harris [117] showed that MUC6 mRNA is expressed transiently in the nephrogenic zone of the kidney in the early mid-trimester of development in humans. The MUC6 transcripts were detected in the epithelium of ureteric buds at 13 weeks. Interestingly, expression of *MUC6* could not be

detected at 14-16 weeks of gestation, but was expressed again between 17 and 23 weeks, albeit at lower levels. Leroy *et al.* [140] also detected expression of the MUC6 mucin early in embryogenesis: it was expressed in the nephrogenic zone of ureteric bud as early as 9.5 weeks of gestation. After this period, MUC6 protein was constantly detected at the tips of the ureteric buds until 24 weeks. From weeks 18 to 24, MUC6 was also expressed in the collecting ducts, but after 30 weeks its expression was weak and focal [140]. These data may be of significance since, as pointed out by Reid and Harris [117], "the pattern of expression of MUC6 is coincident with other genes that are known to be involved in signaling between the collecting duct epithelium and the surrounding mesenchyme". Thus, MUC6 mucin may be directly involved in epithelial/mesenchymal signaling. As for the possible role of MUC6 in epithelial organogenesis, the presence in the MUC6 apomucin molecule of the CK-domain is of great importance, as a highly similar domain in the TGFβ molecule plays an essential role in epithelial differentiation and organogenesis. Although the precise function of the MUC6 CK-domain is unknown, it could be involved in cellular differentiation and epithelial/mesenchymal transition.

Studies conducted by Leroy and co-workers [140] showed that *MUC6* is expressed not only in embryonic and fetal kidney, but also in normal adult kidney, although at a low level.

Seminal vesicles and ejaculatory ducts: Immunohistochemistry and *in situ* hybrydization assays showed that under physiological conditions ejaculatory epithelial ducts and seminal vesicles express MUC6 mRNA and protein at substantial levels, while prostate and bladder are MUC6-negative [124, 134, 135].

Prostate: Little is known about the expression of mucins, including MUC6, in prostate, and the available data are controversial [135]. According to some studies [124, 134], prostate epithelial tissues do not express MUC6 mucin under normal physiological conditions, while others [103, 137] did demonstrate MUC6 expression at substantial levels in normal prostate. These inconsistencies require examination of the issue by more sensitive methods. Of note, Nakajima *et al.* [137] recently made an important observation about expression of MUC6 mucin in prostate. In this organ, they found dissociation in expression of MUC6 apomucin and GlcNAcα1-4Galβ-R antigen linked to MUC6 protein: only MUC6 molecules, which lack the GlcNAcα1-4Galβ-R antigen, are expressed in normal and malignant prostate tissues. Taking into account that MUC6 and MUC5AC mucin molecules expressed in stomach and pancreas do carry the GlcNAcα14Galβ-R epitopes, the dissociation described by Nakajima *et al.* [137] might be prostate-specific. However, according to analysis of mucin gene expression in the human male urogenital tract performed by Russo *et al.*'s [135] and N'Dow *et al.* [138], MUC6 mRNA is not expressed in prostate, as well as in bladder, urethra, *vas deferens* and foreskin.

<u>*Testis, epididymis, seminal vesicles and seminal plasma*</u>: A review of *MUC6* expression in the male urogenital tract would be incompleted without critical analysis of the studies describing expression of the gene in testis, epididymis, seminal vesicles and seminal plasma. However, there are no comprehensive reports of *MUC6* expression in these organs, only sporadic, unsystematic studies on limited numbers of samples are obtainable. Thus, no reliable conclusions can be drawn regarding MUC6 production and secretion in these organs without further research. Nevertheless, the available data do give some idea about the potential of these organs to express MUC6 mucin. Russo *et al.* [135] detected expression of the MUC6 mRNA in 1 out of 3 testicular tissue specimens and in 1 out of 2 samples of epididymis, using RT-PCR. Reid and Harris [117] confirmed the potential of epididymis to express *MUC6* when they found low levels of MUC6 mRNA by *in situ* hybridization in the epididymis epithelium at weeks 16.5, 18 and 19.5 of gestation. Interestingly, no *MUC6* expression was detected elsewhere in the testis at any gestational age in this study. More reliable results were obtained when *MUC6* expression was analyzed in seminal vesicles. According to Russo *et al.* [135], 5 out of 5 seminal vesicles were positive for MUC6 mRNA; and the MUC6 protein was detected throughout the cytoplasm of seminal vesicle epithelial cells by immunohistochemistry as opposed to other types of cells which were MUC6-negative. These results are in agreement with those of Leroy *et al.* [134] who also detected a 100% rate of MUC6 expression (10 MUC6-positive samples out of 10 tested) in seminal vesicles by immunohistochemistry. Western blot analysis performed by Russo *et al.* [135] showed that all studied samples of seminal plasma from fertile men were MUC6-positive, although the levels of MUC6 expression

were different in different samples. Immunoblot analysis showed that MUC6 and MUC5AC glycoproteins were consistently present in the fertile seminal plasma, although individual differences in MUC6 concentration were noted [135]. As emphasized by Russo *et al.* [135], it is not surprising that the gel-forming mucins are constituents of seminal plasma, since they facilitate sperm mobility and protect its viability, especially along the urethra, analogously to the cervical mucins that protect sperm from the vaginal environment and assist their transport through the female reproductive tract [139]. Interestingly, the expression of *MUC6* was shown to coincide with the expression of *MUC5AC* only in the testis, whereas in all other male reproductive organs tested a reciprocal relationship was observed between expressions of these two gel-forming mucin genes [135]. In comparison with the endocervical epithelium, which excessively expresses three gel-forming mucins (MUC5AC, MUC5B and MUC6) [128], no redundancy in production of these glycoprotein was detected in any cells of the male urogenital tract [135]. Thus, it appears that the male reproductive organs express MUC6 in a tissue-specific manner.

Summing up the studies on *MUC6* gene expression in different organs under physiological conditions, one may conclude that MUC6 can reasonably be considered a stomach-specific mucin, although its physiological expression is also observed in other organs. It clearly serves specific functions in the development of stomach, uterus, seminal vesicles, epididymis, testis and kidney, and probably also in airway epithelium. The precise functions of the MUC6 mucin in embryonic and fetal development of these organs as well as in adults remain to be clarified.

7.6. EXPRESSION OF *MUC6* GENE IN PATHOLOGY

Various organs, tissues and cells that are generally MUC6-negative or weakly-positive under physiological conditions often express this mucin at high levels during different stages of the pathological processes.

7.6.1. Expression of *MUC6* Gene in the Gastrointestinal Tract

Expression of *MUC6* has been intensively studied in pathologically changed gastrointestinal organs, where the aberrant expression of the mucin may contribute to the pathogenesis of various diseases.

Salivary glands: As noted above, in contrast to other organs of the gastrointestinal tract, expression of *MUC6* in salivary glands is poorly investigated. There are no publications on *MUC6* expression in the inflammatory diseases of the salivary glands, and only a few papers on the *MUC6* gene expression in malignant salivary glands. Kusafuka *et al.* [141] described *MUC6* expression in low-grade salivary duct carcinoma of the parotid gland. Interestingly, most cases of this type of salivary carcinoma resemble mammary duct carcinoma, which also expresses MUC6 mucin [84]. Salivay duct carcinomas often express MUC6 mucin at a high rate [142], while mucoepidermal carcinomas of salivary glands are MUC6-positive in only about 30% of cases [111]. Expression of the *MUC6* gene in the salivary glands under pathological conditions requires further study.

Esophagus: Among the pathological conditions described in esophagus, Barrett's esophagus attracts special interest because of its relationship to esophageal cancer. It is considered a precancerous lesion. Chronic inflammation of esophageal mucosa mediated by different damaging factors, including gastro-esophageal reflux (GER), may lead to development of esophageal squamous cell carcinoma (SSC) or adenocarcinoma (ADC) [143, 144]. A number of authors [143, 145, 146] pointed out that esophageal ADC usually develops from a premalignant lesion termed Barrett's esophagus, which is defined as columnar metaplasia of distal esophagus most often due to chronic GER-mediated inflammation [147]. Besides GER, other factors such as exogenous luminal nitric oxide may also induce metaplastic transformation of esophageal epithelium [148]. Barrett's esophagus results from metaplastic replacement of the normal squamous epithelium by columnar epithelial cells which may be of intestinal, cardial or fundic types. The intestinal type is characterized by the presence of the goblet cells and is thought to have carcinogenic potential, while two other types of metaplasia are not considered premalignant lesions [143, 149]. Progression from Barrett's esophagus to ADC involves histological transition from intestinal metaplasia to low- and high-grade displasia followed by the architectural structure specific for adenocarcinoma. Alterations in expression of some mucins, including MUC6 [143, 150], are among the molecular events accompaning malignant transformation of esophageal mucosa.

Glickman and co-workers [150] showed that in nondisplastic Barrett's esophagus, *MUC6* is expressed in 90% of cases both in goblet and nongoblet columnar epithelium. With progression from metaplasia to indefinite, low- and high-grade dysplasia and further to adenocarcinoma, there is a gradual but significant decrease in *MUC6* expression from 100% at the early stages of dysplasia to 57% in the high-grade dysplasia and further to 27% in adenocarcinoma. The same dynamics of *MUC6* expression in related studies were observed by Guillem *et al.* [143] and Arul *et al.* [114]. In Guillem *et al.*'s study [143], *MUC6* was expressed in 60% of gastric type metaplasia cases, in 38% of cases with intestinal type metaplasia without dysplasia, in 36% of cases with low- and high-grade dysplasia, and in only 16% of cases with Barrett's adenocarcinoma. Arul *et al.* [114] noted that all specimens with Barrett's adenocarcinomas tested in their study were MUC6-negative. Warson *et al.* [113] also found frequent expression (>90%) of the *MUC6* gene in areas of incomplete intestinal metaplasia in study of a representative group of patients with Barrett's esophagus. These authors emphasized that the levels of *MUC6* expression in the ulcerated highly malignant samples were significantly lower than in metaplastic samples without dysplasia. Low levels of expression in esophageal adenocarcinomas are characteristic for other gel-forming mucins as well.

Thus, according to the cited papers, progression of Barrett's metaplasia to dysplasia and adenocarcinoma is associated with progressive decrease in *MUC6* expression. In contrast, Yamamoto *et al.* [151] observed high levels of *MUC6* and *MUC5AC* expression in 93.3% of patients with Barrett's adenocarcinomas, and Chaves *et al.* [152] observed high frequency of *MUC6* expression (93-100%) in columnar cells independent of development of metaplasia and/or dysplasia. Importantly, expression of MUC6 mucin in the goblet cells was highly dynamic as the disease progressed: from MUC6-negative at the stage preceding dysplastic changes to 60% of the cells expressing MUC6 after progression to the dysplastic stage. Moreover, Barrett's adenocarcinoma cells also expressed MUC6 mucin at high frequency (86.6%). As pointed out by the authors, gastric characteristics, assessed in particular by *MUC6* expression, are present in both the metaplastic and neoplastic cells, strongly suggesting the involvement of the metaplastic elements with gastric phenotype in malignant transformation. This study also showed that the two cellular metaplastic subtypes, columnar nongoblet and goblet cells, are related to malignant transformation of Barrett's esophagus.

Thus, two conflicting groups of results emerged from investigations of *MUC6* expression during progression of Barrett's esophagus to Barrett's adenocarcinoma. One group found negative dynamics of *MUC6* expression along the axis Barrett's esophagus-neoplastic-dysplasia-Barrett's adenocarcinoma [113, 114, 143, 150], while the other group [151, 152] reported the opposite trend. Do these results really contradict each other, or do they just belong to different subsets of patients and therefore represent different subtypes of Barrett's diseaese? Further studies on larger populations of patients using a combination of methods and criteria are needed to answer this question.

Several reports appeared recently describing expression of *MUC6* in some rare diseases associated with Barrett's esophagus. Kuroda *et al.* [153] studied expression of *MUC6* and *MUC5AC* in a patient with an extremely rare disease, hepatoid adenocarcinoma arising in Barrett's esophagus. They found that this tumor, like Barrett's adenocarcinoma, was MUC6- and MUC5AC- negative [114]. Asthana *et al.* [154] described a rare case of esophageal polypoid dysplasia with gastric phenotype and focal intramucosal carcinoma associated with Barrett's esophagus; histopathology of the lesion revealed an exuberant polypoid gastric epithelium with areas of low- and high-grade dysplasia, and focal intramucosal carcinoma. The proximal and distal ends of the lesion contained foci of intestinal metaplasia consistent with Barrett's esophagus without dysplastic changes. This lesion produced MUC6, MUC1 (focally) and MUC5AC (diffusely) mucins, but was MUC2-negative. Such a profile of mucin expression is consistent with a gastric foveolar phenotype previously described as extremely well-differentiated adenocarcinoma [155].

These two examples of rare adenocarcinomas arising in Barrett's esophagus demonstrate opposite expression profiles of the *MUC6* gene in malignancies developed on a cell background similar to that typical of Barrett's esophagus. These examples, combined with the aforementioned studies, point to the histological heterogeneity of Barrett's esophagus that results in variable expression of mucin genes. Better classification of Barrett's disease subtypes would enable studying more homogeneous groups of patients and lead to a better understanding of the mechanisms governing expression of mucin genes in Barrett's disease.

<u>*Stomach*</u>: Gastric mucins are critical cytoprotective proteins synthesized by stomach epithelial cells [77]. Numerous alterations in expression of gastric

mucins have been described in inflammatory, metaplastic and malignant lesions of stomach [156]. Transition from normal epithelium to intestinal metaplasia and further to adenocarcinoma is accompanied by qualitative and quantitative alterations of mucin glycoproteins. These alterations contribute to changes in cell growth regulation, immune recognition and cell adhesion, thereby influencing the invasive and metastatic potentials of cancer cells [12, 136, 157-161].

a) Helicobacter pylori infection: *Helicobacter pylori (H. pylori)* is an ubiquitous Gram-negative bacterium that colonizes stomach and adheres to the mucus and/or the epithelium [162-169]. *H. pylori* causes gastric diseases such as chronic gastritis, peptic ulcer, and intestinal metaplasia, which may evolve to gastric carcinoma [169-172]. More than 50% of the world population is infected by *H. pylori* [169, 173], but only 3% of the infected individuals develop advanced stages of the pathological conditions, such as peptic ulcer and gastric carcinoma [174]. *H. pylori* is largely associated with surface mucus layer and is rarely found in the deeper portion of the gastric mucosa [165].

The adherent gastric mucous layer is composed of alternating layers of MUC5AC and MUC6 mucin glycoproteins [175]. The relative contribution of each mucin to the gastric mucous layer is not equal. The laminated surface mucus is present in both *H. pylori*-infected and noninfected stomachs, indicating that MUC5AC and MUC6 proteins remain segregated within the mucous gel in a laminated linear arrangement independent on *H. pylori* infection [175]. The physico-chemical properties of mucins, and therefore the properties of the gel layer, are largely dependent on the amount of O-glycans attached to the VNTR region of each mucin. Interestingly, there is substantial variation in the number of TRs (VNTR-polymorphism) leading to the individual variation in the number of O-glycan side chains on mucin molecules. MUC6 shows extensive VNTR variation, whereas MUC5AC is characterized by a moderate degree of polymorphism [20, 176]. The polymorphism leads to production of mucin polypeptides of substantially different lengths and degrees of glycosylation [20, 78]. Thus, the VNTR-determined polymorphism may affect the protective properties of the mucins and, in turn, the suspectibility to *H. pylori* infection. Indeed, Nguyen *et al.* [23] found short *MUC6* alleles associated with higher suspectibility to *H. pylori* infection.

Progression of *H. pylori*-induced diseases is restricted by a deeper portion of the gastric mucosa. In many patients, glandular atrophy appears to be prevented by mucins secreted by gland mucous cells, such as mucous neck cells and pyloric gland cells located in a deeper portion of the gastric mucosa. Moreover, inhibitory activity directed against *H. pylori* is associated with α1,4-N-acetylglucosamine residues attached to the MUC6 mucin expressed in the deeper layer of mucosa. This inhibitory activity is regarded as a natural antibiotic of the MUC6 mucin [169, 177]. From these studies, it follows that the increase in expression of MUC6 mucin may protect gastric mucosa from *H. pylori*-induced diseases. Correctness of this assumption was confirmed by several studies [178-180]. Byrd *et al.* [179] reported that *H. pylori*-infected stomachs exhibit reciprocal up-regulation of *MUC6* associated with down-regulation of *MUC5AC*. Matsuzwa *et al.* [180] also observed an increase in MUC6 glycoprotein production in response to *H. pylori* infection. Both groups of authors noted a correlation between the *MUC6* up-regulation and the degree of infection, and a decrease in *MUC6* expression to almost normal levels after eradication of *H. pylori*. Nam *et al.* [181] showed that the protective effect of geranylgeranylacetone, a chemical agent that defends gastric mucosa from different types of damage [182-184], is associated with activation of *MUC6* expression.

However, several studies contradict these findings [167, 185, 186]. Wang and Fang [185] found that both *MUC5AC* and *MUC6* expression was significantly down-regulated in the *H. pylori*-positive pericancerous mucosa. Kim *et al.* [187] reported that *H. pylori* infection induced down-regulation of MUC5AC mucin, while the bacteria had no effect on *MUC6* expression. No differences in *MUC5AC* and *MUC6* expression were observed by Morgenstern *et al.* [186] between *H. pylori*-positive and *H. pylori*-negative patients. Different sensitivities of the methods used in the above studies can explain the conflicting data. To better understand the published results, one must take into account the divergent functions attributed to MUC5AC and MUC6 mucins [164, 165, 188, 189]. MUC5AC allows *H. pylori* colonization, whereas MUC6 is toxic to bacteria [164] – an observation in line with data showing that MUC5AC functions as a receptor molecule for *H. pylori* [188, 189]. These findings suggest that superficial mucin MUC5AC allows bacteria to move, while glandular mucin MUC6 hampers the movement of bacteria and thereby contributes to elimination of bacteria [165]. As shown by Marques *et al.* [162], MUC5AC and MUC6 mucins are expressed differently in different *H. pylori*-induced diseases, and the impact of each mucin in the pathogenesis of *H. pylori*-associated pathological conditions depends on the topography of MUC5AC and MUC6 glycoprotein expression.

b) Gastric cancer: The pathogenesis of gastric cancer is a multifactorial process in which gastric mucins MUC6 and MUC5AC play different but significant and specific roles. The risk of developing gastric cancer is closely related to *H. pylori*-associated progressive gastric inflammation [190-193]. Intestinal metaplasia is one of the lesions identified in the cascade of events that precede development of gastric carcinoma [156]. This type of metaplasia reflects the replacement of the gastric mucosa by intestinal mucosa type epithelium. Intestinal type gastric carcinomas are thought to evolve through a multiple step process that starts with superficial gastritis and progresses through atrophy, dysplasia and finally carcinoma [170]. Analysis of the expression profile of gel-forming mucins in *H. pylori*-infected preneoplastic and neoplastic human gastric epithelium showed that the decreased expression of gastric mucins MUC5AC and MUC6 and *de novo* expression of the colonic mucin MUC2 are characteristic of *H. pylori*-induced intestinal metaplasia [194-196]. A decrease in MUC6, MUC5AC and MUC1 mucin expression in parallel with an increase in expression of MUC2, MUC3 and MUC4, normally not expressed in gastric epithelium, was described by Ho *et al.* [77] as a typical feature of neoplastic transformation in the stomach. Interestingly, not only decreased expression of MUC6, but also its aberrant expression in foveolar cells of antrum and stomach body, were observed in *H. pylori*-positive patients with duodenal and gastric ulcer, dysplasia and stomach cancer [168]. Recently, an interesting observation was made by Linden *et al.* [197]: mucin expression and glycosylation patterns are similar in children and adults infected by *H. pylori,* while pediatric *H. pylori* infection is not accompained by the aberrant expression of MUC6 and MUC2 glycoproteins seen in adults. Zheng *et al.* [161] found a correlation between *MUC6* down-regulation and progression of gastric carcinoma associated with poor prognosis.

Recent data confirmed the previous observations of a reciprocal relationship between decreased expression of *MUC6* and increased risk of gastric adenocarcinoma [198-200], pointing to a possible tumor-suppression function of the MUC6 mucin glycoprotein.

Liver, gallbladder and biliary duct tree: Mucins play an important role in epithelial cytoprotection in the gallbladder. Mucin glycoproteins are strong polyanions and therefore may serve as a barrier for diffusion of biliary solutes and bile acids [121]. Comparison of MUC6 mucin expression with that of other gel-forming and membrane-bound mucins in gallbladder shows that MUC6 mucin is consistently expressed in the majority (86%) of normal gallbladder specimens [120, 121]. These data are in agreement with the findings of strong MUC6 mRNA expression in normal biliary tree epithelium, gallbladder biliary epithelial cells, and intrahepatic bile ducts [122].

a) Inflammatory diseases: In the inflamed galldbladder, MUC6 mucin was highly expressed in 96% of mild chronic cholecystitis specimens, in 90% of chronic and 81% of acute cholecystitis samples [121]. The high level of MUC6 mRNA and protein expression is retained in biliary tree also in hepathlithiasis [122, 201]. The increased expression of MUC6 glycoprotein associated with high expression of other mucins (MUC1-4 and MUC5B) was detected in gallbladder with calcium bilirubinate stones and cholesterol-based stones [120]. Sasaki *et al.* [202] noted that increased MUC6 apomucin expression is characteristic also for reactive biliary epithelium in chronic viral hepatitis.

In summary, the cited studies show that inflammatory diseases of gallbladder are associated with increased expression of mucins, in general, and MUC6, in particular. However, the high level of MUC6 expression in normal gallbladder epithelium [121, 122] makes it difficult to evaluate the precise role of small alterations in MUC6 expression in the pathogenesis of these diseases. Is it a causative factor that directs and accelerates inflammation in the organ, or is it a consequence of the inflammatory process? The third possibility is that inflammation simply does not significantly affect the high level of MUC6 expression normally observed in gallbladder epithelium. More studies are needed to clarify this isssue.

b) Malignant tumors: Although hepato-biliary neoplasias have been extensively studied, only a few types of hepato-biliary benign and malignant tumors have been analyzed with regard to *MUC6* expression [203, 204]. These tumors include pyloric-gland type adenoma of the gallbladder [203, 204], gallbladder adenocarcinoma [205], biliary epithelial metaplasia and dysplasia [206] and extra- and intra-hepatic cholangiocarcinoma [207, 208]. Recently, Nagata *et al.* [209] analyzed expression of several mucin genes, including *MUC6*, in normal gallbladder epithelium and pyloric gland type adenoma of the gallbladder and found that *MUC6* was overexpressed in 100% of adenomas and in only 50% of normal specimens. Of note, the level of *MUC6* expression was low in the rest 50% of normal specimens. High frequency of *MUC6* expression in pylori-gland type adenoma of gallbladder was observed previously by Chang *et al.* [203, 204]. Nakanuma *et al.* [206] found aberrant overexpression of MUC6, MUC2 and MUC5AC mucins in biliary epithelial gastroenteric metaplasia.

Cholangiocarcinoma accounts for about 3% of all gastrointestinal cancers [210]. It arises from ductal epithelium of the biliary tree. Depending on localization, the tumors are classified as an intra- or extrahepatic. Thuwajit *et al.* [208] found high levels of *MUC6* expression in 37% of patients with cholangiocarcinomas, but the expression levels decreased dramatically in the advanced metastatic stage of the disease. The same relationship between *MUC6* expression and metastatic process in cholangicarcinoma was reported by Aishima *et al.* [207]. Importantly, high expression of *MUC6* was significantly correlated with a 5-year survival rate in Thuwajit and colleagues' study [208]. Park *et al.* [205] also detected high levels of *MUC6* expression especially in the well-differentiated cholangiocarcinomas. These data are in line with Sasaki *et al.*'s report [211] of a high level of *MUC6* expression in 89% of patients with well-differentiated cholangiocarcinoma compared with 42% of patients with poorly-differentiated cholangiocarcinoma. In a study of mucin expression in intrahepatic cholangiocarcinoma, Aishima *et al.* [207] came to the conclusion that high level of *MUC5AC* expression correlates with tumor invasiveness, whereas the expression of *MUC6* correlates with the degree of tumor differentiation and is inversely proportional to invasiveness. These findings are in agreement with previous studies on pancreatic adenocarcinoma [212] and gallbladder carcinoma [203]. Chang *et al.* [203] reported that aggressiveness of gallbladder carcinoma was higher in patients with low expression of MUC6 mucin. Matsukita *et al.* [213] consider that MUC6 mucin prevents the spreading of cancer cells, resulting in less aggressive tumor growth.

Although the studies cited above suggest the potential of MUC6 mucin to deter the growth of neoplasms, contradictory data have also been published. For instance, biliary neoplasia with extensive intraductal spread associated with liver cirrhosis did not express MUC6 but did produce the MUC5AC mucin [214]. Of note, this neoplasia also did not express MUC2 glycoprotein, a mucin with tumor-suppressive activity in colon cancer detected in mouse model [215]. In another study [216], MUC6 and MUC2 were expressed in only 3.8% of cholangiocarcinomas, while MUC5AC was expressed in 46.1% of the tumors tested. The coincidence in expression of MUC6 and MUC2 observed in this study is of interest as it points to a possible interdependence or common regulation of expression of the two genes, located "head to head" on the 11p15.5 chromosome.

c) Liver cancer: Hepatocellular carcinoma (HCC), a hepatic primary parenchymal cell cancer, is an aggressive tumor characterized by high dissemination power and low survival rate [217]. HCC is less studied in the context of mucin gene expression than intrahepatic cholangiocarcinoma. This issue is investigated relatively well in regard to *MUC1* expression [218-221], whereas expression of *MUC6* in HCC has been examined in only a few studies [222, 223]. As pointed out by Sasaki *et al.* [222, 223], there is no difference in MUC6 expression profile in HCC and cholangiocarcinomas. Cholangiocarcinomas and hepatocellular cholangiocarcinomas share the same *MUC6, MUC3, MUC5AC* and *MUC7* expression profiles, which may indicate similar or common histogenesis. Clearly, more data are needed before the contribution of MUC6 mucin to pathogenesis of HCC and associated pathologies can be properly evaluated.

Pancreas: Pancreas is one of the few organs that express the *MUC6* gene under physiological conditions [14, 117, 127, 212, 224]. However, the available information regarding *MUC6* expression in the pathological pancreas is controversial and therefore this issue needs further investigation.

a) Inflammatory diseases: No systematic analyses of the role of MUC6 mucin in inflammatory diseases of pancreas have been undertaken. What is known about

MUC6 expression in pancreatitis comes from comparison of its expression in normal and inflamed tissues adjacent to malignant lesions.Very few cases of a ‘pure’ pancreatitis have been examined. Expression of *MUC1-MUC7* mucin genes in pancreatic adenocarcinoma, chronic pancreatitis and normal pancreas were analyzed by Andrianfahanana *et al.* [224]: only *MUC6* and *MUC5B* were overexpressed in pancreatitis specimens, while other mucin genes were expressed at very low levels (*e.g. MUC1*) or strongly repressed (*e.g. MUCMUC2, MUC3A, MUC4, MUC5AC* and *MUC7*). Wada *et al.* [225] also noted the elevated level of *MUC6* expression in pancreatic mucous gland specimens affected by fibrosis or inflammation. More research is needed to determine the functions of the *MUC6* gene in the pathogenesis of acute and chronic pancreatitis.

b) Cystic fibrosis: Pancreas is one of the target organs of cystic fibrosis (CF), which leads to destruction of pancreatic tissue and, as a result, to pancreatic insufficiency. It is generally accepted that obstruction of the pancreatic ducts by condensed secretion is the cause of the tissue destruction. The MUC6 mucin appears to contribute heavily to this destructive process by its over-expression in pancreatic epithelial and centroacinar cells. The *MUC6* gene has been shown to be highly expressed in fetal, postnatal and adult pancreas [117, 124, 226]. The localization and time of the *CFTR* gene expression coincides with that of *MUC6* in fetal and postnatal pancreas [125, 227, 228] – a finding that may explain how the abnormal expression of two genes contributes to the pathophysiology observed in the CF pancreas. As pointed out by Reid and co-workers [125], a key defect in the CF pancreas is deficiency of secreted bicarbonate ions that are essential for the normal flow of pancreatic duct secretion. This deficiency and reduced water content in the CF pancreatic duct fluid increase the density of the duct contents, resulting in deposition of material in the small intralobular ducts. MUC6 mucin is a significant constituent of these ductal deposits [125]. The role of *MUC6* in pathogenesis of the pancreatic form of CF was studied in the *Cftr*-knockout mice model [123]: the expression of *Muc6* was significantly higher in pancreatic tissue of Cftr-/- mice than in their wild-type littermates. A large quantity of Muc6 protein was detected in centroacinar cells in these animals – perhaps explaining why acinar diameters are higher in *Cftr-/-* mice than in *Cftr+/+* mice. Taken together, these data suggest that MUC6 glycoprotein likely participates in the formation of pancreatic plugs in CF patients.

c) Malignant diseases: Pancreatic cancer has a relatively high incidence and an extraordinarily poor prognosis [229, 230]. Approximately 90% of exocrine tumors of pancreas are adenocarcinomas of ductal cell origin [230]. The relationship between hyperplastic and dysplastic epithelial lesions of the pancreatic ducts and invasive ductal carcinomas (IDC) is reflected in a series of lesions termed pancreatic intraepithelial neoplasia (PanIN); according to morphological, cytological and architectural criteria, these can be graded as PanIN-1A, PanI-1B, Pan-2 or PanIN-3 lesions, in increasing levels of severity [231]. The higher neoplastic potential of each of the subsequent grades is associated with an increasing number of genetic abnormalities in the precancerous cells that progress to invasive ductal carcinoma [232].

Numerous studies have shown over-expression of *MUC6* and *de novo* expression of *MUC5AC* in all PanINs and IDCs, both at mRNA and protein levels [212, 224, 233]. Interestingly, like in hepatobiliary cancer, *MUC6* expression in invasive pancreatic adenocarcinoma strongly correlates with the degree of differentiation: *MUC6* is expressed predominantly in well-differentiated tumors and less in moderately differentiated neoplasms (Fig. **4**). Ringel and Lohr [230] reported that only about 9% of poorly differentiated pancreatic carcinomas are MUC6-positive. These data are in line with the results reported by Kim *et al.* [212] and Nagata *et al.* [127] showing that the expression of *MUC6* in PanINs is an early event: it is especially high at the early PanIN-1A and PanIN-1B stages and gradually decreases at the subsequent stages characterized by less differentiated phenotypes. At the same time, Andrianfahanana *et al.* [224] detected expression of *MUC6* gene in 100% (16 out of 16 samples) of pancreatic adenocarcinomas (the stage and location of tumors were not specified) and in 27% (4 out of 15) pancreatic tumor cell lines characterized with different degree of differentiation. The authors noted that the level of *MUC6* expression in most of the examined samples correstonded to that observed in normal pancreatic specimens. Based on these observations the authors concluded that *MUC6* gene did not display a distinguishing feature that might be specific for pancreatic adenocarcinoma.

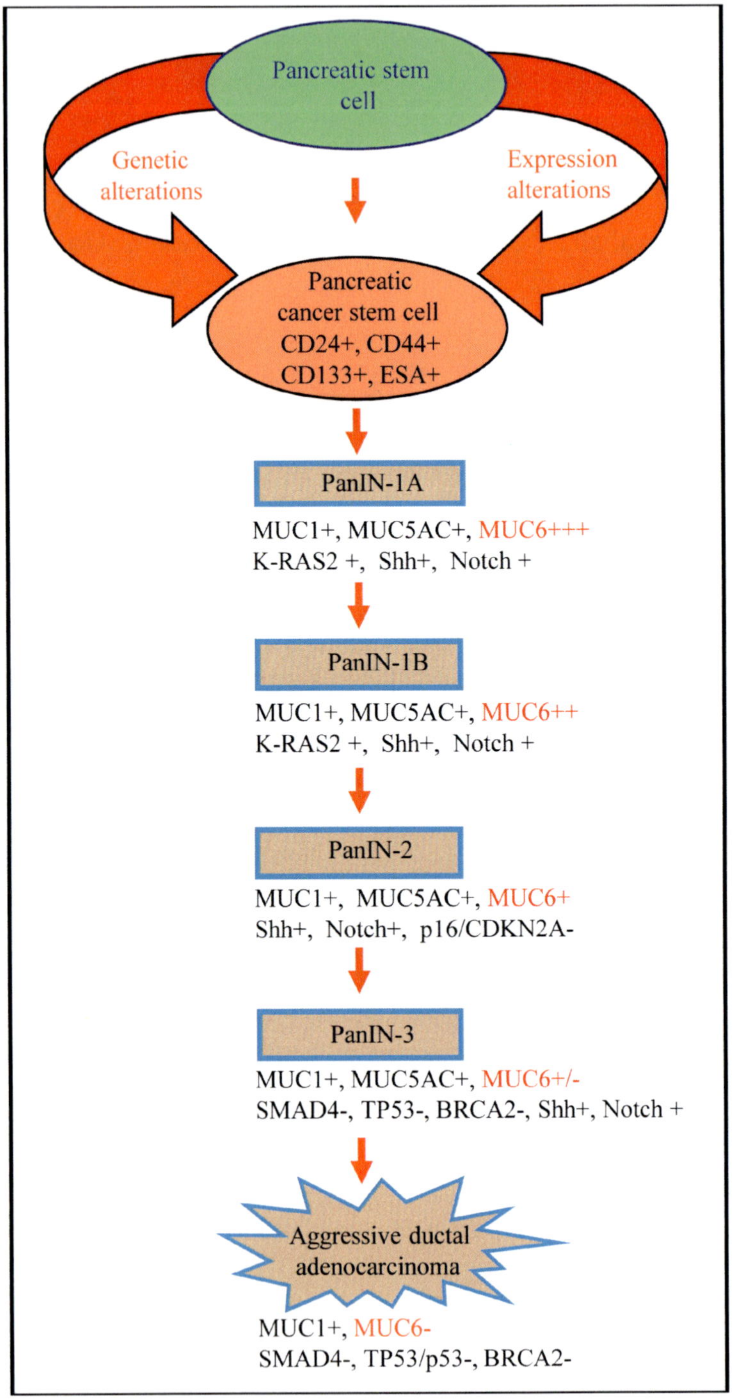

Figure 4: Expression of the *MUC6* gene in PanINs (based on the data reported in [212, 224, 231-233]).

Pancreatic intraductal papillary-mucinous neoplasia (IPMN) is a ductal entity of pancreatic malignancy characterized by a proliferation of the epithelium lining the pancreatic ducts [234]. IPMN may progress stepwise from adenoma to intraductal carcinoma and further to invasive carcinoma. Various studies demonstrate a dysregulation of mucin expression in IPMNs that might contribute to neoplastic progression. IPMNs are classified into two subtypes: IPMN-D type, also called intestinal type, composed of dark columnar cells developing villous architecture morphologically similar to colonic villous adenoma, and IPMN-C type, also called pancreato-biliary type, composed of clear columnar cells forming papillary architecture [235-238]. These two types of IPMNs differ in the expression of *MUC6* (Fig. **5**).

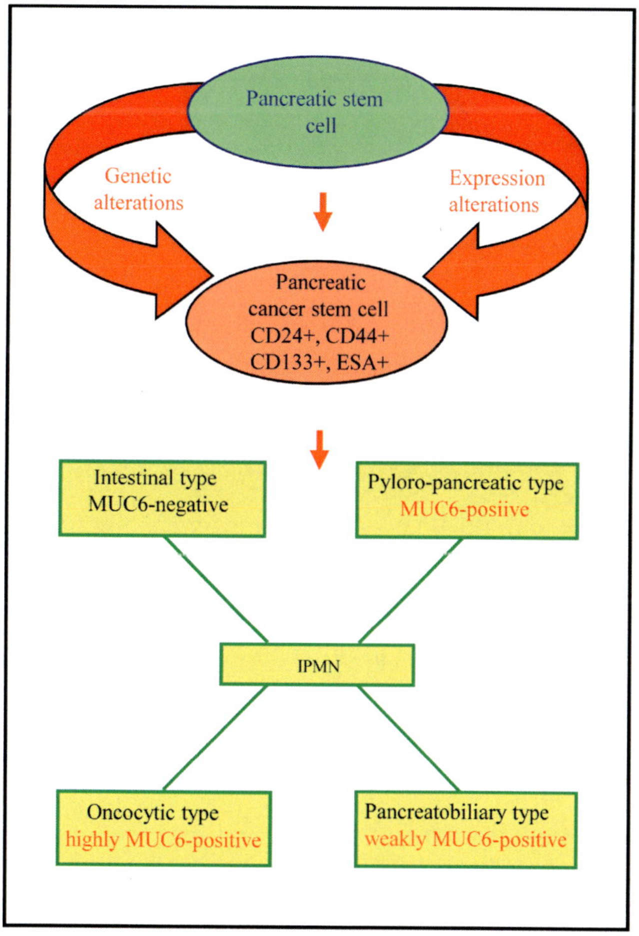

Figure 5: Expression of the *MUC6* gene in IPMNs (based on the data reported in [126, 127, 235-238]).

Nagata *et al.* [127] found the expression rate of *MUC6* to be significantly higher (92%) in IPMN-C type samples than in IPMN-D type samples (37%). Interestingly, there was no significant difference in expression of another gastric mucin, MUC5AC, between the lesions: 92% of IPMN-D tumors and 100% of IPMN-C tumors were MUC5AC-positive. Importantly, MUC5AC is frequently expressed in the papillary or villous lesions of IPMNs in the projected areas, whereas MUC6 is expressed mainly in the basal areas, corresponding to the topography of these mucins in the stomach. Differences in expression and localization of two gastric mucins in two types of IPMNs correlate with the malignant potentials of these types of neoplasia: IPMN-D type tends to have a worse prognosis than IPMN-C type and frequently develops into invasive mucinous carcinoma; IPMN-C type closely resembles the intraductal papillary-mucinous adenoma having better prognosis [239].

Recently, Basturk *et al.* [126] suggested a new classification of IPMNs. They divided these tumors into three main groups based on cell lineage and MUC6 mucin expression: 1) villous or intestinal group that has a "columnar cell" pattern and does not express MUC6; 2) pyloric or pancreatic group characterized by a "cuboidal-cell" pattern and significant MUC6 expression. The later can be further subdivided into two subsets: subgroup 2A composed of oncocytic neoplasms that strongly express MUC6, and subgroup 2B including pancreatobiliary tumors with weak expression of MUC6 mucin.

In addition to PanINs and IPMNs, a third type of pancreatic tumors, mucinous cyctic neoplasms (MCNs), has been also identified. According to the World Health Organization classification [240], the MCNs are divided into adenomas (MCA), borderline tumors, carcinoma *in situ* and invasive carcinoma of pancreas. MCNs are rare tumors that occur exclusively in women. They display ovarian-like stroma and specific localization between the pancreatic body and pancreatic tail [127]. The low incidence of MCNs explains why mucin expression profiles have not yet been established for these tumors. At present, only a few reports are available in which expression of *MUC6* in MCNs has been analyzed. In the study of Luttges *et al.* [241], MUC6 mucin was detected in only a few acinar cells while the most of the specimens were MUC6-negative. These data correlate with the data of Ji *et al.* [242] showing that MUC6 mucin is seldom expressed in mucinous cystic adenomas and can be detected only sporadically in a few cells per tumor. Nearly the same results were reported by Basturk *et al.* [126], who found most of the MCNs studied to be MUC6-negative, with focal and weak expression of MUC6 in rare cells. Nevertheless, critical analysis of their and published data led these authors to conclusion that regardless of the type of tumor (Pan IN, IPMN or MCN), the earliest forms of neoplastic transformation in the pancreatic ductal epithelium have features in common – the earliest metaplastic-like areas are typically MUC6-positive. According to the authors, it is not surprising that *MUC6* expression is manifested by early forms of different types of pancreatic neoplasias, as *MUC6* transcription is controlled by PDX1 transcription factor, an important determinant of pancreatic differentiation. These authors [126] further consider that the expression of *MUC6* in pancreatic papillary neoplasms *in situ* might be a marker of a pyloropancreatic pathway of malignant transformation in this organ. In our opinion, it is too early to draw any conclusions about the specific contribution of *MUC6* expression in pancreas carcinomas, and more studies are needed to clarify the matter.

Colon: Expression of the gel-forming mucin genes in the colon under normal and pathological conditions has been the focus of several studies [137, 243-245]. While *MUC6* is not expressed in normal colon epithelium [243, 245], the pathological processes that take place in colon are often associated with *de novo* expression of the *MUC6* gene.

a) Inflammatory diseases: Ulcerative colitis (UC) and Crohn's disease are two major entities of the inflammatory colon pathology. The role of MUC6 mucin in the pathogenesis of these diseases has been studied, but the results are not uniform. Longman *et al.* [246] could not detect MUC6 mucin in normal and ulcerated colorectal mucosa. Byrd and Bresalier [244] stressed that although *MUC6* was not expressed in most cases of sporadic colorectal cancer, it was substantially expressed in the UC lesions accompanying colorectal neoplasms. Indeed, UC-associated colorectal adenocarcinomas showed significantly higher *MUC6* expression than that observed in the cases of "pure" adenocarcinoma (34% *vs.* 2%) in this study. According to Tatsumi *et al.* [247], 32-44% of UC-associated neoplasms expressed MUC6 mucin, while in neoplasms not associated with UC the rate of the MUC6 expression was very low or not detectable: it reached 4.2%

in tubular adenoma, and was not found at all in sporadic adenocarcinoma. These results point to the role of inflammation in the development of colon cancer and the possible participation of MUC6 mucin in inflammatory processes in colon.

Crohn's disease is a chronic relapsing inflammatory bowel disease of unknown origin. It is characterized by mucosal ulceration that may affect any part of the intestine but is most commonly found in the ileum and proximal colon [248]. Expression of mucins was shown to be deregulated in Crohn's disease and aberrant expression of the mucin-encoding genes contributes to pathogenesis of this nosology [249]. The expression of *MUC6, MUC5AC* and *MUC5B* was never detected in healthy ileal mucosa of patients with Crohn's disease [124, 248, 250]. They were, however, expressed in ileal mucosa adjacent to ulceration area in the so-called ulcer-associated cell lineage (UACL) [248]. The UACL is a unique cell lineage that is developed *de novo* and secretes large amounts of the neutral mucins, in contrast to the intestinal goblet cells which secrete the acid mucins [251]. This lineage is typical of all diseases of the gastrointestinal tract with chronic mucosal ulcerations, especially Crohn's disease and duodenal ulcer. MUC6 and MUC5AC mRNAs and proteins are expressed in acinar cells of the UACL that arise from crypts adjacent to the ulcerations and in the ducts that arise from these acini [248]. These findings suggest a possible role for both MUC6 and MUC5AC mucins in the healing of wounds developed on chronic ulcerative background. Different studies showed that the expression pattern of the *MUC6* and *MUC5AC* genes in UACL is very similar to that observed in stomach and duodenum [252, 253]. Similarity in histology and mucin expression between UACL and stomach was also noted by Longman *et al.* [254], who found that, like in stomach, the expression of MUC6 mucin in UACL is co-localized with trefoil factor family peptide 2 (TFF2), while MUC5AC is co-localized with TFF1. The similarity between UACL and stomach mucosa was also brought out in the study of Kaneko *et al.* [255], who found that histological differentiation of UACL simulates the gastric pylori mucosa, with up-regulation of mucins MUC6 and MUC5AC and transcription factor PDX-1 that regulates expression of the pyloric mucosa specific genes. Collectively, the presented data show that MUC6 plays a definite role in the pathogenesis of Crohn's disease. Future studies are needed to delineate this role more specifically and clarify how MUC6 mucin operates in the development of intestinal inflammation in general and Crohn's disease in particular.

b) Colon cancer: As noted above, sporadic colon adenocarcinomas do not express *MUC6* [247], a finding confirmed by Sylvester *et al.* [245] who could not detect expression of *MUC6* in mucinous and nonmucinous adenocarcinomas of colon at either the mRNA or protein levels. Nakajima *et al.* [137] also did not observe *MUC6* expression in colon adenocarcinomas or in lymph node metastasis, although they did find weak expression of *MUC6* in 3.7% of benign adenomas. In agreement with the aforementioned data are the findings by Levi and Harpaz [256] in a study of an unusual adenocarcinoma, the so-called well-differentiated low-grade tubuloglandular adenocarcinoma (LGTGA), that results from chronic idiopathic inflammatory bowel disease with extensive colon involvement. Immunohistochemistry of LGTGA disclosed expression of MUC2 in 72% of specimens, CK7 in 69%, CK20 in 100%, and MUC6 in no specimen (0%). Two other unusual tumors, colonic clear cell adenocarcinoma and colonic tubular adenoma with clear cell component, were recently studied with regard to mucin expression, including expression of MUC6 [257]. Although clear cell metaplasia is observed in a very small subset of colonic adenoma and adenocarcinoma [258-261], it represents a diagnostic problem for clinical oncology that demands better diagnostic tools. Thus, mucin expression in clear cell adenomas and adenocarcinomas was examined [257]. No expression of the *MUC6* gene was observed in the analyzed specimens, while characteristic changes in expression of *MUC2* and *MUC5AC* were detected in clear cell-containing specimens compared with the background tissues. The available data indicates that most of the malignant lesions in colon studied to date have a MUC6-negative phenotype.

c) Colorectal serrated polyps: Colorectal polyps are heterogeneous lesions that include hyperplastic polyps (HPs), sessile serrated polyp/adenoma (SSA), traditional serrated adenoma (TSA), and mixed polyps [262, 263]. Colorectal polyps are considered to be precursors of colorectal carcinomas [263]. Increasing evidence suggests that a certain proportion of colorectal cancer develops *via* the recently recognized serrated polyp-neoplasia pathway, which differs from the conventional adenoma-carcinoma sequence [262, 264-267]. While there are no strict criteria for reliable differential diagnosis of colorectal polyps at the present time, mucins are emerging as an attractive tool for this purpose. MUC6 mucin expression in particular is considered an appropriate marker for differential diagnosis of HP, SSP and TSA, although conflicting results of studies make it premature to draw conclusions about its usefulness. According to Bartman *et al.* [268], MUC6 mRNA and protein is rarely expressed in normal colon and in hyperplastic polyps, but is present more frequently and at higher levels in polyps of intermediate size and extent of villous histology and dysplasia. On the other hand, Mochizuka *et al.* [269] observed substantial expression of MUC6 mucin both in hyperplastic polyps and in sessile serrated adenomas. Owens *et al.* [270] showed by immunostaining that MUC6 had 100% specificity in distinguishing SSA (MUC6-positive phenotype) from hyperplastic polyp and TSA, which have a MUC6-negative phenotype. The authors note that neither location in the right or left colon nor polyp size accounted for the differences in the MUC6 expression.

In contrast to these data, Bartley *et al.* [271] found that only 53% of SSAs expressed MUC6, and that 17% of hyperplastic polyps and 18% of TSAs also expressed MUC6 mucin. The expression was limited to the lower crypts in all serrated polyps studied. The authors conclude that MUC6 expression is strongly associated with proximal location of serrated polyps, but has only modest utility as a tissue biomarker for SSA. Fujita *et al.* [262] differ with this conclusion, claiming that detection of mucin core protein expression, including MUC6, MUC5AC and MUC2, is insufficient for differentiation of SSA from other types of serrated polyps.

In summary, the data are clearly insufficient for coming to definitive conclusions about the value of MUC6 mucin expression as a discriminating factor in the diagnosis of colorectal polyps.

7.6.2. Expression of *MUC6* Gene in the Female Reproductive Tract

Mucins play an important role in reproductive functions and defense of the female reproductive tract against infections and noxious agents [128]. Alterations in the expression of the several gel-forming and membrane-bound mucins seen in female reproductive organs are related to hormone status associated with the phases of the menstrual cycle [272]. In normal physiological conditions, the *MUC6* gene is expressed in endometrium and endocervix only during the proliferative phase of the menstrual cycle [128], while it is constantly seen in these tissues in the postmenopausal period [129]. The ectocervix, fallopian tubes and vagina do not produce MUC6 mucin [128, 131, 273]. The *MUC6* gene expression has been studied in different types of pathology in the female reproductive tract, including benign and malignant tumors of ovary, hyperplasia and adenocarcinoma of endometrium, and adenocarcinoma of cervix and related glandular lesions. Expression of the gene in other pathological conditions of female reproductive organs has been poorly investigated.

Ovary: The expression of *MUC6* gene in benign and malignant tumours of ovary has been studied by several groups. Boman *et al.* [274] found that *MUC6* was aberrantly expressed in columnar cells of benign lesions in 64% of specimens studied. Activation of gastric MUC6 mucin in ovarian cells that normally do not produce it suggests that gastric type differentiation is an early and frequent event in ovarian mucinous tumorigenesis, at least at the stage of transition from normal tissue to benign lesion. However, as the benign lesion progresses to borderline tumor, the expression of *MUC6* is down-regulated and frequency of MUC6 mRNA positive detection decreases to 10-20%.

Hirabayashi *et al.* [275] observed expression of *MUC6* in three types of ovarian mucinous tumors: mucinous adenoma (MA), mucinous borderline tumors (MB) and mucinous adenocarcinoma (MC). However, the expression was weak and focal in all three types of lesions, and there were no significant differences between them in the proportion of MUC6-positive cells. These results contradict Boman *et al.* [274], who observed a high rate of *MUC6* expression in ovarian benign lesions, which, as mentioned above, dramatically decreased in ovarian adenocarcinomas. The limited number of studies does not permit conclusions regarding the role of MUC6 mucin in pathogenesis of ovarian tumors.

Endometrium: Pathology of endometrium represents a serious diagnostic problem for clinical gynecology. Although expression of mucin genes in endometrial diseases including endometrial cancer has been insufficiently studied, changes in composition of the surface endometrial mucin gel have been detected in various pathological conditions. Alameda *et al.* [131] showed that the expression of mucin genes was higher in hyperplasia than in the normal endometrium. MUC6 mucin was detected in 13% of simple hyperplasia samples and in 20.4% of complex hyperplasia samples. However, the difference between MUC6 expression in normal and hyperplastic endometrium was not statistically significant. On the other hand, the increased expression of MUC6 mucin in endometrial adenocarcinoma (found in 48.5% of the analyzed samples) was statistically significant compared with its expression in normal endometrium. Another study of 310 patients performed by Morrison *et al.* [276], in which only 1.2 % of endometrial adenocarcinomas were MUC6-positive, throws into question the reliability of MUC6 expression as a diagnostic marker for endometrial adenocarcinoma. In general, the role of mucins in pathogenesis of endometrial pathology is poorly studied. Of note, the role of the MUC1 mucin in endometrial pathology has been studied relatively well compared to the scant and sporadic literature on the MUC6 mucin. Nevertheless, the studies do give at least a general idea about the MUC6 mucin's involvement in this malignancy.

Primary mucinous adenocarcinomas of endometrium are typically of low grade. These lesions are frequently associated with endometrial hyperplasia and/or ordinary endometrioid adenocarcinoma and differ from minimal deviation adenocarcinoma (MDA) of the uterine cervix [277-279]. Recently, Abiko *et al.* [280] reported a case of primary mucinous endometrial adenocarcinoma that had many features in common with MDA in terms of morphology and gastric immunophenotype typical for MDA but not for ordinary mucinous endometrial adenocarcinoma [281-283]. The tumor cells were positive for MUC6, HIK1088, CEA and CA19-9, and negative for estrogen receptor and MUC2 mucin. Another rare genital tract malignancy associated with gastric metaplasia was reported by Mikami *et al.* [284] who analyzed six cases of multifocal mucinous metaplastic lesions of endometrium with features of lobular endocervical glandular hyperplasia/pyloric gland metaplasia (LEGH/PGM) and/or adenocarcinoma. In some patients these lesions were combined with mucinous metaplasia and borderline tumors of the fallopian tubes, cervical adenocarcinoma *in situ*, LEGH/PGM lesion of cervix, or MAD. All mucinous lesions were positive for HIK1083 and/or MUC6 regardless of localization [281, 284]. These studies show that gastric type MUC6-positive metaplasia is a stage in tumorigenesis leading to female genital malignancies, and implicate MUC6 in the pathogenesis of these diseases. More studies are needed, however, to determine the precise function(s) of the MUC6 mucin in female genital oncogenesis.

Uterine cervix: In comparison with endometrium, expression of *MUC6* in neoplasms of uterine cervix is studied more thoroughly. Histologically, there are several phenotypes of cervical tumors, including endocervical, mucinous, endometrioid, clear cell, serous and mesonephric phenotypes [285, 286]. MDA of the uterine cervix is a highly differentiated form of adenocarcinoma, regarded by the WHO as a variant of mucinous adenocarcinoma [287-289]. MDA often demonstrates gastric immunophenotype evidenced by expression of MUC6 and positive HIK1083-immunoreactivity [290-293]. Comparison of HIK1083 and MUC6 as gastric phenotype-specific markers showed that, at least for mucinous endocervical adenocarcinomas (ECA), HIK1083 is more specific than MUC6: 75% of ECA related by morphological criteria to gastric phenotype were HIK1083-positive, whereas only 11% of nongastric phenotype ECA were HIK1083-positive. Based on reactivity with anti-MUC6 antibodies, only 31% of gastric phenotype ECA were MUC6-positive, while 16% of nongastric phenotype tumours also expressed MUC6. The available data led Kojima *et al.* [294] to conclusion that the difference in MUC6 reactivity found between cervical adenocarcinomas of gastric and nongastric types is not significant.

Adenoma malignum, a mucinous variant of MAD, is an unusual primary cervical adenocarcinoma of gastric phenotype with extremely differentiated glands [286, 295-297]. Recently, another variant of cervical adenocarcinoma called "gastric type adenocarcinoma" was described, which, in contrast to adenoma malignum, – possesses the highly expressed malignant cytological features [294]. However, the expression of MUC6 appears to be nonspecific in establishing the "gastric phenotype" of this tumor.

An unusual cervical adenocarcinoma with two morphologically distinct and spatially separated components was described in a patient with Peutz Jeghers syndrome [298]. One component was comprised of typical well-differentiated adenoma malignum, whereas the other was a moderately differentiated neoplasm with the gastric phenotype features. Both components were MUC6 and HIK1083-positive. Based on the molecular and cytological data, the authors consider that the gastric type component arose through a process of dedifferentiation within adenoma malignum.

Colloid carcinoma of the cervix – a variant of mucus-producing adenocarcinoma – is characterized by the extracellular accumulation of a large amount of mucus and formation of numerous mucous nodules (mucous lakes) with a relative paucity of neoplastic cells within the nodules [299]. This type of tumor is encountered in the uterine cervix very rarely [285, 300]. It is difficult to differentiate colloid carcinoma of intestinal phenotype from gastric phenotype carcinoma by the histopathological features alone. However, in some cases mucins, including MUC6, are considered possible diagnostic markers and discriminating factors between different types of cervical adenocarcinomas [281, 291, 294, 299].

According to the literature, the detection of MUC6 expression is a helpful diagnostic tool in many cases of malignant cervix; however, the lack of systematic studies of large groups of patients and the absence of animal models diminish the value of the obtained data and preclude drawing conclusions about the efficacy of MUC6 expression for diagnostic purposes.

7.6.3. Expression of *MUC6* Gene in the Male Urogenital Tract

Mucin expression in the male urogenital organs has been poorly studied. This is especially true for MUC6 expression in male urogenital diseases [301, 302].

<u>Kidney</u>: We could find only two publications pertaining to *MUC6* expression in pathology of kidney. By RT-PCR approach, Leroy *et al.* [302] demonstrated expression of *MUC6* in 90% of renal clear cell carcinomas and in 100% of papillary carcinomas, albeit at low levels. No expression of *MUC6* tested by *in situ* hybridization or by immunochemistry was found in malformed cysts of autosomal recessive polycystic kidney disease [140].

Other organs of the human urinary tract, including urethra, bladder and foreskin, appear to be MUC6-negative in both normal physiological conditions [135, 138, 303] and pathological states such as various forms of cystic glandularis [138] and invasive micropapillary urothelial carcinoma of the bladder [304].

<u>Prostate</u>: Study of mucin expression in normal and pathologically altered prostate is of special importance because of the high incidence of prostate cancer in

Western countries [305]. Despite early diagnosis through testing for prostate-specific antigen (PSA) in serum and modern treatment, up to 30% of treated patients suffer relapse. The clinical course of prostate cancer is highly variable due to its biological heterogeneity. Mucins, in general, are widely accepted not only as protectors of epithelial surfaces against damaging factors, but also as active players in epithelial renewal and tissue-specific oncogenesis [306]. This is evidenced by qualitative and quantitative alterations in the expression of mucins in preneoplastic and neoplastic lesions of different organs and tissues [307]. However, little is known about expression of mucins, including MUC6, in prostate. Moreover, conflicting results were obtained on expression of this gene in prostate cancer. Cozzi *et al.* [308] could not detect *MUC6* expression in prostate cancer specimens, while Legrier *et al.*[309] found it in hormone-independent variants of prostate cancer. As previously shown, hormone-independent growth of prostate tumors is associated, among other factors, with mucinous and endocrine differentiation [310-312]. Up to 55% of mucinous and colloid prostate adenocarcinomas were shown to produce mucins [311]. These data are in line with the report that hormone-dependent prostate adenocarcinomas do not express MUC6, while their hormone-independent variants do express large amounts of MUC6, MUC1 and MUC2 mucins and form so-called "mucin lakes" [309]. This observation underscores mucinous differentiation as an important step in the acquisition of hormone-independence in progression of prostate cancer.

As noted above, a unique MUC6 expression was found in normal and malignant prostate tissues in a few studies in recent years [137, 309]: dissociation in expression of MUC6 apomucin and GlcNAcα1-4Galβ-R antigen. That is, only MUC6 molecules that lacked GlcNAcα1-4Galβ-R antigen were expressed in normal prostate and in prostate affected by adenocarcinoma [137, 309]. This is an important finding in view of the known fact that the MUC6 and MUC5AC mucin molecules expressed in the stomach and pancreas carry the specific carbohydrate antigen, GlcNAcα1-4Galβ-R antigen [313].

On the whole, the data from studies of *MUC6* gene expression in normal and malignant prostate are highly inconsistent. Some authors demonstrate expression of *MUC6* in prostate both in physiological and pathological conditions while others assert the opposite. This inconsistency warrants future study of the issue by more sensitive methods.

Testis, epidydimis and seminal vesicle: Analysis of *MUC6* expression in the male urogenital tract must include also studies describing the gene's expression in testis, epididymis, seminal vesicle and seminal plasma. However, the literature contains only sporadic, nonsystematic studies based on a limited number of samples. The available data do, however, give an approximate idea of the expression of MUC6 mucin in the testis and seminal vesicle. Using RT-PCR, Russo *et al.* [135] detected the expression of the MUC6 mRNA in 1 out of 3 samples of testicular tissues, in 1 out of 2 samples of epididymis, and in 5 out of 5 seminal vesicles. On the other hand, Naito *et al.* [301] could not find expression of MUC6 mucin glycoprotein in mucinous cystadenoma of testis assayed by immunohistochemical methods. Expression of MUC6 in pathologically altered epidydimis and seminal vesicle has not been reported.

In conclusion, the data on *MUC6* expression in the male urogenital tract show that some urine-producing and reproductive organs express MUC6 mucin under pathological conditions, whereas others are MUC6-negative. However, the data are sporadic and limited, and in many cases statistically unreliable, highlighting the need for further research.

7.6.4. Expression of *MUC6* Gene in the Respiratory Tract

The role of mucins in the physiology and pathology of the organs comprising the respiratory tract is difficult to overestimate. The epithelial surface of the respiratory tract is coated with mucus that protects the airways from dehydration and damage by inhaled bacteria, viruses, mechanical particles and chemical irritants [107, 314, 315]. Nevertheless, the role of the MUC6 mucin in normal physiological processes appears to be minimal, and MUC6 is not expressed in respiratory epithelium during embryonic and fetal development [106, 107]. Moreover, the literature on its expression in adult airways is controversial. The general opinion is that most regions of the respiratory tract do not express MUC6, or, if at all, express it at very low, almost undetectable levels. In contrast, aberrant expression of the *MUC6* gene in the airway epithelium is often observed in pathological conditions.

a) Inflammatory diseases: Respiratory epithelium affected by inflammation often produces the MUC6 glycoprotein [316, 317]. However, the available information varies in regard different regions of the respiratory tract, and different inflammatory diseases. Surprisingly, just several reports have been published on *MUC6* expression in chronic inflammatory disorders such as asthma, CF and chronic obstructive pulmonary disease (COPD). Caramori *et al.* [318] reported expression of MUC6 in cytoplasm and cilia in all samples of bronchiolar epithelial cells of the COPD patients studied, but could not detect MUC6 mucin in the luminal mucus of these patients. The pathophysiological relationships between the airway mucus secretion and progression of COPD were studied earlier [319-321], but the role of MUC6 mucin in the development of the disease was not clarified in these studies. Although it is difficult to define the contribution of MUC6 mucin to the pathogenesis of COPD on the basis of a single study, Cammori *et al.*'s data [318] implicate MUC6 mucin in the development of this disease. The functional role and significance of the MUC6 glycoprotein in the pathogenesis of COPD await clarification.

MUC6 mucin expression in the upper airways under pathological conditions, particularly in inflammatory diseases of nasal epithelium, is also poorly investigated. Nasal polyp, one of the inflammatory diseases of the nasal cavity, may occur in combination with a variety of other diseases such as asthma, CF, bilateral nasal polyposis, and unilateral antrochoanal polyps (ACP). Although *MUC6* has been shown to be expressed in normal nasal epithelium [317], epithelial cells of a singular nasal polyp as well as polyps combined with asthma, CF and/or ACP were found to be MUC6-negative [317, 322], suggesting down-regulation of *MUC6* expression as one of the events in development of these pathological conditions.

In summary, taking into account the impact of inflammatory respiratory diseases in human pathology, more studies are needed to determine the precise functions of the MUC6 glycoprotein in the pathogenesis of airway imflammatory diseases.

b) Malignant tumors: The rate of lung cancer-associated death has increased in many countries, making early detection of lung neoplasias by specific markers essential [323]. Since mucin expression is organ- and tissue-specific and may be altered during carcinogenesis [324-326], examination of its expression could be informative for diagnostics and follow-up of the disease.

There are several phenotypes of lung mucin-producing carcinomas: mucinous bronchoalveolar carcinoma (m-BAC), solid adenocarcinoma with mucin production (SA), signet-ring cell carcinoma (SRCC), mucinous ("colloid") adenocarcinoma, mucoepidermoid carcinoma, and mixed adenocarcinoma (MX) [327, 328]. Atypical adenomatous hyperplasia (AAH) has been shown to be a precursor of BAC [329-331]. As reported by Awaya *et al.* [323], BAC consists of cancer cells lining the alveolar wall. It is considered a nonaggressive carcinoma, whereas adenocarcinoma with mixed subtypes (MX) usually consists of BAC and elements of invasive carcinoma. Based on these features, a progression from AAH through BAC to MX has been suggested [332, 333]. The study conducted by Awaya *et al.* [323] showed that progression from AAH to MX is associated with an increase in *MUC6* expression, which is higher in mucinous BAC (m-BAC) than in the nonmucinous variant (nm-BAC). The alteration in *MUC6* expression was found to correlate with tumor size and abnormalities in p53 gene, and was associated with dedifferentiation of bronchial epithelium [316]. According to Lopez-Ferrer *et al.* [316] and Barsky *et al.* [334], the higher aggressiveness of m-BAC compared to nonmucinous BAC might stem from the action of secretory mucins including over-expressed MUC6. In support of this speculation, Hamamoto *et al.* [335] observed aberrant expression of *MUC6* in pulmonary adenocarcinoma. Copin *et al.* [307] also demonstrated a specific pattern of mucin gene expression, including *MUC6*, in primary lung adenocarcinomas. Moreover, Honda *et al.* [336] showed that development of m-BAC is associated with organoid differentiation simulating the pyloric mucosa of the stomach. Interestingly, the gastric foveolar cell transformation of alveolar lining cells characterized by strong up-regulation of *MUC6* expression was observed also in aggressive m-BAC, while less aggressive pulmonary tumors, SRCC and SA, kept the alveolar phenotype at much lower levels of *MUC6* expression [328]. These results may indicate that different carcinogenic pathways are exploited by different pulmonary tumors. The pathway activated in the case of m-BAC involves high transcriptional activity of the *MUC6* gene, whereas the participation of *MUC6* in the development of SRCC and SA appears not to be critical. Based on their study of mucin gene expression in primary lung adenocarcinomas, Copin *et al.* [307]

concluded that the aberrant expression of the MUC6 mucin could serve as one of the diagnostic markers in the management of mucinous type BAC. Nishiumi *et al.* [337] confirmed the usefulness of the MUC6 glycoprotein as a lung carcinoma specific marker by showing that expression of MUC6 is related to lymph node metastasis and affects prognosis of small size adenocarcinoma of the lung.

In summary, the participation of MUC6 mucin in carcinogenesis is highly probable in some forms of adenocarcinomas, in which the use of MUC6 glycoprotein as a diagnostic marker appears to be scientifically proven. Its role in tumor development in other forms has not been proven and requires further study. More research is also needed to clarify the role of MUC6 mucin in the pathogenesis of inflammatory and allergic diseases of the human respiratory tract.

7.6.5. Role of *MUC6* gene in Pathology of the Eye, Ear, Thyroid and Mammary Glands

The glycoprotein MUC6 is involved in normal functioning and pathogenesis of diseases in some organs outside the gastrointestinal, urogenital and respiratory tracts. These include the eye, ear and mammary gland, where the impact of the MUC6 mucin in pathology of these organs is substantial.

Eye: The role of MUC6 mucin in the physiology and pathology of eyes is uncertain at the present time. The MUC6 mRNA was expressed in 50% of analyzed normal lacrimal sacs and nasolacrimal duct; the "normal" level of its expression was retained in patients with primary acquired nasolacrimal duct obstruction, while the expression levels of other gel-forming mucin genes, including *MUC2, MUC5AC* and *MUC5B*, were down-regulated [338, 339]. However, at the protein level, MUC6 could not be detected in the specimens tested [338-340]. Can the observed discrepancy be explained by different dynamics of MUC6 mRNA and MUC6 protein synthesis and/or degradation? Future studies should clarify the role of the MUC6 mucin in the normal and pathologically changed eye tissues.

Ear: The reports on the role of *MUC6* in the physiology and pathology of the ear are limited and controversal. According to Takeuchi *et al.* [341], several mucin genes including *MUC6* are expressed by the middle ear epithelium, and their mRNAs could be identified in middle ear effusion. Interestingly, the expression profile of these mucin genes was similar to that of the bronchial epithelial cell line BEAS-2B. In contrast to these data, Kerschner [342] found no expression of the MUC6 mRNA in human middle ear epithelial specimens obtained *in vivo*, or in immortalized human, mouse and chinchilla middle ear epithelial cell cultures *in vitro* [343, 344]. These authors [344] consider that "further investigation of this mucin as a contributor to respiratory epithelial pathology is not warranted". In our opinion, more studies are needed to clarify the possible participation of the MUC6 mucin in the physiology and pathology of the middle ear epithelium. These studies should employ a broad range of *in vitro* and *in vivo* experimental systems under variable conditions simulating the effects of different physiological and pathological signals.

Thyroid gland: Normal thyroid gland, hyperplastic and benign thyroid tissues, and papillary-type thyroid carcinoma have turned up no *MUC6* expression in any analyzed specimens [124, 345, 346]. However, the data come from separate studies with limited numbers of samples, precluding drawing any conclusions regarding activity of the *MUC6* gene in normal and diseased thyroid. It is obviously that more studies are needed to clarify the role MUC mucin plays in pathology of the thyroid.

Mammary gland: Compared to the many studies on the MUC1 mucin in breast cancer [347-352], much less attention has been paid to the MUC6 mucin in the initiation and progression of mammary malignancy [84, 124]. Mammary gland does not express MUC6 mucin under physiological conditions [84, 124], but a substantial level of MUC6 expression was detected in the pathologically altered breast. The MUC6 mucin and Lewis^{y} antigen associated with MUC6-apomucin are commonly expressed in breast cancer [353]. MUC6 was expressed in 41% of fibrocystic mammary glands without atypia, in 100% of atypical fibrocystic glands, and in 95% of breast adenocarcinomas [84]. Conflicting data were reported by Rakha *et al.* [354], who examined expression of several mucins, including MUC6, in 1447 cases of invasive breast carcinoma using tissue microarray technology and immunochemistry, and found the MUC6 mucin expressed in only 20% of tumors. Importantly, the expression rate of MUC6 in mucinous carcinomas was much higher than in other types of breast tumors. The discrepancy between high (95%) [84] and relatively low (20%) [354] levels of

MUC6 expression in breast tumours reported in different studies may be explained by the findings of Matsukita *et al.* [213]: the expression rate of MUC6 mucin was significantly higher in mucinous breast carcinoma (71%) than in invasive ductal carcinoma (15%). Approximately the same low frequency of MUC6 expression (23%) reported by. Rakha *et al.* [354] was observed by Pereira *et al.* [355]. It should be pointed out that these low figures were related exclusively to invasive carcinoma, confirming the data reported by Matsukita *et al.* [213]. Importantly, as noted by Rakha *et al.* [354], MUC6 expression appeared to be a reliable survival factor – a conclusion in agreement with Komaki *et al.*'s suggestion [356] that abundant mucus within the tumors acts as a barrier to cancerous extension in mucinous carcinoma. Hence, the high expression of *MUC6* observed in many mucinous tumors, including mucinous breast carcinoma, determines the nonaggressive behavior of these neoplasms [213].

In summary, the presented data show that the MUC6 mucin obviously plays an important, but as yet unidentified, role in breast cancer pathogenesis. Conflicting results obtained in different laboratories indicate to different functions of the MUC6 mucin in pathology of the mammary gland. Future studies based on usage of the modern molecular biology methods, such as sensetive PCR, DNA and protein sequencing and immunochemical detection of specific MUC6 antigens, will have to elucidate the specific function(s) the MUC6 glycoprotein serves in this process and how they are carried out.

REFERENCES

[1] Pigny P, Guyonnet-Duperat V, Hill AS, *et al.* Human mucin genes assigned to 11p15.5: identification and organization of a cluster of genes. Genomics 1996;38:340-52.

[2] Lang T, Hansson GC, Samuelsson T. Gel-forming mucins appeared early in metazoan evolution. Proc Natl Acad Sci U S A 2007;104:16209-14.

[3] Rousseau K, Byrne C, Kim YS, *et al.* The complete genomic organization of the human MUC6 and MUC2 mucin genes. Genomics 2004;83:936-9.

[4] Rousseau K, Byrne C, Griesinger G, *et al.* Allelic association and recombination hotspots in the mucin gene (MUC) complex on chromosome 11p15.5. Ann Hum Genet 2007;71:561-9.

[5] Desseyn JL, Buisine MP, Porchet N, *et al.* Evolutionary history of the 11p15 human mucin gene family. J Mol Evol 1998;46:102-6.

[6] Meezaman D, Charles P, Daskal E, *et al.* Cloning and analysis of cDNA encoding a major airway glycoprotcin, human tracheobronchial mucin (MUC5). J Biol Chem 1994;269:12932-9.

[7] Kummerfeld SK, Weiss AS, Fekete A, Jermiin LS. AMID: autonomous modeler of intragenic duplication. Appl Bioinformatics 2003;2:169-76.
[8] Desseyn JL, Aubert JP, Porchet N, Laine A. Evolution of the large secreted gel-forming mucins. Mol Biol Evol 2000;17:1175-84.
[9] Toribara NW, Ho SB, Gum E, *et al.* The carboxyl-terminal sequence of the human secretory mucin, MUC6. Analysis Of the primary amino acid sequence. J Biol Chem 1997;272:16398-403.
[10] Herrmann A, Davies JR, Lindell G, *et al.* Studies on the "insoluble" glycoprotein complex from human colon. Identification of reduction-insensitive MUC2 oligomers and C-terminal cleavage. J Biol Chem 1999;274:15828-36.
[11] Wickstrom C, Davies JR, Eriksen GV, Veerman EC, Carlstedt I. MUC5B is a major gel-forming, oligomeric mucin from human salivary gland, respiratory tract and endocervix: identification of glycoforms and C-terminal cleavage. Biochem J 1998;334 (Pt 3):685-93.
[12] Toribara NW, Roberton AM, Ho SB, *et al.* Human gastric mucin. Identification of a unique species by expression cloning. J Biol Chem 1993;268:5879-85.
[13] Audie JP, Janin A, Porchet N, *et al.* Expression of human mucin genes in respiratory, digestive, and reproductive tracts ascertained by *in situ* hybridization. J Histochem Cytochem 1993;41:1479-85.
[14] Balague C, Audie JP, Porchet N, Real FX. *In situ* hybridization shows distinct patterns of mucin gene expression in normal, benign, and malignant pancreas tissues. Gastroenterology 1995;109:953-64.
[15] Carrato C, Balague C, de Bolos C, *et al.* Differential apomucin expression in normal and neoplastic human gastrointestinal tissues. Gastroenterology 1994;107:160-72.
[16] Lesuffleur T, Zweibaum A, Real FX. Mucins in normal and neoplastic human gastrointestinal tissues. Crit Rev Oncol Hematol 1994;17:153-80.
[17] Gehring WJ, Affolter M, Burglin T. Homeodomain proteins. Annu Rev Biochem 1994;63:487-526.
[18] Strous GJ, Dekker J. Mucin-type glycoproteins. Crit Rev Biochem Mol Biol 1992;27:57-92.
[19] Loomes KM, Senior HE, West PM, Roberton AM. Functional protective role for mucin glycosylated repetitive domains. Eur J Biochem 1999;266:105-11.
[20] Vinall LE, Hill AS, Pigny P, *et al.* Variable number tandem repeat polymorphism of the mucin genes located in the complex on 11p15.5. Hum Genet 1998;102:357-66.
[21] Garcia E, Carvalho F, Amorim A, David L. MUC6 gene polymorphism in healthy individuals and in gastric cancer patients from northern Portugal. Cancer Epidemiol Biomarkers Prev 1997;6:1071-4.
[22] Kwon JA, Lee SY, Ahn EK, *et al.* Short rare MUC6 minisatellites-5 alleles influence susceptibility to gastric carcinoma by regulating gene. Hum Mutat 2010;31:942-9.
[23] Nguyen TV, Janssen M, Jr., Gritters P, *et al.* Short mucin 6 alleles are associated with H pylori infection. World J Gastroenterol 2006;12:6021-5.
[24] Parry S, Sutton-Smith M, Heal P, *et al.* Evaluation of MUC6 mucin tandem repeats. Biochim Biophys Acta 2005;1722:77-83.
[25] Gum JR, Jr., Hicks JW, Toribara NW, *et al.* The human MUC2 intestinal mucin has cysteine-rich subdomains located both upstream and downstream of its central repetitive region. J Biol Chem 1992;267:21375-83.
[26] Desseyn JL, Buisine MP, Porchet N, Aubert JP, Laine A. Genomic organization of the human mucin gene MUC5B. cDNA and genomic sequences upstream of the large central exon. J Biol Chem 1998;273:30157-64.

[27] Chen Y, Zhao YH, Kalaslavadi TB, *et al.* Genome-wide search and identification of a novel gel-forming mucin MUC19/Muc19 in glandular tissues. Am J Respir Cell Mol Biol 2004;30:155-65.

[28] Lang T. Evolution of transmembrane and gel-forming mucins studied with bioinformatic methods. Goteborg: Goteborg University Institute of Biomedicine, Sahlgrenska Academy; 2007.

[29] Escande F, Porchet N, Bernigaud A, *et al.* The mouse secreted gel-forming mucin gene cluster. Biochim Biophys Acta 2004;1676:240-50.

[30] Desseyn JL, Laine A. Characterization of mouse muc6 and evidence of conservation of the gel-forming mucin gene cluster between human and mouse. Genomics 2003;81:433-6.

[31] Shekels LL, Lyftogt C, Kieliszewski M, *et al.* Mouse gastric mucin: cloning and chromosomal localization. Biochem J 1995;311 (Pt 3):775-85.

[32] van Klinken BJ, Einerhand AW, Buller HA, Dekker J. The oligomerization of a family of four genetically clustered human gastrointestinal mucins. Glycobiology 1998;8:67-75.

[33] Voorberg J, Fontijn R, Calafat J, *et al.* Assembly and routing of von Willebrand factor variants: the requirements for disulfide-linked dimerization reside within the carboxy-terminal 151 amino acids. J Cell Biol 1991;113:195-205.

[34] Dekker J, Strous GJ. Covalent oligomerization of rat gastric mucin occurs in the rough endoplasmic reticulum, is N-glycosylation-dependent, and precedes initial O-glycosylation. J Biol Chem 1990;265:18116-22.

[35] Klomp LW, van Rens L, Strous GJ. Identification of a human gastric mucin precursor: N-linked glycosylation and oligomerization. Biochem J 1994;304 (Pt 3):693-8.

[36] Tytgat KM, Buller HA, Opdam FJ, *et al.* Biosynthesis of human colonic mucin: Muc2 is the prominent secretory mucin. Gastroenterology 1994;107:1352-63.

[37] Klomp LW, de Lely AJ, Strous GJ. Biosynthesis of a human gall-bladder mucin. Biochem J 1994;304 (Pt 3):737-44.

[38] van Klinken BJ, Oussoren E, Weenink JJ, *et al.* The human intestinal cell lines Caco-2 and LS174T as models to study cell-type specific mucin expression. Glycoconj J 1996;13:757-68.

[39] Dekker J, Van Beurden-Lamers WM, Strous GJ. Biosynthesis of gastric mucus glycoprotein of the rat. J Biol Chem 1989;264:10431-7.

[40] Doms RW, Ruusala A, Machamer C, *et al.* Differential effects of mutations in three domains on folding, quaternary structure, and intracellular transport of vesicular stomatitis virus G protein. J Cell Biol 1988;107:89-99.

[41] Dube S, Fisher JW, Powell JS. Glycosylation at specific sites of erythropoietin is essential for biosynthesis, secretion, and biological function. J Biol Chem 1988;263:17516-21.

[42] Leavitt R, Schlesinger S, Kornfeld S. Impaired intracellular migration and altered solubility of nonglycosylated glycoproteins of vesicular stomatitis virus and Sindbis virus. J Biol Chem 1977;252:9018-23.

[43] Guan JL, Cao H, Rose JK. Cell-surface expression of a membrane-anchored form of the human chorionic gonadotropin alpha subunit. J Biol Chem 1988;263:5306-13.

[44] Miyazaki J, Appella E, Zhao H, Forman J, Ozato K. Expression and function of a nonglycosylated major histocompatibility class I antigen. J Exp Med 1986;163:856-71.

[45] Nordman H, Davies JR, Lindell G, *et al.* Gastric MUC5AC and MUC6 are large oligomeric mucins that differ in size, glycosylation and tissue distribution. Biochem J 2002;364:191-200.

[46] Yakubov GE, Papagiannopoulos A, Rat E, Waigh TA. Charge and interfacial behavior of short side-chain heavily glycosylated porcine stomach mucin. Biomacromolecules 2007;8:3791-9.
[47] Dekker J, van der Ende A, Aelmans PH, Strous GJ. Rat gastric mucin is synthesized and secreted exclusively as filamentous oligomers. Biochem J 1991;279 (Pt 1):251-6.
[48] Yakubov GE, Papagiannopoulos A, Rat E, Easton RL, Waigh TA. Molecular structure and rheological properties of short-side-chain heavily glycosylated porcine stomach mucin. Biomacromolecules 2007;8:3467-77.
[49] Hong Z, Chasan B, Bansil R, *et al.* Atomic force microscopy reveals aggregation of gastric mucin at low pH. Biomacromolecules 2005;6:3458-66.
[50] Di Cola E, Yakubov GE, Waigh TA. Double-globular structure of porcine stomach mucin: a small-angle X-ray scattering study. Biomacromolecules 2008;9:3216-22.
[51] McColl J, Yakubov GE, Ramsden JJ. Temperature dependence of mucin adsorption. Langmuir 2008;24:902-5.
[52] Bansil R, Stanley E, LaMont JT. Mucin biophysics. Annu Rev Physiol 1995;57:635-57.
[53] Griffiths PC, Occhipinti P, Morris C, *et al.* PGSE-NMR and SANS studies of the interaction of model polymer therapeutics with mucin. Biomacromolecules 2010;11:120-5.
[54] Yakubov GE, McColl J, Bongaerts JH, Ramsden JJ. Viscous boundary lubrication of hydrophobic surfaces by mucin. Langmuir 2009;25:2313-21.
[55] Kantor Y, Kardar M. Conformations of randomly linked polymers. Phys Rev E Stat Phys Plasmas Fluids Relat Interdiscip Topics 1996;54:5263-7.
[56] Cao X, Bansil R, Bhaskar KR, *et al.* pH-dependent conformational change of gastric mucin leads to sol-gel transition. Biophys J 1999;76:1250-8.
[57] Strous GJ. Initial glycosylation of proteins with acetylgalactosaminylserine linkages. Proc Natl Acad Sci U S A 1979;76:2694-8.
[58] Abeijon C, Hirschberg CB. Subcellular site of synthesis of the N-acetylgalactosamine (alpha 1-0) serine (or threonine) linkage in rat liver. J Biol Chem 1987;262:4153-9.
[59] Johnson WV, Heath EC. Evidence for posttranslational O-glycosylation of fetuin. Biochemistry 1986;25:5518-25.
[60] Piller V, Piller F, Klier FG, Fukuda M. O-glycosylation of leukosialin in K562 cells. Evidence for initiation and elongation in early Golgi compartments. Eur J Biochem 1989;183:123-35.
[61] Asker N, Baeckstrom D, Axelsson MA, Carlstedt I, Hansson GC. The human MUC2 mucin apoprotein appears to dimerize before O-glycosylation and shares epitopes with the'insoluble' mucin of rat small intestine. Biochem J 1995;308 (Pt 3):873-80.
[62] Freire T, Bay S, von Mensdorff-Pouilly S, Osinaga E. Molecular basis of incomplete O-glycan synthesis in MCF-7 breast cancer cells: putative role of MUC6 in Tn antigen expression. Cancer Res 2005;65:7880-7.
[63] Ten Hagen KG, Fritz TA, Tabak LA. All in the family: the UDP-GalNAc:polypeptide N-acetylgalactosaminyltransferases. Glycobiology 2003;13:1R-16R.
[64] Hanisch FG. O-glycosylation of the mucin type. Biol Chem 2001;382:143-9.
[65] Fukuda M. Roles of mucin-type O-glycans in cell adhesion. Biochim Biophys Acta 2002;1573:394-405.
[66] Ju T, Brewer K, D'Souza A, Cummings RD, Canfield WM. Cloning and expression of human core 1 beta1,3-galactosyltransferase. J Biol Chem 2002;277:178-86.

[67] Iwai T, Inaba N, Naundorf A, *et al.* Molecular cloning and characterization of a novel UDP-GlcNAc:GalNAc-peptide beta1,3-N-acetylglucosaminyltransferase (beta 3Gn-T6), an enzyme synthesizing the core 3 structure of O-glycans. J Biol Chem 2002;277:12802-9.

[68] Mollicone R, Bara J, Le Pendu J, Oriol R. Immunohistologic pattern of type 1 (Lea, Leb) and type 2 (X, Y, H) blood group-related antigens in the human pyloric and duodenal mucosae. Lab Invest 1985;53:219-27.

[69] De Bolos C, Garrido M, Real FX. MUC6 apomucin shows a distinct normal tissue distribution that correlates with Lewis antigen expression in the human stomach. Gastroenterology 1995;109:723-34.

[70] David L, Nesland JM, Clausen H, Carneiro F, Sobrinho-Simoes M. Simple mucin-type carbohydrate antigens (Tn, sialosyl-Tn and T) in gastric mucosa, carcinomas and metastases. APMIS Suppl 1992;27:162-72.

[71] Sakai H, Jinawath A, Yamaoka S, Yuasa Y. Upregulation of MUC6 mucin gene expression by NFkappaB and Sp factors. Biochem Biophys Res Commun 2005;333:1254-60.

[72] Gum JR, Jr., Hicks JW, Toribara NW, Siddiki B, Kim YS. Molecular cloning of human intestinal mucin (MUC2) cDNA. Identification of the amino terminus and overall sequence similarity to prepro-von Willebrand factor. J Biol Chem 1994;269:2440-6.

[73] Dufosse J, Porchet N, Audie JP, *et al.* Degenerate 87-base-pair tandem repeats create hydrophilic/hydrophobic alternating domains in human mucin peptides mapped to 11p15. Biochem J 1993;293 (Pt 2):329-37.

[74] Gum JR, Hicks JW, Swallow DM, *et al.* Molecular cloning of cDNAs derived from a novel human intestinal mucin gene. Biochem Biophys Res Commun 1990;171:407-15.

[75] Klomp LW, Van Rens L, Strous GJ. Cloning and analysis of human gastric mucin cDNA reveals two types of conserved cysteine-rich domains. Biochem J 1995;308 (Pt 3):831-8.

[76] Bresalier RS, Niv Y, Byrd JC, *et al.* Mucin production by human colonic carcinoma cells correlates with their metastatic potential in animal models of colon cancer metastasis. J Clin Invest 1991;87:1037-45.

[77] Ho SB, Shekels LL, Toribara NW, *et al.* Mucin gene expression in normal, preneoplastic, and neoplastic human gastric epithelium. Cancer Res 1995;55:2681-90.

[78] Debailleul V, Laine A, Huet G, *et al.* Human mucin genes MUC2, MUC3, MUC4, MUC5AC, MUC5B, and MUC6 express stable and extremely large mRNAs and exhibit a variable length polymorphism. An improved method to analyze large mRNAs. J Biol Chem 1998;273:881-90.

[79] Tan KO, Sater GR, Myers AM, Robson RM, Huiatt TW. Molecular characterization of avian muscle titin. J Biol Chem 1993;268:22900-7.

[80] Zhang JQ, Luo G, Herrera AH, Paterson B, Horowits R. cDNA cloning of mouse nebulin. Evidence that the nebulin-coding sequence is highly conserved among vertebrates. Eur J Biochem 1996;239:835-41.

[81] Tytgat KM, Bovelander FJ, Opdam FJ, *et al.* Biosynthesis of rat MUC2 in colon and its analogy with human MUC2. Biochem J 1995;309 (Pt 1):221-9.

[82] Inoue D, Yamaya M, Kubo H, *et al.* Mechanisms of mucin production by rhinovirus infection in cultured human airway epithelial cells. Respir Physiol Neurobiol 2006;154:484-99.

[83] Zhu Z, Tang W, Ray A, *et al.* Rhinovirus stimulation of interleukin-6 *in vivo* and *in vitro*. Evidence for nuclear factor kappa B-dependent transcriptional activation. J Clin Invest 1996;97:421-30.

[84] de Bolos C, Guma M, Barranco C, *et al.* MUC6 expression in breast tissues and cultured cells: abnormal expression in tumors and regulation by steroid hormones. Int J Cancer 1998;77:193-9.

[85] Kang W, Rathinavelu S, Samuelson LC, Merchant JL. Interferon gamma induction of gastric mucous neck cell hypertrophy. Lab Invest 2005;85:702-15.

[86] Sekine A, Akiyama Y, Yanagihara K, Yuasa Y. Hath1 up-regulates gastric mucin gene expression in gastric cells. Biochem Biophys Res Commun 2006;344:1166-71.

[87] Akazawa C, Ishibashi M, Shimizu C, Nakanishi S, Kageyama R. A mammalian helix-loop-helix factor structurally related to the product of Drosophila proneural gene atonal is a positive transcriptional regulator expressed in the developing nervous system. J Biol Chem 1995;270:8730-8.

[88] Tsukamoto T, Mizoshita T, Mihara M, *et al.* Sox2 expression in human stomach adenocarcinomas with gastric and gastric-and-intestinal-mixed phenotypes. Histopathology 2005;46:649-58.

[89] Tsukamoto T, Inada K, Tanaka H, *et al.* Down-regulation of a gastric transcription factor, Sox2, and ectopic expression of intestinal homeobox genes, Cdx1 and Cdx2: inverse correlation during progression from gastric/intestinal-mixed to complete intestinal metaplasia. J Cancer Res Clin Oncol 2004;130:135-45.

[90] Ishii Y, Rex M, Scotting PJ, Yasugi S. Region-specific expression of chicken Sox2 in the developing gut and lung epithelium: regulation by epithelial-mesenchymal interactions. Dev Dyn 1998;213:464-75.

[91] Sakamoto N, Fukuda K, Watanuki K, *et al.* Role for cGATA-5 in transcriptional regulation of the embryonic chicken pepsinogen gene by epithelial-mesenchymal interactions in the developing chicken stomach. Dev Biol 2000;223:103-13.

[92] Xiao C, Ogle SA, Schumacher MA, *et al.* Hedgehog signaling regulates E-cadherin expression for the maintenance of the actin cytoskeleton and tight junctions. Am J Physiol Gastrointest Liver Physiol 2010;299:G1252-65.

[93] Shiotani A, Iishi H, Uedo N, *et al.* Evidence that loss of sonic hedgehog is an indicator of Helicobater pylori-induced atrophic gastritis progressing to gastric cancer. Am J Gastroenterol 2005;100:581-7.

[94] Stepan V, Ramamoorthy S, Nitsche H, *et al.* Regulation and function of the sonic hedgehog signal transduction pathway in isolated gastric parietal cells. J Biol Chem 2005;280:15700-8.

[95] Suzuki H, Minegishi Y, Nomoto Y, *et al.* Down-regulation of a morphogen (sonic hedgehog) gradient in the gastric epithelium of Helicobacter pylori-infected Mongolian gerbils. J Pathol 2005;206:186-97.

[96] van den Brink GR, Hardwick JC, Nielsen C, *et al.* Sonic hedgehog expression correlates with fundic gland differentiation in the adult gastrointestinal tract. Gut 2002;51:628-33.

[97] Xiao C, Ogle SA, Schumacher MA, *et al.* Loss of parietal cell expression of Sonic hedgehog induces hypergastrinemia and hyperproliferation of surface mucous cells. Gastroenterology 2010;138:550-61, 61 e1-8.

[98] Oettgen P, Alani RM, Barcinski MA, *et al.* Isolation and characterization of a novel epithelium-specific transcription factor, ESE-1, a member of the ets family. Mol Cell Biol 1997;17:4419-33.

[99] Oettgen P, Finger E, Sun Z, *et al.* PDEF, a novel prostate epithelium-specific ets transcription factor, interacts with the androgen receptor and activates prostate-specific antigen gene expression. J Biol Chem 2000;275:1216-25.

[100] Kas K, Finger E, Grall F, *et al.* ESE-3, a novel member of an epithelium-specific ets transcription factor subfamily, demonstrates different target gene specificity from ESE-1. J Biol Chem 2000;275:2986-98.
[101] Oettgen P, Kas K, Dube A, *et al.* Characterization of ESE-2, a novel ESE-1-related Ets transcription factor that is restricted to glandular epithelium and differentiated keratinocytes. J Biol Chem 1999;274:29439-52.
[102] Horst D, Gu X, Bhasin M, *et al.* Requirement of the epithelium-specific Ets transcription factor Spdef for mucous gland cell function in the gastric antrum. J Biol Chem 2010;285:35047-55.
[103] Ho SB, Roberton AM, Shekels LL, *et al.* Expression cloning of gastric mucin complementary DNA and localization of mucin gene expression. Gastroenterology 1995;109:735-47.
[104] Jeffrey GP, Oates PS, Wang TC, Babyatsky MW, Brand SJ. Spasmolytic polypeptide: a trefoil peptide secreted by rat gastric mucous cells. Gastroenterology 1994;106:336-45.
[105] Vincent A, Perrais M, Desseyn JL, *et al.* Epigenetic regulation (DNA methylation, histone modifications) of the 11p15 mucin genes (MUC2, MUC5AC, MUC5B, MUC6) in epithelial cancer cells. Oncogene 2007;26:6566-76.
[106] Reid CJ, Gould S, Harris A. Developmental expression of mucin genes in the human respiratory tract. Am J Respir Cell Mol Biol 1997;17:592-8.
[107] Buisine MP, Devisme L, Copin MC, *et al.* Developmental mucin gene expression in the human respiratory tract. Am J Respir Cell Mol Biol 1999;20:209-18.
[108] Bernacki SH, Nelson AL, Abdullah L, *et al.* Mucin gene expression during differentiation of human airway epithelia *in vitro*. Muc4 and muc5b are strongly induced. Am J Respir Cell Mol Biol 1999;20:595-604.
[109] Van Klinken BJ, Dekker J, Buller HA, de Bolos C, Einerhand AW. Biosynthesis of mucins (MUC2-6) along the longitudinal axis of the human gastrointestinal tract. Am J Physiol 1997;273:G296-302.
[110] Cornfield AP. The role of mucins in the gastrointestinal tract: an overview with emphasis on nutrition. In: Renner B, Sawatzki G, editors. New perspectives in infant nutrition. New York: Thieme Medical Publication; 1995. p. 57-65.
[111] Alos L, Lujan B, Castillo M, *et al.* Expression of membrane-bound mucins (MUC1 and MUC4) and secreted mucins (MUC2, MUC5AC, MUC5B, MUC6 and MUC7) in mucoepidermoid carcinomas of salivary glands. Am J Surg Pathol 2005;29:806-13.
[112] Kusafuka K, Ueno T, Kurihara K, *et al.* Cystadenoma of the palate: immunohistochemistry of mucins. Pathol Int 2008;58:524-8.
[113] Warson C, Van De Bovenkamp JH, Korteland-Van Male AM, *et al.* Barrett's esophagus is characterized by expression of gastric-type mucins (MUC5AC, MUC6) and TFF peptides (TFF1 and TFF2), but the risk of carcinoma development may be indicated by the intestinal-type mucin, MUC2. Hum Pathol 2002;33:660-8.
[114] Arul GS, Moorghen M, Myerscough N, *et al.* Mucin gene expression in Barrett's oesophagus: an *in situ* hybridisation and immunohistochemical study. Gut 2000;47:753-61.
[115] Seregni E, Botti C, Massaron S, *et al.* Structure, function and gene expression of epithelial mucins. Tumori 1997;83:625-32.
[116] Lopez-Ferrer A, Barranco C, de Bolos C. Apomucin expression and association with Lewis antigens during gastric development. Appl Immunohistochem Mol Morphol 2001;9:42-8.

[117] Reid CJ, Harris A. Developmental expression of mucin genes in the human gastrointestinal system. Gut 1998;42:220-6.
[118] Buisine MP, Devisme L, Savidge TC, *et al.* Mucin gene expression in human embryonic and fetal intestine. Gut 1998;43:519-24.
[119] Kim CH, Oh Y, Ha Y, *et al.* Expression of mucins in the mucosal surface of small intestines in 1 week-old pigs. J Vet Med Sci 2010;72:245-7.
[120] Lee KT, Liu TS. Mucin gene expression in gallbladder epithelium. J Formos Med Assoc 2002;101:762-8.
[121] Ho SB, Shekels LL, Toribara NW, *et al.* Altered mucin core peptide expression in acute and chronic cholecystitis. Dig Dis Sci 2000;45:1061-71.
[122] Vandenhaute B, Buisine MP, Debailleul V, *et al.* Mucin gene expression in biliary epithelial cells. J Hepatol 1997;27:1057-66.
[123] Gouyer V, Leir SH, Tetaert D, *et al.* The characterization of the first anti-mouse Muc6 antibody shows an increased expression of the mucin in pancreatic tissue of Cftr-knockout mice. Histochem Cell Biol 2010;133:517-25.
[124] Bartman AE, Buisine MP, Aubert JP, *et al.* The MUC6 secretory mucin gene is expressed in a wide variety of epithelial tissues. J Pathol 1998;186:398-405.
[125] Reid CJ, Hyde K, Ho SB, Harris A. Cystic fibrosis of the pancreas: involvement of MUC6 mucin in obstruction of pancreatic ducts. Mol Med 1997;3:403-11.
[126] Basturk O, Khayyata S, Klimstra DS, *et al.* Preferential expression of MUC6 in oncocytic and pancreatobiliary types of intraductal papillary neoplasms highlights a pyloropancreatic pathway, distinct from the intestinal pathway, in pancreatic carcinogenesis. Am J Surg Pathol 2010;34:364-70.
[127] Nagata K, Horinouchi M, Saitou M, *et al.* Mucin expression profile in pancreatic cancer and the precursor lesions. J Hepatobiliary Pancreat Surg 2007;14:243-54.
[128] Gipson IK, Ho SB, Spurr-Michaud SJ, *et al.* Mucin genes expressed by human female reproductive tract epithelia. Biol Reprod 1997;56:999-1011.
[129] Andersch-Bjorkman Y, Thomsson KA, Holmen Larsson JM, Ekerhovd E, Hansson GC. Large scale identification of proteins, mucins, and their O-glycosylation in the endocervical mucus during the menstrual cycle. Mol Cell Proteomics 2007;6:708-16.
[130] Zhao S, Hayasaka T, Osakabe M, *et al.* Mucin expression in nonneoplastic and neoplastic glandular epithelia of the uterine cervix. Int J Gynecol Pathol 2003;22:393-7.
[131] Alameda F, Mejias-Luque R, Garrido M, de Bolos C. Mucin genes (MUC2, MUC4, MUC5AC, and MUC6) detection in normal and pathological endometrial tissues. Int J Gynecol Pathol 2007;26:61-5.
[132] Arias EB, Verhage HG, Jaffe RC. Complementary deoxyribonucleic acid cloning and molecular characterization of an estrogen-dependent human oviductal glycoprotein. Biol Reprod 1994;51:685-94.
[133] Lapensee L, Paquette Y, Bleau G. Allelic polymorphism and chromosomal localization of the human oviductin gene (MUC9). Fertil Steril 1997;68:702-8.
[134] Leroy X, Ballereau C, Villers A, *et al.* MUC6 is a marker of seminal vesicle-ejaculatory duct epithelium and is useful for the differential diagnosis with prostate adenocarcinoma. Am J Surg Pathol 2003;27:519-21.
[135] Russo CL, Spurr-Michaud S, Tisdale A, *et al.* Mucin gene expression in human male urogenital tract epithelia. Hum Reprod 2006;21:2783-93.
[136] Stax MJ, van Montfort T, Sprenger RR, *et al.* Mucin 6 in seminal plasma binds DC-SIGN and potently blocks dendritic cell mediated transfer of HIV-1 to CD4(+) T-lymphocytes. Virology 2009;391:203-11.

[137] Nakajima K, Ota H, Zhang MX, *et al.* Expression of gastric gland mucous cell-type mucin in normal and neoplastic human tissues. J Histochem Cytochem 2003;51:1689-98.
[138] N'Dow J, Pearson JP, Bennett MK, Neal DE, Robson CN. Mucin gene expression in human urothelium and in intestinal segments transposed into the urinary tract. J Urol 2000;164:1398-404.
[139] Eriksen GV, Carlstedt I, Uldbjerg N, Ernst E. Cervical mucins affect the motility of human spermatozoa *in vitro*. Fertil Steril 1998;70:350-4.
[140] Leroy X, Devisme L, Buisine MP, *et al.* Expression of human mucin genes during normal and abnormal renal development. Am J Clin Pathol 2003;120:544-50.
[141] Kusafuka K, Itoh H, Sugiyama C, Nakajima T. Low-grade salivary duct carcinoma of the parotid gland: report of a case with immunohistochemical analysis. Med Mol Morphol 2010;43:178-84.
[142] Simpson RH, Prasad AR, Lewis JE, Skalova A, David L. Mucin-rich variant of salivary duct carcinoma: a clinicopathologic and immunohistochemical study of four cases. Am J Surg Pathol 2003;27:1070-9.
[143] Guillem P, Billeret V, Buisine MP, *et al.* Mucin gene expression and cell differentiation in human normal, premalignant and malignant esophagus. Int J Cancer 2000;88:856-61.
[144] Sagara M, Yonezawa S, Nagata K, *et al.* Expression of mucin 1 (MUC1) in esophageal squamous-cell carcinoma: its relationship with prognosis. Int J Cancer 1999;84:251-7.
[145] Endo T, Tamaki K, Arimura Y, *et al.* Expression of sulfated carbohydrate chain and core peptides of mucin detected by monoclonal antibodies in Barrett's esophagus and esophageal adenocarcinoma. J Gastroenterol 1998;33:811-5.
[146] Labouvie C, Machado JC, Carneiro F, *et al.* Differential expression of mucins and trefoil peptides in native epithelium, Barrett's metaplasia and squamous cell carcinoma of the oesophagus. J Cancer Res Clin Oncol 1999;125:71-6.
[147] Spechler SJ. Clinical practice. Barrett's Esophagus. N Engl J Med 2002;346:836-42.
[148] Endo H, Iijima K, Asanuma K, *et al.* Exogenous luminal nitric oxide exposure accelerates columnar transformation of rat esophagus. Int J Cancer 2010;127:2009-19.
[149] Paull A, Trier JS, Dalton MD, *et al.* The histologic spectrum of Barrett's esophagus. N Engl J Med 1976;295:476-80.
[150] Glickman JN, Blount PL, Sanchez CA, *et al.* Mucin core polypeptide expression in the progression of neoplasia in Barrett's esophagus. Hum Pathol 2006;37:1304-15.
[151] Yamamoto S, Kijima H, Hara T, *et al.* Mucin expression and proliferating cell index of esophageal Barrett's adenocarcinoma. Int J Mol Med 2005;16:375-80.
[152] Chaves P, Cruz C, Dias Pereira A, *et al.* Gastric and intestinal differentiation in Barrett's metaplasia and associated adenocarcinoma. Dis Esophagus 2005;18:383-7.
[153] Kuroda N, Onishi K, Lee GH. Combined tubular adenocarcinoma and hepatoid adenocarcinoma arising in Barrett esophagus. Ann Diagn Pathol 2010;15:450-3.
[154] Asthana N, Mandich D, Ligato S. Esophageal polypoid dysplasia of gastric foveolar phenotype with focal intramucosal carcinoma associated with Barrett's esophagus. Am J Surg Pathol 2008;32:1581-5.
[155] Nokubi M, Kawanowa K, Kawata H, Hanatsuka K, Hosoya Y. Extremely well-differentiated adenocarcinoma of the gastric cardia: a unique case with columnar cells and laminated stones. Pathol Int 2004;54:854-60.
[156] Correa P. A human model of gastric carcinogenesis. Cancer Res 1988;48:3554-60.

[157] Rokkas T, Filipe MI, Sladen GE. Detection of an increased incidence of early gastric cancer in patients with intestinal metaplasia type III who are closely followed up. Gut 1991;32:1110-3.

[158] Nakasaki H, Mitomi T, Noto T, *et al.* Mosaicism in the expression of tumor-associated carbohydrate antigens in human colonic and gastric cancers. Cancer Res 1989;49:3662-9.

[159] Dong B, Xie YQ, Chen K, *et al.* Differences in biological features of gastric dysplasia, indefinite dysplasia, reactive hyperplasia and discriminant analysis of these lesions. World J Gastroenterol 2005;11:3595-600.

[160] Mino-Kenudson M, Tomita S, Lauwers GY. Mucin expression in reactive gastropathy: an immunohistochemical analysis. Arch Pathol Lab Med 2007;131:86-90.

[161] Zheng H, Takahashi H, Nakajima T, *et al.* MUC6 down-regulation correlates with gastric carcinoma progression and a poor prognosis: an immunohistochemical study with tissue microarrays. J Cancer Res Clin Oncol 2006;132:817-23.

[162] Marques T, David L, Reis C, Nogueira A. Topographic expression of MUC5AC and MUC6 in the gastric mucosa infected by Helicobacter pylori and in associated diseases. Pathol Res Pract 2005;201:665-72.

[163] Beil W, Enss ML, Muller S, *et al.* Role of vacA and cagA in Helicobacter pylori inhibition of mucin synthesis in gastric mucous cells. J Clin Microbiol 2000;38:2215-8.

[164] Bravo JC, Correa P. Sulphomucins favour adhesion of Helicobacter pylori to metaplastic gastric mucosa. J Clin Pathol 1999;52:137-40.

[165] Hidaka E, Ota H, Hidaka H, *et al.* Helicobacter pylori and two ultrastructurally distinct layers of gastric mucous cell mucins in the surface mucous gel layer. Gut 2001;49:474-80.

[166] Micots I, Augeron C, Laboisse CL, Muzeau F, Megraud F. Mucin exocytosis: a major target for Helicobacter pylori. J Clin Pathol 1993;46:241-5.

[167] Slomiany BL, Slomiany A. Disruption in gastric mucin synthesis by Helicobacter pylori lipopolysaccharide involves ERK and p38 mitogen-activated protein kinase participation. Biochem Biophys Res Commun 2002;294:220-4.

[168] Kang HM, Kim N, Park YS, *et al.* Effects of Helicobacter pylori Infection on gastric mucin expression. J Clin Gastroenterol 2008;42:29-35.

[169] Fukuda M, Kawakubo M, Ito Y, *et al.* Assay of human gastric mucin as a natural antibiotic against Helicobacter pylori. Methods Enzymol 2006;415:164-79.

[170] Correa P. Chronic gastritis as a cancer precursor. Scand J Gastroenterol Suppl 1984;104:131-6.

[171] Correa P. Helicobacter pylori and gastric carcinogenesis. Am J Surg Pathol 1995;19 Suppl 1:S37-43.

[172] Dinis-Ribeiro M, Lopes C, da Costa-Pereira A, *et al.* A follow up model for patients with atrophic chronic gastritis and intestinal metaplasia. J Clin Pathol 2004;57:177-82.

[173] Brown LM. Helicobacter pylori: epidemiology and routes of transmission. Epidemiol Rev 2000;22:283-97.

[174] Peek RM, Jr., Blaser MJ. Helicobacter pylori and gastrointestinal tract adenocarcinomas. Nat Rev Cancer 2002;2:28-37.

[175] Ho SB, Takamura K, Anway R, *et al.* The adherent gastric mucous layer is composed of alternating layers of MUC5AC and MUC6 mucin proteins. Dig Dis Sci 2004;49:1598-606.

[176] Vinall LE, King M, Novelli M, *et al.* Altered expression and allelic association of the hypervariable membrane mucin MUC1 in Helicobacter pylori gastritis. Gastroenterology 2002;123:41-9.

[177] Nakayama J, Yeh JC, Misra AK, *et al.* Expression cloning of a human alpha1, 4-N-acetylglucosaminyltransferase that forms GlcNAcalpha1-->4Galbeta-->R, a glycan specifically expressed in the gastric gland mucous cell-type mucin. Proc Natl Acad Sci U S A 1999;96:8991-6.

[178] Byrd JC, Yan P, Sternberg L, *et al.* Aberrant expression of gland-type gastric mucin in the surface epithelium of Helicobacter pylori-infected patients. Gastroenterology 1997;113:455-64.

[179] Byrd JC, Yunker CK, Xu QS, Sternberg LR, Bresalier RS. Inhibition of gastric mucin synthesis by Helicobacter pylori. Gastroenterology 2000;118:1072-9.

[180] Matsuzwa M, Ota H, Hayama M, *et al.* Helicobacter pylori infection up-regulates gland mucous cell-type mucins in gastric pyloric mucosa. Helicobacter 2003;8:594-600.

[181] Nam SY, Kim N, Lee CS, *et al.* Gastric mucosal protection *via* enhancement of MUC5AC and MUC6 by geranylgeranylacetone. Dig Dis Sci 2005;50:2110-20.

[182] Murakami M, Oketani K, Fujisaki H, *et al.* Effects of the antiulcer drug geranylgeranylacetone on aspirin-induced gastric ulcers in rats. Jpn J Pharmacol 1982;32:299-306.

[183] Terano A, Shiga J, Hiraishi H, Ota S, Sugimoto T. Protective action of tetraprenylacetone against ethanol-induced damage in rat gastric mucosa. Digestion 1986;35:182-8.

[184] Arakawa T, Yamada H, Nakamura A, *et al.* Gastric cytoprotection by tetraprenylacetone in human subjects. Digestion 1988;39:111-7.

[185] Wang RQ, Fang DC. Effects of Helicobacter pylori infection on mucin expression in gastric carcinoma and pericancerous tissues. J Gastroenterol Hepatol 2006;21:425-31.

[186] Morgenstern S, Koren R, Fraser G, Okon E, Niv Y. Gastric corpus mucin expression after partial gastrectomy, in relation to colonization with Helicobacter pylori. J Clin Gastroenterol 2001;32:218-21.

[187] Kim H, Seo JH, Kim KH. The effect of p38 mitogen-activated protein kinase on mucin gene expression and apoptosis in Helicobacter pylori-infected gastric epithelial cells. Ann N Y Acad Sci 2003;1010:90-4.

[188] Van de Bovenkamp JH, Mahdavi J, Korteland-Van Male AM, *et al.* The MUC5AC glycoprotein is the primary receptor for Helicobacter pylori in the human stomach. Helicobacter 2003;8:521-32.

[189] Van den Brink GR, Tytgat KM, Van der Hulst RW, *et al.* H pylori colocalises with MUC5AC in the human stomach. Gut 2000;46:601-7.

[190] Shiotani A, Haruma K, Uedo N, *et al.* Histological risk markers for non-cardia early gastric cancer. Pattern of mucin expression and gastric cancer. Virchows Arch 2006;449:652-9.

[191] Shiotani A, Iishi H, Uedo N, *et al.* Histologic and serum risk markers for noncardia early gastric cancer. Int J Cancer 2005;115:463-9.

[192] Meining A, Morgner A, Miehlke S, Bayerdorffer E, Stolte M. Atrophy-metaplasia-dysplasia-carcinoma sequence in the stomach: a reality or merely an hypothesis? Best Pract Res Clin Gastroenterol 2001;15:983-98.

[193] Dixon MF. Prospects for intervention in gastric carcinogenesis: reversibility of gastric atrophy and intestinal metaplasia. Gut 2001;49:2-4.

[194] Babu SD, Jayanthi V, Devaraj N, Reis CA, Devaraj H. Expression profile of mucins (MUC2, MUC5AC and MUC6) in Helicobacter pylori infected pre-neoplastic and neoplastic human gastric epithelium. Mol Cancer 2006;5:10.

[195] Reis CA, David L, Correa P, *et al.* Intestinal metaplasia of human stomach displays distinct patterns of mucin (MUC1, MUC2, MUC5AC, and MUC6) expression. Cancer Res 1999;59:1003-7.
[196] Uemura N, Okamoto S, Yamamoto S, *et al.* Helicobacter pylori infection and the development of gastric cancer. N Engl J Med 2001;345:784-9.
[197] Linden S, Semino-Mora C, Liu H, Rick J, Dubois A. Role of mucin Lewis status in resistance to Helicobacter pylori infection in pediatric patients. Helicobacter 2010;15:251-8.
[198] Seki T, Ito T, Kawachi H, *et al.* Relationship between mucin expression of gastric intramucosal signet ring cell carcinoma and its background mucosa. J Med Dent Sci 2009;56:25-35.
[199] Wada R, Yamaguchi T, Tanizaki T. Mucin phenotypic expression and p53 gene abnormality of gastric super-minute well-differentiated adenocarcinoma: re-evaluation with relationship between histogenesis of well-differentiated adenocarcinoma and intestinal metaplasia in distal stomach. J Carcinog 2005;4:14.
[200] Choi JS, Kim MA, Lee HE, Lee HS, Kim WH. Mucinous gastric carcinomas: clinicopathologic and molecular analyses. Cancer 2009;115:3581-90.
[201] Sasaki M, Nakanuma Y, Kim YS. Expression of apomucins in the intrahepatic biliary tree in hepatolithiasis differs from that in normal liver and extrahepatic biliary obstruction. Hepatology 1998;27:54-61.
[202] Sasaki M, Nakanuma Y, Ho SB, Kim YS. Increased MUC6 apomucin expression is a characteristic of reactive biliary epithelium in chronic viral hepatitis. J Pathol 1998;185:191-8.
[203] Chang HJ, Kim SW, Lee BL, Hong EK, Kim WH. Phenotypic alterations of mucins and cytokeratins during gallbladder carcinogenesis. Pathol Int 2004;54:576-84.
[204] Chang HJ, Jee CD, Kim WH. Mutation and altered expression of beta-catenin during gallbladder carcinogenesis. Am J Surg Pathol 2002;26:758-66.
[205] Park SY, Roh SJ, Kim YN, *et al.* Expression of MUC1, MUC2, MUC5AC and MUC6 in cholangiocarcinoma: prognostic impact. Oncol Rep 2009;22:649-57.
[206] Nakanuma Y, Sasaki M, Ishikawa A, *et al.* Biliary papillary neoplasm of the liver. Histol Histopathol 2002;17:851-61.
[207] Aishima S, Kuroda Y, Nishihara Y, *et al.* Gastric mucin phenotype defines tumour progression and prognosis of intrahepatic cholangiocarcinoma: gastric foveolar type is associated with aggressive tumour behaviour. Histopathology 2006;49:35-44.
[208] Thuwajit P, Chawengrattanachot W, Thuwajit C, *et al.* Enhanced expression of mucin 6 glycoprotein in cholangiocarcinoma tissue from patients in Thailand as a prognostic marker for survival. J Gastroenterol Hepatol 2008;23:771-8.
[209] Nagata S, Ajioka Y, Nishikura K, *et al.* Co-expression of gastric and biliary phenotype in pyloric-gland type adenoma of the gallbladder: immunohistochemical analysis of mucin profile and CD10. Oncol Rep 2007;17:721-9.
[210] Khan SA, Davidson BR, Goldin R, *et al.* Guidelines for the diagnosis and treatment of cholangiocarcinoma: consensus document. Gut 2002;51 Suppl 6:VI1-9.
[211] Sasaki M, Nakanuma Y, Kim YS. Characterization of apomucin expression in intrahepatic cholangiocarcinomas and their precursor lesions: an immunohistochemical study. Hepatology 1996;24:1074-8.

[212] Kim GE, Bae HI, Park HU, *et al.* Aberrant expression of MUC5AC and MUC6 gastric mucins and sialyl Tn antigen in intraepithelial neoplasms of the pancreas. Gastroenterology 2002;123:1052-60.

[213] Matsukita S, Nomoto M, Kitajima S, *et al.* Expression of mucins (MUC1, MUC2, MUC5AC and MUC6) in mucinous carcinoma of the breast: comparison with invasive ductal carcinoma. Histopathology 2003;42:26-36.

[214] Aishima S, Nishihara Y, Tsujita E, *et al.* Biliary neoplasia with extensive intraductal spread associated with liver cirrhosis: a hitherto unreported variant of biliary intraepithelial neoplasia. Hum Pathol 2008;39:939-47.

[215] Velcich A, Yang W, Heyer J, *et al.* Colorectal cancer in mice genetically deficient in the mucin Muc2. Science 2002;295:1726-9.

[216] Mall AS, Tyler MG, Ho SB, *et al.* The expression of MUC mucin in cholangiocarcinoma. Pathol Res Pract 2010;206:805-9.

[217] Franca AV, Elias Junior J, Lima BL, Martinelli AL, Carrilho FJ. Diagnosis, staging and treatment of hepatocellular carcinoma. Braz J Med Biol Res 2004;37:1689-705.

[218] Ichikawa T, Yamamoto T, Uenishi T, *et al.* Clinicopathological implications of immunohistochemically demonstrated mucin core protein expression in hepatocellular carcinoma. J Hepatobiliary Pancreat Surg 2006;13:245-51.

[219] Cao Y, Blohm D, Ghadimi BM, *et al.* Mucins (MUC1 and MUC3) of gastrointestinal and breast epithelia reveal different and heterogeneous tumor-associated aberrations in glycosylation. J Histochem Cytochem 1997;45:1547-57.

[220] Cao Y, Karsten U, Otto G, Bannasch P. Expression of MUC1, Thomsen-Friedenreich antigen, Tn, sialosyl-Tn, and alpha2,6-linked sialic acid in hepatocellular carcinomas and preneoplastic hepatocellular lesions. Virchows Arch 1999;434:503-9.

[221] Ma CK, Zarbo RJ, Frierson HF, Jr., Lee MW. Comparative immunohistochemical study of primary and metastatic carcinomas of the liver. Am J Clin Pathol 1993;99:551-7.

[222] Sasaki M, Yamato T, Nakanuma Y. Expression of sialyl-Tn, Tn and T antigens in primary liver cancer. Pathol Int 1999;49:325-31.

[223] Sasaki M, Nakanuma Y, Ho SB, Kim YS. Cholangiocarcinomas arising in cirrhosis and combined hepatocellular-cholangiocellular carcinomas share apomucin profiles. Am J Clin Pathol 1998;109:302-8.

[224] Andrianifahanana M, Moniaux N, Schmied BM, *et al.* Mucin (MUC) gene expression in human pancreatic adenocarcinoma and chronic pancreatitis: a potential role of MUC4 as a tumor marker of diagnostic significance. Clin Cancer Res 2001;7:4033-40.

[225] Wada R, Ogawa K, Yamaguchi T, Tanizaki T, Matsumoto M. Intercalated duct cell is starting point in development of pancreatic ductal carcinoma? J Carcinog 2005;4:9.

[226] Leir SH, Harris A. MUC6 mucin expression inhibits tumor cell invasion. Exp Cell Res 2011;317:2408-19.

[227] Trezise AE, Chambers JA, Wardle CJ, Gould S, Harris A. Expression of the cystic fibrosis gene in human foetal tissues. Hum Mol Genet 1993;2:213-8.

[228] Anderson MP, Rich DP, Gregory RJ, Smith AE, Welsh MJ. Generation of cAMP-activated chloride currents by expression of CFTR. Science 1991;251:679-82.

[229] Rosewicz S, Wiedenmann B. Pancreatic carcinoma. Lancet 1997;349:485-9.

[230] Ringel J, Lohr M. The MUC gene family: their role in diagnosis and early detection of pancreatic cancer. Mol Cancer 2003;2:9.

[231] Hruban RH, Adsay NV, Albores-Saavedra J, *et al.* Pancreatic intraepithelial neoplasia: a new nomenclature and classification system for pancreatic duct lesions. Am J Surg Pathol2001;25:579-86.
[232] Biankin AV, Biankin SA, Kench JG, *et al.* Aberrant p16(INK4A) and DPC4/Smad4 expression in intraductal papillary mucinous tumours of the pancreas is associated with invasive ductal adenocarcinoma. Gut 2002;50:861-8.
[233] Prasad NB, Biankin AV, Fukushima N, *et al.* Gene expression profiles in pancreatic intraepithelial neoplasia reflect the effects of Hedgehog signaling on pancreatic ductal epithelial cells. Cancer Res 2005;65:1619-26.
[234] Luttges J, Zamboni G, Longnecker D, Kloppel G. The immunohistochemical mucin expression pattern distinguishes different types of intraductal papillary mucinous neoplasms of the pancreas and determines their relationship to mucinous noncystic carcinoma and ductal adenocarcinoma. Am J Surg Pathol 2001;25:942-8.
[235] Yonezawa S, Horinouchi M, Osako M, *et al.* Gene expression of gastric type mucin (MUC5AC) in pancreatic tumors: its relationship with the biological behavior of the tumor. Pathol Int 1999;49:45-54.
[236] Nakamura A, Horinouchi M, Goto M, *et al.* New classification of pancreatic intraductal papillary-mucinous tumour by mucin expression: its relationship with potential for malignancy. J Pathol 2002;197:201-10.
[237] Fukushima N, Mukai K, Kanai Y, *et al.* Intraductal papillary tumors and mucinous cystic tumors of the pancreas: clinicopathologic study of 38 cases. Hum Pathol 1997;28:1010-7.
[238] Adsay NV, Conlon KC, Zee SY, Brennan MF, Klimstra DS. Intraductal papillary-mucinous neoplasms of the pancreas: an analysis of *in situ* and invasive carcinomas in 28 patients. Cancer 2002;94:62-77.
[239] Thompson R. Cancer: Epithelial subtype influences the prognosis of invasive IPMN. Nat Rev Gastroenterol Hepatol 2011;8:420.
[240] Kloppel G, Luttges J. WHO-classification 2000: exocrine pancreatic tumors. Verh Dtsch Ges Pathol 2001;85:219-28.
[241] Luttges J, Feyerabend B, Buchelt T, Pacena M, Kloppel G. The mucin profile of noninvasive and invasive mucinous cystic neoplasms of the pancreas. Am J Surg Pathol 2002;26:466-71.
[242] Ji Y, Xu JF, Kuang TT, *et al.* Mucin profile of the pancreatic mucinous cystic neoplasms. Chin Med J (Engl) 2006;119:328-30.
[243] Gouyer V, Gottrand F, Desseyn JL. The extraordinarily complex but highly structured organization of intestinal mucus-gel unveiled in multicolor images. PLoS One 2011;6:e18761.
[244] Byrd JC, Bresalier RS. Mucins and mucin binding proteins in colorectal cancer. Cancer Metastasis Rev 2004;23:77-99.
[245] Sylvester PA, Myerscough N, Warren BF, *et al.* Differential expression of the chromosome 11 mucin genes in colorectal cancer. J Pathol 2001;195:327-35.
[246] Longman RJ, Poulsom R, Corfield AP, *et al.* Alterations in the composition of the supramucosal defense barrier in relation to disease severity of ulcerative colitis. J Histochem Cytochem 2006;54:1335-48.
[247] Tatsumi N, Kushima R, Vieth M, *et al.* Cytokeratin 7/20 and mucin core protein expression in ulcerative colitis-associated colorectal neoplasms. Virchows Arch 2006;448:756-62.

[248] Buisine MP, Desreumaux P, Leteurtre E, *et al.* Mucin gene expression in intestinal epithelial cells in Crohn's disease. Gut 2001;49:544-51.

[249] Buisine MP, Desreumaux P, Debailleul V, *et al.* Abnormalities in mucin gene expression in Crohn's disease. Inflamm Bowel Dis 1999;5:24-32.

[250] Buisine MP, Devisme L, Degand P, *et al.* Developmental mucin gene expression in the gastroduodenal tract and accessory digestive glands. II. Duodenum and liver, gallbladder, and pancreas. J Histochem Cytochem 2000;48:1667-76.

[251] Wright NA, Pike C, Elia G. Induction of a novel epidermal growth factor-secreting cell lineage by mucosal ulceration in human gastrointestinal stem cells. Nature 1990;343:82-5.

[252] Buisine MP, Devisme L, Maunoury V, *et al.* Developmental mucin gene expression in the gastroduodenal tract and accessory digestive glands. I. Stomach. A relationship to gastric carcinoma. J Histochem Cytochem 2000;48:1657-66.

[253] Roberts IS, Stoddart RW. Ulcer-associated cell lineage ('pyloric metaplasia') in Crohn's disease: a lectin histochemical study. J Pathol 1993;171:13-9.

[254] Longman RJ, Douthwaite J, Sylvester PA, *et al.* Coordinated localisation of mucins and trefoil peptides in the ulcer associated cell lineage and the gastrointestinal mucosa. Gut 2000;47:792-800.

[255] Kaneko Y, Nakamura T, Hayama M, *et al.* Altered expression of CDX-2, PDX-1 and mucin core proteins in "Ulcer-associated cell lineage (UACL)" in Crohn's disease. J Mol Histol 2008;39:161-8.

[256] Levi GS, Harpaz N. Intestinal low-grade tubuloglandular adenocarcinoma in inflammatory bowel disease. Am J Surg Pathol 2006;30:1022-9.

[257] Shi C, Scudiere JR, Cornish TC, *et al.* Clear cell change in colonic tubular adenoma and corresponding colonic clear cell adenocarcinoma is associated with an altered mucin core protein profile. Am J Surg Pathol 2010;34:1344-50.

[258] Barisella M, Lampis A, Perrone F, Carbone A. Clear cell adenocarcinoma of the colon is a unique morphological variant of intestinal carcinoma: case report with molecular analysis. World J Gastroenterol 2008;14:6575-7.

[259] Braumann C, Schwabe M, Ordemann J, Jacobi CA. The clear cell adenocarcinoma of the colon: case report and review of the literature. Int J Colorectal Dis 2004;19:264-7.

[260] Domoto H, Terahata S, Senoh A, *et al.* Clear cell change in colorectal adenomas: its incidence and histological characteristics. Histopathology 1999;34:250-6.

[261] Hellstrom HR, Fisher ER. Physaliferous Variant of Carcinoma of Colon. Cancer 1964;17:259-63.

[262] Fujita K, Hirahashi M, Yamamoto H, *et al.* Mucin core protein expression in serrated polyps of the large intestine. Virchows Arch 2010;457:443-9.

[263] Snover DC, Jass JR, Fenoglio-Preiser C, Batts KP. Serrated polyps of the large intestine: a morphologic and molecular review of an evolving concept. Am J Clin Pathol 2005;124:380-91.

[264] Kambara T, Simms LA, Whitehall VL, *et al.* BRAF mutation is associated with DNA methylation in serrated polyps and cancers of the colorectum. Gut 2004;53:1137-44.

[265] Yang S, Farraye FA, Mack C, Posnik O, O'Brien MJ. BRAF and KRAS Mutations in hyperplastic polyps and serrated adenomas of the colorectum: relationship to histology and CpG island methylation status. Am J Surg Pathol 2004;28:1452-9.

[266] Goldstein NS, Bhanot P, Odish E, Hunter S. Hyperplastic-like colon polyps that preceded microsatellite-unstable adenocarcinomas. Am J Clin Pathol 2003;119:778-96.

[267] Lu FI, van Niekerk de W, Owen D, *et al.* Longitudinal outcome study of sessile serrated adenomas of the colorectum: an increased risk for subsequent right-sided colorectal carcinoma. Am J Surg Pathol 2010;34:927-34.

[268] Bartman AE, Sanderson SJ, Ewing SL, *et al.* Aberrant expression of MUC5AC and MUC6 gastric mucin genes in colorectal polyps. Int J Cancer 1999;80:210-8.

[269] Mochizuka A, Uehara T, Nakamura T, Kobayashi Y, Ota H. Hyperplastic polyps and sessile serrated'adenomas' of the colon and rectum display gastric pyloric differentiation. Histochem Cell Biol 2007;128:445-55.

[270] Owens SR, Chiosea SI, Kuan SF. Selective expression of gastric mucin MUC6 in colonic sessile serrated adenoma but not in hyperplastic polyp aids in morphological diagnosis of serrated polyps. Mod Pathol 2008;21:660-9.

[271] Bartley AN, Thompson PA, Buckmeier JA, *et al.* Expression of gastric pyloric mucin, MUC6, in colorectal serrated polyps. Mod Pathol 2010;23:169-76.

[272] Hafez ESE. Human Reproduction: Conception and Contraception. Hagerstown: Harper and Row; 1980. p. 221-52.

[273] Aubert S, Van Seuningen I, Leroy X. Mucins in the uro-genital tract. Potential for therapeutic approaches using mucins. In: Van Seuningen I, editor. The Epithelial mucins: Structure/Function> Roles in Cancer and Inflammation Diseases. Kerala, India: Research Singpost 2008. p. 249-72.

[274] Boman F, Buisine MP, Wacrenier A, *et al.* Mucin gene transcripts in benign and borderline mucinous tumours of the ovary: an *in situ* hybridization study. J Pathol 2001;193:339-44.

[275] Hirabayashi K, Yasuda M, Kajiwara H, *et al.* Alterations in mucin expression in ovarian mucinous tumors: immunohistochemical analysis of MUC2, MUC5AC, MUC6, and CD10 expression. Acta Histochem Cytochem 2008;41:15-21.

[276] Morrison C, Merati K, Marsh WL, Jr., *et al.* The mucin expression profile of endometrial carcinoma and correlation with clinical-pathologic parameters. Appl Immunohistochem Mol Morphol 2007;15:426-31.

[277] Longacre TA, Hendrickson MR. Diffusely infiltrative endometrial adenocarcinoma: an adenoma malignum pattern of myoinvasion. Am J Surg Pathol 1999;23:69-78.

[278] Mai KT, Yazdi HM, Boone SA.'Minimal deviation' endometrioid carcinoma with oncocytic change of the endometrium. Arch Pathol Lab Med 1995;119:751-4.

[279] Landry D, Mai KT, Senterman MK, *et al.* Endometrioid adenocarcinoma of the uterus with a minimal deviation invasive pattern. Histopathology 2003;42:77-82.

[280] Abiko K, Baba T, Ogawa M, *et al.* Minimal deviation mucinous adenocarcinoma ('adenoma malignum') of the uterine corpus. Pathol Int 2010;60:42-7.

[281] Mikami Y, Kiyokawa T, Hata S, *et al.* Gastrointestinal immunophenotype in adenocarcinomas of the uterine cervix and related glandular lesions: a possible link between lobular endocervical glandular hyperplasia/pyloric gland metaplasia and'adenoma malignum'. Mod Pathol 2004;17:962-72.

[282] Kondo T, Hashi A, Murata S, *et al.* Endocervical adenocarcinomas associated with lobular endocervical glandular hyperplasia: a report of four cases with histochemical and immunohistochemical analyses. Mod Pathol 2005;18:1199-210.

[283] Nara M, Hashi A, Murata S, *et al.* Lobular endocervical glandular hyperplasia as a presumed precursor of cervical adenocarcinoma independent of human papillomavirus infection. Gynecol Oncol 2007;106:289-98.

[284] Mikami Y, Kiyokawa T, Sasajima Y, *et al.* Reappraisal of synchronous and multifocal mucinous lesions of the female genital tract: a close association with gastric metaplasia. Histopathology 2009;54:184-91.

[285] Walker AN, Mills SE. Unusual variants of uterine cervical carcinoma. Pathol Annu 1987;22 (Pt 1):277-310.

[286] Young RH, Clement PB. Endocervical adenocarcinoma and its variants: their morphology and differential diagnosis. Histopathology 2002;41:185-207.

[287] Nanbu K, Konishi I, Yamamoto S, *et al.* Minimal deviation adenocarcinoma of endometrioid type may arise in the isthmus: clinicopathological and immunohistochemical study of two cases. Gynecol Oncol 1995;58:136-41.

[288] Scully RE, Bonfiglio TA. Histological typing of Female Genital Tract tumours. Berlin: Springer; 1994.

[289] Tavassoli FA, Devilee P. World Health Organization Classification of Tumors, Tumors of the Breast and Female Genital Organs. Lyon: IARC Press; 2003.

[290] Ishihara K, Kurihara M, Goso Y, *et al.* Peripheral alpha-linked N-acetylglucosamine on the carbohydrate moiety of mucin derived from mammalian gastric gland mucous cells: epitope recognized by a newly characterized monoclonal antibody. Biochem J 1996;318 (Pt 2):409-16.

[291] Ishii K, Hosaka N, Toki T, *et al.* A new view of the so-called adenoma malignum of the uterine cervix. Virchows Arch 1998;432:315-22.

[292] Ishii K, Katsuyama T, Ota H, *et al.* Cytologic and cytochemical features of adenoma malignum of the uterine cervix. Cancer 1999;87:245-53.

[293] Tsuda H, Mikami Y, Kaku T, *et al.* Reproducible and clinically meaningful differential diagnosis is possible between lobular endocervical glandular hyperplasia and'adenoma malignum' based on common histopathological criteria. Pathol Int 2005;55:412-8.

[294] Kojima A, Mikami Y, Sudo T, *et al.* Gastric morphology and immunophenotype predict poor outcome in mucinous adenocarcinoma of the uterine cervix. Am J Surg Pathol 2007;31:664-72.

[295] McKelvey JL. Early invasive carcinoma of the cervix. Minn Med 1968;51:1205.

[296] Gilks CB, Young RH, Aguirre P, DeLellis RA, Scully RE. Adenoma malignum (minimal deviation adenocarcinoma) of the uterine cervix. A clinicopathological and immunohistochemical analysis of 26 cases. Am J Surg Pathol 1989;13:717-29.

[297] Young RH, Scully RE. Invasive adenocarcinoma and related tumors of the uterine cervix. Semin Diagn Pathol 1990;7:205-27.

[298] McCluggage WG, Harley I, Houghton JP, *et al.* Composite cervical adenocarcinoma composed of adenoma malignum and gastric type adenocarcinoma (dedifferentiated adenoma malignum) in a patient with Peutz Jeghers syndrome. J Clin Pathol 2010;63:935-41.

[299] Shintaku M, Kushima R, Abiko K. Colloid carcinoma of the intestinal type in the uterine cervix: mucin immunohistochemistry. Pathol Int 2010;60:119-24.

[300] Lewis TL. Colloid (mucus secreting) carcinoma of the cervix. J Obstet Gynaecol Br Commonw 1971;78:1128-32.

[301] Naito S, Yamazumi K, Yakata Y, *et al.* Immunohistochemical examination of mucinous cystadenoma of the testis. Pathol Int 2004;54:355-9.

[302] Leroy X, Copin MC, Devisme L, *et al.* Expression of human mucin genes in normal kidney and renal cell carcinoma. Histopathology 2002;40:450-7.

[303] Jankovic Velickovic L, Katic V, Hattori T, *et al.* Differences in the expression of mucins in various forms of cystitis glandularis. Pathol Res Pract 2007;203:653-8.

[304] Lopez-Beltran A, Montironi R, Blanca A, Cheng L. Invasive micropapillary urothelial carcinoma of the bladder. Hum Pathol 2010;41:1159-64.

[305] Jemal A, Murray T, Samuels A, *et al.* Cancer statistics, 2003. CA Cancer J Clin 2003;53:5-26.

[306] Live DH, Williams LJ, Kuduk SD, *et al.* Probing cell-surface architecture through synthesis: an NMR-determined structural motif for tumor-associated mucins. Proc Natl Acad Sci U S A 1999;96:3489-93.

[307] Copin MC, Buisine MP, Leteurtre E, *et al.* Mucinous bronchioloalveolar carcinomas display a specific pattern of mucin gene expression among primary lung adenocarcinomas. Hum Pathol 2001;32:274-81.

[308] Cozzi PJ, Wang J, Delprado W, *et al.* MUC1, MUC2, MUC4, MUC5AC and MUC6 expression in the progression of prostate cancer. Clin Exp Metastasis 2005;22:565-73.

[309] Legrier ME, de Pinieux G, Boye K, *et al.* Mucinous differentiation features associated with hormonal escape in a human prostate cancer xenograft. Br J Cancer 2004;90:720-7.

[310] Epstein JI, Lieberman PH. Mucinous adenocarcinoma of the prostate gland. Am J Surg Pathol 1985;9:299-308.

[311] Sentinelli S. Mucins in prostatic intra-epithelial neoplasia and prostatic carcinoma. Histopathology 1993;22:271-4.

[312] McWilliam LJ, Manson C, George NJ. Neuroendocrine differentiation and prognosis in prostatic adenocarcinoma. Br J Urol 1997;80:287-90.

[313] Zhang MX, Nakayama J, Hidaka E, *et al.* Immunohistochemical demonstration of alpha1,4-N-acetylglucosaminyltransferase that forms GlcNAcalpha1,4Galbeta residues in human gastrointestinal mucosa. J Histochem Cytochem 2001;49:587-96.

[314] Rose MC. Mucins: structure, function, and role in pulmonary diseases. Am J Physiol 1992;263:L413-29.

[315] Rogers DF. Airway goblet cells: responsive and adaptable front-line defenders. Eur Respir J 1994;7:1690-706.

[316] Lopez-Ferrer A, Curull V, Barranco C, *et al.* Mucins as differentiation markers in bronchial epithelium. Squamous cell carcinoma and adenocarcinoma display similar expression patterns. Am J Respir Cell Mol Biol 2001;24:22-9.

[317] Martinez-Anton A, Debolos C, Garrido M, *et al.* Mucin genes have different expression patterns in healthy and diseased upper airway mucosa. Clin Exp Allergy 2006;36:448-57.

[318] Caramori G, Di Gregorio C, Carlstedt I, *et al.* Mucin expression in peripheral airways of patients with chronic obstructive pulmonary disease. Histopathology 2004;45:477-84.

[319] Vestbo J, Prescott E, Lange P. Association of chronic mucus hypersecretion with FEV1 decline and chronic obstructive pulmonary disease morbidity. Copenhagen City Heart Study Group. Am J Respir Crit Care Med 1996;153:1530-5.

[320] Rose MC, Nickola TJ, Voynow JA. Airway mucus obstruction: mucin glycoproteins, MUC gene regulation and goblet cell hyperplasia. Am J Respir Cell Mol Biol 2001;25:533-7.

[321] Borchers MT, Carty MP, Leikauf GD. Regulation of human airway mucins by acrolein and inflammatory mediators. Am J Physiol 1999;276:L549-55.

[322] Ali MS, Wilson JA, Bennett M, Pearson JP. Mucin gene expression in nasal polyps. Acta Otolaryngol 2005;125:618-24.

[323] Awaya H, Takeshima Y, Yamasaki M, Inai K. Expression of MUC1, MUC2, MUC5AC, and MUC6 in atypical adenomatous hyperplasia, bronchioloalveolar carcinoma,

adenocarcinoma with mixed subtypes, and mucinous bronchioloalveolar carcinoma of the lung. Am J Clin Pathol 2004;121:644-53.

[324] Perrais M, Pigny P, Copin MC, Aubert JP, Van Seuningen I. Induction of MUC2 and MUC5AC mucins by factors of the epidermal growth factor (EGF) family is mediated by EGF receptor/Ras/Raf/extracellular signal-regulated kinase cascade and Sp1. J Biol Chem 2002;277:32258-67.

[325] Yonezawa S, Sato E. Expression of mucin antigens in human cancers and its relationship with malignancy potential. Pathol Int 1997;47:813-30.

[326] Copin MC, Devisme L, Buisine MP, *et al.* From normal respiratory mucosa to epidermoid carcinoma: expression of human mucin genes. Int J Cancer 2000;86:162-8.

[327] Moran CA, Hochholzer L, Fishback N, Travis WD, Koss MN. Mucinous (so-called colloid) carcinomas of lung. Mod Pathol 1992;5:634-8.

[328] Tsuta K, Ishii G, Nitadori J, *et al.* Comparison of the immunophenotypes of signet-ring cell carcinoma, solid adenocarcinoma with mucin production, and mucinous bronchioloalveolar carcinoma of the lung characterized by the presence of cytoplasmic mucin. J Pathol 2006;209:78-87.

[329] Koga T, Hashimoto S, Sugio K, *et al.* Lung adenocarcinoma with bronchioloalveolar carcinoma component is frequently associated with foci of high-grade atypical adenomatous hyperplasia. Am J Clin Pathol 2002;117:464-70.

[330] Mori M, Rao SK, Popper HH, Cagle PT, Fraire AE. Atypical adenomatous hyperplasia of the lung: a probable forerunner in the development of adenocarcinoma of the lung. Mod Pathol 2001;14:72-84.

[331] Travis WD, Colby B, Corrin B. Histological Typing of Lung and Pleural Tumours. World Health Organization International Histological Classification of Tumours. 3rd ed. Berlin: Springer; 1999.

[332] Hayashi H, Ito T, Yazawa T, *et al.* Reduced expression of p27/Kip1 is associated with the development of pulmonary adenocarcinoma. J Pathol 2000;192:26-31.

[333] Seki N, Takasu T, Mandai K, *et al.* Expression of eukaryotic initiation factor 4E in atypical adenomatous hyperplasia and adenocarcinoma of the human peripheral lung. Clin Cancer Res 2002;8:3046-53.

[334] Barsky SH, Cameron R, Osann KE, Tomita D, Holmes EC. Rising incidence of bronchioloalveolar lung carcinoma and its unique clinicopathologic features. Cancer 1994;73:1163-70.

[335] Hamamoto A, Abe Y, Nishi M, *et al.* Aberrant expression of the gastric mucin MUC6 in human pulmonary adenocarcinoma xenografts. Int J Oncol 2005;26:891-6.

[336] Honda T, Ota H, Ishii K, *et al.* Mucinous bronchioloalveolar carcinoma with organoid differentiation simulating the pyloric mucosa of the stomach: clinicopathologic, histochemical, and immunohistochemical analysis. Am J Clin Pathol 1998;109:423-30.

[337] Nishiumi N, Abe Y, Inoue Y, *et al.* Use of 11p15 mucins as prognostic factors in small adenocarcinoma of the lung. Clin Cancer Res 2003;9:5616-9.

[338] Paulsen FP, Corfield AP, Hinz M, *et al.* Characterization of mucins in human lacrimal sac and nasolacrimal duct. Invest Ophthalmol Vis Sci 2003;44:1807-13.

[339] Paulsen F, Langer G, Hoffmann W, Berry M. Human lacrimal gland mucins. Cell Tissue Res 2004;316:167-77.

[340] Schafer G, Hoffmann W, Berry M, Paulsen F. [Lacrimal gland-associated mucins. Age related production and their role in the pathophysiology of dry eye]. Ophthalmologe 2005;102:175-83.

[341] Takeuchi K, Yagawa M, Ishinaga H, *et al.* Mucin gene expression in the effusions of otitis media with effusion. Int J Pediatr Otorhinolaryngol 2003;67:53-8.

[342] Kerschner JE. Mucin gene expression in human middle ear epithelium. Laryngoscope 2007;117:1666-76.

[343] Kerschner JE, Li J, Tsushiya K, Khampang P. Mucin gene expression and mouse middle ear epithelium. Int J Pediatr Otorhino-Laryngol 2010;74::864-8.

[344] Kerschner JE, Khampang P, Samuels T. Extending the chinchilla middle ear epithelial model for mucin gene investigation. Int J Pediatr Otorhinolaryngol 2010;74:980-5.

[345] Magro G, Schiappacassi M, Perissinotto D, *et al.* Differential expression of mucins 1-6 in papillary thyroid carcinoma: evidence for transformation-dependent post-translational modifications of MUC1 *in situ*. J Pathol 2003;200:357-69.

[346] Alves P, Soares P, Fonseca E, Sobrinho-Simoes M. Papillary Thyroid Carcinoma Overexpresses Fully and Underglycosylated Mucins Together with Native and Sialylated Simple Mucin Antigens and Histo-Blood Group Antigens. Endocr Pathol 1999;10:315-24.

[347] Kondo K, Kohno N, Yokoyama A, Hiwada K. Decreased MUC1 expression induces E-cadherin-mediated cell adhesion of breast cancer cell lines. Cancer Res 1998;58:2014-9.

[348] Rahn JJ, Dabbagh L, Pasdar M, Hugh JC. The importance of MUC1 cellular localization in patients with breast carcinoma: an immunohistologic study of 71 patients and review of the literature. Cancer 2001;91:1973-82.

[349] Zaretsky JZ, Barnea I, Aylon Y, *et al.* MUC1 gene overexpressed in breast cancer: structure and transcriptional activity of the MUC1 promoter and role of estrogen receptor alpha (ERalpha) in regulation of the MUC1 gene expression. Mol Cancer 2006;5:57.

[350] Zaretsky JZ, Weiss M, Tsarfaty I, *et al.* Expression of genes coding for pS2, c-erbB2, estrogen receptor and the H23 breast tumor-associated antigen. A comparative analysis in breast cancer. FEBS Lett 1990;265:46-50.

[351] Zaretsky JZ, Sarid R, Aylon Y, *et al.* Analysis of the promoter of the MUC1 gene overexpressed in breast cancer. FEBS Lett 1999;461:189-95.

[352] Wreschner DH, Hareuveni M, Tsarfaty I, *et al.* Human epithelial tumor antigen cDNA sequences. Differential splicing may generate multiple protein forms. Eur J Biochem 1990;189:463-73.

[353] Pastan I, Lovelace ET, Gallo MG, *et al.* Characterization of monoclonal antibodies B1 and B3 that react with mucinous adenocarcinomas. Cancer Res 1991;51:3781-7.

[354] Rakha EA, Boyce RW, Abd El-Rehim D, *et al.* Expression of mucins (MUC1, MUC2, MUC3, MUC4, MUC5AC and MUC6) and their prognostic significance in human breast cancer. Mod Pathol 2005;18:1295-304.

[355] Pereira MB, Dias AJ, Reis CA, Schmitt FC. Immunohistochemical study of the expression of MUC5AC and MUC6 in breast carcinomas and adjacent breast tissues. J Clin Pathol 2001;54:210-3.

[356] Komaki K, Sakamoto G, Sugano H, Morimoto T, Monden Y. Mucinous carcinoma of the breast in Japan. A prognostic analysis based on morphologic features. Cancer 1988;61:989-96.

CHAPTER 8

Gel-Forming Mucin MUC19

Abstract: The mucin MUC19 is a recently discovered member of the gel-forming mucin family. At present, limited information is available regarding its structure and functions. This chapter covers several issues: the evolutionary connection between MUC19 and other gel-forming mucin genes, domain structure of the MUC19 mucin glycoprotein, and specific features of the mouse Smgc/Muc19 mRNA transcription and splicing. Some recent data regarding human *MUC19* and mouse *Muc19* expression in embryonic development and in adult tissues are presented, and the possible role of MUC19 in physiology of the eye and middle ear are discussed.

Keywords: MUC19, Muc19, Smgc/Muc19, domain structure, splicing, expression.

8.1. CLONING, CHROMOSOMAL LOCALIZATION AND EVOLUTION OF THE *MUC19* GENE

The human glycoprotein MUC19 is a gel-forming mucin discovered in 2004 [1, 2]. According to its "birth day", it is the youngest member of the gel-forming mucin family. As noted by Zhu *et al.* [1], "MUC19 has an unusual discovery path that appears different from all others". Although pig Muc19/PSM (porcin submaxillary gland mucin, AF005273) was discovered and cloned in 1997 [3], no attempt was made to identify its human ortholog till 2004. Because of its late discovery, our knowledge about MUC19 mucin structure and functions is scant and superficial. The studied properties of MUC19 mucin and its possible role in cell physiology are discussed in this chapter.

Chen *et al.* [2] was the first to attempt to identify the human and rodent MUC19 mucin. Assuming that all gel-forming mucins were developed by gene duplication followed by evolutional modifications, the authors searched for a new mucin using the "Hidden Marcov Model" (HMM), one of the powerful gene search and identification tools [4-6]. Their search, based on the common properties of the conserved CK-domains present in all gel-forming mucins, together with molecular cloning resulted in isolation of the human and mouse *MUC19/Muc19* genes [2]. At the same time, but using a different approach, Culp *et al.* [7] cloned the mouse *Muc19* gene.

Joseph Z. Zaretsky and Daniel H. Wreschner

The human *MUC19* gene was found on chromosome locus 12q12 while its mouse homolog *Muc19* was detected on chromosome locus 15E3 [2, 7]. Each gene is represented in the genome by a single copy [2, 7]. Interestingly, the gene encoding the von Wilibrand factor (*vWF)*, which evolved from the ancestor common to both *vWF* and gel-forming mucin genes, also resides on chromosome 12 at 12p13 locus, close to *MUC19*. This points to a common evolution of these genes, although the evolutionary relationships of *vWF*, *MUC19* and other gel-forming mucin genes remain unclear. On the one hand, the human *MUC19* and mouse *Muc19* genes have structural features similar to the porcine and bovine submaxillary gel-forming mucin genes, but differ somewhat from the human and mouse gel-forming mucin genes as well as from *vWF* [2, 8-11]. This is especially true with regard to C-terminal regions of these genes. On the other hand, in contrast to *MUC19* resided, as noted above, on chromosome 12, human gel-forming mucin genes *MUC2-MUC6* are located as a cluster on chromosome 11p15 [12]. Additional lack of clarity about the relationship between *MUC19* and *MUC2-MUC6* gel-forming mucin genes emerged from the phylogenetic study of Pigny *et al.* [12], who showed that *MUC19* is much closer on the phylogenic tree to *MUC2, MUC5AC* and *MUC5B* than to *MUC6*, even though *MUC6* is also a member of the 11p15 cluster like *MUC2, MUC5AC* and *MUC5B*. Recently, Kawahara and Nishida [13] found that evolutionary roots of the human *MUC19* and mouse *Muc19* genes are linked to the spiggin multi-gene family, and that these genes are orthologs of the spiggin gene. Further studies are needed to clarify the evolutionary history of the 11p15 cluster genes, spigging genes, *vWF* and *MUC19* genes. Phylogenetic analysis performed by Zhu *et al.* [1] showed that "the *MUC19* gene family appears to be the first gel-forming mucin to branch out from the common ancestor gene with *vWF*. The appearance of four 11p15 gel-forming mucins occurred after separation of *MUC19*, and segregation between *MUC5AC* and *MUC5B* was the most recent event" (see Chapter 6).

8.2. DOMAIN STRUCTURE OF THE MUC19 GLYCOPROTEIN

The methods used by Chen *et al.* [2] to search for the sequence of the *MUC19* gene allowed cloning of only 2.23 kb cDNA corresponding to the 3'-region of the gene. Although the full *MUC19* genomic and cDNA sequences had not been identified at that time, Chen *et al.* [2] could determine the gene structure upstream to the cloned 3'-region by using bioinformatics. The results suggested that both the human *MUC19* and mouse *Muc19* genes share partially the structural domain composition common to the von Willebrand factor coding gene and other gel-forming mucin genes. The domain structure of the MUC19 mucin that emerged from the study of Chen *et al.* [2] is presented in Fig. **1**.

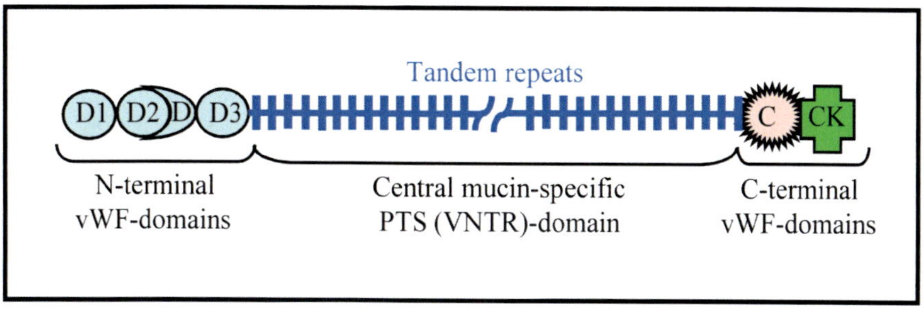

Figure 1: Partial domain structure of the MUC19 mucin (based on the data reported in [2]).

According to this proposed model, the MUC19 glycoprotein does not contain vWD$_4$- and B-domains, in contrast to MUC2, MUC5AC and MUC5B mucins. The absence of these domains makes the MUC19 mucin protein similar to the MUC6 molecule, however, in contrast to MUC19, MUC6 does not contain vWC-domain characteristic for other gel-forming mucins and vWF glycoprotein. Importantly, another structural difference between MUC19 and MUC6 is associated with the mucin-specific tandem repeat (TR)-containing domain, which is much larger in the MUC19 molecule than in the MUC6 polypeptide. According to the predicted peptide sequence, the large mucin domain of the MUC19 molecule must contain more than 7000 amino acids. The predicted human MUC19 genomic sequence appears to have more than 180000 nucleotides [2]. Any discussion of the human MUC19 mucin structure and amino acid sequences (especially those belonged to the N-terminal half of the molecule) must take into account that all predictions made before 2011 were made on data obtained by bioinformatics and not by biochemical or molecular biology methods, as the full human *MUC19* gene had not been cloned yet.

In 2011, Chen's group succeeded in cloning and sequencing the full length human *MUC19* gene [1]. This gene, isolated from human trachea and salivary gland tissues, consists of 185903 bp, distributed among 182 exons (25863 bp) and 181 introns (160040 bp). Interestingly, 56 exons (~31%) are 54 bp each. *MUC19* is the largest of the gel-forming mucin genes. The protein encoded by this gene possesses the typical structural features of the gel-forming mucins, including vWD1-vWD2-vWD'-vWD3-domains, central mucin-specific domain containing threonine/serine-rich TRs, vWC-domain and cysteine knot-like domain (CK), and lacks, as noted above, the vWD4 and vWB domains present in MUC2, MUC5AC and MUC5B mucins. Nevertheless, as noted by the authors [1], "the deduced MUC19 protein has *bona fide* gel-forming mucin structure". However, in addition to the typical gel-forming mucin characteristics, the human MUC19 mucin also has unique features not found in its pig or mouse counterparts or in other gel-forming mucins. The most prominent unique trait is its unusually long N-terminal polypeptide fragment located upstream to the first vWD-domain and encoded by the highly variable region (HVR) of the *MUC19* gene (Fig. **2**).

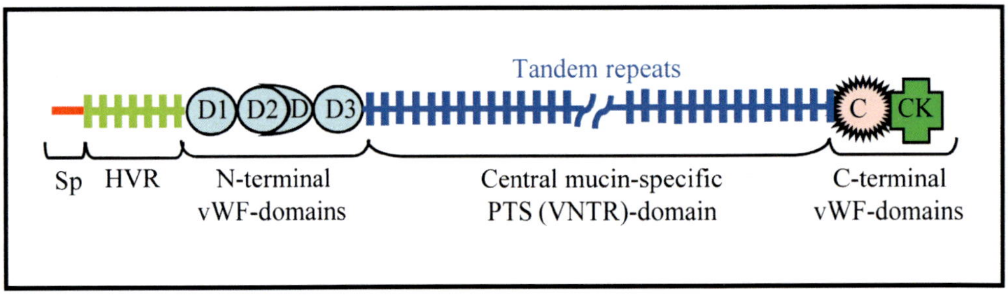

Figure 2: Structure of the full length human MUC19 glycoprotein (based on the data reported in [1]; Sp-signal peptide, HVR- highly variable region).

8.3. TRANSCRIPTION AND SPLICING OF THE PRIMARY MUC19 RNA

This fragment contains serine-rich repetitive structure, which is reminiscent of mouse submandibular gland protein C (Smgc) [8], encoded by the mouse *Smgc* gene fused at its 3'-end with the 5'-end of the mouse *Muc19* (accession number NT_039621.2) [7]. Importantly, this fusion determines the mode of mRNA transcription and splicing of the *Smgc/Muc19* gene (Fig. **3**). Culp and co-workers [7] found that Smgc mRNA and Muc19 mRNA appear to be the splice variants of the same primary transcript transcribed from the fused *Smgc/Muc19* gene. Both mRNAs share exon 1. The fused *Smgc/Muc19* gene contains 60 exons: the first 18 encode mouse Smgc protein, while the Muc19 transcript incorporates exon 1 and exons 19-60 encoding the mouse Muc19 mucin [7, 14]. Given the different tissue, sex and developmental expression patterns of Muc19 and Smgc proteins in rodents [15, 16], the fused gene encoding both mRNAs must be under the control of a complex regulatory mechanism that allows regulated transcription and splicing of the 18 exon-containing Smgc mRNA (1-18 exons) and the 43 exon-containing Muc19 mRNA (1, 19-60 exons).

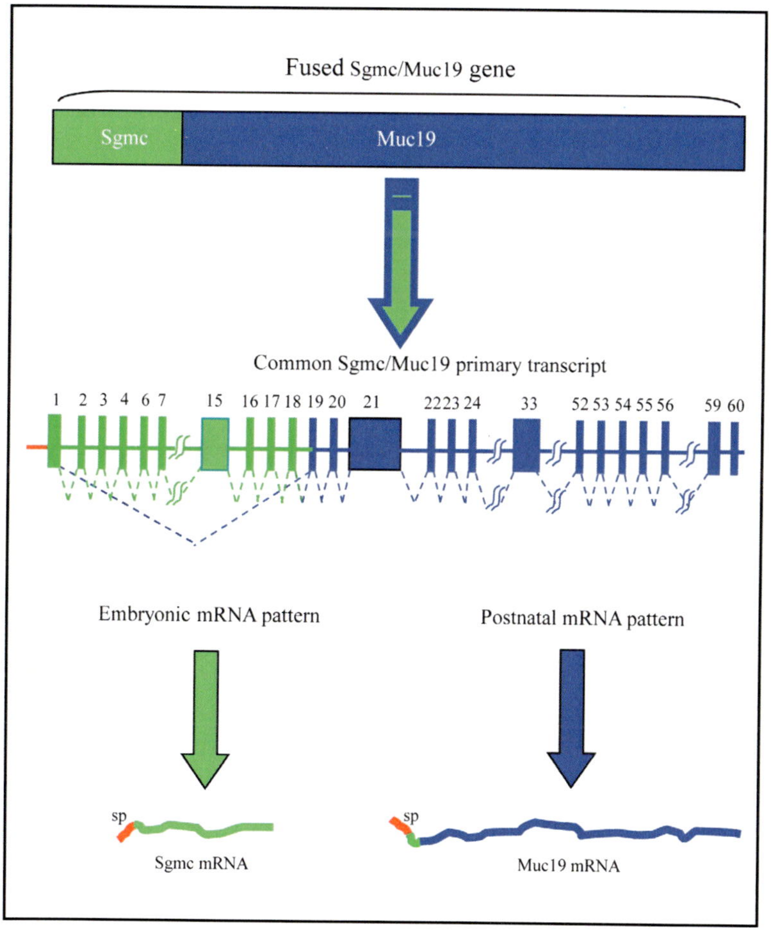

Figure 3: Transcription and splicing of the Sgmc/Muc19 primary transcript (based on the data reported in [7, 14]; Sp – signal peptide).

The exons comprising Muc19 mRNA are distributed over the 106.4 kb genomic sequence. The large TR-containing domain is encoded by exon 33 (17997 bp) that consists of 36 full TR units, each 489 bp in length, plus one partial repeat. The consensus repeat sequence contains 80 potential sites for O-glycosylation and two potential sites for N-glycosylation. Exon 1 starts with the initial A as the transcription start site, followed by 42 nucleotides comprising the 5'-UTR region. The potential translation ATG codon placed within a Kozak consensus sequence [17] is located at position +44. The TATA-box consensus motif (TATAAAAT) is 28 bp upstream to the transcription start site within the genomic sequence [7, 14].

A comparison of the deduced amino acid sequences of the MUC19 mucins of different origins (mouse, rat, porcine and human) revealed a high degree of conservation in sequences located 5' and 3' to the TRs, especially within domains corresponding to the von Willebrand factor type D- and C-domains and to the cysteine knot-like domain. The small cysteine-containing domains described in porcine submaxillary mucin [11] are also found in rodent Muc19, and show a high degree of interspecies similarities. As pointed out by Culp *et al.* [7], the low number of single nucleotide polymorphisms observed between mouse strains, and the absence of detectable splice variants of the corresponding mRNAs, also indicate to conservation of the sequences located 5'-up-stream and 3'- down-stream to the TRs. However, the sequences adjacent to the central TR-containing region appear to be less conserved both in primary amino acid composition and in length. This is especially true for the region 3'-adjacent to the TRs: the 3'-located unique non-repeat sequences of human and porcine MUC19 mucins are larger than the corresponding sequences in rat and mouse mucins. This region of the rat Muc19 mucin is larger than that of the mouse homolog by 5 exons [2, 7].

Collectively, the data of Chen's group [1, 2] and Culp's group [7] lead to the conclusion that MUC19 is the largest gel-forming mucin studied to date. It appears that the very N-terminal end of the human MUC19 mucin located upstream to the vWD1 domain corresponds to the mouse Smgc protein. This observation raises the question of whether the highly variable region of the human *MUC19* gene is a human homolog of the mouse *Smgc* gene fused to the human *MUC19* gel-forming mucin gene, or it is a native integral part of the human *MUC19* gene. Importantly, expression of the mouse *Muc19* suggests splicing of the primary Smgc/Muc19 transcript, an event that is not characteristic of the gel-forming mucin genes [7]. It is noteworthy that the transcription start site and the first splice donor site are located in the *Smgc* gene. In the human *MUC19*, the HVR is responsible for at least 20 different splicing variants of the MUC19 mRNAs, including 18 mRNAs that are translated into proteins [1]. Future studies will have to clarify the nature of the HVR and its role in expression of the human *MUC19* gene.

While the HVR is a unique element of the MUC19 glycoprotein that separates it from the other gel-forming mucins, the presence in the MUC19 protein of the cysteine-rich vWD-domains and CK-domain relates MUC19 to the group of the gel-forming mucins and implies the ability of the MUC19 monomer molecules to develop oligomeric forms. Indeed, Rousseau *et al.* [18] established that Muc19 is a polymeric mucin stabilized by disulfide bonds - a conclusion that followed from analysis of the rate-zonal distribution of the horse and rat salivary mucins, including Muc19, subjected to centrifugation before and after treatment with a reducing agent.

The process of posttranslational glycosylation of MUC19 apomucin has not been studied, although it is known that MUC19 contains multiple O- and two N-glycosylation sites. This and other processes involved in synthesis, maturation and secretion of the MUC19 glycoprotein await clarification.

8.4. EXPRESSION OF *MUC19* GENE

Little is known about the biosynthesis and maturation of functionally active MUC19 glycoprotein, but some progress has been made in the analysis of the *MUC19* gene expression. In a pioneering study, Chen *et al.* [2] observed the expression of human and mouse MUC19 mRNAs in the salivary glands and tracheal tissues. In mouse, *Muc19* is expressed mainly in two major salivary glands, the sublingual and submandibular glands, whereas the parotid gland appears to be Muc19-negative. The highest level of *Muc19* expression found in the sublingual gland is consistent with the highest amount of mucous cell in this gland compared with other salivary glands. The submandibular gland, comprised of a mixture of mucous and serous cells, demonstrates a moderate level of *Muc19* expression, whereas the parotid gland, which lacks mucous cells, does not express the Muc19 mRNA. In trachea, only mucous cells of the submucosal glands express the Muc19 mRNA. The same patterns of expression have been detected in humans. Recently, Kouznetsova *et al.* [19] extended these findings by showing that mucous acini of the labial glands in humans also express MUC19. The expression of MUC19 mRNA in the human salivary glands was also detected by Rousseau *et al.* [18], but the MUC19 mucin was undetectable in human saliva. Using the same methods, the authors did find the Muc19 mucin in the saliva of horse, pig, cow and rat. The biological meaning of these differences in MUC19 glycoprotein expression in different species remains to be elucidated.

A comprehensive study of the *Smgc/Muc19* gene expression during murine sublingual gland development was undertaken by Das *et al.* [15]. As mentioned above, this fused gene encodes two translational products: the submandibular gland protein C (Smgc) and the gel-forming mucin Muc19 [7, 14]. The mouse Muc19 mucin is the major secretion product of the mucous acinar cells of adult sublingual glands [7]. In contrast, during early postnatal development, the Smgc was found first in the submandibular glands and later on also in the sublingual glands [7, 20]. Analysis of the dynamics of the *Smgc/Muc19* gene expression during sublingual gland development shows that expression of *Smgc* directly precedes that of *Muc19* during mucous cell differentiation. The Smgc transcripts can be first detected at embryonic day 17 (E17), after which their expression increases several folds, reaching maximal levels at E19-20, and substantially declines at birth (E20-21). During postnatal development males and females differ in expression of the Smgc transcripts: females display a relatively constant level from birth through postnatal day 14 (P14), while males show transient expression that reaches a peak at P10 and becomes essentially undetectable after P21. The Smgc transcripts and Smgc protein are barely detectable in adults of either gender [21]. The transcripts of the *Muc19* gene are hardly detectable during early embryonic development. However, the Muc19 mRNA is readily expressed at E19, increases more than 2-fold at E20, reaches a plateau from E20 through P21, and sharply increases at P28 to a level that persists into adulthood, equal in both genders. Importantly, the absence of transcripts of several other mucins (Muc2, Muc5AC, Muc6 and Muc16) in murine sublingual glands during the studied developmental stages indicate that Muc19 is the only large secreted mucin in these cells that is expressed during the embryonic and postnatal periods. It has been noted that the level of Muc19 mucin expression correlates with the early differentiation of mucous cell in sublingual glands [15].

Based on the above results, Das *et al.* [15] suggested a model of the key regulatory events of mucous cell development. The initial differentiation of the terminal bulb cells is associated with the initiation of *Smgc/Muc19* transcription and splicing of the primary Smgc/Muc19 mRNA, resulting in production of Smgc protein and conversion of bulb cells to an exocrine phenotype. The next stage involves mucus cell differentiation dictated by production of the Muc19 mucin. Thus, mucus cell differentiation is associated with a switch in the regulation of splicing to produce the Muc19 transcripts rather than the Smgc mRNAs. A simple switch in mRNA splicing to direct translation from Smgc protein to Muc19 mucin during embryonic development may, according to Das *et al.* [15], contribute to early morphological maturation of sublingual glands. By expanding the analysis of the *Smgc/Muc19* expression to other murine cells and tissues, Das *et al.* [15] claimed preferential Muc19 mRNA expression in the major (sublingual) and minor (hard and soft palate, buccal mucosa and posterior tongue) salivary mucous glands as well as in submucosal glands of the tracheo-larynx and bulbourethral glands. Importantly, though hardly detectable, the expression of Muc19 transcripts was observed also in the prostate, testis, ovaries, fallopian tubes and uterus. The Smgc transcripts were detected only in neonatal tracheo-larynx and salivary tissues except buccal mucosa. Mouse bulbourethral glands, ileum, ovaries, fallopian tubes, uterus, cervix, vagina and lacrimal glands are Smgc-negative [15]. Interestingly, Das *et al.* [15] could not detect *Muc19* expression in mouse conjunctiva and lacrimal glands, whereas Yu *et al.* [22] did find MUC19 mRNA and protein in human conjunctiva, lacrimal glands and cornea, suggesting interspecies differences in expression and tissue distribution of MUC19 between man and mice. Yu *et al.* [22] showed that Sjorgen syndrome is associated with substantial decrease in MUC19 expression. Surprisingly, MUC19 mucin as well as MUC2 and MUC6 were found in this study to be expressed in cornea epithelium, which is traditionally classified as squamous epithelium. According to the authors [22], these mucins may play different roles in cornea than in secretory cells, or a part of epithelia that expresses those mucins may be under mucinous *trans*-differentiation as described by Pera *et al.* [23].

Kerschner *et al.* [24, 25] recently observed expression of several mucins, including Muc19, in epithelium of mouse middle ear. Earlier, this group showed that the MUC19 glycoprotein is expressed in both human middle ear epithelium *in vivo* and *in vitro* and in chinchilla middle ear tissues [25, 26]. Expression of MUC19 has been shown to be sensitive to inflammatory cytokines TNFα, IL-1β, IL-6 and IL-8, which sharply up-regulate *MUC19* transcription after a short exposure (1-2 hours) and down-regulate production of its transcripts after long cytokine treatment (more than 6 hours) [26].

In summary, the presence of MUC19 glycoprotein has been demonstrated in epithelial cells of different tissues including salivary and bulbourethral glands, airway tissues, and ocular and middle ear epithelia. Its expression is observed under normal physiological conditions as well as in pathologically changed cells and tissues. It appears that MUC19 plays an important and specific role in basic biological processes, but our knowledge of the molecular biology of this mucin is only in the initial stages, and additional efforts are required to gain insight into the properties and functions of this glycoprotein molecule.

REFERENCES

[1] Zhu L, Lee P, Yu D, Tao S, Chen Y. Cloning and characterization of human MUC19 gene. Am J Respir Cell Mol Biol 2011;45:348-58.

[2] Chen Y, Zhao YH, Kalaslavadi TB, *et al.* Genome-wide search and identification of a novel gel-forming mucin MUC19/Muc19 in glandular tissues. Am J Respir Cell Mol Biol 2004;30:155-65.

[3] Eckhardt AE, Timpte CS, DeLuca AW, Hill RL. The complete cDNA sequence and structural polymorphism of the polypeptide chain of porcine submaxillary mucin. J Biol Chem 1997;272:33204-10.

[4] Bateman A, Birney E, Durbin R, *et al.* Pfam 3.1: 1313 multiple alignments and profile HMMs match the majority of proteins. Nucleic Acids Res 1999;27:260-2.

[5] Sonnhammer EL, Eddy SR, Birney E, Bateman A, Durbin R. Pfam: multiple sequence alignments and HMM-profiles of protein domains. Nucleic Acids Res 1998;26:320-2.

[6] Hofmann K. Sensitive protein comparisons with profiles and hidden Markov models. Brief Bioinform 2000;1:167-78.

[7] Culp DJ, Latchney LR, Fallon MA, *et al.* The gene encoding mouse Muc19: cDNA, genomic organization and relationship to Smgc. Physiol Genomics 2004;19:303-18.

[8] Perez-Vilar J, Eckhardt AE, Hill RL. Porcine submaxillary mucin forms disulfide-bonded dimers between its carboxyl-terminal domains. J Biol Chem 1996;271:9845-50.

[9] Perez-Vilar J, Eckhardt AE, DeLuca A, Hill RL. Porcine submaxillary mucin forms disulfide-linked multimers through its amino-terminal D-domains. J Biol Chem 1998;273:14442-9.

[10] Perez-Vilar J, Hill RL. The carboxyl-terminal 90 residues of porcine submaxillary mucin are sufficient for forming disulfide-bonded dimers. J Biol Chem 1998;273:6982-8.

[11] Perez-Vilar J, Hill RL. Identification of the half-cystine residues in porcine submaxillary mucin critical for multimerization through the D-domains. Roles of the CGLCG motif in the D1- and D3-domains. J Biol Chem 1998;273:34527-34.

[12] Pigny P, Guyonnet-Duperat V, Hill AS, *et al.* Human mucin genes assigned to 11p15.5: identification and organization of a cluster of genes. Genomics 1996;38:340-52.

[13] Kawahara R, Nishida M. Extensive lineage-specific gene duplication and evolution of the spiggin multi-gene family in stickleback. BMC Evol Biol 2007;7:209.

[14] Zinzen KM, Hand AR, Yankova M, Ball WD, Mirels L. Molecular cloning and characterization of the neonatal rat and mouse submandibular gland protein SMGC. Gene 2004;334:23-33.

[15] Das B, Cash MN, Hand AR, Shivazad A, Culp DJ. Expression of Muc19/Smgc gene products during murine sublingual gland development: cytodifferentiation and maturation of salivary mucous cells. J Histochem Cytochem 2009;57:383-96.

[16] Das B, Cash MN, Hand AR, *et al.* Tissue Distibution of Murine Muc19/Smgc Gene Products. J Histochem Cytochem 2010;58:141-56.

[17] Kozak M. An analysis of 5'-noncoding sequences from 699 vertebrate messenger RNAs. Nucleic Acids Res 1987;15:8125-48.

[18] Rousseau K, Kirkham S, Johnson L, *et al.* Proteomic analysis of polymeric salivary mucins: no evidence for MUC19 in human saliva. Biochem J 2008;413:545-52.

[19] Kouznetsova I, Gerlach KL, Zahl C, Hoffmann W. Expression analysis of human salivary glands by laser microdissection: differences between submandibular and labial glands. Cell Physiol Biochem 2010;26:375-82.

[20] Ball WD, Redman RS. Two independently regulated secretory systems within the acini of the submandibular gland of the perinatal rat. Eur J Cell Biol 1984;33:112-22.

[21] Denny PC, Ball WD, Redman RS. Salivary glands: a paradigm for diversity of gland development. Crit Rev Oral Biol Med 1997;8:51-75.

[22] Yu DF, Chen Y, Han JM, *et al.* MUC19 expression in human ocular surface and lacrimal gland and its alteration in Sjogren syndrome patients. Exp Eye Res 2008;86:403-11.

[23] Pera M, Pera M, de Bolos C, *et al.* Duodenal-content reflux into the esophagus leads to expression of Cdx2 and Muc2 in areas of squamous epithelium in rats. J Gastrointest Surg 2007;11:869-74.

[24] Kerschner JE. Mucin gene expression and mouse middle ear epithelium. Int J Pediatr Otorhinolaryngol 2010;74:864-8.

[25] Kerschner JE. Mucin gene expression in human middle ear epithelium. Laryngoscope 2007;117:1666-76.

[26] Kerschner JE, Khampang P, Erbe CB, Kolker A, Cioffi JA. Mucin gene 19 (MUC19) expression and response to inflammatory cytokines in middle ear epithelium. Glycoconj J 2009;26:1275-84.

CHAPTER 9

Mucin MUC7

Abstract: The *MUC7* gene belongs to a group of genes encoding soluble mucin glycoproteins. This chapter describes the structure of the *MUC7* gene and its protein product; the mechanisms responsible for regulation of its expression; important functions associated with the individual domains; and the role of posttranslational modifications of the MUC7 apomucin in expression of the functions embedded in its polypeptide structure.

Keywords: MUC7, structure, regulation, expression, functions, antimicrobial activity.

9.1. GENERAL CHARACTERISTICS OF *MUC7* GENE AND ITS PROTEIN PRODUCT

The *MUC7* gene encoding a small salivary mucin MUC7, previously known as MG2 mucin, was cloned and sequenced and mapped to chromosome 4q13-q21 by Levine's group [1]. It spans ~10.0 kb and comprises three exons and two introns. Exons 1, 2 and 3 are 100, 68 and 2200 bp in length, respectively. Intron 1 is ~1.7 kb long and occupies the region corresponding to the 5'-UTR of the MUC7 cDNA. Intron 2 spans ~6.0 kb and is located between the putative leader peptide and secreted protein. The MUC7 core polypeptide is encoded by exon 3.

The MUC7 cDNA (~2.4 kb), also isolated by Levine's group [2], contains a coding sequence of 1131 nucleotides and 1120 bp of 3'-untranslated region (3'-UTR) without the poly (A) tail. The MUC7 mRNA encodes a polypeptide containing 377 amino acid residues with molecular weight of 39 kDa. The N-terminally located 20 residues are strongly hydrophobic and function as a leader peptide. Human salivary MUC7 mucin consists of a single polypeptide chain primarily made up of Ser, Thr, Pro and Ala [3, 4]. According to biochemical studies, MUC7 mucin isolated from human glandular saliva is a 125-kDa glycoprotein consisting of 30.4% protein, 68% carbohydrate, and 1.6% sulfate [5, 6]. Some 83% of the Ser+Thr residues in MUC7 mucin are O-glycosylated, with the majority located in the tandem repeat (TR) region [7]. Based on the content of fucose and sialic acid, two isoforms of MUC7 mucin, *a* and *b*, have been

Joseph Z. Zaretsky and Daniel H. Wreschner

identified [3]: the first one containing 16% fucose and 14% sialic acid, and another one comprising much less fucose, 7%, and substantially more sialic acid, 26%. The amounts of these saccharides in the MUC7 glycoprotein can vary in different cells and conditions. Shori *et al.* [8] observed a decrease in the content of sialic acid and increased fucosylation in patients with cystic fibrosis (CF).

Up to 80% of carbohydrate chains in MUC7 are rather simply organized: they consist of di- and trisaccharides such as Fucα(1-2) or NeuAcα(2-3) linked to Gal*β*(1-3)-GalNAc and conjugated to Ser/Thr residues. However, longer carbohydrate chains up to seven sugar residues have also been reported [9, 10]. In comparison, the O-glycan chains of another salivary mucin, MUC5B, contain more than 40 carbohydrate residues [11]. The glycosylation pattern of MUC7 differs significantly from that of MUC5B. For instance, neutral oligosaccharides and blood group antigens are less abundant on MUC7 molecules, and glycosylation patterns of MUC7 appear to be more preserved between individuals compared to salivary MUC5B mucin [12]. On the other hand, salivary MUC7 mucin is a major carrier of blood group I type O-linked oligosaccharides serving as a scaffold for sialyl Lewis^{x} antigen [12]. As noted by Karlsson and Thomsson [12], MUC7 mucin expresses a unique profile of O-glycans that is totally different from that of MUC5B, which carries more blood group epitopes and exhibits significant individual variation. A characteristic feature of MUC7 O-glycosylation is branched poly-lactosamine sequence backbones carrying terminal Si-Le^{x} epitopes. These differences in glycosylation of two salivary mucins may stem from different cellular origins [13] that result in different regulation of glycosylation mechanisms [12].

9.2. DOMAIN STRUCTURE OF MUC7 MUCIN GLYCOPROTEIN

According to Liu *et al.* [14], the MUC7apomucin moiety consists of an N-terminal 144-aa fragment containing two cysteines, a central region of 139-aa in length including six nearly identical 23-aa containing tandem repeat (TR) units, and a C-terminal region comprised of 74 amino acids. There are two common allelic forms with 5 or 6 TRs [15-17]. Analysis of the derived MUC7 peptide sequence by Gururaja *et al.* [7] revealed five distinct domains, in addition to a signal peptide encompassing the first 20 amino acids (1-20 aa): 1) N-terminal histatin-like domain D1 (21-71 aa); 2) moderately glycosylated domain D2 (72-164 aa); 3) heavily glycosylated tandem repeat-containing domain D3 (165-303 aa); 4) MUC1/MUC2-like domain D4 (304-355 aa), and 5) C-terminal domain D5 (356-377 aa) including a leucine-zipper module (Fig. **1**).

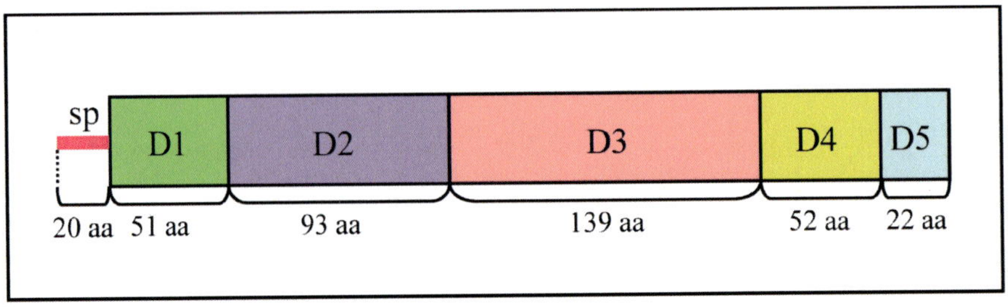

Figure 1: Domain structure of the MUC7 mucin (based on the data from [7]; Sp-signal peptide).

Different domains of the MUC7 mucin are characterized by specific features that make MUC7 unique among mucin glycoproteins. As follows from its name, the N-terminal histatin-like domain contains a histatin-like sequence [18] of 15 amino acids (aa 23-37) and exhibits a high level of candidacidal activity [19]. It also contain a leucine-zipper segment [7]. A distinguish property of the mucin domain D3 is associated with 16 diproline segments non-uniformly distributed among repeat units, an integral parts of this domain. The diproline segment located at positions 5 and 6 from the N-terminus of a repeat unit occurs in all six repeat units; the segment located at position 21 and 22 close to the C-terminus of a repeat unit is present in five repeats; and diproline residues at positions 15 and 16 in the center of repeat sequence are present in only three repeat units. According to Gururaja *et al.* [7], these diproline segments are the key elements that determine the type II polyproline extended conformation [20] of the third domain, which, in turn, contributes to the overall conformational rigidity of MUC7. Interestingly, circular dichroism (CD) spectra of glycosylated and nonglycosylated MUC7 peptides turned up no profound effect of GalNAcα and Galβ(1-3)GalNAcα O-linked to Ser on the diproline-containing peptide backbone conformation. On the other hand, the same carbohydrate moiety O-linked to Thr residues in the same modeling peptides corresponding to a 23-aa TR sequence of MUC7 exerts a strong intramolecular stabilizing effect on polyproline-determined polypeptide structure [21].

Comparison of human MUC7 mucin with ovine and porcine submaxillary mucins shows that humans and animals use different mechanisms to adopt an extended structure. Animal mucin molecules adopt a semi-rigid rod-like structure that can be transformed into a random coil-like structure by sequential removal of carbohydrate units [22]. Thus, in these mucins the O-linked carbohydrate chains play an important role in the adoption of filamentous, nonrandom coil structure and in the stiffening of the mucin conformation. In human MUC7, this function is performed mostly by diprolene segments, the majority of which are located in the TR domain [7]. Thus, polyproline sequences of the second and especially the third domains are responsible for adoption of extended rigid structure by human MUC7 mucin.

The fourth domain of the MUC7 mucin differs from the other MUC7 domains by the presence of sequences homologous to the sequences of mouse Muc1 and human MUC2 mucins. The level of homology with Muc1 mucin is about 27%, and in the case of MUC2 it reaches 43%. Similarity between Muc1, MUC2 and MUC7 is also reflected in a high content of proline residues in TRs of all three molecules [7]. Although these mucins belong to different mucin subfamilies (membrane-bound, gel-forming or soluble mucins), the observed sequence homologies indicate some crossing points of the evolutionary histories of the genes encoding these proteins. One of the characteristic features of the fourth domain is the presence of N-glycosylation site, whose function in the physiology and biochemistry of MUC7 mucin is unknown.

The fifth domain is the smallest one, comprising only 23 amino acids including one serine residue. This domain does not contain proline or threonine residues, which means that proline can not affect the domain structure. On the other hand, glycosylation of a single serine residue by attachment of Galβ(1,3)-α-GalNAc chain does determine transition of the chemically synthesized 23-aa peptide derived from the MUC7 C-terminal domain from the β-strand conformation to the helical structure [23, 24]. This structural transition illustrates that the carbohydrate moiety of the MUC7 C-terminus exerts a significant effect on the peptide backbone conformation.

A part of the polypeptide sequence that makes up domain 5 is represented by the leucine-zipper motif (-LLYMKNLL-) important for stabilization of a polypeptide molecule at the coiled-coil interface *via* hydrophobic interactions [25]. Molecular dynamic simulation analysis revealed specific arrangement of hydrophobic and hydrophilic amino acid residues in the domain 5 polypeptide, indicating the amphipathic character of the domain 5 structure. The two parameters together – leucine-zipper motif and amphipathic nature of the domain 5 polypeptide – suggest that domain 5 is one of the key elements in self-association (di- or oligomerization, aggregation) of MUC7 *in vivo* [23].

However, the process of MUC7self-association is not a prerogative of the domain D5. It appears that the first domain also participates in this process. Electron microscope and light-scattering studies carried out by Mehrotra and co-workers [26] showed that MUC7 might self-associate *via* a protein-protein interaction involving the N-terminus of the molecule.These data are in agreement with the results of Gururaja *et al.* [7] who succeeded in showing that two cysteine residues located at the N-terminus of the domain D1 are responsible for self-association of the MUC7 molecules. This was evidenced by experiments in which reduction of the heterogeneous MUC7 preparation yielded pure monomer species. Moreover, according to mass spectral analysis of the peptides obtained after endoproteinase Lys-C digestion of MUC7 polypeptide, the cysteines in MUC7 exist in the dithiol form. This finding also indicates to involvement of cystein residues of domain D1 in the mechanism of MUC7 self-association. However, the described results were contradicted by Soares *et al.* [27], who could not find intermolecular disulfide bonds between MUC7 molecules by electrophoretic analysis conducted under reducing or nonreducing conditions.

Despite the inconsistent results, it appears that both N- and C- termini of the MUC7 mucin polypeptide are involved in the self-association process. Based on these results, Satyanarayana and co-workers [23] proposed a model for MUC7 self-association depicting head-to-head and tail-to-tail interactions, in which linear polymers are formed *via* cysteine-mediated disulfide linkage between two N-termini on the one side, and leucine-zipper type interactions between two C-termini on the other side. Although this model explains the involvement of the N-terminally located cysteine residues and C-terminally located leucine-zipper sequence in MUC7 polymerization, it does not specify the role of the second leucine-zipper motif found in domain D1.

9.3. REGULATION OF *MUC7* GENE EXPRESSION

Expression of the *MUC7* gene was detected in early studies mainly in the acinar cells of submandibular and sublingual salivary glands [5, 6, 28, 29], and later in palatine and labial salivary glands [30, 31]. The highest production of MUC7 mucin in salivary glands was observed in sublingual gland (980 μg/ml), with much lower amounts in submandibular (220 μg/ml) and palatine glands (100 μg/ml) [30]. Interestingly, during the first year of an infant's life, salivary mucins are expressed differentially, with first MUC7 and later MUC5B being predominant [32].

For a long time *MUC7* expression was thought to be restricted to salivary glands [33]. It was shown both by immunohistochemistry in humans and in transgenic mice [13, 29, 33]. However, *MUC7* expression was detected recently also in submucosal glands in human upper airways and bronchus [34, 35], in bladder tumors [36, 37], in kidneys affected by pyelonephritis [38], in ampullary carcinomas [39], in the middle ear epithelium [40, 41], and in the lacrimal gland and conjunctive mucosa [42-44]. Moreover, *MUC7* over-expression was detected not only in glandular cells, but also in goblet cells [45].

The wide spectrum of *MUC7* expression suggests a complex mechanism of its regulation. These mechanisms have not been well studied, mainly because of the absence of an appropriate cell system (salivary gland acinar cell culture) suitable for *MUC7* transfection assays. Only a few publications have addressed this issue. In an early study by Bobek and colleagues [1], the 5'-regulatory region was identified as *MUC7* promoter and several transcription *cis*-elements were detected (Fig. **2**). A TATA-box and a CAAT-box were identified at positions -24/-19 bp (TATAAAA-box) and -83/-79 bp (CAAT-box) upstream to the MUC7 transcription start site, respectively. Recently, the same research group found in the *MUC7* promoter two more consensus sequences of TATA-box in the -600/-450 bp region, suggesting that *MUC7* gene may have more than one transcription unit [46].

Several potential regulatory elements were also found in the *MUC7* promoter, including a *cis*-AP-1 site, a glucocorticoid regulatory element, and a cAMP

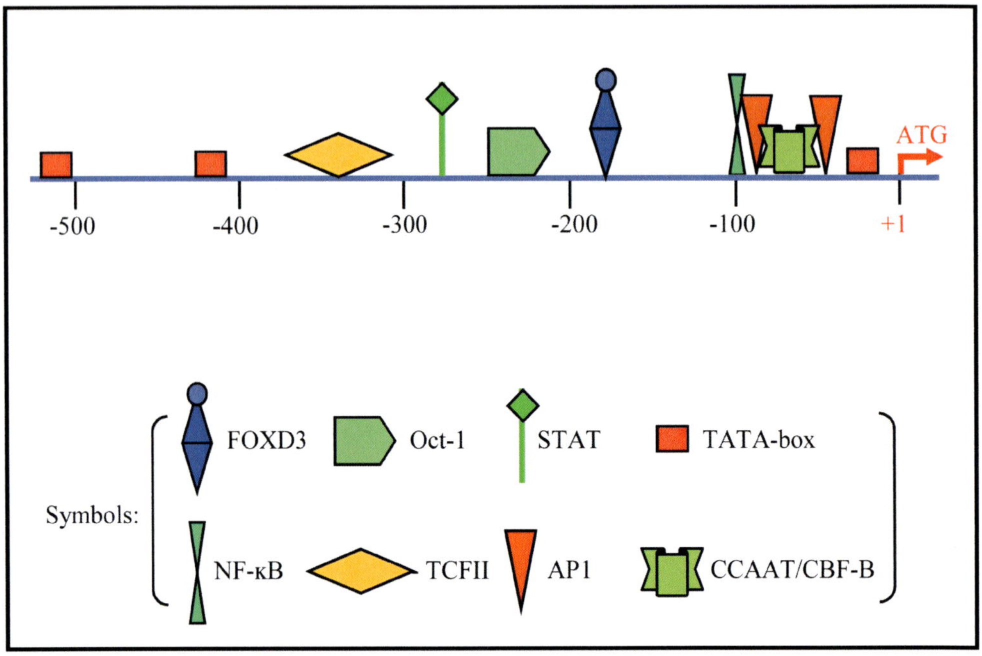

Figure 2: Transcription factor *cis*-elements of *MUC7* promoter (based on the data reported in [1, 46-49]).

response element. In addition to the transcription factor *cis*-elements already mentioned, *MUC7* promoter contains binding sites for transcription factors NF-κB, C/EBPβ, FOXD3, Oct-1, TCF11, and STATs. Moreover, transfection of the *MUC7* promoter-containing construct in a human lung epithelial cell line A549 established the functional role of the NF-κB *cis*-site (-100/-91 bp) in inducible *MUC7*-directed expression. The NF-κB element was shown to play a crucial role in *MUC7* gene response to TNFα stimulation, and the AP-1 sites located at -89/-86 bp and -80/-77 bp were shown to be indispensable for constitutive expression of *MUC7* in lung epithelial cells. It was also found that the minimal MUC7 5'-flanking sequence that possesses promoter activity in A549 cells is located between -138 and +30 bp. Furthermore, analysis of *MUC7* gene expression in differentiated and nondifferentiated normal human tracheo-bronchial epithelial cells, treated with growth factor EGF, retinoic acid (RA), bacterial lipopolysaccharide (LPS) and cytokines IL-1β, IL-4, IL-13 and TNFα showed that *MUC7* gene expression can be stimulated by all aforementioned agents [47], and its transcription is differentiation dependent [47-49]. The ability to stimulate *MUC7* promoter activity by so many different agents implies the involvement of several different signaling pathways in regulation of the *MUC7* expression. Although most of these pathways await identification, activation of *MUC7* expression in human gastric epithelial cell infected with *Helicobacter pylori* appears to be associated with activation of p38 MAPK cascade [50].

Thus, our understanding of the mechanisms regulating activity of the *MUC7* gene is still very elementary. More studies are needed to identify transcription factor binding sites in the *MUC7* promoter, to establish appropriate cell systems for *in vitro* analysis of *MUC7* promoter-directed transcription of a test gene(s), to identify signaling pathways involved in regulation of *MUC7* gene expression, and to elucidate the role, if any, of the intron sequences and 3'-UTR in mechanisms that regulate expression of *MUC7* gene.

9.4. FUNCTIONS OF MUC7 MUCIN

The MUC7 mucin as a component of the mucous layer serves a number of functions analogous to those of other secreted mucins: lubrication and hydration of mucosal and dental surfaces, protection against chemical and mechanical damage, and formation of a permeable barrier. It also performs functions specific to the oral cavity, including, among others, mastication, food bolus formation, speaking, prevention of oral infection, and defense of dental tissues by protection of teeth against decalcification [10]. Clearly, the MUC7 mucin is a multifunctional glycoprotein with unique biological functions associated with each of the MUC7 domains [23].

Theoretically, a protein can realize its functional potential in a number of ways: 1) by expressing function(s) associated exclusively with its native molecular structure; 2) by interacting with other proteins while keeping its own structure intact; 3) by uncovering functional potentials "hidden" in different parts of its molecule by posttranslational proteolysis, which breaks up the native structure of a protein and creates two or more peptides possessing new functional activities; and 4) by acquiring novel functions *via* posttranslational modification such as glycosylation, sulfation, acetylation and others of native polypeptide backbone.

The MUC7 glycoprotein utilizes several mechanisms to express its multiple functions.

The use of the whole native molecule to protect oral epithelial and dental surfaces by pellicle formation is thought to be one of the mechanisms utilized by salivary MUC7 mucin. Indeed, Fisher *et al.* [51] showed that the native MG2 (MUC7) mucin may protect teeth *in vivo* and *in vitro* by formation of tooth pellicle, although investigation of this function in several laboratories yielded conflicting results. Nieuw Amerongen and co-workers [52, 53] found that the mucin layer on human tooth enamel protects tooth against decalcification by organic acids, confirming Fisher *et al.*'s findings [51]. Al-Hashimi and Levine [54], on the other hand, could not detect MG2 (MUC7) in *in vivo* formed pellicle. Jensen *et al.* [55] consider that the failure to detect the native MG2 (MUC7) in tooth pellicle might be associated with proteolytic degradation.

Another function of MUC7 mucin, the opposite of its defense function, is the ability of MUC7 glycoprotein to provide receptors for adherence, and thereby colonization, of a wide spectrum of oral microorganisms [10, 56-58]. This function, which is associated with the native MUC7 molecule, depends on the chemical properties of the polypeptide backbone and/or the type of its posttranslational glycosylation. It also depends on the properties of the bacterial membrane glycoproteins. The binding of bacteria to MUC7 is highly selective; the mucin binds, for example, *Streptococcus gordonii* (*S. gordonii*) and *Streptococcus oralis,* but not *Streptococcus salivarius* or *Streptococcus sobrinus* [10, 59]. The intact MUC7 molecule has several tools for binding bacteria and viruses. For example, the finding that *in vitro* binding of purified MUC7 (MG2) to *Streptococcus mutans* could be abolished by reduction and/or alkylation suggests the involvement of cysteine residues of the MUC7 apomucin in this interaction [60]. The MUC7 molecule possesses a number of receptor sites for binding bacteria, and therefore may interact simultaneously with different bacterial proteins [61]. As shown by Kesimer *et al.* [61], MUC7 specifically interacts with several proteins synthesized by *S. gordonii*, including two elongation factors, EF-Tu and EF-G, a product of the *hppA* gene, the oligopeptide-binding lipoprotein, enolase-α and RNA polymerase. Of note, some of these proteins are located on the bacterial surface [62], whereas others are expressed intracellularly, indicating to ability of MUC7 molecule to enter bacterial intracellular compartments.

The MUC7 mucin can bind bacterial proteins by protein-protein interaction *via* electrostatic contact (*e.g.* binding of *Pseudomonas aeruginosa*) [63, 64], or by using carbohydrate moieties of the mucin and bacterial molecules. An example of the latter is the formation of the oligosaccharide bridge NeuAcα(2-3)Galβ(1-3) GalNAc between MUC7 and *Streptococcus sanguis* binding site [5]. MUC7 glycoprotein also makes use of its oligosaccharide elements to trap or aggregate human immunodeficiency virus [65]. Interestingly, binding of *Helicobacter pylori* to human salivary mucin MUC7 occurs through four modes of adhesion by usage: 1) blood-group antigen-binding adhesin, 2) sialic acid-binding adhesin, 3) novel saliva-binding adhesin, and 4) charge/low pH-dependent mechanism [66]. MUC7 may interact also with neutrophil L-selectin, for which MUC7 oligosaccharides serve as ligands [67]. L-selectin may provide MUC7 with antibacterial, antifungal, antiviral, and anti-inflammatory potentials [68], which are important in view of the mucin's interaction with microorganisms [69]. L-selectin also participates in formation of biofilm covering mucosal surfaces [70] and tooth enamel [51].

Thus, in a general sense, the MUC7 mucin participates in inflammation as both a "pro" and a "con" factor: on the one hand it interacts with bacteria and facilitates their colonization [69, 70], leading to inflammation, and on the other hand it binds to neutrophils, activating anti-inflammatory and antibacterial reactions [68].

As noted above, native MUC7 molecules may interact with other proteins *via* protein-protein interactions. An example of such interaction is the binding of MUC7 with lactoferrin, an iron-binding glycoprotein that bridges innate and adaptive immune functions in mammals. Lactoferrin, one of the MUC7 protein partners, was shown to form a heterotypic complex with MUC7 (MG2) *in vitro* and *in vivo via* polypeptide-polypeptide interaction by usage Ala-Leu-Leu-Cys motif present in the lactoferrin molecule [71]. Formation of this complex may have important functional consequences thanks to a synergistic effect. Because lactoferrin is sensitive to proteolysis, combining with MUC7 may protect it from proteolytic attack. The joining together of the two proteins also influences the rate of their removal from the oral cavity, extending the time they can function in the oral environment [71]. And finally, interaction of MUC7 with lactoferrin enables MUC7 mucin to take part in immune reactions mediated by lactoferrin.

The functions of MUC7 described above are associated with the intact full-length mucin molecule. There are, however, functions that can be activated only by its fragmentation [7, 14]. This property of MUC7 has been thoroughly studied by several laboratories. Gururaja *et al.* [7] found that the N-terminal region of MUC7 apomucin contains a short region of a 15-residue sequence highly homologous to histatine 5, known to possess anti- candidacidal activity – indicating the potential candidacidal activity of the MUC7 molecule. Indeed, such activity was identified by Liu *et al.* [14], who found that the recombinant protein, containing the N-terminal 144 aa-fragment of human MUC7 (MG2) mucin, exhibits candidacidal activity in *in vitro* assay. Subsequent experiments with synthetic cationinc peptides of different lengths, all of which contained the MUC7-derived sequence spanning residues from R^{40} to R^{51} (^{40}RKSYK**C**LHKR**C**^{51}R), confirmed the previous suggestion and showed that even an isolated N-terminal fragment of the MUC7 mucin does possess significant fungicidal activity *in vitro* [72-76]. This property of human MUC7 mucin appears to have been conserved in evolution, as the rat and mouse homologs of human MUC7 display structural organization similar to that of human MUC7 [77, 78]: all three have a similar composition of basic amino acids and a pI greater than 10.0. This suggests that fungicidal activity is dependent on electrostatic interactions between the indicated region of MUC7 and the negatively charged head-groups of phospholipids in the candida cell membrane [19]. Wei *et al.* [76] showed that fungicidal activity of the synthetic 12-mer peptide, corresponding to the ^{40}R-^{51}R sequence of the MUC7 domain D1, is not affected by Na^+, K^+, or Mg^{2+}, but is inhibited by Ca^{2+}. This peptide exerts optimum antifungal activity at neutral or slightly basic pH, which is in accord with the above proposed mechanism. Fluorescence microscopy [72] demonstrated the ability of synthetic MUC7 20-mer peptide to cross the fungal cell membrane and accumulate inside the cell. Cysteine residues (^{45}**C** and ^{50}**C**) present in the MUC7 20-mer and 12-mer synthetic peptides were found to be completely dispensable and not crucial for expression of antifungal potential, as substitution of these two residues with alanine did not alter their antifungal activity [75].

Comparison of MUC7-associated candidacidal activity tested *in vitro* and *in vivo* raises the question of whether results obtained *in vitro* can be extrapolated to *in vivo* systems. On the one hand, *in vitro* killing assays suggest that the N-terminal region of MUC7 possesses candidacidal activity, while purified MUC7 mucin lacked such activity *in vivo* against *Actinobacillus actinomycetemcomitans* [19, 74]. This problem was solved by the finding that the saliva of most subjects tested for candidacidal activity contained a 20 kDa polypeptide immunoreactive with anti-MUC7 antibody [70]. Importantly, this polypeptide molecule was not present in submandibular and sublingual glandular extracts, indicating that the 20 kDa polypeptide is most probably a fragment of the MUC7 molecule generated by a salivary protease following secretion of MUC7 into the oral cavity. Collectively, these data suggest that cleavage of MUC7 *in vivo* may produce fragment(s) that possess microbicidal properties and thereby activate a novel host defense function hidden in the native MUC7 molecule [70]. This suggestion is in agreement with Gururaja and co-workers' observation [19] that incubation of MUC7 with whole saliva leads to fragmentation of the native MUC7 polypeptide. Thus, the available information shows that both antibacterial and antifungal functions of the MUC7 mucin are linked to the same N-terminal region of domain D1. However, despite association with the same polypeptide region, these functions adopt different mechanisms: antibacterial activity requires active participation of the cysteine residues [14], whereas antifungal activity depends on the net positive charge of the specific N-terminal amino acid sequence effective only after disconnection of this sequence from the rest of the MUC7 polypeptide [75].

Analysis of the functions associated with domain D1 of the MUC7 glycoprotein clearly demonstrates the multifunctionality of the MUC7 mucin that is partially dependent on disintegration of the MUC7 molecule. However, as mentioned above, other MUC7 domains also possess multiple and, in a sense, opposite functions, whose manifestation may also be dependent on proteolytic cleavage of the native protein. The MUC7 central domain D3, containing six TRs rich in serine, threonine and proline, determines the stretched out, rigid rod-like structure of the whole molecule necessary for several mucin-specific functions, including lubrication and hydration of mucosal surfaces. It also facilitates binding of MUC7 to bacterial cells. Studies have shown that MUC7 mucin performs two apparently opposite functions: it facilitates bacterial colonization on tooth surface and enhances clearance of the colonized bacteria from tooth surface [10, 79]. How can a mucin molecule perform two contradictory functions? According to Antonyraj *et al.* [79], a native full-length

MUC7 molecule participates in bacterial absorption and colonization, but before the colonized organisms turn into pathogens, native mucin protein undergoes proteolytic degradation by protease(s) present in saliva, resulting in release of proline-rich TR-containing peptides harboring antimicrobial activity.

As noted above, each TR of MUC7 contains 8 proline residues specifically distributed along 23-aa repeat units. The presence of prolines allows the protein molecule to adopt a poly-L-proline II conformation, which can interact with microbial membrane and carry out antimicrobial activity [80, 81]. Like natural bactenecins with poly-L-proline structure, proline-containing TR-derived peptides possess anti-microbial activity. In contrast to naturally occurring cationic peptides, which usually contain arginine and lysine interspersed with proline residues [82], MUC7 TRs contain only serine, threonine, alanine and proline. The fact that MUC7 nevertheless exhibits strong antimicrobial activity may be explained by the ability of the liberated TR-containing fragments to adopt a specific structure – as suggested by Antonyraj *et al.* [79]. These fragments may fold into bundles of trimeric or hexameric poly-L-proline II helices stabilized by hydrophobic effects and hydrogen bonds between the strands. Apparently, such molecular configuration allows the antimicrobial activity embedded in the liberated MUC7 fragments to be displayed. The correctness of this hypothesis was demonstrated in experimental models [79].

Thus, the domains D1 and D3 of the MUC7 mucin possess both overt and covert functions hidden in their structures. Notable, the MUC7 derivatives resulting from posttranslational modifications of the primary apomucin may harbor still more functions yet to be disclosed. One of the most powerful approaches used to search for novel functions and properties of a protein is the yeast two-hybrid system. Using this system, Bruno *et al.* [83] showed that human mucin MUC7 may interact with several structurally diverse proteins secreted into saliva by submandibular gland. Interestingly, the N-terminal region, comprised of domains D1 and D2, could form complexes with such functionally and structurally different proteins as amylase, acidic proline-rich protein 2 (PRP2), basic proline-rich protein 3 (PRP3), lacrimal proline-rich protein 4 (PRP4), statherin, and histatin 1, whereas the C-terminal region, including domains D4 and D5, did not interact with any of the proteins in saliva. Surprised by their failure to detect complexes with the C-terminal region, which shares many other parameters with the N- terminal regions, the authors suggested that specific structural peculiarities of the C-terminal region did not allow its interaction with the proteins they examined, but might be suitable for interactions with proteins produced by other salivary glands, epithelial cells or bacteria.

The functional implications of the protein-protein interactions detected in the yeast two-hybrid system [83] are not known at present, but may be significant. Proline-rich proteins have been shown to interact with a variety of signaling proteins containing SH3 and/or WW domains [84-88]. As proline-rich protein, MUC7 also has the potential to participate in signaling pathways [84-88] probably as proline-rich ligand of signaling proteins [84, 89]. Interaction of MUC7 with different partner proteins [83] implies variability of MUC7-partner combinations and multiple modulations of MUC7 functions (Fig. **3**).

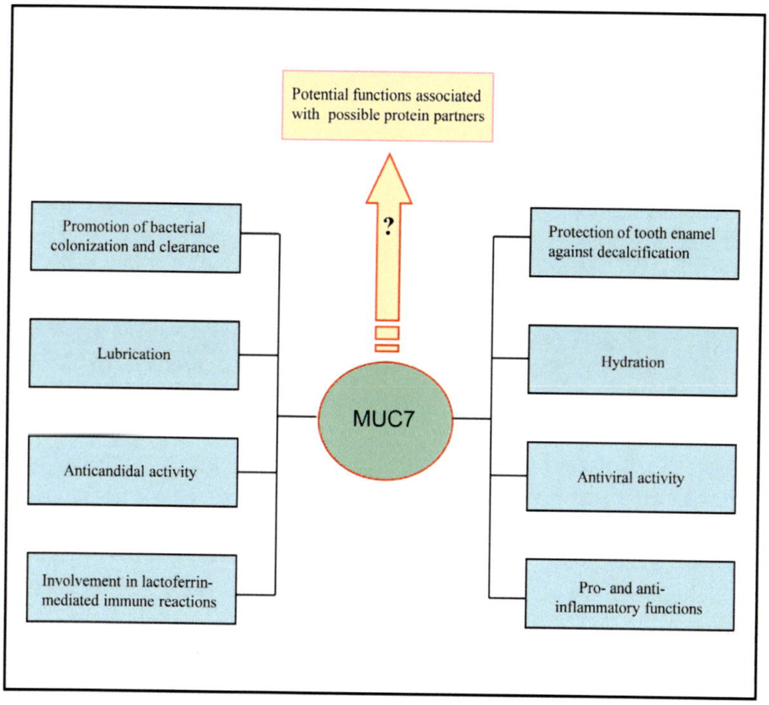

Figure 3: Functions of the MUC7 mucin.

It appears that the functional potentials of the MUC7 mucin concealed in its structure may go far beyond the overt and already described functions of this molecule. Expression of MUC7 in a wide spectrum of different tissue observed in physiological [34, 35] and pathological conditions [15, 17, 38] including malignant transformation [36, 37] suggests many additional yet not identified functions of this mucin.

REFERENCES

[1] Bobek LA, Liu J, Sait SN, *et al.* Structure and chromosomal localization of the human salivary mucin gene, MUC7. Genomics 1996;31:277-82.

[2] Bobek LA, Tsai H, Biesbrock AR, Levine MJ. Molecular cloning, sequence, and specificity of expression of the gene encoding the low molecular weight human salivary mucin (MUC7). J Biol Chem 1993;268:20563-9.

[3] Ramasubbu N, Reddy MS, Bergey EJ, *et al.* Large-scale purification and characterization of the major phosphoproteins and mucins of human submandibular-sublingual saliva. Biochem J 1991;280 (Pt 2):341-52.

[4] Reddy MS, Bobek LA, Haraszthy GG, Biesbrock AR, Levine MJ. Structural features of the low-molecular-mass human salivary mucin. Biochem J 1992;287 (Pt 2):639-43.

[5] Levine MJ, Reddy MS, Tabak LA, *et al.* Structural aspects of salivary glycoproteins. J Dent Res 1987;66:436-41.

[6] Loomis RE, Prakobphol A, Levine MJ, Reddy MS, Jones PC. Biochemical and biophysical comparison of two mucins from human submandibular-sublingual saliva. Arch Biochem Biophys 1987;258:452-64.

[7] Gururaja TL, Ramasubbu N, Venugopalan P, *et al.* Structural features of the human salivary mucin, MUC7. Glycoconj J 1998;15:457-67.

[8] Shori DK, Genter T, Hansen J, *et al.* Altered sialyl- and fucosyl-linkage on mucins in cystic fibrosis patients promotes formation of the sialyl-Lewis X determinant on salivary MUC-5B and MUC-7. Pflugers Arch 2001;443 Suppl 1:S55-61.

[9] Reddy MS, Levine MJ, Prakobphol A. Oligosaccharide structures of the low-molecular-weight salivary mucin from a normal individual and one with cystic fibrosis. J Dent Res 1985;64:33-6.

[10] Amerongen AV, Bolscher JG, Veerman EC. Salivary mucins: protective functions in relation to their diversity. Glycobiology 1995;5:733-40.

[11] Thomsson KA, Prakobphol A, Leffler H, *et al.* The salivary mucin MG1 (MUC5B) carries a repertoire of unique oligosaccharides that is large and diverse. Glycobiology 2002;12:1-14.

[12] Karlsson NG, Thomsson KA. Salivary MUC7 is a major carrier of blood group I type O-linked oligosaccharides serving as the scaffold for sialyl Lewis^x. Glycobiology 2009;19:288-300.

[13] Nielsen PA, Mandel U, Therkildsen MH, Clausen H. Differential expression of human high-molecular-weight salivary mucin (MG1) and low-molecular-weight salivary mucin (MG2). J Dent Res 1996;75:1820-6.

[14] Liu B, Rayment SA, Gyurko C, *et al.* The recombinant N-terminal region of human salivary mucin MG2 (MUC7) contains a binding domain for oral Streptococci and exhibits candidacidal activity. Biochem J 2000;345 (Pt 3):557-64.

[15] Kirkbride HJ, Bolscher JG, Nazmi K, *et al.* Genetic polymorphism of MUC7: allele frequencies and association with asthma. Eur J Hum Genet 2001;9:347-54.
[16] Rousseau K, Vinall LE, Butterworth SL, *et al.* MUC7 haplotype analysis: results from a longitudinal birth cohort support protective effect of the MUC7*5 allele on respiratory function. Ann Hum Genet 2006;70:417-27.
[17] Watson AM, Ngor WM, Gordish-Dressman H, Freishtat RJ, Rose MC. MUC7 polymorphisms are associated with a decreased risk of a diagnosis of asthma in an African American population. J Investig Med 2009;57:882-6.
[18] Oppenheim FG, Xu T, McMillian FM, *et al.* Histatins, a novel family of histidine-rich proteins in human parotid secretion. Isolation, characterization, primary structure, and fungistatic effects on Candida albicans. J Biol Chem 1988;263:7472-7.
[19] Gururaja TL, Levine JH, Tran DT, *et al.* Candidacidal activity prompted by N-terminus histatin-like domain of human salivary mucin (MUC7)1. Biochim Biophys Acta 1999;1431:107-19.
[20] Williamson MP. The structure and function of proline-rich regions in proteins. Biochem J 1994;297 (Pt 2):249-60.
[21] Naganagowda GA, Gururaja TL, Satyanarayana J, Levine MJ. NMR analysis of human salivary mucin (MUC7) derived O-linked model glycopeptides: comparison of structural features and carbohydrate-peptide interactions. J Pept Res 1999;54:290-310.
[22] Butenhof KJ, Gerken TA. Structure and dynamics of mucin-like glycopeptides. Examination of peptide chain expansion and peptide-carbohydrate interactions by stochastic dynamics simulations. Biochemistry 1993;32:2650-63.
[23] Satyanarayana J, Gururaja TL, Narasimhamurthy S, Naganagowda GA, Levine MJ. Synthesis and conformational features of human salivary mucin C-terminal derived peptide epitope carrying Thomsen-Friedenreich antigen: implications for its role in self-association. Biopolymers 2001;58:500-10.
[24] Narasimhamurthy S, Naganagowda GA, Janagani S, Gururaja TL, Levine MJ. Solution structure of O-glycosylated C-terminal leucine zipper domain of human salivary mucin (MUC7). J Biomol Struct Dyn 2000;18:145-54.
[25] Kohn WD, Kay CM, Hodges RS. Protein destabilization by electrostatic repulsions in the two-stranded alpha-helical coiled-coil/leucine zipper. Protein Sci 1995;4:237-50.
[26] Mehrotra R, Thornton DJ, Sheehan JK. Isolation and physical characterization of the MUC7 (MG2) mucin from saliva: evidence for self-association. Biochem J 1998;334 (Pt 2):415-22.
[27] Soares RV, Liu B, Oppenheim FG, Offner GD, Troxler RF. Structural characterisation of cysteines in a bacterial-binding motif of human salivary mucin MG2. Arch Oral Biol 2002;47:591-7.
[28] Prakobphol A, Levine MJ, Tabak LA, Reddy MS. Purification of a low-molecular-weight, mucin-type glycoprotein from human submandibular-sublingual saliva. Carbohydr Res 1982;108:111-22.
[29] Khan SH, Aguirre A, Bobek LA. *In-situ* hybridization localized MUC7 mucin gene expression to the mucous acinar cells of human and MUC7-transgenic mouse salivary glands. Glycoconj J 1998;15:1125-32.
[30] Bolscher JG, Groenink J, van der Kwaak JS, *et al.* Detection and quantification of MUC7 in submandibular, sublingual, palatine, and labial saliva by anti-peptide antiserum. J Dent Res 1999;78:1362-9.

[31] Piras M, Hand AR, Tore G, Ledda GP, Piludu M. Ultrastructural localization of salivary mucins MUC5B and MUC7 in human labial glands. Eur J Oral Sci 2010;118:14-8.
[32] Ruhl S, Rayment SA, Schmalz G, Hiller KA, Troxler RF. Proteins in whole saliva during the first year of infancy. J Dent Res 2005;84:29-34.
[33] Bobek LA, Li H, Rojstaczer N, *et al.* Tissue-specific expression of human salivary mucin gene, MUC7, in transgenic mice. Transgenic Res 1998;7:195-204.
[34] Sharma P, Dudus L, Nielsen PA, *et al.* MUC5B and MUC7 are differentially expressed in mucous and serous cells of submucosal glands in human bronchial airways. Am J Respir Cell Mol Biol 1998;19:30-7.
[35] Martinez-Anton A, Debolos C, Garrido M, *et al.* Mucin genes have different expression patterns in healthy and diseased upper airway mucosa. Clin Exp Allergy 2006;36:448-57.
[36] Okegawa T, Kinjo M, Horie S, Nutahara K, Higashihara E. Detection of mucin 7 gene expression in exfoliated cells in urine from patients with bladder tumor. Urology 2003;62:182-6.
[37] Kinjo M, Okegawa T, Horie S, Nutahara K, Higashihara E. Detection of circulating MUC7-positive cells by reverse transcription-polymerase chain reaction in bladder cancer patients. Int J Urol 2004;11:38-43.
[38] Lehmann J, Suttmann H, Gerber M, *et al.* [Expression of antimicrobial peptide MUC7 in kidneys with pyelonephritis]. Urologe A 2006;45:1300, 2-7.
[39] Paulsen FP, Varoga D, Paulsen AR, Corfield A, Tsokos M. Prognostic value of mucins in the classification of ampullary carcinomas. Hum Pathol 2006;37:160-7.
[40] Kerschner JE. Mucin gene expression in human middle ear epithelium. Laryngoscope 2007;117:1666-76.
[41] Takeuchi K, Yagawa M, Ishinaga H, *et al.* Mucin gene expression in the effusions of otitis media with effusion. Int J Pediatr Otorhinolaryngol 2003;67:53-8.
[42] Gipson IK. Distribution of mucins at the ocular surface. Exp Eye Res 2004;78:379-88.
[43] Jumblatt MM, McKenzie RW, Steele PS, Emberts CG, Jumblatt JE. MUC7 expression in the human lacrimal gland and conjunctiva. Cornea 2003;22:41-5.
[44] Ramamoorthy P, Nichols JJ. Mucins in contact lens wear and dry eye conditions. Optom Vis Sci 2008;85:631-42.
[45] Penia MT, Aujla PK, Zudaire E, *et al.* Localization and expression of MUC5B and MUC7 mucins in pediatric sinus mucosa. Ann Otol Rhinol Laryngol 2007;116:389-97.
[46] Li S, Bobek LA. Functional analysis of human MUC7 mucin gene 5'-flanking region in lung epithelial cells. Am J Respir Cell Mol Biol 2006;35:593-601.
[47] Li S, Intini G, Bobek LA. Modulation of MUC7 mucin expression by exogenous factors in airway cells *in vitro* and *in vivo*. Am J Respir Cell Mol Biol 2006;35:95-102.
[48] Woenckhaus M, Klein-Hitpass L, Grepmeier U, *et al.* Smoking and cancer-related gene expression in bronchial epithelium and non-small-cell lung cancers. J Pathol 2006;210:192-204.
[49] Gray T, Koo JS, Nettesheim P. Regulation of mucous differentiation and mucin gene expression in the tracheobronchial epithelium. Toxicology 2001;160:35-46.
[50] Kim H, Seo JH, Kim KH. The effect of p38 mitogen-activated protein kinase on mucin gene expression and apoptosis in Helicobacter pylori-infected gastric epithelial cells. Ann N Y Acad Sci 2003;1010:90-4.
[51] Fisher SJ, Prakobphol A, Kajisa L, Murray PA. External radiolabelling of components of pellicle on human enamel and cementum. Arch Oral Biol 1987;32:509-17.

[52] Nieuw Amerongen AV, Oderkerk CH, Driessen AA. Role of mucins from human whole saliva in the protection of tooth enamel against demineralization *in vitro*. Caries Res 1987;21:297-309.

[53] Nieuw Amerongen AV, Oderkerk CH, Veerman EC. Influence of phytate on the adsorption of human salivary mucins onto hydroxyapatite. J Biol Buccale 1988;16:203-8.

[54] Al-Hashimi I, Levine MJ. Characterization of *in vivo* salivary-derived enamel pellicle. Arch Oral Biol 1989;34:289-95.

[55] Jensen JL, Lamkin MS, Oppenheim FG. Adsorption of human salivary proteins to hydroxyapatite: a comparison between whole saliva and glandular salivary secretions. J Dent Res 1992;71:1569-76.

[56] Gibbons RJ, Etherden I, Moreno EC. Association of neuraminidase-sensitive receptors and putative hydrophobic interactions with high-affinity binding sites for Streptococcus sanguis C5 in salivary pellicles. Infect Immun 1983;42:1006-12.

[57] Gibbons RJ. Adherent interactions which may affect microbial ecology in the mouth. J Dent Res 1984;63:378-85.

[58] Liljemark WF, Bloomquist CG, Fenner LJ, Antonelli PJ, Coulter MC. Effect of neuraminidase on the adherence to salivary pellicle of Streptococcus sanguis and Streptococcus mitis. Caries Res 1989;23:141-5.

[59] Murray PA, Prakobphol A, Lee T, Hoover CI, Fisher SJ. Adherence of oral streptococci to salivary glycoproteins. Infect Immun 1992;60:31-8.

[60] Liu B, Rayment S, Oppenheim FG, Troxler RF. Isolation of human salivary mucin MG2 by a novel method and characterization of its interactions with oral bacteria. Arch Biochem Biophys 1999;364:286-93.

[61] Kesimer M, Kilic N, Mehrotra R, Thornton DJ, Sheehan JK. Identification of salivary mucin MUC7 binding proteins from Streptococcus gordonii. BMC Microbiol 2009;9:163.

[62] Ge J, Catt DM, Gregory RL. Streptococcus mutans surface alpha-enolase binds salivary mucin MG2 and human plasminogen. Infect Immun 2004;72:6748-52.

[63] Reddy MS, Levine MJ, Paranchych W. Low-molecular-mass human salivary mucin, MG2: structure and binding of Pseudomonas aeruginosa. Crit Rev Oral Biol Med 1993;4:315-23.

[64] Reddy MS. Binding of the pili of Pseudomonas aeruginosa to a low-molecular-weight mucin and neutral cystatin of human submandibular-sublingual saliva. Curr Microbiol 1998;37:395-402.

[65] Habte HH, Mall AS, de Beer C, Lotz ZE, Kahn D. The role of crude human saliva and purified salivary MUC5B and MUC7 mucins in the inhibition of Human Immunodeficiency Virus type 1 in an inhibition assay. Virol J 2006;3:99.

[66] Linden SK, Wickstrom C, Lindell G, Gilshenan K, Carlstedt I. Four modes of adhesion are used during Helicobacter pylori binding to human mucins in the oral and gastric niches. Helicobacter 2008;13:81-93.

[67] Prakobphol A, Tangemann K, Rosen SD, *et al.* Separate oligosaccharide determinants mediate interactions of the low-molecular-weight salivary mucin with neutrophils and bacteria. Biochemistry 1999;38:6817-25.

[68] Soares RV, Siqueira CC, Bruno LS, *et al.* MG2 and lactoferrin form a heterotypic complex in salivary secretions. J Dent Res 2003;82:471-5.

[69] Weinberg ED. Human lactoferrin: a novel therapeutic with broad spectrum potential. J Pharm Pharmacol 2001;53:1303-10.

[70] Liu B, Rayment SA, Soares RV, *et al.* Interaction of human salivary mucin MG2, its recombinant N-terminal region and a synthetic peptide with Actinobacillus actinomycetemcomitans. J Periodontal Res 2002;37:416-24.

[71] Collins LM, Dawes C. The surface area of the adult human mouth and thickness of the salivary film covering the teeth and oral mucosa. J Dent Res 1987;66:1300-2.

[72] Bobek LA, Situ H. MUC7 20-Mer: investigation of antimicrobial activity, secondary structure, and possible mechanism of antifungal action. Antimicrob Agents Chemother 2003;47:643-52.

[73] Satyanarayana J, Situ H, Narasimhamurthy S, *et al.* Divergent solid-phase synthesis and candidacidal activity of MUC7 D1, a 51-residue histidine-rich N-terminal domain of human salivary mucin MUC7. J Pept Res 2000;56:275-82.

[74] Situ H, Bobek LA. *In vitro* assessment of antifungal therapeutic potential of salivary histatin-5, two variants of histatin-5, and salivary mucin (MUC7) domain 1. Antimicrob Agents Chemother 2000;44:1485-93.

[75] Situ H, Wei G, Smith CJ, Mashhoon S, Bobek LA. Human salivary MUC7 mucin peptides: effect of size, charge and cysteine residues on antifungal activity. Biochem J 2003;375:175-82.

[76] Wei GX, Campagna AN, Bobek LA. Factors affecting antimicrobial activity of MUC7 12-mer, a human salivary mucin-derived peptide. Ann Clin Microbiol Antimicrob 2007;6:14.

[77] Albone EF, Hagen FK, VanWuyckhuyse BC, Tabak LA. Molecular cloning of a rat submandibular gland apomucin. J Biol Chem 1994;269:16845-52.

[78] Denny PC, Mirels L, Denny PA. Mouse submandibular gland salivary apomucin contains repeated N-glycosylation sites. Glycobiology 1996;6:43-50.

[79] Antonyraj KJ, Karunakaran T, Raj PA. Bactericidal activity and poly-L-proline II conformation of the tandem repeat sequence of human salivary mucin glycoprotein (MG2). Arch Biochem Biophys 1998;356:197-206.

[80] Raj PA, Marcus E, Edgerton M. Delineation of an active fragment and poly(L-proline) II conformation for candidacidal activity of bactenecin 5. Biochemistry 1996;35:4314-25.

[81] Raj PA, Soni SD, Levine MJ. Membrane-induced helical conformation of an active candidacidal fragment of salivary histatins. J Biol Chem 1994;269:9610-9.

[82] Zasloff M. Antibiotic peptides as mediators of innate immunity. Curr Opin Immunol 1992;4:3-7.

[83] Bruno LS, Li X, Wang L, *et al.* Two-hybrid analysis of human salivary mucin MUC7 interactions. Biochim Biophys Acta 2005;1746:65-72.

[84] Kay BK, Williamson MP, Sudol M. The importance of being proline: the interaction of proline-rich motifs in signaling proteins with their cognate domains. Faseb J 2000;14:231-41.

[85] Feng S, Chen JK, Yu H, Simon JA, Schreiber SL. Two binding orientations for peptides to the Src SH3 domain: development of a general model for SH3-ligand interactions. Science 1994;266:1241-7.

[86] Grabs D, Slepnev VI, Songyang Z, *et al.* The SH3 domain of amphiphysin binds the proline-rich domain of dynamin at a single site that defines a new SH3 binding consensus sequence. J Biol Chem 1997;272:13419-25.

[87] Chen HI, Sudol M. The WW domain of Yes-associated protein binds a proline-rich ligand that differs from the consensus established for Src homology 3-binding modules. Proc Natl Acad Sci U S A 1995;92:7819-23.

[88] Ermekova KS, Zambrano N, Linn H, *et al.* The WW domain of neural protein FE65 interacts with proline-rich motifs in Mena, the mammalian homolog of Drosophila enabled. J Biol Chem 1997;272:32869-77.

[89] Simon C, Barathieu K, Laguerre M, *et al.* Three-dimensional structure and dynamics of wine tannin-saliva protein complexes. A multitechnique approach. Biochemistry 2003;42:10385-95.

Send Orders of Reprints at reprints@benthamscience.net

Mucins – Potential Regulators of Cell Functions, 2013, 418-424

CHAPTER 10

Mucin MUC8

Abstract: Glycoprotein MUC8, one of the less studied mucins, belongs to the group of soluble mucins. Its cDNA is only partially cloned. Like other mucins, MUC8 has a tandem repeat containing domain, but in contrast to other mucins, which have, as a rule, only one type of repeats, the MUC8 mucin has two types of repeats: one of 13 amino acids and another one containing 41 amino acid residues. The *MUC8* gene is localized on human chromosome 12q24.3. MUC8 mucin appears to serve a ciliated cell marker. Regulation of *MUC8* gene expression has been studied and some regulatory pathways have been identified. MUC8 mucin plays a major role in the airway mucosa and its over-expression and hypersecretion are associated with airway inflammatory diseases.

Keywords: MUC8, mucin, glycoprotein, tandem repeats, transcriptional regulation.

10.1. GENERAL PROPERTIES OF *MUC8* GENE AND THE ENCODED MUCIN GLYCOPROTEIN

Little information is available about the structure and functions of the *MUC8* gene and the encoded mucin. The initial cloning of a part of MUC8 cDNA was carried out 17 years ago [1], however, the full cDNA and complete genomic sequences of the *MUC8* gene are still not identified. Shankar *et al.* [1, 2] cloned a 1422 bp sequence of the *MUC8* gene made up of a centrally located 941 bp cDNA fragment encoding a tandem repeat (TR)-containing polypeptide of 313 amino acids. In addition to 941 bp of the cDNA, the cloned sequence also includes a 3'-adjacent 481 nucleotides of non-coding sequence containing a stop codon, a polyadenylate signal with a poly-A tail, a 3'-UTR of 458 bp in length, and the extreme carboxy terminus of the *MUC8* gene. The *MUC8* gene is thought to be approximately 90 kb in length [2].

Despite the meager information available, several unique features of the MUC8 mucin have emerged. It is well known that all mucins have extended arrays of TR sequences within a protein core rich in proline and highly glycosylated serine and threonine residues. The overall composition of the deduced amino acid sequence of MUC8 apomucin corresponds to that expected for an apoprotein with serine, threonine, proline and alanine amino acids comprising ~51% [1]. Although the TR

Joseph Z. Zaretsky and Daniel H. Wreschner

units vary in length from as short as 8 amino acids per repeat unit in the MUC5AC mucin [3] to as long as 169 amino acids in MUC6 [4], there is usually only one type of consensus repeat in a given mucin. In *MUC8*, Shankar *et al.* [1] discovered a novel type of mucin gene organization. The TR-containing region of the *MUC8* gene consists of imperfect tandem repeats each of 41nucleotides (CCAGGAGGGACACCGGGTTCACGAGCTGCCCACGCCCTCT) that encode a unique polypeptide with two types of consensus repeats: one, represented by short amino acid sequence (TSCPRPLQEGTRV), and another one, represented by amino acid chain (TSCPRPLQEGTPGSRAAHALSRRGHRVHELPTSSPGGDTGF) which is 3 time longer that a short one [1].

Another unique feature of the MUC8 mucin is associated with the cysteine residues found in its TR units. While mucins described in the previous chapters do not contain cysteines in the repeat units (cysteine residues, if any, are usually located between TR units), the MUC8 mucin contains at least one cysteine residue per repeat unit. The cysteine residues in the gel-forming mucins usually transform mucin monomers into oligomers *via* formation of disulfide bonds [5, 6]. The role the cysteine residues play in the physiology and biochemistry of the MUC8 glycoprotein is unknown.

The *MUC8* gene differs from other gene encoding gel-forming and soluble mucins also by chromosomal localization. While most of the gel-forming mucin genes are located on chromosome 11p15.5, and *MUC7* and *MUC9* occupy chromosomes 4q13 and 1p13, respectively, the *MUC8* gene resides on chromosome 12q24.3 [2], which contains also *MUC19* mucin gene.

Early studies on expression of *MUC8* gene reported this gene expressed exclusively by the mucous cells of the submucosal glands in the airways [1, 7]. Subsequent investigations showed that it is also expressed in nasal epithelium [8], human middle ear epithelial cells [9], stomach [2], and both the male and female reproductive tracts [10, 11]. According to Hebbar *et al.* [11], MUC8 mucin may be considered a "specific marker for endometrial carcinomas" as its expression in this type of cancer is significantly higher than in normal endometrium. It should be pointed out that the levels of *MUC8* gene expression are very high in the cervix in both cancerous and noncancerous tissues [11].

The MUC8 glycoprotein is considered as a ciliated cell marker [8]. This conclusion came out from Kim *et al.* [8] study of the MUC8 mRNA and MUC8 protein expression patterns during mucociliary differentiation of normal human nasal epithelial cells (NHNE) *in vitro*. An increase in the number of ciliated cells in time correlated with the levels of *MUC8* gene expression in the NHNE cell culture. The number of ciliated cells rose as a function of differentiation, while the number of secretory cells remained relatively constant. *MUC8* was expressed also in the ciliated cells of human nasal polyps. Analogous to ciliogenesis of normal nasal epithelial cells, the ciliogenesis of human middle ear epithelial cells could be promoted by IL-1β which induces in parallel over-expression of the MUC8 mucin [9]. These results suggest that MUC8 mucin might be related to differentiation and/or function of the ciliated cells. For the sake of objectivity, it should be said that Lopez-Ferrer *et al.* [12] observed expression of the MUC8 protein in both goblet and ciliated cells in human bronchial epithelium, while Kim *et al.* [8] could not detect expression of the gene in secretory goblet cells of nasal origin.

10.2. REGULATION OF *MUC8* GENE EXPRESSION

The mechanisms of *MUC8* gene transcriptional regulation are poorly understood. Practically all what is known about these mechanisms is based on the reports from Yoon's group [13-15]. In their initial study Seong *et al.* [13] found that inflammation increases *MUC8* expression in the nasal epithelium. Using a mixture of the inflammatory mediators TNFα, IL-1β, LPS, IL-4 and PAF, they showed that these mediators up-regulate *MUC8* expression in the NHNE cell culture. This group then showed that the induction of *MUC8* gene expression by IL-1β is mediated by a sequential ERK MAPK/RSK1/CREB pathway in human airway epithelial cells [14]. Interestingly, although several reports suggested that more than one MAPK might be necessary for the signal transduction activated by various inflammatory mediators including IL-1β [16-18], Song *et al.* [14] showed that activation of the ERK MAPK pathway is not only required, but is sufficient for IL-1β-induced *MUC8* expression. They also established the role of RSK1 and CREB in the downstream signaling of ERK MAPK that resulted in induction of *MUC8* gene expression. CREB, at least in part, is essential for IL-1β-induced *MUC8* expression through binding to the CRE *cis*-element in the *MUC8* promoter. Indeed, the CRE *cis*-element located at -803 position of the *MUC8* promoter, encompassing the region from -1644 to +87 bp, was found to be critical for the up-regulation of *MUC8* transcriptional activity both by IL-1β, which activates the IL-1β receptor/Ras/Raf/ERK/RSK1/CREB cascade pathway [14], and by prostaglandin E2, which converts its signal through the EP1-4 receptor/MEK1/ERK/RSK1/CREB pathway [18].

Physiological homeostasis supposes the involvement of active mechanisms that depending on various stimuli may up-regulate expression of different genes or inhibit their transcriptional activities. The suppressors of cytokine signaling (SOCS) are elements of the inhibitory mechanisms. SOCS 1-8 are the negative feedback regulators of the Jak/STAT pathways [19, 20]. The main physiological function of the SOCS is the negative regulation of the Jak/STAT-dependent IL-6 signaling [21, 22], which in turn inhibits also IL-1 signaling [23, 24]. The intracellular negative regulatory mechanism that affects IL-1β-induced expression of the *MUC8* gene is not fully understood. Song *et al.* [25] recently examined the inhibitory effects of SOCS3 on IL-1β-induced *MUC8* gene expression, and found that SOCS3 over-expression suppressed the IL-1β-induced *MUC8* gene expression in NCI-H292 cells, while silencing of SOCS3 restored the IL-1β-induced *MUC8* gene transcriptional activity. SOCS3 did not affect the ERK1/2/RSK1/CREB pathway activated by IL-1β, indicating that SOCS3 is directed to another target(s). The authors found that the interaction of SOCS3 with the NonO (non-POU-domain containing octamer-binding protein) attenuates the IL-1β-dependent *MUC8* expression, while silencing of SOCS3 dramatically increases the NonO-mediated *MUC8* expression caused by IL-1β.

Nucleotides, including adenosine-5-triphosphate (ATP) and uridine-5-triphosphate (UTP), play fundamental roles in secretion of mucins in many polarized epithelial cells. Choi *et al.* [26] found that UTP and ATP induce secretion of MUC5AC and MUC5B mucins as well as MUC8 glycoprotein in normal human nasal epithelial cells *via* Ca^{2++}-dependent pathway. However, these nucleotides did not significantly change expression of mucin mRNAs.

AP2α is one of the relatively well studied factors that may up-regulate *MUC8* gene transcription. Several studies have shown that AP2α plays a critical role in transcriptional regulation of epithelial cell specific genes: a number of epithelial cells express AP2α protein, which participates in epithelial gene expression [27-29]; nasal polyp epithelium over-expresses the AP2 protein compared with normal nasal mucosa [30]. Based on an earlier report that expression of *MUC8* was also up-regulated in nasal polyp epithelium [31], one may consider AP2α an important regulator of *MUC8* expression in nasal polyp epithelium. Indeed, Moon *et al.* [30] showed that AP2α is essential for *MUC8* expression in human airway epithelial cells, having demonstrated that phorbol ester PMA up-regulates expression of both AP2α and *MUC8* gene in NCI-H299 cells. Moreover, a critical role of AP2α plays in *MUC8* expression was evidenced by experiments in which silencing of the AP2α-coding gene by siRNA suppressed the PMA-induced *MUC8* expression. The authors found that in human airway epithelial cells PMA induces *MUC8* expression through a mechanism involving protein kinase C (PKC), ERK1/2 and AP2α activation.

In conclusion, although the MUC8 mucin appears to play a major role in the airway mucosa and its over-expression is related to mucus hypersecretion leading to airway inflammatory diseases, there is very little information about the structure of the *MUC8* gene and mechanisms that regulate its transcription. Practically, there are no data about synthesis, structure and functions of the MUC8 mucin. At present, only a small part of the C-terminal domain sequence has been published of what promises to be a multifunctional mucin with many important properties yet to be discovered.

REFERENCES

[1] Shankar V, Gilmore MS, Elkins RC, Sachdev GP. A novel human airway mucin cDNA encodes a protein with unique tandem-repeat organization. Biochem J 1994;300 (Pt 2):295-8.

[2] Shankar V, Pichan P, Eddy RL, Jr., *et al.* Chromosomal localization of a human mucin gene (MUC8) and cloning of the cDNA corresponding to the carboxy terminus. Am J Respir Cell Mol Biol 1997;16:232-41.

[3] Escande F, Aubert JP, Porchet N, Buisine MP. Human mucin gene MUC5AC: organization of its 5'-region and central repetitive region. Biochem J 2001;358:763-72.

[4] Toribara NW, Roberton AM, Ho SB, *et al.* Human gastric mucin. Identification of a unique species by expression cloning. J Biol Chem 1993;268:5879-85.

[5] Gum JR, Jr. Mucin genes and the proteins they encode: structure, diversity, and regulation. Am J Respir Cell Mol Biol 1992;7:557-64.

[6] Bell SL, Xu G, Forstner JF. Role of the cystine-knot motif at the C-terminus of rat mucin protein Muc2 in dimer formation and secretion. Biochem J 2001;357:203-9.
[7] Leikauf GD, Borchers MT, Prows DR, Simpson LG. Mucin apoprotein expression in COPD. Chest 2002;121:166S-82S.
[8] Kim CH, Kim HJ, Song KS, *et al.* MUC8 as a ciliated cell marker in human nasal epithelium. Acta Otolaryngol 2005;125:76-81.
[9] Choi JY, Kim JY, Kim CW, *et al.* IL-1beta promotes the ciliogenesis of human middle ear epithelial cells: possible linkage with the expression of mucin gene 8. Acta Otolaryngol 2005;125:260-5.
[10] D'Cruz OJ, Dunn TS, Pichan P, Hass GG, Jr., Sachdev GP. Antigenic cross-reactivity of human tracheal mucin with human sperm and trophoblasts correlates with the expression of mucin 8 gene messenger ribonucleic acid in reproductive tract tissues. Fertil Steril 1996;66:316-26.
[11] Hebbar V, Damera G, Sachdev GP. Differential expression of MUC genes in endometrial and cervical tissues and tumors. BMC Cancer 2005;5:124.
[12] Lopez-Ferrer A, Curull V, Barranco C, *et al.* Mucins as differentiation markers in bronchial epithelium. Squamous cell carcinoma and adenocarcinoma display similar expression patterns. Am J Respir Cell Mol Biol 2001;24:22-9.
[13] Seong JK, Koo JS, Lee WJ, *et al.* Upregulation of MUC8 and downregulation of MUC5AC by inflammatory mediators in human nasal polyps and cultured nasal epithelium. Acta Otolaryngol 2002;122:401-7.
[14] Song KS, Seong JK, Chung KC, *et al.* Induction of MUC8 gene expression by interleukin-1 beta is mediated by a sequential ERK MAPK/RSK1/CREB cascade pathway in human airway epithelial cells. J Biol Chem 2003;278:34890-6.
[15] Bernatchez PN, Allen BG, Gelinas DS, Guillemette G, Sirois MG. Regulation of VEGF-induced endothelial cell PAF synthesis: role of p42/44 MAPK, p38 MAPK and PI3K pathways. Br J Pharmacol 2001;134:1253-62.
[16] Waetzig GH, Seegert D, Rosenstiel P, Nikolaus S, Schreiber S. p38 mitogen-activated protein kinase is activated and linked to TNF-alpha signaling in inflammatory bowel disease. J Immunol 2002;168:5342-51.
[17] Plath KE, Grabbe J, Gibbs BF. Calcineurin antagonists differentially affect mediator secretion, p38 mitogen-activated protein kinase and extracellular signal-regulated kinases from immunologically activated human basophils. Clin Exp Allergy 2003;33:342-50.
[18] Cho KN, Choi JY, Kim CH, *et al.* Prostaglandin E2 induces MUC8 gene expression *via* a mechanism involving ERK MAPK/RSK1/cAMP response element binding protein activation in human airway epithelial cells. J Biol Chem 2005;280:6676-81.
[19] Hanada T, Kinjyo I, Inagaki-Ohara K, Yoshimura A. Negative regulation of cytokine signaling by CIS/SOCS family proteins and their roles in inflammatory diseases. Rev Physiol Biochem Pharmacol 2003;149:72-86.
[20] Kubo M, Hanada T, Yoshimura A. Suppressors of cytokine signaling and immunity. Nat Immunol 2003;4:1169-76.
[21] Schmitz J, Weissenbach M, Haan S, Heinrich PC, Schaper F. SOCS3 exerts its inhibitory function on interleukin-6 signal transduction through the SHP2 recruitment site of gp130. J Biol Chem 2000;275:12848-56.
[22] Nicholson SE, De Souza D, Fabri LJ, *et al.* Suppressor of cytokine signaling-3 preferentially binds to the SHP 2 binding site on the shared cytokine receptor subunit gp130. Proc Natl Acad Sci U S A 2000;97:6493-8.

[23] Frobose H, Ronn SG, Heding PE, *et al.* Suppressor of cytokine Signaling-3 inhibits interleukin-1 signaling by targeting the TRAF-6/TAK1 complex. Mol Endocrinol 2006;20:1587-96.

[24] Yang XP, Albrecht U, Zakowski V, *et al.* Dual function of interleukin-1beta for the regulation of interleukin-6-induced suppressor of cytokine signaling 3 expression. J Biol Chem 2004;279:45279-89.

[25] Song KS, Kim K, Chung KC, Seol JH, Yoon JH. Interaction of SOCS3 with NonO attenuates IL-1beta-dependent MUC8 gene expression. Biochem Biophys Res Commun 2008;377:946-51.

[26] Choi JY, Namkung W, Shin JH, Yoon JH. Uridine-5'-triphosphate and adenosine triphosphate gammaS induce mucin secretion *via* Ca^{2+}-dependent pathways in human nasal epithelial cells. Acta Otolaryngol 2003;123:1080-6.

[27] Hennig G, Lowrick O, Birchmeier W, Behrens J. Mechanisms identified in the transcriptional control of epithelial gene expression. J Biol Chem 1996;271:595-602.

[28] Batsche E, Muchardt C, Behrens J, Hurst HC, Cremisi C. RB and c-Myc activate expression of the E-cadherin gene in epithelial cells through interaction with transcription factor AP-2. Mol Cell Biol 1998;18:3647-58.

[29] Anttila MA, Kellokoski JK, Moisio KI, *et al.* Expression of transcription factor AP-2alpha predicts survival in epithelial ovarian cancer. Br J Cancer 2000;82:1974-83.

[30] Moon UY, Kim CH, Choi JY, *et al.* AP2alpha is essential for MUC8 gene expression in human airway epithelial cells. J Cell Biochem 2010;110:1386-98.

[31] Martinez-Anton A, de Bolos C, Alobid I, *et al.* Corticosteroid therapy increases membrane-tethered while decreases secreted mucin expression in nasal polyps. Allergy 2008;63:1368-76.

CHAPTER 11

Mucin MUC9

Abstract: The *MUC9/OGP* gene encodes the oviduct-specific secreted glycoprotein, MUC9/OGP. This glycoprotein is a member of two protein superfamilies: the mucin protein superfamily and the glycosyl hydrolase 18 one. It contains 11 exons and 10 introns. Although *MUC9/OGP* genes cloned from different mammals display the same structure, species-specific features have been observed for each gene. The mechanisms of transcriptional regulation of the *MUC9/OGP* gene have been investigated. Comparative studies of the cloned 5'-flanking promoter sequences showed that the homologous *MUC9/OGP* genes demonstrate conservation of transcription factor-specific *cis*-elements, but their individual promoters bear species-specific features that determine species-specific expression of a given *MUC9/OGP* gene. The biosynthetic pathway leading to production of the mature MUC9/OGP glycoprotein has been identified. Overt and covert functional potentials of the MUC9/OGP mucin are discussed.

Keywords: MUC9/OGP, promoter, biosynthesis, domain structure, functions.

11.1. GENERAL CHARACTERISTICS OF *MUC9* GENE

The MUC9 glycoprotein contains structural elements characteristic of two protein superfamilies: in addition to being a mucin, it also belongs to the family 18 glycosylhydrolases [1-4]. Comparison of the genes encoding proteins of both families, as well as the amino acid sequences of the polypeptides encoded by these genes, indicates a common ancestral progenitor [2]. However, during evolution *MUC9* gene apparently lost one or more amino acids critical for hydrolase (chitinase) activity [5]. Consistent with this, all homologs of the human MUC9 mucin isolated from the examined species are lacking an essential glutamic acid residue and, as a result, do not display chitinase activity [2, 3].

MUC9 encodes the oviduct-specific secreted glycoprotein [6, 7]. Initially isolated from the New Zealand white doe [8], this glycoprotein was later identified in many mammalian species and is known by different names: oviduct secreted glycoprotein, pOSP, [1], estrus-associated glycoprotein, EAP, [9, 10], oviduct-specific, estrus-associated glycoprotein, EGP, [11], sheep oviduct glycoprotein, sOP92, [12], oviductin [13], MUC9 [4, 7], glycoprotein GP [14] and oviduct-specific glycoprotein, OGP [15, 16]. Buhi [2] suggested defining these glycoproteins collectively as oviduct-specific estrogen-dependent glycoproteins, OGPs. Since this

Joseph Z. Zaretsky and Daniel H. Wreschner

monograph is aimed to analysis of mucins, we suggest assigning the label MUC9/OGP to this oviduct-specific mucinous glycoprotein, underscoring its membership in the mucin superfamily.

The gene encoding the MUC9/OGP glycoprotein has been cloned from different species, with a high degree of conservation in both nucleotide and amino acid sequences across the species [2]. It was identified in various mammals: monotremes [17], marsupials [18] and placentals [19]. The human genome contains a single copy of the *MUC9/OGP* gene located on chromosome 1p13 [7]. Its porcine homolog is located on chromosome 4q22/q23 [20] and the mouse *Muc9/ogp* is found on chromosome 3 [21].

11.2. *MUC9/OGP* GENE: COMPARISON OF MAMMALIAN HOMOLOGS

The *MUC9/OGP* gene has been cloned from at least eight species [2], documenting its general structure [6, 7, 15, 21]. *MUC9/OGP* gene consists of a 5'-flanking promoter region of about 2-3 kb in length, a coding region of 11 exons and 10 introns spanning about 13 kb, and a relatively short 3'-untranslated region (3'-UTR) consisting of 150-400 bp. Although all the *MUC9/OGP* genes display the same structure, species-specific peculiarities have been observed for each gene (Table **1**).

Rabbit *Muc9/ogp*: The rabbit *Muc9/ogp* gene (GenBank accession number DQ640063) spans 13,042 bp and contains 11 exons and 10 introns [22], showing similarity to human [23], mouse [21], and hamster [24] homologs. However, in contrast to these homologous genes, exon 11 of the rabbit *Muc9/ogp* does not contain the Ser/Thr-rich tandem repeats (TR) common to most mucin glycoproteins and is therefore shorter. Intron 3 is also shorter in the rabbit than in the human, mouse and hamster genes: intron 3 of the rabbit gene contains 84 bp compared to 676 - 997 bp in other species [22]. There are two microsatellite-type repeats in the rabbit *Muc9/ogp* gene: the $(TA)_8$ microsatellite sequence located in intron 8, which is specific to the rabbit gene, and the $(TG)_6$-TA-$(TG)_7$ sequence mapped to intron 9, which is present also in the human, mouse, and hamster genes. Several single nucleotide polymorphic sites specific to the rabbit gene were detected in its promoter region. The 1922 bp of the rabbit *Muc9/ogp* cDNA has been cloned (GenBank accession number AF347052); it encodes a 489-aa protein.

Table 1: Characteristics of *MUC9/OGP* Gene and its Product from Different Species.

Species	Gene	No. of Exons	No. of Introns	cDNA	Protein	GenBank
Rabbit	13042 bp	11	10	1922 bp	489 aa	AF347052
Mouse	13400 kb	11	10	2163 bp	721 aa	D32137
Hamster	13600 kb	11	10	2387 bp	671 aa	AH003613
Simian	13868 kb	11	10	2228 bp	623 aa	AAB39765
Human	13257 bp	11	10	2226 bp	742 aa	AH003613

Mouse Muc9/ogp: The mouse *Muc9/ogp* gene, comprised of 13.4 kb spanning over 11 exons and 10 introns, is larger than the rabbit one by several hundreds of nucleotides [21]. Exons and introns correspond in length to those of the human *MUC9/OGP* gene, even though the 3'-UTR of the mouse gene is almost two times larger than the corresponding region of the human gene. The mouse *Muc9/ogp* cDNA (GenBank accession number D32137) contains a 2163 bp sequence that encodes a 721-aa polypeptide including a 21-aa signal peptide [24]. The cDNA contains a 21-bp unit repeated 21 times. These TRs are located between nucleotides +1471 and +1912 in the species-specific region of exon 11 [21, 25-27]. Northern analysis demonstrated two patterns of mRNA – a main pattern of 2.8 kb and a minor one of 1.4 kb – that may indicate allelic polymorphisms or alternative splicing of the primary transcript [25].

Hamster Muc9/ogp: The hamster *Muc9/ogp* gene, like other homologs, consists of 11 exons distributed over a 13.6 kb genomic sequence [24]. The cDNA is comprised of 2387 bp, including 2013 bp of the coding region. The gene product is a 671-aa polypeptide containing a 21-aa signal peptide [28]. A part of the C-terminal region of the gene, located between nucleotides +1502 and +1862, contains a 45-bp sequence repeated 8 times. Allelic polymorphism resulting from variation in the number of repeats, frequently observed in the hamster *Muc9/ogp* gene [29], may account for the production of two types of Muc9/ogp polypeptide molecules of 650 and 635 amino acids [24, 28, 29].

Baboon and rhesus Muc9/ogp: The complete coding sequence of the baboon (*Papio anubis*) *Muc9/ogp* gene is comprised of 2228 bp that includes an open reading frame that starts at position 13 corresponding to the first methionine and extends to position 1881. This sequence allows translation of a 623-aa polypeptide [30]. The *Muc9/ogp* cDNA derived from the rhesus (*Macaca mulata*) *Muc9/ogp* gene is highly similar to the baboon gene described above. It is only 9 nucleotides longer than the coding sequence of the *Papio anubis Muc9/ogp* (2237 *vs.* 2228 bp) [31]. 98.7% of the rhesus *Muc9/ogp* cDNA nucleotide sequence is identical to the baboon sequence [30]. Slightly less identity (95.2%) is observed between rhesus and human cDNAs [6].

Human *MUC9/OGP*: The human *MUC9/OGP* gene (accession numbers AL390195, AH003613, AB208783) is comprised of 13257 bp that cover 11 exons and 10 introns. According to data obtained in 1994, its cDNA consists of 2216 bp, including an open reading frame of 1962 bp that enables translation of a 654-aa protein [6]. The data submitted to GenBank in 1996 (accession number AH003613) lists mRNA of 2187 bp in length containing an open reading frame of 2034 bp (from 13 bp to 2046 bp) sufficient for translation of a 678-aa protein. According to the data posted in 2005 (accession number AB208783), the *MUC9/OGP* gene can produce mRNA of 2618 bp with an open reading frame encompassing 2226 bp that may direct synthesis of a 742-aa protein. Different sources of genomic clones and different methods of analysis may explain the inconsistent results. More studies are needed to clarify this matter.

Despite the small discrepancies, the presented data show that genes encoding oviduct-specific mucin glycoprotein in different species have many features in common. They appear to be highly conserved, and they share 70-78% identity and 76-87% similarity in the overall nucleotide sequences. Importantly, the degree of evolutionary conservation is very high at the N-terminal region where identity reaches 84% and similarity 90%, and is relatively low at the C-terminal end where identity reaches 37-63% and similarity 50-75% [2]. The most pronounced differences are in the part of the gene that corresponds to exon 11, which contains species-specific sequences, and as such should, by definition, be different in different species. The differences at the protein level are even more pronounced and significant, as they include in addition to differences in amino acid sequences also those associated with posttranslational modifications.

11.3. REGULATION OF *MUC9/OGP* GENE EXPRESSION

The mechanisms of transcriptional regulation that control expression of the *MUC9/OGP* gene have been studied in various animal models. Structural aspects of this regulation were partially clarified by analysis of the promoter regions of the homologous genes isolated from different mammalian species. Comparisons of the cloned and sequenced 5'-flanking regions showed that promoters of the homologous genes demonstrate conservation of transcription factor-specific *cis*-elements [21, 24, 32], however, they also possess species-specific features that determine peculiarities in expression of a given *MUC9/OGP* gene in a given species (Fig. **1**).

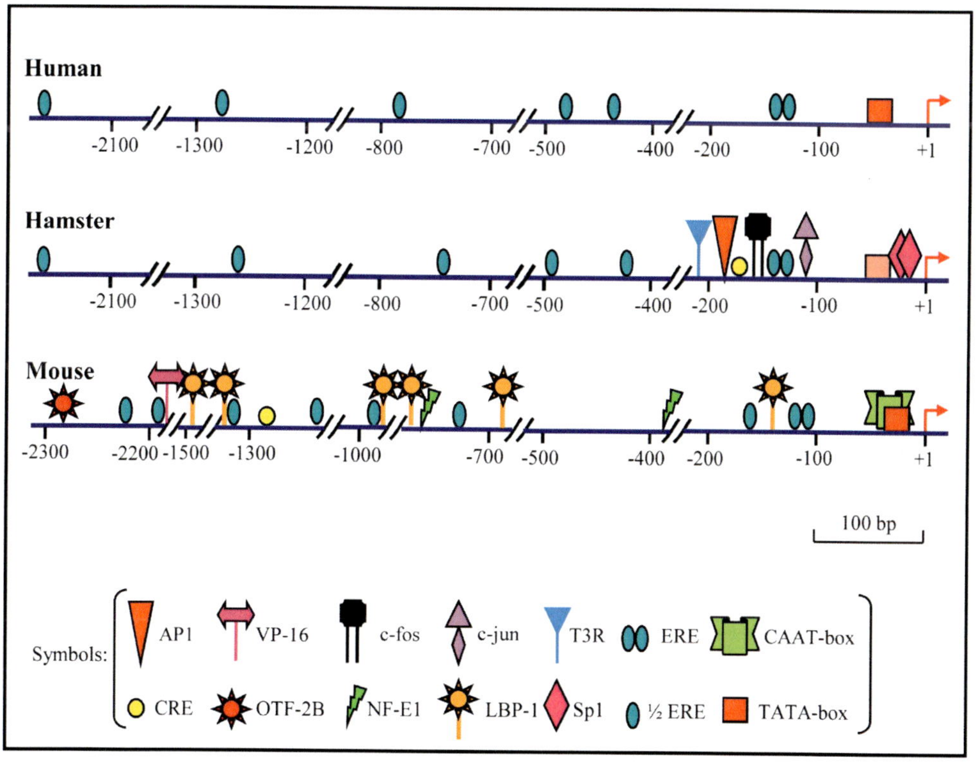

Figure 1: Promoters of human, hamster and mouse *MUC9/OGP* genes (based on the data reported in [21, 24, 32]).

Sequence analysis of the 2633 bp human *MUC9/OGP* promoter revealed a putative TATA-box as a basic regulatory element located 26 bp upstream to the transcription start site [32]. In addition, nine estrogen receptor binding elements (*cis*-ERE), including one imperfect full-length *cis*-ERE and eight 5'-half and 3'-half *cis*-EREs, were found in the promoter region. Two half *cis*-EREs were detected also in the first intron. Functional activity of the imperfect full-length *cis*-ERE was evidenced by electrophoresis mobility shift analysis, while the role of the remaining *cis*-EREs is uncertain. The important functions of two proximal half *cis*-EREs located at positions -491/-487 bp and -464/-460 bp were demonstrated by transient transfection assay. The role of the distal *cis*-EREs is not known. It has been suggested that the activity of these elements may be masked or neutralized by silencing elements found in this region [32]. Besides the regulatory *cis*-elements mentioned above, no other transcription factor binding sites have been described in this promoter at present, although it is hard to believe that the 2.6 kb promoter sequence does not contain other regulatory *cis*-units as well. Then there is the matter of the possible role of epigenetic mechanisms in regulation of the *MUC9/OGP* gene, which is also not clear at the present time. More studies are needed to elucidate the mechanisms regulating human *MUC9/OVGP* gene transcription.

In contrast to the human *MUC9/OGP* promoter, the hamster *hMuc9/ovgp* promoter does not contain a typical TATA-box, having, rather, a nonconsensus TATA-box. It also does not have the typical CAAT-box usually found in the TATA-box-containing promoters [24]. However, like the human *MUC9/OGP* promoter, it contains six half *cis*-EREs and one almost perfect full length *cis*-ERE. In addition, there is a perfect inverted Sp1 binding site located between the transcription start site (TSS) and the atypical TATA box. One more *cis*-Sp1 site was identified downstream to TATA-box at -14/-7 bp relative to TSS. Several other transcription factor binding sites have been identified in the hamster promoter, including AP-1, CREB, c-fos, c-jun and T3R. These elements are compactly distributed between nucleotides -1 and -200. How all these elements cooperate in regulation of *hMuc9/ogp* gene is not clear. Nor is it clear how estradiol regulates expression of the hamster Muc9/ogp mucin glycoprotein. The presence of EREs in hamster promoter supports the view that *hMuc9/ogp* gene is sensitive to estrogen control – a concept supported by the finding that concentration of hamster Muc9/ogp glycoprotein is the highest during estrus and lowest at diestrus [33]. Interestingly, when expression of the hamster Muc9/ogp glycoprotein was examined by antibody specific to nascent polypeptide epitopes, no significant influence of estrogen was observed. Consistent with this finding, several studies showed that hamster Muc9/ogp mRNA expression levels do not vary significantly during the estrous cycle [29, 34]. However, when examined by antibodies specific to hMuc9/ogp glycosidic epitopes, an increasing amount of hMuc9/ogp glycoprotein was detected in the presence of estrogen. Thus, the role of estrogen in regulation of hMuc9/ogp glycoprotein expression appears to be contradictory. It is possible that estrogen controls posttranslational glycosylation of the hMuc9/ogp apomucin rather than transcription of the *hMuc9/ogp* gene.

A study conducted by McBride *et al.* [35] shed some light on this complex issue: indeed, ovarian hormones did regulate glycosylation of hamster oviduct glycoprotein, but not transcription of the *hMuc9/ogp* gene encoding this protein. These results are consistent with the different expression of the *Muc9/ogp* gene in mammals with long duration of the estrous cycle (*i.e.* primate, human) compared to those with a relatively short cycle (*i.e.* rodents) [2, 35, 36]. As pointed out by McBride *et al.* [35], during evolution species developed their own mechanisms for controlling the oviduct mucin glycoprotein expression, whether at the transcriptional, translational, or posttranslational level. Species with long estrous cycle, such as human, adopted transcriptional mechanisms [6, 37], whereas species with short estrous cycle, such as hamster, adopted posttranslational instruments [2, 35, 36]. While McBride *et al.*'s study [35] highlighted important aspects of the hMuc9/ogp glycoprotein expression associated with estrogen-mediated enhancement of the oviduct mucin glycosylation, it did not answer the question of the role of the numerous *cis*-EREs in the promoter region in transcriptional regulation of the *hMuc9/ogp* gene. Future studies will provide the scientific community with an answer to this enigma.

The role of epigenetic mechanisms in regulation of hamster *Muc9/ovgp* gene is also elusive, although some investigators believe these mechanisms may be involved in repression of the gene in tissues outside the oviduct [24]. At least some structural elements important for this type of regulation were found in the *hMuc9/ogp* promoter and in downstream intronic sequences. Sequence analysis revealed 18 CpG and 108 GpC dinucleotides over 2200 bp of the 5'-flanking region, and another 18 CpG sites over 1090 bp located downstream to the transcription start site [24]. If the Sp1 site located between TATA-box and TSS is active in *hMuc9/ogp* gene transcriptional regulation in oviduct, then this region should be protected from CpG-specific methylation, since, being methylated, the Sp1-specific *cis*-element cannot bind Sp1 transcription factor and transcription of the *hMuc9/ogp* gene cannot take place [38-41].

The promoter region of the mouse *Muc9/ogp* (*mMuc9/ogp*) gene cloned by Takahashi *et al.* [21] demonstrates some degree of sequence homology and *cis*-element conservation between mouse and hamster *Muc9/ogp* genes. Like the *hMuc9/ogp* promoter, the promoter of *mMuc9/ogp* gene does not contain typical TATA, CAAT, and GC boxes, but has suboptimal TATA-like (TATTAA) and CAAT-like (CAAC) nucleotide sequences [21, 42] located at -29 bp and -36 bp, respectively. As noted by Takahashi *et al.* [21], the CAAC motif is most probably not functional since its location in the *mMuc9/ogp* promoter differs from that of the functional CAAT boxes in other mammalian genes [43]. The *mMuc9/ogp* gene contains two transcription start sites located 18 and 14 nucleotides upstream to the first ATG initiation codon. What is the physiological import of two transcription initiation sites? Does it influence the level and/or type of mRNA expression? Does the use of one or the other TSS determine the use of the canonical or downstream located ATG codons? Does one transcript initiate translation of the full-length mucin polypeptide while the other one is translated into an N-terminal truncated isoform, and which does which? These questions remain unanswered.

Computer-assisted analysis of the mouse *Muc9/ogp* promoter sequences revealed several consensus motifs specific for the binding of various transcription factors. In particular, one full-length imperfect *cis*-ERE [44] and ten 5'-half and 3'-half palindromic *cis*-ERE sequences [45-47] were found in the mouse *Muc9/ogp* promoter region [21]. Also identified were the binding sites specific for OTF-2B (B cell-specific octamer binding protein) [48, 49], NF-E1 (erythroid-cell-specific nuclear factor [50-52], TGGCA-site (specific for binding of chicken liver protein) [53], VP-16 (protein activating HSV transcription) [54], CACCC-site (erythroid cell-specific transcription factor) [55], several LBP-1- sites (leader-binding protein-1) [56], and CRE site (cAMP response element) [57]. Which of these transcription factors function as transcriptional activators and which as transcriptional repressors has not been established. Nevertheless, transfection assays showed that suppressive effect might be associated with the sequences containing single OTF-2B and VP-16 *cis*-elements, one of five LBP-1 sites and four half-*cis*-ERE units. Importantly, estrogen could stimulate transcription of the test gene containing promoter of the *mMuc9/ogp* gene only when it was transfected into estrogen receptor-containing cells, and was ineffective in cells lacking the receptors [21]. Moreover, transgenic mice carrying a heterologous DNA encoding the SV40 T-antigen under control of 2.2 kb *mMuc9/ogp* promoter sequence expressed T-antigen and developed tumors only in the female reproductive organs, demonstrating tissue specificity of the T-antigen gene expression directed by *mMuc9/ogp* promoter. Estrogen was shown to enhance expression of T-antigen and growth of tumors in both the oviduct and uterus [58]. Taken together, these results provide unambiguous evidence of the positive role estrogen plays in transcriptional regulation of the *mMuc9/ogp* gene.

The above data testify to a principal difference in the regulation of mouse and hamster *Muc9/ogp* genes even though both species are related to rodents and have a short (4-6 day) estrous cycle [59, 60]. It appears that in mouse, estrogen regulates transcription of the *mMuc9/ogp* gene, while in hamster, it influences glycosylation of the *hMuc9/ogp* gene product. Transcription is evidently controlled by direct binding of estrogen receptors to *cis*-EREs in *mMuc9/ovgp* promoter, while enhancement of hamster Muc9/ogp glycoprotein glycosylation may result from estrogen-mediated activation of the genes encoding glycosyltransferases.

The role of epigenetic mechanisms in mouse *Muc9/ovgp* gene regulation has not been investigated.

In summary, although a relatively long time has passed since the mammalian *MUC9/OGP* genes and their promoter regions were cloned and sequenced, limited progress has been made in our knowledge of the mechanisms regulating the activity of these genes. Some basic features of these mechanisms have been identified, and some properties specific to individual genes have been studied. However, more work needs to be done to elucidate many unsolved issues: the role of hormones in transcriptional regulation of these genes, factors that determine stability and degradation of their mRNAs, the ability of these genes to produce various mRNA and protein isoforms, and the involvement of epigenetic mechanisms in transcriptional regulation of these genes.

11.4. BIOSYNTHESIS OF MUC9/OGP GLYCOPROTEIN

Soluble mucins, in general, and MUC9/OGP, in particular, have been less studied compared to the gel-forming mucins with regard to biosynthesis and posttranslational processing. Only a few studies were carried out on the biosynthesis, posttranslational modification and maturation of the MUC9/OGP mucin in different species. Among all mammalian MUC9/OGPs studied, the hamster Muc9/ogp mucin has attracted the most attention. The onset of hamster Muc9/ogp biosynthesis could not be detected in females younger than 7 days old [61], a result in line with data [62] that the rough endoplasmic reticulum (rER) and the Golgi apparatus are undeveloped in the hamster oviductal cells up to postnatal day 5.5. On the other hand, intensive production of Muc9/ogp was observed in hamster oviduct epithelial cells of 10-day-old females, the age at which a well-developed Golgi apparatus, extensive rER, and immature secreted granules appear in hamster [61, 62]. In other words, biosynthesis of hamster Muc9/ogp mucin occurs in parallel with the development and maturation of the biosynthetic machinery of the oviduct epithelial cells, pointing the role of rER and the Golgi in biosynthesis and post-translational modification of the hamster Muc9/ogp glycoprotein. Interestingly, treatment of newborn hamster females with estradiol leads to enhanced oviduct cytodifferentiation [62]. This phenomenon correlates well with the observation that pre-pubertal hamster females treated with estradiol synthesize and secrete Muc9/ogp mucin before the physiological age of sexual maturation [61]. Altogether, these data demonstrate the dependence of the biosynthesis of hamster Muc9/ogp on estrogen. Taking into account that the hamster Muc9/ogp mucin shares many properties with the homologous glycoproteins of other mammalian species, one may assume that dependence of mammalian Muc9/ogp mucin biosynthesis on estrogen is a common feature of the glycoproteins of this subgroup of oviduct mucins. However, it should be pointed out one again that although the final production of mature mucin glycoprotein in most species studied is estrogen dependent, different mechanisms are controlled by steroid hormone in different species: estrogen in human controls transcription of the MUC9/OGP mRNA [6]; estrogen in hamster regulates posttranslational glycosylation [35].

When the dynamics of hamster Muc9/ogp mucin biosynthesis were studied by the pulse-chase method [63], no mucin molecules were detected at pulse times less than 20 min, at which point the first precursor polypeptide with apparent molecular mass of 69 kDa appeared and underwent progressive maturation leading to the consequent appearance of glycoprotein molecules of 71, 73 and 80-90 kDa. These intermediate forms have different degrees of glycosylation and maturation: the completely matured O-glycosylated and sulfated Muc9/ogp mucin molecules of 160-350 kDa are seen after 45 min of pulse-labeling; and at 60 min the mature mucin molecules appear in secretory granules, from which they are secreted into extracellular lumen during several hours [63]. The spacio-temporal parameters of the hamster Muc9/ogp mucin biosynthesis are presented in Fig. **2**.

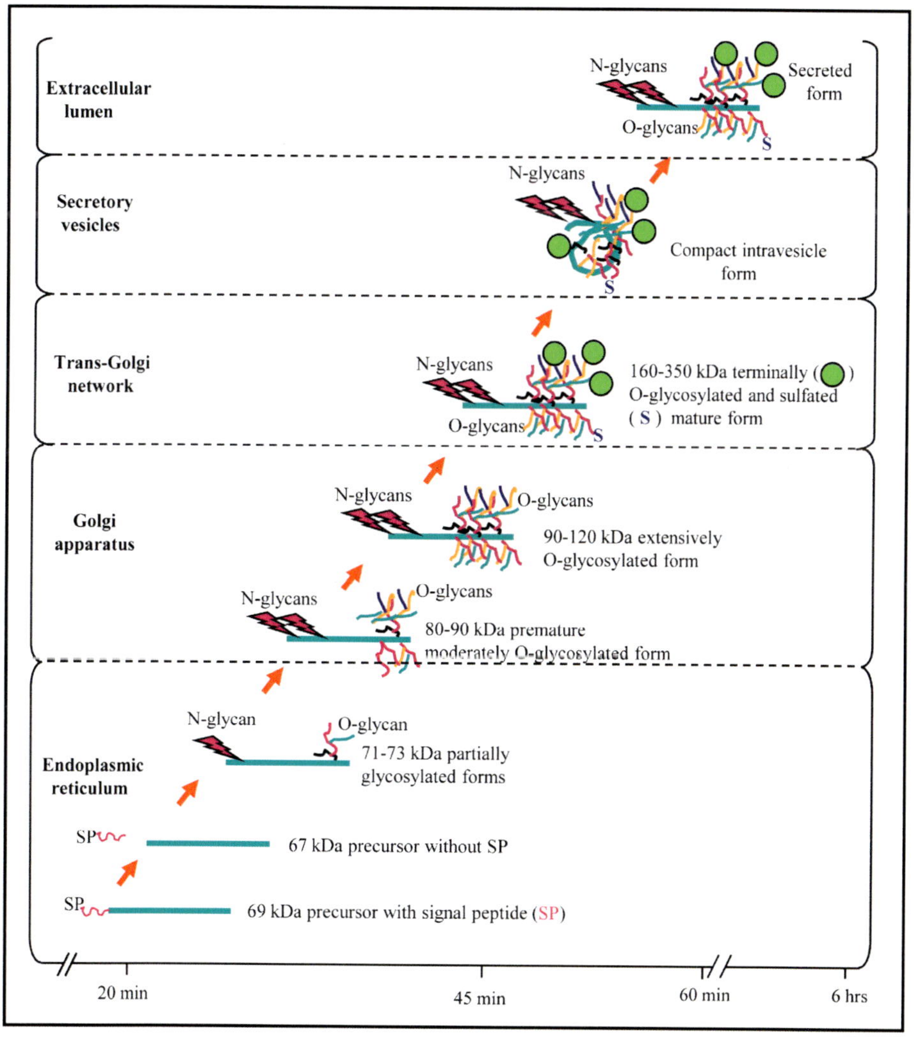

Figure 2: Biosynthesis of the hamster Muc9/ogp glycoprotein (based on the data reported in [63, 64, 66]).

As shown in Fig. **2**, N-glycosylation occurs in the rER. Interestingly, only a limited number (1-2) of 7 potential N-glycosylation sites found in the deduced amino acid sequence of hamster Muc9/ogp [64, 65] could be detected experimentally [64]. Malette *et al.* [63] showed that inhibition of N-glycosylation causes a considerable intracellular accumulation of the processed polypeptides, probably by altering their normal transport to the Golgi. Under physiological conditions, the polypeptide precursors are not retained in the rER, and after being N-glycosylated are rapidly transferred to the Golgi for further O-glycosylation. When the physiological pathway from rER to Golgi and further to the *trans*-Golgi network is blocked by Brefeldin A (BFA), conversion of the premature 80-90 kDA-form into completely glycosylated sulfated mature 160-350kDa-form does not occur. Instead, incompletely glycosylated unsulfated intermediate molecules of 90-120 kDa are observed. These findings show that sulfation as well as terminal O-glycosylation are late maturational processes taking place in the post-Golgi compartments, such as the *trans*-Golgi network and, probably, in secretory granules [63]. The ability of Muc9/ogp-producing cells to synthesize normally sulfated glycoprotein when N-glycosylation is inhibited, and their inability to produce sulfated mucin when O-glycosylation is prevented, indicates that sulfation occurs only on O-linked carbohydrates [64, 66].

Two major immunologically related but differently charged isoforms, α and β, of hamster Muc9/ogp has been discovered [64]. The α isoform (160-210 kDa) is negatively charged and has pI 8.0 [64, 65, 67]. These isoforms result from different glycosylation and sulfation of the same polypeptide precursor. This conclusion is based on the data obtained by N-terminal sequencing, peptide mapping and partial deglycosylation approaches. It was shown that the β isoform can be transformed into the α isoform by partial deglycosylation [64, 66]. What factors determine the occurrence of two Muc9/ovgp isoforms? Are they produced by the same cell or are different isoforms synthesized by different cells belonging to the same epithelial type? These questions are still unanswered. The mechanisms that determine glycosylation of mucin/oviductin, particularly the number and positions of O-glycans, their composition and size have not been delineated. These mechanisms are of great importance since the resulting polymorphism may be associated with different biological functions of each isoform.

Hamster Muc9/ogp mucin is synthesized only in nonciliated secretory cells, as the ciliated cells of the columnar epithelium lining the oviduct lumen do not synthesize this mucin [36]. Interestingly, immunohistochemistry and electron microscopy disclosed a faint amount of the hamster Muc9/ogp mucin in the epithelial lining of the endometrium [68, 69]. However, this glycoprotein was not detected in the Golgi complex of these cells, indicating that the uterine epithelium does not synthesize Muc9/ogp. The current results suggest that Muc9/ogp mucin produced in hamster oviduct is simply taken up by the uterine epithelium, where it may function as a modulator of uterine receptivity [36, 70].

The presented data give an idea of the biosynthetic processes resulting in production of the mature hamster Muc9/ogp glycoprotein. Although very limited, the results obtained in other animal models show that the same mechanisms are mobilized for synthesis of MUC9 mucins in different species. Generally, the polypeptide precursors of all studied members of the MUC9 group are translated from mRNAs of a similar size (2.2-2.9 kb) and having the same exon structure [6, 9, 71-77]. This suggests that the kinetics of their precursor polypeptide translation must be similar. However, closer examination allows separation of all Muc9/ogp mRNAs into two groups according to size: one group is comprised of relatively large (2.4-2.9 kb) mRNAs of human, baboon, hamster and mice origin [6, 76, 77], whereas the other group consists of smaller mRNAs (2.2-2.3 kb) encoding for cow, pig, rabbit, sheep and cat mucinous oviductins [9, 71, 72, 74, 75].

11.5. DOMAIN STRUCTURE OF MUC9/OGP GLYCOPROTEIN

Analysis of the amino acid sequences of the MUC9/OGP apomucins synthesized by different mammalian species enabled construction of a domain map of these molecules (Fig. **3**). As mentioned above, one of the MUC9/OGP specific features is the chitinase-like domain comprising 338 amino acid residues at the N-terminus of the mucin molecule [78, 79]. The similarity between this part of the MUC9/OGP molecule and the native chitinase domain is especially high in the region of the purported active site of the chitinase, with the notable exception of one or two amino acids believed to be essential for enzymatic activity [6]. No chitinase-like activity has been reported for the oviductal MUC9/OGPs [1, 30, 80], but if the chitinase-like domain retains some of its sugar-binding properties it would determine the interaction of MUC9/OGP with other glycoproteins, and therefore affect their functions [66].

As a mucin, the MUC9/OGP glycoprotein, independent of origin, contains a mucin-like domain, a hallmark of these proteins. This domain, comprised of highly glycosylated TRs, is encoded mainly by exon 11 [2]. This is the most variable domain of the MUC9/OGP glycoprotein containing the main species-specific antigenic determinants [22]. It was suggested that different forms of MUC9/OGP, responsible for diverse biological functions, vary in the number and distribution of oligosaccharide side- chains attached to the serine and threonine residues of the TRs [2, 36]. This suggestion was confirmed by analysis of the oligosaccharides in MUC9/OGP glycoproteins of different species.

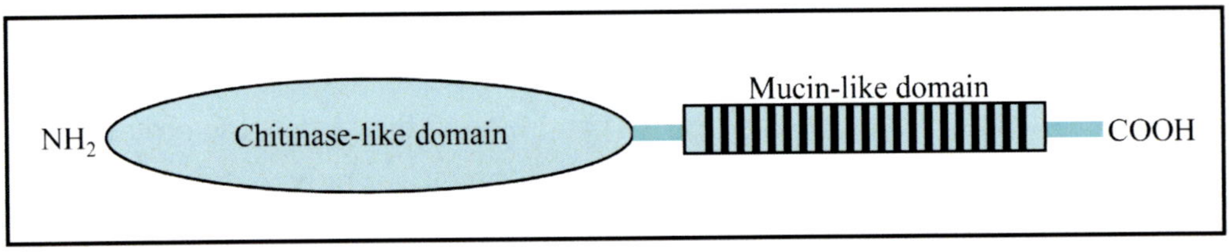

Figure 3: Domain structure of the MUC9/OGP mucin (based on the data reported in [2, 36]).

Hamster Muc9/ogp contains terminal α-D-GalNAc, terminal α-D-NeuAc and/or non-terminal β-D-$(GlcNAc)_2$ residues, but does not contain α-L-fucose and high-mannose/hybrid N-linked oligosaccharide chains [64]. Different types of glycosylation are evidenced also by the existence of two differently charged isoforms, α and β, of hamster Muc9/ogp. The different types and degrees of glycosylation are reflected also in different mobility of the two isoforms in two-dimensional PAG-electrophoresis: α-isoform behaves as monomer, whereas β-isoform demonstrates poor mobility due to extensive glycosylation and/or noncovalent aggregation. Importantly, Muc9/ogp from cattle (97 kDa), sheep (90-92 kDa) and baboon (120 kDa) closely resemble the hamster homolog in that they all resolve into two major isoelectric variants, an acidic isoform and a basic isoform [71, 74, 81]. The mucin-like domain with its variable number of TRs and multiple glycosylation sites determines the mucin-specific properties and associated functions of the MUC9/OGP glycoprotein.

In addition to the mucin-specific and chitinase-like domains, *in silico* analysis of the amino acid sequence of the bonnet monkey Muc9/ogp polypeptide (GenBank Accession No AY341432) revealed a long list of numerous protein domain and amino acid motifs with the potential for various functions [78]. This list includes: 1) signal peptide that determines endoplasmic reticulum localization; 2) clatrin box participating in endocytic vesicle formation; 3) DPF/W motif, important for recruitment of a protein to site of clatrin-coated vesicle formation; 4) KAWTTD motif found in the proteins involved in signal transduction, cell-cycle control, apoptosis, stress response and malignant transformation; 5) several class III PDZ-domain-binding motifs that control plasma membrane and cytosol trafficking; 6) a number of SH2 and SH3-domain-binding motifs usually found in the proteins participating in signal transduction, cytoskeleton organization, organel formation and nucleus and membrane targeting; 7) several class IV WW-ligand motifs important for ubiquitin-mediated degradation and for mitosis; 8) two nardilysine cleavage sites important for processing of secreted proteins; 9) multiple sites specific for posttranslational modifications such as phosphorylation, N- and O-glycosylation, sulfation, and peptide C-terminal amidation, which all together influence processes associated with cell communication, morphogenesis, development, secretory granule formation and secretion.

The *in silico* data regarding involvement of the indicated structural elements in the processes mediated by MUC9/OGP is supported by experimental findings. For instance, the finding that Muc9/ogp mucin molecules are indeed endocytosed by blastocysts of developing hamster embryos and most probably undergo degradation through the ubiquitin pathway, as seen in experiments of Kan *et al.* [82], argues in favor of the active participation of the clatrin box, found by bioinformatic methods in the bonnet monkey Muc9/ogp [78], in edocytosis and ubiquitin-mediated degradation. Another example is demonstration of the requirement of signal peptide, nardilysine cleavage sites, and glycosylation sites for extracellular secretion of MUC9/OGP-containing vesicles [2, 78].

Thus, the multiple motifs disclosed by *in silico* analysis of the bonnet monkey Muc9/ogp are not just traces of evolutionary events, but the structural elements actively involved in the physiology and biochemistry of this mucin. Moreover, these elements, or at least some of them, are present in MUC9/OGP molecules of other species, as evidenced by finding of the highly conserved PDZ-domain binding sequences in human, baboon and porcine MUC9/OGP s [2, 78]. Thus, biochemical, biophysical and *in silico* assays suggest structural complexity of MUC9/OGP glycoproteins with potentially multiple functions.

11.6. FUNCTIONAL POTENTIALS OF MUC9/OGP MUCIN

The main purpose of protein studies is to translate the information enciphered in the protein structures into functional potentials. The structural complexity of the MUC9/OGP molecule suggests its multifunctionality, but at present we are far from complete understanding of its functional potentials. Nevertheless, some important functions of this glycoprotein have been discovered. Many reports show that MUC9/OGP is one of the important participants in biological reproduction [2, 10, 27]. It can directly interact with gametes and/or the early embryo in the oviduct, thereby influencing their properties and functions [2, 27]. Several *in vitro* studies in mammals have shown that MUC9/OGP binds to oocytes, spermatozoa and embryos, and that these interactions positively affect sperm capacitation, motility and viability, sperm-ovum binding, ovum penetration, fertilization and early embryo development. These interactions decrease polyspermy, and increase cleavage rate of embryos and the number of embryos reaching the blastocyst stage [2, 16, 26, 83-86]. Importantly, oviductin (MUC9) behaves differently in different species: it interacts with bovine and hamster sperm [10, 87], but does not bind to ovine or human sperm [16, 88]; the MUC9/OGP glycoprotein is present in the perivitelline space of baboon and human oocytes [71], but oocytes collected directly from the baboon or human ovary during *in vitro* fertilization do not contain this glycoprotein [16]. Association of the purified bovine Muc9/ogp with the bovine sperm surface, which occurs through the midpiece and tail regions, significantly increases the viability of the frozen-thawed sperm, pointing to the importance of Muc9/ogp for sperm function [89, 90]. Indeed, several studies highlighted the importance of MUC9/OGP interaction with spermatozoa and oocytes for *in vivo* fertilization [71, 91-97]. However, its physiological significance remains controversial, as normal fertility was observed in *Muc9/ovp* gene-null mice [98]. Moreover, there is no justification for direct extrapolation of results obtained un mice to other species. Instead, as noted by Merchan *at el.* [22], the role of MUC9/OGPs in fertilization may be different in different species of mammals.

The oviduct consists of two main regions, istmus and ampula, which provide a microenvironment for oocyte and spermatozoa transport, interaction and fertilization, and development of early embryos. The oviductal epithelium, especially cells located at istmus (the sperm reservoir), produce several secreted glycoproteins, including *MUC9/OGP* [92, 93]. Mammalian spermatozoa need to reside a certain amount of time in istmus in order to undergo the modifications required for fertilizing ova [99, 100]. Studies in hamster and other mammals revealed that during the pre-ovulatory phase, the spermatozoa are tightly bound by their heads to the oviductal mucosa of the istmus, where they interact with *MUC9/OGP* and other secreted proteins. Only after interaction with Muc9/ogp are spermatozoa capable of fertilizing ova in hamster [101, 102]. This interaction is highly species-specific and is mediated by the C-terminal region of the oviductin molecule. Interestingly, the C-terminal region of the MUC9/OGP glycoprotein participates not only in fertilization, but also in the release of early embryonic development block observed when the fertilized eggs of mammals are cultured *in vitro* [86]. This region is less conserved between different species both in amino acid sequence and in O-glycosylation patterns, and possesses species-specific antigenic determinants [22]. Because carbohydrate moieties participate in recognition and interaction between proteins, they were suggested to determine the species-specific interaction between the oocyte zona pelucida (ZP) and sperm [16, 25, 26, 103]. MUC9/OGP (oviductin) interacts with ZP and/or with perivitelline (PV) space of the oocyte during its passage through the oviduct, and remains associated with the early embryo until implantation [66]. ZP is composed of three glycoproteins, including primary and secondary sperm receptors [104-106], but exactly how these components of ZP interact with MUC9/OGP is not known. Although most studies point to direct interaction of MUC9/OGP with ZP proteins, any viable hypothesis suggesting such a mechanism must take into account the results of *in vitro* fertilization, a procedure in which these glycoproteins are clearly dispensable [66]. On the other hand, both *in vitro* and *in vivo* studies showed that incubation of ovarian oocytes with purified MUC9/OGP increases sperm penetration [107, 108]. In the hamster model, Muc9/ogp was shown to enhance both sperm binding to ZP and ZP-induced acrosome reaction [109]. Boatman and Magnoni [108] demonstrated that purified hamster Muc9/ogp binds *in vitro* to the homologous sperm on the acrosomal crescent, but in Reuter *et al.*'s experiments

[110], neither hamster nor human homologous MUC9/OGP mucins could associate with human sperm. Taken together, these results suggest the species-specific character of the interaction between the MUC9/OGP molecules, on the one hand, and ZP and spermatozoa, on the other hand. They do not, however, tell how this mechanism operates. Does it correspond to a ligand/receptor mechanism? Does the MUC9/OGP undergo structural modification preceding interaction with ZP, or does it form a complex with other protein(s) prior to interaction with the ZP glycoproteins? Does MUC9/OGP covalently interact with the ZP proteins, or do they associate through the sulfate and/or sialic acid groups of the olygosaccharide side-chains by ionic interactions? More studies of these basic biological processes are needed.

What is the biological meaning of MUC9/OGP's interaction with the gametes in the oviduct? The maternal immune system can react with sperm and with preimplantation embryos, and can destroy them by antibodies and by the complement system. Hence, control over the humoral immune system may play a role in gamete and embryo survival. Oliphant *et al.* [111] claimed that MUC9/OGP may constitute a local barrier at the oocyte/embryo surface to attack by the immune system.

Many studies show that defense is not the only function of the MUC9/OGP glycoprotein in the reproduction process. The fact that MUC9 mucin molecules were found in specific endocytic structures of the fertilized eggs and early embryos in hamster [82], in the cytoplasm of the porcine blastocyst [112], and in the baboon cleavage state embryos [81] shows that these molecules interact not only with the surface of gametes, but enter also into the cells, where they may carry out various tasks. One such task is participation in oviduct tissue remodeling [78].

As mentioned above, the *in silico* analysis of the deduced amino acid sequence of bonnet monkey Muc9/ogp revealed the presence in its apomucin of several conserved motifs that imply multi-protein complex formation between Muc9/ogp and other functionally active proteins [78]. Indeed, Kadam *et al.* [78, 113] showed that nonmuscle myosin IIA, an active element of the cell remodeling apparatus that determines cell shape, polarity and morphogenesis, is a protein partner of Muc9/ogp in gametes. Through interaction with this protein, Muc9/ogp is involved in oviduct epithelial remodeling [78].

The oviduct epithelium is characterized by segmental differentiation [114]. Interestingly, the MUC9/OGP also demonstrates differentiation-dependent patterns of expression, with higher levels in areas of high cell turnover [115]. Indirect evidence has been obtained of MUC9/OGP involvement in tissue remodeling, cell shaping and polarity. *In vitro* studies revealed a close link between MUC9/OGP mRNA levels and the maintenance of polarity and differentiation [116]. Analogous to tubulin up-regulation in ciliogenesis [117], MUC9/OGP is regulated by a number of genes acting in concert in differentiation [118]. The finding of MUC9/OGP molecules inside the cells (in endosomes, multivesicular bodies, blastomere membranes) [82, 112], in the intercellular space at the tight junctions of the developing embryos [112], and in oviduct epithelia [36] indicates possible intra- and intercellular functions other than mechanical lubrication or coating of gametes for the purpose of defending them from immunological or proteolytic attacks. Indeed, during contact with oocyte or spermatozoa, the N-terminus of the MUC9/OGP glycoprotein (amino acid residues 11-137) interacts with the gamete cytoskeleton protein MYH9, a nonmuscle myosin IIA. This is consistent with the data from several laboratories showing that *in vitro* the MUC9/OGP-binding sites present on the sperm can be visualized only following partial permeability of cell membrane, or on the capacitated sperm [10, 119]. In other words, the sperm membrane must be modified to permit interaction of MUC9/OGP with intracellular proteins. This observation is in line with the well established fact that membrane reorganization occurs during sperm capacitation [120, 121]. Moreover, *in vitro* capacitated sperm were reported to have cytoskeleton protein on the surface of the sperm head [120]. MUC9/OGP was shown to co-localize and associate with the filamentous actin at the cleavage furrow of the developing blastocysts [122], where cortical myosin IIA was also found [123]. Kadam *et al.* [78] demonstrated that MUC9/OGP interacts with cortical myosin IIA, a regulator of cell polarity during normal morphogenesis of epithelia [124-126].

Immunoelectron microscopy disclosed the presence of the MUC9/OGP mucin at the tight junctions (TJ) of developing ciliated cells undergoing alteration in shape, and on the adluminal surface of plasma membranes of mature ciliated cells developed from secretory cells [127-129]. Co localization studies of MUC9/OGP and cadherin, a master regulator of the epithelial phenotype, showed them to co-localize at the adluminal face of the plasma membrane [78], suggesting possible interaction of both proteins in oviduct tissue remodeling.

That MUC9/OGP, in addition to a number of different domains and motifs, contains PDZ-binding motifs, implicates it as a PDZ-ligand protein. This property of the MUC9/OGP glycoprotein allows the mucin molecule to be bound through the PDZ-domain-containing proteins to multiprotein complexes, which, as suggested by Giallourakis *et al.* [130], can be further targeted to specific sub-cellular compartments including TJs. This is consistent with the findings of Bauersachs *et al.* [115, 118] who observed coordinated expression of MUC9/OGP mucin with two TJ-proteins, claudin 1 and ZO-1, a PDZ-domain-containing protein. Localization of MUC9/OGP at the tonofilaments of TJs in secretory and differentiating ciliated cells, together with spacio-temporal regulation of the coordinated expression of potential partner-proteins, point to the presence of the intracellular MUC9/OGP isoform. This isoform may interact with the epithelial actin-myosis cytoskeleton, inducing various physiological events such as altering cell shape, sperm motility, acrosome reaction, egg activation and cytokinesis. Hence, MUC9/OGP might be one of the factors governing tissue remodeling and the maintenance of cell shape and polarity [78].

In summary, the MUC9/OGP mucin is a unique polyfunctional glycoprotein molecule containing numerous amino acid motifs and domains, including mucin-specific and chitinase-specific ones. This mucin exists in two isoforms, extra- and intracellular, whose expression and glycosylation are under hormonal control. The presence of multiple functional knots in the MUC9 molecule allows it to interact with multiple protein partners and form different protein complexes. Such protein-protein interactions might trigger numerous extra- and intracellular reactions, placing MUC9/OGP at the center of the multi-chain network.

REFERENCES

[1] Buhi WC, Alvarez IM, Choi I, Cleaver BD, Simmen FA. Molecular cloning and characterization of an estrogen-dependent porcine oviductal secretory glycoprotein. Biol Reprod 1996;55:1305-14.
[2] Buhi WC. Characterization and biological roles of oviduct-specific, oestrogen-dependent glycoprotein. Reproduction 2002;123:355-62.

[3] DeSouza MM, Murray MK. An estrogen-dependent secretory protein, which shares identity with chitinases, is expressed in a temporally and regionally specific manner in the sheep oviduct at the time of fertilization and embryo development. Endocrinology 1995;136:2485-96.

[4] Lagow E, DeSouza MM, Carson DD. Mammalian reproductive tract mucins. Hum Reprod Update 1999;5:280-92.

[5] Watanabe T, Kobori K, Miyashita K, *et al.* Identification of glutamic acid 204 and aspartic acid 200 in chitinase A1 of Bacillus circulans WL-12 as essential residues for chitinase activity. J Biol Chem 1993;268:18567-72.

[6] Arias EB, Verhage HG, Jaffe RC. Complementary deoxyribonucleic acid cloning and molecular characterization of an estrogen-dependent human oviductal glycoprotein. Biol Reprod 1994;51:685-94.

[7] Lapensee L, Paquette Y, Bleau G. Allelic polymorphism and chromosomal localization of the human oviductin gene (MUC9). Fertil Steril 1997;68:702-8.

[8] Oliphant G, Reynolds AB, Smith PF, Ross PR, Marta JS. Immunocytochemical localization and determination of hormone-induced synthesis of the sulfated oviductal glycoproteins. Biol Reprod 1984;31:165-74.

[9] Buhi WC, Alvarez IM, Sudhipong V, Dones-Smith MM. Identification and characterization of *de novo*-synthesized porcine oviductal secretory proteins. Biol Reprod 1990;43:929-38.

[10] King RS, Killian GJ. Purification of bovine estrus-associated protein and localization of binding on sperm. Biol Reprod 1994;51:34-42.

[11] Hill JL, Walker SK, Brown GH, Nancarrow CD. The effects of an ovine oviductal estrus-associated glycoprotein on early embryo development. Theriogenology 1996;46:1367-77.

[12] Gandolfi F, Modina S, Brevini TA, *et al.* Oviduct ampullary epithelium contributes a glycoprotein to the zona pellucida, perivitelline space and blastomeres membrane of sheep embryos. Eur J Basic Appl Histochem 1991;35:383-92.

[13] Robitaille G, St-Jacques S, Potier M, Bleau G. Characterization of an oviductal glycoprotein associated with the ovulated hamster oocyte. Biol Reprod 1988;38:687-94.

[14] Kapur RP, Johnson LV. Ultrastructural evidence that specialized regions of the murine oviduct contribute a glycoprotein to the extracellular matrix of mouse oocytes. Anat Rec 1988;221:720-9.

[15] Sendai Y, Abe H, Kikuchi M, Satoh T, Hoshi H. Purification and molecular cloning of bovine oviduct-specific glycoprotein. Biol Reprod 1994;50:927-34.

[16] O'Day-Bowman MB, Mavrogianis PA, Reuter LM, *et al.* Association of oviduct-specific glycoproteins with human and baboon (Papio anubis) ovarian oocytes and enhancement of human sperm binding to human hemizonae following *in vitro* incubation. Biol Reprod 1996;54:60-9.

[17] Warren WC, Hillier LW, Marshall Graves JA, *et al.* Genome analysis of the platypus reveals unique signatures of evolution. Nature 2008;453:175-83.

[18] Mikkelsen TS, Wakefield MJ, Aken B, *et al.* Genome of the marsupial Monodelphis domestica reveals innovation in non-coding sequences. Nature 2007;447:167-77.

[19] Aviles M, Gutierrez-Adan A, Coy P. Oviductal secretions: will they be key factors for the future ARTs? Mol Hum Reprod 2010;16:896-906.

[20] Merchan M, Rendon M, Folch JM. Assignment of the oviductal glycoprotein 1 gene (OVGP1) to porcine chromosome 4q22-->q23 by radiation hybrid panel mapping. Cytogenet Genome Res 2006;114:93C.

[21] Takahashi K, Sendai Y, Matsuda Y, *et al.* Mouse oviduct-specific glycoprotein gene: genomic organization and structure of the 5'-flanking regulatory region. Biol Reprod 2000;62:217-26.

[22] Merchan M, Peiro R, Santacreu MA, Francino O, Folch JM. Rabbit oviductal glycoprotein 1 gene: genomic organization polymorphism analysis and mRNA expression. Mol Reprod Dev 2007;74:687-93.

[23] Jaffe RC. Characterization of the human oviductal glycoprotein gene. Biol Reprod 1996;54:74.

[24] Merlen Y, Bleau G. Organization of a gene coding for an oviduct-specific glycoprotein (oviductin) in the hamster. Mol Reprod Dev 2000;57:238-46.

[25] Sendai Y, Komiya H, Suzuki K, *et al.* Molecular cloning and characterization of a mouse oviduct-specific glycoprotein. Biol Reprod 1995;53:285-94.

[26] Verhage HG, Mavrogianis PA, O'Day-Bowman MB, *et al.* Characteristics of an oviductal glycoprotein and its potential role in the fertilization process. Biol Reprod 1998;58:1098-101.

[27] Araki Y, Yoshida-Komiya H. Mammalian oviduct-specific glycoprotein: characterization and potential role in fertilization process. J Reprod Dev 1998;44:215-28.

[28] Suzuki K, Sendai Y, Onuma T, *et al.* Molecular characterization of a hamster oviduct-specific glycoprotein. Biol Reprod 1995;53:345-54.

[29] Paquette Y, Merlen Y, Malette B, Bleau G. Allelic polymorphism in the hamster oviductin gene is due to a variable number of mucin-like tandem repeats. Mol Reprod Dev 1995;42:388-96.

[30] Jaffe RC, Arias EB, O'Day-Bowman MB, *et al.* Regional distribution and hormonal control of estrogen-dependent oviduct-specific glycoprotein messenger ribonucleic acid in the baboon (Papio anubis). Biol Reprod 1996;55:421-6.

[31] Verhage HG, Mavrogianis PA, Boomsma RA, *et al.* Immunologic and molecular characterization of an estrogen-dependent glycoprotein in the rhesus (Macaca mulatta) oviduct. Biol Reprod 1997;57:525-31.

[32] Agarwal A, Yeung WS, Lee KF. Cloning and characterization of the human oviduct-specific glycoprotein (HuOGP) gene promoter. Mol Hum Reprod 2002;8:167-75.

[33] Roux E, Kan FW. Stage-specific immunolabeling for oviductin in the secretory granules of the oviductal epithelium of the golden hamster during the estrous cycle. Anat Rec 1995;241:369-76.

[34] Komiya H, Onuma T, Hiroi M, Araki Y. *In situ* localization of messenger ribonucleic acid for an oviduct-specific glycoprotein during various hormonal conditions in the golden hamster. Biol Reprod 1996;55:1107-18.

[35] McBride DS, Boisvert C, Bleau G, Kan FW. Evidence for the regulation of glycosylation of golden hamster (Mesocricetus auratus) oviductin during the estrous cycle. Biol Reprod 2004;70:198-203.

[36] McBride DS, Boisvert C, Bleau G, Kan FW. Detection of nascent and/or mature forms of oviductin in the female reproductive tract and post-ovulatory oocytes by use of a polyclonal antibody against recombinant hamster oviductin. J Histochem Cytochem 2004;52:1001-9.

[37] Lok IH, Briton-Jones CM, Yuen PM, Haines CJ. Variable expression of oviductin mRNA at different stages of human reproductive cycle. J Assist Reprod Genet 2002;19:569-76.

[38] Turker MS, Bestor TH. Formation of methylation patterns in the mammalian genome. Mutat Res 1997;386:119-30.

[39] Jost JP, Bruhat A. The formation of DNA methylation patterns and the silencing of genes. Prog Nucleic Acid Res Mol Biol 1997;57:217-48.

[40] Kass SU, Pruss D, Wolffe AP. How does DNA methylation repress transcription? Trends Genet 1997;13:444-9.

[41] Nan X, Campoy FJ, Bird A. MeCP2 is a transcriptional repressor with abundant binding sites in genomic chromatin. Cell 1997;88:471-81.

[42] Martens JA, Brandl CJ. GCN4p activation of the yeast TRP3 gene is enhanced by ABF1p and uses a suboptimal TATA element. J Biol Chem 1994;269:15661-7.
[43] Efstratiadis A, Posakony JW, Maniatis T, *et al.* The structure and evolution of the human beta-globin gene family. Cell 1980;21:653-68.
[44] Teng CT, Liu Y, Yang N, Walmer D, Panella T. Differential molecular mechanism of the estrogen action that regulates lactoferrin gene in human and mouse. Mol Endocrinol 1992;6:1969-81.
[45] Tora L, Gaub MP, Mader S, *et al.* Cell-specific activity of a GGTCA half-palindromic oestrogen-responsive element in the chicken ovalbumin gene promoter. Embo J 1988;7:3771-8.
[46] Kato S, Sasaki H, Suzawa M, *et al.* Widely spaced, directly repeated PuGGTCA elements act as promiscuous enhancers for different classes of nuclear receptors. Mol Cell Biol 1995;15:5858-67.
[47] Kato S, Tora L, Yamauchi J, *et al.* A far upstream estrogen response element of the ovalbumin gene contains several half-palindromic 5'-TGACC-3' motifs acting synergistically. Cell 1992;68:731-42.
[48] Schreiber E, Matthias P, Muller MM, Schaffner W. Identification of a novel lymphoid specific octamer binding protein (OTF-2B) by proteolytic clipping bandshift assay (PCBA). Embo J 1988;7:4221-9.
[49] Kemler I, Schreiber E, Muller MM, Matthias P, Schaffner W. Octamer transcription factors bind to two different sequence motifs of the immunoglobulin heavy chain promoter. Embo J 1989;8:2001-8.
[50] deBoer E, Antoniou M, Mignotte V, Wall L, Grosveld F. The human beta-globin promoter; nuclear protein factors and erythroid specific induction of transcription. Embo J 1988;7:4203-12.
[51] Martin DI, Orkin SH. Transcriptional activation and DNA binding by the erythroid factor GF-1/NF-E1/Eryf 1. Genes Dev 1990;4:1886-98.
[52] Yamamoto M, Ko LJ, Leonard MW, *et al.* Activity and tissue-specific expression of the transcription factor NF-E1 multigene family. Genes Dev 1990;4:1650-62.
[53] Rupp RA, Sippel AE. Chicken liver TGGCA protein purified by preparative mobility shift electrophoresis (PMSE) shows a 36.8 to 29.8 kd microheterogeneity. Nucleic Acids Res 1987;15:9707-26.
[54] Triezenberg SJ, LaMarco KL, McKnight SL. Evidence of DNA: protein interactions that mediate HSV-1 immediate early gene activation by VP16. Genes Dev 1988;2:730-42.
[55] Miller IJ, Bieker JJ. A novel, erythroid cell-specific murine transcription factor that binds to the CACCC element and is related to the Kruppel family of nuclear proteins. Mol Cell Biol 1993;13:2776-86.
[56] Jones KA, Luciw PA, Duchange N. Structural arrangements of transcription control domains within the 5'-untranslated leader regions of the HIV-1 and HIV-2 promoters. Genes Dev 1988;2:1101-14.
[57] Montminy MR, Sevarino KA, Wagner JA, Mandel G, Goodman RH. Identification of a cyclic-AMP-responsive element within the rat somatostatin gene. Proc Natl Acad Sci USA 1986;83:6682-6.
[58] Miyoshi I, Takahashi K, Kon Y, *et al.* Mouse transgenic for murine oviduct-specific glycoprotein promoter-driven simian virus 40 large T-antigen: tumor formation and its hormonal regulation. Mol Reprod Dev 2002;63:168-76.
[59] Barbacka-Surowiak G, Surowiak J, Stoklosowa S. The involvement of suprachiasmatic nuclei in the regulation of estrous cycles in rodents. Reprod Biol 2003;3:99-129.
[60] Nelson JF, Felicio LS, Randall PK, Sims C, Finch CE. A longitudinal study of estrous cyclicity in aging C57BL/6J mice: I. Cycle frequency, length and vaginal cytology. Biol Reprod 1982;27:327-39.

[61] Malette B, Filion B, St-Jacques S, Kan FW, Bleau G. Hormonal control of the biosynthesis of hamster oviductin. Microsc Res Tech 1995;31:470-7.
[62] Abe H, Oikawa T. Differentiation of the golden hamster oviduct epithelial cells during postnatal development: an electron microscopic study. J Exp Zool 1989;252:43-52.
[63] Malette B, Paquette Y, Bleau G. Size variations in the mucin-type domain of hamster oviductin: identification of the polypeptide precursors and characterization of their biosynthetic maturation. Biol Reprod 1995;53:1311-23.
[64] Malette B, Bleau G. Biochemical characterization of hamster oviductin as a sulphated zona pellucida-binding glycoprotein. Biochem J 1993;295 (Pt 2):437-45.
[65] Suzuki K, Araki Y, Sendai Y. Molecular cloning of an oviduct specific glycoprotein in hamster. Biol Reprod 1994;50:164.
[66] Malette B, Paquette Y, Merlen Y, Bleau G. Oviductins possess chitinase- and mucin-like domains: a lead in the search for the biological function of these oviduct-specific ZP-associating glycoproteins. Mol Reprod Dev 1995;41:384-97.
[67] Merlen Y, Paquette Y, Bleau G. Protein sequence of hamster oviductin deduced from its cloned cDNA. Int J Gynecol Obstet 1994;46:132.
[68] St-Jacques S, Malette B, Chevalier S, Roberts KD, Bleau G. The zona pellucida binds the mature form of an oviductal glycoprotein (oviductin). J Exp Zool 1992;262:97-104.
[69] Martoglio AM, Kan FW. Immunohistochemical localization of oviductin in the endometrial lining of the golden hamster (Mesocricetus auratus) during the estrous cycle and early gestation. Histochem J 1996;28:449-59.
[70] Roux E, Bleau G, Kan FW. Fate of hamster oviductin in the oviduct and uterus during early gestation. Mol Reprod Dev 1997;46:306-17.
[71] Boice ML, McCarthy TJ, Mavrogianis PA, Fazlebas AT, Verhage HG. Localization of oviductal glycoproteins within the zona pellucida and perivitelline space of ovulated ova and early embryos in baboons (Papio anubis). Biol Reprod 1990;43:340-6.
[72] Wegner CC, Killian GJ. Origin of oestrus-associated glycoproteins in bovine oviductal fluid. J Reprod Fertil 1992;95:841-54.
[73] Sendai Y, Suzuki K, Hochi H. Molecular characterization of a mouse oviduct specific glycoprotein. Biol Reprod 1994; 50:164.
[74] Gandolfi F, Brevini TA, Richardson L, Brown CR, Moor RM. Characterization of proteins secreted by sheep oviduct epithelial cells and their function in embryonic development. Development 1989;106:303-12.
[75] Rapisarda J, Mavrogianis PA, O'Day-Bowman MB. Immunological characterization and immunocytochemicallocalization of an oviduct-specific glycoprotein in the human. J Clin Endocrinol &Metab 1993;76:1483-8.
[76] Donnelly KM, Fazleabas AT, Verhage HG, Mavrogianis PA, Jaffe RC. Cloning of a recombinant complementary DNA to a baboon (Papio anubis) estradiol-dependent oviduct-specific glycoprotein. Mol Endocrinol 1991;5:356-64.
[77] Kapur RP, Johnson LV. Selective sequestration of an oviductal fluid glycoprotein in the perivitelline space of mouse oocytes and embryos. J Exp Zool 1986;238:249-60.
[78] Kadam KM, D'Souza SJ, Natraj U. Identification of cellular isoform of oviduct-specific glycoprotein: role in oviduct tissue remodeling? Cell Tissue Res 2007;330:545-56.
[79] Verhage HG, Fazleabas AT, Mavrogianis PA, *et al.* The baboon oviduct: characteristics of an oestradiol-dependent oviduct-specific glycoprotein. Hum Reprod Update 1997;3:541-52.
[80] DeSouza MM, Murray MK. An estrogen-dependent sheep oviductal glycoprotein has glycan linkages typical of sialomucins and does not contain chitinase activity. Biol Reprod 1995;53:1517-26.

[81] Boice ML, Geisert RD, Blair RM, Verhage HG. Identification and characterization of bovine oviductal glycoproteins synthesized at estrus. Biol Reprod 1990;43:457-65.

[82] Kan FW, Roux E, Bleau G. Immunolocalization of oviductin in endocytic compartments in the blastomeres of developing embryos in the golden hamster. Biol Reprod 1993;48:77-88.

[83] Kouba AJ, Abeydeera LR, Alvarez IM, Day BN, Buhi WC. Effects of the porcine oviduct-specific glycoprotein on fertilization, polyspermy, and embryonic development *in vitro*. Biol Reprod 2000;63:242-50.

[84] Killian GJ. Evidence for the role of oviduct secretions in sperm function, fertilization and embryo development. Anim Reprod Sci 2004;82-83:141-53.

[85] McCauley TC, Buhi WC, Wu GM, *et al.* Oviduct-specific glycoprotein modulates sperm-zona binding and improves efficiency of porcine fertilization *in vitro*. Biol Reprod 2003;69:828-34.

[86] Yong P, Gu Z, Luo JP, Wang JR, Tso JK. Antibodies against the C-terminal peptide of rabbit oviductin inhibit mouse early embryo development to pass 2-cell stage. Cell Res 2002;12:69-78.

[87] Anderson SH, Killian GJ. Effect of macromolecules from oviductal conditioned medium on bovine sperm motion and capacitation. Biol Reprod 1994;51:795-9.

[88] Nancarrow CD, Hill JL. Co-culture, oviduct secretion and the function of oviduct-specific glycoproteins. Cell Biol Int 1994;18:1105-14.

[89] Abe H, Sendai Y, Satoh T, Hoshi H. Bovine oviduct-specific glycoprotein: a potent factor for maintenance of viability and motility of bovine spermatozoa *in vitro*. Mol Reprod Dev 1995;42:226-32.

[90] Satoh T, Abe H, Sendai Y, Iwata H, Hoshi H. Biochemical characterization of a bovine oviduct-specific sialo-glycoprotein that sustains sperm viability *in vitro*. Biochim Biophys Acta 1995;1266:117-23.

[91] Kimura H, Matsuda J, Ogura A, Asano T, Naiki M. Affinity binding of hamster oviductin to spermatozoa and its influence on *in vitro* fertilization. Mol Reprod Dev 1994;39:322-7.

[92] Kan FW, St-Jacques S, Bleau G. Immunoelectron microscopic localization of an oviductal antigen in hamster zona pellucida by use of a monoclonal antibody. J Histochem Cytochem 1988;36:1441-7.

[93] Abe H, Oikawa T. Study of the differentiation of secretory cells in the golden hamster oviductal epithelium by use of a monoclonal antibody. J Exp Zool 1990;254:97-106.

[94] Brown CR, Cheng WK. Changes in composition of the porcine zona pellucida during development of the oocyte to the 2- to 4-cell embryo. J Embryol Exp Morphol 1986;92:183-91.

[95] Hedrick JL, Wardrip NJ, Berger T. Differences in the macromolecular composition of the zona pellucida isolated from pig oocytes, eggs, and zygotes. J Exp Zool 1987;241:257-62.

[96] McNutt TL, Killian GJ. Influence of bovine follicular and oviduct fluids on sperm capacitation *in vitro*. J Androl 1991;12:244-52.

[97] Wegner CC, Killian GJ. *In vitro* and *in vivo* association of an oviduct estrus-associated protein with bovine zona pellucida. Mol Reprod Dev 1991;29:77-84.

[98] Araki Y, Nohara M, Yoshida-Komiya H, *et al.* Effect of a null mutation of the oviduct-specific glycoprotein gene on mouse fertilization. Biochem J 2003;374:551-7.

[99] Austin CR. Syngamy in a mammalian egg study by phasecontrast microscopy. Med Biol Illus 1960;10:63.

[100] Yanagimachi R. Sperm capacitation and gamete interaction. J Reprod Fertil Suppl 1989;38:27-33.

[101] Smith TT, Yanagimachi R. The viability of hamster spermatozoa stored in the isthmus of the oviduct: the importance of sperm-epithelium contact for sperm survival. Biol Reprod 1990;42:450-7.

[102] Smith TT, Yanagimachi R. Attachment and release of spermatozoa from the caudal isthmus of the hamster oviduct. J Reprod Fertil 1991;91:567-73.
[103] Boatman DE. Responses of gametes to the oviductal environment. Hum Reprod 1997;12:133-49.
[104] Wassarman PM. Zona pellucida glycoproteins. Annu Rev Biochem 1988;57:415-42.
[105] Shabanowitz RB, O'Rand MG. Characterization of the human zona pellucida from fertilized and unfertilized eggs. J Reprod Fertil 1988;82:151-61.
[106] Moller CC, Bleil JD, Kinloch RA, Wassarman PM. Structural and functional relationships between mouse and hamster zona pellucida glycoproteins. Dev Biol 1990;137:276-86.
[107] Bavister BD, Andrews JC. A rapid sperm motility bioassay procedure for quality-control testing of water and culture media. J *In vitro* Fert Embryo Transf 1988;5:67-75.
[108] Boatman DE, Magnoni GE. Identification of a sperm penetration factor in the oviduct of the golden hamster. Biol Reprod 1995;52:199-207.
[109] Boatman DE, Magnoni GE, Robbins RS. Modulation of spermatozoa and zona pellucida properties by the soluble acrosome reaction-inducing factor of the ovulated egg-cumulus complex. Mol Reprod Dev 1994;38:410-20.
[110] Reuter LM, O'Day-Bowman MB, Mavrogianis PA, Fazleabas AT, Verhage HG. *In vitro* incubation of golden (Syrian) hamster ovarian oocytes and human sperm with a human oviduct specific glycoprotein. Mol Reprod Dev 1994;38:160-9.
[111] Oliphant G, Cabot C, Ross P, Marta J. Control of the humoral immune system within the rabbit oviduct. Biol Reprod 1984;31:205-12.
[112] Buhi WC, O'Brien B, Alvarez IM, Erdos G, Dubois D. Immunogold localization of porcine oviductal secretory proteins within the zona pellucida, perivitelline space, and plasma membrane of oviductal and uterine oocytes and early embryos. Biol Reprod 1993;48:1274-83.
[113] Kadam KM, D'Souza SJ, Bandivdekar AH, Natraj U. Identification and characterization of oviductal glycoprotein-binding protein partner on gametes: epitopic similarity to non-muscle myosin IIA, MYH 9. Mol Hum Reprod 2006;12:275-82.
[114] Abe H. The mammalian oviductal epithelium: regional variations in cytological and functional aspects of the oviductal secretory cells. Histol Histopathol 1996;11:743-68.
[115] Bauersachs S, Blum H, Mallok S, *et al.* Regulation of ipsilateral and contralateral bovine oviduct epithelial cell function in the postovulation period: a transcriptomics approach. Biol Reprod 2003;68:1170-7.
[116] Reischl J, Prelle K, Schol H, *et al.* Factors affecting proliferation and dedifferentiation of primary bovine oviduct epithelial cells *in vitro*. Cell Tissue Res 1999;296:371-83.
[117] Shea DK, Walsh CJ. mRNAs for alpha- and beta-tubulin and flagellar calmodulin are among those coordinately regulated when Naegleria gruberi amebae differentiate into flagellates. J Cell Biol 1987;105:1303-9.
[118] Bauersachs S, Rehfeld S, Ulbrich SE, *et al.* Monitoring gene expression changes in bovine oviduct epithelial cells during the oestrous cycle. J Mol Endocrinol 2004;32:449-66.
[119] Natraj U, Bhatt P, Vanage G, Moodbidri SB. Overexpression of monkey oviductal protein: purification and characterization of recombinant protein and its antibodies. Biol Reprod 2002;67:1897-906.
[120] Liu DY, Clarke GN, Baker HW. Exposure of actin on the surface of the human sperm head during *in vitro* culture relates to sperm morphology, capacitation and zona binding. Hum Reprod 2005;20:999-1005.
[121] McNutt T, Rogowski L, Vasilatos-Younken R, Killian G. Adsorption of oviductal fluid proteins by the bovine sperm membrane during *in vitro* capacitation. Mol Reprod Dev 1992;33:313-23.

[122] Murray M, Messinger SM. Early embryonic development in the Djungarian hamster (Phodopus sungorus) is accompanied by alterations in the distribution and intensity of an estrogen (E2)-dependent oviduct glycoprotein in the blastomere membrane and zona pellucida and in its association with F-actin. Biol Reprod 1994;51:1126-39.

[123] Simerly C, Nowak G, de Lanerolle P, Schatten G. Differential expression and functions of cortical myosin IIA and IIB isotypes during meiotic maturation, fertilization, and mitosis in mouse oocytes and embryos. Mol Biol Cell 1998;9:2509-25.

[124] Auersperg N. Microfilaments in epithelial morphogenesis. J Cell Biol 1972;52:206-11.

[125] Fey EG, Wan KM, Penman S. Epithelial cytoskeletal framework and nuclear matrix-intermediate filament scaffold: three-dimensional organization and protein composition. J Cell Biol 1984;98:1973-84.

[126] Meyer TN, Schwesinger C, Bush KT, *et al.* Spatiotemporal regulation of morphogenetic molecules during *in vitro* branching of the isolated ureteric bud: toward a model of branching through budding in the developing kidney. Dev Biol 2004;275:44-67.

[127] Hagiwara H, Shibasaki S, Ohwada N. Ciliogenesis in the human oviduct epithelium during the normal menstrual cycle. J Electron Microsc (Tokyo) 1992;41:321-9.

[128] Comer MT, Leese HJ, Southgate J. Induction of a differentiated ciliated cell phenotype in primary cultures of Fallopian tube epithelium. Hum Reprod 1998;13:3114-20.

[129] Suuroia T, Aunapuu M, Arend A, Sepp E. ["Light" epithelial cells of swine and bovine oviducts]. Tsitologiia 2002;44:656-60.

[130] Giallourakis C, Cao Z, Green T, *et al.* A molecular-properties-based approach to understanding PDZ domain proteins and PDZ ligands. Genome Res 2006;16:1056-72.

CHAPTER 12

Secreted Mucin Multifunctionality: Overt Functions

Abstract: Multifunctionality is one of the ubiquitous properties of biological macromolecules and is also a feature of the gel-forming and soluble secreted mucin glycoproteins. These mucin molecules participate in various basic cell processes in both physiological and pathological conditions. Their functions appear to be important both for embryonic/fetal development and for adult cells and tissues. They perform such functions as lubrication and hydration of mucosal surfaces, defending them against mechanical and chemical harm and infection. The gel-forming mucins have the potential to modify and catalyze organic chemical reactions. The secreted glycoproteins are active participants in innate immunity. Depending on cell type and type of neoplasm, gel-forming mucins function as tumor suppressors or tumor promoters. The secreted mucins possess antibacterial, antifungal and antiviral activities. In addition, the important processes of biological reproduction are also regulated by mucins, in particular by the soluble MUC9/OGP mucin glycoprotein. This chapter presents the data on the multifunctional potentials of the secreted mucins as regulators of cell functions.

Keywords: Overt functions, secreted mucins, multifunctional potential.

Multifunctionality is one of the fundamental properties of biological macromolecules [1]. The experimentally established functions – called expressed or "overt" functions – often represent only a part of the functional potentials of a protein. Many others occur hidden in the protein's molecular structure and remain undetected. These "covert" functions are embeded in the protein's amino acid sequence and are unveiled by various mechanisms including posttranslational modifications of the primary translated polypeptide. Not only do these mechanisms disclose functions associated with the protein primary structure itself, they also unmask functions whose expression depends on interactions of a given protein or its fragments with other functionally active protein and nonprotein molecules.

12.1. SECRETED MUCINS AS REGULATORS OF CELL FUNCTIONS

Functions of the gel-forming and soluble mucins can be expressed both in the intracellular compartments and the extracellular matrix. Most of the overt functions of these proteins are associated with the extracellular secreted mucin molecules, although theoretically these mucins can interact with multiple intracellular adaptor and effecter molecules, which may involve them in diverse

Joseph Z. Zaretsky and Daniel H. Wreschner

intracellular processes. In this chapter, only experimentally verified functions of the gel-forming and soluble mucins are discussed. The "covert" functional potentials of the secreted mucins will be discussed in chapter 13.

12.1.1. Secreted Mucins as Regulators of Embryonic and Fetal Development

Studies of secreted mucin gene expression during embryonic and fetal development demonstrate specific patterns of expression for each of the tested genes. These different patterns may reflect specific functions of the mucins in cell and tissue differentiation. The expression of a given secreted mucin occurs at a specific stage of the developmental process (Fig. **1**). For example, no secreted mucin genes were observed in human embryonic and fetal specimens before 8 weeks of gestation. Expression of *MUC5AC* is first detected in the primitive gut at 8 weeks of gestation, lasts one week only, after which the MUC5AC mRNA is expressed constantly in the stomach. In the ileum, but not in the colon, MUC5AC transcript is detected between 11-12 weeks. After 12 weeks, the MUC5AC mRNA was not expressed in any region of the fetal or adult intestine [2].

The expressions of *MUC5AC* and *MUC2* differ both spatially and temporally (Fig. **1**). Expression of the *MUC2* gene occurs in the primitive gut at 9 weeks of gestation simultaneous with the cessation of the brief expression of the *MUC5AC* gene. The expression of *MUC2* appears in crypts of the small intestine at the 10th week and continues throughout the rest of fetal development and in adults; it appears in the surface epithelium of small intestine 2.6 weeks later. Interestingly, the MUC2 mucin appears in the colon at the 18th week of gestation simultaneously in the surface and crypt epithelium [2]. Since 9.5 weeks of gestation *MUC5AC* is constantly expressed in the stomach, but only in the surface epithelium and not in glands, while *MUC2* is expressed only in glands of antrum after 26 weeks of gestation [2, 3].

The distinct patterns of expression in embryonic, fetal and adult tissues are characteristic also for other secreted mucin genes [4-6]. For instance, the expression pattern of the *MUC6* gene differs in the developing stomach and in the normal adult gastric mucosa. For 10 weeks, from weeks 8 to 18 of gestation, *MUC6* is expressed in all surface epithelial cells of stomach, however, after this time its expression is restricted to the epithelial folds and to the developing submucosal glands (Fig. **1**) [3]. This observation indicates the association of *MUC6* expression with the cytodifferentiation of the gastric surface cells into submucosal glands [7, 8]. In contrast to MUC6, the MUC2 mucin is not involved in gastric epithelium morphogenesis and cytodifferentiation since its expression occurs late in the fetal stomach, at 26 weeks of gestation, when fully differentiated glands are already present in gastric mucosa.

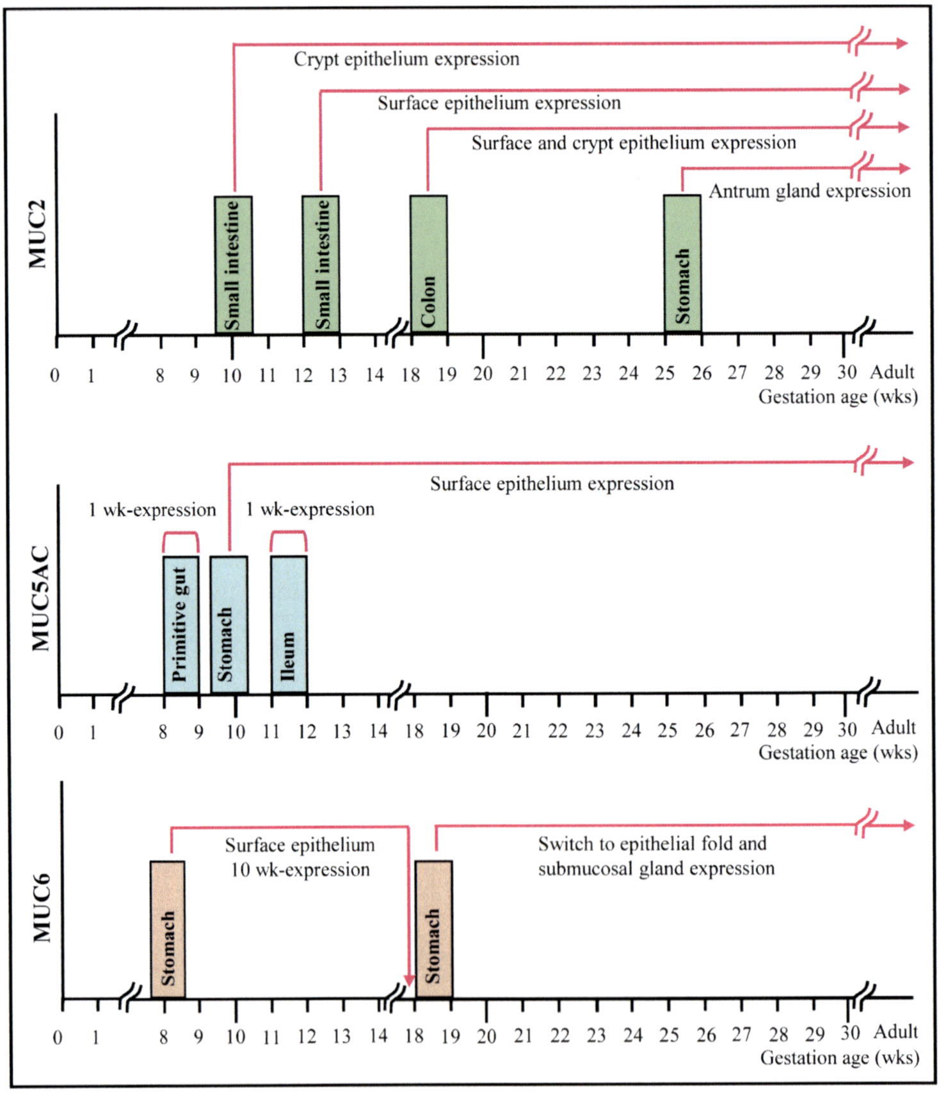

Figure 1: Developmental expression of the *MUC5AC, MUC2* and *MUC6* genes (based on the data reported in [2, 3, 7, 8]).

Like the MUC6 mucin, the MUC5B glycoprotein also participates in the development and differentiation of submucosal glands, but, unlike MUC6, which is expressed in the stomach, MUC5B is expressed in the respiratory tract [9]. The expression of *MUC2, MUC5AC, MUC5B* and *MUC7* in the developing tissues of the respiratory tract was associated with differentiation of specific cell lineages. It appears that *MUC6* and *MUC8* are not involved in development of the respiratory tract [5].

The examples presented – only a few among the many illustrating the participation of the gel-forming and soluble mucins in cell differentiation and tissue morphogenesis – suffice to show their involvement in these processes during embryonic and fetal development. The precise functions of individual mucins in specific developmental processes are not known at present, but the available data suggest that the secreted mucins operate as regulators of the embryonic and developmental processes. This suggestion is strengthened by, as noted above, specific dynamics of expression of a given mucin during embryonic and fetal development. Moreover, some secreted mucin glycoproteins are expressed in embryonic and fetal tissues, which usually do not express a given mucin in the corresponding normal adult tissues. For example, the MUC2 glycoprotein, a typical intestinal mucin, which is not expressed under physiological conditions in human adult airway tissues, is easily detected in embryonic and fetal trachea and terminal sac. Noteworthy, the dynamics of its expression in these tissues is characterized by the waves of activation and suppression of transcription of the MUC2 mucin gene, thereby pointing to the specific phases in human development dependent on the presence of functionally active MUC2 glycoprotein. On the other hand, mucin glycoprotein MUC5AC that is constantly expressed in the airways of normal adults individuals is actively expressed in the embryonic ileum, the organ that never expresses this mucin during lifetime of normal adult organisms. Importantly, its expression, as described above, occurs at the specific point of human development (the 11th week of gestation) and can be detected only for one week, after which the gene encoding MUC5AC mucin is strongly suppressed. Altogether, these data suggest that at the definite stage of development the molecular target(s) specific for a given mucin occur in different embryonic and fetal tissues. Interaction of mucins with these targets may induce the next stage(s) in embryonic and fetal development. It appears that these glycoproteins serve ligands for the receptors participating in different signal transduction pathways important for normal development.

12.1.2. Secreted Mucins – Regulators of Homeostasis in Adult Tissues

Mucus plays a central role in the evolution of multi-cellular organisms, enabling adaptation to constantly changing environment [10]. Mucins as components of mucus participate in "the dynamic, interactive mucosal defensive system active at the mucosal surface" of the respiratory, gastrointestinal and urogenital tracts as well as in the organs of the ocular and audio systems [11]. Mucus is a semi-permeable barrier that allows exchange of nutrients, water, gases, hormones and gametes while being impermeable to most pathogenic microorganisms including bacteria and viruses [12]. The most important functions of the mucus gel are protection, hydration and lubrication of the epithelial surfaces and transport of water, nutrients and gases [13]. These functions of secreted mucins allow regulation of cell and tissue homestasis.

In the airways, mucus entraps inhaled foreign debris and bacteria and clears them by a mucociliary "escalator" [14]. In the gastrointestinal tract, the mucus barrier prevents the epithelial cells from auto-digestion and proteolytic damage [12]. In the reproductive tract, mucus protects the uterine cavity from bacterial invasion and controls the survival and penetrability of the spermatozoa [13]. In the eye, the mucus gel of the tear film is the primary barrier defending the ocular surface from chemical, enzymatic and mechanical attacks [15]. These mucus functions are attributed to and maintained by the multifunctional mucin glycoproteins that provide the structural framework of the mucous gel, prevent barrier dehydration, ensure lubrication, present carbohydrate receptors to immobilize pathogens, and, "*via* binding to other components of the secretion, have the potential to act as a sink for host-protective proteins and peptides" [16-19].

The secreted mucins are characterized by complex multidomain structures with specific functional potential(s) attributed to each domain. Although some functions of some domains have been established, "it is very likely that other, as-yet-unascribed functionalites reside in these domains" [16]. As noted by Thornton *et al.* [16], these domains may be sites of "interaction with other components of mucus, that are important for barrier organization or innate defense". The ability of the gel-forming mucins to build a protective "gel blanket" is associated mainly with their N- and C-termini, which determine the character of the mucin polypeptide polymerization. To fulfill the barrier function noted above, the mucus gel has to have a specific structure and viscoelastic properties, which result from correct polymerization of the mucin filamentous "wormlike" molecules with other secreted mucus components such as peptides, proteins, lipids and enzymes [20-23]. Multiple low-affinity hydrophobic interactions play a major role in the ability of the mucus gel to stick to most particles, bacteria and viruses, preventing their penetration to the epithelial surfaces. Due to biophysical properties of the gel-forming mucins, mucus gel possesses "a shear-thinning non-Newtonian property making it an excellent lubricant that ensures an unstirred layer of mucus remains adherent to the epithelial surfaces" [12].

The ability of the gel-forming mucins to form two-layered mucus gel consisting of an entangled nonadherent luminal "sloppy" layer and a cell-adherent unstirred layer is the basic property allowing realization of the lubrication function. Biological lubrication is known to be associated with the ability of gel-forming mucin molecules "to form a dense hydrated layer at the surfaces of practically any chemistry" [24]. This unique ability to form a gel layer is an intrinsic property of the "block-copolymer structure" of mucins consisting of "glycosylated hydrated comb-brushes" that are polyampholyte domains with both positively and negatively charged amino acids [24-27]. Under adsorbtion, the glycosylated hydrophilic comb sections, which are abundant in negatively charged sialic acid residues and sulfate groups, "stretch out into aqueous media due to favorable interactions with water, thereby adopting a brush-like architecture" [24]. Such a structure of gel-forming mucins and the presence of sialic acid residues and sulfate groups are known to be crucial for biolubrication, a physical process minimizing friction between contacting surfaces that is very important for mechanical defense of the external and internal epithelial surfaces in human and animal organisms [28-30].

Adsorption of highly hydrated hydrophilic and amphiphilic polymers (in our case the gel-forming mucins) onto surfaces reduces friction between most notably hydrophobic surfaces [24, 30, 31]. It is important for defense of the respiratory, gastrointestinal, reproductive, ocular, oral and many other surfaces covered with the mucosal layer [32-34]. Yakubov *et al.* [35] studied the mechanisms of viscous boundary lubrication of hydrophobic surfaces by mucins and showed that "with increasing mucin bulk concentration the adsorbed layer thickness increases, leading to a decrease in local shear rate and thus boundary viscous stress". This, in turn, reduces the friction forces between the lubricating surfaces.

In the airways, the presence on the epithelial surfaces of the double-layered gel, consisting of the cell-adherent nonmovable layer and the nonadherent "slipping" layer, allows not only entrapping of a pathogen by a "slipping" part of the gel, but also permits moving it off of the luminal surface [36]. In the gastrointestinal tract, mucus aids the transport of chime from the gut to the colon by serving as a lubricant during the peristaltic process, while allowing rapid entry and exit of nutrients and waste [37]. Mucus transport depends directly on regulated mucus viscoelasticity, which, in turn, strongly depends on degree and nature of mucin glycosylation and correlates with mucus sialomucin content [37, 38]. It should be noted that the mucus mesh, important for control of the passage of nutrients and waste through the mucus barrier, is composed primarily of entangled mucins and other mucus constituents (lipids, salts, proteins) combined by a reversible linkage between them [37]. The role of these interactions in gel "breathing" (changes in viscoelasticity) cannot be overestimated. Comprehensive analysis of the possible interactions is beyond the scope of this monograph, but some of the most important interactions deserve at least a brief discussion.

As shown by a number of studies, most of the lipids present in mucus are associated with the hydrophobic domains of mucin glycoproteins. Lipids are known to contribute greatly to the viscoelasticity of mucus: extracting lipids from the mucus reduces the steady shear viscosity [39-41], while the increasing total lipid content enhances mucus viscoelasticity [42]. By changing lipid content in mucus, one can regulate the viscoelasticity of the mucin gel and facilitate the normal functioning of this important homeostatic mechanism [41-43]. Although limited, the presented data show that lipids are very powerful constituents of mucus which may significantly change the mucus properties thus influencing the defense function of the mucus gel.

Ionic strength is also an important biophysical property of mucus gel. Alterations in ionic strength can directly lead to shrinkage or swelling of mucus, significantly changing its viscoelasticity [37]. It has been shown that an increase in ion concentration correlates with a decrease in mucus viscoelasticity [44].

High acidity increases viscoelasticity of mucus by the mechanism that reduces the negative charges associated with the sialic acid residues distributed along the glycosylated regions of the mucin molecules. Hong *et al.* [45] analyzed the role of different pH on gastric mucin aggregation, which may affect the viscoelastic status of mucus. It is well known that at low pH, high concentrations of mammalian gastric mucins may lead to gel formation. This molecular transformation is essential to protect the stomach mucosa from auto-digestion. Hong *et al.* [45] established that gastric mucins aggregate at or below pH 4. At pH 5-7, mucin molecules are in the extended fiber-like conformation measuring some 400 nm in length, whereas at pH 4 and below they cluster together. Studies of deglycosylated mucins conducted at pH 5 showed the importance of the sugar side-chains in maintaining the extended structure and solubility of the mucin molecule, since the deglycosylated regions of the molecule tended to re-fold into compact spherical structures. However, it appears that oligosacchride side-chains are involved in the aggregation process in a roundabout manner. Earlier, Cao *et al.* [46] showed that aggregation/gelation of mucins represents the interplay of hydrophobic and electrostatic interactions. In line with these findings, Hong *et al.*'s results [45] suggest that glutamic and aspartic acid residues, with pH 4.1 and 3.9, respectively, play the key role in aggregation of gastric mucins. The authors emphasize the similarity of the gel cluster morphology to the' "pearl necklace" morphology produced by acid-induced gelation of micelles formed by polystyrene-polyacrylic acid di-blocks in aqueous solution. Still other investigators [47] concluded that "although hydrophobic interactions may lead to formation of the micellar cores, ionic interactions of the negatively charged sugar and amino acid residues and other ions present in mucus are contributing factors in determining the morphology of gels". Importantly, there is a clear correlation between the aggregation of gastric mucins at pH 2 and the physiological function of these glycoproteins, which serve a "protective coating for the stomach at the low values of pH at which food must be digested" [45].

Important data regarding the dependence of changes in MUC5B mucin structure and mucus gel permeability in cervical mucus on pH value and the stages of the ovulatory cycle was published by Brunelli *et al.* [48]. Human cervical mucus is a heterogeneous mixture of mucin glycoproteins whose relative concentration and physical, chemical and rheological properties change dramatically during the phases of ovulation [49]. During the normal menstrual cycle but outside the ovulatory phase, the mucus is arranged in a dense filamentous structure impermeable to spermatozoa. Immediately before ovulation, MUC5B becomes the dominant mucin in the cervical mucus, and at the point of ovulation changes in its structure decreases mucus viscosity [48], maximizing the mucus' permeability to sperm [50-52]. Brunelli *et al.* [48] showed that the changes in mucus gel permeability are associated with the transition of the MUC5B molecule from the fiber-like preovulatory structure to the globular conformation characteristic of the ovulatory phase. Importantly, the compact fiber-like arrangement could be easily restored in the ovulatory mucus by lowering the pH, indicating that the "switch from globular to fibrous conformation of the MUC5B mucin largely depends on a pH-driven mechanism" [48].

Depending on the nature, origin and physical location of a mucin molecule, the changes in pH may induce opposite effects on the mucin structure. As noted above, the gastric surface mucins, represented mainly by MUC5AC molecules, tend to aggregate at low pH, whereas the cervical MUC5B mucin aggregates at high pH and reverts to a fiber-like conformation at low pH. It appears that primary amino acid structure of a mucin molecule and its glycosylation pattern, as well as various mucus constituents including ions, lipids and polypeptides, determine the type of mucin structural transformation induced by alterations in pH.

In addition to lipids, ions, salts and pH, various nonmucin proteins in the mucus can also change its properties and mucin functions. According to studies on interactions between mucus mucins and nonmucin proteins [53], mucin-protein interactions are either covalent or noncovalent and cause formation of stable heterotypic complexes that may potentiate functions of the individual molecules [54]. The MG1 salivary mucin (later identified as MUC5B) was found to interact with hydrophobic and hydrophilic domains of protein macromolecules present in glandular secretion or in the whole saliva [55, 56]. Iontcheva *et al.* [53] showed that MG1 (MUC5B) forms complexes with amylase, proline-rich proteins, statherin and histatin in one of two ways: through low-affinity hydrophobic or ionic interactions, including hydrogen bonds and van der Waals forces; or by formation of covalent linkages between the γ-carboxyl group of glutamine and the ε-amino group of lysine present in the interacting protein molecules. Several studies showed that intestinal, salivary and cervical secreted mucins form low-affinity binding with antibodies [57-59], specifically with the Fc moieties of antibodies [59, 60].

In addition to lubrication and hydration of mucosa, mucins represent a first line of defense of mucosal surfaces against bacterial, fungal and viral infections [61]. Interaction of secreted mucins with glycoproteins present in the outer membranes of different microorganisms is a principal link in the mechanism of cell defense. Glycosidic groups of the O-glycosylated domain(s) of the secreted mucins are involved in lectin-like interaction with the corresponding components of microbial walls, thereby mediating entrapment and clearance of microorganisms from the body tracts [61]. As noted by Piludu *et al.* [61], the human mucosal linings produce chemically different mucins depending on the functional necessities of the mucosal surfaces and microorganisms that come into contact with these mucins. During contact, the mucins serve as receptors specific for definite types of microorganisms: for example, O-glycans of salivary mucins function as specific receptors for *Candida albicans* blastospores and *Streptococcus mutans* [62, 63]; the stomach MUC5AC mucin is a receptor for *Helicobacter pylori (H. pylori)* [64, 65]; and both the gel-forming mucin MUC5B and the soluble mucin MUC7 purified from saliva, as well as as gel-forming mucins MUC2, MUC5AC and MUC5B purified from cervical mucus plug developed in pregnant women, bind *in vitro* HIV-1, inhibiting its activity [66, 67]. The authors of these studies [66, 67] suggest that the mechanism of inhibition is "a purely physical phenomenon", associated with aggregation of viral particles mediated by the mucin sugar side-chains. Preliminary incubation of the purified mucins with the permissive cells does not prevent the consequent HIV-1 infection, demonstrating that only direct binding of the mucin molecules to the viral particles results in inhibition of viral infection.

MUC6 mucin present in seminal plasma also inhibits HIV-1 infection [68]. It is known that $CD4^+$ T-lymphocytes, Langerhans cells and dendritic cells (DC) are the primary targets for HIV-1 [69-71]. Many viruses transmitted *via* the genital or oral mucosa, including HIV-1, interact with DC through DC-specific intercellular adhesion molecule-3 grabbing non-integrin (DC-SIGN) expressed on immature DC [70-73]. DC-SIGN is used by DCs to capture viral pathogen or its antigens for subsequent presentation to T-lymphocytes [71, 74]. However, the same mechanism can also be used by viruses to enhance host infection [73]. As shown by Stax *et al.* [68], MUC6 in seminal plasma binds DC-SIGN and potently blocks DC mediated transfer of HIV-1 to $CD4^+$ T-lymphocytes. The results of this study suggest that Lewis sugars are the main groups providing the MUC6 mucin with its DC-SIGN-binding capacity.

The above data present strong evidence that secreted mucins, both gel-forming and soluble, are effective components of the mucosal surface's first line of defense against harmful exogenous agents. Importantly, mucins defend cells against pathogens not only physically through aggregation of the microbial agents, but chemically by antifungal and antibacterial activities, as shown in the cases of MUC7 and MUC6 mucins, respectively [75-77]. Interestingly, the antifungal activity of the MUC7 mucin is seen only after posttranslational proteolysis of the intact MUC7 glycoprotein. The full-length MUC7 apoprotein comprised of 357 amino acid residues does not display any significant candidacidal activity [78], whereas the synthetic 51-aa residue polypeptide corresponding to the N-terminal domain D1 exhibits strong broad-spectrum antifungal activity comparable to the activity of the human salivary histatin-5, a natural 24-residue peptide possessing high antifungal potential [79]. Liu *et al.* [75] succeeded in showing that a 20 kD polypeptide, present in the whole saliva and possessing high microbicidal activity, is a product of *in vivo* proteolytic cleavage of a native soluble MUC7 mucin molecule. This polypeptide corresponds to the N-terminal region of the MUC7 glycoprotein.

Evidence is accumulating that the gel-forming mucins also undergo proteolytic processing at their N- and C-termini during biosynthesis [80-83]. The functional potentials of these modifications are not fully understood. The cleavage in the N-terminal region of MUC5B was proposed to be associated with transition of the mucin from the gel- to the sol-phase, thus influencing viscoelasticity of mucus and lubrication function [82]. The C-terminal 14-aa peptide of the human and rat

MUC2 mucins was found to interact with heparin *in vitro* [84], and later on it was established that endopeptidase furin is responsible for this fragment cut-off *in vivo* [85]. Immunofluorescent localization of the peptide was confined to the basal perinuclear regions of the goblet cells, while the truncated MUC2 molecule was found in the secretory granules [84], indicating different targeting of the two parts of the same mucin molecule and suggesting different functioning. The absence of the C-tail in the MUC2 molecule was shown to lead to formation of abnormally structured dimers [85]. It is not clear whether furin-mediated cleavage of the MUC2 mucin occurs in physiological conditions, and, if so, whether the free 14-aa peptide interacts with heparin *in vivo*. Although the functional consequence of such an interaction, if any, is unknown, it is highly likely that the native MUC2 molecule or its C-terminal MUC2 fragment binds with heparin. Moreover, Borsig *et al.'s* findings [86] suggest that binding of heparin to mucins expressed on cancer cells (membrane-bound mucins) or tightly attached to cancer cells (gel-forming mucins) may interrupt the interaction of P-selectins expressed on the platelet membrane with the mucins associated with cancer cells, thereby inhibiting the metastatic process. If such an interaction does occur *in vivo*, it may contribute to MUC2-mediated tumor suppression [87].

In addition to furin-mediated cleavage, the auto-catalytic proteolysis of MUC2 and MUC5AC mucins occurs during biosynthesis of these glycoproteins. This proteolytic reaction takes place in the GDPH sequence located at the C-terminus of the indicated mucins. It appears to be a natural phenomenon with biological relevance, as it happens at a pH comparable to that of the late secretory pathway [82, 83]. Importantly, despite the nuances specific for cleavage of each mucin, the fragments resulting from proteolysis of MUC2 and MUC5AC molecules possess similar biochemical reactivity. The functional power of the relatively small fragment containing the original C-terminus has not been studied in depth, but at least in the case of the MUC2 mucin, the fragment contains the above-described short 14-aa peptide that has the potential to bind heparin. In addition to binding with heparin, this fragment may interact with other biologically active macromolecules, modifying its own functional potentials or gaining new ones.

Studies of the biochemical reactivity of the large N-terminal fragments resulting from the GDPH cleavage of the MUC2 and/or MUC5AC glycoproteins [82, 83] showed that cleavage generates a reactive Asp residue (anhydride) at the newly formed C-terminal ends of these mucins. High biochemical activity of Asp enables interaction of this residue with the carboxyl groups of any O-glycans attached to various proteins, including MUC2 or MUC5AC mucin molecules, resulting in ester linkage formation. Occurrence of this type of ester bond was described between Asp anhydride of H3 polypeptide chain and chondroitin sulfate chain attached to bikunin, generating functionally active pre-α-inhibitor [88, 89]. As pointed out by the authors [82, 83], such chemical linkage is likely to occur in the cleaved MUC2 and MUC5AC, adding novel functional abilities to the newly generated fragments. Besides, the cleaved fragments with the active C-terminal Asp residue can form an ester bridge with the corresponding reactive groups located on the molecules of other proteins, thereby generating covalently bound protein complexes with new functions. Although it is not known whether the cleavage of the GDPH sequence in the mucin molecule is accompanied by a covalent ester bond formation with another protein molecule *in vivo*, "the potential appearance of such a linkage could be physiologically very important" [82, 83]. This cleavage can also be important in pathological conditions as one of the events in the pathogenesis of mucin-associated diseases.

Different regions of mucin molecules varies considerably in their potential to interact with other protein and nonprotein biomolecules. In some cases, specific experimental conditions are necessary to disclose these interactions. By usage of the yeast two-hybrid system it was established that the Cys-domains of the MUC5B mucin interact with salivary histatins [90] – hinting at another defense function(s) of the MUC5B mucin hidden in its structure. Exploration of other mucins in the yeast two-hybrid system may uncover still more mucin-protein interactions and as-yet-unknown but intrinsic mucin functions. Tomasetto *et al.* [91] supported this supposition with the finding that in the yeast two-hybrid system "cysteine-rich vWF C-domains at the C-termini of MUC2 and MUC5AC interact with TFF1 peptide". Further experiments showed that the trefoil peptides interact with each of mucins tested in mucin- and TFF-specific manner, and that these interactions modulate the rheological properties of the mucus gel [92-94].

The important role of mucus constituents in the proper functioning of mucins and mucus gel was highlighted in a study by Raynal *et al.* [95]. These investigators showed that gel formed by guanidine chloride-purified MUC5B at concentrations similar to those found in natural saliva was approximately 20 times more permeable than native mucus; this suggested that the native mucus contained components that are absent in the gel composed of the purified mucin. Indeed, further analysis implicated Ca^{2+} as a key regulator of MUC5B mucin supra-molecular conformation in salivary mucus [96]. Later on, the same group established that MUC5B interacts with gp-340 glycoprotein, a component of the airway mucus possessing antibacterial properties [17]. This interaction increases MUC5B functionality by promoting its bacteriocidal activity.

In summary, the biochemical and biophysical properties of the secreted mucins, together with their ability to interact with different components of mucus secretion, enable them to defend tissues against various infections, making them participants in innate immunity.

12.2. ROLE OF SECRETED MUCINS IN INNATE IMMUNITY

Several excellent reviews were published recently that discussed in detail the mechanisms of innate immunity and the role of mucins in this type of cell defense [19, 97-106]. We will focus here only on a specific issue of the mucin functioning in innate immunity: the ambiguous role of the gel-forming mucins in the cell's defense against bacterial infections. Indeed, the gel-forming mucins behave in this mechanism as a double-edged sword. On the one hand, they build a defense barrier for epithelial cells, while on the other hand, they serve as sources of energy for bacteria and supply them with "comfortable" niches for colonization. The relationship between the MUC2 or MUC5AC mucins and bacteria during intestinal or stomach infections, respectively, are examples of the complex relationships between gel-forming mucins and microorganisms. It must be noted that the functions MUC2 and MUC5AC perform against bacteria can be carried out by other gel-forming mucins as well, and in other locations including oral cavity, airways and ocular surface.

12.2.1. MUC2-Mediated Defense of the Intestine Against Infection

The mucus gel covering the intestinal epithelium is "the anatomical site at which the host first encounters gut bacteria" [105]. For a long time, the protective functions of the mucus gel layer were considered static and constitutive. However, recent studies changed that concept and now mucus gel is viewed as a dynamic defense barrier [102, 103, 106]. The first indications of the dynamic nature of mucus gel were brought by Atuma *et al.* [107], who found that the mucus gel in the stomach and colon consists of two layers of different thicknesses and mobility: the outer layer is a thick, "loose" and disorganized structure that is easily moved by peristalsis, while the inner one is a relatively thin and stratified mucus layer "firmly" attached to the epithelium. In the stomach, the outer layer consists of the MUC5AC mucin molecules, while the inner layer contains mainly the MUC6 mucin [108]. In the colon, both layers contain only MUC2 mucin [106]. In general, the outer layer is estimated to be twice as thick as the inner layer. Bacteria are not distributed uniformly in these layers: the outer intestinal layer in the healthy adult individual contains up to 10^{13} - 10^{14} commensal bacteria representing at least 160 species [109], while the inner layer normally is totally free of bacteria [106].

The two differently organized gel layers permit different functioning of the gel-forming mucins in each layer. The outer layer offers numerous ecological advantages to both commensal bacteria and some types of pathogenic bacteria in the lumen. A mucin of the outer layer serves as a source of nutrients (*e.g.* mucin saccharides) for bacteria, thus supporting their growth. It also provides physiologically relevant oligosaccharide ligands for bacterial adhesins, thereby promoting intestinal colonization by the adhering microorganisms [110-113]. The commensal bacteria exist in symbiosis with their host as the host provides "good conditions" for bacterial growth and colonization. On the other hand, bacteria help the host extract energy from the indigestible glycoconjugates and supply vitamins to the host [114]. In contrast to commensal bacteria, pathogenic bacteria may destroy the mucus barrier by moving efficiently through the outer "loose" layer and by making use of the enzymes that degrade mucin carbohydrates and/or mucin polypeptide backbone [113].

It would appear that in the outer layer the MUC2 mucin ensures bacterial survival and colonization, while in the inner layer it builds a dense impervious gel meant to prevent contact of the bacteria with the underlying epithelial cells. However, the actual relationship between MUC2 and bacteria is much more complicated. As noted by Deplancke and Gaskins [105], "the defensive nature of mucins lies in their capacity to entrap microbes". This supposes adhesion of a bacterium to a mucin molecule as the first step of entrapping. The enteric pathogens have on their surfaces specific molecules, bacterial adhesins, that bind the corresponding receptor on the host epithelial cell membrane [115-117]. Importantly, because some of the mucin epitopes are equivalent to the corresponding epithelial cell receptors, mucins (MUC2 in particular) bind a bacterium through interaction of these epitopes with bacterial adhesins, thereby blocking a pathogen on its way to the host epithelial cells [118-120] (Fig. **2**). These events occur in the outer "movable" layer of the intestinal mucosal barrier, activating the pathogen elimination mechanism that provides removal of the outer bacteria-containing mucus layer by gut peristalsis.

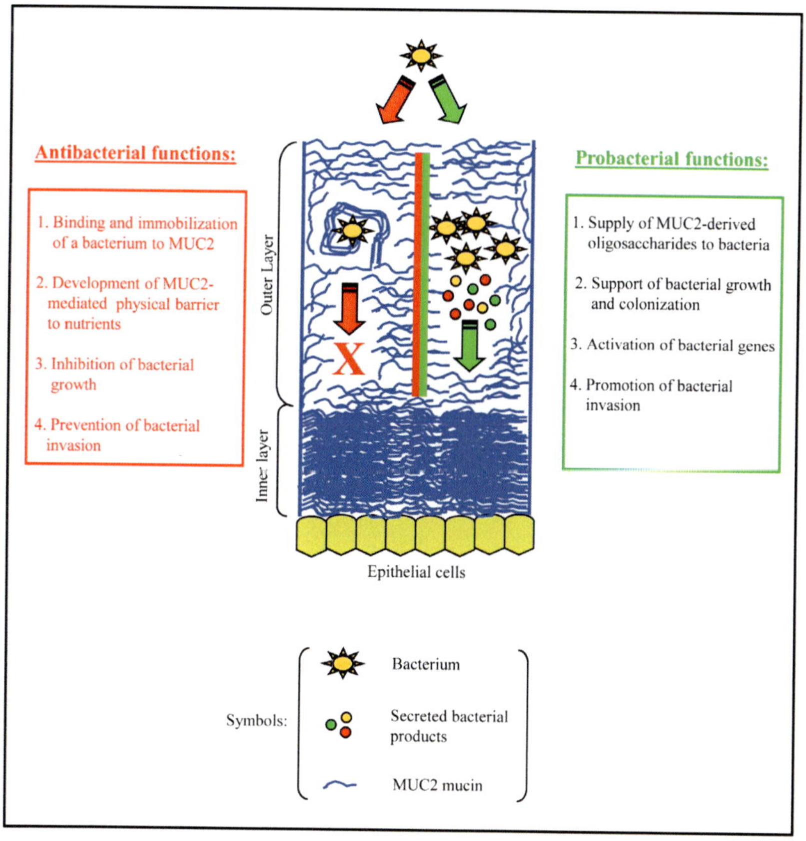

Figure 2: Interactions of MUC2 and bacteria in the intestine (based on the data reported in [113, 118-120]).

The described reactions represent only a part of the complex mechanism by which the MUC2 mucin carries out its defense functions. Recently, Tu *et al.* [113] reported that the MUC2 mucin added to the culture of *Campylobacter jejuni* (*C. jejuni*) inhibited bacterial growth. Two possible explanations for this phenomenon were suggested.

The observed inhibition of bacterial growth might result from the binding of mucin oligosaccharide ligands to the bacterial adhesins, creating a physical barrier to nutrients necessary for bacterial growth. Another explanation might be associated with the possible MUC2 antibacterial activity analogous to the antibacterial activity exhibited by the MUC6 [77] and MUC7 mucins [121]. Further research on the MUC2 inhibitory activity is needed to clarify this dilemma. However, it is important to note that the phenomenon was observed *in vitro* under conditions that prevented interference by other components present in the natural intestinal mucus, implying that the ability to inhibit bacterial growth is an intrinsic function of the MUC2 mucin.

In addition to the defensive antibacterial functions MUC2 performs in the outer layer of the mucus gel, it also performs the opposite probacterial functions (Fig. **2**) that promote survival of a pathogen and increase its pathogenecity [113]. These tasks are carried out not only by supplying nutrients to an invading microbe, but also by changing its transcriptional activity in a way that gives advantages to the pathogen. Indeed, interaction of MUC2 with *C. jejuni* activates transcription of a battery of bacterial genes important for colonization and pathogenecity, including genes encoding cytotoxins, invasion antigen, flagellin A, and rod-shape-determining proteins, as well as four gene encoding putative mucin-degrading enzymes [113]. These data raise the question of how the external mucin molecule up-regulates activity of the transcriptional machinery of a bacterium. At least two hypotheses can be proposed for explaination of the underlying mechanism. The MUC2 mucin may function as a ligand with the potential to interact with the bacterial membrane receptor(s) followed by activation of an intracellular signaling pathway(s) that stimulates the transcription process. This explanation is consistent with the fact that the cell membrane receptors represented by membrane-bound mucins such as MUC3, MUC4 and MUC12 do interact with gel-forming mucin (*e.g.* MUC5AC) by formation of both covalent and noncovalent bonds [104, 122].

Another hypothesis suggests that the products of MUC2 degradation resulted from enzymatic activity of the invaded bacteria are "swallowed" by microorganisms followed by interaction of these products with the bacterial transcription factor(s), thus modifying their specificity and activity and changing the transcriptional profile of the invaded bacterium. The suggested hypotheses require examination, but whatever the mechanisms, the ability of MUC2 mucin to change transcription of the bacterial genes unequivocally indicates its potential to function as a transcriptional regulator. Importantly, the observed transcriptional activation cannot be attributed to any contaminants since thoroughly purified MUC2 mucin was used in the reported experiments [113].

The ability of MUC2 to activate bacterial genes important for bacterial colonization and pathogenecity – namely those encoding cytotoxins, the invasion promoting antigen and putative mucin-degrading enzymes [113] – shows that MUC2 behaves, at least partially, as a "suicide protein" that promotes self-destruction and bacterial invasion, thus functioning as an anti-defensive factor. On the other hand, the ability of the MUC2 mucin to inhibit bacterial growth [113] underscores its active participation in host protection as a component of the host innate defense mechanisms.

It is important to emphasize that although the MUC2 functions described by Tu *et al.* [113] were observed *in vitro,* they could be performed *in vivo* as well, since the experimental conditions employed by the authors corresponded to those of the outer layer of the natural mucus gel. As noted by Johansson *et al.* [106], transition of the firmly adherent stratified nonsoluble gel into a "loose" nonattached soluble mucus outer layer is associated with a large expansion in gel volume "owing to proteolysis within the cysteine-rich regions of MUC2 in a way allowing maintenance of the polymeric network". In addition to MUC2, a substantial contribution to this network is made by the trefoil factor 3 (TFF3) and the IgGFcγ-binding protein (IgGFcγBP) present in both intestinal mucus layers. The recent study of Yu *et al.* [123] showed that the rat goblet cell-secreted TFF3, IgGFcγBP and Muc2 are bound together "by covalent disulfide bond-based interaction in the soluble fraction of intestinal mucus, and form heteropolymers", suggesting that this interaction is one of the biochemical bases of the net-like molecular structure of mucus.

The described properties of the intestinal outer friable gel layer permit both the motility of a bacterium and realization of MUC2's controversial functions, part of which supports bacterial survival and colonization, while another part inhibits growth of intestinal pathogens. In contrast, the highly dense inner mucus gel containing tightly connected MUC2 polymer molecules represents an impenetrable barrier to a bacterium on its way to the underlying epithelium [106].

12.2.2. Roles of MUC5B and MUC7 Mucins in Defense of the Oral Mucosa

Saliva represents a substratum that performs defense and perception functions in the oral cavity. MUC5B and MUC7 are two major mucins secreted into saliva [124]. It has been suggested that the mucus glycoprotein layer on the oral mucosa surfaces serves as a matrix for recruitment of other protective proteins. Recently, Amado *et al.* [125] reported that saliva contains more than 2000(!) proteins/peptides, the main constituens of which are mucins, immunoglobulins, enzymes (*e.g.* α–amylase, lysozyme and carbonic anhydrase), lactoferrin, and a number of small proteins such as cystatins, histatins, defensins, statherin, and proline-rich proteins [126, 127]. Analogous to the intestinal defense barrier, the whole saliva also consists of two layers, or phases: soluble (sol) and gel-like (gel) [128]. The MUC7 mucin is present mainly in the sol phase, although some amount is consistently found in the gel phase, suggesting interaction of MUC7 with salivary gel matrix. The majority of MUC5B is present in the gel phase in the "insoluble" state, but a substantial amount is usually detected in the sol state as well. It is represented in the gel phase by two differently charged isoforms [129, 130]. The ratio of "soluble" to "insoluble" fractions of MUC5B mucin in the whole saliva depends on the "proteolytic processing" of mucin molecules in the oral cavity that leads to release of a C-terminal fragment, a phenomenon previously observed during biosynthesis of the human and rat MUC2 mucins in the colon [131, 132].

Similarity between the MUC5B and MUC2 mucins was found not only in formation of the defensive mucus barriers, but also in their interactions with pathogens. Like MUC2, the MUC5B mucin serves as a source of nutrients for the oral microbiata. Dental plaque is capable of utilizing salivary MUC5B as a source of carbohydrates, a process that requires synergistic degradation of the complex oligosaccharide side-chains of the mucin by a consortium of oral bacteria in the plaque community [133, 134]. Salivary glycoproteins, especially MUC5B, can influence the establishment and selection of the oral microflora, promoting adhesion of certain bacteria and facilitating clearance of other species by aggregation [134, 135].

The interactions between microorganisms, salivary mucins and oral tissues occur through specific and nonspecific physico-chemical forces [136]. The binding of some bacteria to MUC5B is a pH-dependent process that results from a nonspecific interaction between positively charged bacterial surfaces and a negatively charged mucin molecule. MUC5B may interact with other bacteria through specific receptor-mediated binding. As shown by Walz *et al.* [137], the interaction of the salivary MUC5B mucin with *H. pylori* is mediated mainly by the BabA adhesin of *H. pylori* and ABO/Lewis^b blood group antigens of the MUC5B mucin, and to a lesser degree by the SabA bacterial adhesin and MUC5B receptors [138, 139]. In contrast, the binding of MUC7 with *Actinobacillus actinomycetemcomitans* is pH-independent and occurs through the specific interaction of bacterial wall receptors and sialic acid residues on the MUC7 molecule [136, 140]. Interaction of *H. pylori* and MUC7 is mediated exclusively by SabA adhesin and sialyl-Lewis^x antigen of MUC7 [137, 141]. Importantly, the presented data show that MUC5B and MUC7 use different mechanisms for binding bacteria. Since different cells are responsible for secretion of MUC5B and MUC7, the difference in glycosylation may explain the difference in the functions of each mucin, and their interactions with specific bacteria [142].

In addition to antibacterial activity, MUC5B and MUC7 also possess antiviral and antifungal capabilities [66, 79, 143]. Both mucins possess anti-HIV-1 activity, however, this activity is associated only with MUC5B and MUC7 mucins isolated from HIV-negative individuals [66], while mucins from HIV-positive patients almost did not inhibit the HIV particles [144]. Compared to HIV-negative individuals, saliva from the HIV-positive patients possesses also considerably lower anti-candicidal activity [145]. In immunocompromised individuals, MUC7 lost its carbohydrate sialyl-Lewis^x receptor. The alteration in carbohydrate content results in higher suspectibility to oral diseases [141]. Because the side-chain oligosaccharides of MUC5B and MUC7 are the binding sites for bacteria and viruses [146, 147], the change in the glycosylation pattern associated with HIV infection decreases the ability of MUC5B and MUC7 mucins to aggregate the virus [144].

Thus, the salivary mucins MUC5B and MUC7, like the intestinal MUC2 mucin, perform two opposite functions: on the one hand, they provide nutrients to the pathogens, promoting bacterial colonization, but, on the other hand, they aggregate microorganisms, facilitating clearance of bacteria, viruses or fungi. For these purposes, salivary mucins form two phase-containing defense barrier that can be considered an analog of the two-layer barrier defending epithelial surface in the intestine [128].

In addition to defense oral cavity, saliva performs important functions in mastication. This can explain the increasing interest in the effect of saliva and its multiple components, mucins in particular, on perception and interaction with food products [148, 149]. The interaction between emulsion droplets and salivary proteins, including MUC5B and MUC7 mucins, might be important for the sensory perception of food emulsions [150]. Silletti *et al.* [150] studied the interaction between salivary proteins and the emulsifiers at the oil-water interfaces stabilized by lysozyme and β-lactoglobin. They found that MUC5B and MUC7 interacted different ways with the oil-water surfaces: MUC5B was strongly bound to the lysozyme-stabilized emulsions and moderately to the β-lactoglobin-stabilized emulsion, while MUC7 associated only with the β-lactoglobin-containing emulsions. The interactions of the MUC5B and MUC7 mucins with the oil-water interfaces are probably emulsifier-dependent, in which case the physico-chemical properties of each component would determine the interaction [149, 150]. Following this line of thought, the different nature and content of O-glycan in different species of MUC5B [129, 151], and the different sialic acid and fucose content of the MUC7 isoforms [152] may determine how salivary mucins interact with emulsifiers and other proteins and peptides in saliva [127].

12.2.3. Defensive Functions of MUC5AC and MUC5B Mucins in the Airways

The airway mucus is a complex mixture of water, lipids, glycoconjugates and proteins, including mucins represented mainly by the MUC5AC and MUC5B glycoproteins [153]. It appears that the airway mucus barrier covering epithelial surfaces of the respiratory tract is based on the same principle as that of the gastrointestinal tract and the oral cavity. Like the intestinal mucus gel and saliva, "airway mucus is believed to form a liquid bi-layer: an upper gel layer floats above a lower, more watery sol, or periciliary liquid, layer" [153]. The two layers are probably separated by a thin layer of surfactant that could facilitate the spreading of mucus over the epithelial surface [154]. The functions of each layer are different and specific. The sol layer lubricates the beating cilia; the sticky gel layer traps the inhaled mechanical particles and/or microorganisms and moves them "on the tips of the beating cilia" out of the airways by the mucociliary clearance mechanism [153].

Viscoelasticity of the airway mucus is a crucial property affecting the efficiency of ciliary clearance [155]. The respiratory tract mucus gel is a dynamic structure replaced every 10-20 minutes by newly synthesized components [13]. Several secreted mucins, including MUC5B, MUC5AC, MUC2, MUC7 and MUC8, were detected in airway mucus, with MUC5AC and MUC5B the predominant ones [156]. In normal airways, synthesis of these two mucins is spatially separated [129, 157], and the MUC5B mucin is produced in two differently charged isoforms [130]. These are key factors affecting the physical properties of the respiratory mucus gel. Variations in concentrations of these mucins and their isoforms affect the efficiency of mucociliary clearance in both physiological and pathological conditions. However, conflicting data are reported on this issue by different groups. For example, according to Evans *et al.* [158], the amount of MUC5B in the normal airway secretion is approximately 20% that of MUC5AC mucin, whereas according to Kesimer *et al.* [159] and Ross *et al.* [160], MUC5B is the predominant mucin in the airway secretion (45%) and amount of MUC5AC is much less (21%) [159]. In the asthmatic exudate, the MUC5B and MUC5AC mucins accounted for approximately 86% and 14% the total mucin content, respectively [161]. Furthermore, the low-charged isoform of the MUC5B mucin was the predominant type and comprised 79% of the total gel-forming mucins present in the specimens tested. Interestingly, female steroid hormones were shown to affect lung function in chronic lung diseases associated with abnormal mucociliary clearance and mucus obstruction of the pathological airways [162]. In this context, it is noteworthy that estrogen can up-regulate *MUC5B* gene expression in normal human airway epithelial cells [163]. Taken together, these studies show that estrogen has the potential to increase production of the MUC5B mucin, which, in turn, would regulate ciliary clearance in normal and pathological airways.

The mechanism of mucociliary "escalator" has been thoroughly studied [164, 165], although not on the scale of that of cilia [166, 167]. It has generally been assumed that "mucus flows freely above cilia as a homogeneous layer without any attachment" [168]. However, recent study by Sears and co-workers [168] of mucus dynamics and mucociliary interactions in ciliated human bronchial epithelial cell culture changed this assumption. Using live-cell confocal microscopy in combination with fluorescent beads and exogeneous MUC5B, the authors showed that mucus forms discontinuous layers with temporary attachments to the cilia surface. Importantly, mucus and MUC5B used in this study as a model mucin interacted exclusively with the cilia but not with nonciliated surfaces. The attachments were dynamic in that they were short-lived in the well-ciliated culture. According to the mechanism proposed by the authors for mucociliary clearance, the gel-forming mucins are secreted into the mucus layer, where they display their intrinsic ability to form attachments to cilia (Fig. **3**).

Upon contact of the mucus layer with contaminants (inhaled particle, dust, microorganisms), the mucus collapses around the contaminants followed by "the binding of the collapsed mucus into strands by the cilia and their (the contaminants) transport in a packaged form" [168]. Repeated mucociliary interactions provoked by the collapse of the mucus cause additional secretion of mucus, which, in turn helps "bundle the contaminants and regenerate the layer of mucus in its less sticky form" [168].

Thus, the two layers of the airway mucus work in concert to modulate the mucociliary interactions in response to the inhaled agents. Both airway mucins, MUC5B and MUC5AC, are apparently important for these interactions, although the precise role of each mucin is not known. The proposed new mechanism of mucociliary clearance is considerably simplified as it does not take into account possible interactions of the airway gel-forming mucins with the discarded membrane-bond mucins [159] and many nonmucin proteins in the airway secretion, including trefoil family factors, inflammatory, anti-inflammatory, anti-oxidative and antimicrobial proteins and peptides [17, 169]. Nevertheless, Sears *et al.*'s data

[168] bring a new intrinsic function of the gel-forming mucins: to form attachments to the cilia. The proposed model presumes appropriate modulation of the mucociliary interactions, which may be distorted in pathological conditions, such as cystic fibrosis and asthma, by a permanent shift to more vigorous interaction resulting in abnormal functioning of the mucociliary escalator.

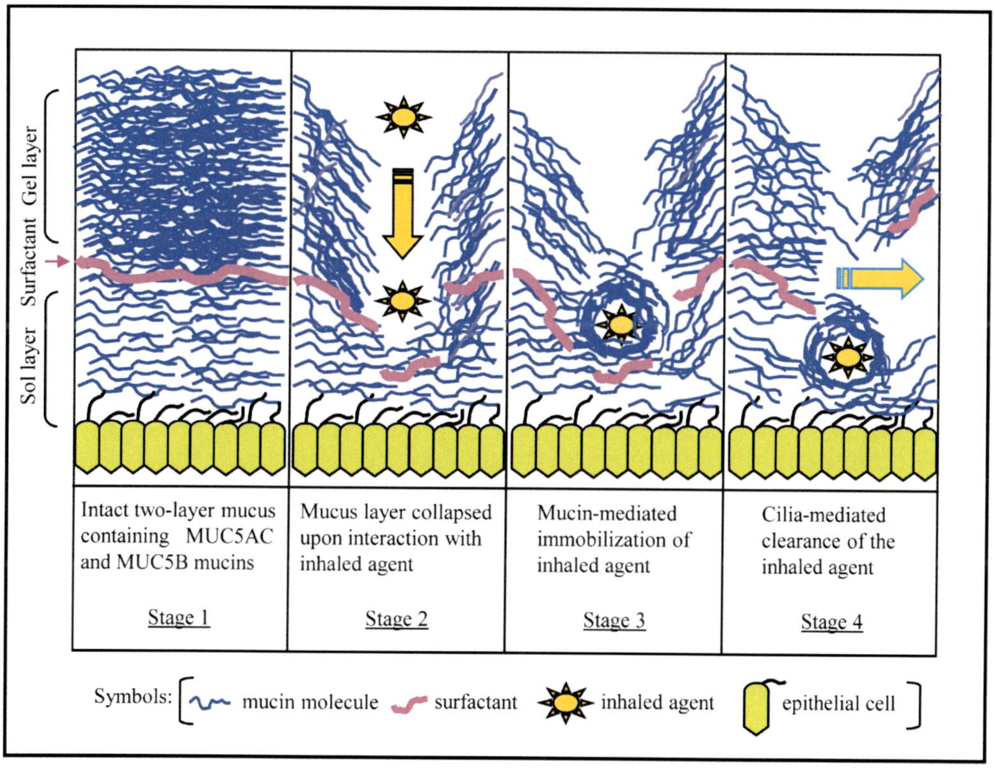

Figure 3: Mechanism of the airway mucociliary "escalator" (based on the data reported in [168]).

This new mechanism of mucociliary clearance [168] attributes the trapping of a microbe solely to collapse of mucus around an inhaled bacterium and the consequent bundling of mucus and the entrapped agent. However, it appears that interaction between the active chemical groups on the bacterial wall and the mucin molecule is necessary for strong attachment of a bacterial particle to the airway mucus. Numerous studies have shown that most of the tested bacteria and viruses express on their surfaces various adhesins and/or lectins that can specifically bind to the carbohydrate moieties attached to the mucin polypeptide backbones [20, 170, 171]. Thus, the carbohydrate epitopes distributed on the airway mucin molecules function as the binding sites for microbes; this would retain bacterial particles in the mucus, which undergoes further collapse around an individual bacterium or bacterial colony and transports the mucus-microbial conglomerate for removal *via* the mucociliary escalator.

12.2.4. MUC5AC and MUC6 - Regulators of Mucus Gel Functions in the Stomach

As follows from the studies discussed in the previous sections, the mucosal defensive barrier covering the intestinal, oral and airway epithelial surfaces is made up of two layers: an inner layer that is firmly adherent to the epithelial surface and an outer layer that is a "loose" movable soft gel floating on the inner layer. However, depending on localization, each gel has a specific mucin composition and posttranslational modifications (O-glycosylation, in particular) that strongly affect the properties of the gel layer. In addition to serving as a physical barrier to the gastric epithelium, a mucus gel in stomach is also a selective molecular filter for nutrients, chemicals, enzymes (*e.g.* hydrolases) and luminal noxious agents. The molecules secreted into stomach mucus gel, including enzymes and antibacterial peptides, serve as "decoys" for different bacterial and viral pathogens [172, 173]. The mucus gel in the stomach is also important for establishing and maintaining a pH gradient from an acidic pH in the gastric lumen to a neutral pH in the mucus closest to the epithelium [174].

The gastric mucus layer is constantly renewed. Its functional efficiency depends on its thickness and stability, which in turn depend on the physical and chemical properties of the gel and secretion of the gel-forming mucins by the underlying cells. The gel thickness depends on both secretion of mucins and degree of mechanical erosion and proteolytic degradation. Thickness of the mucus gel differs in different anatomical parts of the stomach. The thickness of the inner firmly adherent layer in the corpus of the rat stomach is estimated to be about 80 μm and reaches 154 μm in the antrum [107]. The thickness of the outer loosely adherent layer is 109 μm in the corpus and 120 μm in the antrum. Interestingly, mechanical removal of the loosely adherent mucus layer stimulates the accumulation of new mucus in the stomach – a renewal restricted solely to the outer loosely adherent layer. This renewal of the mucus layer is associated with an increase in its thickness, while the thickness of the inner layer is not changed by removal of the outer layer [107]. The data obtained in the numerous studies of human gastric gel thickness – most of which are controversial – usually relate to the inner adherent gel layer. According to Al-Marhoon *et al.* [175], the mean human gastric mucus layer thickness is about 51 μm. Allen *et al.* [176] reported a thickness of 180 μm of the mucus layer adherent to human antral mucosa. The conflicting measurements of gel layer thickness may be due to the difficulty of accurately measuring. As emphasized by Atuma *et al.* [107], who analyzed, among others, the problem of gel measuring, "the possibilities range from a continuum of the same secretion, which reflects a gradual diluting out of the gel the further one proceeds from the mucosa, to two different mucus secretion".

In the stomach, two main gel-forming mucins, MUC5AC and MUC6, are synthesized and secreted by the surface foveolar or pit cells (MUC5AC) and by gland cells (MUC6) [177, 178]; however, the contribution of each mucin to the different mucus layers is not clear. As both mucins are present in gastric mucus, "they may segregate or may combine to form non-covalent mixtures or combine to form covalent oligomers" [108]. According to Ho *et al.* [108], the mucus layer on the gastric surface consists primarily of the MUC5AC mucin molecules extending in layered sheets interspersed with layers of MUC6 mucin molecules. The authors believe that the segregation of this laminated linear arrangement of the two mucin glycoproteins confers increased strength to the mucous layer. Atuma *et al.* [107] consider two options for MUC5AC and MUC6 participation in the mucus gel structure. The first option assumes that the outer and inner layers of the gastric mucus are comprised of different mucins. The second option suggests that the mucus gel layers are comprised of different mixtures of MUC5AC and MUC6, perhaps with different glycosylation patterns in each layer. The latter suggestion is supported by Sawaguchi *et al.* [179], who showed that the rat gastric mucins Muc5ac and Muc6 are indeed differently glycosylated and form laminated gel structure with alternating layers of Muc5ac and Muc6. The alternating laminated array of two mucins in the human gastric mucus layers was demonstrated earlier by Ota and Katsuyama [180]. Functional studies by Ichikawa *et al.* [181, 182] brought indirect evidence of the different roles of MUC5AC and MUC6 in structure formation and functioning of the two mucus gel layers in the stomach. They found that mucins synthesized in corpus surface mucous cells (MUC5AC) and antral gland cells (MUC6) are differently secreted into the mucus gel. The differential regulation of the two mucus layers was recently demonstrated also by Phillipson *et al.* [183], who showed that the luminal administration of prostaglandin E_2 increased the thickness of both layers, while the luminal nitric oxide stimulated accumulation of mucus only in the firmly adherent layer. It is important to emphasize that this type of differential regulation of mucin accumulation by two layers relates to the same mucin, as only MUC5AC was detected in both layers. Interestingly, the *MUC1* gene, which encodes the membrane-bound mucin, also participates in regulation of mucus accumulation in both gastric mucus layers. It is evidenced by the finding that *MUC1* knock-out leads to down-regulation of gastric mucus accumulation. However, Phillipson *et al.* [174] pointed out that the MUC1 mucin does not contribute to the attachment of the inner mucus layer to the epithelial surface. Their research also identified MUC5AC as the only mucin glycoprotein present in both mucus layers, which do not contain even a trace of MUC6. These authors do not, however, rule out the participation of MUC6 in gastric mucus gel formation and functioning, claiming that the inability to detect this mucin in the studied model system may resulted from the specific type of MUC6 posttranslational glycosylation that compromises detection of MUC6 by the antibodies used in the study.

At present, the mechanism of two gel layer formation is not known. What determines transition of the firmly adherent gel layer into a loosely adherent layer? What determines the thickness of each mucus layer? Meanwhile, as stated by Phillipson *et al.* [174], "the thickness of the firmly adherent mucus layer appears preset and variations between individuals are minimal". The proteomic analysis turned up a smaller amount of Muc5ac-derived peptides in the loosely adherent layer in the mouse stomach, suggesting that processing or proteolytic degradation of the inner firmly adherent mucus layer might trigger transition from the firmly adherent layer to the loosely adherent mucus gel [174].

The two mucus layers of the gastric defensive barrier perform different functions. For example, while both layers play an important role in establishing pH gradient from acid pH in the gastric lumen to neutral pH at the epithelial surface (so-called juxtamucosal pH), the loosely adherent mucus layer does not participate in maintaining the neutral juxtamucosal pH [174]. This clearly indicates the importance of the inner mucus layer in protecting the gastric mucosa from damaging acid content in the stomach. On the other hand, the loosely adherent mucus layer is responsible for such important functions as binding of luminal noxious agents, including microorganisms, and defense against chemicals like exogenous food nitrite and endogenous nitric oxide [184].

Multiple studies have been undertaken to determine the role gastric defense barrier plays in *H. pylori* infection, since this Gram-negative bacterium infects more than 50% of the world's population [185] and can cause atrophic gastritis, dysplasia and gastric carcinoma [186]. *H. pylori* colonizes mainly the gastric mucus gel and is rarely detected in the deeper parts of the gastric mucosa. On the other hand, adhesion to the gastric epithelium is a crucial step of a successful *H. pylori* infection, since it provides protection to a bacterium from clearance by liquid flow, peristalsis or shedding of the mucus layer [187]. Adhesion of *H. pylori* to gastric mucus gel is mediated by glycan receptors expressed on the gastric mucins and depends on adhesion molecules present on the bacterial wall [187]. It has been shown that *H. pylori* strains expressing the BabA adhesin bind to the gastric mucus layer through interaction with the Lewis^b (Le^b) antigen expressed on the MUC5AC backbone [64, 65]. *H. pylori* strains that express sialic acid-binding adhesin, SabA, interact with the host mucus gel through binding to another MUC5AC epitope, sialyl-Le^x [188]. However, not every *H. pylori* strain expresses functional BabA or SabA adhesins, suggesting that other bacterial proteins such as the adherence-associated lipoproteins and HopZ protein accomplish the binding functions [189, 190]. The mechanisms of this interaction and corresponding receptors are unknown and remain to be determined.

A fundamental study of the mechanisms mediating *H. pylori* interactions with gastric mucins and the role of pH in these interactions, carried out by Linden *et al.* [191], identified three glycoforms of MUC5AC in gastric mucus gel: neutral mucins, sialomucins, and a mixture of sialo- and sulfomucins [65]. At pH 3, all *H. pylori* strains demonstrated binding to the most charged MUC5AC glycoforms irrespective of host blood group. In contrast, at pH 7.4, only Le^b-binding BabA-positive strains bound to Le^b-positive MUC5AC molecules. Thus, *H. pylori* binding to human gastric mucins occurs by two different mechanisms: one is Le^b-

blood-group-dependent and active at neutral pH, and the other which is based on a charge-dependent interaction with sulfated or sialated mucins at pH 3-4. Interestingly, at pH 7.4, the BabA-positive *H. pylori* bound also to the ectodomain of membrane-bound MUC1 mucin through the H-type-1 structure on the MUC1 molecule, suggesting a possibility for *H. pylory* to activate intracellular processes through signal transduction pathways associated with the intracytoplasmic domain of the MUC1 molecule [191]. This type of signal transduction across the epithelial barrier was previously identified in a cell culture model of *Pseudomonas aeruginosa* (*P. aeruginosa*) infection. What type of interaction (bacteria-MUC5AC or bacteria-MUC1) is preferable depends both on the glycosylation patterns of the respective mucins and on the adhesins present on bacteria. As noted by the authors [65], BabA appears to have a higher avidity for Le^{b}-antigen present on MUC5AC than for the H-type-1 epitope on the MUC1 molecule. In addition, the MUC5AC mucin is present in gastric mucus layer in much larger quantity than MUC1. These two factors may serve as the key triggers of *H. pylori*-MUC5AC interaction. The resultant secreted gel-forming mucins may function as decoys for a microbe, preventing its binding to the membrane-bound mucin molecules. In other words, competition between secreted and cell-associated mucins for binding to *H. pylori* is likely to influence the host-microbe cross-talk.

The ability of *H. pylori* to influence intracellular processes by interaction with MUC1 (and probably with other membrane-bound mucins as well) could explain the aberrant expression of MUC5AC and MUC6 in gastric epithelial cells associated with *H. pylori* infection. Expressions of the MUC5AC and MUC6 mucins in the human stomach are reciprocal processes: gastric epithelial cells specifically express only one of the two mucins. Moreover, *H. pylori* colonizes only that part of mucus gel layer that overlies the cells producing MUC5AC, and never associates with the cells producing MUC6 [192]. Importantly, *H. pylori* can cause a shift in the proportion of cell populations expressing either MUC5AC or MUC6. *H. pylori* infection was shown to affect all major secretory cell populations in the human antrum by down-regulation of *MUC5AC* in parallel with up-regulation of *MUC6* expression [193]. The presented data show that *H. pylori* indeed may change expression of the gastric-specific mucins MUC5AC and MUC6, apparently through MUC1-mediated intracellular signaling. This statement was confirmed by Byrd *et al.* [194] when they found dramatic changes in expression of the *MUC6* gene induced by *H.pylori* infection in humans; the changes were associated with *de novo* expression of the gland type MUC6 mucin in the surface epithelium normally expressing only *MUC5AC*. Kang *et al.* [195], on the other hand, found no difference in *MUC5AC* expression between *H. pylori*-positive and *H. pylori*-negative individuals. Moreover, *MUC6* expression was significantly lower in the *H. pylori*-positive patients than in the *H. pylori*-negative ones. Aberrant expression of MUC6 mucin in foveolar cells was observed in both antrum and body only in the *H. pylori*-positive individuals, however, as underlined by the authors [195], it reverted to normal level following *H. pylori* eradication.

Kang *et al.*'s results [195] were confirmed by some studies [196], and contradicted by others that showed *MUC5AC* down-regulation in *H. pylori*-infected gastric mucosa [193, 194, 197]. Controversial results were reported also regarding *MUC6* down- and up-regulation in *H. pylori*-infected gastric mucosa [192, 198, 199]. These inconsistent results may reflect differences in methodology and/or cohorts of the various studies. Despite these differences, it is clear that *H. pylori* induces alterations in expression of the gastric-specific genes in the epithelial cells of stomach.

The mucus barrier of the stomach is considered the first line of mucosal defense against bacterial invasion. However, at the same time it provides an environment for *H. pylori* colonization [173, 200]. Despite extensive research, it remains a matter of debate whether or not *H. pylori* colonization results in changes in mucus gel structure [201-203]. According to Newton *et al.* [173], *H. pylori* infection in human causes structural changes in the adherent gastric mucus layer, which, however, do not affect the thickness of the mucus gel. The authors noted a significant 18% reduction in the proportion of polymeric gel-forming mucins in the adherent mucus layer in *H. pylori*-positive compared with negative subjects, but these changes were not sufficient to cause a collapse of the mucus barrier. Shimizu *et al.* [204] also found that the gastric surface mucosa gel layer inhabited by *H. pylori* exhibited marked derangement of the multilaminated structure, with fragmentation of both mucin-containing layers. Importantly, after eradication of *H. pylori,* the mucus layers regained the laminated structure. According to Dekker [205], the "shift in expression levels of individual mucins that constitute the mucus layer may easily explain the disturbance of the gastric mucus layer" and the mucin functions associated with the physiology of the defensive mucus gel barrier.

Thus, the gel-forming gastric mucins MUC5AC and MUC6 as well as nongastric MUC2 and MUC5B mucins and the soluble mucin MUC7 perform several important functions in the gastrointestinal tract. They determine properties of the mucus gel and regulate processes associated with the defense of epithelial surfaces against multiple noxious agents, including infectious microorganisms and chemicals. Further studies are needed to establish the precise role of each mucins noted above as regulators of the defensive processes occurred in the stomach and intestine.

12.2.5. Functions of the Secereted Mucins in the Eye

The tear film plays an important role in protecting the underlying cells and tissues from damage by microbial, chemical, enzymatic and mechanical attacks [15]. Located on the external ocular surface, it provides primary defense against external stress factors [206-209]. The nonhomogeneous structure of the tear film relates it to the mucus gel lining the epithelial surfaces in the stomach, intestine and airways, while it has its own specific features and properties. The unique structure of the tear film is composed of three layers: the outermost lipid layer, the middle aqueous layer, and the innermost mucus layer containing ocular mucins [206] (Fig. **4**). According to recent estimations, the tear film is approximately 35-45 μm thick [206-209].

The lipid layer, composed of the oil produced by the meibomian glands, provides a smooth optical surface for light refraction, reduces the evaporation of tears, and prevents contamination of the underlying layers by debris [208]. Spreading of the lipids over the ocular surface decreases the surface tension of the tear film, which in turn draws water into the tear film, increasing the film thickness and allowing the spreading of lipids during blinking [208].

In addition to water, the aqueous layer contains antibacterial factors – including immunoglobulins, lysozyme, lactoferrin – that participate in defense of the ocular surface [210, 211]. It also contains a soluble MUC7 mucin, which decreases the surface tension, enhances the spread and coherence of the aqueous layer, and contributes to the viscosity of the tear film [206, 212]. The aqueous layer participates in lubrication and protection of the ocular surface and provides nutrients to the cornea [210, 211].

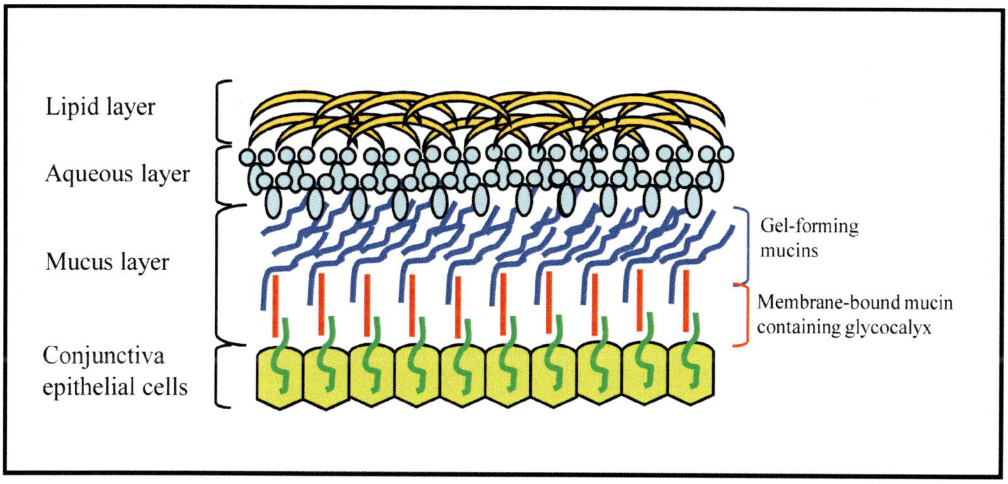

Figure 4: Composition of tear film (based on the data reported in [206-209])

The innermost mucus layer is composed of mucins, immunoglobulins, urea, salt, glucose, leukocytes, enzymes and cellular debris [213]. This layer lubricates and protects the cornea by specific interaction with the corneal epithelium. On the one hand, the hydrophilic mucus layer, which contains mucin molecules densely decorated by the highly hydrophilic O-glycans, neutralizes the repulsion forces on the surface of the hydrophobic corneal epithelium, facilitating the spread of the aqueous layer evenly over the ocular surface. On the other hand, with its mucin content, the mucus layer is not tightly attached to the epithelial surface but rather attaches to the glycocalyx. This type of interaction allows the mucus layer to move freely across the cornea, to spread evenly over the epithelial cells, and to protect them from damage during blinking [214].

The secretory mucins in the mucus layer cause an increase in surface pressure *via* lateral reorganization of the lipids and alteration of surface viscoelastic properties [207, 215]. All gel-forming mucins (MUC2-MUC6, MUC19), two soluble mucins (MUC7, MUC8), and several membrane-bound mucins (MUC1, MUC4, MUC16) are expressed in the eye by conjunctival goblet cells, lacrimal gland and nasolacrimal ducts [216-221]. Among the gel-forming mucins expressed in the ocular tissues, the MUC5AC glycoprotein is predominant [222]. Watanabe [223] made an important observation about cooperation between the gel-forming and membrane-bound mucins in facilitating physiological functions of the tear film. He showed that the mucus layer containing gel-forming mucins spreads over the glycocalyx containing membrane-bound mucins derived from the corneal epithelial cells and conjunctival epithelium, and the interaction between the two types of mucins promotes generation of the overlying aqueous layer of the tear film. This cooperation was demonstrated by comparing of tear film spreading in normal individuals and patients with corneal ulcer: the tear film could not spread over the ulcer site because the area was short of the transmembrane mucins required for spreading of the gel-forming mucins. Thus, cooperation of the transmembrane mucins with the gel-forming ocular mucins is necessary for the spread of the tear film [223].

Secreted and membrane-bound mucins perform different functions in the context of ocular physiology. The secreted mucins perform clearance of allergens, pathogens and debris; lubrication; and antimicrobial tasks [216]. The membrane-bound mucins participate in boundary lubrication; formation of apical cell surface barrier; and osmosensing [216]. However, there is evidence that such a strict "division of labour" between the two groups of mucins is not justified. In the first place, both gel-forming and membrane-bound mucins cooperate in maintaining the hydrophilic character of the wet-surfaced epithelia by utilizing their carbohydrate moieties [216]. In addition, gel-forming mucins may also contribute to osmosensing, since the expression of the rat ocular surface gel-forming Muc5ac mucin is down-regulated by hyper-osmotic stress [224], implicating the Muc5ac mucin in the maintenance of physiologically optimal osmotic pressure.

According to Davidson and Kuonen [206], the tear film performs seven main functions based largely on the properties of the ocular mucins: 1) maintaining a smooth surface for light refraction; 2) lubricating the eyelids, conjunctiva and cornea; 3) supplying the cornea with nutrients; 4) transporting metabolic by-products from the corneal surface; 5) providing white blood cells to the cornea and conjunctiva; 6) removing foreign materials from the cornea and conjunctiva; and 7) defending the ocular surface from pathogens *via* specific and nonspecific antimicrobial substances. As noted by Mantelli and Argueso [216], "traditionally, mucin function at the ocular surface has been ascribed to secreted gel-forming mucins acting as lubricating agents and clearing molecules". Indeed, the ability of a gel-forming mucin to retain water through multiple O-glycans densely distributed on the mucin backbone allows mucin-mediated "lubrication of the ocular surface and reduction of shear stress during blinking or rubbing" [216].

In this context, it is of interest to look at how the different MUC5AC glycoforms [207, 225] contribute to properties of the lubricating gel. As known, MUC5AC molecules are present as glycoforms of low, medium and high charge species. Sialylation predominates in the moderately charged and sulfation in the highly charged species. Sialylated structure can modulate immune cell adhesion to the lubricating gel and can interact with adhesion sites on ocular pathogen. Sulfation may have a role in blocking bacterial degradation of the oligosaccharide side-chain. Each glycoform is characterised by different conformational properties and polymer diameters, both of which can determine the properties of the lubricating tear film gel [207, 225]. Of note, the O-glycans of the tear film mucins are relatively small compared with O-glycans found in gastrointestinal and airway secretions, perhaps reflecting the specific requirements of the ocular surface [226-229]. In contrast to the highly polymeric mucin molecules forming lubricating gels in the gastrointestinal tract and airways, the lubricating gel component of the tear film is comprised of the low molecular weight mucin monomers [230, 231]. This reflects the unique requirements of the ocular surface for a transparent refracting medium enabling vision. The presented data show that MUC5AC present in the ocular lubrication gel acts as a regulator of tear film functions.

The ability of ocular secreted mucins to carry out clearance is based mainly on the hydrodynamic properties of the tear film and the ability of a mucin carbohydrate moiety to bind specific epitopes expressed by a pathogen on its outer membrane [232, 233]. It is increasingly evident that the relationships between host mucins and bacteria are both active and antagonistic: on the one hand, ocular mucins (MUC5AC, in particular) inhibit growth of commensal bacteria on the ocular surface, while on the other hand, bacteria display considerable mucolytic activity, including both proteolysis and glycolysis [231]. Fleiszig *et al.* [232, 234] showed that the ocular mucin binds *P. aeruginosa* in a specific manner by modulating accessibility to the epithelial glycocalyx and protecting corneal epithelial cells from bacterial invasion. Further studies showed that *P. aeruginosa* binds preferentially to the tear film gel-forming mucin containing α2-6 sialic acids [235]. The sialic acid-containing glycans expressed on the mucin backbone may function in two ways: they may serve as decoys for bacteria by binding to the bacterial receptors, thereby inhibiting ability of bacteria to interact with epithelial cells; or the glycans may competitively block the attachment of bacterial pathogens by binding to specific adhesins expressed on the epithelial cells, thereby preventing bacteria from binding to the cell. Ocular mucins also bind to viral particles through sialic acid, providing antiviral defense [206].

In addition to protection of the epithelial cells from bacteria and viruses, the tear film also defends the underlying epithelium and tissues from noxious chemicals that may come into contact with the ocular surface. However, the resistance to chemicals has a definite limit and can be broken by toxic compounds. Cigarette smoke is one of the toxic agents whose effect on the ocular surface and tear film has been thoroughly studied [236, 237]. It was shown that even brief passive exposure (5 min) to cigarette smoke in healthy nonsmokers adversely affects ocular surface physiology, as evidenced by an increase in tear inflammatory cytokines and lipid peroxidation products and a decrease in mucosal defense resulting in tear instability and damage to the ocular surface epithelia [236].

Apparently, different toxic chemicals activate different defense mechanisms operating on the ocular surfaces. Lipids, which easily bind to mucins, may be one of the effective instruments protecting the corneal and conjuctival epithelium from aggressive chemicals. More than 30 years ago Gong *et al.* [238] showed that the binding of lipids to gastric mucins shields the mucin molecules from attack by oxygen radicals. Recently, Setala *et al.* [239] reported that MUC5AC mucin in human tear fluid directly interacts with the phospholipid transfer protein (PLTP) secreted from the lacrimal gland, resulting in formation of the PLTP/MUC5AC complex, which in turn helps maintain homeostasis on the anterior tear lipid film. These results suggest a significant role for PLTP in lipid transfer in human tears and in preventing instability of the lipid film. Apparently, PLTP functions as a vehicle for lipid molecules on their way to shield mucin polypeptides [238], thereby protecting the tear film mucins and the underlying epithelial layer from external noxious agents.

It was recently shown that the human tear fluid contains a large number (about 50 reported by Li *et al.* [240] or even 500 identified by De Souza *et al.* [241]) of different proteins. The intimate contact of these proteins with ocular mucins suggests multiple interactions between them. Future studies should reveal new functionally active complexes between mucins and various proteins and extend our understanding of the functional potentials of the ocular mucin glycoproteins.

12.3. MUCIN-MEDIATED PROTECTION AGAINST CHEMICAL INJURY

The role of the gel-forming mucins in protecting the epithelial cells is not limited to lubrication, hydration and antimicrobial activity. These mucins exhibit other important defense functions, including protection of the underlying epithelium from chemical assaults associated with acidic conditions and proteolytic enzymes in the stomach and intestine [242], inhaled irritants in airways [243, 244], and toxic compounds on the ocular surface [236, 237]. The studies of Loomes *et al.* [242] showed that glycosylated tandem repeat-containing domain of the MUC6 mucin possesses anti-acidic properties that protect the gastric epithelium from low pH in the stomach. Moreover, the carbohydrate side-chains of this domain defend the MUC6 molecule itself against proteolytic enzyme attack in the gastric environment.

The main defensive actions of cells exposed to chemicals entail activation of mucin secretion, which, owing to its hydrophilic nature, increases the volume of fluid at the site of exposure. Borchers *et al.* [243] reported that rats exposed to acrolein developed increased numbers of the airway epithelial mucus cells synthesizing and secreting increased amounts of the Muc5ac mucin. Interestingly, the *MUC5AC* and *MUC5B* genes encoding the main airway mucins react differently to the chemical "aggression": the *MUC5AC* gene exhibits enhanced transcription, while *MUC5B* expression remains unchanged [245]. It seems that MUC5AC mucin responds to acute assault of irritants in the airway, while MUC5B mucin participates mainly in chronic airway diseases. Irritant gases, including cigarette smoke, sulfur dioxide and ammonia vapor, are notable in inducing secretion of mucin by goblet cells [244, 246, 247]. A rapid secretory response to inhaled irritants assists airway defense by providing a protective mucus/fluid layer against further damage [247]. Recently, the damaging effect on mucus epithelia of medications containing irritants such as camphor, eucalyptus oil and menthol was reported [248]. The combined effect of these chemicals tested in ferret trachea specimens induced 63% increase in mucin secretion and 72% increase in water content. The presented data show that the secreted mucins possess properties that enable a rapid defensive reaction on the epithelial surface to damaging effects of chemicals.

However, secretory gel-forming mucins were recently shown to actually promote the negative effects of some chemical substances [249]. This ability of the gel-forming mucins is associated with the new property of these glycoproteins, discovered only two years ago, to function as catalyst of organic chemical reactions [250]. It appears that a new page in the study of functional potentials of the gel-forming mucins has been opened by discoveries made by M. Gozin and his colleagues from Tel Aviv University [249, 250]. These investigators showed for the first time that mucin glycoproteins can promote organic chemical reactions: under physiological conditions two animal gel-forming mucins, bovine submandibular mucin (Muc5b) and porcine gastric mucin (Muc5ac), accelerated the rate of fatty acid ester hydrolysis up to 377 times relative to the control reaction [250]. The "remarkable and unprecedented" property of the mucin to promote Diels-Adler reaction (DA-cycloaddition) was also discovered [250]. As noted by the authors, this ability of mucins to accelerate organic chemical reactions (carboxylic ester hydrolysis and DA carbon-carbon bond-forming reactions) provides a new and unique example of natural nonenzymatic proteins capable of promoting reactions of hydrophobic compounds in aqueous solution.

The bovine Muc5b gel-forming mucin possesses one more remarkable property: the ability to bind and solubilize highly hydrophobic materials in physiological solution [249]. While the mucus gel covering all mucosal surfaces in organisms provides an effective physical and chemical shield against a wide range of toxic materials [251], mucins may act also as factors that strengthen natural toxicity of some hydrophobic materials. Drug *et al.* [249] showed for the first time that complexation of polyaromatic toxic hydrocarbons with gel-forming mucins (bovine submandibular mucin Muc5b) enhanced bioavailability of the hydrophobic materials and led to formation of previously unknown materials with enhanced toxicity. The authors [249] concluded that the studied mucin has an impressive and unique capability to bind and solubilize water-insoluble materials in physiological solution. Using an internalization assay, they demonstrated the bioavailability, toxicity and membrane-penetration capability of the solubilized chemical agents *in vitro* and *in vivo*. This assay unambiguously showed that the enhanced bioavailability of the solubilized hydrocarbon results directly from using mucin as a carrier agent. The exploited mucin increases the membrane-penetrating ability of the insoluble agent, resulting in better accumulation of the toxic compound in the cytoplasm of the tested cells [249]. These studies [249, 250] describe the new unexpected functions of the gel-forming mucins, ability to regulate chemical reactions, thereby highlighting once again the multifunctional nature of the mucin glycoproteins.

12.4. FUNCTIONS OF SECRETED MUCINS IN TUMORIGENESIS

12.4.1. General Characteristics of Tumor Suppressors and Tumor Promoters

In addition to functions described above, mucins also play an important role in tumorigenesis. Recent studies showed that mucin expression profiles and tumor phenotypes are tightly associated with the clinico-pathological findings and pathogenesis of a particular malignancy [252-255]. Although the importance of mucins in carcinogenesis is indisputable, the precise role of a specific mucin in development of a specific tumor is often difficult to determine. This difficulty stems from the bivalent nature of the mucins: the same mucin glycoprotein promotes development of a tumor in one tissue and hampers initiation and/or progression of a neoplasm in another [87, 254, 256, 257]. To better understand the role of a protein in tumorigenesis, one must to find out whether the protein is a tumor suppressor or a tumor promoter. Theoretically, any gene can be considered a tumor suppressor gene (TSG), if its expression is associated with suppression of tumor development [258, 259], or it may be defined as an oncogene, if its product activates and/or promote tumorigenesis [260]. TSGs and oncogenes perform antagonistic functions: members of the first group inhibit cell growth and survival, while members of the second group promote cell growth and survival [261]. Knowledge of the properties of tumor suppressors and oncoproteins is important for evaluation of the possible oncogenic and tumor suppressing potentials of mucins.

One of the fundamental properties of tumor suppressors and tumor promoters (oncoproteins) is their multifunctionality. In addition to functions associated with suppression or initiation and progression of carcinogenic process, tumor suppressors and oncoproteins may perform other functions as well. For example, they may function as enzymes (*e.g.* phosphatase), as regulators of proteolysis, as elements of the ubiquitin degradation system (*e.g.* E3 ligase), and more [262, 263]. Numerous studies have demonstrated that, depending on cell type and conditions, the same protein can function as a tumor suppressor or an oncoprotein [262, 264-266]. The tumor suppressors and oncoproteins belong to various protein families inluding secreted and membrane-bound mucins [87, 267, 268]. TSGs and oncogenes are located on practically the entire spectrum of human chromosomes, including the 11p15 locus that contains a cluster of gel-forming mucin genes [259, 269-272]. It appears that TSGs and cellular proto-oncogenes are critical "gatekeepers" of tissue differentiation [273]. Mutations in these genes associated with the loss of functions lead to significant dedifferentiation, a typical feature of tumor development.

TSGs and oncogenes can affect different links in the tumorigenic pathways. According to Kinzler and Vogelstein [274-276], TSGs and oncogenes can be classified into three groups based on their functions: "gatekeepers", "caretakers" and "landscapers" (Fig. **5**). The "gatekeeper" genes control cell cycle and proliferation. The "caretaker" genes contribute to maintaining genome integrity, thereby preventing aberrations in the "gatekeepers" and the consequent development of cancer. The "landscaper" genes modulate the microenvironment of tumor cells by direct or indirect regulation of extracellular matrix proteins, adhesion proteins and secreted growth factors. According to this clasification, tumor suppressive or oncogenic activity of a protein is directed to a specific link in a tumorigenic pathway.

Proteins in each group are equally important for cell homeostasis and for tumor development since they participate both in physiological processes and in tumor development. If the "gatekeeper" genes specific for a particular cell lineage function according to the normal cell program, the cells and tissues exhibit normal patterns of differentiation specific for a given tissue. Structural (mutation) or functional (expression) changes in the "gatekeeper" genes may initiate the latent tumorigenic process, which, however, can result in tumor development only if the "caretaker" genes function abnormally. If they function in accord with the normal the cell program, the potential tumor cells will remain in a latent phase of tumorigenesis without progression to a growing tumor. Even if a tumor occurs, its progression, invasiveness and metastatic dissemination will not be possible if the "landscaper" genes function normally. However, aberrations in expression of these genes can switch on the "green light" to tumor progression. Defining a given protein – in our case the mucins – according to Kinzler's and Vogelstein's classification [274-276], might clarify the role of a given protein in the carcinogenic process and indicate its possible target in a tumorigenic pathway. According to Shrestha-Bhattarai and Rangneker [277], the tumor suppressors belonging to "gatekeepers" are responsible for regulation of cell cycle, cell proliferation, differentiation and apoptosis, and those of "caretaker" group - for preservation of the genome integrity.

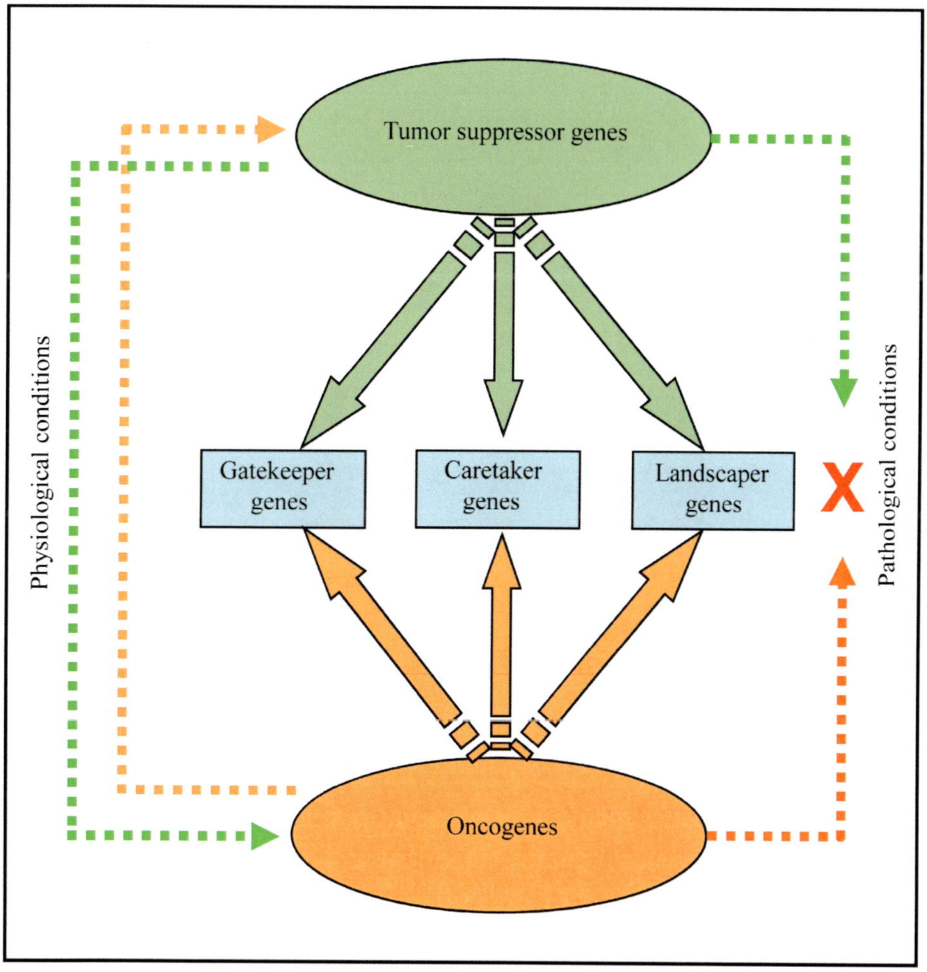

Figure 5: Relationship between tumor suppressor genes and oncogenes (based on the data reported in [274-276])

Several mechanisms may disturb the functions of TSGs and proto-oncogens. The abnormalities of tumor suppressor genes can be characterized by a "loss-of-function" mediated by *cis*-inactivation (gene mutation) and/or by *trans*-inactivation (abnormal expression), resulting in tumor initiation, growth and progression [277, 278]. According to Knudson's "two hit hypothesis", the germline mutation on one allele of a TSG may destroy the antitumor potential of a tumor suppressor gene and predispose to tumor formation, although, the somatic mutation in the second allele must occur during an individual's lifetime for initiation of tumor alterations [258]. As mentioned above, mutations in genes responsible for cell proliferation and survival are not the only factors that may activate tumorigenic process; changes in expression of these genes may also initiate and advance tumor development. Among the mechanisms that may change gene expression, "epigenetic alterations, which, by definition, comprise mitotically and meiotically heritable changes in gene expression that are not caused by changes in the primary DNA sequence, are increasingly being recognized for their role in carcinogenesis" [261].

Although gene amplification, translocation, loss of the whole or part of a chromosome, an intragenic deletion, and point mutations can inactivate TSGs and/or activate oncogenes, it is clear that epigenetic silencing of TSGs also plays an important role in tumorigenesis, being considered the functional equivalent of mutations and deletions [261, 279, 280]. Epigenetic modifications, including DNA methylation and covalent modification of histones, are reversible but stable alterations that dramatically change gene expression and consequently the cell properties and functions. Hypermethylation of the promoter regions of various TSGs causes transcriptional silencing of these genes and, as a result, loss of tumor suppression functions and appearance of potential tumor cells (changes directed against "gatekeeper" function). On the other hand, hypomethylation of the regulatory DNA sequences may activate transcription of proto-oncogenes leading to genomic instability and tumor development (changes directed against "caretaker" function) followed by malignant cell metastasis (changes directed against "landscaper" function) [281]. With these considerations in mind, we will analyze the functions of the gel-forming mucins in tumor development.

12.4.2. MUC2 Mucin as Regulator of Tumorigenesis

As known, expression of the gel-forming mucins is tissue specific [2-5, 282-284]. For example, under physiological conditions, MUC2 is expressed mainly in the intestinal epithelium and not in the stomach and pancreas, while MUC5AC is expressed in the surface epithelium of the stomach and airway epithelium but not in the intestine and pancreas [2-5, 282-284]. The MUC6 mucin is expressed in the submucosal glands in the stomach [285] but not in the normal prostate, breast and airway tissues [9, 286, 287]. It appears that these mucins contribute to preservation of a specific differentiation pattern in the tissues where they are normally expressed. This supposition is strengthened by the fact that when a given mucin gene is expressed *de novo* in the tissue where it is normally not expressed, the tissue acquires features of the tissues normally expressing that mucin gene. For example, when the *MUC2* gene is expressed in gastric or breast carcinomas, the tumors display the mucinous type of differentiation specific to the intestinal epithelium normally expressing MUC2 mucin [256, 286, 288, 289]. Thus, in terms of their ability to preserve a specific type of differentiation, the gel-forming mucins may be related to the group of "gatekeeper" tumor suppressors.

Can a gel-forming mucin suppress development of malignant tumors? This critical question received an affirmative answer, at least with regard to the mouse Muc2 mucin [87]. Velcich and co-workers showed that *Muc2*$^{-/-}$ knock-out mouse develops intestinal carcinoma at a rate much higher than that of the wild type animals [87]. The loss of *Muc2* expression in these mice was associated with increased proliferation and survival of intestinal epithelial cells. Notably, the loss of *Muc2* expression was accompanied by significant up-regulation of the proto-oncogene *c-Myc*. Moreover, these mice developed adenomas that progressed to invasive adenocarcinomas in all parts of the intestine. These results indicate that *Muc2* behaves like a tumor suppressor gene. In a recent paper from Velcich's group [290], the authors noted that the "lack of Muc2 can be an independent initiating event". Moreover, they demonstrated that the "tumors from *Muc2*$^{-/-}$ mice do not have alterations of Wnt/β-catenin/Tcf4 signaling" which is involved in the intestinal tumorigenesis initiated by mutations in *APC* gene, another tumor suppressor gene active in intestinal epithelial cells. The authors concluded that tumors detected in *Muc2*$^{-/-}$ mice developed through an inflammation-related pathway that is distinct from the mechanisms of tumorigenesis in mice carrying *APC* mutations.

In line with Velcich *et al.*'s findings in intestine [87], Iwase *et al.* [291] reported a correlation between reduced *MUC2* expression and development and progression of colorectal neoplasms. Other investigators [292, 293] found that the decrease in the *MUC2* expression in colorectal cancer occurred in parallel with the decrease or loss of the expression of other tumor suppressor genes including *p53* and *E-cadherin* [292, 293]. Further studies showed that expression of the MUC2 mucin is significantly decreased in hyperplastic polyp - low grade dysplasia - high grade dysplasia - colorectal adeno-carcinoma sequence [294], and is associated with increased risk of invasion to mucosa [295]. Importantly, these data were obtained not only in the artificial knock-out model system, but also in patients with different types and stages of colorectal neoplasms.

Additional evidence of the active role of the *Muc2* gene in suppression of intestinal tumorigenesis was obtained by Fijneman *et al.* [296], who found that over-expression of the secretory phosphatase *Pla2g2a* gene, one of the intestinal TSGs, prevented carcinogenesis in the *Muc2*-deficient mice. Deficiency of both Muc2 mucin and Pla2g2a phosphatase caused up-regulation of genes involved in cell cycle regulation, inflammation and oncogenicity. Importantly, prevention of tumor development in *Muc2*$^{-/-}$ mice by over-expression of Pla2g2a phosphtase was associated with significant modulation of a number of genes involved in intestinal lipid and energy metabolism, cell signaling, nuclear transactivation, apoptosis, immune responses and inflammation. Modulation in expression of these genes could compensate for and correct the abnormal processes originally induced by deficiency of Muc2 mucin. Alterations in expression of 99 genes, including *Nfkbil1, Reg2, Limk1, Hoxb9, Ctrb1* and *Ela2a*, were shown to be unique to the *Muc2*$^{-/-}$ mice background [296]. Notably, the proto-oncogene *c-Myc*, whose expression is significantly up-regulated by *Muc2* deficiency [87], is not changed by expression of *Pla2g2a*, indicating that not all alterations induced by cancellation of *Muc2* expression in the *Muc2*-knock-out mice can be corrected by *Pla2g2a* gene expression. Of note, the relatively low level of *c-Myc* expression in the wild type mice and its up-regulation in the *Muc2*$^{-/-}$ mice may indicate a certain balance between the tumor suppression activity of the MUC2 mucin and the oncogenic potential of the c-Myc oncoprotein in the wild type mice, and the disturbance of this balance in the *Muc2*-deficient animals.

Collectively, the presented data suggest that the mouse Muc2 mucin, and apparently the human MUC2 as well, direct their tumor suppression activities against the genes responsible for initiation of tumorigenic cascade. Hence, *MUC2* and *Muc2* genes share common functions with the "gatekeeper" tumor suppression genes. The relationship between the *MUC2* gene and the "caretaker" and "landscaper" genes in intestinal carcinogenesis is not known and requires further study.

In addition to the already discussed evidence of the tumor suppression potential of the MUC2 glycoprotein [87, 290, 291], other characteristics common to the MUC2 mucin and known tumor suppressors also indicate a link between the genes encoding gel-forming mucins and tumor suppressor genes. One of the mechanisms by which tumor suppressor genes keep tumor growth under control is apoptosis. TSGs induce apoptosis in response to different stress factors that may activate oncogenic processes. Many tumor suppressors, including p53, PTEN, APC and PML, exert tumor suppressor activity by inducing apoptosis [297-300]. Several mechanisms can be utilized by tumor suppressors to induce apoptosis: the tumor suppressor p53 activates the pro-apoptotic regulator Bax [301, 302], induces the apoptosis-specific Fas receptor [303], and/or promotes cytochrome C release [259]; the c-Myc protein also expresses its antitumorigenic potential through regulation of cytochrome C release [304].

Like tumor suppressor p53, MUC2 also uses several ways to induce apoptosis. One is by activation of the c-Myc-mediated apoptotic pathway [87], and another way is by direct interaction with the cells condemned to apoptosis. Ishida *et al.* [305] recently showed that MUC2 mucin induces apoptosis of monocyte-derived dendritic cells through direct interaction of α2,6-sialic acid-containing glycans attached to the MUC2 tandem repeats and sialic acid-binding Ig-like lectin (Siglec 3) expressed on the dendritic cells [306]. Although this study does not identify all events occurred after the indicated MUC2-Siglec 3 interaction leading to apoptosis of malignant cells, it brings evidence of the ability of the MUC2 mucin to induce cell death, a property characteristic of many tumor suppressors. Is the ability to induce apoptosis a common feature of the gel-forming mucin? Recently, a link was found between apoptosis and MUC5AC mucin expression [307]: *MUC5AC* knock-down induced by the MUC5AC si-RNA protected airway epithelial cells against LPS-induced apoptosis [307]. Further studies are needed to determine the connection between other gel-forming as well as soluble mucins and apoptosis in the context of tumorigenesis.

Consistent with the concept of protein multifunctionality [1], the MUC2 mucin possesses both ability to induce and to prevent apoptosis. The latter property was evidenced by the ability of the MUC2 mucin to prevent the profound effects of rotaviruses on intestinal homeostasis which they exert by inducing apoptosis and increasing epithelial cell turnover [308, 309]. Interestingly, MUC2, secreted by the intestinal goblet cells, inhibits rotavirus infection through direct interaction of the MUC2-associated glycan containing sialic acid with the epitopes expressed on the viral particle, thereby preventing virus-cell interaction [310] and, as a result, virus-induced apoptosis [309].

Epigenetic mode of transcriptional regulation is another common feature of tumor suppressor genes and genes encoding gel-forming mucins [311-317]. Transcription of the *MUC2* gene is highly regulated by epigenetic mechanisms [311-314]. It has been shown that the decreased or completely blocked expression of *MUC2* in colorectal carcinoma is strictly dependent on the level of *MUC2* promoter methylation in human colorectal carcinoma cells and metastasis [311]. In contrast, the *de novo* expression of *MUC2* in cells that do not normally express the MUC2 mucin is triggered by promoter demethylation [311, 318]. Thus, while the epigenetic regulation of *MUC2* gene expression does not by itself support its belonging to TSGs, it does point to their possible relationship.

As noted above, depending on cell type, tissue and conditions, tumor suppressor genes may function as oncogenes [262, 266, 319-324]. As underscored by Paige [278], TSGs "are genes whose function is not always tumour suppressive, but which can also act to promote tumorigenesis". This bivalent nature of tumor suppressor genes and oncogenes is seen, for example, in *TGFβ1*, a recognized tumor suppressor gene that is typically expressed in normal cells and is often over-expressed in tumor cells [325-327]. TGFβ1 inhibits formation of benign skin tumors (tumor suppressive effect), but enhances progression to the invasive stage of spindle carcinomas (oncogenic effect) in transgenic mice [328]. The TGFβ1-mediated pathway was shown to be necessary for carcinoma cell invasiveness and metastasis [329]. Numerous studies showed that TGFβ is able to promote and to suppress tumorigenesis [327, 330, 331]. Apparently, the overall effect of TGFβ on a tumor reflects "the imbalance between these two opposing influences" [278]. It seems that the products of TSGs have different effects on different neoplasms at different phases of tumor progression, inhibiting tumor growth in one type of tumors but promoting it in others. Since most of the studied proteins are known to be multifunctional, "genes that can be both tumor suppressive and tumor promoting may be more common" than one could expect [278].

Like many TSGs, the *MUC2* gene has also the oncogenic potential usually displayed in organs that normally do not express the MUC2 mucin, namely in pancreas and stomach. However, in some cases the tumor-promoting properties of MUC2 may be realized in the intestinal epithelium as well. Mucinous colorectal carcinoma, a relatively rare type of colorectal cancer, follows a distinct tumorigenic pathway that includes over-expression of *MUC2* and development of the aggressive metastatic variant of carcinoma, suggesting a partial or total loss of the tumor suppressive properties of the MUC2 mucin and activation of its tumor-promoting functions [332-335]. Interestingly, increased expression of MUC2 mucin in the appendiceal mucinous adenocarcinoma correlates with increased expression of proteins with oncogenic potentials (β-catenin, cyclinD1, Ki67, NFκB, VEGF). Of note, the more aggressive behavior of malignant neoplasms in the appendix was associated with a relative increase in expression of the TSGs including *p53* and *E-cadherin*, which might indicate re-programming of their functions from tumor suppression to tumor promotion [336].

The relationships between the tumor suppression and tumor promotion potentials of MUC2 mucin in most cases of colorectal tumorigenesis can not be characterized as'black and white'. Carcinogenesis in the colorectal epithelial cells is a complex process including both genetic and epigenetic alterations often associated with activation of oncogenes or oncogenic potentials of TSGs and inactivation of tumor suppression genes [337, 338]. It appears that these relationships are reciprocal: the gradual weakening of the tumor suppressive activity is associated with the gradual activation of the onco-promotive functions, and *vice versa*. In line with the tumor suppressor effects on the different links of the tumorigenic cascade, the oncogenic activity of the MUC2 mucin also may affect different stages of colorectal carcinogenesis. At present, there are no data indicating the direct targets attacted by MUC2 oncomucin in colorectal carcinogenesis. Their discovery will hopefully point the way to new therapeutic modalities.

The link between the *de novo* expression of *MUC2* and development of malignant tumors in the stomach is well documented. As noted by Leteurte *et al.* [256], the expression of *MUC2* was frequently observed in all subtypes of gastric carcinoma whatever the classification; higher expression of this mucin was significantly correlated with the mucinous subtype and was associated with poor prognosis. Interestingly, in this type of gastric carcinoma the expression of MUC2 mucin, which normally is not expressed by gastric mucosa, reached 81-83%. At the same time, the expression of the MUC5AC and MUC6 glycoproteins, the native markers of normal gastric differentiation that may be considered as the "gatekeepers" of the gastric type differentiation, was only about 33% and 19%, respectively [256]. Such dynamics in expression of the "oncogenic" *MUC2* and "tumor suppressive" *MUC5AC* and *MUC6* reflects the reciprocal relationships between pro- and anti-tumorigenic activities of mucins in gastric epithelium during carcinogenesis. Our suggestion that MUC5AC may function as a tumor suppressor in the gastric epithelium is supported by findings of Baldus *et al.* [257] who showed that high expression of *MUC5AC* is significantly associated with a better prognosis, while loss of *MUC5AC* expression correlates with dedifferentiation phenomena [289].

The high expression of *MUC2* in the intestinal (mucinous) type gastric epithelial neoplasias correlates well with the data of Park *et al.* [339] showing up-regulation of *CDX2* in this type of gastric neoplasia. CDX2 is a homeobox transcription factor that controls intestinal differentiation by transcriptional activation of the intestine-specific genes including *MUC2* [340, 341]. However, some reports describing *CDX2* expression in gastric neoplasias contradict the results reported by Park *et al.* [339]. For example, Liu *et al.* [342] observed a lower *CDX2* expression in the high-grade gastric adenomas compared with the low-grade adenomas and suggested that mucin expression in gastric epithelial displasias is not associated with *CDX2* expression. On the other hand, Rugge *et al.* [343] found *CDX2* expressed in all types of gastric epithelial dysplasia irrespective of the grade. In support of Park *et al.'* [339] findings, Kim *et al.* [344] showed that increased ectopic expression of intestinal mucin MUC2 and decreased expression of gastric mucin MUC5AC were associated with increased expression of CDX2 in gastric cancers characterized by intestinal mucin phenotype. The large number of reports with conflicting results reflects the debate still in progress about whether

MUC2 and other gel-forming mucins can function as both tumor suppressors and oncopromoters depending on the type of tumor. Future studies must answer this important question. At this point, the available data allows suggestion that secreted mucins possess both anti- and pro-oncogenic potentials.

Besides the stomach, ectopic expression of *MUC2* has been observed in malignant tumors of several other organs including pancreas, breast, prostate and ovary [345-348]. In pancreas, *de novo* expression of *MUC2* was reported only in the intraductal papillary mucinous neoplasia (IPMN) [349, 350]. Neither normal pancreas, pancreatic intraepithelial neoplasias (PanIN) nor ductal adenocarcinomas (DAC) express *MUC2* [255, 351]. Importantly, only the intestinal type of IPMNs expresses MUC2, whereas the gastric type does not express the mucin. Various studies revealed deregulation of the mucin genes in IPMNs, which might contribute to neoplastic progression of these tumors [352]. However, it is difficult to evaluate the impact of MUC2 on IPMN development, as many conflicting results have been published. In some studies the high expression of *MUC2* in IPMNs was associated with low invasion and metastatic potential [345, 350, 353], whereas in other investigations the highly MUC2-positive IPMNs demonstrated a high malignant index and high level of invasiveness [354]. Further studies are needed to understand the relationship between tumor suppressive and tumor promotive potentials of MUC2 glycoprotein in pancreatic neoplasms.

The MUC2 mucin is also expressed in prostate neoplasms, but the data relating to its functions are not uniform. According to Ho *et al.* [355] and Cozzi *et al.* [287], MUC2 is not expressed by normal or malignant prostate, while Zhang *et al.* [356] observed MUC2 in 72% of primary prostate carcinoma specimens and 100% of metastatic prostate carcinoma biopsy specimens. As noted by Osunkoya *et al.* [357] "mucinous adenocarcinoma of the prostate shows diffuse expression of MUC2, a known tumor suppressor, which is not present either in normal prostate or the majority of conventional adenocarcinomas of this organ". Adsay *et al.* [358] consider MUC2 expression highly specific for mucinous (colloid) carcinomas in general, and in the pancreas and breast in particular, with "a key role in the morphogenesis" of these tumors. In prostate tumors, it serves as a biological factor that determines a relatively slow growth of colloid carcinomas [357],. In accordance with its tumor suppressive function documented in colon

[87], “MUC2 may have a role in diverting the process of carcinogenesis toward more favorable prognostic pathways” also in other organs, including prostate [357]. In other words, even being ectopically *de novo* expressed, MUC2 mucin preserves, at least in some tumors, its native tumor suppressive activity that weakens the carcinogenic process induced by yet unidentified pathways.

In addition to the tumor suppression function displayed by MUC2 mucin in prostate cancer, MUC2 exhibits an important function(s) associated with acquisition of androgen independence by prostate cancer cells. Legrier *et al.* [347] showed that transition from the androgen-dependent prostate carcinoma cells to the hormone-independent state is associated with increased expression of MUC2 mucin. The precise role of MUC2 in this process is not known but recent data point to several possible scenarios of the mucin's participation in this process. Schroder [359] noted that among the several mechanisms that mediate androgen independence of prostate cancer cells, activation of oncogenes and inactivation of tumor suppression genes are of special significance. The mechanisms regulating apoptosis are also involved in the transition of prostate cancer cells from the hormone-dependent to the hormone-independent state. The MUC2 mucin appears to have all the functions necessary for participation in such a transition: tumor suppressive activity, oncogenic activity and pro-apoptotic properties [87, 305]. Besides, MUC2 induces over-expression of *c-Myc* oncogene, which may participate in development of androgen-independence as an activated oncogene. Although these considerations do not explain how MUC2 glycoprotein operates in hormone escape by prostate cancer cells, they may help design future investigations of the phenomenon.

The expression of MUC2 in the mucinous carcinomas of pancreas, prostate, ovary and breast suggests a common function of the MUC2 mucin in these tumors [286, 345, 346, 360]. It has been shown that MUC2 expressed in mucinous breast cancer prevents tumor invasion by acting as a barrier that interferes with the spread of the malignant cells [346, 360, 361]. These observations, which could explain the lower aggressiveness of the mucinous carcinomas of the breast compared with the intraductal carcinomas [286, 346], point to a connection between the barrier function and the tumor suppressive activity of the MUC2 mucin; by operating as a component of the mucus barrier that surrounds a tumor, MUC2 behaves in breast carcinogenesis as a tumor suppressive "landscaper". By surrounding of the tumor, MUC2 mucin diminishes the access of the immunocompetent cells and molecules to mucinous carcinoma, therefore functioning as an immunoblocker that allows tumor growth. Thus, one of the possibility for MUC2 to fulfill its oncogenic potential is interfering with the immune system.

It appears that whether MUC2 will operate as a tumor suppressor or a tumor promoter depends on whether it is secreted into extracellular space or distributed intracellularly, although other factors are also important. It is known that oncogenic functions of the MUC2 mucin are intensively displayed in the invasive intraductal carcinomas of the breast. As noted by Walsh *et al.* [362], all breast ductal carcinomas *in situ* and the adjacent invasive carcinoma samples demonstrated MUC2 expression only in the cytoplasm with no membrane or extracellular distribution of the mucin, although MUC2 is known to be a secreted protein. It is possible that intracellular expression of MUC2 in ductal breast carcinoma is determined by alterations in its structure and biochemical properties. If this is the case, then not only localization but also biochemical parameters regulate expression of the functional potentials of the MUC2 molecule in the context of tumor development. The presence of MUC2 in intraductal breast carcinomas is significantly associated with shorter disease-free intervals and correlates with lymph node metastasis, suggesting a connection between expression of the MUC2 oncogenic potentials and intracellular expression of the MUC2 glycoprotein [362, 363]. These considerations may apply not only to breast cancer, but also to malignant tumors of other organs and tissues ectopically expressing MUC2.

12.4.3. Oncogenic and Anti-Oncogenic Functions of MUC5AC Mucin

Like the MUC2 glycoprotein, the MUC5AC mucin also possesses both tumor suppressive and oncogenic activities. In many cases of gastric cancer, MUC5AC, a native gastric mucosa-specific mucin and one of the key factors responsible for the gastric-type differentiation, demonstrates decrease of its tumor suppressor activity. This was evidenced by its significant down-regulation in the gastric carcinoma accompanied by reduction in gastric differentiation, positive correlation with poor prognosis [257, 364], and correlation with the loss of expression of other tumor suppression genes including *Smad4*, *E-cadherin*, *MGMT*, *FHIT* and *PTEN* [365]. At the same time, other studies showed that gastric carcinogenesis was often associated with over-expression of the MUC5AC mucin [366]. Interestingly, under physiological conditions the MUC5AC mucin is expressed exclusively by the goblet cells, while in gastric carcinogenesis there is significant expression of the mucin by nongoblet cells as well [366]. As noted by Zoghbi *et al.* [366], the *de novo* expression of *MUC5AC* in nongoblet cells predicts histological progression of gastric pre-neoplastic lesions to the advanced stage of gastric cancer, suggesting expression of the MUC5AC oncogenic potential at the very early stage of gastric tumorigenesis. Strictly speaking, these data do not contradict the view that MUC5AC mucin, being expressed under natural conditions in goblet cells, fulfills its tumor suppressive functions, whereas its expression in nongoblet cells (unnatural environment) activates the oncogenic potential of the mucin. These results do conflict, however, with those reported by Baldus *et al.* [289], who stressed that although some gastric cancers do indeed display high expression of MUC5AC, this over-expression is associated with better prognosis, demonstrating the tumor suppression and not the tumor oncogenic potential of MUC5AC mucin.Thus, MUC5AC mucin functions as a tumor suppressor in some gastric carcinomas and a tumor promoter in others. It appears that gastric carcinomas are a heterogeneous group of neoplasms that needs to be subdivided into more homogeneous subgroups.

The relationship between the tumor suppressive and tumor promotive activities of the *MUC5AC* gene is apparently subject to myriad influences. Such a situation is typical for many TSGs and oncogenes [272, 367]. Tumor development is a complex process of cooperation and counteraction of numerous different genes [367]. Recent studies showed that expression of cellular mucins and tumor phenotype-specific proteins are associated with the coordinated expression and/or inhibition of different genes, including tissue-specific genes and cell-cycle regulating genes such as *cyclin D1*, *cyclin E*, *p53*, *p21* and *p27* [319, 320, 322, 368-370]. In the gastric cancers associated with the gastric phenotype, poor outcomes, distinct histological type of differentiation and specific genetic alterations [371, 372], *MUC5AC* over-expression correlates with increased expression of *cyclin A* and *p53*, whereas in the intestinal phenotype *MUC2* over-expression correlates with *p27* over-expression and decreased expression of *cyclin A* [322]. Interestingly, both tumor phenotypes demonstrate high expression of tumor suppressors p53 and p27, which in the gastric carcinoma specimens display their oncogenic rather than tumor suppressive potential. An analogous transition of oncogenic and tumor suppressive potentials of MUC5AC and MUC2 mucins might occur in the gastric cancers specimens examined in the studies cited above [322]. Results conflicting with this view have also been reported [373-375] – not surprising in light of the multifactorial character of tumor development and differences between individuals. In some patients, the resulting vector of the multiple gene activities tilts the balance in favor of tumor suppression while in others - in favor of tumor promotion. Importantly, however, both the tumor suppressive and oncogenic functions are associated with the same mucin molecule, and, depending on the specific conditions, one or the other will come to the fore.

Generally, normal gastric mucosa expresses native gastric mucin MUC5AC at a high level that gradually decreases along the adenoma - early cancer - advanced carcinoma sequence [374, 376, 377]. In developing tumors with such dynamics of MUC5AC expression, the declining activity of the mucin's tumor suppressive function is apparent. In tumors with aberrant or *de novo* expression of *MUC5AC* correlated with tumor progression, activation of the MUC5AC oncogenic function is evident. The most representative examples of tumors demonstrating activation of the MUC5AC oncogenic potential are the malignant neoplasms of pancreas. High *de novo* expression of *MUC5AC* is observed in practically all types of pancreatic cancer, including all grades of intraepithelial lesions (PanIN), all types of IPNMs, pancreatic ductal carcinomas (PDC), and intrahepatic cholangiocarcinomas (HCC), while the normal pancreas does not express MUC5AC [255, 345, 378, 379]. Importantly, MUC5AC expression was found to be an early and *de novo* event in pre-cancerous PanIN lesions [255, 352]. As noted by Yonezawa *et al.* [345], "MUC5AC is useful in detecting pancreatic neoplastic lesions from the early stage, but it is not effective in differentiating the different histological types". This observation suggests that *MUC5AC* may function in pancreatic carcinogenesis as a "gatekeeper" oncogene.

The strongest evidence of the MUC5AC oncogenic potential comes from knock-down experiments conducted by Hoshi *et al.* [254] who showed that MUC5AC si-RNA significantly reduces the tumorigenicity and tumor growth of si-MUC5AC cells compared with si-mock cells. Knock-down of *MUC5AC* reduces the capability of pancreatic cancer cells for adhesion and invasion. Moreover, expression of genes associated with adhesion to and invasion in the extracellular matrix, including genes encoding integrins, matrix metalloproteinase (MMP)-3 and vascular endothelial growth factor (VEGF), are down-regulated in the *MUC5AC* knock-down cells. These data suggest that MUC5AC mucin contributes to the invasive potential of the cancer cells by enhancing expression of the integrin, MMP-3 and VEGF, and by activating the Erk pathway. Other studies [292, 380] showed that the up-regulated MUC5AC mucin is expressed in the intercellular junction between cultured pancreatic ductal carcinoma cells and interferes with the membrane localization of E-cadherin, leading to a decrease of E-cadherin-dependent cell-cell adhesion and promoting migration and invasion of cancer cells. These data show that MUC5AC is involved not only in initiation of pancreatic carcinogenesis as a "gatekeeper", but contributes also to the late stages of tumor progression, functioning as a "landscaper".

In agreement with these data, Aloysius *et al.* [381] found a correlation between over-expression of *MUC5AC* in periampullary cancer and tumor recurrence. The results of Yamada *et al.* [382] also suggest possession of tumor suppressive and tumor promotive functions by the MUC5AC mucin. These authors showed that, like many TSGs and oncogenes, the MUC5AC gene is regulated by epigenetic mechanisms including CpG methylation of the *MUC5AC* promoter and histone H3-K9 modification. *MUC5AC* transcriptional activity is also controlled by transcription factors, in particular by the Kruppel-like zinc-finger transcription factor GL11, which up-regulates expression of *MUC5AC* in pancreatic cancer cells (PCC), thereby promoting migration and invasion of PCC by MUC5AC-mediated attenuation of E-cadherin [380].

Interaction of the MUC5AC mucin with E-cadherin appears to play an important and specific role in carcinogenesis. The mechanism of this interaction was disclosed in a study by Truant *et al.* [122]. These investigators found that expression of MUC5AC in colon carcinoma HT-29 cells is associated with the loss of E-cadherin function and the acquisition of invasive behavior. It has been established that the gel-forming mucin MUC5AC cooperates with the membrane-bound mucin MUC1 and heparin sulfate proteoglycans to form a biological complex that inhibits E-cadherin-mediated cell-cell adhesion, thereby promoting cancer cell invasion. Apparently, the newly formed complex combines the oncogenic potentials of each component: MUC1 mucin is a well known oncoprotein [253, 383, 384], and MUC5AC displays tumor promoting activity at least at the late invasive stage of colon carcinogenesis [122]. Importantly, *de novo* expression of the *MUC5AC* gene was also found at early stages of colon cancer development [294, 385, 386]. In summary, the MUC5AC mucin appears to participate in different stages of colorectal tumor development through activation of mucin's oncogenic potential.

The airway epithelium is a native "dwelling place" for the MUC5AC glycoprotein, as this gel-forming mucin is expressed mainly by airway goblet cells and serves as a goblet cell-specific differentiation marker [9]. In the other words, it functions as a "gatekeeper" of airway specific cell differentiation, and as such it may possess tumor suppressive activity. Yu *et al.* [387] presented the data showing that this is the case. It was shown that 74% of the non-small cell lung adenocarcinoma (NSCLA) cells demonstrated complete block of MUC5AC expression, implicating the role of MUC5AC glycoprotein in tumor suppression. Only 26% of the NSCLA samples studied expressed MUC5AC, although each MUC5AC-positive cell expressed the mucin at a high level. These 26% of MUC5AC-positive NSCLA samples apparently represent a separate population of NSCLAs in which the carcinogenic process is associated with MUC5AC's oncogenic functions. This conjecture is supported by finding that expression of MUC5AC highly correlates with sialyl-Le^x antigen expression, post-operative distant metastasis, post-operative recurrence, and shorter overall survival [388, 389]. Altogether, these data underscore the bivalent potential of the MUC5AC mucin, however, further studies of the role MUC5AC plays in lung carcinogenesis are needed.

Several mechanisms are involved in metastatic process. Selectins expressed on the surfaces of platelets, leukocytes and vascular endothelial cells are the elements of one of them [390-392]. As noted above, another mechanism entails the ability of MUC5AC mucin to interact with E-cadherin with the consequent decrease in intercellular adhesion of lung tumor cells resulting in an increase in cell migration and invasiveness [122]. A new mechanism has been recently discovered by showing that the MUC5AC glycoprotein may be involved in the lung cancer metastastatic process through the aquaporin 5 (AQP5)-mediated pathway [393-396]. It has been established that in addition to its osmotic fluid transport function [393], aquaporins participate also in migration and metastasis of tumor cells of different origins including lung cancer cells [394-396]. Over-expression of AQP5 induces up-regulation of MUC5AC production with a parallel increase in the malignant potential of AQP5-transfected cells. On the other hand, silencing of the *AQP5* gene is induces a decrease of *MUC5AC* expression associated with significant decrease in the migration and invasion capabilities of the lung cancer cells [397-399]. Interestingly, in parallel with *MUC5AC* over-expression, AQP5 up-regulates also the oncogenes *PCNA* and *c-myc*, in part by the EGFR/ERK/p38/MAPK signaling pathway [398-400]. The participation of the MUC5AC mucin in metastatic process evidenced by the mechanism that suggests direct non-covalent (electrostatic) interaction of the sialyl-Le^{x} antigen attached to the MUC5AC polypeptide expressed by the tumor cells with selectins expressed on the surface of platelets, leukocytes and vascular endothelial cells [390-392].

According to these and other data [400-402], whatever the activating mechanism, the MUC5AC mucin glycoprotein bears a definite oncogenic potential and promotes tumor progression by participating in the metastatic process and invasion. The MUC5AC mucin most probably participates in lung tumorigenesis as a "landscaper", although other functions must be also considered. For example, according to Evans and Koo [158], the up-regulation of *MUC5AC* in lung cancer occurs "in order to clear cells that accumulate in hyper-proliferative and stressed regions of tumors". They base this view on the fact that mucous cells localize to the hypoxic cores of many tumors, including lung tumors, where HIF-1, one of the critical mediators of *MUC5AC* gene expression [403], is up-regulated [158]. More research is needed to clarify the role of the MUC5AC mucin in lung carcinogenesis.

New and important functions of the mucins, in general, and of the MUC5AC gel-forming mucin, in particular, have emerged from recent studies showing possible cooperation of the mucins with the trefoil factor family proteins: TFF1, TFF2 and TFF3. These proteins, which play an active role in signal transduction and regulation of gene transcription [404-407], are expressed almost exclusively by the cells producing secretory mucins [92, 408-411], and their coordinated expression with mucins has been documented [411]. Moreover, direct intermolecular interactions have been established [92, 412, 413] between trefoil proteins and mucins that may result in alterations of the physical properties of mucins in solution, affecting the viscosity and elasticity of mucin solution and ability of the solution to be transformed into a gel-like structure [414]. All these changes affect lubrication and the protective function of mucins on different epithelial surfaces including intestinal mucosa, airway epithelial lining and ocular conjunctiva. Trefoil proteins interact with mucins through vWF-like domains of mucin proteins [91]. The trefoil and mucin glycoproteins appear to be conformationally suitable to each other, allowing interaction between them. As noted by Paulsen and Berry [415], "the dimeric structure of TFF1 and TFF3 is ideally suited to form an entangled network with MUC5AC". Such an intimate connection of MUC5AC and other gel-forming mucins with the trefoil proteins [192, 411] suggests that mucins may participate in processes regulated by trefoil proteins through activation of multiple signaling pathways [404-407]. Among these processes are cell migration, cell restitution, morphogenesis, apoptosis and immune responses [416-419]. Obviously, interactions of the mucin glycoproteins with the trefoil proteins significantly extend the functional potentials of mucin glycoprotein, including those associated with tumorigenesis.

12.4.4. MUC6 Mucin as Oncopromoter and Tumor Suppressor

As follows from the previous sections, MUC2 and MUC5AC mucins display tumor suppressive and tumor promotive functions. Do other gel-forming mucins also possess anti-tumorigenic and oncogenic properties? Despite the importance of the question, limited information is available on this issue. Nevertheless, study of other gel-forming mucins suggests that, like MUC2 and MUC5AC, they also exhibit anti-tumorogenic and tumor promoting activities. Several studies demonstrated the tumor suppressive potential of the MUC6 mucin [420-422]. *MUC6* expression is typical of gastric submucosal glands under physiological conditions. Some studies claim that the tight association of *MUC6* and *MUC5AC* expression with differentiation of the normal gastric epithelium implies alterations in their expression in neoplasms as proof of their role as tumor suppressors in regard to malignant transformation [423, 424]. Zheng *et al.*'s comprehensive study [420] showed that expression of MUC6 is significantly lower in gastric carcinoma than in normal gastric epithelium or benign gastric adenomas. The down-regulation of *MUC6* was related to malignant transformation and correlated with tumor size, degree of lymphatic and venous invasion, and lymph node metastasis, suggesting that down-regulation of *MUC6* expression was closely linked to gastric carcinoma progression. In agreement with Zheng *et al.* [420], Seki *et al.* [421]

found lack of MUC6 mucin expression in highly malignant, gastric intramucosal signet ring cell carcinoma compared to some levels of MUC6 expression in the less malignant carcinoma cells.

Taking into account that gastric adenoma expresses the MUC6 mucin at a rate comparable to that in the normal gastric mucosa, and that the decrease in MUC6 expression is observed only in the advanced gastric carcinomas, one may assume that MUC6 glycoprotein functions as a tumor suppressor of the late molecular events in gastric carcinogenesis. This places *MUC6* in the category of "landscaper" tumor suppressor genes, whose decreased expression determines the late events in the tumor progression pathway. Recently, this point of view was confirmed *in vitro* by experiments conducted by Leir and Harris [422] who showed that over-expression of MUC6 glycoprotein significantly inhibited adhesion of pancreatic, colorectal and breast tumor cells to matrix proteins. These results suggest that MUC6 may inhibit invasion of tumor cells through the basement membrane and slow the development of infiltrating tumor.

In addition to gastric cancer, MUC6 also has an important impact on lung carcinogenesis. The level of *MUC6* expression increases significantly with the progression from lung atypical adenomatous hyperplasia (9%) through nonmucinous bronchioloalveolar carcinoma (48%) to adenocarcinoma with mixed subtypes (52-71%) and mucinous bronchioloalveolar carcinoma (100%), with significant correlation between up-regulation of MUC6 expression and tumor size [425]. Although the precise role of the MUC6 mucin in lung carcinogenesis is unknown, its association with tumor development is obvious. The *MUC6* knock-out experiments are needed to evaluate the role MUC6 mucin plays in carcinogenesis of the lung.

12.5. FUNCTIONS OF SOLUBLE MUCINS

The group of soluble mucins consists of three glycoproteins that differ from each other both structurally and functionally. Two members of this group, MUC7 and MUC9, have been relatively well studied and important information regarding their functional potentials has been obtained. In contrast, the MUC8 mucin has been less studied. Its full-length primary amino acid sequence has not been established, its domain structure has not been fully evaluated, and its functional potentials have only been scantily explored. The expression of MUC8 has been detected in some cells and tissues both in physiological and pathological conditions, implying functions in various basic cell processes. However, more studies are clearly needed on this glycoprotein before discussion of its role in regulation of cell functions will be possible.

12.5.1. MUC7 Mucin Glycoprotein - Multifunctional Cell Regulator

The MUC7 mucin performs a number of functions analogous to those of other secreted mucins: permeable barrier function, lubrication of mucosal surfaces, and protection against chemical and mechanical damage and micrrobial invasion. It performs specific functions in the oral cavity by participation in mastication, speaking, food bolus formation, and prevention of oral microbial infections [426]. Like other secreted mucins, MUC7 is a multidomain glycoprotein. Each domain has its own unique biological function(s) [427], reflecting the multifunctional nature of the MUC7 mucin.

Different mechanisms are involved in the expression of the multiple functions of the MUC7 mucin in the oral cavity. Protection of oral epithelial and dental surfaces by pellicle formation is thought to be one of the main functions of the salivary MUC7 mucins. Indeed, Fisher *et al.* [428] showed that MG2 (MUC7) is a major component of tooth pellicle formed *in vivo* and *in vitro*. However, Al-Hashimi and Levine [429] could not detect MG2 in any pellicle produced *in vivo*. Failure to detect the native MG2 (MUC7) in tooth pellicle may be due to its proteolytical degradation [430]. According to Nieuw Amerongen and co-workers [431, 432], the role of MUC7 in tooth physiology is to regulate metabolic homeostasis and protect the tooth against decalcification by organic acids.

Besides defense against chemical insults, MUC7 mucin provides defense against microbial infection [426, 433-435]. Importantly, MUC7 interacts with a variety of bacteria *via* carbohydrate receptors and its binding with bacteria is highly selective. For example, MUC7 binds to *Streptococcus gordonii* (*S. gordonii*) and *Streptococcus oralis*, but not to *Streptococcus salivarius* or *Streptococcus sobrinus* [426, 436]. The intact MUC7 molecule makes use of several tools to bind bacteria or viruses, one of which is cysteine residues. *In vitro* binding of purified MUC7 (MG2) to *Streptococcus mutans* could be abolished by reduction and/or alkylation, indicating that cysteine residues are involved in this interaction [143]. The MUC7 mucin can interact simultaneously with different bacteria thanks to the multiple binding sites on its molecule specific for different bacterial proteins. MUC7 interacts with several proteins expressed by *S. gordonii* [437], including the two elongation factors EF-Tu and EF-G, the product of the *hppA* gene, the oligopeptide-binding lipoprotein, enolase-α and RNA polymerase. Some of these proteins are expressed while others (*e.g.* enolase-α) are located on the bacterial surface [438] pointing to ability of MUC7 to function both intra- and extracellularly. Interestingly, binding of *Helicobacter pylori* to human salivary mucins including MUC7 occurs through four adhesion routes: 1) the blood-group antigen-binding adhesin; 2) the sialic acid-binding adhesin; 3) a novel saliva-binding adhesin, and 4) a charge/low pH-dependent mechanism [439]. The MUC7 mucin can bind bacterial proteins by a protein-to-protein interaction (*e.g.* binding of *Pseudomonas aeruginosa*) [440, 441], or through carbohydrate moieties of the mucin molecule and bacterial receptors. For instance, binding of the MUC7 glycoprotein to *Streptococcus sanguis* occurs through the oligosaccharide bridge NeuAcα(2-3)Galβ(1-3)GalNAc [442]. MUC7 also makes use of its oligosaccharide elements for aggregation of the human HIV-1 virus [66]. MUC7 interacts with L-selectin expressed on the neutrophil leukocytes, thereby taking part in inflammation [147].

As the above studies clearly show, carbohydrates are of great importance in the interaction of the MUC7 mucin with bacteria, viruses [66], and neutophils [147], but of less or no importance in the binding of MUC7 to other partners such as lactoferrin. MUC7 (MG2) and lactoferrin form a heterotypic complex *in vitro* and *in vivo* through polypeptide-polypeptide interaction in which the Ala-Leu-Leu-Cys motif present in lactoferrin molecule plays a major role [443]. The formation of complexes between different proteins and MUC7 may have significant functional consequences that lead to modulation of the functions associated with both proteins. For example, MUC7 modifies such properties of lactoferrin as ferric ion binding and antineoplastic and anti-inflammatory potentials [444]. It also protects lactoferrin from proteolytic attack. On the other hand, lactoferrin affects the antibacterial and tooth protective functions of MUC7 [75, 428, 445]. The MUC7-lactoferrin complex also modulates turnover of these proteins, extending the time the two proteins are active in the oral environment [443].

The aforementioned functions of the MUC7 mucin are associated with the native MUC7 molecule; other functions are displayed only after its fragmentation. The autonomic functioning of various parts of the MUC7 molecule has been extensively studied by several laboratories. Gururaja *et al.* [446] were the first to discover that the N-terminal region of the MUC7 apomucin contains a short region of a 15-residue sequence highly homologous to histatine 5 known as an anticandidal agent. This finding suggested potential candidacidal activity associated with the MUC7 molecule, and indeed such activity was detected by Liu *et al.* [447] and confirmed by others [79, 121, 448, 449]. This property of the MUC7 mucin appears to be conserved in evolution, as its structure is similar to that of rat and mouse homologs of human MUC7 [450, 451]. All three mucins have a similar composition of basic amino acids and high pI (greater than 10.0), suggesting that the fungicidal activity is dependent on electrostatic interactions between the N-terminal region of the MUC7 molecule and the negatively charged groups of phospholipids in the candidal cell membrane [78]. Wei *et al.* [449] showed that fungicidal activity of synthetic 12-mer peptide, corresponding to the sequence present in the MUC7 domain D1, is inhibited by Ca^{2+}, but is not affected by Na^{+}, K^{+}, or Mg^{2+}. This peptide exerts optimum antifungal activity at neutral or slightly basic pH, in line with the proposed mechanism [78]. Fluorescence microscopy conducted by the same group [121] revealed the ability of the synthetic MUC7 20-mer peptide to cross the fungal cell membrane and accumulate inside the cell. In addition, cysteine residues present in the MUC7 derived peptides could be substituted with alanine withou affecting the antifungal activity of the peptides [76].

Comparison of the data obtained in *in vitro* and *in vivo* studies of the MUC7 candicidal potentials disclosed some contradictions. *In vitro* assays suggested that the N-terminal region of MUC7 possesses candidacidal activity, while according to *in vivo* studies, purified native MUC7 mucin lacked such activity [78, 79]. This contradiction was resolved when the whole saliva of most subjects was found to contain` a 20 kDa polypeptide immunoreactive with anti-MUC7 antibody, indicating the MUC7 origin of the polypeptide [75]. Importantly, this polypeptide molecule was not present in the submandibular and sublingual glandular extracts, suggesting generation of the MUC7-derived 20 kDa fragment after secretion of the native MUC7 mucin into the oral cavity. Collectively, these data suggest that cleavage of the MUC7 mucin *in vivo* produces fragment(s) with microbicidal properties. According to Liu *et al.* [75], such cleavage may represent a novel mechanism of host defense. This suggestion is in agreement with the findings by Gururaja and co-workers [78] that incubation of the MUC7 mucin with the whole saliva leads to fragmentation of the native MUC7 molecule. It must be noted that while both the antibacterial and antifungal properties of MUC7 are associated with the same N-terminal region of the mucin molecule, they apparently adopt different mechanisms: the antibacterial activity requires active participation of the cysteine residues [447], whereas the antifungal activity depends on the net positive charge of the specific region in domain-1 [76].

Analysis of the functions associated with domain D1 of the MUC7 molecule clearly demonstrates the multifunctionality of this mucin. However, functional potentials of this mucin are not restricted only to this N-terminal domain. MUC7 also carries out lubrication and hydration functions associated mainly with the central domain (domain D3) of the MUC7 glycoprotein containing six highly glycosylated tandem repeats that determine the hydrophilic properties of the MUC7 mucin. This domain also facilitates binding of MUC7 to bacterial cells and promotes their colonization. Numerous studies showed that the MUC7 mucin performs two opposite functions simultaneously: it facilitates bacterial colonization on tooth surface and enhances clearance of the colonized bacteria from this surface [426, 452]. The mechanism by which the same mucin molecule performs two different functions has been investigated. According to Antonyraj *et al.* [452], a native full-length MUC7 mucin molecule participates in bacterial adsorbtion and colonization, but before the colonized organisms turn into pathogens, native mucin protein undergoes proteolytic degradation by saliva protease(s), resulting in release of proline-rich tandem repeat-containing peptides which, like other natural bactenecins with poly-L-proline structure, possess antimicrobial activity [453]. In contrast to the naturally occurring cationic peptides, which usually contain arginine and lysine amino acids interspersed with proline residues [454], the MUC7 tandem repeats comprise only serine, threonine, alanine and proline. Despite the absence of cationic amino acids, tandem repeats do exhibit strong antimicrobial activity. According to Antonyraj *et al.* [452], this bactericidal activity results from the ability of the liberated tandem repeat-containing fragment to adopt a specific structure, which folds into bundles of trimeric or hexameric poly-L-proline II helices stabilized by hydrophobic effects and hydrogen bonds between the strands. Apparently, such molecular configuration allows *in vivo* expression of antimicrobal activity embedded in the structure of the liberated MUC7 fragment, observed earlier in the experimental systems [452].

The presented examples demonstrate both the overt and covert functional potentials of the MUC7 mucin. Importantly, the MUC7 derivatives resulting from posttranslational modifications of the primary apomucin, in general, and by proteolysis, in particular, may harbor other hidden functions. It appears that MUC7 multifunctionality has not been fully evaluated and many functions of this mucin remain to be discovered. Using the yeast two-hybrid system, Bruno *et al.* [455] showed that human mucin MUC7 may interact with several structurally diverse proteins secreted into saliva by submandibular gland. Interestingly, the N-terminal region, comprised of domains D1 and D2, could form complexes with such diverse proteins as amylase, acidic proline-rich protein 2 (PRP2), basic proline-rich protein 3 (PRP3), lacrimal proline-rich protein 4 (PRP4), statherin and histatin 1, whereas the C-terminal region, including domains D4 and D5, did not interact with any of the proteins found in saliva. It is possible that the specific structure of the C-terminal region hampers interaction of this part of MUC7 molecule with the proteins secreted by submandibular gland [455]. Further studies are needed to clarify this issue. The functional implications of the detected protein-protein interactions [455] are not known at present, but may be significant. Many proline-rich proteins have been shown to interact with a variety of signaling proteins containing SH3 and/or WW domains. Interaction of MUC7 with proline-rich proteins may connect the MUC7 mucin to numerous signaling pathways [456-461]. Moreover, proline-rich ligands of the signaling proteins frequently adopt a polyproline type II helix [456], a structure predicted to occur in salivary proline-rich proteins [462] and found in domains D2 and D3 of the MUC7 mucin [447]. Interaction of MUC7 with structurally and functionally different proteins (*e.g.* amylase, statherin, PRPs and histatin-1) implies modulations of functions specific for each participant.

In conclusion, it appears that the number of functional potentials of the MUC7 mucin hidden in its chemical structure is higher than the number of overt functions tested and proven experimentally. The available information shows that MUC7 may regulate many cell functions by itself and *via* interaction with protein-partners. The expression of MUC7 observed in different tissues in physiological as well as in pathological conditions, including malignant transformation, suggests additional functions. Investigations based on the new methodologies may uncover those functional potentials of the MUC7 mucin.

12.5.2. Functional Potentials of MUC9/OGP Mucin

Numerous reports have demonstrated the importance of the MUC9/OGP mucin functional activity, especially in processes of biological reproduction. It interacts directly with gametes and with the early embryo in the oviduct, thereby modulating their properties and functions [463, 464]. Several *in vitro* studies have shown that in mammals MUC9/OGP glycoprotein binds to oocytes, spermatozoa and embryos, and that these interactions have positive effects on sperm capacitation, motility and viability, sperm-ovum binding, ovum penetration, fertilization, and early embryo development. By these interactions, the MUC9/OGP mucin decreases polyspermy and increases cleavage rate of embryos and the number of embryos reaching the blastocyst stage [464-470]. Importantly, studies of fertilization and early embryo development showed that MUC9/OGP glycoprotein (oviductin) may function differently in different species. For instance, it binds to both bovine and hamster sperm [471, 472], but not to ovine or human sperm [465, 473]. It is present in the perivitelline space of baboon and human oocytes and binds to zona pelucida [474], but oocytes collected directly from the baboon or human ovary during *in vitro* fertilization do not contain this glycoprotein [465]. Purified bovine Muc9/ogp interacts with bovine sperm surface through the midpiece and tail regions, leading to significant increase of the frozen-thawed sperm viability and indicating the importance of Muc9/ogp to sperm function [475, 476]. The importance of MUC9/OGP interaction with spermatozoa and oocytes for *in vivo* fertilization was shown in a number of studies [477-483]. While these findings indicate the crucial role of MUC9/OVGP in fertilization, its physiological significance remains controversial in light of the normal fertility of *Muc9/ogp* null mice [484]. However, the validity of extrapolating the findings in mice to other species is not proven. It must be taken into account that the role of the MUC9/OGPs in fertilization as well as the mechanisms of fertilization themselves may be different in various mammalian species [485].

The oviduct consists of two main regions, istmus and ampula, which provide a microenvironment for the oocyte and spermatozoa transport, fertilization, and development of early embryos. The oviductal epithelium, especially cells located in the istmus, which serves as sperm reservoir, produce several secreted glycoproteins, including MUC9/OGP [478, 479]. Mammalian spermatozoa have to reside in the istmus for a certain time in order to undergo the modifications required for their capacity to fertilize ova [486, 487]. Studies in hamster and other mammals revealed that during the pre-ovulatory phase, the spermatozoa are tightly bound by their heads to the oviductal mucosa of the istmus, where they interact with MUC9/OGP and other secreted proteins. This interaction is of great importance as only after interaction with MUC9/OGP mucin spermatozoa are capable of fertilizing ova [488, 489]. Of note, interaction between spermatozoa and MUC9/OGP is highly species-specific. It is mediated by the C-terminal region of the MUC9/OGP molecule – a region that is less conserved in different species both in amino acid sequence and in O-glycosylation patterns, and contains the species-specific antigenic determinants [485]. This suggests that carbohydrate moiety determines the species-specific interaction between zona pelucida (ZP) of an oocyte and sperm [465, 466, 490, 491], an interaction that is crucial for successful fertilization.

Numerous studies have shown that the MUC9/OGP mucin (oviductin) interacts with ZP and/or the perivitelline (PV) space of an oocyte during its passage through the oviduct, and that the two remain associated with the early embryo until implantation [492]. ZP is known to be composed of three glycoproteins, including primary and secondary sperm receptors [493-495], but the precise mechanism of their interaction with MUC9/OGP is not known. On the one hand, *in vitro* fertilization studies clearly showed the dispensability of these glycoproteins in fertilization [492], despite numerous indications of their direct interaction with MUC9/OGP [465, 466, 490, 491]. On the other hand, *in vitro* and *in vivo* studies showed that incubation of ovarian oocytes with purified MUC9/OGP increases sperm penetration [496, 497]. There is convincing evidence that hamster Muc9/ogp enhances both sperm binding to ZP and ZP-induced acrosome reaction [498]. Boatman and Magnoni [497] found that purified hamster Muc9/ogp bound *in vitro* to the homologous sperm on the acrosomal crescent, in contrast to Reuter *et al.* [499] who reported that neither hamster nor human homologous MUC9 mucins were associated with human sperm. At the same time, incubation of the bovine Muc9/ogp with bull spermatozoa resulted in association of the Muc9/ogp glycoprotein with the sperm membranes [472]. Taken together, these results suggest species-specific interaction of MUC9/OGP molecules with ZP and spermatozoa. However, these results do not clarify the mechanism of Muc9/ogp-ZP-spermatozoa interaction. Does it correspond to a mechanism of ligant/receptor interaction? Does the MUC9/OGP undergo structural modification prior to interaction with ZP, or does it form a complex with other protein(s) before its interaction with the ZP glycoproteins? Does MUC9/OGP covalently interact with the ZP proteins, or do they associate through the sulfate and/or sialic acid groups of the olygosaccharide side-chains by ionic interactions? More studies are needed to ferret out these basic biological processes.

The significance of the MUC9/OGP mucin interaction with gametes in the oviduct is associated with the ability of MUC9/OGP to form a local barrier at the oocyte/embryo surface to attack by the immune system [500]. The maternal immune system can attack the sperm on its way to fertilize the oocyte. The preimplantation embryo can also serve as a target for the immune system. Sperm, oocyte, and embryo can be destroyed by antibodies and by the complement system. Hence, control over the humoral immune system may play a role in gamete and embryo survival, highlighting defense of gametes and embryo against the maternal immune system as one of the important functions of the MUC9/OGP protein.

Defense of embryo is not the only function of the MUC9/OGP glycoprotein in the process of reproduction. The findings of MUC9/OGP mucin molecules in specific endocytic structures of the fertilized eggs and early embryos in hamster [501], in cytoplasm of the porcine blastocyst [502], and in the baboon cleavage state embryos [503] show that these molecules interact not only with the surface of gametes, but penetrate also into the cells, where they may carry out other tasks among which is participation in oviduct tissue remodeling [504].

The *in silico* analysis of the deduced MUC9/OGP amino acid sequence revealed the presence in the MUC9/OGP apomucin of several conserved motifs, suggesting formation of multi-protein complexes between MUC9/OGP and various functionally active proteins [504]. Indeed, Kadam *et al.* [504, 505] showed that nonmuscle myosin IIA, an active element of the cell remodeling apparatus determining cell shape, polarity, and morphogenesis, is a protein partner of MUC9/OGP in gametes. Through interaction with this protein, MUC9/OGP may regulate oviduct epithelial remodeling [504].

The *MUC9*/OGP exhibits differential expression along the oviduct, with higher levels in areas of high cell turnover [506]. Importantly, differential expression of *MUC9*/OGP corresponds to segmental differentiation of the oviduct epithelium [507]. Several studies showed the involvement of the MUC9/OGP glycoprotein in tissue remodeling, cell shape and polarity, and probably in other intracellular processes as well [501, 502, 507, 508]. *In vitro* studies revealed a close link between the level of MUC9/OGP expression and the maintenance of polarity and differentiation [508]. Like genes involved in tubulin up-regulation in ciliogenesis [509], genes employed in differentiation of oviduct epithelial cells function in concert to regulate MUC9/OGP expression [510]. The presence of the MUC9/OGP molecules in endosomes, multivesicular bodies and blastomere membranes [501], as well as in the intercellular space at the tight junctions of the developing embryos [502] and oviduct epithelia [511], point to intra- and intercellular functions of the MUC9/OGP mucin differed from those associated with mechanical lubrication or defense from immunological or proteolytic attacks. Indeed, during the contact of the oocyte or spermatozoa with MUC9/OGP mucin, the N-terminus of the MUC9/OGP glycoprotein (amino acid residues 11-137) interacts with the gamete cytoskeleton protein MYH9, a nonmuscle myosin IIA. This finding is consistent with data showing that the part of the MUC9/OGP molecule that directly interacts with the sperm can be visualized only after partial permeablization of cell membrane or in the capacitated sperm [472, 512]. In the other words, the sperm membrane must be modified to allow interaction of MUC9/OGP with intracellular proteins. Really, reorganization of sperm membrane occurs during sperm capacitation [513, 514]. Moreover, *in vitro* capacıtated sperm was reported to have cytoskeleton protein on the surface of the sperm head [513]. MUC9/OGP was shown to co-localize and to associate with the filamentous actin at the cleavage furrow of the developing blastocysts [515], where cortical myosin IIA was also found [516]. These data are consistent with Kadam *et al.*'s recent findings [504] showing interaction of MUC9/OGP with cortical myosin IIA, a regulator of cell polarity during normal epithelial morphogenesis [517-519].

Immunoelectron microscopy revealed the presence of the MUC9/OGP mucin at the tight junctions (TJ) of developing ciliated cells, which are undergoing change in shape, and at the adluminal surface of plasma membranes of mature ciliated cells [520-522]. Co-localization studies of MUC9/OGP and cadherin, a master regulator of the epithelial phenotype, demonstrated co-localization of MUC9/OGP and cadherin at the adluminal face of the plasma membrane [504], suggesting possible interaction of the two proteins in oviduct tissue remodeling.

For evaluation of the MUC9/OGP potential to interact with the various proteins inside the cell one must take into account that MUC9/OGP contains a number of different domains and motifs, including PDZ-binding motifs that implicate MUC9/OGP as a PDZ-ligand protein. This property of the MUC9/OGP glycoprotein allows the mucin molecule to be bound through the PDZ-domain-containing proteins to multiprotein complexes, which, as suggested by Giallourakis *et al.* [523], can be further targeted to specific subcellular compartments including TJs. This is consistent with the observation of Bauersachs *et al.* [506, 510] of the coordinated expression of the MUC9/OGP mucin with two TJ-proteins, claudin 1 and ZO-1, a PDZ domain-containing protein. Localization of the MUC9/OGP molecules at the tonofilaments of TJs in the secretory and differentiating ciliated cells, on the one hand, and spacio-temporal regulation of the potential partner-protein(s) expression, on the other hand, suggest that the intracellular MUC9/OGP isoform(s) may interact with the epithelial actin-myosin cytoskeleton. Such an interaction would induce a number of physiological events, such as changing of cell shape, sperm motility, acrosome reaction, egg activation and cytokinesis. Hence, MUC9/OGP might be one of the important regulators of tissue remodeling [504].

In summary, the MUC9 mucin is a unique polyfunctional glycoprotein molecule containing multiple amino acid motifs and domains, including mucin-specific one. The multiple functional knots in the MUC9/OGP molecule permit it to interact with multiple protein partners and form protein complexes. Such protein-protein interactions might trigger numerous extra- and intracellular reactions, placing MUC9/OGP in the center of the multichain network.

12.6. GENERAL CONCLUSION

In this chapter the experimental evidence-based functions of the secreted mucins has been described. The available data demonstrate their multifunctionality. Some functions such as lubrication and protection against mechanical and chemical injuries are common to all members of this group. Other functions such as interaction between ova and spermatozoids are specific to MUC9/OGP. MUC7 possesses antifungial potential, and MUC6 demonstrates antibacterial activity. Practically all of the discussed mucins participate in innate immunity. Importantly, all secreted mucins take part in cell morphogenesis and differentiation during embryonic and fetal development. The dynamics of the expressions of individual mucins point to cooperative and coordinated functions of the secreted mucins in the developmental processes. Importantly, the gel-forming mucins are involved in tumorigenic processes in epithelial tissues, where they may function as tumor suppressors or tumor promoters. These polyfunctional molecules involved in multiple physiological and pathological processes are attractive targets for future molecular therapy.

REFERENCES

[1] Zaretsky JZ, Wreschner DH. Protein multifunctionality: principles and mechanisms. Transl Oncogenomics 2008;3:99-136.

[2] Buisine MP, Devisme L, Savidge TC, *et al.* Mucin gene expression in human embryonic and fetal intestine. Gut 1998;43:519-24.

[3] Buisine MP, Devisme L, Maunoury V, *et al.* Developmental mucin gene expression in the gastroduodenal tract and accessory digestive glands. I. Stomach. A relationship to gastric carcinoma. J Histochem Cytochem 2000;48:1657-66.

[4] Reid CJ, Gould S, Harris A. Developmental expression of mucin genes in the human respiratory tract. Am J Respir Cell Mol Biol 1997;17:592-8.

[5] Reid CJ, Harris A. Developmental expression of mucin genes in the human gastrointestinal system. Gut 1998;42:220-6.

[6] Kan FW, Roux E. Elaboration of an oviductin by the oviductal epithelium in relation to embryo development as visualized by immunocytochemistry. Microsc Res Tech 1995;31:478-87.

[7] Deren JS. Development of structure and function in the fetal and newborn stomach. Am J Clin Nutr 1971;24:144-59.

[8] Grand RJ, Watkins JB, Torti FM. Development of the human gastrointestinal tract. A review. Gastroenterology 1976;70:790-810.

[9] Buisine MP, Devisme L, Copin MC, *et al.* Developmental mucin gene expression in the human respiratory tract. Am J Respir Cell Mol Biol 1999;20:209-18.

[10] Perez-Vilar J. Mucin granule intraluminal organization. Am J Respir Cell Mol Biol 2007;36:183-90.

[11] Corfield AP, Carroll D, Myerscough N, Probert CS. Mucins in the gastrointestinal tract in health and disease. Front Biosci 2001;6:D1321-57.

[12] Cone RA. Barrier properties of mucus. Adv Drug Deliv Rev 2009;61:75-85.

[13] Ali MS, Pearson JP. Upper airway mucin gene expression: a review. Laryngoscope 2007;117:932-8.

[14] Thornton DJ, Sheehan JK. From mucins to mucus: toward a more coherent understanding of this essential barrier. Proc Am Thorac Soc 2004;1:54-61.

[15] Corfield AP, Carrington SD, Hicks SJ. Ocular mucins: purification, metabolism and functions. Prog Retinal Eye Res 1997;16: 627-56.
[16] Thornton DJ, Rousseau K, McGuckin MA. Structure and function of the polymeric mucins in airways mucus. Annu Rev Physiol 2008;70:459-86.
[17] Thornton DJ, Davies JR, Kirkham S, *et al.* Identification of a nonmucin glycoprotein (gp-340) from a purified respiratory mucin preparation: evidence for an association involving the MUC5B mucin. Glycobiology 2001;11:969-77.
[18] Felgentreff K, Beisswenger C, Griese M, *et al.* The antimicrobial peptide cathelicidin interacts with airway mucus. Peptides 2006;27:3100-6.
[19] Sheehan JK, Kesimer M, Pickles R. Innate immunity and mucus structure and function. Novartis Found Symp 2006;279:155-66; discussion 67-9, 216-9.
[20] Lamblin G, Degroote S, Perini JM, *et al.* Human airway mucin glycosylation: a combinatory of carbohydrate determinants which vary in cystic fibrosis. Glycoconj J 2001;18:661-84.
[21] Creeth JM, Bridge JL, Horton JR. An interaction between lysozyme and mucus glycoproteins. Implications for density-gradient separations. Biochem J 1979;181:717-24.
[22] Van-Seuningen I, Houdret N, Hayem A, Davril M. Strong ionic interactions between mucins and two basic proteins, mucus proteinase inhibitor and lysozyme, in human bronchial secretions. Int J Biochem 1992;24:303-11.
[23] Houdret N, Perini JM, Galabert C, *et al.* The high lipid content of respiratory mucins in cystic fibrosis is related to infection. Biochim Biophys Acta 1986;880:54-61.
[24] Yakubov GE, Papagiannopoulos A, Rat E, Waigh TA. Charge and interfacial behavior of short side-chain heavily glycosylated porcine stomach mucin. Biomacromolecules 2007;8:3791-9.
[25] Cleary J, Bromberg L, Magner E. Adhesion of polyether-modified poly(acrylic acid) to mucin. Langmuir 2004;20:9755-62.
[26] Alexandridis P, Andersson K. Effect of Solvent Quality on Reverse Micelle Formation and Water Solubilization by Poly(ethylene oxide)/Poly(propylene oxide) and Poly(ethylene oxide)/Poly(butylene oxide) Block Copolymers in Xylene. J Colloid Interface Sci 1997;194:166-73.
[27] Celli JP, Turner BS, Afdhal NH, *et al.* Rheology of gastric mucin exhibits a pH-dependent sol-gel transition. Biomacromolecules 2007;8:1580-6.
[28] Lee S, Muller M, Rezwan K, Spencer ND. Porcine gastric mucin (PGM) at the water/poly(dimethylsiloxane) (PDMS) interface: influence of pH and ionic strength on its conformation, adsorption, and aqueous lubrication properties. Langmuir 2005;21:8344-53.
[29] Zhu Y, Granick S. Superlubricity: a paradox about confined fluids resolved. Phys Rev Lett 2004;93:096101.
[30] Raviv U, Giasson S, Kampf N, *et al.* Lubrication by charged polymers. Nature 2003;425:163-5.
[31] Yakubov GE, Papagiannopoulos A, Rat E, Easton RL, Waigh TA. Molecular structure and rheological properties of short-side-chain heavily glycosylated porcine stomach mucin. Biomacromolecules 2007;8:3467-77.
[32] Lamblin G, Roussel P. Airway mucins and their role in defence against micro-organisms. Respir Med 1993;87:421-6.
[33] Nayeb-Hashemi H, Kaunitz JD. Gastroduodenal mucosal defense. Curr Opin Gastroenterol 2009;25:537-43.

[34] Stokes JR, Davies GA. Viscoelasticity of human whole saliva collected after acid and mechanical stimulation. Biorheology 2007;44:141-60.

[35] Yakubov GE, McColl J, Bongaerts JH, Ramsden JJ. Viscous boundary lubrication of hydrophobic surfaces by mucin. Langmuir 2009;25:2313-21.

[36] Chilvers MA, O'Callaghan C. Local mucociliary defence mechanisms. Paediatr Respir Rev 2000;1:27-34.

[37] Lai SK, Wang YY, Wirtz D, Hanes J. Micro- and macrorheology of mucus. Adv Drug Deliv Rev 2009;61:86-100.

[38] Puchelle E, Zahm JM, Havez R. [Biochemical and rheological data in sputum. 3. Relationship between the biochemical constituents and the rheological properties of sputum (author's transl)]. Bull Physiopathol Respir (Nancy) 1973;9:237-57.

[39] Nadziejko CE, Slomiany BL, Slomiany A. Most of the lipid in purulent sputum is bound to mucus glycoprotein. Exp Lung Res 1993;19:671-84.

[40] Slomiany BL, Tsukada H, Slomiany A. Cotranslational attachment of fatty acids to nascent peptides in gastric mucus glycoprotein. Biochem Biophys Res Commun 1986;141:387-93.

[41] Murty VL, Sarosiek J, Slomiany A, Slomiany BL. Effect of lipids and proteins on the viscosity of gastric mucus glycoprotein. Biochem Biophys Res Commun 1984;121:521-9.

[42] Galabert C, Jacquot J, Zahm JM, Puchelle E. Relationships between the lipid content and the rheological properties of airway secretions in cystic fibrosis. Clin Chim Acta 1987;164:139-49.

[43] Girod S, Galabert C, Lecuire A, Zahm JM, Puchelle E. Phospholipid composition and surface-active properties of tracheobronchial secretions from patients with cystic fibrosis and chronic obstructive pulmonary diseases. Pediatr Pulmonol 1992;13:22-7.

[44] Yeates DB, Besseris GJ, Wang LB. Physicochemical properties of mucus and its propulsion. In: Crystal RG, West JB, Weibel ER, Barnes PJ, editors. The Lung: Scientific Foundations. Philadelphia: Lippincott-Raven Publishers; 1977. p. 487-503.

[45] Hong Z, Chasan B, Bansil R, *et al.* Atomic force microscopy reveals aggregation of gastric mucin at low pH. Biomacromolecules 2005;6:3458-66.

[46] Cao X, Bansil R, Bhaskar KR, *et al.* pH-dependent conformational change of gastric mucin leads to sol-gel transition. Biophys J 1999;76:1250-8.

[47] Zhang L, Yu K, Eisenberg A. Ion-Induced Morphological Changes in "Crew-Cut" Aggregates of Amphiphilic Block Copolymers. Science 1996;272:1777-9.

[48] Brunelli R, Papi M, Arcovito G, *et al.* Globular structure of human ovulatory cervical mucus. Faseb J 2007;21:3872-6.

[49] Lagow E, DeSouza MM, Carson DD. Mammalian reproductive tract mucins. Hum Reprod Update 1999;5:280-92.

[50] Moghissi KS. The function of the cervix in fertility. Fertil Steril 1972;23:295-306.

[51] Wolf DP, Blasco L, Khan MA, Litt M. Human cervical mucus. IV. Viscoelasticity and sperm penetrability during the ovulatory menstrual cycle. Fertil Steril 1978;30:163-9.

[52] Gipson IK, Moccia R, Spurr-Michaud S, *et al.* The Amount of MUC5B mucin in cervical mucus peaks at midcycle. J Clin Endocrinol Metab 2001;86:594-600.

[53] Iontcheva I, Oppenheim FG, Troxler RF. Human salivary mucin MG1 selectively forms heterotypic complexes with amylase, proline-rich proteins, statherin, and histatins. J Dent Res 1997;76:734-43.

[54] Aguirre A, Testa-Weintraub LA, Banderas JA, *et al.* Sialochemistry: a diagnostic tool? Crit Rev Oral Biol Med 1993;4:343-50.

[55] Kawagishi S, Fahim RE, Wong KH, Bennick A. Purification and characterization of subunits of a high molecular weight human salivary mucin. Arch Oral Biol 1990;35:265-72.

[56] Loomis RE, Prakobphol A, Levine MJ, Reddy MS, Jones PC. Biochemical and biophysical comparison of two mucins from human submandibular-sublingual saliva. Arch Biochem Biophys 1987;258:452-64.

[57] Biesbrock AR, Reddy MS, Levine MJ. Interaction of a salivary mucin-secretory immunoglobulin A complex with mucosal pathogens. Infect Immun 1991;59:3492-7.

[58] McSweegan E, Burr DH, Walker RI. Intestinal mucus gel and secretory antibody are barriers to Campylobacter jejuni adherence to INT 407 cells. Infect Immun 1987;55:1431-5.

[59] Jager S, Kremer J, Kuiken J, Mulder I. The significance of the Fc part of antispermatozoal antibodies for the shaking phenomenon in the sperm-cervical mucus contact test. Fertil Steril 1981;36:792-7.

[60] Olmsted SS, Padgett JL, Yudin AI, *et al.* Diffusion of macromolecules and virus-like particles in human cervical mucus. Biophys J 2001;81:1930-7.

[61] Piludu M, Hand AR, Cossu M, Piras M. Immunocytochemical localization of MG1 mucin in human bulbourethral glands. J Anat 2009;214:179-82.

[62] May SJ, Blackwell CC, Weir DM. Lewis a blood group antigen of non-secretors: a receptor for candida blastospores. FEMS Microbiol Immunol 1989;1:407-9.

[63] Ligtenberg AJ, Veerman EC, Nieuw Amerongen AV. A role for Lewis a antigens on salivary agglutinin in binding to Streptococcus mutans. Antonie Van Leeuwenhoek 2000;77:21-30.

[64] Van de Bovenkamp JH, Mahdavi J, Korteland-Van Male AM, *et al.* The MUC5AC glycoprotein is the primary receptor for Helicobacter pylori in the human stomach. Helicobacter 2003;8:521-32.

[65] Linden S, Nordman H, Hedenbro J, *et al.* Strain- and blood group-dependent binding of Helicobacter pylori to human gastric MUC5AC glycoforms. Gastroenterology 2002;123:1923-30.

[66] Habte HH, Mall AS, de Beer C, Lotz ZE, Kahn D. The role of crude human saliva and purified salivary MUC5B and MUC7 mucins in the inhibition of Human Immunodeficiency Virus type 1 in an inhibition assay. Virol J 2006;3:99.

[67] Habte HH, de Beer C, Lotz ZE, *et al.* The inhibition of the Human Immunodeficiency Virus type 1 activity by crude and purified human pregnancy plug mucus and mucins in an inhibition assay. Virol J 2008;5:59.

[68] Stax MJ, van Montfort T, Sprenger RR, *et al.* Mucin 6 in seminal plasma binds DC-SIGN and potently blocks dendritic cell mediated transfer of HIV-1 to CD4(+) T-lymphocytes. Virology 2009;391:203-11.

[69] Haase AT. Perils at mucosal front lines for HIV and SIV and their hosts. Nat Rev Immunol 2005;5:783-92.

[70] de Witte L, Nabatov A, Pion M, *et al.* Langerin is a natural barrier to HIV-1 transmission by Langerhans cells. Nat Med 2007;13:367-71.

[71] de Jong MA, de Witte L, Bolmstedt A, van Kooyk Y, Geijtenbeek TB. Dendritic cells mediate herpes simplex virus infection and transmission through the C-type lectin DC-SIGN. J Gen Virol 2008;89:2398-409.

[72] de Witte L, Abt M, Schneider-Schaulies S, van Kooyk Y, Geijtenbeek TB. Measles virus targets DC-SIGN to enhance dendritic cell infection. J Virol 2006;80:3477-86.

[73] Geijtenbeek TB, Kwon DS, Torensma R, *et al.* DC-SIGN, a dendritic cell-specific HIV-1-binding protein that enhances trans-infection of T cells. Cell 2000;100:587-97.
[74] Halary F, Amara A, Lortat-Jacob H, *et al.* Human cytomegalovirus binding to DC-SIGN is required for dendritic cell infection and target cell trans-infection. Immunity 2002;17:653-64.
[75] Liu B, Rayment SA, Soares RV, *et al.* Interaction of human salivary mucin MG2, its recombinant N-terminal region and a synthetic peptide with Actinobacillus actinomycetemcomitans. J Periodontal Res 2002;37:416-24.
[76] Situ H, Wei G, Smith CJ, Mashhoon S, Bobek LA. Human salivary MUC7 mucin peptides: effect of size, charge and cysteine residues on antifungal activity. Biochem J 2003;375:175-82.
[77] Kawakubo M, Ito Y, Okimura Y, *et al.* Natural antibiotic function of a human gastric mucin against Helicobacter pylori infection. Science 2004;305:1003-6.
[78] Gururaja TL, Levine JH, Tran DT, *et al.* Candidacidal activity prompted by N-terminus histatin-like domain of human salivary mucin (MUC7)1. Biochim Biophys Acta 1999;1431:107-19.
[79] Situ H, Bobek LA. *In vitro* assessment of antifungal therapeutic potential of salivary histatin-5, two variants of histatin-5, and salivary mucin (MUC7) domain 1. Antimicrob Agents Chemother 2000;44:1485-93.
[80] Thornton DJ, Gray T, Nettesheim P, *et al.* Characterization of mucins from cultured normal human tracheobronchial epithelial cells. Am J Physiol Lung Cell Mol Physiol 2000;278:L1118-28.
[81] Wickstrom C, Carlstedt I. N-terminal cleavage of the salivary MUC5B mucin. Analogy with the Van Willebrand propolypeptide? J Biol Chem 2001;276:47116-21.
[82] Lidell ME, Hansson GC. Cleavage in the GDPH sequence of the C-terminal cysteine-rich part of the human MUC5AC mucin. Biochem J 2006;399:121-9.
[83] Lidell ME, Johansson ME, Hansson GC. An autocatalytic cleavage in the C terminus of the human MUC2 mucin occurs at the low pH of the late secretory pathway. J Biol Chem 2003;278:13944-51.
[84] Xu G, Forstner GG, Forstner JF. Interaction of heparin with synthetic peptides corresponding to the C-terminal domain of intestinal mucins. Glycoconj J 1996;13:81-90.
[85] Xu G, Bell SL, McCool D, Forstner JF. The cationic C-terminus of rat Muc2 facilitates dimer formation post translationally and is subsequently removed by furin. Eur J Biochem 2000;267:2998-3004.
[86] Borsig L, Wong R, Feramisco J, *et al.* Heparin and cancer revisited: mechanistic connections involving platelets, P-selectin, carcinoma mucins, and tumor metastasis. Proc Natl Acad Sci USA 2001;98:3352-7.
[87] Velcich A, Yang W, Heyer J, *et al.* Colorectal cancer in mice genetically deficient in the mucin Muc2. Science 2002;295:1726-9.
[88] Enghild JJ, Salvesen G, Hefta SA, *et al.* Chondroitin 4-sulfate covalently cross-links the chains of the human blood protein pre-alpha-inhibitor. J Biol Chem 1991;266:747-51.
[89] Chan P, Risler JL, Raguenez G, Salier JP. The three heavy-chain precursors for the inter-alpha-inhibitor family in mouse: new members of the multicopper oxidase protein group with differential transcription in liver and brain. Biochem J 1995;306 (Pt 2):505-12.

[90] Iontcheva I, Oppenheim FG, Offner GD, Troxler RF. Molecular mapping of statherin- and histatin-binding domains in human salivary mucin MG1 (MUC5B) by the yeast two-hybrid system. J Dent Res 2000;79:732-9.
[91] Tomasetto C, Masson R, Linares JL, *et al.* pS2/TFF1 interacts directly with the VWFC cysteine-rich domains of mucins. Gastroenterology 2000;118:70-80.
[92] Ruchaud-Sparagano MH, Westley BR, May FE. The trefoil protein TFF1 is bound to MUC5AC in human gastric mucosa. Cell Mol Life Sci 2004;61:1946-54.
[93] Wright NA, Hoffmann W, Otto WR, Rio MC, Thim L. Rolling in the clover: trefoil factor family (TFF)-domain peptides, cell migration and cancer. FEBS Lett 1997;408:121-3.
[94] Kjellev S, Nexo E, Thim L, Poulsen SS. Systemically administered trefoil factors are secreted into the gastric lumen and increase the viscosity of gastric contents. Br J Pharmacol 2006;149:92-9.
[95] Raynal BD, Hardingham TE, Thornton DJ, Sheehan JK. Concentrated solutions of salivary MUC5B mucin do not replicate the gel-forming properties of saliva. Biochem J 2002;362:289-96.
[96] Raynal BD, Hardingham TE, Sheehan JK, Thornton DJ. Calcium-dependent protein interactions in MUC5B provide reversible cross-links in salivary mucus. J Biol Chem 2003;278:28703-10.
[97] Hasnain SZ, Wang H, Ghia JE, *et al.* Mucin gene deficiency in mice impairs host resistance to an enteric parasitic infection. Gastroenterology 2010;138:1763-71.
[98] Ogra PL, Welliver RC, Sr. Effects of early environment on mucosal immunologic homeostasis, subsequent immune responses and disease outcome. Nestle Nutr Workshop Ser Pediatr Program 2008;61:145-81.
[99] Bosch JA, de Geus EJ, Veerman EC, Hoogstraten J, Nieuw Amerongen AV. Innate secretory immunity in response to laboratory stressors that evoke distinct patterns of cardiac autonomic activity. Psychosom Med 2003;65:245-58.
[100] Linden SK, Sutton P, Karlsson NG, Korolik V, McGuckin MA. Mucins in the mucosal barrier to infection. Mucosal Immunol 2008;1:183-97.
[101] Linden SK, Florin TH, McGuckin MA. Mucin dynamics in intestinal bacterial infection. PLoS One 2008;3:e3952.
[102] McGuckin MA, Linden SK, Sutton P, Florin TH. Mucin dynamics and enteric pathogens. Nat Rev Microbiol 2011;9:265-78.
[103] Dharmani P, Srivastava V, Kissoon-Singh V, Chadee K. Role of intestinal mucins in innate host defense mechanisms against pathogens. J Innate Immun 2009;1:123-35.
[104] Lievin-Le Moal V, Servin AL. The front line of enteric host defense against unwelcome intrusion of harmful microorganisms: mucins, antimicrobial peptides, and microbiota. Clin Microbiol Rev 2006;19: 315-37.
[105] Deplancke B, Gaskins HR. Microbial modulation of innate defense: goblet cells and the intestinal mucus layer. Am J Clin Nutr 2001;73:1131S-41S.
[106] Johansson ME, Larsson JM, Hansson GC. The two mucus layers of colon are organized by the MUC2 mucin, whereas the outer layer is a legislator of host-microbial interactions. Proc Natl Acad Sci USA 2011;108 (Suppl 1):4659-65.
[107] Atuma C, Strugala V, Allen A, Holm L. The adherent gastrointestinal mucus gel layer: thickness and physical state *in vivo*. Am J Physiol Gastrointest Liver Physiol 2001;280:G922-9.

[108] Ho SB, Takamura K, Anway R, *et al.* The adherent gastric mucous layer is composed of alternating layers of MUC5AC and MUC6 mucin proteins. Dig Dis Sci 2004;49:1598-606.

[109] Qin J, Li R, Raes J, *et al.* A human gut microbial gene catalogue established by metagenomic sequencing. Nature 2010;464:59-65.

[110] Aristoteli LP, Willcox MD. Mucin degradation mechanisms by distinct Pseudomonas aeruginosa isolates *in vitro*. Infect Immun 2003;71:5565-75.

[111] Helander A, Hansson GC, Svennerholm AM. Binding of enterotoxigenic Escherichia coli to isolated enterocytes and intestinal mucus. Microb Pathog 1997;23:335-46.

[112] Kerneis S, Bernet MF, Coconnier MH, Servin AL. Adhesion of human enterotoxigenic Escherichia coli to human mucus secreting HT-29 cell subpopulations in culture. Gut 1994;35:1449-54.

[113] Tu QV, McGuckin MA, Mendz GL. Campylobacter jejuni response to human mucin MUC2: modulation of colonization and pathogenicity determinants. J Med Microbiol 2008;57:795-802.

[114] Backhed F, Ley RE, Sonnenburg JL, Peterson DA, Gordon JI. Host-bacterial mutualism in the human intestine. Science 2005;307:1915-20.

[115] Chu SH, Walker WA. Bacterial toxin interaction with the developing intestine. Gastroenterology 1993;104:916-25.

[116] Shah U, Walker WA. Adverse host responses to bacterial toxins in human infants. J Nutr 2000;130:420S-5S.

[117] Belley A, Keller K, Gottke M, Chadee K. Intestinal mucins in colonization and host defense against pathogens. Am J Trop Med Hyg 1999;60:10-5.

[118] Gottke MU, Keller K, Belley A, *et al.* Functional heterogeneity of colonic adenocarcinoma mucins for inhibition of Entamoeba histolytica adherence to target cells. J Eukaryot Microbiol 1998;45:17S-23S.

[119] Vimal DB, Khullar M, Gupta S, Ganguly NK. Intestinal mucins: the binding sites for Salmonella typhimurium. Mol Cell Biochem 2000;204:107-17.

[120] de Repentigny L, Aumont F, Bernard K, Belhumeur P. Characterization of binding of Candida albicans to small intestinal mucin and its role in adherence to mucosal epithelial cells. Infect Immun 2000;68:3172-9.

[121] Bobek LA, Situ H. MUC7 20-Mer: investigation of antimicrobial activity, secondary structure, and possible mechanism of antifungal action. Antimicrob Agents Chemother 2003;47:643-52.

[122] Truant S, Bruyneel E, Gouyer V, *et al.* Requirement of both mucins and proteoglycans in cell-cell dissociation and invasiveness of colon carcinoma HT-29 cells. Int J Cancer 2003;104:683-94.

[123] Yu H, He Y, Zhang X, *et al.* The rat IgGFcgammaBP and Muc2 C-terminal domains and TFF3 in two intestinal mucus layers bind together by covalent interaction. PLoS One 2011;6:e20334.

[124] Prakobphol A, Levine MJ, Tabak LA, Reddy MS. Purification of a low-molecular-weight, mucin-type glycoprotein from human submandibular-sublingual saliva. Carbohydr Res 1982;108:111-22.

[125] Amado F, Lobo MJ, Domingues P, Duarte JA, Vitorino R. Salivary peptidomics. Expert Rev Proteomics 2010;7:709-21.

[126] De Smet K, Contreras R. Human antimicrobial peptides: defensins, cathelicidins and histatins. Biotechnol Lett 2005;27:1337-47.

[127] Humphrey SP, Williamson RT. A review of saliva: normal composition, flow, and function. J Prosthet Dent 2001;85:162-9.
[128] Wickstrom C, Christersson C, Davies JR, Carlstedt I. Macromolecular organization of saliva: identification of'insoluble' MUC5B assemblies and non-mucin proteins in the gel phase. Biochem J 2000;351 (Pt 2):421-8.
[129] Wickstrom C, Davies JR, Eriksen GV, Veerman EC, Carlstedt I. MUC5B is a major gel-forming, oligomeric mucin from human salivary gland, respiratory tract and endocervix: identification of glycoforms and C-terminal cleavage. Biochem J 1998;334 (Pt 3):685-93.
[130] Thornton DJ, Howard M, Khan N, Sheehan JK. Identification of two glycoforms of the MUC5B mucin in human respiratory mucus. Evidence for a cysteine-rich sequence repeated within the molecule. J Biol Chem 1997;272:9561-6.
[131] Herrmann A, Davies JR, Lindell G, *et al.* Studies on the "insoluble" glycoprotein complex from human colon. Identification of reduction-insensitive MUC2 oligomers and C-terminal cleavage. J Biol Chem 1999;274:15828-36.
[132] Xu G, Huan L, Khatri I, *et al.* Human intestinal mucin-like protein (MLP) is homologous with rat MLP in the C-terminal region, and is encoded by a gene on chromosome 11 p 15.5. Biochem Biophys Res Commun 1992;183:821-8.
[133] Wickstrom C, Svensater G. Salivary gel-forming mucin MUC5B-a nutrient for dental plaque bacteria. Oral Microbiol Immunol 2008;23:177-82.
[134] Marsh PD. Are dental diseases examples of ecological catastrophes? Microbiology 2003;149:279-94.
[135] Scannapieco FA. Saliva-bacterium interactions in oral microbial ecology. Crit Rev Oral Biol Med 1994;5:203-48.
[136] Groenink J, Veerman EC, Zandvoort MS, *et al.* The interaction between saliva and Actinobacillus actinomycetemcomitans influenced by the zeta potential. Antonie Van Leeuwenhoek 1998;73:279-88.
[137] Walz A, Odenbreit S, Stuhler K, *et al.* Identification of glycoprotein receptors within the human salivary proteome for the lectin-like BabA and SabA adhesins of Helicobacter pylori by fluorescence-based 2-D bacterial overlay. Proteomics 2009;9:1582-92.
[138] Prakobphol A, Leffler H, Fisher SJ. The high-molecular-weight human mucin is the primary salivary carrier of ABH, Le(a), and Le(b) blood group antigens. Crit Rev Oral Biol Med 1993;4:325-33.
[139] Thomsson KA, Prakobphol A, Leffler H, *et al.* The salivary mucin MG1 (MUC5B) carries a repertoire of unique oligosaccharides that is large and diverse. Glycobiology 2002;12:1-14.
[140] Groenink J, Ligtenberg AJ, Veerman EC, Bolscher JG, Nieuw Amerongen AV. Interaction of the salivary low-molecular-weight mucin (MG2) with Actinobacillus actinomycetemcomitans. Antonie Van Leeuwenhoek 1996;70:79-87.
[141] Prakobphol A, Thomsson KA, Hansson GC, *et al.* Human low-molecular-weight salivary mucin expresses the sialyl lewisx determinant and has L-selectin ligand activity. Biochemistry 1998;37:4916-27.
[142] Nielsen PA, Mandel U, Therkildsen MH, Clausen H. Differential expression of human high-molecular-weight salivary mucin (MG1) and low-molecular-weight salivary mucin (MG2). J Dent Res 1996;75:1820-6.
[143] Liu B, Rayment S, Oppenheim FG, Troxler RF. Isolation of human salivary mucin MG2 by a novel method and characterization of its interactions with oral bacteria. Arch Biochem Biophys 1999;364:286-93.

[144] Habte HH, de Beer C, Lotz ZE, Roux P, Mall AS. Anti-HIV-1 activity of salivary MUC5B and MUC7 mucins from HIV patients with different CD4 counts. Virol J 2010;7:269.
[145] Lal K, Pollock JJ, Santarpia RP, 3rd, *et al.* Pilot study comparing the salivary cationic protein concentrations in healthy adults and AIDS patients: correlation with antifungal activity. J Acquir Immune Defic Syndr 1992;5:904-14.
[146] Bosch JA, de Geus EJ, Ligtenberg TJ, *et al.* Salivary MUC5B-mediated adherence (*ex vivo*) of Helicobacter pylori during acute stress. Psychosom Med 2000;62:40-9.
[147] Prakobphol A, Tangemann K, Rosen SD, *et al.* Separate oligosaccharide determinants mediate interactions of the low-molecular-weight salivary mucin with neutrophils and bacteria. Biochemistry 1999;38:6817-25.
[148] Engelen L, de Wijk RA, Prinz JF, *et al.* A comparison of the effects of added saliva, alpha-amylase and water on texture perception in semisolids. Physiol Behav 2003;78:805-11.
[149] Silletti E, Vingerhoeds MH, Norde W, van Aken GA. Complex formation in mixtures of lysozyme-stabilized emulsions and human saliva. J Colloid Interface Sci 2007;313:485-93.
[150] Silletti E, Vitorino RM, Schipper R, Amado FM, Vingerhoeds MH. Identification of salivary proteins at oil-water interfaces stabilized by lysozyme and beta-lactoglobulin. Arch Oral Biol 2010;55:268-78.
[151] Thornton DJ, Khan N, Mehrotra R, *et al.* Salivary mucin MG1 is comprised almost entirely of different glycosylated forms of the MUC5B gene product. Glycobiology 1999;9:293-302.
[152] Reddy MS, Bobek LA, Haraszthy GG, Biesbrock AR, Levine MJ. Structural features of the low-molecular-mass human salivary mucin. Biochem J 1992;287 (Pt 2):639-43.
[153] Rogers DF. Physiology of airway mucus secretion and pathophysiology of hypersecretion. Respir Care 2007;52:1134-46; discussion 46-9.
[154] Morgenroth K, Bolz J. Morphological features of the interaction between mucus and surfactant on the bronchial mucosa. Respiration 1985;47:225-31.
[155] Houtmeyers E, Gosselink R, Gayan-Ramirez G, Decramer M. Regulation of mucociliary clearance in health and disease. Eur Respir J 1999;13:1177-88.
[156] Rose MC, Voynow JA. Respiratory tract mucin genes and mucin glycoproteins in health and disease. Physiol Rev 2006;86:245-78.
[157] Hovenberg HW, Davies JR, Carlstedt I. Different mucins are produced by the surface epithelium and the submucosa in human trachea: identification of MUC5AC as a major mucin from the goblet cells. Biochem J 1996;318 (Pt 1):319-24.
[158] Evans CM, Koo JS. Airway mucus: the good, the bad, the sticky. Pharmacol Ther 2009;121:332-48.
[159] Kesimer M, Kirkham S, Pickles RJ, *et al.* Tracheobronchial air-liquid interface cell culture: a model for innate mucosal defense of the upper airways? Am J Physiol Lung Cell Mol Physiol 2009;296:L92-L100.
[160] Ross AJ, Dailey LA, Brighton LE, Devlin RB. Transcriptional profiling of mucociliary differentiation in human airway epithelial cells. Am J Respir Cell Mol Biol 2007;37:169-85.
[161] Sheehan JK, Howard M, Richardson PS, Longwill T, Thornton DJ. Physical characterization of a low-charge glycoform of the MUC5B mucin comprising the gel-phase of an asthmatic respiratory mucous plug. Biochem J 1999;338 (Pt 2):507-13.
[162] Tam A, Morrish D, Wadsworth S, *et al.* The role of female hormones on lung function in chronic lung diseases. BMC Womens Health 2011;11:24.

[163] Choi HJ, Chung YS, Kim HJ, *et al.* Signal pathway of 17beta-estradiol-induced MUC5B expression in human airway epithelial cells. Am J Respir Cell Mol Biol 2009;40:168-78.
[164] Knowles MR, Boucher RC. Mucus clearance as a primary innate defense mechanism for mammalian airways. J Clin Invest 2002;109:571-7.
[165] Kurosawa H, Wang CG, Dandurand RJ, King M, Eidelman DH. Mucociliary function in the mouse measured in explanted lung tissue. J Appl Physiol 1995;79:41-6.
[166] Matsui H, Randell SH, Peretti SW, Davis CW, Boucher RC. Coordinated clearance of periciliary liquid and mucus from airway surfaces. J Clin Invest 1998;102:1125-31.
[167] Smith DJ, Gaffney EA, Blake JR. A viscoelastic traction layer model of muco-ciliary transport. Bull Math Biol 2007;69:289-327.
[168] Sears PR, Davis CW, Chua M, Sheehan JK. Mucociliary interactions and mucus dynamics in ciliated human bronchial epithelial cell cultures. Am J Physiol Lung Cell Mol Physiol 2011;301:L181-6.
[169] Ali M, Lillehoj EP, Park Y, Kyo Y, Kim KC. Analysis of the proteome of human airway epithelial secretions. Proteome Sci 2011;9:4.
[170] Arora SK, Ritchings BW, Almira EC, Lory S, Ramphal R. The Pseudomonas aeruginosa flagellar cap protein, FliD, is responsible for mucin adhesion. Infect Immun 1998;66:1000-7.
[171] Reddy MS. Binding between Pseudomonas aeruginosa adhesins and human salivary, tracheobronchial and nasopharyngeal mucins. Biochem Mol Biol Int 1996;40:403-8.
[172] Patsos G, Corfield A. Management of the human mucosal defensive barrier: evidence for glycan legislation. Biol Chem 2009;390:581-90.
[173] Newton JL, Jordan N, Oliver L, *et al.* Helicobacter pylori *in vivo* causes structural changes in the adherent gastric mucus layer but barrier thickness is not compromised. Gut 1998;43:470-5.
[174] Phillipson M, Johansson ME, Henriksnas J, *et al.* The gastric mucus layers: constituents and regulation of accumulation. Am J Physiol Gastrointest Liver Physiol 2008;295:G806-12.
[175] Al-Marhoon MS, Nunn S, Soames RW. Effects of cagA+ and cagA- strains of Helicobacter pylori on the human gastric mucus layer thickness. J Gastroenterol Hepatol 2005;20:1246-52.
[176] Allen A, Cunliffe WJ, Pearson JP, Venables CW. The adherent gastric mucus gel barrier in man and changes in peptic ulceration. J Intern Med Suppl 1990;732:83-90.
[177] Ota H, Katsuyama T, Ishii K, *et al.* A dual staining method for identifying mucins of different gastric epithelial mucous cells. Histochem J 1991;23:22-8.
[178] Van Klinken BJ, Dekker J, Buller HA, de Bolos C, Einerhand AW. Biosynthesis of mucins (MUC2-6) along the longitudinal axis of the human gastrointestinal tract. Am J Physiol 1997;273:G296-302.
[179] Sawaguchi A, Ishihara K, Kawano Ji J, *et al.* Fluid dynamics of the excretory flow of zymogenic and mucin contents in rat gastric gland processed by high-pressure freezing/freeze substitution. J Histochem Cytochem 2002;50:223-34.
[180] Ota H, Katsuyama T. Alternating laminated array of two types of mucin in the human gastric surface mucous layer. Histochem J 1992;24:86-92.
[181] Ichikawa T, Ishihara K, Kusakabe T, *et al.* Distinct effects of tetragastrin, histamine, and CCh on rat gastric mucin synthesis and contribution of NO. Am J Physiol 1998;274:G138-46.
[182] Ichikawa T, Ishihara K, Kusakabe T, *et al.* CGRP modulates mucin synthesis in surface mucus cells of rat gastric oxyntic mucosa. Am J Physiol Gastrointest Liver Physiol 2000;279:G82-9.

[183] Phillipson M, Atuma C, Henriksnas J, Holm L. The importance of mucus layers and bicarbonate transport in preservation of gastric juxtamucosal pH. Am J Physiol Gastrointest Liver Physiol 2002;282:G211-9.

[184] Petersson J, Phillipson M, Jansson EA, *et al.* Dietary nitrate increases gastric mucosal blood flow and mucosal defense. Am J Physiol Gastrointest Liver Physiol 2007;292:G718-24.

[185] Atherton JC. The pathogenesis of Helicobacter pylori-induced gastro-duodenal diseases. Annu Rev Pathol 2006;1:63-96.

[186] Correa P, Houghton J. Carcinogenesis of Helicobacter pylori. Gastroenterology 2007;133:659-72.

[187] Magalhaes A, Reis CA. Helicobacter pylori adhesion to gastric epithelial cells is mediated by glycan receptors. Braz J Med Biol Res 2010;43:611-8.

[188] Mahdavi J, Sonden B, Hurtig M, *et al.* Helicobacter pylori SabA adhesin in persistent infection and chronic inflammation. Science 2002;297:573-8.

[189] Odenbreit S, Till M, Hofreuter D, Faller G, Haas R. Genetic and functional characterization of the alpAB gene locus essential for the adhesion of Helicobacter pylori to human gastric tissue. Mol Microbiol 1999;31:1537-48.

[190] Peck B, Ortkamp M, Diehl KD, Hundt E, Knapp B. Conservation, localization and expression of HopZ, a protein involved in adhesion of Helicobacter pylori. Nucleic Acids Res 1999;27:3325-33.

[191] Linden S, Mahdavi J, Hedenbro J, Boren T, Carlstedt I. Effects of pH on Helicobacter pylori binding to human gastric mucins: identification of binding to non-MUC5AC mucins. Biochem J 2004;384:263-70.

[192] Van den Brink GR, Tytgat KM, Van der Hulst RW, *et al.* H pylori colocalises with MUC5AC in the human stomach. Gut 2000;46:601-7.

[193] Van De Bovenkamp JH, Korteland-Van Male AM, Buller HA, Einerhand AW, Dekker J. Infection with Helicobacter pylori affects all major secretory cell populations in the human antrum. Dig Dis Sci 2005;50:1078-86.

[194] Byrd JC, Yan P, Sternberg L, *et al.* Aberrant expression of gland-type gastric mucin in the surface epithelium of Helicobacter pylori-infected patients. Gastroenterology 1997;113:455-64.

[195] Kang HM, Kim N, Park YS, *et al.* Effects of Helicobacter pylori Infection on gastric mucin expression. J Clin Gastroenterol 2008;42:29-35.

[196] Marques T, David L, Reis C, Nogueira A. Topographic expression of MUC5AC and MUC6 in the gastric mucosa infected by Helicobacter pylori and in associated diseases. Pathol Res Pract 2005;201:665-72.

[197] Beil W, Enss ML, Muller S, *et al.* Role of vacA and cagA in Helicobacter pylori inhibition of mucin synthesis in gastric mucous cells. J Clin Microbiol 2000;38:2215-8.

[198] Matsuzwa M, Ota H, Hayama M, *et al.* Helicobacter pylori infection up-regulates gland mucous cell-type mucins in gastric pyloric mucosa. Helicobacter 2003;8:594-600.

[199] Morgenstern S, Koren R, Moss SF, *et al.* Does Helicobacter pylori affect gastric mucin expression? Relationship between gastric antral mucin expression and H. pylori colonization. Eur J Gastroenterol Hepatol 2001;13:19-23.

[200] Marshall BJ, Barrett LJ, Prakash C, McCallum RW, Guerrant RL. Urea protects Helicobacter (Campylobacter) pylori from the bactericidal effect of acid. Gastroenterology 1990;99:697-702.

[201] Slomiany BL, Bilski J, Sarosiek J, *et al.* Campylobacter pyloridis degrades mucin and undermines gastric mucosal integrity. Biochem Biophys Res Commun 1987;144:307-14.
[202] Sidebotham RL, Batten JJ, Karim QN, Spencer J, Baron JH. Breakdown of gastric mucus in presence of Helicobacter pylori. J Clin Pathol 1991;44:52-7.
[203] Markesich DC, Anand BS, Lew GM, Graham DY. Helicobacter pylori infection does not reduce the viscosity of human gastric mucus gel. Gut 1995;36:327-9.
[204] Shimizu T, Akamatsu T, Sugiyama A, Ota H, Katsuyama T. Helicobacter pylori and the surface mucous gel layer of the human stomach. Helicobacter 1996;1:207-18.
[205] Dekker J. Gut. 2001;48: 868-71.
[206] Davidson HJ, Kuonen VJ. The tear film and ocular mucins. Vet Ophthalmol 2004;7:71-7.
[207] Round AN, Berry M, McMaster TJ, Corfield AP, Miles MJ. Glycopolymer charge density determines conformation in human ocular mucin gene products: an atomic force microscope study. J Struct Biol 2004;145:246-53.
[208] Driver PJ, Lemp MA. Meibomian gland dysfunction. Surv Ophthalmol 1996;40:343-67.
[209] King-Smith PE, Fink BA, Fogt N, *et al.* The thickness of the human precorneal tear film: evidence from reflection spectra. Invest Ophthalmol Vis Sci 2000;41:3348-59.
[210] Shamsi FA, Chen Z, Liang J, *et al.* Analysis and comparison of proteomic profiles of tear fluid from human, cow, sheep, and camel eyes. Invest Ophthalmol Vis Sci 2011;52:9156-65.
[211] German AJ, Hall EJ, Day MJ. Measurement of IgG, IgM and IgA concentrations in canine serum, saliva, tears and bile. Vet Immunol Immunopathol 1998;64:107-21.
[212] Kaura R, Tiffany JM. The role of mucus glycoproteins in the tear film. In: Holly FJ, editor. The Preocular Tear Film in Health, Diseases, and Contact Lens Wear. Lubbock: Dry Eye Institute Inc; 1986. p. 728-31.
[213] Nichols BA, Chiappino ML, Dawson CR. Demonstration of the mucous layer of the tear film by electron microscopy. Invest Ophthalmol Vis Sci 1985;26:464-73.
[214] Gamache DA, Wei ZY, Weimer LK, *et al.* Corneal protection by the ocular mucin secretagogue 15(S)-HETE in a rabbit model of desiccation-induced corneal defect. J Ocul Pharmacol Ther 2002;18:349-61.
[215] Millar TJ, Tragoulias ST, Anderton PJ, *et al.* The surface activity of purified ocular mucin at the air-liquid interface and interactions with meibomian lipids. Cornea 2006;25:91-100.
[216] Mantelli F, Argueso P. Functions of ocular surface mucins in health and disease. Curr Opin Allergy Clin Immunol 2008;8:477-83.
[217] Inatomi T, Spurr-Michaud S, Tisdale AS, *et al.* Expression of secretory mucin genes by human conjunctival epithelia. Invest Ophthalmol Vis Sci 1996;37:1684-92.
[218] Yu DF, Chen Y, Han JM, *et al.* MUC19 expression in human ocular surface and lacrimal gland and its alteration in Sjogren syndrome patients. Exp Eye Res 2008;86:403-11.
[219] McKenzie RW, Jumblatt JE, Jumblatt MM. Quantification of MUC2 and MUC5AC transcripts in human conjunctiva. Invest Ophthalmol Vis Sci 2000;41:703-8.
[220] Paulsen F, Langer G, Hoffmann W, Berry M. Human lacrimal gland mucins. Cell Tissue Res 2004;316:167-77.
[221] Paulsen FP, Schaudig U, Fabian A, Ehrich D, Sel S. TFF peptides and mucins are major components of dacryoliths. Graefes Arch Clin Exp Ophthalmol 2006;244:1160-70.
[222] Gipson IK, Hori Y, Argueso P. Character of ocular surface mucins and their alteration in dry eye disease. Ocul Surf 2004;2:131-48.
[223] Watanabe H. Significance of mucin on the ocular surface. Cornea 2002;21:S17-22.

[224] Cai RR, Wang Y, Xu JJ, Zhang CR. [The effects of hyperosmotic stress on rabbit ocular surface and mucin 5AC expression]. Zhonghua Yan Ke Za Zhi 2011;47:252-9.

[225] Ellingham RB, Berry M, Stevenson D, Corfield AP. Secreted human conjunctival mucus contains MUC5AC glycoforms. Glycobiology 1999;9:1181-9.

[226] Xia B, Royall JA, Damera G, Sachdev GP, Cummings RD. Altered O-glycosylation and sulfation of airway mucins associated with cystic fibrosis. Glycobiology 2005;15:747-75.

[227] Podolsky DK. Oligosaccharide structures of human colonic mucin. J Biol Chem 1985;260:8262-71.

[228] Berry M, Ellingham RB, Corfield AP. Polydispersity of normal human conjunctival mucins. Invest Ophthalmol Vis Sci 1996;37:2559-71.

[229] McMaster TJ, Berry M, Corfield AP, Miles MJ. Atomic force microscopy of the submolecular architecture of hydrated ocular mucins. Biophys J 1999;77:533-41.

[230] Spurr-Michaud S, Argueso P, Gipson I. Assay of mucins in human tear fluid. Exp Eye Res 2007;84:939-50.

[231] Berry M, Ellingham RB, Corfield AP. Human preocular mucins reflect changes in surface physiology. Br J Ophthalmol 2004;88:377-83.

[232] Fleiszig SM, Zaidi TS, Ramphal R, Pier GB. Modulation of Pseudomonas aeruginosa adherence to the corneal surface by mucus. Infect Immun 1994;62:1799-804.

[233] Berry M, Harris A, Lumb R, Powell K. Commensal ocular bacteria degrade mucins. Br J Ophthalmol 2002;86:1412-6.

[234] Fleiszig SM, Kwong MS, Evans DJ. Modification of Pseudomonas aeruginosa interactions with corneal epithelial cells by human tear fluid. Infect Immun 2003;71:3866-74.

[235] Aristoteli LP, Willcox MD. The adhesion of Pseudomonas aeruginosa to high molecular weight human tear film species corresponds to glycoproteins reactive with Sambucus nigra lectin. Exp Eye Res 2006;83:1146-53.

[236] Rummenie VT, Matsumoto Y, Dogru M, *et al.* Tear cytokine and ocular surface alterations following brief passive cigarette smoke exposure. Cytokine 2008;43:200-8.

[237] Yoon KC, Song BY, Seo MS. Effects of smoking on tear film and ocular surface. Korean J Ophthalmol 2005;19:18-22.

[238] Gong DH, Turner B, Bhaskar KR, Lamont JT. Lipid binding to gastric mucin: protective effect against oxygen radicals. Am J Physiol 1990;259:G681-6.

[239] Setala NL, Holopainen JM, Metso J, *et al.* Interaction of phospholipid transfer protein with human tear fluid mucins. J Lipid Res 2010;51:3126-34.

[240] Li N, Wang N, Zheng J, *et al.* Characterization of human tear proteome using multiple proteomic analysis techniques. J Proteome Res 2005;4:2052-61.

[241] de Souza GA, Godoy LM, Mann M. Identification of 491 proteins in the tear fluid proteome reveals a large number of proteases and protease inhibitors. Genome Biol 2006;7:R72.

[242] Loomes KM, Senior HE, West PM, Roberton AM. Functional protective role for mucin glycosylated repetitive domains. Eur J Biochem 1999;266:105-11.

[243] Borchers MT, Wert SE, Leikauf GD. Acrolein-induced MUC5ac expression in rat airways. Am J Physiol 1998;274:L573-81.

[244] Spicer SS, Chakrin LW, Wardell JR, Jr. Effect of chronic sulfur dioxide inhalation on the carbohydrate histochemistry and histology of the canine respiratory tract. Am Rev Respir Dis 1974;110:13-24.

[245] Borchers MT, Carty MP, Leikauf GD. Regulation of human airway mucins by acrolein and inflammatory mediators. Am J Physiol 1999;276:L549-55.
[246] Kuo HP, Rohde JA, Barnes PJ, Rogers DF. Cigarette smoke-induced airway goblet cell secretion: dose-dependent differential nerve activation. Am J Physiol 1992;263:L161-7.
[247] Rogers DF. Airway goblet cells: responsive and adaptable front-line defenders. Eur Respir J 1994;7:1690-706.
[248] Abanses JC, Arima S, Rubin BK. Vicks VapoRub induces mucin secretion, decreases ciliary beat frequency, and increases tracheal mucus transport in the ferret trachea. Chest 2009;135:143-8.
[249] Drug E, Landesman-Milo D, Belgorodsky B, *et al.* Enhanced bioavailability of polyaromatic hydrocarbons in the form of mucin complexes. Chem Res Toxicol 2011;24:314-20.
[250] Shraga N, Belgorodsky B, Gozin M. Organic reactions promoted by mucin glycoproteins. J Am Chem Soc 2009;131:12074-5.
[251] Holzer P. Gastroduodenal mucosal defense. Curr Opin Gastroenterol 2000;16:469-78.
[252] Mukhopadhyay P, Chakraborty S, Ponnusamy MP, *et al.* Mucins in the pathogenesis of breast cancer: implications in diagnosis, prognosis and therapy. Biochim Biophys Acta 2011;1815:224-40.
[253] Kufe DW. Mucins in cancer: function, prognosis and therapy. Nat Rev Cancer 2009;9:874-85.
[254] Hoshi H, Saito H, Iijima H, *et al.* Anti-protein-bound polysaccharide-K monoclonal antibody binds the active structure and neutralizes direct antitumor action of the compound. Oncol Rep 2011;25:905-13.
[255] Nagata K, Horinouchi M, Saitou M, *et al.* Mucin expression profile in pancreatic cancer and the precursor lesions. J Hepatobiliary Pancreat Surg 2007;14:243-54.
[256] Leteurtre E, Zerimech F, Piessen G, *et al.* Relationships between mucinous gastric carcinoma, MUC2 expression and survival. World J Gastroenterol 2006;12:3324-31.
[257] Baldus SE, Monig SP, Arkenau V, *et al.* Correlation of MUC5AC immunoreactivity with histopathological subtypes and prognosis of gastric carcinoma. Ann Surg Oncol 2002;9:887-93.
[258] Knudson AG. Antioncogenes and human cancer. Proc Natl Acad Sci USA 1993;90:10914-21.
[259] Macleod K. Tumor suppressor genes. Curr Opin Genet Dev 2000;10:81-93.
[260] Willecke K, Schafer R. Human oncogenes. Hum Genet 1984;66:132-42.
[261] Gronbaek K, Hother C, Jones PA. Epigenetic changes in cancer. Apmis 2007;115:1039-59.
[262] Manfredi JJ. The Mdm2-p53 relationship evolves: Mdm2 swings both ways as an oncogene and a tumor suppressor. Genes Dev 2010;24:1580-9.
[263] Wang Z, Yang B, Dong L, *et al.* A novel oncoprotein Pirh2: rising from the shadow of MDM2. Cancer Sci 2011;102:909-17.
[264] Ferbeyre G, Moriggl R. The role of Stat5 transcription factors as tumor suppressors or oncogenes. Biochim Biophys Acta 2011;1815:104-14.
[265] Rodier F, Campisi J, Bhaumik D. Two faces of p53: aging and tumor suppression. Nucleic Acids Res 2007;35:7475-84.
[266] Chuang LS, Ito Y. RUNX3 is multifunctional in carcinogenesis of multiple solid tumors. Oncogene 2010;29:2605-15.
[267] Zhou Y, Rajabi H, Kufe D. Mucin 1 C-terminal subunit oncoprotein is a target for small-molecule inhibitors. Mol Pharmacol 2011;79:886-93.

[268] Raina D, Ahmad R, Chen D, *et al.* MUC1 oncoprotein suppresses activation of the ARF-MDM2-p53 pathway. Cancer Biol Ther 2008;7:1959-67.

[269] Forgacs E, Biesterveld EJ, Sekido Y, *et al.* Mutation analysis of the PTEN/MMAC1 gene in lung cancer. Oncogene 1998;17:1557-65.

[270] Bepler G, Garcia-Blanco MA. Three tumor-suppressor regions on chromosome 11p identified by high-resolution deletion mapping in human non-small-cell lung cancer. Proc Natl Acad Sci USA 1994;91:5513-7.

[271] Schwienbacher C, Sabbioni S, Campi M, *et al.* Transcriptional map of 170-kb region at chromosome 11p15.5: identification and mutational analysis of the BWR1A gene reveals the presence of mutations in tumor samples. Proc Natl Acad Sci USA 1998;95:3873-8.

[272] Kohno T, Yokota J. How many tumor suppressor genes are involved in human lung carcinogenesis? Carcinogenesis 1999;20:1403-10.

[273] Kaelin WG, Jr., Maher ER. The VHL tumour-suppressor gene paradigm. Trends Genet 1998;14:423-6.

[274] Kinzler KW, Vogelstein B. Breast cancer. What's mice got to do with it? Nature 1996;382:672.

[275] Kinzler KW, Vogelstein B. Landscaping the cancer terrain. Science 1998;280:1036-7.

[276] Kinzler KW, Vogelstein B. Lessons from hereditary colorectal cancer. Cell 1996;87:159-70.

[277] Shrestha-Bhattarai T, Rangnekar VM. Cancer-selective apoptotic effects of extracellular and intracellular Par-4. Oncogene 2010;29:3873-80.

[278] Paige AJ. Redefining tumour suppressor genes: exceptions to the two-hit hypothesis. Cell Mol Life Sci 2003;60:2147-63.

[279] Jones PA, Baylin SB. The fundamental role of epigenetic events in cancer. Nat Rev Genet 2002;3:415-28.

[280] Herman JG, Baylin SB. Gene silencing in cancer in association with promoter hypermethylation. N Engl J Med 2003;349:2042-54.

[281] Luczak MW, Jagodzinski PP. The role of DNA methylation in cancer development. Folia Histochem Cytobiol 2006;44:143-54.

[282] Buisine MP, Devisme L, Degand P, *et al.* Developmental mucin gene expression in the gastroduodenal tract and accessory digestive glands. II. Duodenum and liver, gallbladder, and pancreas. J Histochem Cytochem 2000;48:1667-76.

[283] McIntire MG, Soucy G, Vaughan TL, Shahsafaei A, Odze RD. MUC2 is a highly specific marker of goblet cell metaplasia in the distal esophagus and gastroesophageal junction. Am J Surg Pathol 2011;35:1007-13.

[284] Yonezawa S, Horinouchi M, Osako M, *et al.* Gene expression of gastric type mucin (MUC5AC) in pancreatic tumors: its relationship with the biological behavior of the tumor. Pathol Int 1999;49:45-54.

[285] Ho SB, Shekels LL, Toribara NW, *et al.* Mucin gene expression in normal, preneoplastic, and neoplastic human gastric epithelium. Cancer Res 1995;55:2681-90.

[286] Matsukita S, Nomoto M, Kitajima S, *et al.* Expression of mucins (MUC1, MUC2, MUC5AC and MUC6) in mucinous carcinoma of the breast: comparison with invasive ductal carcinoma. Histopathology 2003;42:26-36.

[287] Cozzi PJ, Wang J, Delprado W, *et al.* MUC1, MUC2, MUC4, MUC5AC and MUC6 expression in the progression of prostate cancer. Clin Exp Metastasis 2005;22:565-73.

[288] Baldus SE, Zirbes TK, Engel S, *et al.* Correlation of the immunohistochemical reactivity of mucin peptide cores MUC1 and MUC2 with the histopathological subtype and prognosis of gastric carcinomas. Int J Cancer 1998;79:133-8.
[289] Baldus SE, Engelmann K, Hanisch FG. MUC1 and the MUCs: a family of human mucins with impact in cancer biology. Crit Rev Clin Lab Sci 2004;41:189-231.
[290] Yang K, Popova NV, Yang WC, *et al.* Interaction of Muc2 and Apc on Wnt signaling and in intestinal tumorigenesis: potential role of chronic inflammation. Cancer Res 2008;68:7313-22.
[291] Iwase T, Kushima R, Mukaisho K, *et al.* Overexpression of CD10 and reduced MUC2 expression correlate with the development and progression of colorectal neoplasms. Pathol Res Pract 2005;201:83-91.
[292] Yu XW, Rong W, Xu FL, *et al.* [Expression and clinical significance of Mucin and E-cadherin in colorectal tumors]. Ai Zheng 2007;26:1204-10.
[293] Kang H, Min BS, Lee KY, *et al.* Loss of E-cadherin and MUC2 expressions correlated with poor survival in patients with stages II and III colorectal carcinoma. Ann Surg Oncol 2011;18:711-9.
[294] Bu XD, Li N, Tian XQ, *et al.* Altered expression of MUC2 and MUC5AC in progression of colorectal carcinoma. World J Gastroenterol 2010;16:4089-94.
[295] Molaei M, Mansoori BK, Mashayekhi R, *et al.* Mucins in neoplastic spectrum of colorectal polyps: can they provide predictions? BMC Cancer 2010;10:537.
[296] Fijneman RJ, Peham JR, van de Wiel MA, *et al.* Expression of Pla2g2a prevents carcinogenesis in Muc2-deficient mice. Cancer Sci 2008;99:2113-9.
[297] Symonds H, Krall L, Remington L, *et al.* p53-dependent apoptosis suppresses tumor growth and progression *in vivo*. Cell 1994;78:703-11.
[298] Stambolic V, Suzuki A, de la Pompa JL, *et al.* Negative regulation of PKB/Akt-dependent cell survival by the tumor suppressor PTEN. Cell 1998;95:29-39.
[299] Morin PJ, Vogelstein B, Kinzler KW. Apoptosis and APC in colorectal tumorigenesis. Proc Natl Acad Sci USA 1996;93:7950-4.
[300] Wang ZG, Ruggero D, Ronchetti S, *et al.* PML is essential for multiple apoptotic pathways. Nat Genet 1998;20:266-72.
[301] McCurrach ME, Connor TM, Knudson CM, Korsmeyer SJ, Lowe SW. bax-deficiency promotes drug resistance and oncogenic transformation by attenuating p53-dependent apoptosis. Proc Natl Acad Sci USA 1997;94:2345-9.
[302] Yin C, Knudson CM, Korsmeyer SJ, Van Dyke T. Bax suppresses tumorigenesis and stimulates apoptosis *in vivo*. Nature 1997;385:637-40.
[303] Bennett M, Macdonald K, Chan SW, *et al.* Cell surface trafficking of Fas: a rapid mechanism of p53-mediated apoptosis. Science 1998;282:290-3.
[304] Juin P, Hueber AO, Littlewood T, Evan G. c-Myc-induced sensitization to apoptosis is mediated through cytochrome c release. Genes Dev 1999;13:1367-81.
[305] Ishida A, Ohta M, Toda M, *et al.* Mucin-induced apoptosis of monocyte-derived dendritic cells during maturation. Proteomics 2008;8:3342-9.
[306] Crocker PR, Paulson JC, Varki A. Siglecs and their roles in the immune system. Nat Rev Immunol 2007;7:255-66.
[307] Shang Y, Wang F, Bai C, *et al.* Dexamethasone protects airway epithelial cell line NCI-H292 against lipopolysaccharide induced endoplasmic reticulum stress and apoptosis. Chin Med J (Engl) 2011;124:38-44.

[308] Boshuizen JA, Reimerink JH, Korteland-van Male AM, *et al.* Changes in small intestinal homeostasis, morphology, and gene expression during rotavirus infection of infant mice. J Virol 2003;77:13005-16.

[309] Boshuizen JA, Reimerink JH, Korteland-van Male AM, *et al.* Homeostasis and function of goblet cells during rotavirus infection in mice. Virology 2005;337:210-21.

[310] Chen CC, Baylor M, Bass DM. Murine intestinal mucins inhibit rotavirus infection. Gastroenterology 1993;105:84-92.

[311] Gratchev A, Siedow A, Bumke-Vogt C, *et al.* Regulation of the intestinal mucin MUC2 gene expression *in vivo*: evidence for the role of promoter methylation. Cancer Lett 2001;168:71-80.

[312] Siedow A, Szyf M, Gratchev A, *et al. De novo* expression of the Muc2 gene in pancreas carcinoma cells is triggered by promoter demethylation. Tumour Biol 2002;23:54-60.

[313] Vincent A, Perrais M, Desseyn JL, *et al.* Epigenetic regulation (DNA methylation, histone modifications) of the 11p15 mucin genes (MUC2, MUC5AC, MUC5B, MUC6) in epithelial cancer cells. Oncogene 2007;26:6566-76.

[314] Yamada N, Hamada T, Goto M, *et al.* MUC2 expression is regulated by histone H3 modification and DNA methylation in pancreatic cancer. Int J Cancer 2006;119:1850-7.

[315] Ye W, Xu P, Jen R, *et al.* Zeranol Down-Regulates p53 Expression in Primary Cultured Human Breast Cancer Epithelial Cells through Epigenetic Modification. Int J Mol Sci 2011;12:1519-32.

[316] Teng IW, Hou PC, Lee KD, *et al.* Targeted methylation of two tumor suppressor genes is sufficient to transform mesenchymal stem cells into cancer stem/initiating cells. Cancer Res 2011;71:4653-63.

[317] Radpour R, Barekati Z, Kohler C, *et al.* Hypermethylation of tumor suppressor genes involved in critical regulatory pathways for developing a blood-based test in breast cancer. PLoS One 2011;6:e16080.

[318] Mesquita P, Peixoto AJ, Seruca R, *et al.* Role of site-specific promoter hypomethylation in aberrant MUC2 mucin expression in mucinous gastric carcinomas. Cancer Lett 2003;189:129-36.

[319] Ioachim E. Expression patterns of cyclins D1, E and cyclin-dependent kinase inhibitors p21waf1/cip1, p27kip1 in colorectal carcinoma: correlation with other cell cycle regulators (pRb, p53 and Ki-67 and PCNA) and clinicopathological features. Int J Clin Pract 2008;62:1736-43.

[320] Handa K, Yamakawa M, Takeda H, Kimura S, Takahashi T. Expression of cell cycle markers in colorectal carcinoma: superiority of cyclin A as an indicator of poor prognosis. Int J Cancer 1999;84:225-33.

[321] Doki Y, Imoto M, Han EK, Sgambato A, Weinstein IB. Increased expression of the P27KIP1 protein in human esophageal cancer cell lines that over-express cyclin D1. Carcinogenesis 1997;18:1139-48.

[322] Sugai T, Tsukahara M, Endoh M, *et al.* Analysis of cell cycle-related proteins in gastric intramucosal differentiated-type cancers based on mucin phenotypes: a novel hypothesis of early gastric carcinogenesis based on mucin phenotype. BMC Gastroenterol 2010;10:55.

[323] Huff V. Wilms' tumours: about tumour suppressor genes, an oncogene and a chameleon gene. Nat Rev Cancer 2011;11:111-21.

[324] Lee EY, Muller WJ. Oncogenes and tumor suppressor genes. Cold Spring Harb Perspect Biol 2010;2:a003236.

[325] Derynck R, Goeddel DV, Ullrich A, *et al.* Synthesis of messenger RNAs for transforming growth factors alpha and beta and the epidermal growth factor receptor by human tumors. Cancer Res 1987;47:707-12.

[326] Dickson RB, Kasid A, Huff KK, *et al.* Activation of growth factor secretion in tumorigenic states of breast cancer induced by 17 beta-estradiol or v-Ha-ras oncogene. Proc Natl Acad Sci USA 1987;84:837-41.

[327] Derynck R, Akhurst RJ, Balmain A. TGF-beta signaling in tumor suppression and cancer progression. Nat Genet 2001;29:117-29.

[328] Cui W, Fowlis DJ, Bryson S, *et al.* TGFbeta1 inhibits the formation of benign skin tumors, but enhances progression to invasive spindle carcinomas in transgenic mice. Cell 1996;86:531-42.

[329] Oft M, Heider KH, Beug H. TGFbeta signaling is necessary for carcinoma cell invasiveness and metastasis. Curr Biol 1998;8:1243-52.

[330] Engle SJ, Hoying JB, Boivin GP, *et al.* Transforming growth factor beta1 suppresses nonmetastatic colon cancer at an early stage of tumorigenesis. Cancer Res 1999;59:3379-86.

[331] Markowitz S, Wang J, Myeroff L, *et al.* Inactivation of the type II TGF-beta receptor in colon cancer cells with microsatellite instability. Science 1995;268:1336-8.

[332] Kim DH, Kim JW, Cho JH, *et al.* Expression of mucin core proteins, trefoil factors, APC and p21 in subsets of colorectal polyps and cancers suggests a distinct pathway of pathogenesis of mucinous carcinoma of the colorectum. Int J Oncol 2005;27:957-64.

[333] Melis M, Hernandez J, Siegel EM, *et al.* Gene expression profiling of colorectal mucinous adenocarcinomas. Dis Colon Rectum 2010;53:936-43.

[334] Baldus SE, Monig SP, Hanisch FG, *et al.* Comparative evaluation of the prognostic value of MUC1, MUC2, sialyl-Lewis^a and sialyl-Lewis^x antigens in colorectal adenocarcinoma. Histopathology 2002;40:440-9.

[335] Blank M, Klussmann E, Kruger-Krasagakes S, *et al.* Expression of MUC2-mucin in colorectal adenomas and carcinomas of different histological types. Int J Cancer 1994;59:301-6.

[336] Yoon SO, Kim BH, Lee HS, *et al.* Differential protein immunoexpression profiles in appendiceal mucinous neoplasms: a special reference to classification and predictive factors. Mod Pathol 2009;22:1102-12.

[337] Kodama M, Kodama T, Murakami M. Oncogene activation and tumor suppressor gene inactivation find their sites of expression in the changes in time and space of the age-adjusted cancer incidence rate. *In vivo* 2000;14:725-34.

[338] Zheng HC, Xia P, Xu XY, Takahashi H, Takano Y. The nuclear to cytoplasmic shift of ING5 protein during colorectal carcinogenesis with their distinct links to pathologic behaviors of carcinomas. Hum Pathol 2011;42:424-33.

[339] Park do Y, Srivastava A, Kim GH, *et al.* CDX2 expression in the intestinal-type gastric epithelial neoplasia: frequency and significance. Mod Pathol 2010;23:54-61.

[340] Beck F, Chawengsaksophak K, Waring P, Playford RJ, Furness JB. Reprogramming of intestinal differentiation and intercalary regeneration in Cdx2 mutant mice. Proc Natl Acad Sci USA 1999;96:7318-23.

[341] Silberg DG, Swain GP, Suh ER, Traber PG. Cdx1 and cdx2 expression during intestinal development. Gastroenterology 2000;119:961-71.

[342] Liu Q, Teh M, Ito K, *et al.* CDX2 expression is progressively decreased in human gastric intestinal metaplasia, dysplasia and cancer. Mod Pathol 2007;20:1286-97.

[343] Rugge M, Ingravallo G, Farinati F, *et al.* Re: CDX2 homeotic gene expression in gastric noninvasive neoplasia. Am J Surg Pathol 2004;28:834-5; author reply 5.

[344] Ha Kim G, Am Song G, Youn Park D, *et al.* CDX2 expression is increased in gastric cancers with less invasiveness and intestinal mucin phenotype. Scand J Gastroenterol 2006;41:880-6.

[345] Yonezawa S, Higashi M, Yamada N, Yokoyama S, Goto M. Significance of mucin expression in pancreatobiliary neoplasms. J Hepatobiliary Pancreat Sci 2010;17:108-24.

[346] Rakha EA, Boyce RW, Abd El-Rehim D, *et al.* Expression of mucins (MUC1, MUC2, MUC3, MUC4, MUC5AC and MUC6) and their prognostic significance in human breast cancer. Mod Pathol 2005;18:1295-304.

[347] Legrier ME, de Pinieux G, Boye K, *et al.* Mucinous differentiation features associated with hormonal escape in a human prostate cancer xenograft. Br J Cancer 2004;90:720-7.

[348] Hanski C, Hofmeier M, Schmitt-Graff A, *et al.* Overexpression or ectopic expression of MUC2 is the common property of mucinous carcinomas of the colon, pancreas, breast, and ovary. J Pathol 1997;182:385-91.

[349] Luttges J, Zamboni G, Longnecker D, Kloppel G. The immunohistochemical mucin expression pattern distinguishes different types of intraductal papillary mucinous neoplasms of the pancreas and determines their relationship to mucinous noncystic carcinoma and ductal adenocarcinoma. Am J Surg Pathol 2001;25:942-8.

[350] Terris B, Dubois S, Buisine MP, *et al.* Mucin gene expression in intraductal papillary-mucinous pancreatic tumours and related lesions. J Pathol 2002;197:632-7.

[351] Moniaux N, Andrianifahanana M, Brand RE, Batra SK. Multiple roles of mucins in pancreatic cancer, a lethal and challenging malignancy. Br J Cancer 2004;91:1633-8.

[352] Ringel J, Lohr M. The MUC gene family: their role in diagnosis and early detection of pancreatic cancer. Mol Cancer 2003;2:9.

[353] Yonezawa S, Sueyoshi K, Nomoto M, *et al.* MUC2 gene expression is found in noninvasive tumors but not in invasive tumors of the pancreas and liver: its close relationship with prognosis of the patients. Hum Pathol 1997;28:344-52.

[354] Nakamura A, Horinouchi M, Goto M, *et al.* New classification of pancreatic intraductal papillary-mucinous tumour by mucin expression: its relationship with potential for malignancy. J Pathol 2002;197:201-10.

[355] Ho SB, Niehans GA, Lyftogt C, *et al.* Heterogeneity of mucin gene expression in normal and neoplastic tissues. Cancer Res 1993;53:641-51.

[356] Zhang S, Zhang HS, Reuter VE, *et al.* Expression of potential target antigens for immunotherapy on primary and metastatic prostate cancers. Clin Cancer Res 1998;4:295-302.

[357] Osunkoya AO, Adsay NV, Cohen C, Epstein JI, Smith SL. MUC2 expression in primary mucinous and nonmucinous adenocarcinoma of the prostate: an analysis of 50 cases on radical prostatectomy. Mod Pathol 2008;21:789-94.

[358] Adsay NV, Merati K, Nassar H, *et al.* Pathogenesis of colloid (pure mucinous) carcinoma of exocrine organs: Coupling of gel-forming mucin (MUC2) production with altered cell polarity and abnormal cell-stroma interaction may be the key factor in the morphogenesis and indolent behavior of colloid carcinoma in the breast and pancreas. Am J Surg Pathol 2003;27:571-8.

[359] Schroder FH. Progress in understanding androgen-independent prostate cancer (AIPC): a review of potential endocrine-mediated mechanisms. Eur Urol 2008;53:1129-37.

[360] Clayton F. Pure mucinous carcinomas of breast: morphologic features and prognostic correlates. Hum Pathol 1986;17:34-8.

[361] Tashiro Y, Yonezawa S, Kim YS, Sato E. Immunohistochemical study of mucin carbohydrates and core proteins in human ovarian tumors. Hum Pathol 1994;25:364-72.

[362] Walsh MD, McGuckin MA, Devine PL, Hohn BG, Wright RG. Expression of MUC2 epithelial mucin in breast carcinoma. J Clin Pathol 1993;46:922-5.

[363] Xu Y, Kimura N, Yoshida R, Lin H, Yoshinaga K. Immunohistochemical study of Muc1, Muc2 and human gastric mucin in breast carcinoma: relationship with prognostic factors. Oncol Rep 2001;8:1177-82.

[364] Mizoshita T, Tsukamoto T, Inada KI, *et al.* Loss of MUC2 expression correlates with progression along the adenoma-carcinoma sequence pathway as well as *de novo* carcinogenesis in the colon. Histol Histopathol 2007;22:251-60.

[365] Lee HS, Lee HK, Kim HS, Yang HK, Kim WH. Tumour suppressor gene expression correlates with gastric cancer prognosis. J Pathol 2003;200:39-46.

[366] Zoghbi S, Drouin E, Claustre J, *et al.* Intestinal MUC2 and gastric M1/MUC5AC in preneoplastic lesions induced by 1,2-dimethylhydrazine in rat: a sequential analysis. Int J Oncol 2007;30:489-97.

[367] Dutrillaux B. Pathways of chromosome alteration in human epithelial cancers. Adv Cancer Res 1995;67:59-82.

[368] Liu XP, Kawauchi S, Oga A, *et al.* Combined examination of p27(Kip1), p21(Waf1/Cip1) and p53 expression allows precise estimation of prognosis in patients with gastric carcinoma. Histopathology 2001;39:603-10.

[369] Sugai T, Inomata M, Uesugi N, *et al.* Analysis of mucin, p53 protein and Ki-67 expressions in gastric differentiated-type intramucosal neoplastic lesions obtained from endoscopic mucosal resection samples: a proposal for a new classification of intramucosal neoplastic lesions based on nuclear atypia. Pathol Int 2004;54:425-35.

[370] Mrena J, Wiksten JP, Kokkola A, *et al.* Prognostic significance of cyclin A in gastric cancer. Int J Cancer 2006;119:1897-901.

[371] Sugai T, Habano W, Uesugi N, *et al.* Three independent genetic profiles based on mucin expression in early differentiated-type gastric cancers--a new concept of genetic carcinogenesis of early differentiated-type adenocarcinomas. Mod Pathol 2004;17:1223-34.

[372] Endoh Y, Tamura G, Ajioka Y, Watanabe H, Motoyama T. Frequent hypermethylation of the hMLH1 gene promoter in differentiated-type tumors of the stomach with the gastric foveolar phenotype. Am J Pathol 2000;157:717-22.

[373] Kocer B, Soran A, Kiyak G, *et al.* Prognostic significance of mucin expression in gastric carcinoma. Dig Dis Sci 2004;49:954-64.

[374] Reis CA, David L, Nielsen PA, *et al.* Immunohistochemical study of MUC5AC expression in human gastric carcinomas using a novel monoclonal antibody. Int J Cancer 1997;74:112-21.

[375] Taylor KL, Mall AS, Barnard RA, Ho SB, Cruse JP. Immunohistochemical detection of gastric mucin in normal and disease states. Oncol Res 1998;10:465-73.

[376] Zhang HK, Zhang QM, Zhao TH, Li YY, Yi YF. Expression of mucins and E-cadherin in gastric carcinoma and their clinical significance. World J Gastroenterol 2004;10:3044-7.

[377] Reis CA, David L, Carvalho F, *et al.* Immunohistochemical study of the expression of MUC6 mucin and co-expression of other secreted mucins (MUC5AC and MUC2) in human gastric carcinomas. J Histochem Cytochem 2000;48:377-88.

[378] Kim GE, Bae HI, Park HU, *et al.* Aberrant expression of MUC5AC and MUC6 gastric mucins and sialyl Tn antigen in intraepithelial neoplasms of the pancreas. Gastroenterology 2002;123:1052-60.

[379] Zen Y, Fujii T, Itatsu K, *et al.* Biliary papillary tumors share pathological features with intraductal papillary mucinous neoplasm of the pancreas. Hepatology 2006;44:1333-43.

[380] Inaguma S, Kasai K, Ikeda H. GLI1 facilitates the migration and invasion of pancreatic cancer cells through MUC5AC-mediated attenuation of E-cadherin. Oncogene 2010;30:714-23.

[381] Aloysius MM, Zaitoun AM, Awad S, *et al.* Mucins and CD56 as markers of tumour invasion and prognosis in periampullary cancer. Br J Surg 2010;97:1269-78.

[382] Yamada N, Nishida Y, Yokoyama S, *et al.* Expression of MUC5AC, an early marker of pancreatobiliary cancer, is regulated by DNA methylation in the distal promoter region in cancer cells. J Hepatobiliary Pancreat Sci 2010;17:844-54.

[383] Ren J, Bharti A, Raina D, *et al.* MUC1 oncoprotein is targeted to mitochondria by heregulin-induced activation of c-Src and the molecular chaperone HSP90. Oncogene 2006;25:20-31.

[384] Huang L, Chen D, Liu D, *et al.* MUC1 oncoprotein blocks glycogen synthase kinase 3beta-mediated phosphorylation and degradation of beta-catenin. Cancer Res 2005;65:10413-22.

[385] Bara J, Chastre E, Mahiou J, *et al.* Gastric M1 mucin, an early oncofetal marker of colon carcinogenesis, is encoded by the MUC5AC gene. Int J Cancer 1998;75:767-73.

[386] Sylvester PA, Myerscough N, Warren BF, *et al.* Differential expression of the chromosome 11 mucin genes in colorectal cancer. J Pathol 2001;195:327-35.

[387] Yu CJ, Yang PC, Shun CT, *et al.* Overexpression of MUC5 genes is associated with early post-operative metastasis in non-small-cell lung cancer. Int J Cancer 1996;69:457-65.

[388] Yu CJ, Shih JY, Lee YC, *et al.* Sialyl Lewis antigens: association with MUC5AC protein and correlation with post-operative recurrence of non-small cell lung cancer. Lung Cancer 2005;47:59-67.

[389] Nishiumi N, Abe Y, Inoue Y, *et al.* Use of 11p15 mucins as prognostic factors in small adenocarcinoma of the lung. Clin Cancer Res 2003;9:5616-9.

[390] Dabelsteen E. Cell surface carbohydrates as prognostic markers in human carcinomas. J Pathol 1996;179:358-69.

[391] Paavonen T, Renkonen R. Selective expression of sialyl-Lewis x and Lewis a epitopes, putative ligands for L-selectin, on peripheral lymph-node high endothelial venules. Am J Pathol 1992;141:1259-64.

[392] Takada A, Ohmori K, Yoneda T, *et al.* Contribution of carbohydrate antigens sialyl Lewis^{A} and sialyl Lewis^{X} to adhesion of human cancer cells to vascular endothelium. Cancer Res 1993;53:354-61.

[393] Verkman AS, Matthay MA, Song Y. Aquaporin water channels and lung physiology. Am J Physiol Lung Cell Mol Physiol 2000;278:L867-79.

[394] Hu J, Verkman AS. Increased migration and metastatic potential of tumor cells expressing aquaporin water channels. Faseb J 2006;20:1892-4.

[395] Moon C, Soria JC, Jang SJ, *et al.* Involvement of aquaporins in colorectal carcinogenesis. Oncogene 2003;22:6699-703.

[396] Chae YK, Woo J, Kim MJ, *et al.* Expression of aquaporin 5 (AQP5) promotes tumor invasion in human non small cell lung cancer. PLoS One 2008;3:e2162.

[397] Chen Z, Wang X, Gao L, *et al.* Regulation of MUC5AC mucin secretion by depletion of AQP5 in SPC-A1 cells. Biochem Biophys Res Commun 2006;342:775-81.

[398] Zhang Z, Chen Z, Song Y, *et al.* Expression of aquaporin 5 increases proliferation and metastasis potential of lung cancer. J Pathol 2010;221:210-20.

[399] Zhang ZQ, Zhu ZX, Bai CX, Chen ZH. Aquaporin 5 expression increases mucin production in lung adenocarcinoma. Oncol Rep 2011;25:1645-50.

[400] Inaba T, Sano H, Kawahito Y, *et al.* Induction of cyclooxygenase-2 in monocyte/macrophage by mucins secreted from colon cancer cells. Proc Natl Acad Sci USA 2003;100:2736-41.

[401] Kitazaki T, Soda H, Doi S, *et al.* Gefitinib inhibits MUC5AC synthesis in mucin-secreting non-small cell lung cancer cells. Lung Cancer 2005;50:19-24.

[402] Kang SK, Chae YK, Woo J, *et al.* Role of human aquaporin 5 in colorectal carcinogenesis. Am J Pathol 2008;173:518-25.

[403] Young HW, Williams OW, Chandra D, *et al.* Central role of Muc5ac expression in mucous metaplasia and its regulation by conserved 5' elements. Am J Respir Cell Mol Biol 2007;37:273-90.

[404] Hoffmann W, Jagla W. Cell type specific expression of secretory TFF peptides: colocalization with mucins and synthesis in the brain. Int Rev Cytol 2002;213:147-81.

[405] Thim L, May FE. Structure of mammalian trefoil factors and functional insights. Cell Mol Life Sci 2005;62:2956-73.

[406] Oertel M, Graness A, Thim L, *et al.* Trefoil factor family-peptides promote migration of human bronchial epithelial cells: synergistic effect with epidermal growth factor. Am J Respir Cell Mol Biol 2001;25:418-24.

[407] Storesund T, Schenck K, Osmundsen H, *et al.* Signal transduction and gene transcription induced by TFF3 in oral keratinocytes. Eur J Oral Sci 2009;117:511-7.

[408] Rio MC, Bellocq JP, Daniel JY, *et al.* Breast cancer-associated pS2 protein: synthesis and secretion by normal stomach mucosa. Science 1988;241:705-8.

[409] Piggott NH, Henry JA, May FE, Westley BR. Antipeptide antibodies against the pNR-2 oestrogen-regulated protein of human breast cancer cells and detection of pNR-2 expression in normal tissues by immunohistochemistry. J Pathol 1991;163:95-104.

[410] Lefebvre O, Chenard MP, Masson R, *et al.* Gastric mucosa abnormalities and tumorigenesis in mice lacking the pS2 trefoil protein. Science 1996;274:259-62.

[411] Longman RJ, Douthwaite J, Sylvester PA, *et al.* Coordinated localisation of mucins and trefoil peptides in the ulcer associated cell lineage and the gastrointestinal mucosa. Gut 2000;47:792-800.

[412] Poulsom R, Begos DE, Modlin IM. Molecular aspects of restitution: functions of trefoil peptides. Yale J Biol Med 1996;69:137-46.

[413] Podolsky DK. Mucosal immunity and inflammation. V. Innate mechanisms of mucosal defense and repair: the best offense is a good defense. Am J Physiol 1999;277:G495-9.

[414] Thim L, Madsen F, Poulsen SS. Effect of trefoil factors on the viscoelastic properties of mucus gels. Eur J Clin Invest 2002;32:519-27.

[415] Paulsen FP, Berry MS. Mucins and TFF peptides of the tear film and lacrimal apparatus. Prog Histochem Cytochem 2006;41:1-53.

[416] Graness A, Chwieralski CE, Reinhold D, Thim L, Hoffmann W. Protein kinase C and ERK activation are required for TFF-peptide-stimulated bronchial epithelial cell migration and tumor necrosis factor-alpha-induced interleukin-6 (IL-6) and IL-8 secretion. J Biol Chem 2002;277:18440-6.

[417] Goke MN, Cook JR, Kunert KS, *et al.* Trefoil peptides promote restitution of wounded corneal epithelial cells. Exp Cell Res 2001;264:337-44.

[418] Taupin DR, Kinoshita K, Podolsky DK. Intestinal trefoil factor confers colonic epithelial resistance to apoptosis. Proc Natl Acad Sci USA 2000;97:799-804.

[419] Baus-Loncar M, Kayademir T, Takaishi S, Wang T. Trefoil factor family 2 deficiency and immune response. Cell Mol Life Sci 2005;62:2947-55.

[420] Zheng H, Takahashi H, Nakajima T, *et al.* MUC6 down-regulation correlates with gastric carcinoma progression and a poor prognosis: an immunohistochemical study with tissue microarrays. J Cancer Res Clin Oncol 2006;132:817-23.

[421] Seki T, Ito T, Kawachi H, *et al.* Relationship between mucin expression of gastric intramucosal signet ring cell carcinoma and its background mucosa. J Med Dent Sci 2009;56:25-35.

[422] Leir SH, Harris A. MUC6 mucin expression inhibits tumor cell invasion. Exp Cell Res 2011;317:2408-19.

[423] Pinto-de-Sousa J, David L, Reis CA, *et al.* Mucins MUC1, MUC2, MUC5AC and MUC6 expression in the evaluation of differentiation and clinico-biological behaviour of gastric carcinoma. Virchows Arch 2002;440:304-10.

[424] Pinto-de-Sousa J, Reis CA, David L, Pimenta A, Cardoso-de-Oliveira M. MUC5B expression in gastric carcinoma: relationship with clinico-pathological parameters and with expression of mucins MUC1, MUC2, MUC5AC and MUC6. Virchows Arch 2004;444:224-30.

[425] Awaya H, Takeshima Y, Yamasaki M, Inai K. Expression of MUC1, MUC2, MUC5AC, and MUC6 in atypical adenomatous hyperplasia, bronchioloalveolar carcinoma, adenocarcinoma with mixed subtypes, and mucinous bronchioloalveolar carcinoma of the lung. Am J Clin Pathol 2004;121:644-53.

[426] Amerongen AV, Bolscher JG, Veerman EC. Salivary mucins: protective functions in relation to their diversity. Glycobiology 1995;5:733-40.

[427] Satyanarayana J, Gururaja TL, Narasimhamurthy S, Naganagowda GA, Levine MJ. Synthesis and conformational features of human salivary mucin C-terminal derived peptide epitope carrying Thomsen-Friedenreich antigen: implications for its role in self-association. Biopolymers 2001;58:500-10.

[428] Fisher SJ, Prakobphol A, Kajisa L, Murray PA. External radiolabelling of components of pellicle on human enamel and cementum. Arch Oral Biol 1987;32:509-17.

[429] Al-Hashimi I, Levine MJ. Characterization of *in vivo* salivary-derived enamel pellicle. Arch Oral Biol 1989;34:289-95.

[430] Jensen JL, Lamkin MS, Oppenheim FG. Adsorption of human salivary proteins to hydroxyapatite: a comparison between whole saliva and glandular salivary secretions. J Dent Res 1992;71:1569-76.

[431] Nieuw Amerongen AV, Oderkerk CH, Driessen AA. Role of mucins from human whole saliva in the protection of tooth enamel against demineralization *in vitro*. Caries Res 1987;21:297-309.

[432] Nieuw Amerongen AV, Oderkerk CH, Veerman EC. Interaction of human salivary mucins with hydroxyapatite. J Biol Buccale 1989;17:85-92.
[433] Gibbons RJ, Etherden I, Moreno EC. Association of neuraminidase-sensitive receptors and putative hydrophobic interactions with high-affinity binding sites for Streptococcus sanguis C5 in salivary pellicles. Infect Immun 1983;42:1006-12.
[434] Gibbons RJ. Adherent interactions which may affect microbial ecology in the mouth. J Dent Res 1984;63:378-85.
[435] Liljemark WF, Bloomquist CG, Fenner LJ, Antonelli PJ, Coulter MC. Effect of neuraminidase on the adherence to salivary pellicle of Streptococcus sanguis and Streptococcus mitis. Caries Res 1989;23:141-5.
[436] Murray PA, Prakobphol A, Lee T, Hoover CI, Fisher SJ. Adherence of oral streptococci to salivary glycoproteins. Infect Immun 1992;60:31-8.
[437] Kesimer M, Kilic N, Mehrotra R, Thornton DJ, Sheehan JK. Identification of salivary mucin MUC7 binding proteins from Streptococcus gordonii. BMC Microbiol 2009;9:163.
[438] Ge J, Catt DM, Gregory RL. Streptococcus mutans surface alpha-enolase binds salivary mucin MG2 and human plasminogen. Infect Immun 2004;72:6748-52.
[439] Linden SK, Wickstrom C, Lindell G, Gilshenan K, Carlstedt I. Four modes of adhesion are used during Helicobacter pylori binding to human mucins in the oral and gastric niches. Helicobacter 2008;13:81-93.
[440] Reddy MS, Levine MJ, Paranchych W. Low-molecular-mass human salivary mucin, MG2: structure and binding of Pseudomonas aeruginosa. Crit Rev Oral Biol Med 1993;4:315-23.
[441] Reddy MS. Binding of the pili of Pseudomonas aeruginosa to a low-molecular-weight mucin and neutral cystatin of human submandibular-sublingual saliva. Curr Microbiol 1998;37:395-402.
[442] Levine MJ, Reddy MS, Tabak LA, *et al.* Structural aspects of salivary glycoproteins. J Dent Res 1987;66:436-41.
[443] Soares RV, Siqueira CC, Bruno LS, *et al.* MG2 and lactoferrin form a heterotypic complex in salivary secretions. J Dent Res 2003;82:471-5.
[444] Weinberg ED. Human lactoferrin: a novel therapeutic with broad spectrum potential. J Pharm Pharmacol 2001;53:1303-10.
[445] Collins LM, Dawes C. The surface area of the adult human mouth and thickness of the salivary film covering the teeth and oral mucosa. J Dent Res 1987;66:1300-2.
[446] Gururaja TL, Ramasubbu N, Venugopalan P, *et al.* Structural features of the human salivary mucin, MUC7. Glycoconj J 1998;15:457-67.
[447] Liu B, Rayment SA, Gyurko C, *et al.* The recombinant N-terminal region of human salivary mucin MG2 (MUC7) contains a binding domain for oral Streptococci and exhibits candidacidal activity. Biochem J 2000;345 Pt 3:557-64.
[448] Satyanarayana J, Situ H, Narasimhamurthy S, *et al.* Divergent solid-phase synthesis and candidacidal activity of MUC7 D1, a 51-residue histidine-rich N-terminal domain of human salivary mucin MUC7. J Pept Res 2000;56:275-82.
[449] Wei GX, Campagna AN, Bobek LA. Factors affecting antimicrobial activity of MUC7 12-mer, a human salivary mucin-derived peptide. Ann Clin Microbiol Antimicrob 2007;6:14.
[450] Albone EF, Hagen FK, VanWuyckhuyse BC, Tabak LA. Molecular cloning of a rat submandibular gland apomucin. J Biol Chem 1994;269:16845-52.
[451] Denny PC, Mirels L, Denny PA. Mouse submandibular gland salivary apomucin contains repeated N-glycosylation sites. Glycobiology 1996;6:43-50.

[452] Antonyraj KJ, Karunakaran T, Raj PA. Bactericidal activity and poly-L-proline II conformation of the tandem repeat sequence of human salivary mucin glycoprotein (MG2). Arch Biochem Biophys 1998;356:197-206.

[453] Raj PA, Marcus E, Edgerton M. Delineation of an active fragment and poly(L-proline) II conformation for candidacidal activity of bactenecin 5. Biochemistry 1996;35:4314-25.

[454] Zasloff M. Antibiotic peptides as mediators of innate immunity. Curr Opin Immunol 1992;4:3-7.

[455] Bruno LS, Li X, Wang L, *et al.* Two-hybrid analysis of human salivary mucin MUC7 interactions. Biochim Biophys Acta 2005;1746:65-72.

[456] Kay BK, Williamson MP, Sudol M. The importance of being proline: the interaction of proline-rich motifs in signaling proteins with their cognate domains. Faseb J 2000;14:231-41.

[457] Feng S, Chen JK, Yu H, Simon JA, Schreiber SL. Two binding orientations for peptides to the Src SH3 domain: development of a general model for SH3-ligand interactions. Science 1994;266:1241-7.

[458] Rickles RJ, Botfield MC, Weng Z, *et al.* Identification of Src, Fyn, Lyn, PI3K and Abl SH3 domain ligands using phage display libraries. Embo J 1994;13:5598-604.

[459] Grabs D, Slepnev VI, Songyang Z, *et al.* The SH3 domain of amphiphysin binds the proline-rich domain of dynamin at a single site that defines a new SH3 binding consensus sequence. J Biol Chem 1997;272:13419-25.

[460] Chen HI, Sudol M. The WW domain of Yes-associated protein binds a proline-rich ligand that differs from the consensus established for Src homology 3-binding modules. Proc Natl Acad Sci USA 1995;92:7819-23.

[461] Ermekova KS, Zambrano N, Linn H, *et al.* The WW domain of neural protein FE65 interacts with proline-rich motifs in Mena, the mammalian homolog of Drosophila enabled. J Biol Chem 1997;272:32869-77.

[462] Simon C, Barathieu K, Laguerre M, *et al.* Three-dimensional structure and dynamics of wine tannin-saliva protein complexes. A multitechnique approach. Biochemistry 2003;42:10385-95.

[463] Akatsuka K, Yoshida-Komiya H, Tulsiani DR, *et al.* Rat zona pellucida glycoproteins: molecular cloning and characterization of the three major components. Mol Reprod Dev 1998;51:454-67.

[464] Buhi WC. Characterization and biological roles of oviduct-specific, oestrogen-dependent glycoprotein. Reproduction 2002;123:355-62.

[465] O'Day-Bowman MB, Mavrogianis PA, Reuter LM, *et al.* Association of oviduct-specific glycoproteins with human and baboon (Papio anubis) ovarian oocytes and enhancement of human sperm binding to human hemizonae following *in vitro* incubation. Biol Reprod 1996;54:60-9.

[466] Verhage HG, Mavrogianis PA, O'Day-Bowman MB, *et al.* Characteristics of an oviductal glycoprotein and its potential role in the fertilization process. Biol Reprod 1998;58:1098-101.

[467] Kouba AJ, Abeydeera LR, Alvarez IM, Day BN, Buhi WC. Effects of the porcine oviduct-specific glycoprotein on fertilization, polyspermy, and embryonic development *in vitro*. Biol Reprod 2000;63:242-50.

[468] Killian GJ. Evidence for the role of oviduct secretions in sperm function, fertilization and embryo development. Anim Reprod Sci 2004;82-83:141-53.

[469] McCauley TC, Buhi WC, Wu GM, *et al.* Oviduct-specific glycoprotein modulates sperm-zona binding and improves efficiency of porcine fertilization *in vitro*. Biol Reprod 2003;69:828-34.
[470] Yong P, Gu Z, Luo JP, Wang JR, Tso JK. Antibodies against the C-terminal peptide of rabbit oviductin inhibit mouse early embryo development to pass 2-cell stage. Cell Res 2002;12:69-78.
[471] Anderson SH, Killian GJ. Effect of macromolecules from oviductal conditioned medium on bovine sperm motion and capacitation. Biol Reprod 1994;51:795-9.
[472] King RS, Killian GJ. Purification of bovine estrus-associated protein and localization of binding on sperm. Biol Reprod 1994;51:34-42.
[473] Nancarrow CD, Hill JL. Co-culture, oviduct secretion and the function of oviduct-specific glycoproteins. Cell Biol Int 1994;18:1105-14.
[474] Boice ML, McCarthy TJ, Mavrogianis PA, Fazlebas AT, Verhage HG. Localization of oviductal glycoproteins within the zona pellucida and perivitelline space of ovulated ova and early embryos in baboons (Papio anubis). Biol Reprod 1990;43:340-6.
[475] Abe H, Sendai Y, Satoh T, Hoshi HA. Japanese Association for Immunology of Reproduction; Proceeding of the 8th annual Meeting; 1993; 1993. p. pp71-4.
[476] Satoh T, Abe H, Sendai Y, Iwata H, Hoshi H. Biochemical characterization of a bovine oviduct-specific sialo-glycoprotein that sustains sperm viability *in vitro*. Biochim Biophys Acta 1995;1266:117-23.
[477] Kimura H, Matsuda J, Ogura A, Asano T, Naiki M. Affinity binding of hamster oviductin to spermatozoa and its influence on *in vitro* fertilization. Mol Reprod Dev 1994;39:322-7.
[478] Kan FW, St-Jacques S, Bleau G. Immunoelectron microscopic localization of an oviductal antigen in hamster zona pellucida by use of a monoclonal antibody. J Histochem Cytochem 1988;36:1441-7.
[479] Abe H, Oikawa T. Study of the differentiation of secretory cells in the golden hamster oviductal epithelium by use of a monoclonal antibody. J Exp Zool 1990;254:97-106.
[480] Brown CR, Cheng WK. Changes in composition of the porcine zona pellucida during development of the oocyte to the 2- to 4-cell embryo. J Embryol Exp Morphol 1986;92:183-91.
[481] Hedrick JL, Wardrip NJ, Berger T. Differences in the macromolecular composition of the zona pellucida isolated from pig oocytes, eggs, and zygotes. J Exp Zool 1987;241:257-62.
[482] McNutt TL, Killian GJ. Influence of bovine follicular and oviduct fluids on sperm capacitation *in vitro*. J Androl 1991;12:244-52.
[483] Wegner CC, Killian GJ. *In vitro* and *in vivo* association of an oviduct estrus-associated protein with bovine zona pellucida. Mol Reprod Dev 1991;29:77-84.
[484] Araki Y, Nohara M, Yoshida-Komiya H, *et al.* Effect of a null mutation of the oviduct-specific glycoprotein gene on mouse fertilization. Biochem J 2003;374:551-7.
[485] Merchan M, Peiro R, Santacreu MA, Francino O, Folch JM. Rabbit oviductal glycoprotein 1 gene: genomic organization polymorphism analysis and mRNA expression. Mol Reprod Dev 2007;74:687-93.
[486] Austin CR. Anomalies of fertilization leading to triploidy. J Cell Comp Physiol 1960;56(Suppl 1):1-15.
[487] Yanagimachi R. Sperm capacitation and gamete interaction. J Reprod Fertil Suppl 1989;38:27-33.

[488] Smith TT, Yanagimachi R. The viability of hamster spermatozoa stored in the isthmus of the oviduct: the importance of sperm-epithelium contact for sperm survival. Biol Reprod 1990;42:450-7.

[489] Smith TT, Yanagimachi R. Attachment and release of spermatozoa from the caudal isthmus of the hamster oviduct. J Reprod Fertil 1991;91:567-73.

[490] Boatman DE. Responses of gametes to the oviductal environment. Hum Reprod 1997;12:133-49.

[491] Sendai Y, Komiya H, Suzuki K, *et al.* Molecular cloning and characterization of a mouse oviduct-specific glycoprotein. Biol Reprod 1995;53:285-94.

[492] Malette B, Paquette Y, Merlen Y, Bleau G. Oviductins possess chitinase- and mucin-like domains: a lead in the search for the biological function of these oviduct-specific ZP-associating glycoproteins. Mol Reprod Dev 1995;41:384-97.

[493] Wassarman PM. Zona pellucida glycoproteins. Annu Rev Biochem 1988;57:415-42.

[494] Shabanowitz RB, O'Rand MG. Characterization of the human zona pellucida from fertilized and unfertilized eggs. J Reprod Fertil 1988;82:151-61.

[495] Moller CC, Bleil JD, Kinloch RA, Wassarman PM. Structural and functional relationships between mouse and hamster zona pellucida glycoproteins. Dev Biol 1990;137:276-86.

[496] Boatman DE, Andrews JC, Bavister BD. A quantitative assay for capacitation: evaluation of multiple sperm penetration through the zona pellucida of salt-stored hamster eggs. Gamete Res 1988;19:19-29.

[497] Boatman DE, Magnoni GE. Identification of a sperm penetration factor in the oviduct of the golden hamster. Biol Reprod 1995;52:199-207.

[498] Boatman DE, Magnoni GE, Robbins RS. Modulation of spermatozoa and zona pellucida properties by the soluble acrosome reaction-inducing factor of the ovulated egg-cumulus complex. Mol Reprod Dev 1994;38:410-20.

[499] Reuter LM, O'Day-Bowman MB, Mavrogianis PA, Fazleabas AT, Verhage HG. *In vitro* incubation of golden (Syrian) hamster ovarian oocytes and human sperm with a human oviduct specific glycoprotein. Mol Reprod Dev 1994;38:160-9.

[500] Oliphant G, Cabot C, Ross P, Marta J. Control of the humoral immune system within the rabbit oviduct. Biol Reprod 1984;31:205-12.

[501] Kan FW, Roux E, Bleau G. Immunolocalization of oviductin in endocytic compartments in the blastomeres of developing embryos in the golden hamster. Biol Reprod 1993;48:77-88.

[502] Buhi WC, O'Brien B, Alvarez IM, Erdos G, Dubois D. Immunogold localization of porcine oviductal secretory proteins within the zona pellucida, perivitelline space, and plasma membrane of oviductal and uterine oocytes and early embryos. Biol Reprod 1993;48:1274-83.

[503] Boice ML, Geisert RD, Blair RM, Verhage HG. Identification and characterization of bovine oviductal glycoproteins synthesized at estrus. Biol Reprod 1990;43:457-65.

[504] Kadam KM, D'Souza SJ, Natraj U. Identification of cellular isoform of oviduct-specific glycoprotein: role in oviduct tissue remodeling? Cell Tissue Res 2007;330:545-56.

[505] Kadam KM, D'Souza SJ, Bandivdekar AH, Natraj U. Identification and characterization of oviductal glycoprotein-binding protein partner on gametes: epitopic similarity to non-muscle myosin IIA, MYH 9. Mol Hum Reprod 2006;12:275-82.

[506] Bauersachs S, Blum H, Mallok S, *et al.* Regulation of ipsilateral and contralateral bovine oviduct epithelial cell function in the postovulation period: a transcriptomics approach. Biol Reprod 2003;68:1170-7.

[507] Abe H. The mammalian oviductal epithelium: regional variations in cytological and functional aspects of the oviductal secretory cells. Histol Histopathol 1996;11:743-68.
[508] Reischl J, Prelle K, Schol H, *et al.* Factors affecting proliferation and dedifferentiation of primary bovine oviduct epithelial cells *in vitro*. Cell Tissue Res 1999;296:371-83.
[509] Shea DK, Walsh CJ. mRNAs for alpha- and beta-tubulin and flagellar calmodulin are among those coordinately regulated when Naegleria gruberi amebae differentiate into flagellates. J Cell Biol 1987;105:1303-9.
[510] Bauersachs S, Rehfeld S, Ulbrich SE, *et al.* Monitoring gene expression changes in bovine oviduct epithelial cells during the oestrous cycle. J Mol Endocrinol 2004;32:449-66.
[511] McBride DS, Boisvert C, Bleau G, Kan FW. Detection of nascent and/or mature forms of oviductin in the female reproductive tract and post-ovulatory oocytes by use of a polyclonal antibody against recombinant hamster oviductin. J Histochem Cytochem 2004;52:1001-9.
[512] Natraj U, Bhatt P, Vanage G, Moodbidri SB. Overexpression of monkey oviductal protein: purification and characterization of recombinant protein and its antibodies. Biol Reprod 2002;67:1897-906.
[513] Liu DY, Clarke GN, Baker HW. Exposure of actin on the surface of the human sperm head during *in vitro* culture relates to sperm morphology, capacitation and zona binding. Hum Reprod 2005;20:999-1005.
[514] McNutt T, Rogowski L, Vasilatos-Younken R, Killian G. Adsorption of oviductal fluid proteins by the bovine sperm membrane during *in vitro* capacitation. Mol Reprod Dev 1992;33:313-23.
[515] Murray M, Messinger SM. Early embryonic development in the Djungarian hamster (Phodopus sungorus) is accompanied by alterations in the distribution and intensity of an estrogen (E2)-dependent oviduct glycoprotein in the blastomere membrane and zona pellucida and in its association with F-actin. Biol Reprod 1994;51:1126-39.
[516] Simerly C, Nowak G, de Lanerolle P, Schatten G. Differential expression and functions of cortical myosin IIA and IIB isotypes during meiotic maturation, fertilization, and mitosis in mouse oocytes and embryos. Mol Biol Cell 1998;9:2509-25.
[517] Auersperg N. Microfilaments in epithelial morphogenesis. J Cell Biol 1972;52:206-11.
[518] Fey EG, Wan KM, Penman S. Epithelial cytoskeletal framework and nuclear matrix-intermediate filament scaffold: three-dimensional organization and protein composition. J Cell Biol 1984;98:1973-84.
[519] Meyer TN, Schwesinger C, Bush KT, *et al.* Spatiotemporal regulation of morphogenetic molecules during *in vitro* branching of the isolated ureteric bud: toward a model of branching through budding in the developing kidney. Dev Biol 2004;275:44-67.
[520] Hagiwara H, Shibasaki S, Ohwada N. Ciliogenesis in the human oviduct epithelium during the normal menstrual cycle. J Electron Microsc (Tokyo) 1992;41:321-9.
[521] Comer MT, Leese HJ, Southgate J. Induction of a differentiated ciliated cell phenotype in primary cultures of Fallopian tube epithelium. Hum Reprod 1998;13:3114-20.
[522] Suuroia T, Aunapuu M, Arend A, Sepp E. ["Light" epithelial cells of swine and bovine oviducts]. Tsitologiia. 2002;44:656-60.
[523] Giallourakis C, Cao Z, Green T, *et al.* A molecular-properties-based approach to understanding PDZ domain proteins and PDZ ligands. Genome Res 2006;16:1056-72.

CHAPTER 13

Secreted Mucins: Covert Functions

Abstract: Functions that potentially can be performed by secreted mucins have been analyzed by bioinformatics using the Motifscan and Eukaryotic Linear Motif (ELM) tools and STRING 9.0 interaction network. A number of potentially important motifs were discovered and possible functional partners of the secreted mucin glycoproteins were predicted. A new role of galactin-3 in physiological transformation of the intestinal mucus gel layers is proposed.

Keywords: Secreted mucins, galectin-3, bioinformatics.

The previous chapter summarized the functions of the gel-forming and soluble mucins that have been verified experimentally. The reader is referred to several recent reviews [1-8] for further information on this subject. This chapter addresses mucin functions and properties that have not yet been directly tested in experimental models but show high probability according to *in silico* bioinformatics analysis and indirect experimental data.

13.1. REVERSIBLE MUC2/GALECTIN-3 INTERACTION AS REGULATOR OF PHYSIOLOGICAL TRANSFORMATION OF MUCUS GEL (HYPOTHESIS)

Galectin-3 (Gal-3), a member of the growing family of β–galactoside binding lectin proteins [9], is a small 32-kDa soluble cytosolic protein [10]. Although Gal-3 does not have a signal peptide, it may be targeted to the nucleus, enter intracellular vesicles, and be secreted by a nonclassical pathway [11]. It exhibits a surprising array of functional activities whose expression depends on its targeting. Secreted extracellular Gal-3 mediates cell migration, adhesion, and cell-cell interactions [12]. In the cytoplasm, Gal-3 displays anti-apoptotic function and actively participates in signal transduction pathways [13, 14]. In the nucleus, it is involved in pre-mRNA splicing, transcriptional regulation and signal transduction [15-18], playing a key role in normal cell proliferation and development. Depending on cell type, Gal-3 has been reported to be exclusively cytoplasmic or predominantly nuclear; or, it may shuttle between the cytoplasm and the nucleus. As noted by Haudek *et al.* [19], a number of intracellular ligands interact with Gal-3 in the cytoplasm and in the

Joseph Z. Zaretsky and Daniel H. Wreschner

nucleus, determining its intracellular activity. Apparently, the extracellular ligands may also influence the cell behavior and intracellular processes *via* interaction with the secreted Gal-3 molecule followed by formation of a Gal-3/ligand complex and its consequent internalization back to the cytoplasm or nucleus, thereby transferring an extracellular ligand inside the cell [20]. Thus, this highly multifunctional protein may combine extra- and intracellular processes into a common network that may operate both inside and outside the cell.

Gal-3 was shown to interact with MUC2 mucin in two ways: as a transcription factor that modulates transcription of the *MUC2* gene [18], and as a partner of the MUC2 glycoprotein, which serves, in this case, as an extracellular ligand for Gal-3 [21]. The role of Gal-3 in transcription of the *MUC2* gene is well established [18], whereas the role of the MUC2-Gal-3 complex is presently unknown. Interaction between these two glycoproteins *in vitro* is specific, depends on peripheral carbohydrate structures, and can be completely abrogated by lactose [21]. This type of binding between the rat colonic mucin (Muc2) and the adherent lectin of *Entamoeba histolytica* (Gal-3) followed by *in vivo* internalization was described by Chadee *et al.* [22]. Still other studies point to a specific relationship between MUC2 and Gal-3. It was shown that alterations in the production of Gal-3 and MUC2 independently correlate with the malignant behavior and the metastatic capacity of human colon cancer cells [23-27]. Importantly, the modulation of either *Gal-3* or *MUC2* expression alters the metastatic potential of colon cancer cells. Dudas *et al.* [28] and later on Song *et al.* [18] showed that Gal-3 directly regulates *MUC2* expression *via* activation of AP-1 and formation of a complex with c-Jun and Fra-1 at the AP-1 site of the *MUC2* promoter. Taken together, these results show that Gal-3 regulates expression of its own ligand (MUC2 !). This mechanism is of great importance as it may explain the coordinated expression of two proteins and perhaps also their functions. It would be interesting to examine the expression and functions of the *Gal-3* and MUC2 genes in experiments using anti-sense knock-down technology. Thus, the data presented above point to the potential function(s) of the MUC2-Gal-3 complex in normal and pathological conditions both inside and outside the cell.

Theoretically, the cooperation between the MUC2 and Gal-3 glycoproteins can take place both in the extracellular space and in intracellular compartments. The extracellular cooperation of these proteins may be associated with normal cell migration or with spreading of the metastatic cells through interaction of the MUC2-Gal-3 complex with the extracellular matrix proteins and the endothelial cell receptors. These aspects of MUC2 and Gal-3 interaction have been intensively studied [24, 28-33]. Both glycoproteins may also cooperate in the formation and functioning of the mucus gel layer – a suggestion supported by the recent discovery of Gal-3's contribution to stabilizing and maintaining the ocular defense barrier function [34]. Two distinct membrane-bound mucins, MUC1 an MUC16, were found to interact with Gal-3 on the apical surface of the ocular epithelial cells through binding of lectin and mucin carbohydrate moieties, determining the stability and permeability of the ocular defense barrier [34].

Thus, Gal-3 has bivalent potential to interact with both gel-forming (MUC2) and membrane-bound (MUC1) mucins. Based on these data we hypothesize that the simultaneous binding of the Gal-3 molecule to MUC1 associated with cell membrane, on the one hand, and to the MUC2 comprising the inner mucus layer of the intestine, on the other hand, is the key mechanism that fulfills tight binding of the innermost gel layer to the epithelial cell surface (Fig. **1**). Moreover, dynamic interaction of Gal-3 and MUC2 in the inner gel layer may explain the physiological alterations of the defense mucus barrier in the intestinal tract.

According to our hypothesis, interaction of MUC2 with Gal-3 forms a regular MUC2-Gal-3 net throughout the inner gel layer that retains tight connection with the apical surface of the epithelial cells *via* the Gal-3-MUC1 "bridge". (Fig. **1**, phase I). The mechanical pressure of the outer layer is important for the integrity of the MUC2-Gal-3 net. Removal of the outer layer by peristalsis or food movement changes the mechanical pressure over the inner layer. This will induce disconnection of the MUC2 from the Gal-3 molecules, destroying the regular structure of the MUC2-Gal-3 net (Fig. **1**, phase II) and facilitating transfer of the MUC2 molecules from the inner gel layer to the outer loosely attached mucus layer (Fig. **1**, phase III). The transfer of the liberated MUC2 molecules toward the lumen initiates renewal of the outer mucus layer (Fig. **1**, phase III). Once the outer layer is formed, the mechanical pressure over the inner layer is restored, inducing the rebinding of the MUC2 and Gal-3 molecules in the inner layer (Fig. **1**, phase IV). The rebinding of the MUC2 and Gal-3 glycoprotein molecules to each other restores the regular polymeric net structure, stiffening the inner layer and rendering it impermeable to a pathogen.

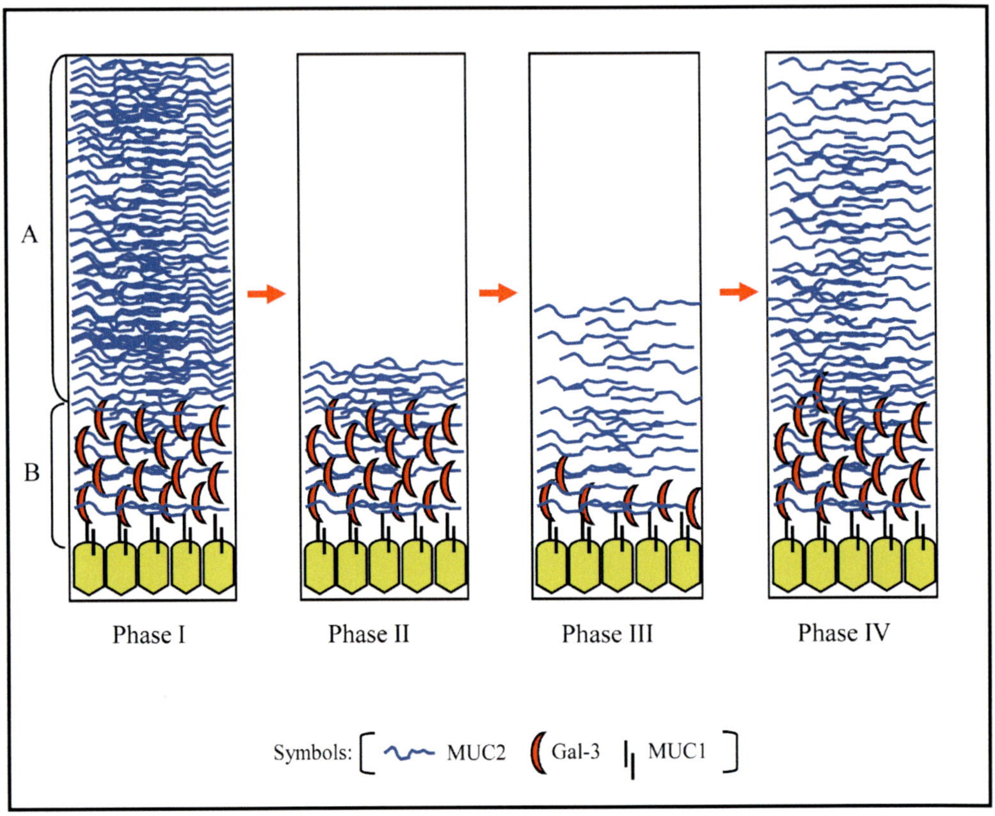

Figure 1: Structure and physiological transformation of the intestinal mucus gel layers. A. Outer layer. B. Inner layer. Phase I – normal double-layered intestinal gel; Phase II – the outer layer is removed as the result of peristalsis or food movement; Phase III – destabilization of the MUC2-Gal-3 net in the inner layer and transfer of the MUC2 molecules to the outer layer; Phase IV-restoration of the outer and inner layer structure.

The proposed hypothesis offers an explanation for the physiological and biochemical reactions occurring in the mucus barrier, its cyclic transformations and the role of MUC2 and Gal-3 proteins in these processes. The hypothesis is purposely simplified to focus on the role of MUC2-Gal-3 binding in the gel barrier functions. We are aware that, in reality, numerous other proteins and functionally active molecules including trefoil proteins, immunoglobulins and metals [35] may take part in the complex process of the gel layer formation and functioning. Nevertheless, mucins in general, and MUC2 in particular, constitute the major potential binding partners of animal lectins as up to 80% of a mucin mass is represented by carbohydrates [34] having the potential to interact with lectins. Future studies will clearify this issue.

13.2. *IN SILICO* DETECTED POTENTIALS OF SECRETED MUCINS TO INTERACT WITH PROTEIN PARTNERS

It is generally accepted that the gel-forming and soluble mucins display their functions in the extracellular space (mucus gel, extracellular matrix, intercellular interactions). Nevertheless, they apparently may function in the intracellular compartments as well. Theoretically, this can occur during the intracellular period of the mucin life cycle, or when (and if) the mucin molecule is transported from outside into the cell compartments. Apparently, both options can be utilized by secreted mucins. As shown above, the Gal-3 protein may function as a suitable vehicle for transportation of mucin molecules (*e.g.* MUC2) into the cytoplasm or nucleus, similar to what happens with other glycoproteins [20, 36]. To be functionally active inside the cell, a secreted mucin glycoprotein would have to possess the potentially functional sites or motifs that allow it to interact with other intracellular molecules. Such interactions may result in engagement of a mucin molecule in various intracellular processes, including signal transductions. To better understand the potentials of the secreted mucins to interact with other proteins, we conducted an *in silico* search in the mucin amino acid sequences for the functional motifs and binding sites specific for various functionally important proteins. For this purpose, we employed the Motifscan and Eukaryotic Linear Motif (ELM) tools at www.Expasy.com. The relevance of these softwares for such a purpose was demonstrated recently by Kadam *et al.* [37].

A number of potentially important motifs were discovered by our analysis in the secreted mucin glycoproteins:

1) Sequences specific for enzymes involved in processing, maturation and posttranslational modifications are found in all of the mucin precursors. All secreted mucin polypeptides contain the nardilysine cleavage sites specific for cleavage by N-arginine dibasic convertase [38].

2) All secreted mucins except MUC8 contain the subtilisin-like protein convertase (SLPC)-specific cleavage sites [39]. This finding is consistent with the furin-mediated proteolysis of the MUC2 precursor [40]. Furin is a member of the SLPC subfamily [41]. The presence of the SLPC specific cleavage sites in all secretory mucins reflects a common proteolytic reaction(s) undergone by the mucin precursors during processing. The MUC2 cleavage by furin is an example of such proteolysis. This proteolytic step may be important for expression of as yet unidentified mucin functions.

3) Several motifs specific for posttranslational glycosylation such as N- and O-glycosylation specific sites [42, 43] as well as C-mannosylation specific motifs [44] are found in all secreted mucins.

4) A large group of binding motifs specific for proteins involved in basic cell processes such as signal transduction, transcription and apoptosis are found in the mucins tested. All secreted mucins except MUC7 contain binding sites specific for the 14-3-3 proteins known to act as adaptor molecules to mediate protein-protein interactions, enzyme activities and subcellular localization [45]. The general mechanisms of 14-3-3 action include changes in activity of the bound ligands and changes in intracellular localization of 14-3-3-bound cargo [46, 47]. Interestingly, the number of the 14-3-3-specific sites and their precise sequence and score are different for each mucin. For example, MUC8 contains seven sites (two specific for 14-3-3-1 ligand and five specific for 14-3-3-3 ligand) characterized by high scores, while the MUC19 mucin contains two sites specific for 14-3-3-2 ligand, only one with a high score. Mucins MUC2, MUC5AC and MUC19 contain multiple sites specific for all three types of 14-3-3 ligands, while MUC5B and MUC6 contain sites specific only for two 14-3-3 ligands. These data indicate that mucins differ in their ability to be bound to different 14-3-3 proteins and therefore to be engaged in processes mediated by this group of adaptor signaling molecules.

5) The BRCT-domain (C-terminal domain of a breast cancer susceptibility protein) binding sequence is found in all secreted mucins except MUC8

and MUC9. Since the BRCT domains are present in proteins associated with DNA damage response and cell-cycle control [48, 49], finding of these motifs in the secreted mucins implicates the secreted mucin glycoproteins in the basic cell processes.

6) Multiple copies of the forkhead-associated (FHA)-domain ligand binding sites are detected in all secreted mucins. The FHA domain is a signal transduction module that recognizes phosphothreonine-containing peptides as the ligand proteins. FHA partakes in many signaling processes including cell cycle checkpoint, DNA repair and transcriptional regulation [50]. The largest number of FHA-binding sites is found in the gel-forming mucins, compared to a smaller number found in the soluble mucins, implicating the gel-forming mucins in the processes mediated by the FHA-containing proteins.

7) All secreted mucins except of MUC8 contain variable numbers of the FBW7 motif specific for binding to WD40 domains of the F-box containing proteins. These proteins are associated with ubiquitin-mediated proteolysis, which in turn controls equilibrium between oncogenes and tumor suppressors and regulates cell cycle and other basic biological processes [51, 52]. Interestingly, all mucins except MUC8 also contain USP7-specific motifs, which can potentially engage mucins in the opposite processes associated with deubiqiutination. USP7 is a deubiquitinating enzyme that cleaves ubiquitin moieties from its substrate, thus participating in regulation of stress response, epigenetic silencing and cell survival pathways [53]. Importantly, the same mucin molecule contains two different motifs, one specific for protein degradation and one indispensable for escape from protein degradation. Such a combination of FBW7 and USP7 motifs in one molecule gives mucins the potential to serve as regulators of two opposite processes associated with degradation and preservation of proteins, thereby supporting intracellular homeostasis. Interestingly, the MUC8 molecule displayed neither pro-degradation FBW7 motif nor anti-degradation USP7 sequence.

8) The PDZ domain binding motif is present in all secreted mucins except MUC8, in a copy number as small as 2 in MUC7 and as large as 62 in MUC5AC. Although high affinity PDZ sites are almost always located at the C-terminus, lower affinity internal sites were also reported [54, 55]. Importantly, all secreted mucins contain both the C-terminal PDZ-binding motifs and the internal sites. In this context, an interesting observation was made by Wang *et al.* [56] who found that various domains, including PDZ domain, "have regions outside of their canonical definition that affect their structure and function". The PDZ domains are found in various proteins in humans and are characterized by a high degree of plasticity in terms of ligand preferences [57]. The secreted PDZ domain containing protein 2 induces senescence or quiescence of prostate, breast and liver cancer cells *via* transcriptional activation of p53 [58], and may function as a potential prostate tumor suppressor [59]. The PDZ-containing proteins affect also the growth and differentiation of human fetal pancreatic stem cells [60]. Thus, potential interaction of MUC2 with PDZ-domain containing proteins could explain many functions of the MUC2 mucin, including the tumor suppression properties of the MUC2 described in mice [61].

9) All gel-forming mucins and the soluble MUC9 mucin contain the binding motifs specific for Src homology 2 (SH2) domains. These domains are found in a large number of proteins involved in different signaling pathways [62]. In addition, all secreted mucins except MUC8 contain the binding sites for SH3-domains present in the proteins involved in biological processes as diverse as signal transduction, cytoskeleton organization, membrane traffic and organelle assembly [63].

10) All secreted mucins with SH2- and SH3-domain specific binding sites possess also the MAPK docking motifs. MAPK-interacting molecules carry docking motifs that facilitate specific interactions in the MAPK-mediated multiple signal transduction pathways [64, 65]. In addition to the MAPK kinases-specific sites, all secreted mucins contain sites

specific for Polo-like-kinase, PKA-type AGC kinase, PIK kinase and GSK3 kinase, all of which conduct phosphorylation of the substrate molecules involved in signal transduction cascades [66-70].

11) Four gel-forming mucins, including MUC2, MUC5B, MUC6 and MUC19, and soluble mucin MUC9 contain motifs specific for interaction with cyclins and cyclin-dependent protein kinase 1 (CDK1) [71-73], which may link these mucins to cell cycle processes. Interestingly, the MUC5AC and MUC7 mucins do not contain CDK1-specific motifs, but contain cyclin-binding sites, whereas MUC8 contains CDK1-specific motifs but lacks the cyclin-binding sequence [74].

12) All secreted mucins are positive for multiple sites specific for interaction with WW domains, which are known to be involved in a number of cellular processes including ubiquitin-mediated degradation and mitotic regulation [75, 76]. These domains mediate protein-protein interaction through binding of short proline-rich regions within proteins. The multiple representations of these motifs in mucins indicate a possible interaction of mucins with proline-rich proteins.

13) All secreted mucins contain the TNF receptor-associated factor (TRAF) binding sites. TRAFs are known to be the main mediators of cell activation, cell survival and anti-apoptotic functions of the TNF receptor superfamily [77]. Thus, binding of TRAFs to mucins may link the mucin glycoproteins to the fundamental processes associated with cell survival.

14) A single copy of a motif specific for the V-domain of the Alix protein associated with ESCRT (Endosomal Sorting Complex Required for Transport) is found only in the MUC2 and MUC5AC mucins. The Alix protein and ESCRT are known mainly as functional complexes important for retroviral budding [78], although human homolog active in human cells has been also described [79]. The fact that only two gel-forming mucins contain the Alix-specific motif, while other

secreted mucins do not have it, may indicate a unique function of MUC2 and MUC5AC in the cell in the context of endosomal sorting and/or retroviral budding.

In addition to the motifs and binding sites mentioned above, manual inspection of the secreted mucin amino acid sequences revealed several motifs specific for interaction with integrin receptors expressed on the cell surface. The presence of these binding sites in mucins suggests interaction of the mucins and integrins in the ligand-receptor reactions. The integrin specific sites found in the secreted mucins represent a wide range of short (three amino acids) sequences characteristic of different groups of integrin receptors [80-85]. The MUC2 mucin contains two DGR sites, two KGD sites (one in direct and one in reverse orientation), and one RHD site in reverse orientation. MUC5AC contains two NGR sites and one copy of RLP, LDV and RQR sequences. Interestingly, integrin-specific triplets are typically located at the N-termini of integrin ligands as well as in mucin molecules, whereas RQR site, unique to the Tat protein of HIV [86], is located at the very C-terminal end of the MUC5AC molecule. The MUC5B mucin contains two LDV sites and one RLD site. The MUC6 mucin contains a single NGR site, whereas MUC19 and all soluble mucins contain highly homologous sites with one nonconsensus amino acid substitution.

Integrins are cell adhesion receptors that mediate cell-extracellular matrix and cell-cell interactions *via* binding of the specific ligands to the specific binding sites, most of which are composed, as noted above, of only three amino acids. Through interaction with the corresponding integrin receptor, a ligand may regulate intracellular processes. For example, cysteine proteinase 5 expressed on the surface of *E. histolytica* binds *via* RGD motif to α(V)β(3) integrin on Caco-2 colonic cells and stimulates intracellular NFκB-mediated pro-inflammatory reactions. The presence of integrin-specific binding sites on the secreted mucin molecules indicate that these glycoproteins may also activate and regulate intracellular processes through interactions with integrins.

Collectively, the presented data show that the gel-forming and soluble mucins contain multiple motifs and binding sites that may allow interaction with the extra- and intracellular protein partners. If such interactions do take place, they could influence biochemical and biophysical processes both in the extracellular matrix and inside the cell.

To verify whether the discovered motifs can determine interactions between the secreted mucins and protein partners, we used the STRING 9.0 interaction network (http://string.embl.de/newstring_cgi/show_network_section.pl) for prediction of the possible interactive partners. Table **1** summarizes the predicted partners for each secreted mucin, with the exception of MUC8, for which the STRING 9.0 interaction network does not have enough information. The interaction scores signify the probability of a particular mucin-protein binding.

Table 1: Predicted Functional Partners of the Secreted Mucin.

Partners	MUC2	MUC5AC	MUC5B	MUC6	MUC19	MUC7	MUC9
CHPF	0.978		0.870				
ATOH1	0.973	0.952	0.954	0.732			
CDX2	0.965			0.784			
MUC7	0.925		0.927	0.876	0.598		
CDX1	0.923						
MUC1	0.922	0.825	0.874	0.894		0.906	
NFκB	0.871		0.849	0.815			
MUC12	0.856		0.863				
TFF3	0.854		0.850	0.834			
SOX9	0.850						
MUC3A	0.841		0.842	0.854		0.860	
IL4	0.839						
TNF	0.837				0.836	0.832	
SP1	0.836		0.865				
HLA-A	0.835						
CD33	0.816						
VIP	0.815						
MUC17	0.804						
MMP9	0.800						
MME	0.786			0.784			
KRT7	0.786			0.784			

Table 1: cont….

KRT20	0.786			0.782			
GALNT14	0.784						
TFF1	0.782		0.954	0.911	0.577		
Tp53	0.780			0.769			
GBGT1	0.780						
SMAD4	0.760			0.706			
AGR2	0.757						
CDH1	0.756			0.730			
TFF2	0.754			0.780			
GALNT12	0.750						
IL8					0.846		
IL6					0.830		
KIF21B					0.704		
ELANE			0.943				
NKX2-1	0.696		0.925				
LYZ			0.911				
SOX2			0.898				
AHR			0.883				
STAT6			0.879				
MUC13			0.863			0.826	
ZFHX3			0.850				
IL17A		0.847	0.848				
GATA6			0.823				
INS							0.786
GCG							0.780
GPR83							0.764
GP6							0.750
RBBP9							0.720
GCK							0.692
ZP3							0.677
SST							0.671
MYH9							0.668
DPF1							0.651
MUC5						0.927	
MUC2						0.925	

Table 1: cont….

MUC6						0.876	
LTF						0.873	
ENO1						0.832	
MUC4		0.866				0.831	
CLDN18	0.740			0.756			
Slc26a4		0.828					
Csf3		0.782					
Clca3		0.770					
Ifitm2		0.694					
IL13		0.637					
Kcnk		0.601					
MUC16	0.694						
MUC15	0.685						
ERBB2	0.683						
APC	0.657						
MYL4	0.649						

Legend: CHPF – chondroitin polymerizing factor; ATOH1 – atonal homolog (E-box transcriptional regulation); CDX2 – caudal type homeobox (transcriptional regulation); MUC7 – soluble mucin; CDX1 – caudal type homeobox (intestinal differentiation); MUC1 – membrane-bound mucin; NF*k*B – transcription factor; MUC12 – membrane-bound mucin; TFF3 – trefoil factor 3; SOX9 – regulator of skeletal development; MUC3A –membrane-bound mucin; IL4 – interleukin 4 (B-cell activation); TNF – cytokine; SP1 – transcription factor; HLA-A – histocompatibility complex; CD33 - adhesion molecule myelomonocytic origin; VIP - vasoactive intestinal peptide; MUC17 – membrane-bound mucin; MMP9 – matrix metallopeptidase 9; MME – membrane metallo-endopeptidase; KRT7 – keratin 7 (stimulation of DNA synthesis); KRT20 – keratin 20 (intestinal differentiation); GALNT14 –GalNAc-14 transferase; TFF1 – trefoil factor 1 (stabilizer of mucus gel); Tp53 – tumor suppressor protein; GBGT1 – globoside-NAc1 transferase; SMAD4 – mediator of signal transduction by TGFβ; AGR2 - factor required for MUC2 synthesis; CDH1 – cadherin 1; TFF2 – trefoil factor 2 (inhibitor of gastric acid secretion); GALNT12 – GalNAc-12 transferase; CLDN18 – claudin 18; NKX2 – NK2 homeobox 1 (transcription factor); MUC15 - membrane-bound mucin; MUC16 – membrane-bound mucin; ERBB2 – v-erb-b2 oncogene coding protein; APC – adenomatous polyposis coli tumor suppressor; MYL4 – myosin, light chain 4; IL17a – interleukin 17A; Slc26a4 – sodium-independent transporter of chloride and iodide; Csf2 – colony stimulating factor 2 (cytokine); IL13 – interferon-induced transmembrane protein 2; Kcnk7 – potassium channel subunit; ELANE – neutrophil elastase; LYZ – lysozyme; SOX2 – SRY-box 2 (transcription factor); AHR – aryl hydrocarbon receptor; STAT6 – signal transducer and activator of transcription; INS – insulin; GCG - glucagon; GPR83 – G protein-coupled receptor; GP6 – collagen receptor; RBBP9 – retinoblastoma binding protein 9; GCK – glucokinase; ZP3 – zona pellucida glycoprotein 3 (sperm receptor); SST – somatostatin; MYH9 – myosin, heavy chain 9; DPF1 – zinc and double PHD finger family 1 protein; IL6 – interleukin 6; IL8 – interleukin 8; TNF – tumor necrosis factor (cytokine); KIF21B – kinesin family member 21B (motor protein); red figures - high confidence – 0.700 and high score (0.978-0.800); black figures - high confidence = 0.700, but relatively low score (0.800- 0.600).

As follows from the data presented in the Table **1**, the predicted functional partners of the secreted mucins represent a wide range of proteins involved in various basic cell processes. The potential partners include transcription factors and cytokines, enzymes and receptors, tumor suppressors and oncoproteins, and more. While the probabilities of interactions between mucins and potential protein partners are different for different pairs of mucins and interacting proteins, a large number of pairs have a high probability (scores of 0.994-0.800) of functionally important interactions. Further experimental studies are expected to identify partners of the gel-forming and soluble mucins and demonstrate the participation of the mucins in basic extra- and intracellular processes.

Figs. **2-8** illustrate possible interactions of 10 most reliable predicted protein partners of the MUC2, MUC5AC, MUC5B, MUC6, MUC19, MUC7 and MUC9 mucins, respectively.

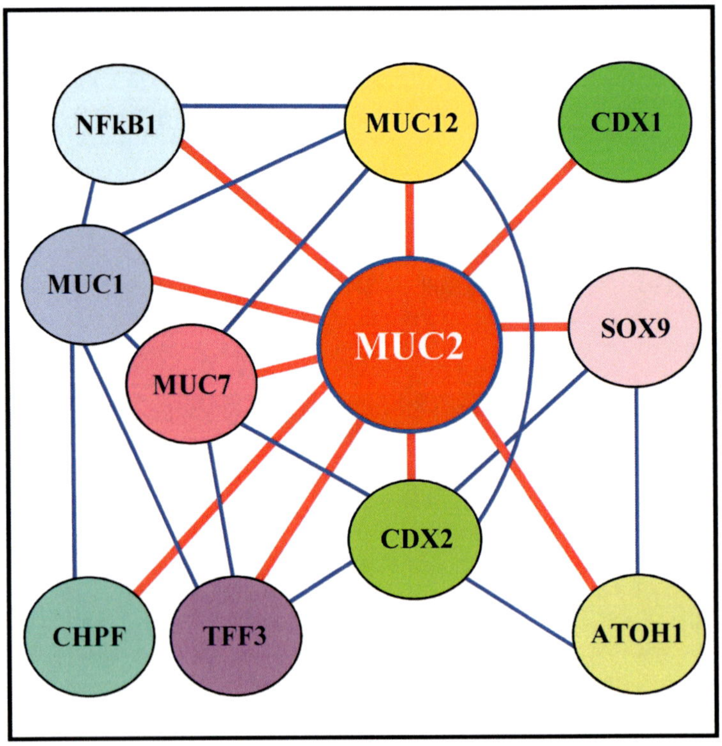

Figure 2: Predicted functional partners of the human MUC2 mucin.

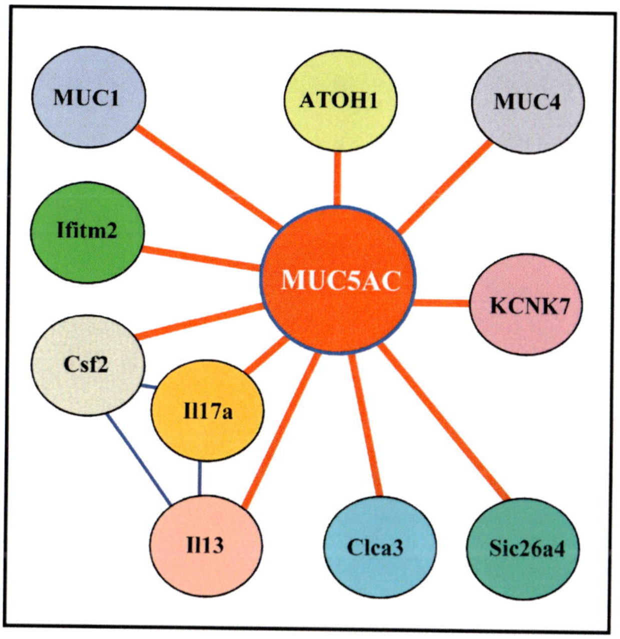

Figure 3: Predicted functional partners of the mouse MUC5AC mucin.

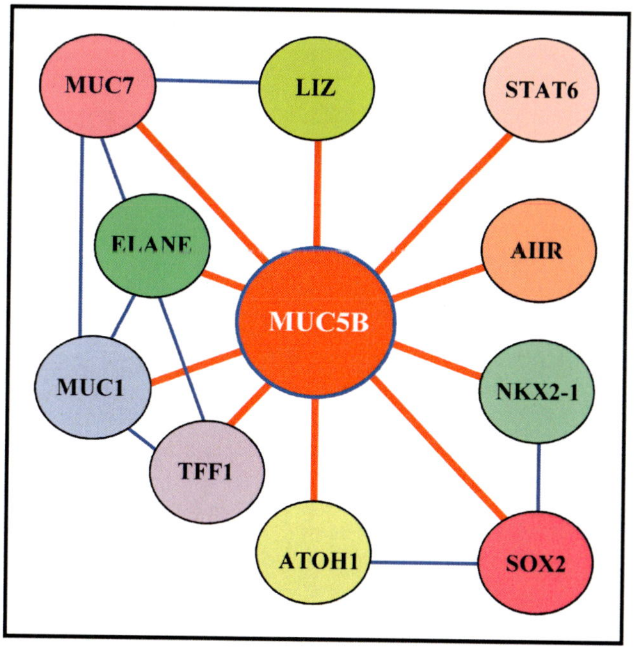

Figure 4: Predicted functional partners of the human MUC5B mucin.

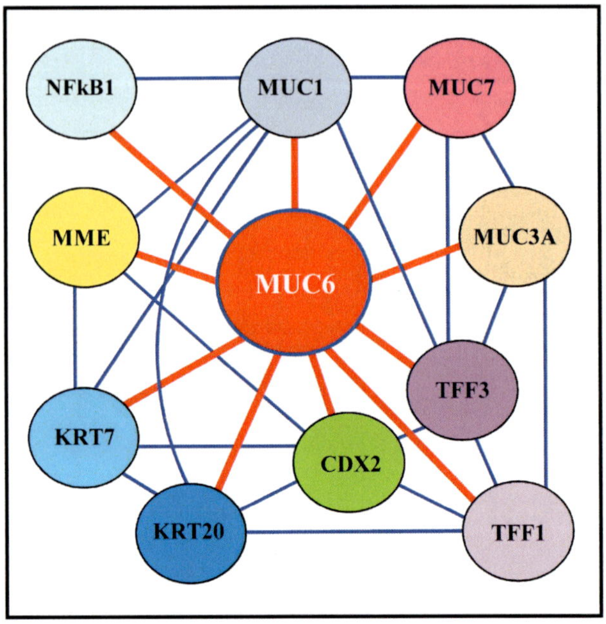

Figure 5: Predicted functional partners of the human MUC6 mucin.

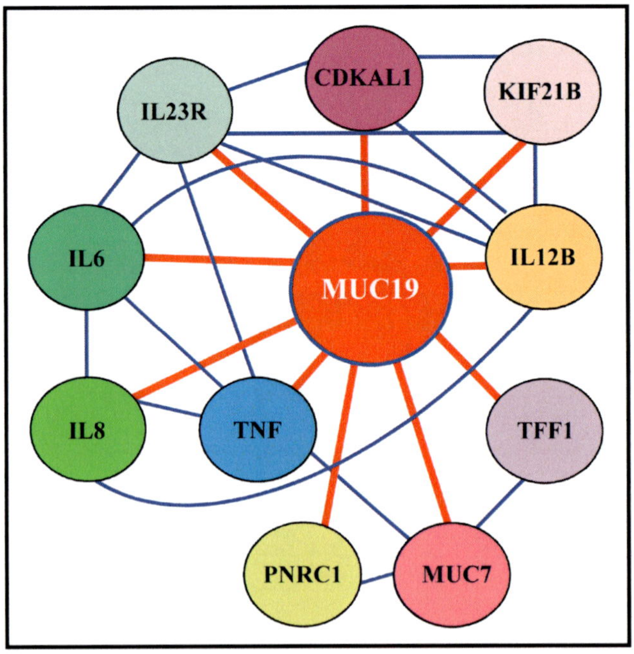

Figure 6: Predicted functional partners of the human MUC19 mucin.

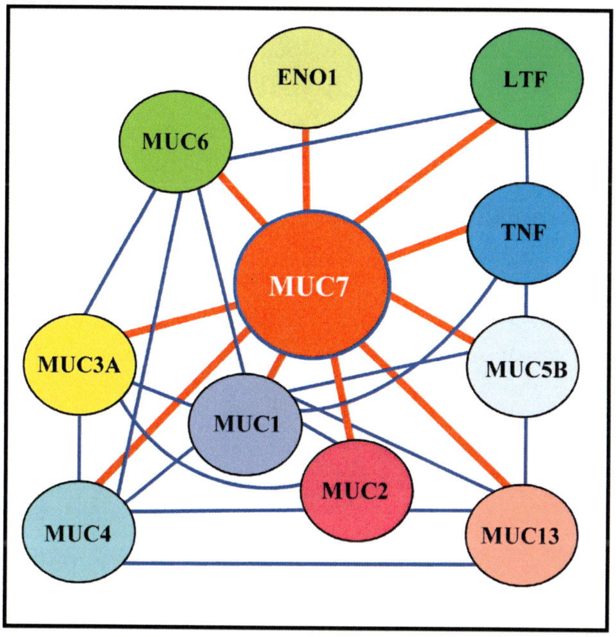

Figure 7: Predicted functional partners of the human MUC7 mucin.

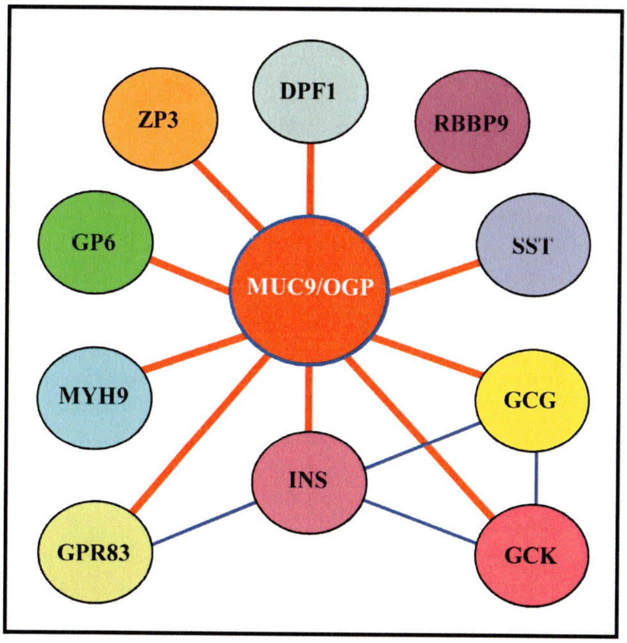

Figure 8: Predicted functional partners of the human MUC9/OGP mucin.

Taken together, the results obtained by experimental approaches and bioinformatic analyses suggest that secreted mucin glycoproteins are multifunctional proteins that have the potentials to be involved in various biological processes. Some of the possible mucin-partner interactions have already been proven experimentally, others await verification. Future studies will extend our knowledge of the functional potentials of mucin glycoproteins.

REFERENCES

[1] Zaretsky JZ, Wreschner DH. Protein multifunctionality: principles and mechanisms. Transl Oncogenomics 2008;3:99-136.

[2] Cone RA. Barrier properties of mucus. Adv Drug Deliv Rev 2009;61:75-85.

[3] Thornton DJ, Rousseau K, McGuckin MA. Structure and function of the polymeric mucins in airways mucus. Annu Rev Physiol 2008;70:459-86.

[4] McGuckin MA, Linden SK, Sutton P, Florin TH. Mucin dynamics and enteric pathogens. Nat Rev Microbiol 9:265-78.

[5] Deplancke B, Gaskins HR. Microbial modulation of innate defense: goblet cells and the intestinal mucus layer. Am J Clin Nutr 2001;73:1131S-41S.

[6] Davidson HJ, Kuonen VJ. The tear film and ocular mucins. Vet Ophthalmol 2004;7:71-7.

[7] Kufe DW. Mucins in cancer: function, prognosis and therapy. Nat Rev Cancer 2009;9:874-85.

[8] Van Seuningen I. The Epithelial Mucins: Structure/Function. Roles in Cancer and Inflammatory Diseases. Kerala: Research Signpost; 2008.

[9] Barondes SH, Castronovo V, Cooper DN, *et al.* Galectins: a family of animal beta-galactoside-binding lectins. Cell 1994;76:597-8.

[10] Ho MK, Springer TA. Mac-2, a novel 32,000 Mr mouse macrophage subpopulation-specific antigen defined by monoclonal antibodies. J Immunol 1982;128:1221-8.

[11] Paz I, Sachse M, Dupont N, *et al.* Galectin-3, a marker for vacuole lysis by invasive pathogens. Cell Microbiol 2010;12:530-44.

[12] Ochieng J, Furtak V, Lukyanov P. Extracellular functions of galectin-3. Glycoconj J 2004;19:527-35.

[13] Nakahara S, Oka N, Raz A. On the role of galectin-3 in cancer apoptosis. Apoptosis 2005;10:267-75.

[14] Ashery U, Yizhar O, Rotblat B, *et al.* Spatiotemporal organization of Ras signaling: rasosomes and the galectin switch. Cell Mol Neurobiol 2006;26:471-95.

[15] Park JW, Voss PG, Grabski S, Wang JL, Patterson RJ. Association of galectin-1 and galectin-3 with Gemin4 in complexes containing the SMN protein. Nucleic Acids Res 2001;29:3595-602.

[16] Wang JL, Gray RM, Haudek KC, Patterson RJ. Nucleocytoplasmic lectins. Biochim Biophys Acta 2004;1673:75-93.

[17] Lin HM, Pestell RG, Raz A, Kim HR. Galectin-3 enhances cyclin D(1) promoter activity through SP1 and a cAMP-responsive element in human breast epithelial cells. Oncogene 2002;21:8001-10.

[18] Song S, Byrd JC, Mazurek N, *et al.* Galectin-3 modulates MUC2 mucin expression in human colon cancer cells at the level of transcription *via* AP-1 activation. Gastroenterology 2005;129:1581-91.

[19] Haudek KC, Spronk KJ, Voss PG, *et al.* Dynamics of galectin-3 in the nucleus and cytoplasm. Biochim Biophys Acta 2010;1800:181-9.

[20] Furtak V, Hatcher F, Ochieng J. Galectin-3 mediates the endocytosis of beta-1 integrins by breast carcinoma cells. Biochem Biophys Res Commun 2001;289:845-50.

[21] Bresalier RS, Byrd JC, Wang L, Raz A. Colon cancer mucin: a new ligand for the beta-galactoside-binding protein galectin-3. Cancer Res 1996;56:4354-7.

[22] Chadee K, Johnson ML, Orozco E, Petri WA, Jr., Ravdin JI. Binding and internalization of rat colonic mucins by the galactose/N-acetyl-D-galactosamine adherence lectin of Entamoeba histolytica. J Infect Dis 1988;158:398-406.

[23] Bresalier RS, Niv Y, Byrd JC, *et al.* Mucin production by human colonic carcinoma cells correlates with their metastatic potential in animal models of colon cancer metastasis. J Clin Invest 1991;87:1037-45.

[24] Schoeppner HL, Raz A, Ho SB, Bresalier RS. Expression of an endogenous galactose-binding lectin correlates with neoplastic progression in the colon. Cancer 1995;75:2818-26.

[25] Bresalier RS, Ho SB, Schoeppner HL, *et al.* Enhanced sialylation of mucin-associated carbohydrate structures in human colon cancer metastasis. Gastroenterology 1996;110:1354-67.

[26] Byrd JC, Bresalier RS. Mucins and mucin binding proteins in colorectal cancer. Cancer Metastasis Rev 2004;23:77-99.

[27] Bresalier RS, Mazurek N, Sternberg LR, *et al.* Metastasis of human colon cancer is altered by modifying expression of the beta-galactoside-binding protein galectin 3. Gastroenterology 1998;115:287-96.

[28] Dudas SP, Yunker CK, Sternberg LR, Byrd JC, Bresalier RS. Expression of human intestinal mucin is modulated by the beta-galactoside binding protein galectin-3 in colon cancer. Gastroenterology 2002;123:817-26.

[29] Bresalier RS. The biology of colorectal cancer metastasis. Gastroenterol Clin North Am 1996;25:805-20.

[30] Yu LG, Andrews N, Zhao Q, *et al.* Galectin-3 interaction with Thomsen-Friedenreich disaccharide on cancer-associated MUC1 causes increased cancer cell endothelial adhesion. J Biol Chem 2007;282:773-81.

[31] Bu X, Li L, Li N, Tian X, Huang P. Suppression of mucin 2 enhances the proliferation and invasion of LS174T human colorectal cancer cells. Cell Biol Int 2011;35:1121-9.

[32] Gibson JA, Hahn HP, Shahsafaei A, Odze RD. MUC expression in hyperplastic and serrated colonic polyps: lack of specificity of MUC6. Am J Surg Pathol 2011;35:742-9.

[33] Bu X, Li N, Tian X, Huang P. Gamma-secretase inhibitor, a potential target therapy for MUC2-positive colorectal carcinoma. Neoplasma 2011;58:343-7.

[34] Argueso P, Guzman-Aranguez A, Mantelli F, *et al.* Association of cell surface mucins with galectin-3 contributes to the ocular surface epithelial barrier. J Biol Chem 2009;284:23037-45.

[35] Yu H, He Y, Zhang X, *et al.* The rat IgGFcgammaBP and Muc2 C-terminal domains and TFF3 in two intestinal mucus layers bind together by covalent interaction. PLoS One 2011;6:e20334.

[36] Pais-Correia AM, Sachse M, Guadagnini S, *et al.* Biofilm-like extracellular viral assemblies mediate HTLV-1 cell-to-cell transmission at virological synapses. Nat Med 2010;16:83-9.
[37] Kadam KM, D'Souza SJ, Natraj U. Identification of cellular isoform of oviduct-specific glycoprotein: role in oviduct tissue remodeling? Cell Tissue Res 2007;330:545-56.
[38] Hospital V, Chesneau V, Balogh A, *et al.* N-arginine dibasic convertase (nardilysin) isoforms are soluble dibasic-specific metalloendopeptidases that localize in the cytoplasm and at the cell surface. Biochem J 2000;349:587-97.
[39] Seidah NG, Mowla SJ, Hamelin J, *et al.* Mammalian subtilisin/kexin isozyme SKI-1: A widely expressed proprotein convertase with a unique cleavage specificity and cellular localization. Proc Natl Acad Sci USA 1999;96:1321-6.
[40] Xu G, Bell SL, McCool D, Forstner JF. The cationic C-terminus of rat Muc2 facilitates dimer formation post translationally and is subsequently removed by furin. Eur J Biochem 2000;267:2998-3004.
[41] Nakayama K. Furin: a mammalian subtilisin/Kex2p-like endoprotease involved in processing of a wide variety of precursor proteins. Biochem J 1997;327 (Pt 3):625-35.
[42] Shakin-Eshleman SH, Spitalnik SL, Kasturi L. The amino acid at the X position of an Asn-X-Ser sequon is an important determinant of N-linked core-glycosylation efficiency. J Biol Chem 1996;271:6363-6.
[43] Shao L, Haltiwanger RS. O-fucose modifications of epidermal growth factor-like repeats and thrombospondin type 1 repeats: unusual modifications in unusual places. Cell Mol Life Sci 2003;60:241-50.
[44] Furmanek A, Hofsteenge J. Protein C-mannosylation: facts and questions. Acta Biochim Pol 2000;47:781-9.
[45] Morrison DK. The 14-3-3 proteins: integrators of diverse signaling cues that impact cell fate and cancer development. Trends Cell Biol 2009;19:16-23.
[46] Gardino AK, Smerdon SJ, Yaffe MB. Structural determinants of 14-3-3 binding specificities and regulation of subcellular localization of 14-3-3-ligand complexes: a comparison of the X-ray crystal structures of all human 14-3-3 isoforms. Semin Cancer Biol 2006;16:173-82.
[47] Fanger GR, Widmann C, Porter AC, *et al.* 14-3-3 proteins interact with specific MEK kinases. J Biol Chem 1998;273:3476-83.
[48] Caldecott KW. Cell signaling. The BRCT domain: signaling with friends? Science 2003;302:579-80.
[49] Varma AK, Brown RS, Birrane G, Ladias JA. Structural basis for cell cycle checkpoint control by the BRCA1-CtIP complex. Biochemistry 2005;44:10941-6.
[50] Durocher D, Henckel J, Fersht AR, Jackson SP. The FHA domain is a modular phosphopeptide recognition motif. Mol Cell 1999;4:387-94.
[51] Kitagawa K, Kotake Y, Kitagawa M. Ubiquitin-mediated control of oncogene and tumor suppressor gene products. Cancer Sci 2009;100:1374-81.
[52] Nakayama KI, Nakayama K. Regulation of the cell cycle by SCF-type ubiquitin ligases. Semin Cell Dev Biol 2005;16:323-33.
[53] Amerik AY, Hochstrasser M. Mechanism and function of deubiquitinating enzymes. Biochim Biophys Acta 2004;1695:189-207.
[54] Hung AY, Sheng M. PDZ domains: structural modules for protein complex assembly. J Biol Chem 2002;277:5699-702.

[55] Sheng M, Sala C. PDZ domains and the organization of supramolecular complexes. Annu Rev Neurosci 2001;24:1-29.

[56] Wang CK, Pan L, Chen J, Zhang M. Extensions of PDZ domains as important structural and functional elements. Protein Cell 2010;1:737-51.

[57] Jemth P, Gianni S. PDZ domains: folding and binding. Biochemistry 2007;46:8701-8.

[58] Tam CW, Liu VW, Leung WY, Yao KM, Shiu SY. The autocrine human secreted PDZ domain-containing protein 2 (sPDZD2) induces senescence or quiescence of prostate, breast and liver cancer cells *via* transcriptional activation of p53. Cancer Lett 2008;271:64-80.

[59] Tam CW, Cheng AS, Ma RY, Yao KM, Shiu SY. Inhibition of prostate cancer cell growth by human secreted PDZ domain-containing protein 2, a potential autocrine prostate tumor suppressor. Endocrinology 2006;147:5023-33.

[60] Suen PM, Zou C, Zhang YA, *et al.* PDZ-domain containing-2 (PDZD2) is a novel factor that affects the growth and differentiation of human fetal pancreatic progenitor cells. Int J Biochem Cell Biol 2008;40:789-803.

[61] Velcich A, Yang W, Heyer J, *et al.* Colorectal cancer in mice genetically deficient in the mucin Muc2. Science 2002;295:1726-9.

[62] Pawson T, Gish GD, Nash P. SH2 domains, interaction modules and cellular wiring. Trends Cell Biol 2001;11:504-11.

[63] Musacchio A, Wilmanns M, Saraste M. Structure and function of the SH3 domain. Prog Biophys Mol Biol 1994;61:283-97.

[64] Bardwell AJ, Flatauer LJ, Matsukuma K, Thorner J, Bardwell L. A conserved docking site in MEKs mediates high-affinity binding to MAP kinases and cooperates with a scaffold protein to enhance signal transmission. J Biol Chem 2001;276:10374-86.

[65] Sharrocks AD, Yang SH, Galanis A. Docking domains and substrate-specificity determination for MAP kinases. Trends Biochem Sci 2000;25:448-53.

[66] Nakajima H, Toyoshima-Morimoto F, Taniguchi E, Nishida E. Identification of a consensus motif for Plk (Polo-like kinase) phosphorylation reveals Myt1 as a Plk1 substrate. J Biol Chem 2003;278:25277-80.

[67] Fujii K, Zhu G, Liu Y, *et al.* Kinase peptide specificity: improved determination and relevance to protein phosphorylation. Proc Natl Acad Sci U S A 2004;101:13744-9.

[68] Kim ST, Lim DS, Canman CE, Kastan MB. Substrate specificities and identification of putative substrates of ATM kinase family members. J Biol Chem 1999;274:37538-43.

[69] Hur EM, Zhou FQ. GSK3 signaling in neural development. Nat Rev Neurosci 2010;11:539-51.

[70] Wu D, Pan W. GSK3: a multifaceted kinase in Wnt signaling. Trends Biochem Sci 2010;35:161-8.

[71] Moseley JB, Nurse P. Cdk1 and cell morphology: connections and directions. Curr Opin Cell Biol 2009;21:82-8.

[72] Porter LA, Donoghue DJ. Cyclin B1 and CDK1: nuclear localization and upstream regulators. Prog Cell Cycle Res 2003;5:335-47.

[73] Klein EA, Assoian RK. Transcriptional regulation of the cyclin D1 gene at a glance. J Cell Sci 2008;121:3853-7.

[74] Pines J. Cyclins and cyclin-dependent kinases: a biochemical view. Biochem J 1995;308 (Pt 3):697-711.

[75] Sudol M, Sliwa K, Russo T. Functions of WW domains in the nucleus. FEBS Lett 2001;490:190-5.

[76] Verdecia MA, Bowman ME, Lu KP, Hunter T, Noel JP. Structural basis for phosphoserine-proline recognition by group IV WW domains. Nat Struct Biol 2000;7:639-43.

[77] Bradley JR, Pober JS. Tumor necrosis factor receptor-associated factors (TRAFs). Oncogene 2001;20:6482-91.

[78] Odorizzi G. The multiple personalities of Alix. J Cell Sci 2006;119:3025-32.

[79] Vincent O, Rainbow L, Tilburn J, Arst HN, Jr., Penalva MA. YPXL/I is a protein interaction motif recognized by aspergillus PalA and its human homologue, AIP1/Alix. Mol Cell Biol 2003;23:1647-55.

[80] Corti A, Curnis F. Isoaspartate-dependent molecular switches for integrin-ligand recognition. J Cell Sci 124:515-22.

[81] Ruoslahti E. RGD and other recognition sequences for integrins. Annu Rev Cell Dev Biol 1996;12:697-715.

[82] Plow EF, Pierschbacher MD, Ruoslahti E, Marguerie GA, Ginsberg MH. The effect of Arg-Gly-Asp-containing peptides on fibrinogen and von Willebrand factor binding to platelets. Proc Natl Acad Sci U S A 1985;82:8057-61.

[83] Ghiso J, Rostagno A, Gardella JE, *et al.* A 109-amino-acid C-terminal fragment of Alzheimer's-disease amyloid precursor protein contains a sequence, -RHDS-, that promotes cell adhesion. Biochem J 1992;288 (Pt 3):1053-9.

[84] Guan JL, Hynes RO. Lymphoid cells recognize an alternatively spliced segment of fibronectin *via* the integrin receptor alpha 4 beta 1. Cell 1990;60:53-61.

[85] Altieri DC, Plescia J, Plow EF. The structural motif glycine 190-valine 202 of the fibrinogen gamma chain interacts with CD11b/CD18 integrin (alpha M beta 2, Mac-1) and promotes leukocyte adhesion. J Biol Chem 1993;268:1847-53.

[86] Vogel BE, Lee SJ, Hildebrand A, *et al.* A novel integrin specificity exemplified by binding of the alpha v beta 5 integrin to the basic domain of the HIV Tat protein and vitronectin. J Cell Biol 1993;121:461-8.

INDEX

A

A-domains 30
Aberrant expression 85, 96, 98, 102, 108, 114, 191, 203, 338, 344, 354, 362, 364-5, 480-1
ABO/Lewis[b] 471
Accessory glands 173, 274-5
Acetylation 58, 405
Acidic isoform 438
Acidic proline-rich protein 2 (PRP2) 410, 513
Acinar cells 88, 193, 274, 352, 354, 394, 403
Acrolein 161, 487
Acrosomal crescent 441, 515
Activation of:
- cryptic splice sites 60
- gastric MUC6 mucin 357
- MUC2 promoter 50
- MUC5B gene 252
- MUC6 expression 343
- MUC7 expression 405
- oncogenes 497, 500
- p38 MAPK cascade 405
- p53 554
- P2Y2 receptor 156, 257
- protein kinase C 156
- the Cdx1 gene 81
- the CDX2 and MUC2 genes 85, 86
- the CREB protein 160
- the EGFR gene 81
- the MUC5AC gene 159, 198
- the Ras/Raf/ERK-signaling pathway 152

Adaptation 456
Adaptor 53, 452, 552

Joseph Z. Zaretsky and Daniel H. Wreschner

Adenocarcinoma:
aggressive 79
breast 212, 366
cervical 109-10, 209, 358-60
clear cell 78, 291, 355
colon 64, 81, 100, 206, 288, 355
colorectal 59, 99, 353
ductal 79, 89, 91-2, 94, 193, 198, 499
endocervical 109, 209
endometrial 110, 208, 358
esophageal 79, 186, 340
gastric 68, 78
invasive 88-9, 91, 99, 110, 193, 209, 493
mixed 182, 364
mucinous 107, 109, 111, 210, 212, 357-9, 497
nonmucinous 107, 355
ovarian 357
pancreatic 196, 197, 287, 346, 348-9
poorly differentiated 181, 195, 200
prostate 107, 108, 176, 290, 361
pulmonary 77-8, 160, 364
solid 182, 289, 364
well-differentiated 188, 341
Adenomas 84, 91, 99, 205-7, 209-10, 289, 346, 352-3, 355-6, 493, 498, 507
Adenomatous hyperplasia 76, 182, 364, 508
Adenomatous polyps 206, 559
Adherent layer 101, 476, 478
Adhesins 170, 407, 466-8, 471, 475, 479-80, 486, 510
bacterial 466-8, 471
Adhesion 5-6, 101, 197, 283, 342, 407, 462, 467, 471, 479, 485, 490, 503-5, 508, 510, 547, 556, 559
Adult duodenum 274
Adult intestine 274, 453
Adult tissues 173, 273, 331, 387, 453, 455-6

AGC kinase 555
Aggregation 267, 323-4, 402, 438, 459, 461-2, 471, 510
Airway epithelium 48, 151-3, 157, 160, 162, 172, 181, 256, 331, 338, 362, 493, 505
Airway mucins 157, 183, 474, 487
Airway mucus 172, 185, 363, 465, 472-5
Airway secretion 156, 171, 183, 473-4, 485
Akt 159-60
Ala-Leu-Leu-Cys motif 407, 510
Alix Protein 555
Allelic polymorphism 248, 427
Allergens 151, 213-14, 484
Allergic rhinitis 75
Alternating layer 342, 477
Alternative splicing 14, 59-60, 164-6, 260-1, 326-7, 427
Alveolar lining cells 364
Amino acid sequences 8, 35, 63, 147, 168, 249, 261, 264-5, 269, 389, 392, 409, 418-9, 425-6, 428, 436-7, 439, 441-2, 452, 508, 515-6, 551, 556
Amino acids 16, 36, 44-5, 145-7, 248-9, 389, 399-401, 408, 418-19, 425, 427, 437, 511-2, 556
Ampullary cancer 94
Ampullary carcinoma 36, 94, 403
Amylase 410, 461, 470, 513
Ancestors 7, 14, 33, 388
Ancestral gene 8, 33, 246
Ancestral progenitor 425
Androgen-dependent prostate carcinoma 500
Animal lectins 551
Anti-apoptotic function 547, 555
Anti-defensive factor 469
Anti-HIV-1 activity 471
Anti-inflammatory potential 407, 510
Anti-MUC7 antibody 409, 511
Anti-VNTR antibodies 206

Antibacterial activity 409, 468, 471, 512, 519
Antibacterial factors 482
Antibacterial peptides 71, 476
Antibacterial properties 465
Antibodies 36, 59, 72, 102, 175, 182, 192, 195, 197, 206, 211, 289, 359, 431, 442, 461, 478, 516
Anticandidal agent 511
Antifungal activity 408-9, 462, 511-12
Antigenic landscape 199
Antigens 71, 170, 192, 195, 265, 325, 367, 399, 462, 471
Antimicrobial activity 71, 398, 410, 487, 512
Antiviral activities 452
Antrum 174, 189, 191, 274, 283, 322, 327, 330, 344, 453, 476, 480-1
AP-1 53-5, 148, 154, 162-3, 251, 254, 403-4, 430, 548
AP-2 56, 148, 258
APC 84, 191-2, 493, 495, 559
Apomucin 20-1, 35, 59, 61-3, 69, 72, 95-6, 99, 102, 166, 168, 172, 186, 188, 190, 195, 203, 260, 262-3, 286, 291, 321, 324-7, 336-7, 345, 366, 393, 398-9, 406, 408, 410, 418, 431, 437, 442, 511, 513, 516
Apomucin expression 69, 188, 195, 291, 345
Apoptosis 104, 153, 439, 491, 494-6, 500, 507, 552
Aquaporin 5 63, 505
Aqueous layer 482-4
Asthma 75-6, 149-51, 158, 162, 183-4, 276, 363, 475
Asthmatic exudate 473
ATF 54, 152, 251-2
ATG codon 320, 392, 432
ATP 155-6, 160, 171, 256-7, 421
Atypical adenomatous hyperplasia (AAH) 182, 364
Atypical fibrocystic glands 366
Atypical TATA box 430
AU-rich sequence 165
Audio system 113-14, 456
Auto-catalytic cleavage 269

Auto-cleavage 22-3, 66
Auto-cleavage of MUC5AC
  mucin glycoprotein 171
Auto-digestion 323, 456, 459
Auto-proteolysis 19, 166
Autocrine/paracrine loop 255
Autosomal recessive polycystic kidney disease 106, 290, 360

B

BabA adhesin 471, 479
Baboon 427-8, 437-8, 440, 442, 514, 516
Baboon gene 428
Bactenecins 410, 512
Bacteria 4, 71, 154, 164, 184, 189, 343-4, 406-7, 411, 456-7, 465-7, 471, 480, 485-6, 509-10
  colonized 409, 512
  commensal 466, 485
  pathogenic 184, 466
Bacterial adsorption 410
Bacterial adhesins 466-8, 471
Bacterial colonization 409, 469, 472, 512
Bacterial genes 468-9
Bacterial growth 466, 468-9
Bacterial infection 52, 97, 265, 268, 465
Bacterial lipoproteins 158
Bacterial LPS 151, 255-6
Bacterial pathogens 148, 486
Bacterial proteins 406-7, 479, 510
Bacterial receptors 486, 510
Bacterial stress 51, 53
Bacterial transcription factor 469
Bacterial toxins 154
Bacterial wall receptors 471
Bacterium 52, 170, 282, 342, 467-70, 475-6, 479

Barrett's adenocarcinoma 187-8, 282, 340-1
Barrett's adenocarcinoma cells 340
Barrett's cancer 187
Barrett's disease 186, 281, 341
Barrett's epithelium 80
Barrett's esophagus 78-81, 92, 186-8, 281-2, 339-41
Barrett's metaplasia 186, 340
Basal promoter 47, 252
Basic isoform 438
Basic proline-rich protein 3 (PRP3) 410, 513
Benign and borderline ovarian tumors 179
Benign gastric adenoma 84, 507
Benign lesions 109, 199, 208, 357
Bikunin 464
Bile acids 53, 81-2, 85, 345
Biliary epithelial metaplasia 346
Biliary tract 96, 275, 286
Biochemical reactivity 463
Biogenesis 321, 326
Bioinformatics 12, 148, 389, 547
Biological heterogeneity 361
Biological reproduction 37, 440, 452, 514
Biosynthesis 11, 20-2, 44, 60-1, 63-5, 67, 74, 103, 145, 166-7, 170, 206-7, 246, 261-3, 268-9, 316, 321, 325, 331, 393, 425, 434-5, 462-3
Biosynthetic pathway 3, 425
Bip 95, 168
BKLF 148
Bladder 36, 73, 106-7, 175, 211, 289, 335, 337, 360, 403
Bladder adenocarcinoma 106
Bladder cancer 36
Bladder tumors 403
Bladder urothelium 106, 175
Blastocyst 439-40, 442-3, 514, 516-17
Blood group 264-5, 399, 471, 479

Blood group epitopes 264, 399
Bonnet monkey Muc9/ogp 439
Bone marrow stem cells (BMSC) 87
Bone morphgenetic protein (BMP) 86
Borderline ovarian tumors 179, 291
Boundary lubrication 458, 484
Bovine Muc9/ogp 440, 514-15
Bovine sperm 440, 514
BRCT domain 553
Breast 51, 74, 114-15, 145, 212, 246, 291-3, 366-7, 493, 499-501, 508, 552, 554
Breast adenocarcinoma 212, 366
Breast cancer 51, 114-15, 145, 212, 291-3, 366-7, 500-1, 552
Breast carcinogenesis 115, 292
Breast carcinomas 114-15, 212, 493, 501
Breast ductal carcinoma 501
Brefeldin A 67, 436
Bronchial cells 54, 153
Bronchioles 68, 75, 173, 184, 278
Bronchioloalveolar carcinoma 76, 508
Brunner glands 79, 274
Bulb cells 395
Bulbourethral gland 289, 395-6

C

C-domain 30, 250, 317, 392, 464
C/EBP 35, 404
c-Ets1 160
C-Fos 55, 430
C-jun 54-5, 154, 430, 548
c-Jun/Fra 1 54
C-mannosylation 20, 166, 170, 268, 271, 552
c-Myc 493-5, 500, 506
c-Src 148
C-terminal CK-domain 317, 319

C-terminal cleavage 22, 269
C-terminal domain 35, 65, 248, 261, 317, 400-1, 422, 552
C-terminal regions 16-17, 22, 30, 32, 45, 147, 250, 254, 388, 399, 410-11, 427, 441, 513, 515
C-terminus 30, 65-6, 166, 172, 269, 400-1, 463, 554
CA^{2+}/Na^{+} exchange 272
CA19-9 358
CAAT-box 403, 430
CACCC box 258
Cadherin 196-7, 444, 494, 497, 501, 504-5, 517-18, 559
Cadherin 1 559
Calnexin 20, 61, 167
Calreticulin 20, 61, 167
cAMP 54, 159, 403, 432, 468, 487 cAMP
    response element 432
Cancer 5, 68, 170, 185, 205, 207, 419, 490
Cancer cells 55, 57, 59, 87, 164, 197, 259-60, 325, 342, 347, 463, 504
    human colon 288, 548
    pancreatic 164, 196-7, 199, 503-4
Cancer stem cells 87-8, 193
Candidacidal activity 35, 400, 408-9, 462, 511
Carbohydrate epitope 265, 476
Carbohydrate receptors 456, 509
Caretaker genes 490
Carcinogenesis 84-5, 87-8, 93-5, 97, 100, 110-11, 191-2, 201, 209-10, 292, 364-5, 494-5, 497-8, 500-6, 508
    breast 115, 292
    colon 100, 207, 505
    colorectal 497
    gastric 84-5, 191-2, 284, 502, 508
    pancreatic 88, 91, 93, 503-4
Carcinomas 78-9, 83-4, 92, 94, 97, 189-92, 199, 202-3, 210, 344, 497, 499
    bladder 106
    colloid 91-3, 96, 115, 198, 209, 212, 291, 360, 499

endometrial 36, 110, 178, 208, 419
gastric 82-4, 87, 189-92, 282-4, 342, 344, 479, 498, 501-2, 507
intestinal cell 82
intraductal 91, 198, 212, 351, 500, 501
invasive ductal 114, 292, 349, 367
lung 77-8, 181-2, 365
mucin-producing 182, 364
renal cell 106, 211, 290
Cascade 47, 52, 154-5, 190, 256, 344
Catenin 493, 497
Cationic peptides 410, 512
cDNA 38, 319, 418, 422, 426-8
Cdx1 49, 81-2, 557, 559
CDX2 49-50, 78, 80-2, 84-6, 93-4, 106, 109, 188, 498
CEA 358
Cell adhesion 5, 342, 485, 504, 556
Cell-cell interactions 254, 547, 556
Cell communication 11, 439
Cell cycle 53, 490-1, 494, 553, 555
Cell differentiation 36, 49-51, 57, 70, 105, 153, 259, 394-5, 455, 505
Cell growth 53, 342, 489
Cell migration 505, 507, 547, 549
Cell proliferation 57, 93, 259, 491-2, 547
Cell remodeling 442, 516
Cell signaling 48, 494
Cell systems 58, 254-5, 403, 405
Cells 5, 56-60, 69-70, 166, 169-70, 172-3, 179-80, 258-62, 270, 272-5, 288, 337-8, 436-7, 441-3, 548
bacterial 409, 512
cancer stem 87-8, 193
centroacinar 88, 193, 334, 348
ciliated cells 153, 277, 420, 437, 444, 518
colon adenocarcinoma 81
colorectal adenocarcinoma 59

human adenocarcinoma 81
inflammatory 86
intraductal carcinoma 198, 351
mucin-producing 259
mucin-secreting 209
nongoblet 79, 105, 502
oval 95
pancreatic adenocarcinoma 196-7
precursor 86-7
vascular endothelial 505-6
Centroacinar cells 88, 193, 334, 348
Cervical adenocarcinoma 109-10, 209, 358-60
Cervical canal 266, 209, 290-1
Cervical mucus 17, 177, 266-7, 275, 460-1
Cervix 45, 74, 109-10, 177-9, 207-9, 265-6, 269, 275, 334-5, 356-60, 395, 419
Cftr-knockout mice 348
Chaperones 20, 167-8, 262
Characterization 38, 264, 444
Checkpoint 553
Chemical irritants 362
Chemical reactions 488
Chemicals 3, 164, 273, 476, 479, 482, 486-8
Chemoresistance 288, 291
Chitinase activity 37, 425
Chitinase-like domain 437, 439
Cholangiocarcinoma 88, 94-6, 192, 198, 200-2, 285-6, 346-7, 503
Cholecystitis 98-9, 200, 202-3, 285, 345
Chromosomal clustering of mucin genes 14
Chromosomal localization 11, 15, 29-30, 32, 316, 387, 419, 422
Chromosome 14, 16, 29, 32, 34, 246, 316, 320, 347, 388, 398, 419, 426, 492
human 14, 16, 316, 418, 490
syntenic mouse 14, 16
Chromosome locus 29, 32-4, 44, 388
Chronic bronchitis 75-6, 184, 276

Chronic gastritis 83, 189, 342
Chronic airway inflammatory diseases 36
Chronic inflammation 86, 106, 203, 211, 339
Chronic obstructive pulmonary disease 75, 154, 256, 277, 363
Chronic pancreatitis 287, 348
Cigarette smoke 151, 161-2, 184, 252, 256, 486-7
Cilia 4, 36, 73, 153, 277, 363, 418, 420, 437, 443-4, 456, 473-6, 517-18
Ciliated cell marker 418, 420
Ciliated cells 153, 277, 420, 437, 444, 518
    developing 443, 517
Ciliogenesis 420, 443, 517
cis-CRE 159-61, 253
cis-elements 35, 46, 48-52, 54-5, 148, 150-3, 159-60, 252-6, 258, 326-8, 330, 403-4, 425, 429-30, 432-3
    cognate 50-2, 148, 155
cis-ERE 429-33
cis-Golgi 21, 169, 261, 264, 325
cis-GRE 163, 253-4
cis-NF*k*B 255, 330
cis-Sp1 252, 430
cis-spdef 330
cis-STAT1 328
cis-STAT6 150
CK-domains 17-18, 45, 147, 247-8, 250, 317, 319, 336, 387, 393
CK7 78, 94, 96, 355
CK20 94, 96, 106, 199, 355
Class III PDZ-domain-binding motifs 439
Class IV WW-ligand 439
Classification 3, 6-8, 29, 76, 82-4, 95, 115, 200, 283-4, 341, 352, 491, 498
    Goseki 82-4, 283-4
    HUGO 8
    Lauren's 83, 283-4
    WHO 76, 82-4, 284
Clatrin box 439

Claudin 18, 444, 518, 559
Clear cell metaplasia 355
Clear columnar cell 351
Cleavage 22-3, 65, 171-2, 269-70, 409, 462-4, 512, 551
CLN 167-8
Cloning 145-6, 387-9, 422
Cluster 14, 16, 32, 34, 109, 174-5, 246, 248, 259, 316, 319-20, 324, 331, 388, 459
Cluster distribution 14, 16
Cluster genes 317, 320, 388
Collapsed mucus 474
Colloid carcinoma 91-3, 96, 115, 198, 209, 212, 291, 360, 499
Colon 56-7, 63-5, 68-9, 99-104, 184-5, 204-7, 287-9, 342-4, 353-6, 409-10, 465-6, 468-70, 479-81, 504-5, 512
Colon adenocarcinomas 100, 206, 355
Colon apomucins 327
Colon cancer 45, 84, 94, 99-100, 288, 347, 354-5
Colon cancer cells 54, 548
Colon carcinogenesis 100, 207, 505
Colon crypts 56
Colon epithelium 45, 68, 353
Colon goblet cells 69
Colon inflammation 100
Colonization 406-7, 410, 465-6, 468, 470, 512
bacterial 409, 469, 472, 512
Colorectal adenocarcinoma 59, 99, 353
Colorectal cancer 59, 99, 205, 288-9, 353, 355, 494, 497
Colorectal neoplasm 353, 494
Colorectal polyp 355-6
Columnar cells 79, 280, 293, 340, 351, 357
Columnar metaplasia 79, 339
Commensal bacteria 466, 485
Common ancestor 33, 37, 316-17, 388
Complement system 442, 516
Complete androgen insensitivity syndrome (CAIS) 213

Components of tobacco smoke 161-2
Congenital cystic adenomatoid malformation (CCAM) 77
Conjunctiva 112, 213, 395, 484, 507
Conjunctival epithelial surface 112
Conjunctival goblet cells 213, 483
Cooperation 49-50, 52, 162, 484, 502, 506, 548
COPD 75-6, 154, 158, 183-4, 256, 276-7, 363
Core peptide 320
Cornea 112, 213, 395, 483-6
Corneal ulcer 213, 484
Cortical myosin IIA 443, 517
Covalent ester bond 464
CpG methylation 48, 56, 504
CpG sites 55-8, 331, 431
CRE *cis*-element 53-4, 420-1
CREB 36, 53-4, 159-61, 251-3, 420-1, 430
CREB protein 159-60
CREB transcription factor 53-4, 159
Crohn's disease 78, 99, 101, 204, 288, 353-4
Crypt epithelium 178, 453
Cryptic epithelial cells 70
Cryptic splice sites 60
Crypts 70, 72, 174, 205, 274, 333, 354, 453
Csf2 559
Cyclic transformation 550
Cyclooxygenase 2 158
Cyclin 502, 555
Cyclin A 502
Cyclin D1 502
Cyclin E 502
Cys-domains 17-18, 147, 464
Cys-subdomain 170, 248-50, 268
Cystadenoma 108, 199, 212, 333, 362
Cystatins 470

Cysteine residues 5, 17, 44, 62, 67, 146-7, 166, 248-9, 402, 406, 408-9, 419, 509-12
Cysteines 62, 249-50, 390, 392, 399, 402, 419
Cystine-knot (CK) 17-18, 390
Cystic glandularis 360
Cystitis glandularis 106, 211
Cytodifferentiation 172, 454
Cytokines 35, 46, 53, 86, 148-50, 158-9, 162, 164, 183, 254-5, 273, 295, 327, 396, 559-60
Cytokinesis 444, 518
Cytoprotection 319, 345
Cytoskeleton 439, 443-4, 517-18, 554
Cytotoxins 468-9

D

Damage 343, 362, 482-3, 486-7
Dark columnar cells 351
De novo expression of MUC2 98, 496, 498-9
De novo expression of MUC5AC 198, 349, 502-3
De novo expression of the MUC6 gene 353
Decalcification 405-6, 509
Dedifferentiation 77, 359, 364, 490, 498
Defense barrier 66, 71, 465-6, 470, 472, 479, 549
Degradation 4, 102, 207, 277, 294, 365, 406, 410, 433, 439, 469-70, 476, 478, 485, 553
Dehydration 362, 456
Dendritic cells 461, 495
Deregulation 88, 499
Development 50-1, 67-71, 84-6, 98-100, 104-6, 172-5, 273-5, 338-40, 362-5, 394-5, 439-41, 452-6, 489-90, 492-4, 497-503
Developmental processes 453, 455, 519
Developmental signaling pathways 91
Dexamethasone 163-4, 253-4

Diagnostic marker 93, 179, 358, 360, 365
Differentiation 5, 36, 46, 53, 57, 78, 84, 259, 332, 420, 443, 455, 490-1, 493, 517
Differentiation patterns 88, 284
Digestive system 185, 193, 273-5, 280, 332
Dimer formations 21, 166-7
Dimerization 17, 20, 61, 67, 166-7, 261-2
Dimers 20, 62, 64, 66, 166, 168-9, 262, 323
Dimerization of secreted mucins 262
Discriminating factors 360
Disease 5, 23, 36, 75-6, 79, 88, 99, 102, 114-15, 184-5, 187-8, 214, 345-6, 353-4, 363-5
  Crohn's 78, 99, 101, 204, 288, 353-4
  gallbladder stone 185, 200
  pylori-induced 343-4
  inflammatory bowel 101, 104, 203, 287, 354-5
Dissociation 336-7, 361
Distal promoter 251-2, 259
Disulfide bond 17, 62, 64-5, 67, 171, 261, 263, 270-1, 322-3, 393, 402, 419, 469
DNA 55, 148, 165, 246-7, 249, 258, 260, 326, 367, 432, 492
DNA methylation 57-8, 164, 198, 259, 331, 492
DNA repair 553
Domain 6-7, 17-19, 23, 29-30, 35-7, 145-7, 170-1, 248-50, 270-1, 319-20, 389-92, 400-2
  D1 100, 102, 109 10, 512 13
  D2 17, 30, 247, 249, 263, 269-70, 319-20, 390, 400, 410, 513
  D3 17, 23, 30, 35, 44, 67, 103, 247, 263, 270, 319-20, 400, 409-10, 427, 512-13
  D4 22, 30-1, 45, 52-3, 103, 147, 163, 171, 247, 250, 256, 295, 317, 389-90, 461-2
  cysteine-rich 5, 146-7
  cytoplasmic 19-20
Domain structure 3, 17, 19, 31, 44-5, 146, 247-8, 317, 387-9, 399-401, 425, 437-8, 456
Domain structure of MUC5B mucin 247

Domain structure of MUC7 mucin glycoprotein 399
Domain structure of MUC9/OGP glycoprotein 437
Down-regulation 53, 113, 187, 189-90, 197, 253, 287, 329, 478, 501
Ducts 95-6, 204, 285, 294, 334, 336, 354
Ductal adenocarcinomas 79, 89, 91-2, 94, 193, 198, 499
Duodenal goblet cells 79
Duodenal mucosa 94
Duodenum 69, 72, 79, 94, 174, 199, 204, 261, 274, 283, 354
Dysplasia 68, 85, 177, 187, 189-91, 203, 205-6, 278, 282, 340, 344, 346, 356, 479
    high-grade 79, 188, 340-1

E

E-box 329, 559
E-cadherin 196, 494, 497, 501, 504-5
E3 ligase 489
Early embryo 440-2, 514-6
ECC 95-6, 200-1
Ectocervix 177, 207, 335, 356
Ectodomain 19, 22, 480
Ectopic expression 288, 498-9
EF-G 406, 510
EF-Tu 406, 510
Effecter molecule 452
Efferent tear duct 293-4
EGF 7-8, 18, 47-8, 81-2, 85, 148-9, 151-3, 155-8, 161-2, 184, 192, 197, 254, 256, 504
EGFR 48, 81-2, 85, 148-9, 151-3, 155-8, 161-2, 192, 256, 506
EGFR activation 151-3, 155
Elastase 151, 156, 164, 184, 559
Electrostatic interaction 324, 408, 459, 506, 511
Elongation factors 406, 510
Embryo 37, 50-1, 67-70, 172-4, 176-7, 273-5, 287, 331-3, 335-6, 394-5, 439-43, 452-3, 455-6, 514-17, 519
Embryo development 37, 440, 514

Embryogenesis 44, 67-70, 72, 172-3, 275, 333, 336
Embryonic colon 287
Embryonic development 50, 84, 387, 395
Embryonic differentiation 81
Embryonic intestine 203
Embryonic liver 174
Embryonic lung 68-9
Embryonic stomach 174, 274
Encoded mucin glycoprotein 418
Endocervical adenocarcinoma 109, 209, 359
Endocervical epithelium 109, 275, 335, 338
Endocervical glands 109
Endocervix 45, 74, 177-9, 207-8, 266, 269, 275, 334-5, 356
Endocrine differentiation 361
Endocytic structures 442, 516
Endocytic vesicle 439
Endometrial adenocarcinomas 110, 208, 358
  ordinary mucinous 358
  primary mucinous 358
Endometrial cancer 37, 357
Endometrial carcinoma 36, 110, 178, 208, 419
Endometrial cycle 74
Endometrial hyperplasia 358
Endomctrium 74, 110, 178, 207-8, 334-5, 356-9, 419, 437
Endoplasmic reticulum 20-1, 60-1, 66-7, 104, 166-70, 172, 261-2, 268, 270-1, 321-2, 328, 434
Endothelial cell receptors 549
Endotoxin LPS 154
Enolase-α 406, 510
Epidermal growth factor receptor 48, 81-2, 148-9, 151-3, 155-8, 161-2, 192, 256, 506
Epidermoid carcinoma 181-2, 278, 280, 364
Epididymis 176, 212, 289, 335, 337-8, 362
Epigenetic mechanisms 44-5, 55, 57-8, 164, 259, 330-1, 430-1, 433, 496, 504

Epigenetic regulation 45, 55-6, 58, 259, 330, 496
Epigenetic silencing 91, 260, 492, 553
Epithelial cell membrane 467
Epithelial cell receptors 467
Epithelial cells 4, 30, 37, 44, 53, 68, 78, 156-7, 174, 204, 274-6, 335, 422, 465-6, 486-7
  airway 51-3, 155-6, 158-9, 162, 181, 183, 332
  biliary 202-3, 285
  cultured human gallbladder 202
  gallbladder biliary 345
  human airway 75, 161, 172, 258, 420, 422
  human middle ear 419-20
  mucin-producing 272
  normal human gallbladder 285
  normal human nasal 420-1
Epithelial folds 173, 175, 274-5, 453
Epithelial organogenesis 335-6
Epithelial renewal 361
Epithelial surfaces 3-5, 35, 72, 172, 213, 324, 334, 361-2, 456-8, 472-3, 476, 478, 482-3, 488, 507
Epithelium 45, 71, 75, 80, 94, 111, 113, 174-5, 187, 199, 208, 210, 285, 334-5, 486-7
Epstein-Barr virus (EBV) 86
ErbB family 254
ERK 47-8, 52-4, 152-5, 158-60, 163, 197, 253-6, 258, 328, 412, 420-2, 504, 506
Erk Pathway 153, 163, 197, 255, 504
Esophageal adenocarcinoma 79, 186, 340
Esophageal cancer 339
Esophageal epithelium 81, 186-7
Esophageal polypoid dysplasia 188, 341
Esophagus 45, 72, 78-9, 84, 185, 188, 273-4, 281-2, 333, 339
Esophagus adenocarcinoma 78, 282
Estradiol 430, 434
Estrogen 178, 266, 431, 433-4, 473

Estrogen receptor 328, 358, 429, 433
Estrous cycle 431, 433
Ets protein family 330
Eukaryotic Linear Motif (ELM) 547, 551
Evolution 3, 7, 11-14, 30-1, 33-4, 37, 166, 246, 250, 316-17, 320, 387, 408, 456, 511
Evolutionary histories 7, 11, 14, 33, 35, 247, 316-17, 331, 388, 401
Exocrine phenotype 395
Exonic minisatellite sequences 60
Exons 33, 44, 145, 147, 165, 246-50, 260-1, 319-20, 390-2, 398, 425-8, 438
Expressed MUC5AC 187, 189, 201-2, 206, 208
Expression of MUC1-MUC7 mucin genes 348
Expression of MUC5AC gene in pathology 179, 185, 207, 210, 212-13
Extracellular ligand 548
Extracellular matrix 18, 452, 490, 504, 549, 551, 556-7

F

Fallopian tube 74, 178, 207, 335, 356, 358, 395
Farnesoid X receptor (FXR) 85
Female reproductive tract 36, 74, 108, 175-7, 179, 207, 210, 246, 273, 275, 290-1, 334-5, 338, 356-7, 419
Fertilization 37, 440-1, 514-15
Fertilizing ova 441, 515
Fetal development 44, 50, 106, 172-3, 273, 275, 284, 289, 333, 335, 338, 362, 453, 455-6, 519
Fetal duodenum 69, 274
Fetal gallbladder 69, 275
Fetal intestine 333
Fetal liver 275
Fetal lungs 69
Fetal pancreatic stem cells 554
Fiber-like structure 266
Film 213, 294, 456, 482-6
Flagellin A 468

Floating globules 266
Folding 20-1, 66, 166, 170, 262, 268, 271, 322
Foreskin 175, 212, 289, 337, 360
Forkhead-associated (FHA)-domain 553
Formation 17-18, 35, 66-7, 93, 159, 165, 207, 269, 271-2, 325, 348, 360, 405-7, 459-61, 548-9
stone 98, 202
Foveolar cells 182, 344, 481
Foveolar type 201
Foxa 50-1, 148, 150-1, 153, 157
Foxa 1 50-1
Foxa 2 50-1, 148, 150-1, 153, 157
FOXD3 35, 404
Fragmentation 408-9, 481, 511-12
Fragments 269, 319, 390, 409-10, 452, 463, 512-13
Friction force 458
Fucose 35, 169, 264-6, 398-9, 438, 472
Fucosylation 63, 265, 268, 325, 399
Functional partners 547, 557, 560-3
Functional potentials 327, 405, 411, 425, 440, 452-3, 462-3, 487-8, 501, 507-9, 512-14, 564
Functions 3-5, 18-19, 67-9, 114-16, 398, 405-12, 438-40, 452-3, 455-6, 468-73, 484-7, 489-92, 499-514, 516-19, 547-8
Functions of MUC5B mucin 282, 287
Functions of secreted mucins 5, 456, 489
Functions of soluble mucins 508
Fundus 174, 274, 322, 327
Fungicidal activity 408, 511
Furin 66, 463, 552

G

GA boxes 258
GABA 153, 162
Gal-3 547-51

Gal-3/ligand complex 548
Galactoside binding lectin protein 547
Galactosylation 325
Galectin-3 54, 547
Gallbladder 45, 87, 94, 96-9, 173-5, 185, 200, 202-3, 273, 285-6, 318, 331, 334, 345-6
Gallbladder adenocarcinoma 346
Gallbladder bile 98, 203
Gallbladder biliary epithelial cells 345
Gallbladder carcinoma 203, 346
Gallbladder epithelium 98, 202, 345-6
Gallbladder folds 334
Gallbladder stone disease 185, 200
GalNAc 65, 169, 264, 325, 399-400, 407, 510
GalNAc residues 62, 169, 325
GalNac-transferases 264
Gamete cytoskeleton protein MYH9 443, 517
Gametes 3, 37, 440, 442-3, 456, 514, 516
Gastric adenocarcinoma 68, 78, 192
Gastric adenomas 84, 508
Gastric atrophy 85, 92, 282
Gastric cancer 82-3, 85, 87, 190-3, 282, 284, 320, 344, 498, 501-2, 508
Gastric carcinogenesis 84-5, 191-2, 284, 502, 508
Gastric carcinoma 82-4, 87, 189-92, 282-4, 342, 344, 479, 498, 501-2, 507
Gastric carcinoma cells 49, 192
Gastric cDNA library 146
Gastric cells 209, 329
Gastric CSCs 87
Gastric differentiation 209, 329, 498, 501
Gastric epithelial cells 85, 330, 480
Gastric epithelium 80, 85-6, 174, 188-9, 191-2, 283, 319, 341, 344, 454, 476, 479, 487, 498, 507
Gastric foveolar cell transformation 364
Gastric foveolar phenotype 188, 341

Gastric genes 329
Gastric immunophenotype 358-9
Gastric metaplasia 187-8, 283, 358
Gastric morphogenetic landscape 86
Gastric MUC6 mucin 318, 357
Gastric mucins 189, 206, 208, 324, 326, 341, 352, 459, 479
    human 82, 422, 479
    mammalian 323, 459
Gastric mucosa 86, 188, 190-1, 203, 283, 342-4, 454, 479, 498
Gastric mucus 477
Gastric mucus gel 479
Gastric mucus layer 188, 476, 478-80, 482
Gastric phenotype 209, 340-1, 359, 502
Gastric tumorigenesis 502
Gastric tumors 83, 86
Gastritis 83, 85, 92, 185, 190, 282, 342, 344, 479
    atrophic 189-90, 479
    chronic 83, 189, 342
    progressive 190
    superficial 190, 344
Gastrointestinal organs 69-70, 338
Gastrointestinal tract 72, 77-8, 174, 185, 289, 316, 318, 332-3, 338-9, 354, 456, 458, 473, 482, 485
GATA-4 50-1
Gatekeeper genes 490
Gatekeeper oncogene 503
GC box 47, 432
GC-rich intronic sequences 250
GDPH sequence 22-3, 66, 171, 463-4
GDPH sequence-associated cleavage 22
GDPH site 22-3, 171, 269
Gel formation 68, 261, 324, 459, 478
Gel-forming genes 14, 31, 33-4, 253, 316, 327
    human 317

Gel-forming mucin biosynthesis 20
Gel-forming mucin characteristics 390
Gel-forming mucin family 261, 387
Gel-forming mucin expression 57, 164, 173
Gel-forming mucin gene expression 58
Gel-forming mucin genes 29, 31-4, 44, 55, 108, 112, 148, 164, 166, 246, 248, 316-18, 326-7, 387-90, 392
Gel-forming mucin glycoprotein 246, 250, 318, 321
Gel-forming mucin molecules 18, 457
Gel-forming mucin MUC2 44
Gel-forming mucin MUC5AC 145, 504
Gel-Forming mucin MUC5B 14, 185, 246, 461
Gel-forming mucin MUC6 316
Gel-forming mucin MUC19 12, 387, 394
Gel-forming mucin O-glycosylation 263
Gel-forming mucin transcript 165
Gel-forming mucins 11-12, 17-23, 29-31, 33-4, 44, 175-7, 247-50, 261-3, 387-90, 457, 465-6, 473-7, 483-9, 505-7, 553-5
  insoluble 3, 5, 11
  secreted 21, 317, 480, 485
Gel layer 101, 342, 457, 466, 470, 473, 476-8, 480, 547, 549-50
  adherent 477-8
  friable 470
  inner 549
  organized 466
  sticky 473
  upper 473
Gene amplification 492
Gene duplication 33, 316, 387
Gene expression 55, 88, 90, 148, 151, 160-1, 163, 183, 194, 258, 393-4, 492
Gene regulation 36, 49, 53, 55, 57, 157, 163, 433
Gene transcription 45-6, 51-3, 55, 148, 151, 160, 190, 198, 421, 430, 506
Genes 31-3, 50-1, 145-7, 151-2, 159-61, 246-8, 253-5, 257-60, 316-20, 329-32, 387-90, 392-4, 425-8, 430-3, 494-8

ancestral 33
bacterial 468-9
common ancestral 8, 246
fish spiggin 33
gatekeeper 490
homologous 426, 429
hppA 406, 510
human 16, 427
membrane-bound mucin Muc1 35
mucin-encoding 354
mucin-like 8 rabbit 426
target 161, 163, 329
test 405, 433
Genes encoding integrins 197, 504
Genetic abnormalities 88, 91, 193, 349
Genetic alteration 84, 91, 502
Genome integrity 490-1
Genomic organization 33, 38, 317, 319-20
Germline mutation 492
Gestation 68-70, 75, 172-5, 203, 273-5, 333-4, 336-7, 453-5
GGGCGG-box 258
GKLF/NRF2 251
Gland 45, 80, 175, 210, 213, 264, 273-4, 277, 282, 333, 393, 453
lacrimal 112, 275, 294, 395, 403, 483, 486
mammary 23, 74, 114-15, 212, 291, 365-7
parotid 339, 393
salivary 261, 263, 269, 273-4, 280-1, 332-3, 339, 389, 393, 403, 411
submandibular 393-4, 410, 513
thyroid 291-2, 366
Glandular lesion 178, 208-9, 357
Glandular secretion 460
Glandular structure 186, 204, 273, 288
GlcNAc 337, 361, 438
GlcNAcα1-4Galβ-R 337, 361

GLI1 196
Globular conformation 460
Glucocorticoid receptor 163, 253-4
Glucocorticoids 163, 253, 403
Glycan receptor 479
Glycan shift 102
Glycans 62, 102, 195, 486, 495
Glycocalyx 483-5
Glycoconjugates 466, 472
Glycolysis 485
Glycoprotein 3, 6, 29, 74-5, 388-90, 393, 396, 425, 430-1, 434-5, 437-8, 440-1, 508-9, 514-16, 548-9
  estrus-associated 425
  multidomain 509
  non-mucin 66
  secreted 425, 441, 452, 515
Glycoprotein expression 344, 394, 431
  oviduct mucin 431
Glycoprotein molecules 4, 6, 396, 435
Glycoprotein MUC5AC 455
Glycoprotein MUC5B 246
Glycoprotein MUC6 365
Glycoprotein MUC8 418
Glycoprotein MUC9/OGP 425
Glycoprotein MUC19 387
Glycosylation 21, 60, 67, 102-3, 105, 169-70, 203, 206, 263, 265, 268, 281, 321-2, 436, 438
Glycosyltransferase 168-9, 268, 433
Goblet cell differentiation 49, 105, 153
Goblet cell hyperplasia 149, 184-5
Goblet cell metaplasia 150-1, 153, 156
Goblet cells 45, 47-8, 50-1, 66, 68-72, 75-6, 79, 80, 83, 101-5, 108, 112, 149-51, 153, 166, 171-3, 179, 182, 183, 186, 191, 203-4, 210, 213-4, 276-9, 291, 293, 339-40, 354, 403, 420, 463, 483, 487, 496, 502, 505

colon 69
conjunctival 213, 483
duodenal 79
intestinal 48, 50, 80, 354, 496
secretory 183, 420
tracheal 172
Golgi 20, 62, 64, 67, 166-8, 261, 263, 271, 321-2, 434, 436-7
GP6 - collagen receptor 559
GpC dinucleotide 431
Granules 21, 66, 102, 271-2, 434-6, 463
secretory 21, 102, 272, 435-6, 463
GRE 148, 163, 251, 253-4
Growth factors 35, 46-7, 148-9, 151, 157, 159, 254
GSK3 kinase 555

H

H. pylori colonization 343, 481
H. pilory infection 189-90
H. pilory receptor 189
Hamster genes 426
Hamster Muc9/ogp gene 427, 432-3
Hamster Muc9/ogp mucin 430, 434-5, 437
Hamster promoter 430
Hamster sperm 440, 514
Hand1/E47 260
Hath1 105, 328-30
HCC 94-5, 97, 347, 503
Hedgehog signaling 329
Helical structure 401
Helicobacter pylori 53, 82, 85, 170, 282, 320, 342, 405, 407, 461, 510
Heparin 45, 463, 504
Hepatic progenitor cells 94-5
Hepatobiliary diseases 285
Hepatobiliary malignancies 95

Hepatoblast 174, 275
Hepatocellular carcinoma 94, 97, 202-3, 347
Hepatocystes 275
Hepatolithiasis 88, 97-8, 200, 202-3, 285-6
Heterodimer 19, 22, 52
Heteropolymers 469
Hidden Marcov model 387
HIF-1 148, 151-2, 506
High grade dysplasia 79, 187-8, 282, 340-1, 494
Highly variable region 390, 392
High-mobility group (HMG) 329
HIK1083 358-9
Histatin 35, 271, 400, 408, 410, 461-2, 464, 470, 511, 513
Histogenesis 88, 95, 182
Histological types 82-3, 209-10, 212, 284, 503
Histone code 55, 58, 331
Histone H3 58, 164, 198, 259, 504
Histone H3 deacetylation 58, 259
Histone H3-K9 modification 198, 504
HIV-1 462
HIV-positive patients 471
HLH 329
hMuc9/ogp 430-3
HNF 50-1
HNF-3*a* 50
HNF-3β 251
Homeostasis 5, 36, 91, 145, 214, 246, 294, 316, 421, 456, 486, 490, 496, 509, 553
Homologs 33, 158, 408, 425-7, 511
HUGO classification 8
Human chromosome 11p15.5 316
Human gastric mucin 82, 479
Human MUC2 gene 44, 50
Human MUC5AC gene 172
Human MUC5B gene 260-1

Human MUC5B mucin 262, 264, 561
Human MUC6 gene 320, 330
Human MUC7 gene 35
Human MUC9/OGP gene 427-8
Human MUC19 gene 32, 388-9, 392-3
Human MUC19 glycoprotein 390
Human mucins 6, 169, 283
Human neutrophil elastase (HNE) 156
Human ortholog 387
Human sperm 440, 442, 514-15
Hydration 35, 68, 266, 405, 409, 452, 456, 461, 487, 512
Hydrogen bond 410, 461, 513
Hydrophilic comb 457
Hydrophilic domains 460
Hydrophilic glycoprotein 213
Hydrophobic domain 458
Hydrophobic interaction 323, 402, 457, 459
Hypermethylation 100, 331, 492
Hyperplasia 71, 76-7, 109-10, 149, 182, 184-5, 191, 208, 212, 278, 286-7, 357-8, 364, 508
Hyperplastic polyps 205, 355-6, 494
Hypoglycosylation 102

I

Identification 145, 148, 158, 331, 405, 422
IgGFcγBP 469
IL-1β 36, 53, 86, 149, 158-9, 254-5, 396, 404, 420, 421
IL-1 36, 53, 86, 149-53, 157-9, 162, 184, 254-5, 396, 404, 420-1
IL-4 36, 53, 77, 85-6, 149-51, 159-60, 162, 255, 404, 420
IL-4Rα 77
IL-5 149-51
IL-6 149, 255, 396
IL-8 165, 256, 396
IL-9 149-50

IL-10 149-50
IL-13 149-53, 157, 162, 184, 254-5, 404
IL-13/IL-4 Rα 149
IL-17 151, 255
IL-17A 158-9, 255
Ileum 69-70, 174, 354, 395, 453, 455
Immune cell adhesion 485
Immune reaction 86, 158, 407
Immune recognition 342
Immune response 53, 85, 157, 494, 507
Immune system 71, 442, 501, 516
Immunochemistry 74, 360, 366
Immunocompetent cells 501
Immunofluorescent localization 463
Immunoglobulins 470, 482-3, 550
Immunohistochemical methods 332, 362
Immunohistochemistry 178, 274, 280, 289, 333, 336-7, 403, 437
Implantation 441, 515
In situ hybridization 177, 202, 211, 274, 285, 332, 337, 360
In vitro fertilization 440-1, 514-15
Inactivation of tumor suppression genes 497, 500
Inflamed gallbladder 98-9, 202-3
Inflamed intestine 100
Inflammation 5, 52, 68, 86, 99, 104-5, 151, 170, 200, 202, 207, 282, 345, 407, 494
Inflammatory bowel disease 101, 104, 203-4, 287, 354-5
Inflammatory diseases 36, 75, 78, 97, 101, 149, 151, 158, 179, 183, 202-3, 207, 276, 278, 285, 291, 339, 345, 347, 353, 363, 418, 422
Inhibition 49, 62, 85, 151, 153, 156-7, 160, 164, 461, 502
Inhibitors 48, 55-6, 61, 154, 158, 257, 322, 464, 559
Initial O-glycosylation 22, 64, 325
Innate immunity 68, 70-1, 259, 452, 465, 519
Inner layer 71, 466, 476-7, 549-50
Insulin 559
Integrin 197, 462, 504, 556

Interaction 3, 17, 37, 48, 53-5, 61, 66, 82, 84-5, 100, 147, 152, 154, 156, 158-63, 165, 167, 177, 183, 254, 257, 262, 266, 269, 271, 323-4, 326, 402, 406-8, 411, 421, 438, 440-4, 452, 455, 457-61, 463-5, 467-72, 474-75, 479-80, 483-4, 487, 495-6, 504, 506-7, 510-11, 513-18, 547-9, 551-2, 554-7, 560, 564
cell-cell 254, 547, 556
electrostatic 324, 408, 459, 506, 511
hydrophobic 323, 402, 457, 459
protein-protein 152, 402, 407, 411, 513, 518, 552, 555
Interaction network 547, 557
Intestinal carcinoma 493
Intestinal crypt 99
Intestinal development 51
Intestinal differentiation 80, 86, 92, 188, 498, 559
Intestinal epithelium 46, 49, 71, 207, 283, 465, 493, 497
Intestinal gastric carcinoma (IGC) 191
Intestinal goblet cells 48, 50, 80, 354, 496
Intestinal metaplasia 68, 81-6, 109, 111, 186-90, 210, 282, 339-42, 344
Intestinal mucin 67, 116, 455, 498
Intestinal mucus gel 473, 547, 550
Intestinal phenotype 84, 87, 96, 111, 360, 502
Intestinal tumorigenesis 100, 493-4
Intestinal type 87, 91, 94, 106, 109-10, 209, 279, 339, 351
Intestinal type mucinous adenocarcinomas 109
Intestine 45, 47, 50-1, 68, 71-2, 75, 81, 84, 99, 100, 105, 174, 203, 205, 274, 287, 289, 318, 332-3, 354, 453, 466-7, 472, 482, 487, 493, 549
large 45, 81, 205, 287, 332
small 51, 99, 333, 453
Intestine-specific genes 50, 498
Intestine-specific homeodomain protein 49
Intracellular ligands 547
Intracellular vesicles 547
Intraductal carcinoma 91, 198, 212, 351, 500, 501
Intraductal papillary mucinous neoplasms (IPMNs) 89-94, 193-4, 198-9, 351-3, 499
Intraductal papillary neoplasia 200

Intrahepatic cholangio-carcinoma (ICCs) 94-7, 200-1
Introns 165, 246-50, 260, 326, 390, 398, 425-8
Invasion 196-7, 469, 494, 503-4, 506
Invasive ductal carcinomas (IDCs) 114, 292, 349
Isoforms 153, 254, 398, 436, 438, 444, 473

J

Jak2 157, 255
Jak/STAT 421
Jejunum goblet cells 69
K9 modification 198, 504
K27H3 trimethylation 259
Key regulator 329, 465
Ki67 497
Kidney 36, 70, 73, 106, 175, 211, 289-90, 335-6, 338, 360, 403
Kozak consensus sequence 392
KRT7 557, 559
KRT20 558-9

L

L-Selectin 407, 510
Labial glands 394
Lacrimal gland 112, 275, 294, 395, 403, 483, 486
Lacrimal proline-rich protein 4 (PRP4) 410, 513
Lacrimal sac 112, 293-4
Lactoferrin 407, 470, 482, 510
Landscaper genes 490, 495
Landscaper tumor suppressor genes 508
Langerhans cells 461
Laryngeal carcinoma 113
Larynx surface epithelium 332
Lauren's classification 83, 283-4
Layer 59, 71, 80, 101, 103, 343, 405-6, 457-8, 466, 468-70, 473-4, 476-8, 481-2, 549-50

adherent 101, 476, 478
inner 71, 466, 476-7, 549-50
Leader peptide 398
Lectin 461, 495, 547-9
LEF 148
Leptin 257-8
Leucine zipper 35, 400-2
Leukocytes 483, 505-6, 510
Leukotriene D4 52, 163
Lewis^{a} 274, 281-2, 325
Lewis^{b} 325, 471, 479
Lewis^{x} 325, 399, 471
Lewisy325, 366
Ligand-activated glucocorticoid receptor 163
Ligands 54, 152, 155, 170, 407, 456, 468, 548, 552, 556
Light refraction 482, 484
Link 22, 65, 172, 257, 461, 490, 495, 497-8, 555
Lipid layer 482
Lipids 171, 457-8, 460, 472, 482-3, 486
Lithogenesis 98, 203
Liver 69, 87, 94, 96-7, 99, 173-4, 185, 200, 275, 285, 345
Lobular carcinoma 114-15
Low grade adenoma 498
Low grade dysplasia 494
LPS 36, 51-2, 76, 154-7, 161, 255-7, 404, 420
LPS-induced apoptosis 495
LPS-induced MUC5AC expression 155-6
LPS-induced MUC5B expression 256
LS174T 58, 288
LTD4 52-3
Lubrication 5, 18, 35, 68, 214, 405, 409, 452, 456, 461, 483-5, 487, 507, 509, 512
Luciferase assays 327
Lung 76-8, 152, 173, 182, 184, 276, 332, 363-5, 508
Lung adenocarcinomas 77, 181, 278, 364, 505

Lung cancer 76, 181, 276, 278, 506
Lung cancer cells 51, 506
Lung cancer pathogenesis 76, 280
Lung carcinogenesis 182, 505-6, 508
Lung carcinoma specific marker 365
Lung carcinoma 77-8, 181-2, 278, 365
Lung tissues 277
Lung tumors 181, 278-9, 506
Lymph node 115, 355, 365, 501, 507
Lymph node metastasis 355, 365, 501, 507
Lysozyme 470, 472, 482, 559

M

M1 antigen 209, 211
Macrophages 176, 295
Male urogenital tract 105, 175, 210, 289-90, 335, 337-8, 360, 362
Malignancies 76, 91, 94-5, 193, 205, 208-9, 289, 341, 358
  cervical 208-9
  colon 289
  colorectal 205
  ductal 95
  endometrial 208
  hepatobiliary 95
  liver 94
  lung 76
Malignant breast lesions 114
Malignant index 499
Malignant intestine 287
Malignant prostate 107, 176, 337, 361, 499
Malignant transformation 68, 89, 92-3, 111, 115, 179, 181, 188, 192, 202, 207, 325, 340, 353, 507
Malignant tumors 179-80, 199-200, 208, 280, 286, 345, 357, 363, 493, 498-9, 501
Mammary gland 23, 74, 114-15, 212, 291-2, 365-7
MAPK 53, 81, 86, 148, 151-2, 154-9, 163-4, 256-8, 328, 405, 420, 506, 554

MAPK kinase pathways 148
MAPK phospatase-1 163
MAPK signaling pathways 154, 157
Markers 36, 51, 57, 78, 80, 88, 92, 94-5, 107, 111, 253, 330, 353, 355, 359
diagnostic 93, 179, 358, 360, 365
Master regulator 444
Mastocytes 176
Maternal immune system 442, 516
Matrix metallopeptidase 559
Mature ciliated cells 443, 517
Mature mucin molecules 21-2, 63-4, 435
Mechanical pressure 549
Mechanical stress 19, 101, 157
Meibomian gland 482
Membrane-bound mucin apoproteins 20
Membrane-bound mucins 3-8, 11-15, 17-23, 59, 68, 177, 185, 285, 327, 345, 356, 463, 468, 483-4, 559
Menstrual cycle 17, 74, 110, 177, 207, 266-7, 275, 290, 335, 356, 460
Mesenchyme 336
Metabolic homeostasis 509
Metalloproteinase TACE 153
Metaplasia 79, 149, 186-7, 190, 339-40, 344
gastric-type 186-7
goblet cell 150-1, 153, 156
Metaplasia-dysplasia-adenocarcinoma sequence 188, 281
Metaplastic cells 49, 83
Metaplastic epithelium 186-7
Metaplastic lesion 180, 186-7, 191, 211, 358
Metaplastic pathway 93
Metaplastic process 109
Metaplastic transformation 339
Metastatic potential 92, 499, 548
Metastatic process 346, 463, 505-6
Metastatic prostate cancer 211, 290

Methotrexate (MTX) 288
Methylation 55-8, 164, 259, 326, 331, 432
MG1/MUC5B mucin 265
MG2 280, 406-8, 509-11
MG2 (MUC7) mucin 406
Microorganisms 71, 170, 190, 319, 406-7, 456, 461, 465, 4696 469, 471-4, 479, 482
Microsatellite instability 100
Microsatellite stable cancers 207
Middle ear epithelium 36-7, 74, 113, 252, 275, 294-5, 365-6, 396, 403
Migration 99, 100, 196-7, 504-7, 547, 549
Minimal deviation adenocarcinoma (MDA) 358
Mixed adenocarcinoma 182, 364
Mixed polyp 205, 355-6
MMP3 197
MMP9 557, 559
Modification 11, 44, 55, 57-8, 60, 65-7, 145, 246, 259, 268-70, 290, 316, 321, 398, 405, 410, 428, 434, 439, 452, 476, 513, 551
  histone 57-8, 259
  posttranslational 11, 44, 55, 66-7, 145, 246, 268, 290, 316, 321, 398, 405, 410, 428, 434, 439, 452, 476, 513, 551
  proteolytic 11, 22-3, 60, 65-6. 268-70
Modulator of uterine receptivity 437
Molecular mechanisms 5, 67, 163, 263
Monomer 17, 61, 64-6, 166, 261, 264, 269, 271, 321-3, 325, 393, 402, 419, 438, 485
Morphogenesis 51, 439, 442-3, 454-5, 499, 507, 516-7, 519
Motifs 67, 168, 263, 268, 444, 518, 547, 551-3, 555-7
Mouse Muc2 mucin 493-4
Mouse *Muc5b* 250, 260-1, 292
Mouse *Muc6* 320
Mouse *Muc9/ogp* 426-7, 429, 432
Mouse *Muc9/ogp* promoter 432
Mouse *Muc19 gene* 387-389

Mouse mucin genes 33
Mouse orthologs 14, 16
Mouse *Smgc* gene 390, 392
Mouse Smgc protein 391-2
mRNAs 59, 145, 165, 179, 210, 260-1, 282, 326-7, 349, 355, 365, 390-5, 427-8, 433, 437
MUC1 molecule 19, 480
MUC1 mucin 14, 95, 288, 358, 366, 401, 478, 504
Muc2 knock-out mouse 478
MUC2 mucin biosynthesis 60
MUC2 apomucin 59, 62-3, 69, 72, 102, 166
MUC2 biosynthesis 64, 67, 103
Muc2-deficient mice 100-1, 494
MUC2 expression 45, 47-9, 53-5, 57-9, 68-74,77-80, 83-6, 88, 91-3, 97, 100, 105-15, 186, 198-9, 284, 493-4, 548
  absence of 71, 73, 112
  activation of 186
  cancellation of 494
  decrease in 79, 100
  dynamics in 80
  frequency of 92, 198
  increase in 111
  inhibitor of 48
  level of 58-9, 72, 74, 83, 86, 110-13
  loss of 493
  pattern of 69
  rate of 79, 114
  regulation of 47, 54
  repression of 55
MUC2-Gal-3 complex 548-9
MUC2-Gal-3 net 549-50
MUC2 gene 35, 44-59, 68-70, 72-6, 78, 81-2, 85-8, 93, 99-100, 103-8, 110-15, 154, 179, 493-7, 548

MUC2 gene expression 46, 48, 51, 54, 56-7, 68, 70, 75, 78, 85-6, 93, 105-6, 108, 113, 496
MUC2 gene promoter 50, 56
MUC2 gene transcription 45-6, 51-2, 55
MUC2 glycoprotein 44-5, 59, 67-8, 76, 78, 100-3, 111, 114-15, 319, 344, 347, 455, 495, 499-501, 548
MUC2 molecule 61, 64-6, 463, 501, 549-50
MUC2 mRNA 59-60, 68-70, 72, 75, 82-3, 98, 102-3, 112-13, 294
  amount 103
  concentration 102
  expression of 69, 103
  differently spliced 59
  full-length 60
MUC2.1 mRNA 60
MUC2 mucin 44-5, 47, 50-62, 64-79, 81-8, 91-3, 99-103, 105, 107-16, 155, 453-5, 463, 466, 468-9, 492-501
  human 65-7, 171, 401, 560
  mouse 493-4
  rat 62, 65, 470
MUC2 mucin expression 51, 52, 54-5, 67, 71, 77-8, 83-6, 100, 105, 108-9
MUC2 mucin expression profile 91
MUC2 mucin gene 52, 60, 455
MUC2 mucin glycoprotein 45
MUC2-negative cells 56, 58
MUC2-negative tumors 92-3
MUC2-non-producing cells 56-7
MUC2-positive cells 79, 99, 109-10
MUC2 precursor 61, 63-5, 103-5, 262, 552
MUC2 precursor biosynthesis 64, 103
MUC2 promoter 45-54, 56-8, 84, 100, 155, 496, 548
  basic 48
  human 46-7, 54
  mouse 51
MUC2 promoter methylation 57, 496

MUC2 promoter region 46, 56, 58
MUC2 protein isoforms 59-60
MUC2 transcription 47-8, 50-3
MUC3 14, 16, 95-6, 98-9, 106, 175, 179, 190-1, 202, 285, 344, 347, 468
MUC3 mucin 204
MUC3A 11, 15-6, 348, 557, 559
MUC3B 11, 15-16
MUC4 11, 16, 19, 21, 69, 106, 164, 177, 179, 185, 190-1, 204, 258, 283, 294-5
MUC4 gene 7, 98
MUC4 mucin 8, 14, 18, 23, 164
  membrane-bound 8
  rat 23
MUC5AC 94, 111, 145-6, 148-50, 152-5, 160-92, 194-208, 211-13, 253-4, 256-7, 341-4, 354-6, 453-5, 477-81, 501-7
MUC5AC apomucin 166, 186, 190, 195
MUC5AC C-terminal region 147
MUC5AC expression 113, 150-6, 158-60, 162-5, 172-6, 178, 180-5, 187, 189-93, 195-6, 200-1, 203-11, 256-7, 503, 505-7
  activation of 204
  dynamics of 180, 187, 207, 503
  EGFR–mediated 152
  induction of 158, 189, 207
  level of 173, 180-1, 189, 195, 201, 208, 213, 346
  LPS-induced 155-6
  mechanism of 158, 257
  pattern of 172
  regulation of 163-4
  suppression of 158
  up-regulation of 151-2, 155
MUC5AC gene 112, 145-6, 148-52, 154-6, 158-64, 172-7, 179-81, 184-5, 187-90, 193, 197-8, 207-14, 253, 255, 504-6
MUC5AC gene expression 151, 154-6, 159-64, 172-3, 175-6, 179, 181, 184-5, 189, 193, 203, 207-8, 210, 212-14, 506

MUC5AC glycoprotein 145, 151, 170, 188, 202, 213, 257, 280, 283, 326, 463, 484, 505
secretion of 170
MUC5AC knockdown 197
MUC5AC mRNA 145, 162, 164-5, 172-5, 180-1, 184-5, 192, 203-4, 208, 210-11, 453
MUC5AC mRNA expression 152, 165, 173-4, 213
MUC5AC mRNA stability 165
MUC5AC mucin 145-7, 167, 173-4, 176-9, 185, 189, 192-3, 197-9, 203-4, 206-9, 211-14, 268-9, 486-7, 501-2, 504-6
MUC5AC mucin expression 155, 162, 176, 178, 193, 197, 209, 211, 213-14, 495
MUC5AC mucin glycoprotein 166, 171, 197, 506
MUC5AC mucin molecules 464, 466, 477
MUC5AC over-expression 150, 155-8, 256-7, 506
MUC5AC polypeptide 168-9, 172, 506
MUC5AC-positive cells 187, 211
MUC5AC-positive tumors 205-6
MUC5AC precursor 166, 168
MUC5AC promoter 148-9, 152-5, 159-64, 183, 198, 253, 504
MUC5AC protein 145-6, 208, 280
MUC5AC transcription 148-54, 157, 160, 162, 164, 183, 253
MUC5AC up-regulation 158, 162-3
MUC5B 11, 29, 145-6, 177-9, 184-5, 246, 248-50, 252-63, 266-70, 273-88, 290-5, 320, 399, 460, 470-4
gel-forming mucin glycoprotein 246
human gel-forming mucins 250
salivary mucin 399
MUC5B apomucin 260, 262, 286, 291
MUC5B biosynthesis 262
MUC5B cDNA 145
MUC5B dimers 261-2
MUC5B expression 250, 252, 254-8, 261, 273-8, 280-6, 288-93, 487
activation of 257
CREB-conducted 252

IL-17-mediated 255
LPS-induced 256
patterns of 273, 277
regulation of 252, 256, 257
MUC5B gene 76, 246-8, 250, 252-8, 260, 263, 273, 275-6, 279-80, 282, 284-5, 287, 289-91, 293-4
MUC5B gene expression 250-1, 273-4, 276, 282-3, 289-91, 293-4, 473
MUC5B glycoprotein 246, 249-50, 259, 261, 266, 274-6, 282, 285, 288-90, 295, 455
MUC5B molecule 264, 266, 460
MUC5B mRNA 256, 260, 273-5, 278, 281, 286-8, 291-2
MUC5B mucin 20, 246, 248, 250, 255, 261-2, 266-7, 269-71, 273-8, 280-4, 286-92, 294-5, 464-5, 470-1, 473-4
MUC5B mucin expression 280, 287, 290
MUC5B mucin molecule 267
MUC5B mucin structure 275, 460
MUC5B promoter 251-6, 258-9, 288
MUC5B transcription 252-3, 255, 259, 261, 283
MUC5B transcriptional regulation 252, 254-5, 258
MUC6 33, 97, 106, 111, 204, 259, 282-3, 316-20, 325-49, 352-67, 388-9, 453-5, 477-8, 480, 507-8
MUC6 apomucin 321, 336, 345, 361
MUC6 expression 110, 186, 190, 328-33, 335-41, 343, 345-9, 352-3, 355-8, 360-4, 366-7, 454, 481, 507-8
regulation of 326-8, 330
MUC6 gene 32, 316, 319, 326-7, 329-32, 334-5, 338-41, 347-9, 353, 355-7, 360, 362, 364-6, 453, 481
MUC6 gene expression 326-7, 331-2, 334, 338-9, 356-7, 360-2
MUC6 glycoprotein 209, 316, 322, 331, 335, 345, 348, 363, 365, 367, 508
MUC6 mRNA 321, 326-7, 333-7, 345, 356-7, 362, 365-6
MUC6 mucin 12, 77, 280, 316, 319-2, 325-6, 331-40, 343, 345-8, 352-4, 356-8, 362-3, 365-7, 461-2, 507-8
biosynthesis of 321
gastric 318, 357

gland type 481
human 321-2, 324, 326, 562
O-glycosylation of 325
oligosaccharide chains of 325
MUC6 mucin expression 97, 283, 328, 345, 352, 355-6, 363, 508
MUC6 mucin glycoprotein 331, 333, 342, 345, 362
MUC6 mucin molecules 322, 477
MUC6 mucin protein 319
MUC6 polypeptide 324, 389
MUC6 precursor 321
MUC6 promoter 327-31
MUC6 protein 324, 336-7, 342, 348, 365
MUC6 transcription 326-7, 330, 353
MUC7 11, 16, 29, 34-5, 37-8, 204, 280, 348, 398-412, 419, 470-3, 483, 508-14, 518, 552
MUC7 apomucin 398, 399, 406, 408, 511
MUC7 cDNA 398
MUC7 expression 35, 347, 403, 405
MUC7 gene 34-5, 258, 398, 403, 404, 405
MUC7 gene expression 403-5
MUC7 glycoprotein 294, 399, 406-7, 409, 462, 510, 512
MUC7 molecule 399, 402, 406-10, 471, 509, 511-13
native 406-7, 409, 511-12
MUC7 mRNA 398
MUC7 mucin 35-6, 294, 398-403, 405-11, 462, 468, 470-2, 482, 509-14, 555, 563
human 401, 408, 563
salivary 398, 406
isoforms of 398
MUC7 mucin glycoprotein 399, 509
MUC7 promoter 403-5
MUC8 11, 16, 29, 34, 36-8, 177, 418-22, 455, 473, 483, 508-9, 552-5, 557
MUC8 cDNA 418
MUC8 expression 36, 420-2
MUC8 gene 36, 418-19, 421-2

MUC8 gene expression 418-20
MUC8 glycoprotein 36, 419-21
MUC8 mRNA 420
MUC8 mucin 36-7, 418-20, 422, 508
MUC8 promoter 421
MUC9 11, 16, 29, 34, 37-8, 178, 334, 419, 425, 437, 440, 442, 444, 508, 515, 518, 553-5, 557, 560, 563
MUC9 glycoprotein 37, 334, 425
MUC9 mucin 37-8, 425, 437, 442, 515, 518, 554, 560
MUC9/OGP 425-44, 514-18, 563
MUC9/OGP apomucin 437, 516
MUC9/OGP biosynthesis 434
MUC9/OGP expression 517
MUC9/OGP gene 425-33
MUC9/OGP glycoprotein 425-6, 430, 433-8, 440-4, 514, 516-18
MUC9/OGP molecules 437, 439-40, 442-3, 515-18
MUC9/OGP mRNA 431, 434, 437, 443
MUC9/OGP mucin 425, 430, 434-5, 437-40, 442-4, 452, 514-18, 563
MUC16 8, 14, 16, 18, 19, 294, 394, 483, 549, 559
MUC19 11, 12, 14, 29, 31-2, 44, 246, 317, 320, 387-396, 419, 483, 555-7, 560, 562
MUC19 gene 11, 29, 31-3, 317, 387-90, 392-4
MUC19 gene expression 393
MUC19 glycoprotein 44, 388-90, 393-4, 396
MUC19 mucin 387, 389-92, 394-5, 552, 562
MUC19 mucin expression 395
MUC19 mRNA 390-1, 393-5
Mucin adsorption 324
Mucin amino acid sequences 551, 556
Mucin backbone 102, 169, 485-6
Mucin biosynthesis 20, 170, 207, 434-5
Mucin carbohydrate moieties 549
Mucin classification 6-7
Mucin dimerization 61-2, 263

Mucin dimers 66-7, 263
Mucin domains 17, 30, 38, 322
Mucin expression patterns 96, 98-9, 201
Mucin expression profiles 94, 97, 175, 177, 185, 209-10, 281, 341, 352, 489
Mucin family 6-8, 37
    gel-forming 261, 387
Mucin functions 460, 482, 485, 547
Mucin gel 66-7, 176, 270, 272, 324, 357, 458
Mucin gene expression 106, 110-12, 115, 154, 176, 179, 183, 191, 193, 255, 257, 286, 332, 337, 364
Mucin gene products 166, 179
Mucin gene transcription 148, 252
Mucin genes 8, 14-16, 69, 88, 99, 106, 108-11, 153-4, 156-7, 174-9, 202-3, 252-3, 258-9, 294-5, 319
    activity of 176, 208, 258
    clustered gel-forming 331
    human gel-forming 317
    membrane-bound 14, 16, 164
    secreted 178, 453
    soluble 16
Mucin glycoprotein maturation 263
Mucin glycoproteins 3, 5-6, 11, 16, 23, 37, 98, 191, 276, 295, 342, 345, 477-8, 488-9, 507
    alterations of 342
    classifications of the 3
    domains of 458
    features of 11
    gel-forming 250, 318, 321
    group of 5
    membrane-bound 12
    multifunctional 456
    ocular 487
    over-production of 98
    secreted 3, 452, 455, 547, 551, 553, 564

soluble 398
structure of 16
Mucin glycosylation 199, 431, 458
Mucin granules 272
Mucin hyper-production 149, 295
Mucin hyper-secretion 185, 202, 277
Mucin incorporation 170, 271
Mucin molecules 5, 7, 16, 18, 22, 64, 169-72, 263-5, 270-2, 322-4, 434-5, 459-61, 463-4, 512, 551
condensation of 271
diversity in 264
gel-forming 18, 30, 457
gigantic 168, 270
full-length 408
high molecular weight 65
immature 20
incorporation of 270
mature 21-2, 63-4, 435
membrane-tethered 19
monomeric 17, 321
multimerization of 263
negatively charged 471
O-glycosylation of 265
packaging of 271
polymeric 485
population of 64
precursor 171
properties of 265, 270
secreted 452, 556
storage of 66
structure of 324
synthesis of 169
unstable 62
Mucin mRNAs 145, 321, 421

Mucin MUC2 44, 344, 498
Mucin MUC5AC 145, 326, 343, 352, 498, 503-4
Mucin MUC5B 14, 185, 246, 399, 461, 488
Mucin MUC6 316, 319, 343
Mucin MUC7 34, 398-9, 401, 403, 405, 407, 409, 411
Mucin MUC8 36, 418-19, 421
Mucin MUC9 37, 425, 427, 429, 431, 433, 435, 437, 439, 441, 443
Mucin MUC19 12, 387, 394
Mucin polymers 170-1
Mucin polypeptide backbone 21, 466
Mucin polypeptides 35, 166, 204, 342, 475
Mucin precursors 20, 168, 270-1, 551-2
Mucin production 5, 48, 151, 157, 161, 171, 182, 206, 276, 294, 364
Mucin profiles 106, 109
Mucin-protein binding 557
Mucin protein coil 266
Mucin-protein interactions 460, 464
Mucin protein superfamily 425
Mucin proteins 6, 295, 389, 410, 507, 512
Mucin secretion 98, 101, 171, 273, 277, 421, 476, 487-8
Mucin-specific domain 36-7, 248-9, 319, 390
Mucinous adenocarcinoma 106-7, 109, 111, 210, 212, 355, 357-9, 497, 499
Mucinous adenomas 111, 210, 357
Mucinous BAC 77, 181, 278, 364
Mucinous borderline tumor 111, 357
Mucinous breast carcinoma 115, 367
Mucinous carcinoma cells 59-60
Mucinous carcinoma 59, 79, 100, 114-15, 207, 352, 366-7, 500-1
Mucinous colorectal carcinoma 79, 497
Mucinous cystadenocarcinoma 77, 182
Mucinous cyctic neoplasms 89-91, 93, 352
Mucinous differentiation 108, 290, 361
Mucinous lung tumors 77
Mucinous metaplasia 358

Mucinous subtype 83, 498
Mucinous *trans*-differentiation 395
Mucinous tumors 84-5, 100, 111, 115, 209-10, 289, 357, 367
borderline 209
ovarian 111, 209-10, 357
Mucins 3-8, 70-7, 96-102, 181-7, 200-4, 284-95, 332-6, 341-5, 389-92, 394-6, 455-67, 470-4, 476-8, 486-9, 505-7
cervical 177, 265, 338
cleaved 22, 66
epithelial 101
expressed 177, 288
intestinal 67, 455
membrane-tethered 7, 19, 22
modified 206, 269
neutral 354, 479
polymeric 38, 393, 485
polymerized 170
porcine submaxillary 262, 392, 396, 401
rat colonic (Muc2) 548
transmembrane 484
Mucociliary clearance 473-5
Mucociliary differentiation 420
Mucociliary escalator 475-6
Mucociliary interactions 474-5
Mucolytic activity 485
Mucosal defensive barrier 476
Mucous cells 204, 273, 278, 393, 395, 419, 506
Mucous cell hyperplasia 278
Mucous glands 73, 112, 274, 293, 395
Mucous metaplasia 150-1, 153, 183
Mucus 3-5, 7, 75, 176, 183-4, 266, 284, 289, 332, 342, 360, 362, 367, 456-60, 473-6
components of 3, 456-7
native 465

Mucus barrier 71, 456, 458, 466, 481, 500, 550
Mucus cell differentiation 395
Mucus constituents 458, 460, 464
Mucus gel 3, 18, 30, 68, 71, 171-2, 183, 185, 456-60, 464-6, 468-70, 473, 476-82, 547, 549-51
Mucus gel layer 466, 477, 480, 547, 549, 550
Mucus gel permeability 460
Mucus layer 101, 103, 105, 170, 467, 474, 476-9, 481-4, 549
  adherent 101, 478-9, 481
  inner 478-9, 549
  innermost 482-3
Mucus permeability 266, 460
Mucus plug 276, 461
Mucus-secreting cells 248, 288
Mucus secretion 4, 74, 465, 474, 477
Mucus viscoelasticity 4, 458-9, 462
Multidomain glycoprotein 509
Multifunctional glycoproteins 34, 405
Multifunctional potential 452
Multifunctionality 19-20, 23, 35, 55, 409, 440, 452-3, 455, 457, 489, 495-7, 511-3, 517-9
Multilaminated structure 481
Multimerization 60, 66-7, 262-3
Multivesicular bodies 443, 517
Murine asthma model 276
Mutations 37, 61-2, 84, 90, 152, 160, 168, 490, 492-3
MYL4 559
Myosin 442-3, 516-17, 559

N

N-acetylgalactosamine 168-9
N-acetyllactosamine 63, 266
N-arginine dibasic convertase 551
N-glycans 16, 20-1, 61-2, 262-3, 321

N-glycosylation 20-1, 60-1, 146-7, 166-7, 249-50, 261-3, 321-2, 392, 401, 436
N-glycosylation sites 147, 249-50, 401, 436
N-Myc/Max 251
N-terminal amino acid sequence 409
N-terminal cleavage 262, 268-9
N-terminal D-domain 67, 261, 263
N-terminal end 392
N-terminal extracellular domain 5
N-terminal histatin-like domain 35, 400
N-terminal polypeptide fragment 390
N-terminal proteolysis 269
N-terminal region 17, 23, 30, 65, 146, 249-50, 268, 320, 408-11, 428, 462, 511-13
N-terminal sequencing 436
N-terminal truncated isoform 432
Napsin A 78
Nardilysine cleavage sites 439, 551
Nasal epithelium 184-5, 363, 419-20
Nasal mucosa 72, 76, 113, 422
Nasal polyposis 276, 363
Nasal polyps 113, 164, 185, 276, 420, 422
Nascent peptide 262
Nasolacrimal duct 112, 275, 293-4, 365, 484
Native markers 498
Negatively charged carbohydrates 266
Negatively charged mucin 471
Negative feedback regulators 421
Neoplasia 79, 89, 91-2, 96, 191, 193, 198, 200, 286, 333, 347, 349, 351-2, 363, 498-9
    biliary 347
    gastric 498
    intraductal papillary mucinous 96, 193, 200, 286, 351, 499
    intraepithelial 89, 96, 191, 193, 198, 200, 349, 499
    lung 363
    mucinous 96, 193, 198, 200, 286, 499

non-invasive 191
pancreatic 91-2, 353
salivary 333
Neoplastic glandular lesions 178, 209
Neoplastic transformation 77, 110, 191, 195, 199, 344, 353
Nephrogenic zone 335-6
Neuregulin 1β1 157, 254, 276
Neutral sugar 265
Neutrophil elastase 151, 156, 164-5, 184, 559
Neutrophils 407
NF1-MUC5B 248, 259-60
NFkB 52, 494, 559
Nitric oxide 81-2, 339, 478-9
NK2 homeobox 1 559
NKX2 148, 558-9
Non-mucinous BAC 77, 181, 278
Non-muscle myosin IIA 442-3, 516-7
Non-small cell lung adenocarcinoma (NSCLA) 505
Non-VNTR antibodies 206
Non-VNTR MUC5B specific antibodies 289
Nonadherent luminal "sloopy" layer 457
Nonciliated secretory cells 437
Noncovalent bonds 22, 468
Normal gallbladder specimens 286, 334, 345
Northern blot 177-8, 202, 285
Noxious agent 331, 335, 356, 476, 479, 482, 486
Nrf2 156, 251
Nuclear factor 50, 148, 248, 259-60, 432
Nucleotides 8, 46, 48, 50-1, 54, 56-7, 249, 258, 273, 389, 392, 398, 418, 421, 426-8
Nucleus 52, 160, 163, 253, 258, 439, 547-8, 551

O

O-glycans 6, 16, 18, 63, 168, 170, 265, 268, 271, 342, 399, 436, 461, 464, 485

O-glycosylation 20, 22, 44, 60-4, 147, 166, 168-70, 264, 268, 324-5, 392, 436, 439, 476
initial 22, 64, 325
limited 65
primary 264
site-directed 63
terminal 436
O-Glycosylation of MUC2 apomucin 62
O-Glycosylation of MUC5AC precursor 168
O-Glycosylation of MUC5B precursor 263
O-Glycosylation of MUC6 polypeptide precursor 324
O-glycosylation of mucin molecules 265
O-glycosylation sites 249
O-linked oligosaccharides 21, 399
Obstruction 184, 294, 348, 365, 473
Oct-1 404
Ocular defense barrier 549
Ocular mucins 482, 484-7
Ocular surface 74, 112, 213-14, 456, 465, 482-7
Oligomerization 20, 62, 65, 103, 105, 147, 261-3, 269-70, 321-2, 325, 402
Oligomers 64, 322-3, 419, 477
Oligosaccharide ligand 466, 468
Oncofetal M1 antigen 209, 284
Oncogenes 88, 489-92, 496-7, 500, 502, 504, 506, 553, 559
Oncogenesis 358, 361
Oncogenic potential 68, 100-1, 494, 497, 499, 501-6
Oncoprotein 489-90, 505, 560
Oocyte zona pelucida 441
Oocytes 37, 440-1, 443, 514-17
Oral cavity 35, 281-3, 405, 407, 409, 465, 470, 472-3, 509, 512
Oral infection 35, 405
Oral mucosa 462, 470
Organogenesis 335-6
Organoid differentiation 364

Osmotic pressure 484
Otitis media 113, 154, 294-5
Outer layer 71, 466, 468-9, 476-7, 549-50
Oval cells 95
Ovarian adenocarcinoma 357
Ovarian cancer 209, 291
Ovarian carcinogenesis 111
Ovarian mucinous cysts 209
Ovarian mucinous tumor 111, 209-10, 357
Ovarian teratoma 209, 291
Ovary 74, 111, 115, 178, 209-10, 289, 291, 357, 395, 440, 499-500, 514
Overt functions 452, 513
Oviduct 37, 334, 425, 431-3, 440-2, 514-17
Oviduct tissue remodeling 442, 444, 516, 518
Oviductin 37, 425, 436-7, 440-1, 514-15
Ovulation 335, 460
Ovulatory cycle 207, 460

P

P. aeruginosa 51-2, 76, 154-5, 480, 485-6
P-selectins 463
P2Y2 receptor 156, 257
P13K 157-8, 161, 254
P21 394, 502
P27 502
p38 54, 158-60, 164
p38 MAPK 157, 164, 254, 258
p53 51, 93, 100, 192, 495, 502, 554
p53 gene 84, 93, 364, 494, 497, 502
PA-LPS 155
Pancreas 69, 72, 79, 87-9, 91-4, 96, 173-4, 193-5, 199-200, 275, 287, 333-4, 347-9, 352-3, 493, 499-500
  normal 92, 195, 334, 348, 499, 503
  pathological 287, 347

Pancreatic adenocarcinoma 196-7, 287, 346, 348-9
Pancreatic cancer 56, 88, 91, 94, 96, 164, 193, 195-9, 287, 349, 503-4
Pancreatic cancer cells 56, 164, 196-7, 199, 503-4
Pancreatic carcinogenesis 88, 91, 93, 503-4
Pancreatic centroacinar cells 334
Pancreatic duct 72, 91, 94, 195, 197-9, 287, 334, 348-9, 351, 353, 503-4
Pancreatic ductal carcinomas 197, 503
Pancreatic epithelium 195
Pancreatic intraepithelial neoplasia 89, 193, 349, 499
Pancreatic neoplasia 91-2, 353
Pancreatic neoplasms 88, 90-2, 96, 194, 499
Pancreatic stem cells 88-9, 193, 554
Pancreatic tumors 93, 199, 352
Pancreatitis 287, 348
    chronic 287, 348
Pancreatobiliary subtypes 94, 198-9
Pancreatobiliary type 91, 94
PanIN 89-92, 96, 193-6, 198, 349-50, 352, 499, 503
PanIN-1A 89, 90, 193, 195
PanIN-1B 89, 193, 195
PanIN-2 89-90, 193, 195
PanIN-3 89-91, 193, 195
Papillary carcinoma 360
Papillary ductal hyperplasia 287
Papillary renal cell carcinoma 290
Parenchymal cells 275
Parotid gland 339, 393
Partner protein 37, 411, 444, 518
Pathogen 3-5, 36, 76, 97-9, 104-5, 107-8, 148-9, 183-6, 211-14, 285-6, 344-5, 347-8, 353-4, 466-70, 484-6, 512, 550
Pathogenesis 36, 52, 76, 78, 88, 97-9, 104-5, 107-8, 184-6, 211-13, 294, 344-5, 347-8, 353-4, 357-8
Pathogenesis of airway inflammatory diseases 278
Pathogenesis of Barrett's esophagus 80-1

Pathogenesis of bowel diseases 289
Pathogenesis of CF 76, 184
Pathogenesis of cholesterol gallstones 98
Pathogenesis of colorectal adenocarcinoma 99
Pathogenesis of COPD 363
Pathogenesis of Crohn's disease 354
Pathogenesis of gallstone disease 202, 286
Pathogenesis of gastric cancer 190, 344
Pathogenesis of hepatobiliary diseases 285
Pathogenesis of hepatolithiasis 97
Pathogenesis of lung cancer 76
Pathogenesis of mucus hypersecretion 36
Pathogenesis of prostate cancer 108
Pathogenesis of the urinary bladder diseases 107
Pathology 29, 34-8, 44, 47, 65-6, 75, 108, 179, 185, 207, 210-13, 286-7,289-94, 332-3, 365-7
Pathways 20, 47, 52, 65, 85-6, 96, 100, 148, 154, 156, 158, 160-2, 188, 253-4, 256-7
PDZ-binding motif 444, 518, 554
PDZ domain 554
PDZ-domain binding sequences 440
PDZ ligand protein 444, 518
Pellicle 406, 509
Penetration 37, 176, 440-1, 457, 488, 514-15
Peptic ulcer 83, 342
Peptide C-terminal amidation 439
Peptides 3, 4, 20-1, 54, 63, 65-6, 71, 98, 146, 169, 186-7, 204, 249-50, 262, 265, 268, 283, 320, 354, 389-402, 405, 408, 410, 427, 436, 439, 456-7,462-4, 470, 472, 474, 476, 478, 511-12, 547, 553, 559
Periciliary liquid layer 473
Perinuclear region 107, 463
Peristalsis 466-7, 479, 549-50
Perivitelline 440-1, 514-15
Peroxidation 486

Peroxysome proliferater-activated receptor 161
pH-driven mechanism 267, 460
Phospholipid transfer protein 486
Phospholipids 408, 511
Phosphorylation 48, 52, 155, 160, 439
Phylogenic tree 388
Physical barrier 468, 476
Physiological transformation 547, 550
PIK kinase 555
PKCα 54, 160, 163, 253
PKCδ 153, 163
Pla2g2a gene 494
Pla2g2a phosphatase 494
Plasma 176, 289, 337-8, 362, 461-2
Plasma membranes 443-4, 517-18
Platelets 505-6
PLCβ3-mediated activation 160
PLTP/MUC5AC complex 486
PMA 151, 153, 157, 252, 422
Pneumolysin 154
Point mutations 492
Polarity 442-4, 516-17
Poly-L-proline 410, 512-13
Polyadenylate signal 418
Polyampholyte domain 457
Polylactosamine 169
Polymeric mucin 393, 485
Polymerization 17, 67, 168, 262, 402, 457
Polymorphisms 18, 258, 342, 436
Polypeptide 249, 398, 401-2, 418, 425, 439, 460, 462, 511
Polypeptide backbone 16, 21, 168, 263-4, 267, 405-6, 466
Polypeptide Gal-NAc-transferase 63
Polypeptide precursors 67, 436-7
Polypeptide structure 398, 400

Polypoid dysplasia 188, 341
Polyproline sequences 401
Polyps 113, 164, 185, 205, 276, 355, 363, 420, 422
  adenomatous 206, 559
  colorectal 355
  hyperplastic 205, 355, 356, 494
  mixed 205, 355, 356
  serrated 205, 355-6
Polyspermy 440, 514
Porcine submaxillary mucin 262, 317, 392, 396, 401
Postmenopausal period 74, 356
Postnatal development 394
Posttranslational glycosylation 184, 393, 406, 431, 434, 478, 552
Posttranslational modification 11, 44, 55, 66-7, 145, 246, 268, 290, 316, 321, 398, 405, 410, 428, 434, 439, 452, 476, 513, 551
Posttranslational processing 434
Posttranslational proteolysis 44, 66, 405, 462
Pre-mRNA splicing 547
Pre-neoplastic lesions 502
Pre-ovulatory phase 441, 515
Precancerous lesion 281, 339
Precancerous stages 190
Precursor 11, 20-2, 44, 60-1, 63-5, 86-7, 92-3, 103-5, 166-8, 170-1, 261-3, 270-1, 321-2, 436-7, 551-2
Precursor cells 86-7
Precursor monomers 61, 261, 321
Predicted functional partners 557, 560-3
Preimplantation embryo 442, 516
Premalignant lesion 76, 79, 206, 278, 339
Primary acquired nasolacrimal duct obstruction 294, 365
Primary mucinous adenocarcinoma of the bladder 106
Primary mucinous adenocarcinomas of endometrium 358
Primary prostate carcinoma 211, 499
Primary transcript 59-60, 260-1, 327, 390-1, 427

Primitive gut 68, 174, 333, 453
Progenitor 87, 89, 94-5, 246, 277, 317, 425
Progenitor cells 89, 94-5, 277
Progenitor genes 246, 317
Prognosis 77, 92, 97, 111, 115, 181, 191, 202, 206, 260, 344, 349, 352, 498, 501-2
Progression 76-7, 79, 88, 91, 98, 109, 182, 184, 187, 195-6, 209, 339-40, 364, 489-90, 492-3
Proliferative phase 110, 356
Proline 6, 18, 21, 30, 63, 250, 264, 401, 409-10, 418, 512
Proline-rich ligand 411, 513
Proline-rich proteins 410-11, 461, 470, 513, 555
Promoter 35, 44-58, 148-55, 159-64, 251-6, 258-9, 288, 326-31, 403-5, 421, 425-6, 429-33, 489, 496, 501-2
Promoter regions 35, 46, 56-7, 148, 152, 163, 258, 326, 328-9, 331, 426, 429-33, 492
Promoter regulatory elements 44
Promoter sequences 47, 148, 249, 329-30, 425, 430, 432-3
Properties 3, 5, 7, 38, 74, 88, 101, 265-6, 326, 408, 433-4, 482, 484-5, 488-9, 510-11
Prostaglandin E2 158, 421, 478
Prostaglandin F2α 161
Prostate 73, 107, 176, 211, 289, 330, 335-7, 360-1, 395, 493, 499-500, 554
    normal 73, 336, 361, 493, 499
Prostate adenocarcinoma 107-8, 176, 290, 361
Prostate cancer 107-8, 211, 290, 360-1, 500
Prostate cancer cells 500
Prostate-specific antigen (PSA) 361
Protection 112, 187, 204, 214, 318-19, 323, 335, 405, 456, 479, 483, 486-7, 509, 518
Protective barrier 35, 290
Protein complexes 257, 269, 444, 464, 516, 518
Protein degradation 553
Protein domain 319, 439
Protein expression 72, 205, 292, 345, 356, 420

mucin core 205, 356
Protein families 8, 37, 490
Protein kinase C 148, 153, 156, 422
Protein levels 103, 177, 212, 281-2, 295, 349, 355, 365, 428
Protein molecules 6, 461
Protein multifunctionality 496
Protein product 398
Protein-protein interaction 152, 402, 407, 411, 513, 518, 552, 555
Protein superfamilies 425
Proteins 17-19, 171-2, 249-50, 405-7, 410-11, 422, 438-9, 441-2, 444, 452, 489-91, 510, 513, 518, 548-55
bacterial 406-7, 479, 510
c-Ets1 160
genes encoding 425
homeodomain 49
human mucin-associated 98
multifunctional 11, 548, 564
mucin core 201, 205, 356
mucin-related 12
mucin type 63
non-mucin 7, 22, 460, 474
nonmucin 460, 474
PDGF-like 17
secreted 398, 439, 441, 501, 515
signaling 254, 411, 513
trefoil 4, 85, 506-7, 550
Proteolysis 19-23, 44, 65-6, 166, 269-70, 320, 405, 407, 462-3, 469, 485, 489, 513, 552-3
Proteolytic cleavage 21, 23, 65, 171, 409, 462
Proteolytic degradation 406, 410, 476, 478, 512
Proteolytic modifications 11, 22-3, 60, 65-6, 268-70
Proteolytic processing 23, 171, 268, 270, 462, 470
Proteolytic reactions 22, 66, 463, 552
Proximal promoter 252, 259

Pseudocysts 199
Pseudomyxoma 289
Pseudomonas aeruginosa 51, 154, 407, 480, 510
PTEN 495, 501
PTS-domain 18-19, 45, 147
Pulmonary adenocarcinoma 77-8, 160, 364
Pulmonary adenocarcinoma with enteric differentiation (PAED) 78
Pulmonary fibrosis 76
Pulmonary inflammation 151
Pulmonary tumors 364
Pulse labeling 321
Pulse time 434
Pulse-chase approach 321
Pulse-chase assay 60
Pulse-chase experiments 172, 268, 321
Pulse-chase method 434
Pulsed-gradient spin-echo nuclear magnetic resonance 323

R

R-end subdomains 249
RA regulation of mucin genes 253
Rabbit *Muc9/ogp* gene 426
Rat colon 63
Rat colonic mucin (Muc2) 548
Rat gastric mucin 64, 321
Rat submandibular gland mucin 262
Reactive hyperplasia (RH) 191
Reactive oxygen species 155
Receptor-mediated binding 471
Receptors 5, 52, 54, 158, 162, 253-4, 257, 433, 456, 461, 467-8, 471, 479, 509-10, 556
Recombinant Cys-subdomains 268
Reduction 105, 187, 203, 213-14, 402, 406, 481, 485, 501, 510
Regulated secretion 171, 261

Regulation 35-6, 46-50, 58-9, 81, 85, 151, 157, 164, 250, 257, 326, 330, 398, 405, 429-31
differential 478
epigenetic 45, 55-6, 58, 259, 330, 496
gene 36, 49, 53, 55, 57, 157, 163, 433
spacio-temporal 444, 518
transcriptional 16, 44, 50, 58, 258, 327, 329, 418, 422, 425, 429, 431, 433, 496, 547
Regulation of *MUC2* gene expression 57, 496
Regulation of *MUC5AC* gene expression 148-9, 151, 159-60, 162, 164
Regulation of *MUC5B* gene expression 250-1
Regulation of *MUC5B* transcription 252-3, 283
Regulation of *MUC6* gene expression 326-7
Regulation of *MUC7* gene expression 403, 405
Regulation of the *MUC2* gene 47-8, 50-1, 56, 58-9, 75-6
Regulators 20, 48-9, 68, 84, 92-3, 158-9, 329, 422, 455, 482, 485, 489, 518, 553
Regulatory elements 44, 148, 316, 403
Regulatory mechanism 30, 58, 75, 154, 157-8, 160, 165, 391, 421
Regulatory region 46, 403
Regulatory sequence 316
Renal clear cell carcinoma 360
Reproduction 3, 37, 440, 442, 452, 514, 516
Reproductive organs 73, 108, 177, 210, 246, 275, 291, 356-7, 362, 433
Reproductive tract 36, 73-4, 175-7, 179, 207, 210, 246, 273, 334-5, 338, 356-7, 419, 456
Repulsion forces 483
Residues 5, 6, 17, 22, 37, 44, 62, 146-7, 168, 249, 264-5, 325, 401-2, 406, 410, 509-12
amino acid 34, 44, 147, 249, 269, 398, 402, 418, 437, 443, 459, 462, 517
cysteine 5, 17, 44, 62, 67, 146-7, 166, 248-9, 402, 406, 408-9, 419, 509-12
proline 168, 249, 401, 410, 512
serine 6, 18, 37
sugar 264-5, 325, 399
sialic acid 22, 457, 459, 471

threonine 6, 16, 18, 21, 30, 45, 63, 65, 168, 264, 325, 401, 418, 438
Respiratory tract 23, 67-9, 72-3, 75-6, 172-3, 183-4, 269, 273, 276, 278, 331-2, 362-3,365 455, 472-3
Restriction fragment length polymorphism 248
Retinoic acid (RA) 54, 160, 253, 404
Retinoic acid signaling pathway 160
Retinoic acid receptors (RAR) 160
RNA-binding proteins 165
RNA polymerase 406, 510
Rodents 12, 391, 431, 433
Role of secreted mucins 465
RORA1 251
ROS 155-7, 161
RSK1 159-60, 420-1

S

SabA adhesin 471, 479
Saliva 72, 265, 273-5, 279-83, 332-3, 339, 393-6, 398-9, 403, 406-7, 409-12, 460-2, 464-5, 470-3, 509-13
Saliva-binding adhesin 407, 510
Salivary duct carcinoma 280, 339
Salivary gel matrix 470
Salivary glands 261, 263, 269, 273-4, 280-1, 332-3, 339, 389, 393, 403, 411
Salivary MUC5B mucin 399, 471
Salivary MUC7 mucins 399, 406, 509
Salivary mucins 263, 265, 399, 403, 471
Salivary proteins 281, 472
SEA-domain 7, 19, 21-3
Secondary sperm receptors 441, 515
Secreted mucin amino acid sequences 556
Secreted mucin gene expression 453
Secreted mucin multifunctionality 453, 455, 457, 459, 461, 463, 465, 467, 469, 471, 473, 475, 477, 479, 481

Secreted mucins 3-6, 11, 29-31, 177-8, 452-3, 455-6, 461-2, 465, 484-5, 488-9, 509, 518-19, 547, 551-7, 559-61
Secretion 4, 60-2, 72-5, 101-3, 156-7, 169-71, 183-5, 257, 261-2, 272-3, 276-7, 294-5, 473-4, 476-7, 487-8
Secretory cells 171, 395, 420, 437, 443
Secretory granules 21, 102, 272, 435-6, 463
Secretory mucins 76-7, 108, 186, 210, 290, 333, 364, 483, 552
Secretory phosphatase 494
Secretory vesicles 170-1, 261, 270, 272, 288, 326
Segmental differentiation 443, 517
Segregation 388, 477
Self-association 35, 402
Self-cleavage 21
Semi-permeable barrier 3, 456
Seminal plasma 176, 289, 337-8, 362, 461-2
Seminal vesicle 176, 212, 289, 335-8, 362
Sequence similarity 316
Sequences 16, 21, 35, 66-7, 145, 147, 166, 168, 249, 319, 392, 400-1, 408-9, 428, 551-2
Ser/Thr-rich domain 324
Ser/Thr-rich tandem repeats 426
Serine 6, 16, 18, 21, 30, 45, 62, 65, 168, 249-50, 264, 325, 409-10, 418, 438
Serous cells 72, 393
Serous cystadenocarcinomas 111, 291
Serous cystadenomas 199
Serrated adenoma 205, 355-6
Serrated polyp 205, 355-6
Serrated polyp-neoplasia pathway 355
SH2 439, 554
SH3 411, 439, 513, 554
SH3-domain 439, 554
SH3-domain-binding motifs 439
SH3-domain specific binding sites 554
SHH 84, 329-30

Short-chain fatty acids 54
SHP-1 157
Sialic acid 22, 35, 168-9, 264, 268, 271, 398-9, 407, 442, 457, 459, 471-2, 479, 486, 495-6
Sialic acid-binding adhesin 407, 479, 510
Sialic acid residues 22, 457, 459, 471
Sialomucin 458, 479
Sialyl-Lewis^x antigen 471
Sialyl-Lewis^x receptor 471
Sialylation 44, 63, 102, 264-5, 268, 271, 325, 485
Signal peptide 146, 390-1, 399-400, 427, 439, 547
Signal transduction 5, 11, 22, 48, 52, 148, 152, 160-1, 164, 420, 439, 456, 480, 547, 551-5
Signaling pathways 35, 52-4, 66, 80, 148-9, 151-2, 154, 156-7, 160-1, 257, 405, 411, 513, 554
Signet-ring cell adenocarcinoma (SRCA) 279
Signet-ring cell carcinoma 106, 182, 364
Silencing elements 430
Single nucleotide polymorphism 104, 258, 392
Sjogren's syndrome 213
Smad 52, 88, 90, 148, 150-2, 501
SMAD3 52, 148, 152
SMAD4 52, 88, 90, 148, 150-2, 501, 558-9
SMAD/Sp1 152
Smad-binding *cis*-elements 52
SMAD-mediated transcriptional activation 152
Small intestine 51, 99, 333, 453
Smgc 32-3, 38, 390, 394
Smgc mRNA 387, 390-1, 395
Smgc/Muc19 33, 387, 390, 394-5
Smgc/Muc19 expression 395
Smgc/Muc19 gene 33, 390, 394
Smgc/Muc19 transcript 392, 395
Smgc/Muc19 transcription 395

Smgc protein 33, 391-2, 394-5
Suppressors of cytokine signaling (SOCS) 36, 421
Solid adenocarcinoma 182, 289, 364
Soluble cytosolic protein 547
Soluble fraction 171, 469
Soluble mucin MUC7 461, 482
Soluble mucin MUC9 555
Soluble MUC9/OGP mucin 452
Soluble Mucins 3-8, 11-23, 29-38, 45-116, 146-214, 247-95, 317-67, 388-96, 399-412, 419-22, 426-44, 452-93, 495-509, 511-19, 547-64
  genes encoding 34
  group of 5, 34, 38, 418, 508
  functions of 508
Soluble mucus outer layer 469
Somatic mutation 492
Somatostatin 559
Sox2 53, 85-6, 329-30, 558-9
Sox9 100, 557, 559
Sp1 46-9, 56-7, 148, 150-3, 161, 250-2, 258-9, 327-8, 330, 430-2, 557, 559
Sp1 protein 48-9, 153
Sp1 transcription factor 48, 57, 252, 259, 432
Spacio-temporal regulation 444, 518
SPDEF 153, 330
Species 6, 18, 175, 249, 320, 394, 426-9, 431, 433-4, 437-8, 440-1, 466, 471-2, 514-15
Species-specific expression 425
Species-specific interaction 441, 515-16
Specimens 80, 83-4, 99, 108, 110, 113, 174, 202, 208, 212, 276, 285-6, 289, 340, 355
Sperm 74, 275, 335, 338, 440-3, 460, 514-17
  bovine sperm 440, 514
  frozen-thawed 440
  hamster sperm 440, 514
  human 440, 442, 514-15

Sperm capacitation 440, 443, 514, 517
Sperm function 440, 514
Sperm membrane 443, 516-17
Sperm mobility 37, 338
Sperm motility 444, 518
Sperm-mucus interaction 266
Sperm-ovum binding 37, 440, 514
Sperm penetration 176, 441, 515
Sperm penetrability 266
Sperm receptors 441, 515, 559
Sperm transition 275
Sperm viability 514
Spermatozoa 37, 177, 440-3, 456, 460, 514-17
Spiggin gene 33, 388
Splicing 14, 33, 59-60, 164-6, 247-8, 260-1, 326-7, 387, 390-3, 395, 427, 547
Splicing donor sequence 260
Sporadic colon adenocarcinomas 355
Squamous cell carcinoma 180, 339
Squamous cell metaplasia 278
Squamous metaplastic lesions 180
Sry-like high-mobility group (HMG) box 329
SSA 205, 355-6
STAT 35, 150-3, 251, 328, 404, 421
STAT1 protein 328
STAT6 150-3, 558-9
Statherin 271, 410, 461, 470, 513
Steroid hormone 110, 177, 266, 328, 434, 473
Stimuli 148, 157-8, 161, 163, 169, 254, 274, 421
Stomach 36, 45, 68, 70, 72, 78, 82-7, 174, 185, 188-93, 261, 274, 282-3, 317, 322-4, 327-35, 337-8, 341-4, 352, 354, 361, 364, 419, 453-5, 459-61, 465-6, 476-82, 487-93,497-9
  adult 174, 274, 329, 333
  developing 70, 453
  embryonic 174, 274

fetal 70, 274, 454
Stomach antrum 189
Stomach cancer 82-4, 87, 192-3, 344
Stomach epithelial cells 333, 341
Stomach fundus 274
Stomach surface epithelium 69
Stomach tumors 84, 87
Stomach ulcer 185
Stop codon 418
Stratified epithelium 186, 333
Stratified mucus layer 466
Streptococcus gordonii 406, 509
Streptococcus oralis 406, 509
Streptococcus salivarius 406, 509
Streptococcus sobrinus 406, 509
Stress factors 51, 53, 495
Stress response 93, 152, 439, 553
Structural complexity 440
Structural features 6, 23, 30, 322, 388, 390
Structural transformations 321, 460
Structure 3-5, 16-19, 145-6, 168-71, 246-9, 265-7, 270-3, 319-21, 323-6, 387-90, 398-402, 409-11, 459-60, 480-2, 511-13
Structure of mucin glycoproteins 16
Subcellular localization 206-7, 552
Subdomains 248-9
Subfamilies 3, 11, 44, 316-17
Subgroups 14, 82, 110-11, 352, 434
Sublingual glands 393-5, 403, 409, 511
maturation of 395
Sublingual glandular extracts 409, 511
Submandibular gland protein C 390, 394
Submucosal glands 68-9, 173, 185, 262, 273-8, 283, 291, 328, 393, 395, 403, 419, 454-5, 493
Substrates 63-4, 169, 264, 325, 553

Subtilisin-like protein convertase (SLPC) 552
Subtypes 83, 181, 198-9, 201, 278, 284, 286, 341, 351, 498
Subunits 19, 21, 23, 52, 159, 272
Suicide protein 469
Sulfate groups 457
Sulfated Muc9/ogp mucin 435
Sulfation 44, 63, 102, 105, 264-5, 268, 271, 281, 322, 325, 405, 436, 439, 485
Sulfur dioxide 487
Sulfomucins 479
Super-repeats 249
Supramolecular conformation 465
Supramolecular nematic crystalline structures 272
Surface epithelial cells 285, 328, 453
Surface epithelium 69-70, 72, 74, 173-5, 178, 186, 207, 273, 275, 278, 318, 332, 453, 481, 493
Surface-free energy 323
Surface tension 482-3
Survival factor 367
SV40 T-antigen 433
Synaptotagmin-2 171
Synthesis 20, 53, 59-60, 62, 67, 72-3, 75, 102-3, 105, 158, 166-7, 169-70, 172, 179, 189, 206, 261-2, 283, 285, 321, 325-6, 365, 393, 422, 428, 437, 473, 559

T

T3R 148, 430
Tandem repeats 6, 11, 18, 30, 36-7, 59, 145, 147, 247-8, 258, 319, 389, 398-400, 418-19, 426, 495, 512
TATA box 46, 57, 148, 150, 251-2, 327, 392, 403, 430-1
Tear film 112, 213, 294, 456, 482-6
Terminal carbohydrate 266
Testicular tissues 108, 176, 362
Testis 70, 73, 108, 176, 212, 289, 337-8, 362, 395
TFF 98, 186, 204, 282-3, 330, 354, 464, 469, 557-9
TFF1 186-7, 203-4, 283, 354, 464, 506-7, 558-9

TFF2 186-7, 204, 283, 330, 354, 506, 558-9
TFF3 186, 282, 469, 506-7, 557, 559
TGFα 47,148-9, 151-3, 155-6, 256
TGFβ 17-8, 52, 149, 151-2, 157, 161, 250, 336, 496, 559
TGFβ2/SMAD 150
TGT3 251
Th2 lymphocytes 150
Thickness 4, 101, 105, 458, 466, 476-8, 481-2
Threonine 6, 16, 18, 21, 30, 37, 45, 63, 65, 168, 249-50, 264, 325, 390, 401, 409-10, 418, 438, 512
Threonine residues 6, 16, 18, 21, 30, 45, 63, 65, 168, 264, 325, 401, 418, 438
Thyroid 246, 292, 366
Thyroid gland 291-2, 366
Tight junctions (TJ) 443, 517
Tissue morphogenesis 455
Tissue remodeling 442-4, 516-18
Tissue-specific genes 502
TJ-protein 444, 518
TLR4 receptor 154, 156, 257
TNF 36, 53, 86, 149, 161-2, 165, 169, 184, 254, 327, 404, 555, 559
TNFα 36, 149, 158, 161-2, 165, 169, 184, 254, 327, 396, 407, 420
TNF receptor-associated factor 555
Tonofilaments 444, 518
Tooth enamel 406-7
Tooth pellicle 406, 509
Toxic agents 486
TR-containing domain 270, 317, 319-20, 326, 391
TR domain 147, 401
TR1 147
TR2 147
TR3 147
TR4 147
Trachea 149, 173, 183, 273, 276, 278, 393
Tracheal goblet cells 172

Traditional serrated adenomas 205, 355
Trans-activation 49, 51
Trans-Golgi 21-2, 62, 67, 169, 261, 263, 271, 436
Trans-Golgi compartments 67, 261, 263, 271
Transcription 33, 35, 44, 48-50, 81, 151, 160, 162, 252, 288, 328, 390-1, 395-6, 431-4, 548
Transcription factor 36, 46-53, 55, 80, 84-5, 148, 150-2, 154, 158-9, 251-2, 254, 259-60, 326-30, 432, 559-60
 AP-1 54, 154
 ATF1 252
 CREB 53, 160-1
 p53 51
 Sp1 252
Transcription start site 46, 148, 151, 249, 251, 327, 392, 402, 429-32
Transcription unit 403
Transcriptional activation 76, 150, 152, 469, 498, 554
Transcriptional regulation 16, 44, 50, 58, 258, 327, 329-30, 418, 422, 425, 429, 431, 433, 496, 547
Transcripts 33, 73, 260, 278, 327, 391-2, 394-6, 432
 primary 59-60, 260-1, 327, 390-1, 427
Transfection assay 252-3, 258, 403, 430, 432
Transgenic mice 150, 255, 403, 433, 496
Translation 59-61, 77, 160, 321, 428, 432, 492
Translocation 61, 77, 160, 492
Trefoil proteins 506-7
Trefoil factors TTF1-3 98
Tubular adenoma 206, 355
Tubular metaplasia 109
Tubulovillous adenoma 289
Tumor cells 107, 109, 115, 188, 286, 289, 358, 490, 496, 505-6, 508
Tumor development 365, 489-90, 492, 494, 501-3, 508
Tumor differentiation 84, 346
Tumor initiation 492
Tumor progression 23, 91, 97, 198, 490, 497, 503-4, 506

Tumor promoters 452, 489, 501-2, 519
Tumor promotion 497, 503
Tumor recurrence 504
Tumor stimulating factor 68
Tumor suppression 91, 100, 329, 345, 463, 492, 494-5, 497, 500-3, 505, 554
  mucin-mediated 100
Tumor suppression activity 100, 494
Tumor suppression potential 495
Tumor suppressor genes 88, 91, 489, 491-7, 500-1, 508
Tumor suppressors 68, 77, 84, 93, 100-1, 329, 452, 489-93, 495-9, 501-2, 507-8, 519, 554, 559, 560
Tumorigenesis 80, 100, 205, 209, 357-8, 489-90, 492-7, 502, 506-7
  colorectal 497
  gastric 502
  intestinal 100, 493, 494
  lung 506
  mechanisms of 493
  ovarian mucinous 209, 357
  regulator of 492
Tumorigenic pathway 490-1, 497
Tumors 77-8, 86-8, 100, 107-8, 181-3, 199-201, 206, 209-12, 278-80, 284, 346-7, 352-3, 489-90, 496-7, 499-507
  borderline 352, 357-8
  intraductal 96
  mucin-producing 107
  primary liver 94-5
Two-hybrid system 410-11, 464, 513
Two-layered mucus gel 457
Tyrosine phosphotase 157

U

Ubiquitin degradation system 489
Ubiquitin-mediated degradation 439, 555
Ubiquitin pathway 439

UC-associated neoplasm 205, 353
UCAN 205
Ulcer associated cell lineage (UACL) 105, 204, 288, 354
Ulceration 103, 105, 204, 288, 354
Ulcerative colitis 78, 99, 101, 185, 204, 288, 353
Up-regulation 76, 81, 85, 98, 114, 153-8, 162, 257, 295, 354, 421, 481, 493-4, 506
Up-regulation of MUC2 gene expression 54, 113
Up-regulation of MUC2 transcription 51-2
Up-regulation of mucin genes 258
Ureteric bud 335-6
Urethra 175, 211-12, 289, 337-8, 360
Uridine-5-triphosphate (UTP) 171, 256-7, 421
Urinary bladder 107, 175
Urogenital organs 67, 73, 105, 145, 212, 290, 335, 360
Urogenital tract 73, 105, 175-6, 289-90, 335, 337-8, 360, 362
Urothelial carcinoma 106, 360
Urothelium 73, 106, 211
Uterus 176-7, 275, 334-5, 338, 395, 433
UTP 171, 256-7, 421
UTP- and ATP- stimulated secretion 257
3'-UTR 165, 398, 405, 418, 426-7
5'-UTR 165, 249, 252, 316, 392, 398

V

V-domain 555
Vagina 74, 108, 177, 207, 335, 357, 395
Vaginal secretion 335
Van der Waals forces 461
Variable number of tandem repeats 6, 18, 36
Vas deferens 176, 212, 289, 337
Vascular endothelial cells 505-6
Vascular endothelial growth factor 197, 504
Vasoactive intestinal peptide 54
Vater's valve 94, 199

VEGF 197, 497, 504
Vernal keratoconjunctivitis 213
Vesicles 166, 170-2, 176, 212, 261, 270-2, 288-9, 326, 335-8, 362, 439, 547
Villin 80, 188
Villous adenoma 206, 289, 351
VIP 54, 557, 559
Viral hepatitis 97, 345
Viral particles 461, 486
Viruses 4, 35, 148, 151, 362, 406, 456-7, 462, 471-2, 475, 486, 509-10
Viscoelasticity 4, 266, 458-9, 462, 473
Visual system 112
In vivo fertilization 440, 514
VNTR 6, 18, 30, 45, 63, 206, 260, 289, 319-20, 327, 342
VNTR polymorphism 319, 327
von Willibrand Factor 23, 29, 246
vWC 7, 389-90
vWD 7-8, 389-90, 392-3
vWF gene 32-3, 317
vWF glycoprotein 248, 250, 269, 389

W

WD40 domain 553
WHO classification 76, 82-3, 284
WXXW motif 170

Y

Yeast two-hybrid system 410-11, 464, 513
YY1 148

Z

Zona pelucida 441, 514-15
ZP glycoproteins 442, 516
ZP proteins 441-2, 516

Made in United States
Orlando, FL
04 November 2026

90205882R00385